Introduction to Genetics
A MOLECULAR APPROACH

Nowadays, genetics focuses on DNA. Just like the first edition, the theme of this new edition, *Introduction to Genetics: A Molecular Approach,* is therefore the progression from molecules (DNA and genes) to processes (gene expression and DNA replication) to systems (cells, organisms and populations). This progression reflects both the basic logic of life and the way in which modern biological research is structured. The molecular approach is particularly suitable for students for whom genetics is part of a broader program in biology, biochemistry, the biomedical sciences or biotechnology. This book presents the basic facts and concepts with enough depth of knowledge to stimulate students to move on to more advanced aspects of the subject.

This second edition has been thoroughly updated to cover new discoveries and developments in genetics from the last ten years. There are new chapters that introduce important techniques such as DNA sequencing and gene editing, and the applications of genetics in our modern world are covered in chapters describing topics as diverse as gene therapy and the use of ancient DNA to study prehistoric ecosystems.

KEY FEATURES:

- This book provides a molecular approach to the study of genetics.
- It is a highly accessible and well-structured book with chapters organized into four parts to aid navigation.
- It presents high-quality illustrations to elucidate the various concepts and mechanisms.
- Each chapter ends with a Key Concepts section, which serves to summarize the most essential points.
- Self-study questions enable the reader to assess their comprehension of chapter content, and discussion topics facilitate a deeper understanding of the material by encouraging conversation and critical evaluation.
- Key terms are emboldened throughout the text and are listed at the end of each chapter, and definitions can be found in the Glossary.
- For instructors who adopt the book, an affiliated question bank is free to download.

Terry Brown is Emeritus Professor of Biomolecular Archaeology at the University of Manchester. He has extensive experience in teaching genetics and is the author of *Genomes*, Fifth Edition.

Introduction to Genetics

A MOLECULAR APPROACH

Second Edition

Terry Brown

CRC Press is an imprint of the
Taylor & Francis Group, an **informa** business

A GARLAND SCIENCE BOOK

Designed cover image: Shutterstock/Billion Photos

Second edition published 2025
by CRC Press
2385 NW Executive Center Drive, Suite 320, Boca Raton FL 33431

and by CRC Press
4 Park Square, Milton Park, Abingdon, Oxon, OX14 4RN

CRC Press is an imprint of Taylor & Francis Group, LLC

© 2025 Terry Brown

Reasonable efforts have been made to publish reliable data and information, but the author and publisher cannot assume responsibility for the validity of all materials or the consequences of their use. The authors and publishers have attempted to trace the copyright holders of all material reproduced in this publication and apologize to copyright holders if permission to publish in this form has not been obtained. If any copyright material has not been acknowledged please write and let us know so we may rectify in any future reprint.

Except as permitted under U.S. Copyright Law, no part of this book may be reprinted, reproduced, transmitted, or utilized in any form by any electronic, mechanical, or other means, now known or hereafter invented, including photocopying, microfilming, and recording, or in any information storage or retrieval system, without written permission from the publishers.

For permission to photocopy or use material electronically from this work, access www.copyright.com or contact the Copyright Clearance Center, Inc. (CCC), 222 Rosewood Drive, Danvers, MA 01923, 978-750-8400. For works that are not available on CCC please contact mpkbookspermissions@tandf.co.uk

Trademark notice: Product or corporate names may be trademarks or registered trademarks and are used only for identification and explanation without intent to infringe.

ISBN: 978-1-032-75413-0 (hbk)
ISBN: 978-1-032-74353-0 (pbk)
ISBN: 978-1-003-47386-2 (ebk)

DOI: 10.1201/9781003473862

Typeset in Leawood Std
by codeMantra

Access the Instructor and Student Resources:
www.routledge.com/9781032754130

CONTENTS

Chapter 1	What Is Genetics and Why Is It Important?	1

PART I GENES AS UNITS OF BIOLOGICAL INFORMATION — 17

Chapter 2	DNA	19
Chapter 3	Genes	37
Chapter 4	Transcription of DNA to RNA	53
Chapter 5	Types of RNA Molecule: Messenger RNA	75
Chapter 6	Types of RNA Molecule: Ribosomal and Transfer RNA	93
Chapter 7	The Genetic Code	111
Chapter 8	Protein Synthesis	129
Chapter 9	Control of Gene Expression	149
Chapter 10	DNA Replication	173
Chapter 11	Mutation and DNA Repair	195

PART II GENES AS UNITS OF INHERITANCE — 217

Chapter 12	Inheritance of Genes During Virus Infection Cycles	219
Chapter 13	Inheritance of Genes in Bacteria	237
Chapter 14	Inheritance of Genes During Eukaryotic Cell Division	255
Chapter 15	Inheritance of Genes During Eukaryotic Sexual Reproduction	279
Chapter 16	Inheritance of Genes in Populations	301

PART III HOW GENES ARE STUDIED — 325

Chapter 17	Mapping the Positions of Genes in Chromosomes	327
Chapter 18	Sequencing Genes and Genomes	345

PART IV GENETICS IN OUR MODERN WORLD — 363

Chapter 19	Genes in Differentiation and Development	365
Chapter 20	The Human Genome	387

Chapter 21	Genes and Medicine	405
Chapter 22	DNA in Forensic Genetics and Archaeology	431
Chapter 23	Genes in Industry and Agriculture	451
Chapter 24	The Ethical Issues Raised by Modern Genetics	471

DETAILED CONTENTS

Preface — xv
Acknowledgments — xvii
About the Author — xix
Organization of the Book — xxi
Learning Aids — xxiii
Note to the Reader — xxv

Chapter 1 What Is Genetics and Why Is It Important? — 1

1.1 THE ORIGINS OF GENETICS — 1
Genetics begins with Mendel — 1
From genes to chromosomes — 3
Genes are made of DNA — 5
The gene as a unit of biological information — 6
Parallel developments in evolutionary genetics — 8
Genetic engineering and the new biotechnology — 9
From genes to genomes — 11
Full circle: the molecular approach to genetic variation — 12

1.2 HOW THIS BOOK IS ORGANIZED — 13
KEY CONCEPTS — 16
FURTHER READING — 16

PART I GENES AS UNITS OF BIOLOGICAL INFORMATION — 17

Chapter 2 DNA — 19

2.1 GENES ARE MADE OF DNA — 19
Biologists once thought that genes must be made of protein — 19
The discovery that the transforming principle is DNA — 20
Bacteriophage genes are made of DNA — 22

2.2 THE STRUCTURE OF DNA — 23
Nucleotides are the basic units of a DNA molecule — 24
Nucleotides join together to make a polynucleotide — 25
In living cells, DNA is a double helix — 26
The key features of the double helix — 28
The double helix exists in several different forms — 29

2.3 THE MOLECULAR EXPLANATION OF THE BIOLOGICAL ROLE OF DNA — 31
Biological information is contained in the nucleotide sequence of a DNA molecule — 31
Complementary base pairing enables DNA molecules to replicate — 31
KEY CONCEPTS — 33
QUESTIONS AND PROBLEMS — 33
FURTHER READING — 35

Chapter 3 Genes — 37

3.1 THE NATURE OF THE INFORMATION CONTAINED IN GENES — 37
Genes are segments of DNA molecules — 37
Genes contain instructions for making RNA and protein molecules — 38
Protein synthesis is the key to the expression of biological information — 38
The RNA molecules that are not translated into protein are also important — 41

3.2 VARIATIONS IN THE INFORMATION CONTENT OF INDIVIDUAL GENES — 43
Genetic variants are called alleles — 43
There may be many variants of the same gene — 44

3.3 FAMILIES OF GENES — 46
Multiple gene copies enable large amounts of RNA to be synthesized rapidly — 46
In some multigene families, the genes are active at different stages of development — 46
The globin families reveal how genes evolve — 47
KEY CONCEPTS — 49
QUESTIONS AND PROBLEMS — 50
FURTHER READING — 51

Chapter 4 Transcription of DNA to RNA — 53

4.1 GENE EXPRESSION IN DIFFERENT TYPES OF ORGANISM — 53
There are three major groups of organism — 53
Gene expression is more than simply 'DNA makes RNA makes protein' — 55

4.2 ENZYMES FOR MAKING RNA — 56
The RNA polymerase of *Escherichia coli* comprises six subunits — 56
Eukaryotes possess more complex RNA polymerases — 57

4.3 RECOGNITION SEQUENCES FOR TRANSCRIPTION INITIATION — 58
Bacterial RNA polymerases bind to promoter sequences — 58

Eukaryotic promoters are more complex	59
Some eukaryotic genes have more than one promoter	60

4.4 INITIATION OF TRANSCRIPTION IN BACTERIA AND EUKARYOTES — 61

The σ subunit recognizes the bacterial promoter — 61
Formation of the RNA polymerase II initiation complex — 62
Initiation of transcription by RNA polymerases I and III — 63

4.5 THE ELONGATION PHASE OF TRANSCRIPTION — 63

Bacterial transcripts are synthesized by the RNA polymerase core enzyme — 64
All eukaryotic mRNAs have a modified 5′ end — 66
Some eukaryotic mRNAs take hours to synthesize — 68

4.6 TERMINATION OF TRANSCRIPTION — 68

Hairpin loops in the RNA are involved in the termination of transcription in bacteria — 68
Eukaryotes use diverse mechanisms for the termination of transcription — 70

KEY CONCEPTS — 71
QUESTIONS AND PROBLEMS — 72
FURTHER READING — 74

Chapter 5 Types of RNA Molecule: Messenger RNA — 75

5.1 THE mRNA COMPONENT OF A TRANSCRIPTOME — 76

The mRNA component of a transcriptome is small but complex — 76
Transcriptome studies are important in cancer research — 77
The mRNA composition of a transcriptome can change over time — 78
There are various pathways for nonspecific mRNA turnover — 79
Individual mRNAs in eukaryotes are degraded by the Dicer protein — 81

5.2 REMOVAL OF INTRONS FROM EUKARYOTIC mRNAs — 82

Intron–exon boundaries are marked by special sequences — 82
The splicing pathway — 83
Enhancer and silencer sequences specify alternative splicing pathways — 85

5.3 CHEMICAL MODIFICATION OF mRNAs — 87

There are various types of RNA editing — 87
Chemical modifications that do not affect the sequence of an mRNA — 89

KEY CONCEPTS — 89
QUESTIONS AND PROBLEMS — 90
FURTHER READING — 92

Chapter 6 Types of RNA Molecule: Ribosomal and Transfer RNA — 93

6.1 RIBOSOMAL RNA — 93

Ribosomes and their components were first studied by density gradient centrifugation — 93
Understanding the fine structure of the ribosome — 95
Electron microscopy and X-ray crystallography have revealed the three-dimensional structure of the ribosome — 96

6.2 TRANSFER RNA — 97

All tRNAs have a similar structure — 97

6.3 PROCESSING OF PRECURSOR rRNA AND tRNA MOLECULES — 99

Ribosomal RNAs are transcribed as long precursor molecules — 99
Transfer RNAs are also cut out of longer transcription units — 101
Transfer RNAs display a diverse range of chemical modifications — 102
Ribosomal RNAs are also modified, but less extensively — 103

6.4 REMOVAL OF INTRONS FROM PRE-rRNAs AND PRE-tRNAs — 104

Some rRNA introns are enzymes — 104
Some eukaryotic pre-tRNAs contain introns — 105

KEY CONCEPTS — 107
QUESTIONS AND PROBLEMS — 107
FURTHER READING — 109

Chapter 7 The Genetic Code — 111

7.1 PROTEIN STRUCTURE — 111

Amino acids are linked by peptide bonds — 111
There are four levels of protein structure — 113
The amino acid sequence is the key to protein structure — 114
Amino acid sequence also determines protein function — 115

7.2 THE GENETIC CODE — 116

There is a colinear relationship between a gene and its protein — 116
Each codeword is a triplet of nucleotides — 117
Elucidation of the genetic code — 118
The genetic code is degenerate and includes punctuation codons — 119
The genetic code is not universal — 120

7.3 THE ROLE OF tRNAs IN PROTEIN SYNTHESIS — 122

Aminoacyl-tRNA synthetases attach amino acids to tRNAs — 122

Unusual types of aminoacylation — 123
The mRNA sequence is read by base pairing between the codon and anticodon — 123
KEY CONCEPTS — 125
QUESTIONS AND PROBLEMS — 126
FURTHER READING — 128

Chapter 8 Protein Synthesis — 129

8.1 THE ROLE OF RIBOSOMES IN PROTEIN SYNTHESIS — 130
Initiation in bacteria requires an internal ribosome binding site — 130
Initiation in eukaryotes is mediated by the cap structure and poly(A) tail — 131
Translation of a few eukaryotic mRNAs initiates without scanning — 133
Elongation of the polypeptide begins with formation of the first peptide bond — 134
Elongation continues until a termination codon is reached — 135
Termination requires special release factors — 136

8.2 POST-TRANSLATIONAL PROCESSING OF PROTEINS — 136
Some proteins fold spontaneously in the test tube — 137
Inside cells, protein folding is aided by molecular chaperones — 138
Some proteins are chemically modified — 140
Some proteins are processed by proteolytic cleavage — 142

8.3 PROTEIN DEGRADATION — 143
Degradation of eukaryotic proteins involves ubiquitin and the proteasome — 144

KEY CONCEPTS — 145
QUESTIONS AND PROBLEMS — 145
FURTHER READING — 147

Chapter 9 Control of Gene Expression — 149

9.1 THE IMPORTANCE OF GENE REGULATION — 149
Gene regulation enables bacteria to respond to changes in their environment — 150
Gene regulation in eukaryotes must be responsive to more sophisticated demands — 151
The underlying principles of gene regulation are the same in all organisms — 152

9.2 REGULATION OF TRANSCRIPTION INITIATION IN BACTERIA — 154
Four genes are involved in lactose utilization by *E. coli* — 154
The regulatory gene codes for a repressor protein — 156
Glucose also regulates the lactose operon — 156
Operons are common in prokaryotes — 158

9.3 REGULATION OF TRANSCRIPTION INITIATION IN EUKARYOTES — 159
RNA polymerase II promoters are controlled by a variety of regulatory sequences — 159
The RNA polymerase II initiation complex is activated via a mediator protein — 161
Signals from outside of the cell must be transmitted to the nucleus in order to influence gene expression — 161

9.4 OTHER STRATEGIES FOR REGULATING GENE EXPRESSION — 163
Modification of the bacterial RNA polymerase enables different sets of genes to be expressed — 163
Transcription termination signals are sometimes ignored — 164
Attenuation is a second control strategy targeting transcription termination — 166
Bacteria and eukaryotes are both able to regulate the initiation of translation — 168

KEY CONCEPTS — 169
QUESTIONS AND PROBLEMS — 170
FURTHER READING — 172

Chapter 10 DNA Replication — 173

10.1 THE OVERALL PATTERN OF DNA REPLICATION — 173
DNA replicates semiconservatively, but this causes topological problems — 173
The Meselson–Stahl experiment proved that replication is semiconservative — 175
DNA topoisomerases solve the topological problem — 176
Variations on the semiconservative theme — 178

10.2 DNA POLYMERASES — 180
DNA polymerases synthesize DNA but can also degrade it — 180
Bacteria and eukaryotes possess several types of DNA polymerase — 182
The limitations of DNA polymerase activity cause problems during replication — 182

10.3 DNA REPLICATION IN BACTERIA — 184
E. coli has a single origin of replication — 184
The elongation phase of bacterial DNA replication — 185
Replication of the *E. coli* DNA molecule terminates within a defined region — 186

10.4 DNA REPLICATION IN EUKARYOTES — 187
Eukaryotic DNA has multiple replication origins — 187
The eukaryotic replication fork: variations on the bacterial theme — 188
Termination of replication in eukaryotes — 189
DNA replication in eukaryotes must be coordinated with the cell cycle — 190

KEY CONCEPTS	191
QUESTIONS AND PROBLEMS	192
FURTHER READING	193

Chapter 11 Mutation and DNA Repair 195

11.1 THE CAUSES OF MUTATIONS 196

Errors in replication are a source of point mutations	196
Replication errors can also lead to insertion and deletion mutations	197
Mutagens are one type of environmental agent that causes damage to cells	198
There are many types of chemical mutagen	199
There are also several types of physical mutagen	201

11.2 DNA REPAIR 203

Direct repair systems fill in nicks and correct some types of nucleotide modification	204
Many types of damaged nucleotide can be repaired by base excision	204
Nucleotide excision repair is used to correct more extensive types of damage	205
Mismatch repair corrects errors in replication	206
DNA breaks can also be repaired	208
In an emergency, damaged nucleotides can be bypassed during DNA replication	209
Defects in DNA repair underlie human disorders, including cancers	209

11.3 THE EFFECTS OF MUTATIONS ON GENES, CELLS AND ORGANISMS 210

Mutations have various effects on the biological information contained in a gene	210
Mutations have various effects on multicellular organisms	211
A second mutation may reverse the phenotypic effect of an earlier mutation	212

KEY CONCEPTS	214
QUESTIONS AND PROBLEMS	214
FURTHER READING	216

PART II GENES AS UNITS OF INHERITANCE 217

Chapter 12 Inheritance of Genes During Virus Infection Cycles 219

12.1 BACTERIOPHAGES 220

Bacteriophages have diverse structures and equally diverse genomes	220
The lytic infection cycle of bacteriophage T4	222
The lytic infection cycle is regulated by the expression of early and late genes	223
The lysogenic infection cycle of bacteriophage λ	224

Many genes are involved in establishment and maintenance of lysogeny	225
There are some unusual bacteriophage life cycles	226

12.2 VIRUSES OF EUKARYOTES 227

Eukaryotic viruses have diverse structures and infection strategies	227
Viral retroelements integrate into the host cell DNA	228
Some retroviruses cause cancer	229

12.3 TOWARD AND BEYOND THE EDGE OF LIFE 231

There are cellular versions of viral retroelements	231
Satellite RNAs, virusoids, viroids and prions are all probably beyond the edge of life	232

KEY CONCEPTS	233
QUESTIONS AND PROBLEMS	233
FURTHER READING	235

Chapter 13 Inheritance of Genes in Bacteria 237

13.1 BACTERIAL GENOMES 237

Bacterial genomes vary greatly in size and some are linear DNA molecules	238
Plasmids are independent DNA molecules within a bacterial cell	239
Some bacteria have multipartite genomes	240
Genome sizes and gene numbers vary within individual species	242

13.2 INHERITANCE OF CHROMOSOMES AND PLASMIDS DURING BACTERIAL CELL DIVISION 243

The E. coli nucleoid contains supercoiled DNA attached to a protein core	243
Partitioning systems control the inheritance of bacterial chromosomes and plasmids	244

13.3 TRANSFER OF GENES BETWEEN BACTERIA 245

Plasmids can be transferred between bacteria by conjugation	246
Chromosomal genes can also be transferred during conjugation	247
Bacterial genes can also be transferred without contact between donor and recipient	248
Transferred bacterial DNA can become a permanent feature of the recipient's genome	249

KEY CONCEPTS	251
QUESTIONS AND PROBLEMS	251
FURTHER READING	253

Chapter 14 Inheritance of Genes During Eukaryotic Cell Division 255

14.1 EUKARYOTIC GENOMES AND CHROMOSOMES 255

Eukaryotic nuclear genomes are contained in chromosomes — 255
Chromosomes contain DNA and proteins — 256
Histones are constituents of the nucleosome — 258
The 10 nm fiber is present in nondividing cells — 259
Higher levels of DNA packaging are needed when a cell divides — 259

14.2 THE SPECIAL FEATURES OF METAPHASE CHROMOSOMES — 261

Individual metaphase chromosomes have distinct morphologies — 261
Centromeres contain repetitive DNA and modified nucleosomes — 261
Telomeres protect chromosome ends — 263
Chromosomes should get progressively shorter during multiple rounds of replication — 263

14.3 PARTITIONING OF CHROMOSOMES DURING EUKARYOTIC CELL DIVISION — 265

Partitioning of chromosomes during mitosis — 265
Meiosis involves two successive nuclear divisions — 267
During meiosis I, bivalents are formed between homologous chromosomes — 268
Formation of bivalents ensures that siblings are not identical to one another — 269
Recombination occurs between homologous chromosomes within a bivalent — 270
Recombination involves a DNA heteroduplex — 271
The biochemical pathway for homologous recombination — 273

KEY CONCEPTS — 275
QUESTIONS AND PROBLEMS — 276
FURTHER READING — 278

Chapter 15 Inheritance of Genes During Eukaryotic Sexual Reproduction — 279

15.1 WHAT MENDEL DISCOVERED — 280

Mendel's experiments were carefully planned — 280
Mendel's crosses revealed a regular pattern in the inheritance of characteristics — 281
Mendel's interpretation of the results of the monohybrid crosses — 283
Why is there a 3:1 ratio in the F_2 generation? — 283
Crosses involving two pairs of characteristics — 284
Mendel pre-empted the discovery of meiosis, but did he cheat? — 285

15.2 RELATIONSHIPS BETWEEN PAIRS OF ALLELES — 286

Recessive genes are often nonfunctional — 286
Some pairs of alleles display incomplete dominance — 288
Lethal alleles result in the death of a homozygote — 289
Some alleles are codominant — 290

15.3 INTERACTIONS BETWEEN ALLELES OF DIFFERENT GENES — 292

Functional alleles of interacting genes can have additive effects — 292
Important interactions occur between genes controlling different steps in a biochemical pathway — 294
Epistasis occurs when alleles of one gene mask the effect of a second gene — 295
Interactions between multiple genes result in quantitative traits — 295

15.4 A SINGLE GENE CAN GIVE RISE TO A COMPLEX PHENOTYPE — 296

Pleiotropic genes affect multiple characteristics — 296
Not all deleterious alleles have an immediate effect — 297

KEY CONCEPTS — 298
QUESTIONS AND PROBLEMS — 299
FURTHER READING — 300

Chapter 16 Inheritance of Genes in Populations — 301

16.1 FACTORS AFFECTING THE FREQUENCIES OF ALLELES IN POPULATIONS — 302

Allele frequencies do not change simply as a result of mating — 302
Genetic drift occurs because populations are not of infinite size — 303
Alleles become fixed at a rate determined by the effective population size — 304
New alleles are continually created by mutation — 306
Natural selection favors individuals with high fitness for their environment — 306
Natural selection acts on phenotypes but causes changes in the frequencies of the underlying genotypes — 308
There are different types of natural selection, with different outcomes — 309
Heterozygotes sometimes have a selective advantage — 310
Alleles not subject to selection can hitchhike with ones that are — 311
What are the relative impacts of drift and selection on allele frequencies? — 312

16.2 THE EFFECTS OF POPULATION BIOLOGY ON ALLELE FREQUENCIES — 313

Environmental variations can result in gradations in allele frequencies — 314
There may be internal barriers to the movement of alleles within a population — 315
Transient reductions in effective population size can result in sharp changes in allele frequencies — 317

A population split can be the precursor to speciation	319	Repetitive DNA complicates shotgun assembly of eukaryotic genomes	350
KEY CONCEPTS	**322**	What is a 'genome sequence'	352
QUESTIONS AND PROBLEMS	**322**	**18.3 ANNOTATING A DNA SEQUENCE**	**353**
FURTHER READING	**324**	Genes can be located by searching for open reading frames	353

PART III HOW GENES ARE STUDIED 325

Genes can be located by searching for open reading frames — 353
ORF scans are less effective with eukaryotic genomes — 354
Locating genes by RNA sequencing — 355
Deducing the function of genes identified in a DNA sequence — 356
Gene editing can be used to assign functions to genes — 357
Genome browsers — 358

Chapter 17 Mapping the Positions of Genes in Chromosomes 327

17.1 PARTIAL LINKAGE PROVIDES THE BASIS TO GENE MAPPING WITH EUKARYOTES 327

Linkage is not absolute, it is only partial — 328
Partial linkage is explained by crossing over — 329
The frequency of recombinants enables the map positions of genes to be worked out — 330
The test cross is central to gene mapping when breeding experiments are possible — 331
Multipoint crosses – mapping the positions of more than two genes at once — 332
Gene mapping in humans is carried out by pedigree analysis — 333
Practicalities of pedigree analysis — 335

17.2 GENE MAPPING WITH BACTERIA 336

Basic features of gene mapping in bacteria — 336
Mapping by conjugation – the interrupted mating technique — 337
Transduction and transformation can provide more detailed bacterial gene maps — 338

17.3 PHYSICAL MAPPING OF DNA MOLECULES 339

Optical mapping can locate short sequence motifs in longer DNA molecules — 340
Optical mapping with fluorescent probes — 340

KEY CONCEPTS 341
QUESTIONS AND PROBLEMS 342
FURTHER READING 343

Chapter 18 Sequencing Genes and Genomes 345

18.1 METHODS FOR DNA SEQUENCING 345

DNA sequencing methods have gradually become more powerful — 345
Reversible terminator sequencing is the most popular short-read method — 346
There are different approaches to long-read sequencing — 347

18.2 ASSEMBLING A GENOME SEQUENCE 349

Small bacterial genomes can be sequenced by the shotgun method — 349

KEY CONCEPTS 359
QUESTIONS AND PROBLEMS 360
FURTHER READING 361

PART IV GENETICS IN OUR MODERN WORLD 363

Chapter 19 Genes in Differentiation and Development 365

19.1 CHANGES IN GENE EXPRESSION DURING CELLULAR DIFFERENTIATION 365

Differentiation requires permanent and heritable changes in gene activity — 366
Immunoglobulin and T-cell diversity in vertebrates results from genome rearrangement — 367
A genetic feedback loop can ensure permanent differentiation of a cell lineage — 369
Chemical modification of nucleosomes leads to heritable changes in chromatin structure — 370
The fruit fly proteins called Polycomb and trithorax influence chromatin packaging — 371
DNA methylation can silence regions of a genome — 372
Genomic imprinting and X inactivation result from DNA methylation — 373

19.2 COORDINATION OF GENE ACTIVITY DURING DEVELOPMENT 375

Research into developmental genetics is underpinned by the use of model organisms — 375
C. elegans reveals how cell-to-cell signaling can confer positional information — 377
Positional information in the fruit fly embryo is established by a cascade of gene activity — 379
Segment identity is determined by the homeotic selector genes — 380
Homeotic selector genes are also involved in vertebrate development — 381
Homeotic genes also underlie plant development — 383

KEY CONCEPTS 384
QUESTIONS AND PROBLEMS 385
FURTHER READING 386

Chapter 20 The Human Genome — 387

20.1 THE GENETIC ORGANIZATION OF THE HUMAN NUCLEAR GENOME — 387

The human genome has fewer protein-coding genes than we expected — 387

The arrangement of genes in the human genome is largely random — 389

The genes make up only a small part of the human genome — 389

Genes are more densely packed in the genomes of lower eukaryotes — 391

20.2 THE REPETITIVE DNA CONTENT OF THE HUMAN NUCLEAR GENOME — 392

Interspersed repeats include RNA transposons of various types — 393

Tandemly repeated DNA forms satellite bands in density gradients — 394

Minisatellites and microsatellites are special types of tandemly repeated DNA — 395

20.3 HOW DIFFERENT IS THE HUMAN GENOME FROM THAT OF OTHER ANIMALS? — 396

Chimpanzees are our closest relatives in the animal kingdom — 396

The human genome is very similar to that of the chimpanzee — 397

20.4 THE MITOCHONDRIAL GENOME — 398

The human mitochondrial genome is packed full of genes — 398

Mitochondrial genomes as a whole display a great deal of variability — 399

Mitochondria and chloroplasts were once free-living prokaryotes — 400

KEY CONCEPTS — 401

QUESTIONS AND PROBLEMS — 402

FURTHER READING — 403

Chapter 21 Genes and Medicine — 405

21.1 THE GENETIC BASIS TO INHERITED DISORDERS — 405

Many inherited disorders are caused by loss-of-function mutations — 406

Inherited disorders resulting from a gain of function are much less common — 407

Trinucleotide repeat expansions can cause inherited disorders — 408

There are dominant and recessive inherited disorders — 410

The dominant–recessive relationship can tell us something about the basis of a disorder — 411

Inherited disorders can also be caused by large deletions and chromosome abnormalities — 413

21.2 IDENTIFYING A GENE FOR AN INHERITED DISORDER — 414

DNA markers are often used in pedigree analysis — 414

How the human breast cancer susceptibility gene *BRCA1* was identified — 417

Genome-wide association studies can also identify genes for genetic disorders — 418

21.3 THE GENETIC BASIS TO CANCER — 419

Cancers begin with the activation of proto-oncogenes — 419

Tumor suppressor genes inhibit cell transformation — 420

The development of cancer is a multistep process — 422

21.4 GENE THERAPY — 422

There are two approaches to gene therapy — 423

Gene therapy can also be used to treat cancer — 425

KEY CONCEPTS — 426

QUESTIONS AND PROBLEMS — 427

FURTHER READING — 429

Chapter 22 DNA in Forensic Genetics and Archaeology — 431

22.1 DNA AND FORENSIC GENETICS — 431

The first genetic fingerprints were based on minisatellite variability — 432

Short tandem repeats are used to obtain DNA profiles — 433

A DNA profile is obtained by multiplex PCR — 434

DNA profiling can solve cold cases — 436

Familial DNA testing extends the range of DNA profiling — 437

DNA testing is also used to identify human remains — 437

DNA testing can be used in kinship analysis — 438

22.2 ANCIENT DNA EXTENDS GENETICS INTO ARCHAEOLOGY — 439

Ancient DNA is degraded and prone to contamination with modern DNA — 439

Ancient DNA typing provides a means of studying kinship in past societies — 440

Ancient DNA has identified family groups among archaeological skeletons — 441

Ancient DNA can reveal more distant relationships between groups of prehistoric people — 442

The Neanderthal genome has been sequenced — 444

Our ancestors interbred with both Neanderthals and Denisovans — 445

Ancient DNA can be obtained from many species other than humans — 446

KEY CONCEPTS — 448

QUESTIONS AND PROBLEMS — 449

FURTHER READING — 450

Chapter 23 Genes in Industry and Agriculture — 451

23.1 PRODUCTION OF RECOMBINANT PROTEIN BY MICROORGANISMS — 451

DNA cloning is used to transfer foreign genes into a bacterium — 452

Special cloning vectors are needed for the synthesis of recombinant protein — 453

Insulin was one of the first human proteins to be synthesized in *E. coli* — 454

E. coli is not an ideal host for recombinant protein production — 456

Recombinant protein can also be synthesized in eukaryotic cells — 458

Factor VIII is an example of a protein that is difficult to produce by recombinant means — 459

Vaccines can also be produced by recombinant technology — 460

23.2 GENETIC ENGINEERING OF PLANTS AND ANIMALS — 461

Crops can be engineered to become resistant to herbicides — 461

Various GM crops have been produced by gene addition — 462

Plant genes can also be silenced by genetic modification — 464

Gene editing has many applications in plant genetic engineering — 466

Engineered plants can be used for the synthesis of recombinant proteins, including vaccines — 467

Recombinant proteins can be produced in animals by pharming — 467

KEY CONCEPTS — 468

QUESTIONS AND PROBLEMS — 469

FURTHER READING — 470

Chapter 24 The Ethical Issues Raised by Modern Genetics — 471

24.1 AREAS OF CONCERN RELATING TO GM CROPS — 472

There have been concerns that GM crops might be harmful to human health — 472

There are also concerns that GM crops might be harmful to the environment — 473

The safety or otherwise of GM crops remains a contentious subject — 475

GM technology could be used to ensure that farmers have to buy new seed every year — 477

24.2 ETHICAL ISSUES RELATING TO THE USE OF PERSONAL DNA DATA — 478

DNA profiling alerted the public to the concept of personal DNA data — 479

Nonforensic DNA databases complicate issues regarding personal DNA data — 480

24.3 ETHICAL ISSUES RAISED BY GENE THERAPY AND PHARMING — 481

Germline gene therapy has the potential to make permanent changes to a human lineage — 481

The use of SCNT in pharming raises additional issues — 482

24.4 REACHING A CONSENSUS ON ETHICAL ISSUES RELATING TO MODERN GENETICS — 483

Ethical debates must be based on science fact not science fiction — 483

Few geneticists, probably none at all, are mad or evil — 484

KEY CONCEPTS — 486

QUESTIONS AND PROBLEMS — 486

FURTHER READING — 487

Glossary — 489

Index — 511

PREFACE

Thirteen years have passed since the publication of the first edition of Introduction to *Genetics: A Molecular Approach*. During that period, there has been significant progress in genetics. In particular, the introduction of 'next-generation' sequencing methods has transformed our knowledge of the genomes of many species, revealing genetic features and patterns of organization that were unsuspected ten years ago. This knowledge extends to prehistoric humans and to extinct animals, thanks to improvements in the methodology for sequencing ancient DNA from archaeological and palaeontological remains. Other technical breakthroughs, such as the development of gene editing, have led to new innovations in genetic engineering and in the experimental analysis of gene function. Even some of the very basic 'facts' of genetics have changed during the last decade, such as the structure of the bacterial RNA polymerase and the identities of the polymerases involved in eukaryotic DNA replication.

All of these recent advances are covered in the second edition of Introduction to *Genetics: A Molecular Approach*. In particular, Chapter 20 on the human genome has been completely revised to reflect new information on the genetic organization of our genome and its relationship with the genomes of other primates. There is a new chapter devoted to modern sequencing technology and the methods used to annotate a genome sequence and another new chapter that brings together the methods used in gene mapping, both by linkage analysis and physical examination of DNA molecules. Chapter 22 on forensic genetics and archaeology has also undergone complete revision to bring it up to date.

I have also made other structural changes to the book to address the many helpful points made by users of the first edition who were kind enough to email me. In the first edition, I experimented with some novel ways of presenting information; some of these innovations were well received and others were less successful. In the second edition, I have given greater emphasis to the historical foundations of modern genetics, including a more detailed description of Mendel's experiments. I have also weaved the information previously contained in standalone Research Briefings into the text, in the hope that this provides a better flow to the narrative.

None of these changes affect the underlying philosophy of this book that genetics is today inexorably centered on DNA and that the teaching of genetics should reflect this fact. The theme of this book remains the progression from *molecules* (DNA and genes) to processes (gene expression and DNA replication) to systems (cells, organisms and populations), a progression that reflects both the basic logic of life and the way in which modern biological research is structured.

ACKNOWLEDGMENTS

I would like to thank the reviewers who provided helpful comments on the first edition of *Introduction to Genetics: A Molecular Approach* and who gave detailed feedback on the proposal for the second edition.

I am also grateful to the Garland staff who have contributed to the creation of the second edition of *Introduction to Genetics: A Molecular Approach* and who have helped convert my own contribution – words not necessarily in the right order and scribbled diagrams – into an actual textbook.

Finally, but not least, I would like to thank my wife Keri for her constant support and encouragement in my endeavors to write useful textbooks.

ABOUT THE AUTHOR

Terry Brown is Emeritus Professor of Biomolecular Archaeology at the University of Manchester. He has extensive experience in teaching genetics and is the author of *Genomes*, Fifth Edition.

ORGANIZATION OF THE BOOK

Introduction to Genetics: A Molecular Approach is divided into four parts:

PART I – GENES AS UNITS OF BIOLOGICAL INFORMATION

First, we will become familiar with the structure of DNA (Chapter 2) and the way in which genes are organized within DNA molecules (Chapter 3). Then, we will be ready to follow the process of gene expression, which results in the information contained in a gene being utilized by the cell. We will study the way in which DNA is copied into RNA (Chapter 4), and we will look in detail at the roles in the cell of the different types of RNA molecule that are made (Chapters 5 and 6). We will then discover how the genetic code is used to direct the synthesis of protein molecules whose structures and functions are specified by the information contained in the genes (Chapters 7 and 8). We will examine how all of these events are controlled so that only those genes whose information is needed are active at a particular time (Chapter 9). Finally, we will study how DNA is replicated and how mutations can arise and be repaired (Chapters 10 and 11).

PART II – GENES AS UNITS OF INHERITANCE

In Part II, we will ask three questions. The first is how a complete set of genes is passed to the daughter cells when the parent cell divides. We will examine this process during the infection cycles of viruses (Chapter 12), when a bacterium divides (Chapter 13) and when an animal cell or a plant cell divides (Chapter 14). The second question concerns the inheritance of genes during sexual reproduction. We will study how DNA molecules are passed from parents to their offspring, and we will investigate how the genes on these DNA molecules specify the biological characteristics of the offspring in such a way that they resemble, but are not identical to, their parents (Chapter 15). The third question concerns the link between the inheritance of genes and the evolution of species. To answer this question, we will examine how the new gene variants that are created by mutation can spread through a population (Chapter 16).

PART III – HOW GENES ARE STUDIED

Part III is devoted to the important research techniques that are used in modern genetics. In Chapter 17, we will explore the ways in which the positions of genes are mapped on the chromosomes of animals, plants and bacteria, including the special approaches that must be taken with humans. Then, in Chapter 18, we will study how DNA molecules are sequenced.

PART IV – GENETICS IN OUR MODERN WORLD

In Part IV, we will explore some of the areas of research that are responsible for the high profile that genetics has in our modern world. The first topic that we will study is the role of genes in development. How do genes control the pathway that begins with a fertilized egg cell and ends with an adult organism? Finding the answer to this question is one of the biggest challenges in all of genetics (Chapter 19). Then, we will devote three chapters to our own species. In Chapter 20, we will look closely at the human genome, and in particular, we will ask what it is about our genome that makes us special. Then, we will look at the ways in which alterations to genes can give rise to cancer and to inherited disorders such as cystic fibrosis (Chapter 21). In Chapter 22, we will examine how DNA profiles are obtained and why these have become so important in forensic biology. We will also learn how the presence of ancient DNA in some archaeological materials is enabling genetics to be used to study the prehistory of our species. We will then look at the applications of genetics in industry and agriculture, in the production of important pharmaceuticals and in the design of genetically modified crops (Chapter 23). This will lead us to some of the controversial aspects of modern genetics. We must not ignore these controversies and so in Chapter 24, we will examine some of the ethical issues raised by genetics, and we will ask how these issues should be debated so that the controversies can be resolved.

LEARNING AIDS

I have tried to make *Introduction to Genetics: A Molecular Approach* as user-friendly as possible. The book therefore includes a number of devices intended to help the reader and to make the book an effective learning aid.

QUESTIONS AND PROBLEMS

Each set of questions and problems is divided into three sections, designed to help you in different ways in individual and group study programs.

Key terms

The first section asks you to define the key terms encountered in a chapter. All the terms are defined in the Glossary and you should check that you can remember the definitions yourself. If you can remember the definitions, then you have an excellent grasp of the main facts for that particular chapter.

Self-study questions

The aim of the self-study questions is to test not just your recall of the facts but also your understanding of the concepts behind those facts. Each question can be answered in 50–100 words, or possibly by a table or annotated diagram. The questions cover the entire content of each chapter in a fairly straightforward manner. You can check your answers by comparing them with the relevant parts of the text. You can use the self-study questions to work systematically through a chapter, or you can select individual ones in order to confirm that you have the correct understanding of a specific topic.

Discussion topics

These vary in nature and difficulty. The simplest can be answered by a well-directed search of the genetics literature, the intention being that you advance your learning a few stages from where the book leaves off. In some cases, the questions point you forward to issues that will be discussed later in the book, to help you see how the basic information that we deal with in the early chapters is relevant to the more complex topics that we study later on. Other problems require you to evaluate a statement or a hypothesis, based on your understanding of the material in the book, possibly supplemented by reading around the subject. These problems are intended to make you think carefully about the subject and perhaps to realize for yourself that often there are hidden complexities that are not immediately apparent. A few problems are very difficult, in some cases to the extent that there is no solid answer to the question posed. These are designed to stimulate debate and speculation, stretching your knowledge and that of your colleagues with whom you discuss the problems. If you find these discussions stimulating, then you will know that you have become a real geneticist.

FURTHER READING

The reading lists at the end of each chapter are intended to help you obtain further information, for example, when writing extended essays or dissertations on particular topics. In some cases, I have appended a few words summarizing the particular value of each entry, to help you decide which ones you wish to seek out. The lists are not all-inclusive, and I encourage you to spend some time searching your library and the internet for other books and articles. Browsing is an excellent way to discover interests that you never realized you had!

GLOSSARY

I am very much in favor of glossaries as learning aids and I have provided an extensive one for *Introduction to Genetics: A Molecular Approach*. Every term that is highlighted in bold in the text is defined in the Glossary. The glossary therefore provides a quick and convenient means by which the reader can remind themselves of the technical terms relevant to the study of genomes and also acts as a revision aid to make sure those definitions are clearly understood during the minutes of uncertainty that many students experience immediately before a test.

Terry Brown
Manchester, United Kingdom

NOTE TO THE READER

Available online for download are learning and teaching resources created for *Introduction to Genetics: A Molecular Approach*.

The following resource is available to students and can be downloaded from the book's product page: www.routledge.com/9781032743530

- **Hints and Answers to In-book Questions**

    Freely available for students are the answers to, or hints on how to answer, the Self-study Questions and Discussion Topics found at the end of each chapter in the book.

The following resources are available only to qualified instructors and can be accessed at the following link, where instructors can request access: https://routledgetextbooks.com/textbooks/instructor_downloads/

- **Test Bank of Multiple Choice Questions**

    New to this edition, the author has written a series of MCQs that relate to the book's content. MCQs have been written for all chapters except Chapter 1, which is more introductory in nature. Instructors may choose to use these MCQs as assessments of the courses they teach.

- **Lecture Outlines**

    The section headings, concept subheadings, and figures from the book have all been integrated into PowerPoint presentations.

- **Figure Slides**

    The high-resolution images are available to download in two convenient formats: PowerPoint and JPEG. Instructors may wish to make use of these in their lecturers.

What Is Genetics and Why Is It Important?

CHAPTER 1

Genetics is the study of **genes**. The first geneticists looked on genes as **units of inheritance**. They studied the way in which genes are transmitted from parents to their offspring during reproduction and also explored how different genes act together to control characteristics such as height and eye color.

Today, we look on genes as **units of biological information**. The entire set of genes possessed by an organism contains the total amount of information needed to construct a living, functioning example of that organism. Geneticists in the 21st century study the way in which information is stored in genes and how that information is utilized by cells and organisms.

Genetics has also become an **applied science** that is important in areas as diverse as industry, agriculture and medicine. **Genetic engineering** is being used to produce pharmaceuticals and vaccines and to improve the nutritional value and productivity of the world's crops. **Gene therapy** is being developed as a means of treating diseases, and **DNA profiling** is used to catch criminals. Genetics is even being applied to archaeology, enabling us to study human evolution and the migrations of prehistoric people.

1.1 THE ORIGINS OF GENETICS

Geneticists, unlike scientists in some other disciplines, have a keen awareness of the history of their subject. This is for two reasons. First, the history of genetics is inherently interesting, especially during the second half of the 20th century when a series of key discoveries drove a rapid advance in our knowledge of the structure and activity of genes. The history of genetics is also important because the key discoveries followed a logical progression, and by following that progression we can see how the strands that make up genetics today are linked together. The diverse roles of genes in inheritance, cellular function, and applied science can be bewildering if these topics are simply studied one after the other without any reference to how the topics are interconnected. The history of genetics reveals those connections and makes an understanding of modern genetics much easier to achieve. We will therefore begin by examining the historical and intellectual backdrop against which our more recent endeavors to understand the gene have been set.

Genetics begins with Mendel

The origin of genetics is linked inexorably with Gregor Mendel (Figure 1.1), in particular the experiments he carried out in 1856–1864 on the inheritance of variable features such as height, flower color and seed shape in garden peas. Mendel was not the first person to study the inheritance of variable features. It is an obvious fact of nature that children are not exact images of their mother or father, but instead are a composite of their two parents. This means that siblings look similar to one another but are not identical unless they are twins (Figure 1.2). Scientists and naturalists going back as far as the Ancient Greeks had proposed theories to explain how characteristics are inherited and how these characteristics blend together in different ways in the offspring. However, these theories imagined vague processes such as the mixing of fluids. Mendel is hailed as a scientific hero because he was the first person to state that inheritance is controlled

FIGURE 1.1 Gregor Mendel, the founder of genetics.

FIGURE 1.2 **Members of a single family share biological characteristics.** The American rock band Haim is made up of three sisters: Este, Danielle and Alana. Their family resemblance is clear. (From Raph_PH, on Wikimedia Commons, published under CC BY 2.0 license.)

by genes. He never used the word 'gene', but the '**unit factors**' that he described in the two papers he published in 1866 are the same things that today we call genes. The way in which these unit factors are inherited and expressed in offspring, which Mendel deduced from his experiments, is precisely the way in which genes are inherited and expressed. Mendel's deductions were so correct that we still study his experiments as the best way of understanding the gene as a unit of inheritance.

Gregor Mendel was an Augustinian friar who eventually became abbot of St Thomas' Abbey in Brno, which is now a city in the Czech Republic but in the mid-19th century was part of the Austrian Empire. This does not mean that Mendel had no scientific training. Although his parents were peasant farmers, he displayed a keen intelligence at an early age. His village schoolmaster arranged a place for him at the Gymnasium in Troppau (now Opava) and one of his three sisters used part of her dowry to allow him to continue his studies at the Olmutz Philosophical Institute. He became a friar in 1843, but continued his education by attending Vienna University for four terms between October 1851 and August 1853. He carried out his breeding experiments in the Abbey gardens and greenhouses (Figure 1.3) and continued his work until 1871 when his duties as abbot effectively curtailed his scientific career. Unfortunately, his later experiments were with hawkweed, which proved much less easy to study than the pea.

FIGURE 1.3 **Mendel's garden.** Mendel conducted his experiments in a greenhouse and in these outdoor plots at St Thomas' Abbey in Brno. (From Science Photo Library, with permission.)

It is often suggested that Mendel's paper was 'lost' because it was published in an obscure journal by an unknown monk in a remote monastery, but this is not really consistent with the facts. The *Proceedings of the Society of Natural Sciences in Brno* was sent to at least 55 European libraries and learned societies, including the Royal Society and Linnean Society in London. Mendel also sent reprints of his paper to many of the leading botanists of the day, including Professor Kerner of the University of Innsbruck (who evidently did not read it as the reprint was found unopened after his death) and Professor Nageli of Munich, with whom Mendel corresponded regularly between 1866 and 1873. Mendel's work was noticed to the extent that he or his paper was mentioned in 16 publications in the late 19th century, including the 9th edition of the *Encyclopaedia Britannica*. However, Mendel remained until his death the only person who understood and appreciated the true significance of his work. He was simply too far ahead of his time and other biologists needed another 35 years to catch up with his way of thinking. Eventually, in 1900, Hugo De Vries, Carl Correns and Erich von Tschermak each conceived the idea of experiments similar to Mendel's to test their own theories of heredity. Each performed their experiments and then, when studying the literature before publishing their results, each independently discovered Mendel's paper. After a lengthy gestation, the science of genetics was finally born.

From genes to chromosomes

Mendel's experiments, and those of biologists such as De Vries, Correns and Tschermak, removed the mystique from heredity and showed that the process follows predictable rules. These rules describe the passage of physical factors, each controlling a separate heritable trait, from the parents to offspring during reproduction. These factors went under a variety of names until 1909 when W. Johannsen suggested that they be called 'genes'. By this time, it was understood that the genes were located on **chromosomes** (Figure 1.4). This became clear when it was shown that the transmission of chromosomes during cell division and reproduction exactly parallels the behavior of genes during these events. The chromosome theory was stated in its most convincing form in 1903 by W.S. Sutton, who at that time was a graduate student at Columbia University in New York and who subsequently became a surgeon before his early death in 1916.

Once the experimental basis of heredity had been established and the chromosome theory accepted, the way was open for a rapid advance in the understanding of genetics. That this advance occurred as rapidly as it did was mainly the result of the intuition and imagination of Thomas Hunt Morgan and the members of his research group at Columbia University, New York, notably Calvin Bridges, Arthur Sturtevant and Hermann Muller. Morgan and his colleagues achieved something that many biologists dream about: they discovered an organism that was ideally suited for the particular research program that they wished to carry out. The organism was *Drosophila melanogaster*, the fruit fly (Figure 1.5). *Drosophila* possesses several features that make it very suitable for genetic analysis, but most important from Morgan's point of view was the fact that a large number of stable variant forms of the fly could be obtained. The differences between these variants involve features such as wing shape and eye color, some traits having many different varieties (Figure 1.6). Morgan showed that each of these varieties is specified by a different gene. Between 1911 and 1929, his group developed many of the techniques that have now become standard methods in genetic analysis, including those that map the relative positions of different genes on a chromosome.

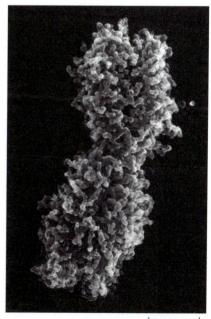

FIGURE 1.4 A scanning micrograph of a typical human chromosome.
(From S. Inaga, K. Tanaka & T. Ushiki (eds), *Chromosome Nanoscience and Technology*, 2007. With permission from CRC Press.)

FIGURE 1.5 The fruit fly *Drosophila melanogaster*. The fruit fly is an important model organism in many areas of biology, including genetics. (Courtesy of Nicolas Gompel, University of Bonn.)

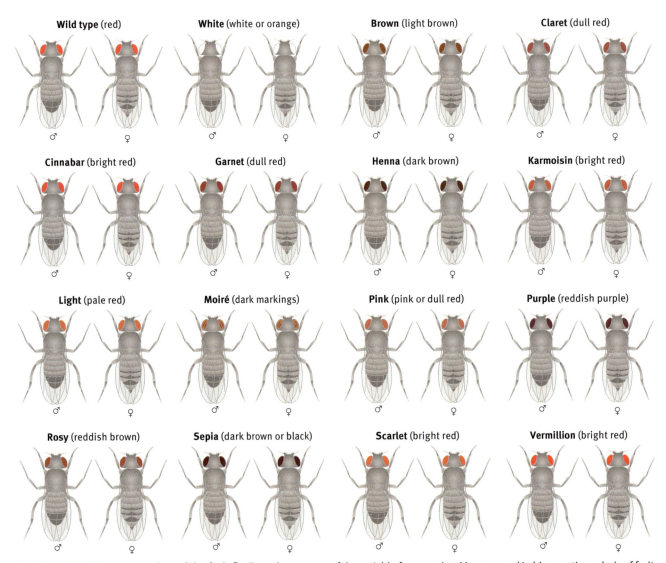

FIGURE 1.6 Different eye colors of the fruit fly. Eye color was one of the variable features that Morgan used in his genetic analysis of fruit flies.

Genes are made of DNA

The rediscovery of Mendel's work and the remarkable advances made by Morgan and his colleagues attracted the attention of many scientists not actively engaged in genetic research. These included a group of physicists who became fascinated by the possibility that by studying biological processes they might be able to describe life itself in purely physical terms. Prominent in this group was Max Delbrück, who decided that the life cycle of **bacteriophages** (**viruses** that attack **bacteria**) could be used as a simple experimental system with which to tackle the question of what genes are and how they work (Figure 1.7). In 1940, he set up the 'Phage Group', an informal association of physicists, biologists and chemists, working in different laboratories but all with a common interest in the gene.

The most important contribution that the Phage Group made to genetics was the discovery that genes are made of DNA. This conclusion had already been reached by Oswald Avery and his colleagues Colin MacLeod and Maclyn McCarty, at the Rockefeller Institute in New York, following a lengthy research project with *Streptomyces pneumoniae*, a bacterium that causes one type of pneumonia. Avery's group showed that an extract prepared from virulent, disease-causing bacteria was capable of converting harmless versions of *S. pneumonia* into the virulent form. The extract, which Avery called the **transforming principle**, must contain the genes for virulence, which when taken up by the harmless bacteria transform these into virulent ones. When Avery's group did biochemical tests on the transforming principle, they found that the active component – the part containing the genes – was DNA. This was a surprising discovery as DNA was not looked on as a strong candidate for the genetic material: it was thought that genes were much more likely to be made of protein. There are some suggestions that the Phage Group intended to show that Avery had got the wrong result, but the key experiments, carried out by Alfred Hershey and Martha Chase (Figure 1.8) at the Cold Spring Harbor Laboratory in 1952, confirmed that Avery was in fact correct and genes are indeed made of DNA.

In 1952, scientists only had a partial understanding of DNA structure. Chemical studies had shown that DNA is a **polymer** of four different **nucleotides**, which we refer to as A, C, G and T, the abbreviations of their full names. A single DNA molecule can contain thousands or millions of nucleotides joined end to end, in any order. If genes are made of DNA, then biological information must be coded in DNA molecules in the form of a four-letter language made up of A, C, G and T. But it was very difficult to imagine how DNA molecules could make exact replicates of themselves, as required in order for a parent's genes to be copied before being passed to offspring. There must be something not yet discovered about DNA structure that would explain how the nucleotide polymers could replicate. It was thought that the answer might be found by carefully extracting DNA from living cells and then studying purified samples by **X-ray diffraction** analysis. This method had originally been developed for working out the structures of crystallized substances, but had been adapted for use with fibers such as silk and wool. It was first used successfully with DNA fibers in 1939 by Florence Bell, a PhD student at the University of Leeds, UK. Florence Bell was not the first important female scientist, but in 1939 the fact that women could be scientists was still a novel idea for many men. There is a famous photograph from the Yorkshire Evening News of 23rd March 1939 in which 'Miss Florence Bell, a 25-year-old graduate of Cambridge University' is shown describing her work at a conference on industrial physics to a group of male onlookers, under the banner headline 'Woman scientist explains' (Figure 1.9).

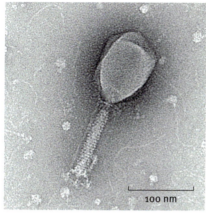

FIGURE 1.7 T2, a typical bacteriophage. T2 is a bacteriophage that infects the bacterium *Escherichia coli*. The bacteriophage was first isolated from sewage in 1932. As well as the head and tube structures seen in this electron micrograph, T2 also has a set of fibers or 'legs' attached to the base of the tube. (Courtesy of SnaxMikn, on Wikimedia Commons, published under CC BY-SA 4.0 license.)

FIGURE 1.8 Martha Chase and Alfred Hershey. This photo was taken in 1953, shortly after Chase and Hershey had shown that genes are made of DNA. Alfred Hershey (but not Martha Chase) was awarded the 1969 Nobel Prize in Physiology or Medicine. (Courtesy of Karl Maramorosch.)

Although Bell's X-ray diffraction results were the clearest that had been obtained with DNA at that time, they were not detailed enough to enable her to deduce a definite structure. In fact, she had not known that the DNA sample she had been given was a mixture of two types of DNA, called the A- and B-forms, and her diffraction results combined the features of both types. It was not until 1952, when Rosalind Franklin and Raymond Gosling, at King's College London, subjected a pure sample of the B-form to X-ray analysis, that it became possible to deduce the structure of this type of DNA. However, these deductions require a complicated analysis of the patterns of spots and swirls present in the diffraction pattern, and although Franklin came very close to working out the structure, it was James Watson and Francis Crick of the University of Cambridge, who took the final steps and discovered the famous double helix. In the double helix, two strands of DNA are wound around one another, with A nucleotides in one strand pairing only with Ts in the other strand, and Cs with Gs. This pairing provides an obvious means by which DNA molecules can be copied into exact replicates (Figure 1.10). The lingering doubts were finally settled and it was accepted that genes are indeed made of DNA.

The gene as a unit of biological information

The discovery of the double helix structure was followed by a period of intensive research, involving many geneticists in all parts of the world, directed at understanding how biological information is stored in genes and how that information is made available to the cell. It was quickly established that **gene expression** is a two-step process, with the gene

FIGURE 1.9 Florence Bell. From the Yorkshire Evening News of 23rd March 1939. In a report published on the previous day, as well as her age the newspaper felt it important to mention that Miss Bell was slim. (Reproduced with the permission of Special Collections, Leeds University Library, Brotherton Collection, MS 419/A.1.)

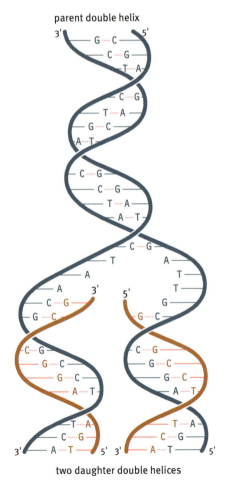

FIGURE 1.10 The DNA double helix. In the double helix structure, two strands of DNA are wound around one another, with A nucleotides in one strand pairing only with Ts in the other strand, and Cs with Gs. This pairing provides a means for replicating the DNA molecule. As shown here, the separation of the strands of the parent DNA (shown in blue), followed by the synthesis of new strands that follow the pairing rules (shown in orange), results in two identical double helices.

first copied into an **RNA** molecule (Figure 1.11). DNA and RNA are very similar types of molecule and this step of gene expression, called **transcription**, is quite straightforward in chemical terms. In the second step, the nucleotide sequence of the gene, now contained in its RNA copy, is **translated** into the amino acid sequence of a protein. Deciphering the **genetic code** that is used during this **translation** process was the major achievement of genetics research during the 1960s. By 1980, most of what we now know about gene expression had been discovered, although even today important new details are still being added.

An important part of this huge research effort was focused on understanding how individual genes are switched on and off. It was recognized that even the simplest organisms must be able to control the activity of their genes in order to respond to changes in the environment. Many bacteria, for example, deal with sudden increases in temperature by switching on a set of genes that code for proteins that help to protect the cell from damage (Figure 1.12). In multicellular organisms, individual cells change their gene activity patterns in response to hormones, growth factors and other regulatory molecules. These chemical signals are therefore able to coordinate the activities of groups of cells in a manner that is beneficial to the organism as a whole. The foundation of our understanding of gene regulation was laid by François Jacob and Jacques Monod, at the Institut Pasteur in Paris, in a paper that they published in the *Journal of Molecular Biology* in 1961. Jacob and Monod carried out an intricate genetic analysis of the way in which lactose is used as a source of energy by the bacterium *Escherichia coli*. They identified three genes that code for the enzymes needed to import lactose molecules into the bacterium and convert these into two other sugars, glucose and galactose (Figure 1.13). Jacob and

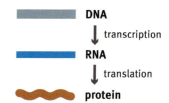

FIGURE 1.11 Gene expression. In its simplest form, gene expression can be looked on as a two-step process. The first step is the transcription of DNA into RNA, and the second step is the translation of some of the RNA molecules into protein.

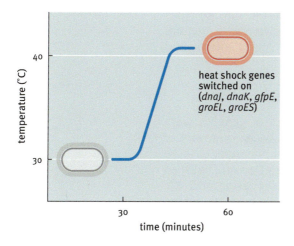

FIGURE 1.12 The heat shock response of a bacterium such as *Escherichia coli*. An increase in temperature from 30°C to 42°C results in genes such as *dnaJ* and *dnaK* being switched on. The proteins specified by these genes help to protect the cell from damage.

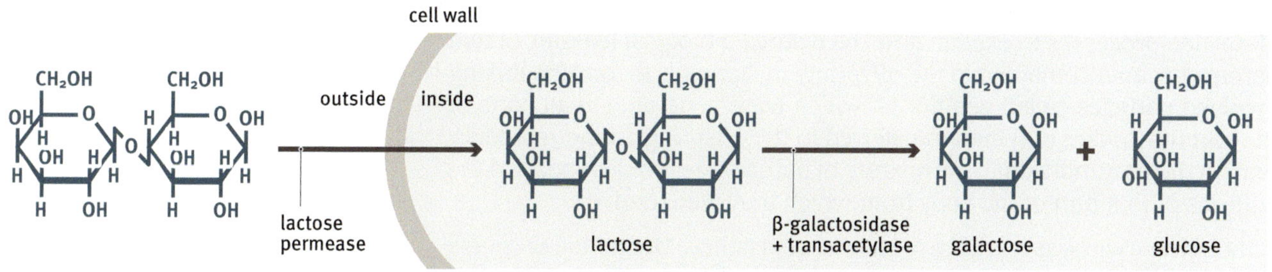

FIGURE 1.13 The utilization of lactose by *E. coli*. In order to utilize lactose as an energy source, the *E. coli* cell must import lactose molecules from the environment and then split the molecules into their glucose and galactose components. Import requires the enzyme called lactose permease, and splitting the lactose molecules involves the combined activity of β-galactosidase and β-galactosidase transacetylase.

Monod also identified a fourth gene that regulates the activity of the other three and deduced exactly how this gene responds to the amount of lactose that is available so that the genes can be switched off when no lactose is present. Although lactose utilization by *E. coli* is a relatively simple process, it illustrates the basic principles of gene regulation in all organisms including the most complex ones such as humans.

The final question that was studied in the years following the discovery of the double helix is how DNA replication is carried out. The copying process that appeared obvious from the structure of the double helix was confirmed by a single experiment carried out in 1958, and the various enzymes responsible for DNA replication were isolated during the 1950s and 1960s. These included the **DNA polymerases**, the enzymes that make the new DNA copies. The first of these enzymes to be discovered was **DNA polymerase I** of *E. coli*, which was purified by Arthur Kornberg of Washington University, St Louis in 1957. Naturally, it was assumed that this enzyme was responsible for DNA replication in the bacterial cell. Surprisingly, as the enzyme was studied in greater detail facts inconsistent with this role began to emerge. The most problematic of these arose with the discovery in 1969 of a type of *E. coli* in which the gene specifying DNA polymerase I was inactive. If this enzyme is responsible for DNA replication, then we would not expect these bacteria to be able to divide, but they appeared to be more or less normal and certainly were able to replicate their DNA. We know now that a different enzyme, called **DNA polymerase III**, which was not isolated until 1972, is primarily responsible for DNA replication in *E. coli*. So what is the role of DNA polymerase I? This enzyme has the equally important task of repairing DNA molecules that contain structural alterations called **mutations**. These are caused by various chemicals present in the environment and are also brought about by physical agents such as ultraviolet radiation from the sun. All types of organism have repair processes, involving DNA polymerases, that are able to correct most of these mutations so that the biological information contained in the genes can still be read.

Parallel developments in evolutionary genetics

During the period when the fundamental principles of genetics were being explored, other biologists were examining how the discovery of genes impacted their own areas of interest. The most important of these areas concerns the role of genes in evolution. Charles Darwin published his theory of evolution by natural selection in *Origin of Species* in 1859, a few years before Mendel presented the results of his work describing the role of genes in inheritance. It appears that Darwin never became aware of Mendel's work, and until his death in 1882 favored a theory of inheritance that he referred to as 'pangenesis'. This was another of the schemes involving mixing that was put forward by 19th-century biologists who, without the benefit of Mendel's insights, had to resort to rather ill-defined processes to explain how the distinct biological features of two parents became combined in the offspring. In Darwin's theory, the mixing involved particles called gemmules, which were generated in all parts of the parents' bodies and then transferred to the offspring during reproduction. In the new individual, the mixture of parental gemmules specified the features of that part of the body from which they were derived.

Although Darwin was unaware of Mendel's existence, Mendel did of course know about Darwin's theories. As Darwin was the most talked about biologist of his time, Mendel could hardly fail to hear about the new theory of evolution. There has been a lot of speculation about Mendel's views on evolution: a part of his faith as a friar would be a belief in divine creation.

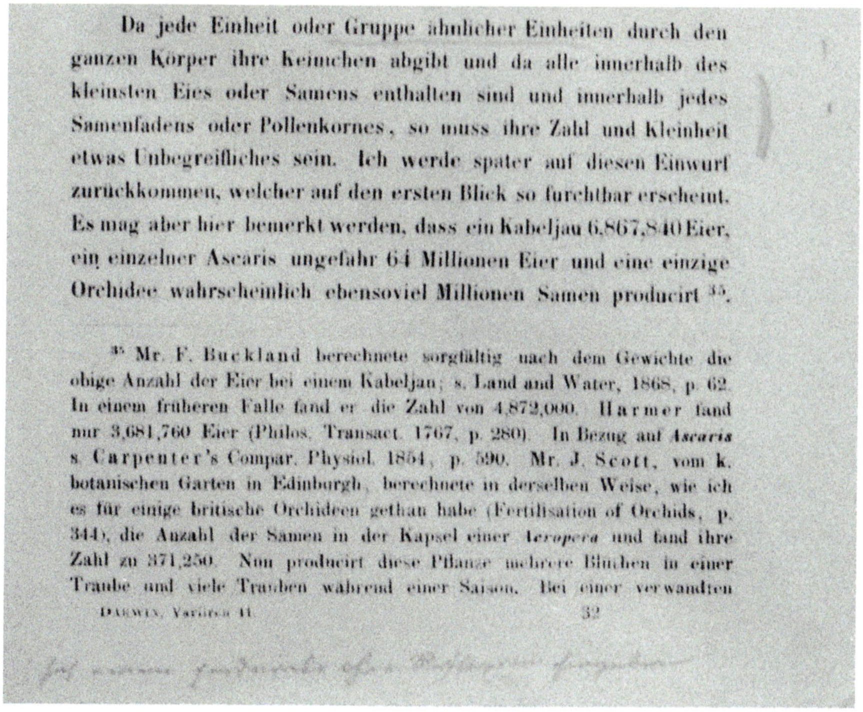

Da jede Einheit oder Gruppe ähnlicher Einheiten durch den ganzen Körper ihre Keimchen abgibt und da alle innerhalb des kleinsten Eies oder Samens enthalten sind und innerhalb jedes Samenfadens oder Pollenkornes, so muss ihre Zahl und Kleinheit etwas Unbegreifliches sein. Ich werde später auf diesen Einwurf zurückkommen, welcher auf den ersten Blick so furchtbar erscheint. Es mag aber hier bemerkt werden, dass ein Kabeljau 6,867,840 Eier, ein einzelner Ascaris ungefähr 64 Millionen Eier und eine einzige Orchidee wahrscheinlich ebensoviel Millionen Samen producirt [35].

[35] Mr. F. Buckland berechnete sorgfältig nach dem Gewichte die obige Anzahl der Eier bei einem Kabeljau; s. Land and Water, 1868, p. 62. In einem früheren Falle fand er die Zahl von 4,872,000. Harmer fand nur 3,681,760 Eier (Philos. Transact. 1767, p. 280). In Bezug auf *Ascaris* s. Carpenter's Compar. Physiol. 1854, p. 590. Mr. J. Scott, vom k. botanischen Garten in Edinburgh, berechnete in derselben Weise, wie ich es für einige britische Orchideen gethan habe (Fertilisation of Orchids, p. 344), die Anzahl der Samen in der Kapsel einer *Acropera* und fand ihre Zahl zu 371,250. Nun producirt diese Pflanze mehrere Blüthen in einer Traube und viele Trauben während einer Saison. Bei einer verwandten

DARWIN, Variiren II. 32

FIGURE 1.14 Mendel's thoughts on Darwin's theory of pangenesis. This is page 497 of Mendel's copy of *The Variation of Animals and Plants Under Domestication* by Charles Darwin. The text marked by Mendel describes part of Darwin's pangenesis theory. Mendel's handwritten note in German, in the bottom margin, is "sich einem Eindrucke ohne Reflexion hingeben", which in English means "to surrender to an impression without reflection". (From D. J. Fairbanks, *Heredity* 124:263–273, 2020. Springer Nature, published under a CC BY 4.0 license.)

What is clear is that Mendel made a detailed study of Darwin's theory, and in particular owned copies of both *Origin of Species* and Darwin's second book, *The Variation of Animals and Plants Under Domestication*. In the margins of these books, he made various notes, some of these critical of the pangenesis theory and others pointing out where aspects of inheritance that Darwin considered intractable were easily explained by what Mendel knew about genes (Figure 1.14).

When Mendel's results were rediscovered, evolutionary biologists at first struggled to reconcile how inheritance controlled by genes could explain the continuous patterns of variation seen in natural populations. It was not until the 1940s that a group of biologists proposed what is called the **Modern Synthesis**, which linked together Mendelian genetics, natural selection, and population variation into a single consistent explanation of evolutionary genetics. Even then the controversies were not over, and as recently as 1980 the influential American biologist Stephen Jay Gould declared that the Modern Synthesis is 'effectively dead'. Other biologists leapt to its defense but the debate continues today.

Genetic engineering and the new biotechnology

The research that was carried out in the 1950s and 1960s identified a number of enzymes that are involved in gene expression and in the replication and repair of DNA molecules. In addition to the DNA polymerases that make copies of DNA molecules, there are also enzymes that cut DNA into segments and others that can join those pieces together again. Once these enzymes had been purified, the possibility that DNA molecules could be restructured and copied in the test tube became a possibility. This methodology, which we call **recombinant DNA technology**, provided an entirely new way of studying genes and gave a further boost to efforts to understand the gene expression pathway.

Recombinant DNA technology also made genetic engineering possible. Genetic engineering involves changing an organism's DNA in order to alter its biological information in some way. In an early example, the

FIGURE 1.15 **Genetic engineering at the industrial scale.** Part of a biotechnology complex for the production of human insulin from genetically engineered *Escherichia coli* bacteria. (Photo by Denis Félix. With permission from Sanofi-Aventis.)

human gene that specifies synthesis of the growth hormone somatotropin was transferred to *E. coli*, so that the bacterium now made the growth hormone. The hormone is used to treat growth disorders, especially in children. Originally, it was obtained from the pituitary glands of dead people, which meant that only limited amounts were available. By engineering bacteria to make somatotropin, supplies could be increased and growth hormone treatment made more readily available. This, project, completed in 1979, was an example of **biotechnology**, the industrial use of biological processes (Figure 1.15). The company that carried out the project, Genentech, was one of the first to make use of recombinant DNA techniques.

During the 1990s, more ambitious genetic engineering projects were attempted, including several aimed at the improvement of crop plants by transferring new genes into their chromosomes. One example of what we called a **genetically modified** (**GM**) **plant** is Golden rice (Figure 1.16). Golden rice has been genetically modified to synthesize increased amounts of β-carotene, a precursor of vitamin A, and is designed for consumption in parts of the world where people suffer from vitamin A deficiency. The modification was achieved by transferring two genes, one from daffodil and the second from the soil bacterium *Erwinia uredovora*, into rice.

FIGURE 1.16 **Golden rice.** Golden rice, on the right, compared with unmodified white rice. Golden rice has been genetically modified to synthesize increased amounts of β-carotene, a precursor of vitamin A. Golden rice is designed for consumption in parts of the world where people suffer from vitamin A deficiency. (Courtesy of Golden Rice Humanitarian Board [www.goldenrice.org].)

FIGURE 1.17 Emmanuelle Charpentier (left) and Jennifer Doudna (right), the inventors of gene editing. (Courtesy of Hallbauer & Floretti.)

More recent applications of genetic engineering have taken a new approach, one that avoids the need to transfer one or more new genes into the organism that is being modified. In this new technique, called **gene editing**, the modification is brought about by making one or more very specific changes to the nucleotide sequence of a single gene, the changes intended to alter the function of the gene so the characteristic it specifies is modified in the desired manner. Gene editing was invented in 2012 by Emmanuelle Charpentier of the Max Planck Institute, Berlin and Jennifer Doudna of the University of California, Berkeley (Figure 1.17).

Genetic engineering has been a controversial aspect of genetics. In fact, the first genetic engineers were so concerned about the possible dangers of an engineered bacterium escaping into the environment that they voluntarily halted their work until a scientific conference could be held to discuss the hazards and how these ought to be avoided. That conference, held in Asilomar, California, in February 1975, was important not just because it established a set of guidelines that enabled the new technology to continue to develop in a safe way. It was also important because it stimulated public debate about the social and environmental issues relating to genetics. Those debates have broadened and continued to the present day, to the extent that **bioethics** has emerged as a distinct subdiscipline of biology.

From genes to genomes

The most important recombinant DNA technique is **DNA sequencing**, in which the nucleotide sequence of a piece of DNA is worked out. In the early 1970s, several possible ways of sequencing DNA were explored, but one method stood out as much more efficient than all the others. This was called **chain termination sequencing** and was invented by Frederick Sanger and colleagues at the University of Cambridge, UK. After its introduction in 1977, chain termination sequencing was quickly adopted by genetics researchers around the world. Among the first pieces of DNA to be sequenced was the segment of human chromosome 11 containing the genes for the β-globin proteins (Figure 1.18). These are the proteins that combine with α-globins (whose genes are on chromosome 16) to make hemoglobin, which is present in red blood cells and carries oxygen around the bloodstream.

```
ATGGTGCATCTGACTCCTGAGGAGAAGTCTGCCGTTACTGCCCTGTGGGGCAAGGTGAAC
GTGGATGAAGTTGGTGGTGAGGCCCTGGGCAGGTTGGTATCAAGGTTACAAGACAGGTTT
AAGGAGACCAATAGAAACTGGGCATGTGGAGACAGAGAAGACTCTTGGGTTTCTGATAGG
CACTGACTCTCTCTGCCTATTGGTCTATTTTCCCACCCTTAGGCTGCTGGTGGTCTACCC
TTGGACCCAGAGGTTCTTTGAGTCCTTTGGGGATCTGTCCACTCCTGATGCTGTTATGGG
CAACCCTAAGGTGAAGGCTCATGGCAAGAAAGTGCTCGGTGCCTTTAGTGATGGCCTGGC
TCACCTGGACAACCTCAAGGGCACCTTTGCCACACTGAGTGAGCTGCACTGTGACAAGCT
GCACGTGGATCCTGAGAACTTCAGGGTGAGTCTATGGGACGCTTGATGTTTTCTTTCCCC
TTCTTTTCTATGGTTAAGTTCATGTCATAGGAAGGGGATAAGTAACAGGGTACAGTTTAG
AATGGGAAACAGACGAATGATTGCATCAGTGTGGAAGTCTCAGGATCGTTTTAGTTTCTT
```

FIGURE 1.18 Part of the DNA sequence for the human β-globin gene. The entire β-globin gene is 1,423 nucleotides in length, of which 600 are shown here.

The segment of DNA containing the human β-globin genes is about 45,000 nucleotides in length. A single chain termination sequencing experiment can only generate a maximum of 800 nucleotides of sequence, so to sequence the entire β-globin region it is necessary to break the DNA into overlapping fragments and sequence each one in a separate experiment. This requires a huge amount of work, but the development of robotic systems for preparing the DNA and performing the sequencing meant that by the 1990s, it was possible for a single laboratory to generate millions of nucleotides of DNA sequence in a single day. This factory-like approach made it feasible to sequence the **genomes** of different organisms. The genome is the entire collection of DNA molecules that an organism possesses and hence includes a copy of every single gene. A few genomes had been sequenced in the 1970s and 1980s, but these were virus genomes, which are much smaller than the genomes of organisms such as bacteria, animals and plants. Sanger's group, for example, sequenced the genome of bacteriophage λ in 1982, but this DNA molecule is only 48,502 nucleotides in length. The simplest bacteria, in comparison, have genomes made up of over a million nucleotides. Despite their size, bacterial genomes proved to be relatively easy to sequence using the factory approach, and once the feasibility of the method had been demonstrated with *Haemophilus influenzae*, whose 1,830,137 nucleotide genome was completed in 1995, a rapidly increasing number of other bacterial genomes were published: today, we have genome sequences for over 17,000 species.

A much greater challenge is provided by the genomes of animals and plants, as these are much larger than the genomes of bacteria. But factory sequencing was so powerful an approach that the genome of the nematode worm, *Caenorhabditis elegans* (Figure 1.19) was published in 1998, despite this genome being just over 100 million nucleotides in length. The first plant genome, for *Arabidopsis thaliana*, with 115 million nucleotides, was completed in 2000. But the most heroic of these endeavors was the Human Genome Project, which was conceived in the late 1980s and reached its initial goal – a sequence covering 90% of the 3,100 million nucleotides that make up the complete genome – in 2000.

FIGURE 1.19 The nematode worm *Caenorhabditis elegans*. This was the first animal whose genome was completely sequenced. The adult worm is about 1 mm in length and lives in soil and rotting vegetation in temperate regions of the world. (From Steve Gschmeissner/Science Photo Library, with permission.)

Full circle: the molecular approach to genetic variation

During the 2000s, new methods were developed that enabled much greater amounts of DNA sequence to be obtained than is possible with the chain termination technique (Figure 1.20). As a result, sequencing a human genome is now a trivial undertaking, so much so that thousands of genomes have been sequenced from people living all over the world. These genomes are not identical: the sequences have differences that underlie the variations in physical and biochemical features possessed by individual people. Genetics has therefore come full circle. Mendel's objective was to understand the biological basis of the variable features displayed by peas and other plants that he studied. Those features were

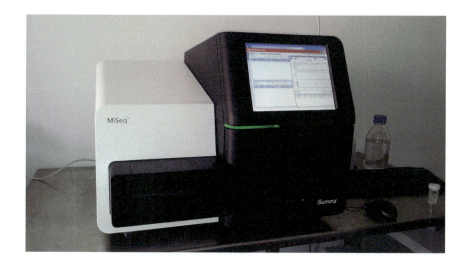

FIGURE 1.20 **A modern DNA sequencing device.** The MiSeq sequencer, marketed by Illumina, is able to generate up to 15×10^9 nucleotides of sequence (equivalent to five entire human genomes) in a 56-hour operating period. In today's genetics, that is mid-range performance: the Illumina NovaSeq, for example, can sequence 128 human genomes in a single run. (Courtesy of Oxford Nanopore Technologies.)

characteristics such as the height of the plants and the color of their flowers. Today's geneticists are similarly interested in genetic variations and seek to link the sequence variations in the human genome with features such as susceptibility to disease. In that way, we hope to develop therapies for cancers and for the many inherited disorders that afflict the human population.

The variations in the human genome are also the key to the use of DNA in forensic science. As no two human genomes are identical, unless they belong to identical twins, the variable features of a genome are characteristic of the person who possesses that genome. Traces of DNA obtained from a crime scene can therefore be used to identify individuals who were present when the crime was committed. The original method used for this purpose was called **genetic fingerprinting** and was invented by Alec Jeffreys of the University of Leicester, UK, in 1984. During the 1990s, this was superseded by a more powerful technique, **DNA profiling**, which is able to work with smaller amounts of DNA and hence provides a better chance of obtaining results with the traces present at a crime scene. This is partly because DNA profiling makes use of the **polymerase chain reaction** (**PCR**), a remarkable procedure that enables many identical copies to be made of a single DNA molecule. PCR was invented by Kary Mullis of the Cetus biotechnology company, one Friday evening in 1983 during a drive along California State Route 128 from Berkeley to Mendocino (Figure 1.21), and has become a central component of many areas of genetics research. In particular, it has made possible a new discipline called **archaeogenetics**, in which traces of **ancient DNA** preserved in bones and other archaeological specimens are used to study prehistoric human populations. Now the variations in the human genome are used to identify the relationships between different groups of people living thousands of years ago and to follow the movements of prehistoric populations as humans left Africa and gradually spread through Asia, Europe and the Americas.

FIGURE 1.21 **California State Route 128 in Mendocino County, where PCR was invented.**

1.2 HOW THIS BOOK IS ORGANIZED

We have completed our historical survey of the way in which genetics developed into the sophisticated and multifaceted science that it has become today. Now we are ready to study the subject in depth. How to begin? The traditional way to learn genetics is to follow the historical journey, by first studying everything there is to know about the gene as a unit of

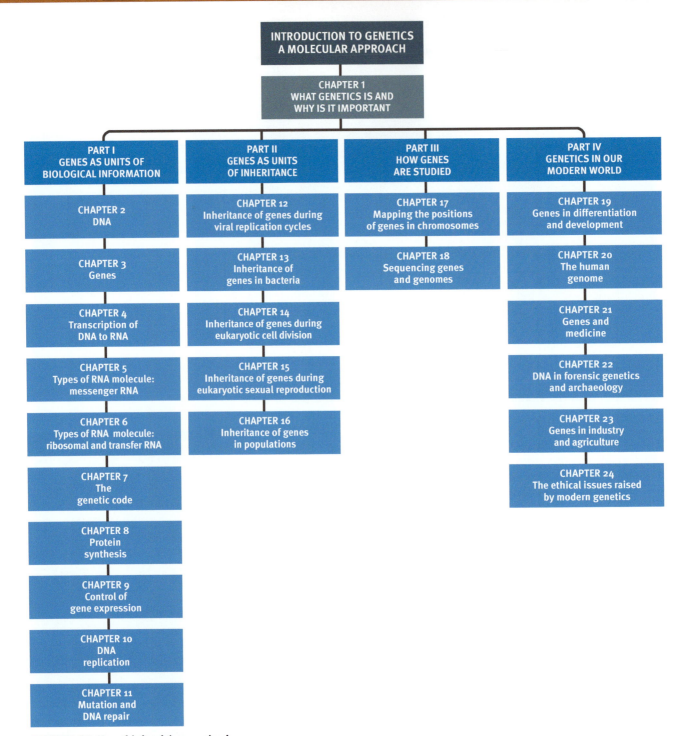

FIGURE 1.22 How this book is organized.

inheritance, and only then moving on to the modern interpretation of the gene as a unit of biological information. In this book, we take a different approach. Genetics today is centered on DNA, and most geneticists study DNA rather than heredity. We will therefore begin our study of genetics with DNA and gene expression, and only after those topics have been thoroughly explored will we move on to consider how genes act in inheritance. We will discover that this molecular approach makes it much easier to understand the complexities of inheritance, not just the way in which the characteristics coded by genes are passed from parents to offspring, but also the way in which different genes interact with one another to specify the complex biological features of humans and other organisms.

This book is divided into four parts (Figure 1.22). In Part I, we examine genes as units of biological information. First, we will become familiar with the structure of DNA (Chapter 2) and the way in which genes are organized within DNA molecules (Chapter 3). Then, we will be ready to follow the process of gene expression, which results in the information contained in a gene being utilized by the cell. We will study the way in which DNA is copied into RNA (Chapter 4), and we will look in detail at the roles in the cell of the different types of RNA molecule that are made (Chapters 5 and 6). We will then discover how the genetic code is used to direct the synthesis of protein molecules whose structures and functions are specified by the information contained in the genes (Chapters 7 and 8). We will examine how all of these events are controlled so that only those genes whose information is needed are active at a particular time (Chapter 9). Finally, in Chapters 10 and 11, we will study how DNA is replicated and how mutations can arise and be repaired.

In Part II, we will study genes as units of inheritance. To do this, we will ask three questions. The first is how a complete set of genes is passed to the daughter cells when the parent cell divides. We will examine this process during the infection cycles of viruses (Chapter 12), when a bacterium divides (Chapter 13), and when an animal or plant cell divides (Chapter 14). The second question concerns the inheritance of genes during sexual reproduction. We will study how DNA molecules are passed from parents to their offspring, and we will investigate how the genes on these DNA molecules specify the biological characteristics of the offspring in such a way that they resemble, but are not identical to, their parents (Chapter 15). The third question concerns the link between the inheritance of genes and the evolution of species. To answer this question, we will examine how the new gene variants that are created by mutation can spread through a population (Chapter 16).

Part III is devoted to the important research techniques that are used in modern genetics. In Chapter 17, we will explore the ways in which the positions of genes are mapped on the chromosomes of animals, plants and bacteria, including the special approaches that must be taken with humans. Then, in Chapter 18, we will study how DNA molecules are sequenced.

In Part IV, we will explore some of the areas of research that are responsible for the high profile that genetics has in our modern world. The first topic that we will study is the role of genes in development. How do genes control the pathway that begins with a fertilized egg cell and ends with an adult organism? Finding the answer to this question is one of the biggest challenges in all of genetics (Chapter 19). Then, we will devote three chapters to our own species. In Chapter 20, we will look closely at the human genome, and in particular we will ask what it is about our genome that makes us special. Then, we will look at the ways in which alterations to genes can give rise to cancer and to inherited disorders such as cystic fibrosis (Chapter 21). In Chapter 22, we will examine how DNA profiles are obtained and why these have become so important in forensic biology. We will also learn how the presence of ancient DNA in some archaeological materials is enabling genetics to be used to study the prehistory of our species. We will then look at the applications of genetics in industry and agriculture, in the production of important pharmaceuticals and in the design of GM crops (Chapter 23). This will lead us to some of the controversial aspects of modern genetics. We must not ignore these controversies, and so in Chapter 24 we will examine some of the ethical issues raised by genetics, and we will ask how these issues should be debated so that the controversies can be resolved.

KEY CONCEPTS

- Genetics is the study of genes. Genes act as units of inheritance and units of biological information.

- Gregor Mendel was the first scientist to explain how inheritance is controlled by genes, following experiments he carried out with pea plants in 1856–1864.

- Many of the important methods used to study genes as units of inheritance were developed by Thomas Hunt Morgan and his colleagues between 1911 and 1928.

- In 1952, it was shown that genes are made of DNA. The structure of DNA was studied by X-ray diffraction analysis by Florence Bell, Rosalind Franklin and Raymond Gosling. The resulting data enabled James Watson and Francis Crick to deduce that DNA is a double helix.

- Extensive research between 1953 and 1980 led to an understanding of gene expression, including how the activities of individual genes are regulated.

- Geneticists in the early part of the 20th century struggled to reconcile how inheritance controlled by genes could explain the continuous patterns of variation seen in natural populations. A solution, called the Modern Synthesis, was reached in the 1940s.

- Studies of DNA replication and repair during the 1950s and 1960s identified enzymes that are able to copy, cut and join DNA molecules. Purification of these enzymes made genetic engineering possible.

- A new version of genetic engineering, called gene editing, was invented in 2012 by Emmanuelle Charpentier and Jennifer Doudna.

- The first efficient DNA sequencing method was developed by Frederick Sanger and colleagues in 1977. Using this method, the human genome sequence was completed in 2000.

- New DNA sequencing methods invented during the 2000s make it possible to sequence multiple copies of the human genome, revealing individual variations whose association with features such as disease susceptibility is being studied.

- Variations in genome sequences are used in DNA profiling and in studies of prehistoric humans.

FURTHER READING

Backman K (2001) The advent of genetic engineering. *Trends in Biochemical Science* 26, 268–270.

Cherfas J (1982) *Man-Made Life*. New York: Pantheon. *A history of the early years of genetic engineering.*

Cobb M (2015) *Life's Greatest Secret: The Race to Crack the Genetic Code*. London: Profile Books.

Fairbanks DJ (2020) Mendel and Darwin: untangling a persistent enigma. *Heredity* 124, 263–273.

Judson HF (1996) *The Eighth Day of Creation: Makers of the Revolution in Biology*. New York: Cold Spring Harbor Laboratory Press. *A highly readable account of the research on DNA and gene expression carried from the 1930s to the 1990s.*

Lander ES & Weinberg RA (2000) Genomics: journey to the center of biology. *Science* 287, 1777–1782. *A brief description of from Mendel to the human genome sequence.*

Maddox B (2002) *Rosalind Franklin: The Dark Lady of DNA*. London: HarperCollins. *A biography of one of the key people involved in discovery of the double helix structure.*

Mullis KB (1990) The unusual origins of the polymerase chain reaction. *Scientific American* 262(4), 56–65.

Olby R (1974) *The Path to the Double Helix*. London: Macmillan. *A scholarly account of the research that led to the discovery of the double helix.*

Orel V (1995) *Gregor Mendel: The First Geneticist*. Oxford: Oxford University Press.

Shenure J, Balasubramanian S, Church GM, et al. (2017) DNA sequencing at 40: past, present and future. *Nature* 550, 345–353.

Shine I & Wrobel S (1976) *Thomas Hunt Morgan: Pioneer of Genetics*. Lexington: University Press of Kentucky.

Watson JD (1968) *The Double Helix*. London: Atheneum. *The most important discovery of 20th century biology, written as a soap opera.*

GENES AS UNITS OF BIOLOGICAL INFORMATION

2 **DNA**
3 **Genes**
4 **Transcription of DNA to RNA**
5 **Types of RNA Molecule: Messenger RNA**
6 **Types of RNA Molecule: Ribosomal and Transfer RNA**
7 **The Genetic Code**
8 **Protein Synthesis**
9 **Control of Gene Expression**
10 **DNA Replication**
11 **Mutation and DNA Repair**

CHAPTER 2

DNA

We begin our study of **genetics** with **DNA**. We start with DNA because genes are made of DNA. By studying the structure of DNA, we can immediately understand how genes are able to fulfill their two related functions in living organisms, as **units of biological information** and as **units of inheritance**. This is the **molecular approach** to genetics, which we will be following in this book. The molecular approach enables us to make a logical and step-by-step progression from the structure of DNA to the process by which the biological information contained in DNA is released to the cell when it is needed. An understanding of DNA structure is also the best starting point for investigating how DNA molecules are copied and passed to offspring during reproduction and for understanding how DNA molecules can change over time enabling evolution to take place. The most accessible route into the more complex intricacies of genetics therefore starts with the structure of DNA.

In this chapter, we will study the structure of DNA. We will learn how geneticists first came to understand that DNA is the genetic material and then examine how biological information is encoded in DNA and how copies can be made of DNA molecules.

2.1 GENES ARE MADE OF DNA

Today, we are so familiar with the role of DNA as the **genetic material** that it comes as quite a surprise to learn that this idea was considered ridiculous by most biologists until the 1940s and that experimental proof that genes are made of DNA was not obtained until 1952. Why did it take so long to establish this fundamental fact of genetics?

Biologists once thought that genes must be made of protein

The first speculations about the chemical nature of genes were prompted by the discovery in the very early years of the 20th century that genes are contained in **chromosomes** (Figure 2.1). Cell biologists had established that chromosomes are made of DNA and **protein**, in roughly equal amounts. One or the other must therefore be the genetic material.

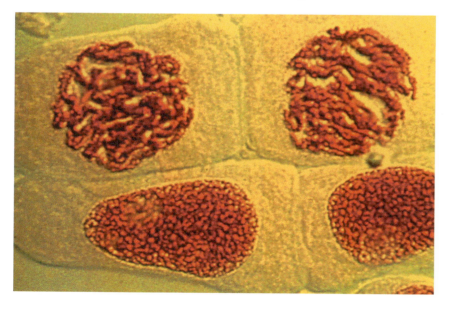

FIGURE 2.1 Genes are contained in chromosomes. Light micrograph of cells of the common bluebell stained so that the DNA content of the nuclei appears red. In the upper two cells, the chromosomes are beginning to condense prior to cell division. Cell biologists in the early 20th century realized that the distribution of chromosomes to new cells during cell division and reproduction exactly parallels the behavior of genes during these events. This led to the hypothesis that genes are located on chromosomes. The use of stains specific for DNA, as shown here, showed that chromosomes contain DNA as well as protein. (Courtesy of G. Giminez-Martin/Science Photo Library.)

DOI: 10.1201/9781003473862-3

In deciding between protein and DNA, biologists considered the properties of genes and how these properties might be provided for by the two types of compound. The most fundamental requirement of the genetic material is that it must be able to exist in an almost infinite variety of forms. Each cell contains a large number of genes, several thousand in the simplest **bacteria**, and tens of thousands in higher organisms. Each gene specifies a different biological characteristic and each presumably has a different structure. The genetic material must therefore have a great deal of chemical variability.

This requirement did not appear to be satisfied by DNA, because in the early part of the 20th century it was thought that all DNA molecules were the same. On the other hand, it was known, correctly, that proteins are polymers made up of different combinations of **amino acids**, with 20 different amino acids and apparently no restrictions on the order in which these could be linked together. This means there can be many types of protein, different from one another by virtue of their amino acid sequences. Proteins therefore possess the variability required by the genetic material. Not surprisingly, biologists during the first half of the 20th century concluded that genes were made of protein and looked on the DNA component of chromosomes as much less important – perhaps a structural material, needed to hold the protein genes together.

Although the hypothesis that genes are made of protein was very widely held in the first half of the 20th century, its supporters were aware that they had no solid evidence to back their view. The necessity for an experimental identification of the genetic material became pressing during the late 1930s, when it gradually became apparent that DNA, rather than being a simple molecule unsuitable as the genetic material, was in fact a long polymer and, like protein, could exist in an almost infinite number of variable forms. If both protein and DNA satisfy the fundamental requirement of the genetic material, and both are present in chromosomes, then which are genes made of? Two critical experiments, radically different in their design, eventually led to the conclusion that genes are made of DNA and not protein. The first of these experiments was the identification of the chemical nature of the **transforming principle**, a substance that can change the bacterium *Streptococcus pneumoniae* from one form to another.

The discovery that the transforming principle is DNA

Streptococcus pneumonia is the causative agent of one type of pneumonia. Virulence is associated with the presence of a **capsule**, which is a coating of polysaccharides that surrounds some cells (Figure 2.2). The capsule helps the bacterium evade the host immune system, so encapsulated bacteria are virulent, and those lacking a capsule are harmless. There are over 100 strains, or **serotypes**, of *S. pneumonia*, different from one another in the composition of the polysaccharide making up the capsule.

Pneumonia was a serious problem in the early part of the 20th century and a great deal of research was devoted to understanding the disease and how to treat it. One of these researchers was a London medical officer, Frederick Griffith. He noted that patients recovering from pneumonia often had avirulent, nonencapsulated bacteria in their saliva. He thought perhaps the bacteria lost their capsules as part of the patient's recovery from the illness, and to test this idea he carried out a series of experiments in which he injected various combinations of bacteria into mice. In one of these experiments, Griffith mixed a sample an extract from heat-killed, virulent Type I bacteria with live, nonencapsulated Type II bacteria and

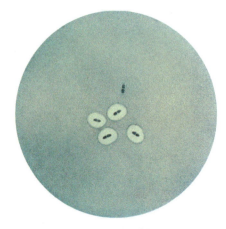

FIGURE 2.2 The capsule of *Streptococcus pneumoniae*. The bacteria in this light micrograph have been treated so their capsules become visible as the pale circles surrounding the cells. The capsule is made of polysaccharide, its composition determining the serotype of the bacterium. The capsule provides a defense against the immune system by preventing the bacteria from being attacked by phagocytes in the host tissues. (Image #2113 from CDC Public Health Image Library.)

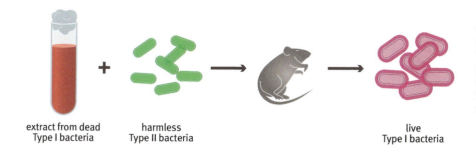

FIGURE 2.3 Bacterial transformation. Griffith mixed an extract from heat-killed, virulent Type I bacteria with live, harmless Type II bacteria and injected this mixture into the bloodstream of a mouse. The mouse developed pneumonia and died. Samples of live Type I bacteria were obtained from the animal's lungs.

injected this mixture into the bloodstream of a mouse (Figure 2.3). As the only live bacteria is the mixture were harmless ones, we might expect the mouse to remain healthy. Instead, the mouse developed pneumonia and died. When its lungs were examined, they were found to contain live Type I bacteria. These live bacteria had not only acquired the ability to cause disease, they had also been converted into the Type I serotype – the serotype of the dead bacteria from which the transforming extract had been obtained.

The change in serotype is critical for understanding the biological significance of this transformation. The explanation cannot be that the live bacteria in the original preparation simply regained their ability to synthesize their capsular polysaccharides. If this had happened then the resulting bacteria would still have been Type II. The fact that they had been transformed into a different serotype showed that some components of the extract prepared from the dead bacteria contained the biological information for the synthesis of the Type I capsule. The extract must therefore contain genes.

Bacteriologists in the 1930s referred to the critical component of Griffith's cell extracts as the **transforming principle**. Griffith did not himself attempt to identify what it was made of. Instead, this work was carried out by Oswald Avery and his colleagues Colin MacLeod and Maclyn McCarty at the Rockefeller Institute in New York. Their strategy was to digest the individual components of the extract with specific degradative enzymes (Figure 2.4). A **protease** was used to degrade all the protein in the extract, and a **ribonuclease** to degrade the **ribonucleic acid** (**RNA**), the second type of nucleic acid present in living cells (although, unlike DNA, not a major constituent of chromosomes). After enzymatic digestion of one or more components, the extract was tested for retention of its transforming ability. Surprisingly (as genes were still thought to be made of protein), treatment with protease had no effect on the extract. Neither did ribonuclease treatment. However, digestion of the DNA with **deoxyribonuclease** totally destroyed the transforming ability so that the extract was no longer able to convert one bacterial serotype into the other. The transforming principle must be DNA.

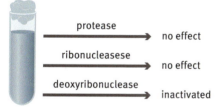

FIGURE 2.4 The transforming principle is DNA. Avery and his colleagues treated extracts of *S. pneumoniae* cells, containing the transforming principle, with enzymes that specifically degrade protein, RNA, or DNA. Only the deoxyribonuclease inactivated the transforming principle.

The notion that genes were made of protein was so widely held that it was not going to be overturned by a single experiment. When Avery published his results in 1944, skeptics examined the paper in detail to try to find loopholes. Doubts were raised about the efficiency of the enzymes used to digest the protein component of the transforming extract. It was suggested that a small amount of protein might remain after enzyme treatment and be responsible for the transformation – if genes were made of protein. These nagging doubts, although unfounded, prevented the results of Avery's group from being completely accepted by the scientific community. A second, independent, experimental identification of the genetic material was required.

Bacteriophage genes are made of DNA

The skeptics included various members of the Phage Group, the association of geneticists who were using bacteriophages as a model system with which to study genes. An experiment carried out by the Phage Group, which we now look back on as confirming that genes are made of DNA, was in fact carried out with the expectation that the results would prove that Avery was wrong.

Bacteriophages, or **phages** as they are commonly known, are **viruses** that specifically infect bacteria. Phage T2 (see Figure 1.7), for example, is one of several types of phage that are specific to the bacterium *Escherichia coli*. If T2 bacteriophages are introduced into a culture of *E. coli*, the cells will become infected and will produce large numbers of new phage particles. Details of how the infection cycle proceeds were not worked out until several years later (Section 12.1), but it was hypothesized in the early 1950s that phages are not able to replicate on their own. Instead, the new phage particles must be synthesized by the bacteria. To do this, the bacteria would need to make use of the information carried by the phage genes, and probably (so it was argued), these genes would have to enter the bacterial cells in order for their information to be used (Figure 2.5). Should it be possible to identify a component of the infecting phage particles that enters the bacterial cell then the chemical nature of the genetic material would be known.

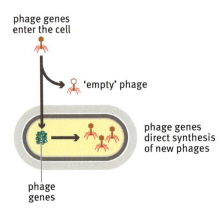

FIGURE 2.5 The role of bacteriophage genes during the infection cycle. In order to direct the synthesis of new bacteriophages, the phage genes must enter the bacterial cell.

Identifying the substance injected by phages into bacterial cells is simplified by the fact that phage particles contain only protein and DNA. In 1951, a new method that would allow protein and DNA to be distinguished had just been developed. This procedure is called **radiolabeling** and involves attachment of a radioactive atom (or '**marker**') to the molecule in question. Protein and DNA can be distinguished because protein can be labeled with ^{35}S, a radioactive isotope of sulfur, which is incorporated into the sulfur-containing amino acids cysteine and methionine. DNA contains no sulfur and cannot be labeled with ^{35}S. Conversely, DNA can be labeled with ^{32}P, which will not be incorporated into proteins because proteins do not contain phosphorus (Figure 2.6). Once labeled, samples of DNA and protein can be distinguished from each other because the two radioisotopes, ^{32}P and ^{35}S, emit radiation of different characteristic energies.

The critical experiment was carried out by Alfred Hershey and Martha Chase at the Cold Spring Harbor Laboratory on Long Island, New York. They prepared a radioactive sample of T2 phage, one in which the protein was labeled with ^{35}S and the DNA with ^{32}P, from a T2-infected culture of *E. coli* that had been grown with ^{35}S- and ^{32}P-labeled nutrients. The labeled phages produced by this culture were used to infect a new, nonradioactive culture of *E. coli* (Figure 2.7). However, this time the infection process was interrupted a few minutes after inoculation by shaking the cells in

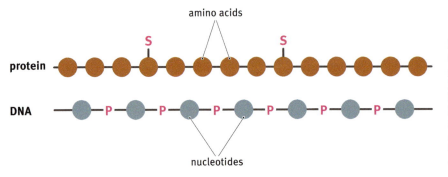

FIGURE 2.6 Radiolabeling of DNA and protein. Protein and DNA can be distinguished by differential radiolabeling. Proteins do not contain phosphorus, but can be labeled with ^{35}S which is incorporated into the sulfur-containing amino acids cysteine and methionine. DNA contains no sulfur but can be labeled with ^{32}P, which is incorporated into the bonds that link nucleotides together.

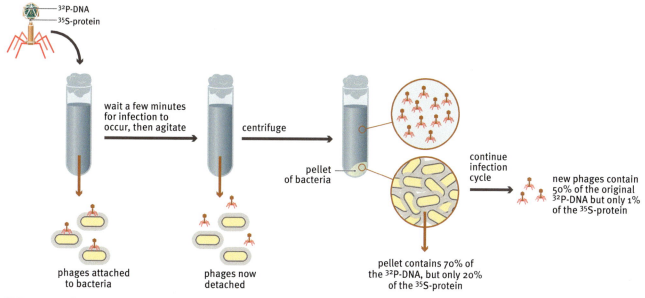

FIGURE 2.7 The Hershey–Chase experiment. Hershey and Chase infected *E. coli* with a sample of T2 phage in which the protein was labeled with ^{35}S and the DNA with ^{32}P. After a few minutes, the cells were shaken in a blender to remove the phage material attached to the outside of the cell. Over 70% of the ^{32}P remained in the bacterial pellet. The bacteria were then allowed to continue through the infection process and produce new phages. Almost half of the ^{32}P, but less than 1% of the ^{35}S, was present in these new phage particles.

a blender. These few minutes were long enough for the phage genes to enter the bacteria but not enough time for new bacteriophages to be synthesized and the bacteria killed. Hershey and Chase believed that agitation would remove the phage material attached to the outside of the cell so that the only component retained by the bacteria would be the injected substance, the phage genes. The culture was then centrifuged so that the relatively heavy bacterial cells, containing the phage genes, were collected at the bottom of the tube, leaving the empty phage particles in suspension. Hershey and Chase discovered that over 70% of the ^{32}P was present in the bacterial pellet. The bacteria were then allowed to continue through the infection process and produce new phages. Almost half of the ^{32}P, but less than 1% of the ^{35}S, was present in these new phage particles. Clearly, only the phage DNA and not its protein enters the bacterial cell.

Although very few doubts could be raised about the interpretation and validity of the Hershey–Chase experiment, several biologists still remained unconvinced that DNA was the genetic material in all organisms. Strictly speaking, all that the Hershey–Chase experiment demonstrated was that the genes of T2 bacteriophages are made of DNA. Phages are very unusual organisms, if indeed they can be called organisms at all, so perhaps the results of experiments with phages cannot be extrapolated to animals and plants. However, from 1953 onwards, it was universally accepted that the genes of all organisms are made of DNA. This was because of the discovery of the structure of the **double helix**, and the realization that the structure was absolutely compatible with the requirements of the genetic material.

2.2 THE STRUCTURE OF DNA

Deoxyribonucleic acid (DNA) is a **polymer**, a long chain-like molecule made up of subunits called **nucleotides** (Figure 2.8). These are linked together to form a **polynucleotide** chain that can be hundreds, thousands or even millions of nucleotides in length. First, we will study the structure

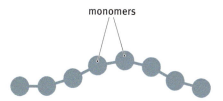

FIGURE 2.8 DNA is a linear polymer. In this depiction, each bead in the chain is an individual nucleotide.

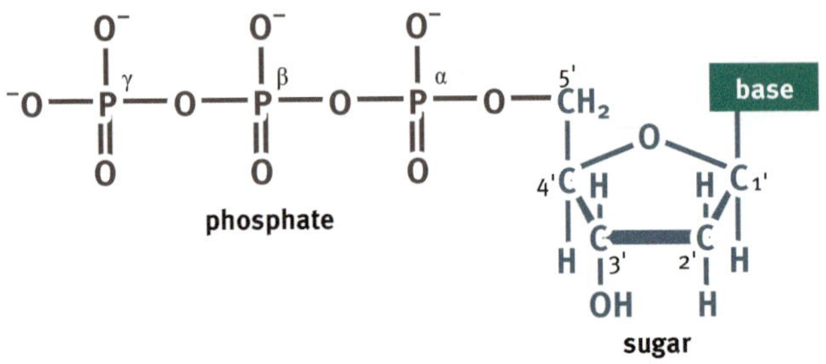

FIGURE 2.9 The components of a deoxyribonucleotide.

of a nucleotide, and then, we will examine how nucleotides are joined together to form a polynucleotide.

Nucleotides are the basic units of a DNA molecule

The basic unit of the DNA molecule is the nucleotide. Nucleotides are found in the cell either as components of nucleic acids or as individual molecules. Nucleotides have several different roles and are not just used to make DNA. For example, some nucleotides are important in the cell as carriers of energy used to power enzymatic reactions.

The nucleotide is itself quite a complex molecule, being made up of three distinct components. These are a sugar, a nitrogenous base and a phosphate group (Figure 2.9). We will look at each of these in turn.

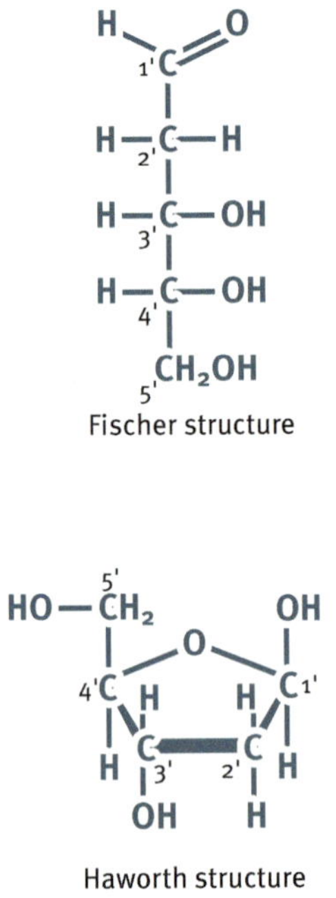

The sugar component of the nucleotide is a **pentose**. A pentose is a sugar that contains five carbon atoms. The particular type of pentose present in the nucleotides found in DNA is called 2′-deoxyribose. Pentose sugars can exist in two forms, the straight chain or Fischer structure and the ring or Haworth structure (Figure 2.10). It is the ring form of 2′-deoxyribose that occurs in the nucleotide. The name 2′-deoxyribose indicates that the standard ribose structure has been altered by the replacement of the hydroxyl group (–OH) attached to carbon atom number 2′ with a hydrogen group (–H). The carbon atoms are always numbered in the same way, with the carbon of the carbonyl group (–C=O), which occurs at one end of the chain structure, numbered 1′. It is important to remember the numbering of the carbons because it is used to indicate the positions at which other components of the nucleotide are attached to the sugar. It is also important to realize that the numbers are not just 1, 2, 3, 4 and 5, but 1′, 2′, 3′, 4′ and 5′. The dash is called the 'prime' and the numbers are called 'one-prime', 'two-prime' and so on. The prime is used to distinguish the carbon atoms in the sugar from the carbon and nitrogen atoms in the nitrogenous base, which are numbered 1, 2, 3 and so on.

The nitrogenous bases are single- or double-ring structures that are attached to the 1′-carbon of the sugar. In DNA, any one of four different nitrogenous bases can be attached at this position. These are called **adenine** and **guanine**, which are double-ring **purines**, and **cytosine** and **thymine**, which are single-ring **pyrimidines**. Their structures are shown in Figure 2.11. The base is attached to the sugar by a **β-N-glycosidic bond** attached to nitrogen number 1 of the pyrimidine or number 9 of the purine.

A molecule comprising the sugar joined to a base is called a **nucleoside**. This is converted into a nucleotide by attachment of a phosphate group to the 5′-carbon of the sugar. Up to three individual phosphates can be attached in series. The individual phosphate groups are designated α, β and γ, with the α-phosphate being the one attached directly to the sugar.

FIGURE 2.10 The two structural forms of 2′ deoxyribose.

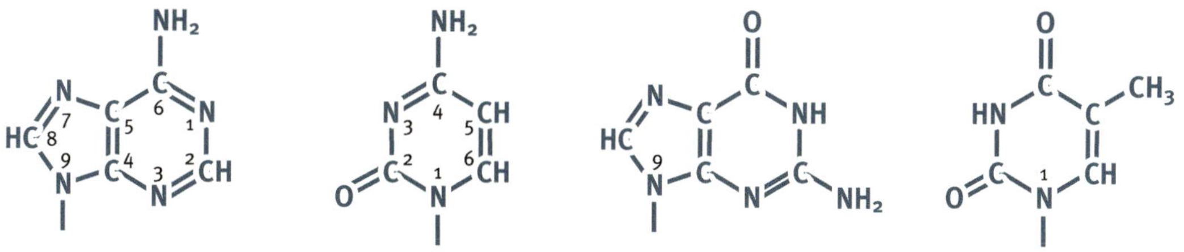

THE STRUCTURE OF DNA

FIGURE 2.11 The structures of the four bases that occur in deoxyribonucleotides.

The full names of the four different nucleotides that polymerize to form DNA are as follows:

- 2′-deoxyadenosine 5′-triphosphate
- 2′-deoxycytidine 5′-triphosphate
- 2′-deoxyguanosine 5′-triphosphate
- 2′-deoxythymidine 5′-triphosphate

Normally, however, we abbreviate these to dATP, dCTP, dGTP and dTTP, or even just to A, C, G and T, especially when writing out the sequence of nucleotides found in a particular DNA molecule.

Nucleotides join together to make a polynucleotide

The next stage in building up the structure of a DNA molecule is to link the individual nucleotides together to form a polymer. This polymer is called a polynucleotide and is formed by attaching one nucleotide to another through the phosphate groups.

The structure of a trinucleotide, a short DNA molecule comprising three individual nucleotides, is shown in Figure 2.12. The nucleotide monomers are linked together by joining the α-phosphate group, attached to the 5′-carbon of one nucleotide, to the 3′-carbon of the next nucleotide in the chain. Normally, a polynucleotide is built up from nucleoside triphosphate subunits, so during polymerization the β- and γ-phosphates are cleaved off. The hydroxyl group attached to the 3′-carbon of the second nucleotide is also lost. The linkage between the nucleotides in a polynucleotide is called a **phosphodiester bond**, 'phospho' indicating the presence of

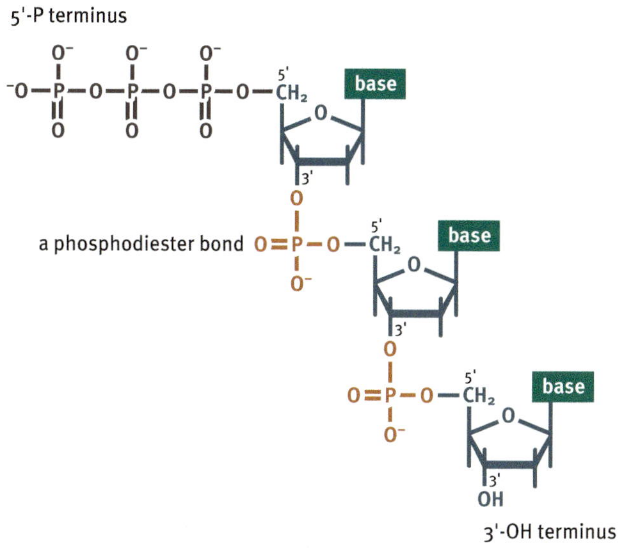

FIGURE 2.12 The structure of a trinucleotide. A trinucleotide is a DNA molecule comprising three individual nucleotides.

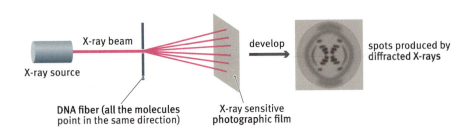

FIGURE 2.13 X-ray diffraction analysis.

a phosphorus atom and 'diester' referring to the two ester bonds (C–O–P) in each linkage. To be precise, we should call this a 3'–5' phosphodiester bond so that there is no confusion about which carbon atoms in the sugar participate in the bond.

An important feature of the polynucleotide is that the two ends of the molecule are not the same. This is clear from an examination of Figure 2.12. The top of this polynucleotide ends with a nucleotide in which the triphosphate group attached to the 5'-carbon has not participated in a phosphodiester bond and the β- and γ-phosphates are still in place. This end is called the **5'** or **5'–P terminus**. At the other end of the molecule, the unreacted group is not the phosphate but the 3'-hydroxyl. This end is called the **3'** or **3'–OH terminus**. The chemical distinction between the two ends means that polynucleotides have a direction, which can be looked as 5' → 3' (down in Figure 2.12) or 3' → 5' (up in Figure 2.12). An important consequence of the polarity of the phosphodiester bond is that the chemical reaction needed to extend a DNA polymer in the 5' → 3' direction is different from that needed to make a 3' → 5' extension. All of the enzymes that make new DNA polynucleotides in living cells carry out 5' → 3 synthesis. No enzymes that are capable of catalyzing the chemical reaction needed to make DNA in the opposite direction, 3' → 5', have ever been discovered.

There is apparently no limitation to the number of nucleotides that can be joined together to form an individual DNA polynucleotide. Molecules containing several thousand nucleotides are frequently handled in the laboratory and the DNA molecules in chromosomes are much longer, sometimes several million nucleotides in length. In addition, there are no chemical restrictions on the order in which the nucleotides can join together. At any point in the chain, the nucleotide could be A, C, G or T.

In living cells, DNA is a double helix

The polynucleotide structure had been fully worked out by the early 1950s, when geneticists were first starting to question whether genes are made of DNA rather than protein. It had also become clear from density measurements of the fibers that form in solutions of DNA extracts that the DNA molecules present in living cells consist of two polynucleotides assembled together in some way. The possibility that unraveling the nature of this assembly might provide insights into how genes function prompted several scientists to attempt to solve its structure.

Then, as now, **X-ray diffraction** analysis was the most important technique used to study the structures of complex biological molecules. An X-ray diffraction pattern for DNA can be obtained by bombarding a DNA fiber with X-rays. X-rays have very short wavelengths – between 0.01 and 10 nm – comparable with the spacings between atoms in chemical structures. When a beam of X-rays is directed onto a DNA fiber some of the X-rays pass straight through, but others are diffracted and emerge at a different angle (see Figure 2.13). As the fiber is made up of many DNA molecules, all positioned in a regular array, the individual X-rays are diffracted in similar ways, resulting in overlapping circles of diffracted waves that interfere with one another. An X-ray-sensitive photographic film placed across the beam reveals a series of spots and smears, called the X-ray diffraction pattern.

X-ray diffraction analysis was first applied to DNA by Florence Bell of the University of Leeds, UK in 1939. However, her sample of DNA was a mixture of the **A-** and **B-forms**, which have slightly different structures. As a result, her X-ray diffraction patterns were difficult to interpret. Rosalind Franklin and her PhD student Raymond Gosling, at King's College, London obtained a pure sample of the B-form, and in 1952 achieved a much clearer diffraction pattern (Figure 2.14). This famous **photo 51** showed that DNA is a helix, and mathematical calculations based on the pattern enabled dimensions such as the diameter, distance between base pairs, and the pitch (the distance taken by a complete turn of the helix) to be calculated.

From her notebooks, it is clear that Rosalind Franklin came very close to solving the double helix structure solely by analysis of her X-ray diffraction data. The final steps, however, were taken by James Watson and Francis Crick, of the University of Cambridge, UK. As well as Franklin's X-ray photo, Watson and Crick were also aware of biochemical studies carried out by Erwin Chargaff of Columbia University, New York. Chargaff analyzed DNA samples from various sources and showed that, although the values are different in different organisms, the amount of adenine is always the same as the amount of thymine and the amount of guanine always equals the amount of cytosine (Figure 2.15). Watson and Crick realized that these **base ratios** meant that an adenine in one polynucleotide must be paired with a thymine in the second polynucleotide, and similarly guanines must pair with cytosines.

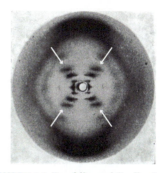

FIGURE 2.14 Franklin and Gosling's photo 51, showing the diffraction pattern obtained with a fiber of DNA. The cross shape indicates that DNA has a helical structure, and the extent of the shadowing within the 'diamond' spaces above, below and to either side of the cross shows that the sugar–phosphate backbone is on the outside of the helix. The positions of the various smears that make up the arms of the cross enable dimensions such as the diameter, distance between base pairs and the distance taken by a complete turn of the helix to be calculated. The 'missing smears' (the gap in each arm of the cross, marked by the arrows) indicate the relative positioning of the two polynucleotides. (From R. Franklin & R. G. Gosling, *Nature* 171: 740–741, 1953. With permission from Springer Nature.)

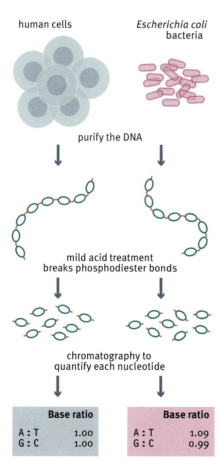

FIGURE 2.15 The base ratio experiments performed by Chargaff. DNA was extracted from various organisms and treated with acid to break the phosphodiester bonds and release the individual nucleotides. Each nucleotide was then quantified by chromatography. The data show some of the actual results obtained by Chargaff. These indicate that, within experimental error, the amount of adenine is the same as that of thymine, and the amount of guanine is the same as that of cytosine.

CHAPTER 2: DNA

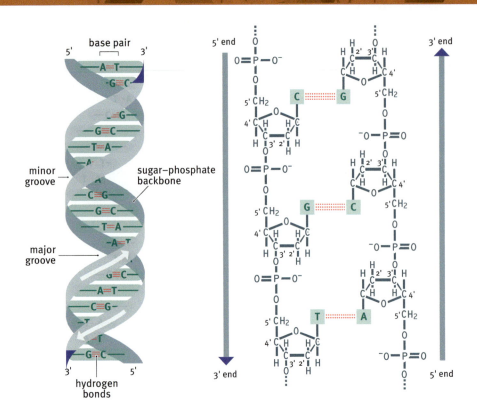

FIGURE 2.16 The double helix structure of DNA. On the left, the double helix is drawn with the sugar–phosphate backbones of each polynucleotide shown as a gray ribbon with the base pairs in green. On the right, the chemical structure for three base pairs is given.

Watson and Crick made scale models of possible DNA structures whose internal dimensions agreed with the X-ray data and which involved A–T and G–C pairings between two polynucleotides that spiraled around one another. The models enabled the relative positioning of the various atoms to be checked, to ensure that pairs of atoms that formed bonds were not too far apart and that other atoms were not so close together as to interfere with one another. The result was the famous double helix structure (Figure 2.16).

The key features of the double helix

The double helix is quite a complicated structure but the key facts about it are not too difficult to understand. The two polynucleotides in the helix are arranged in such a way that their sugar–phosphate 'backbones' are on the outside of the helix, and their bases are on the inside. The bases are stacked on top of each other rather like a pile of plates. The two polynucleotides are **antiparallel**, meaning that they run in different directions, one being orientated in the 5′ → 3′ direction and the other in the 3′ → 5′ direction. The polynucleotides must be antiparallel in order to form a stable helix and molecules in which the two polynucleotides run in the same direction are unknown in nature. The double helix is right-handed, so if it were a spiral staircase then the banister (that is, the sugar–phosphate backbone) would be on your right-hand side as you were climbing upwards. The final point is that the helix is not absolutely regular. Instead, on the outside of the molecule, we can distinguish a **major** and a **minor groove**. These two grooves are clearly visible in Figure 2.16.

Those are the key facts about the double helix. There remains just one additional feature to consider, but this one is the most important of all. Within the helix, an adenine in one polynucleotide is always adjacent to a thymine in the other strand, and similarly, guanine is always adjacent to cytosine. This is called **base pairing** and involves the formation of

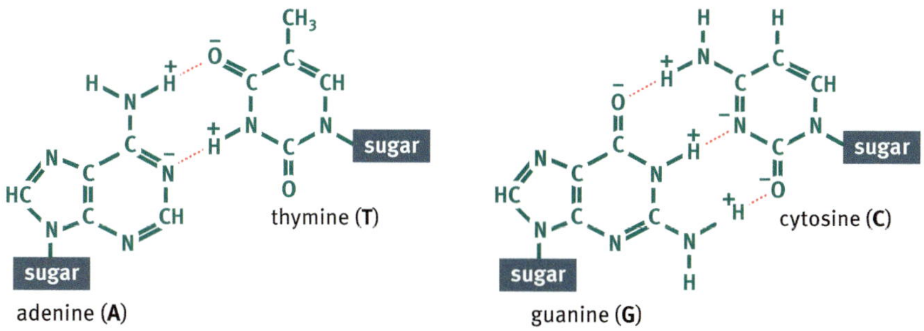

FIGURE 2.17 Base pairing. A base pairs with T and G base pairs with C. The bases are drawn in outline, with the hydrogen bonds indicated by dotted lines. Note that a G–C base pair has three hydrogen bonds, whereas an A–T base pair has just two.

hydrogen bonds between an adenine and a thymine, or between a guanine and a cytosine. A hydrogen bond is a weak electrostatic attraction between an electronegative atom (such as oxygen or nitrogen) and a hydrogen atom attached to a second electronegative atom. The base pairing between adenine and thymine involves two hydrogen bonds and that between guanine and cytosine involves three hydrogen bonds (Figure 2.17).

The two base-pair combinations – A base-paired with T and G base-paired with C – are the only ones that are permissible. This is partly because of the geometries of the nucleotide bases and the relative positions of the atoms that are able to participate in hydrogen bonds and partly because the pair must be between a purine and a pyrimidine. A purine–purine pair would be too big to fit within the helix, and a pyrimidine–pyrimidine pair would be too small. Because of the base pairing, the sequences of the two polynucleotides in the helix are **complementary** – the sequence of one polynucleotide determines the sequence of the other.

The double helix exists in several different forms

The double helix shown in Figure 2.17 is the B-form of DNA. This is by far the commonest type of DNA in living cells, but it is not the only version that is known. Other versions are possible because the three-dimensional structure of a nucleotide is not entirely rigid. For example, the orientation of the base relative to the sugar can be changed by rotation around the β-*N*-glycosidic bond, to give the *anti* and *syn* conformations (Figure 2.18). This has a significant effect on the double helix as it alters the relative positioning of the two polynucleotides. The relative positions of the carbons in the sugar can also be changed slightly, affecting the conformation of the sugar–phosphate backbone.

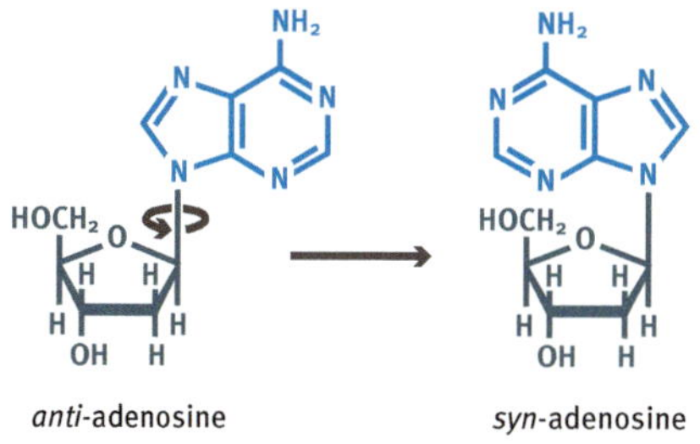

The common, B-form of DNA has ten **base pairs** (abbreviated to 10 bp) per turn of the helix, with 0.34 nm between adjacent base pairs and hence a pitch (the distance needed for one complete turn) of 3.40 nm. The diameter of the helix is 2.37 nm. The second type of DNA double helix, called the A-form, is more compact with 11 bp per turn, 0.23 nm between each base pair, and a diameter of 2.55 nm (Table 2.1). But comparing these dimensions does not reveal what is probably the most significant difference between the two conformations of the double helix. This is the extent to which the internal regions of the DNA molecule are accessible from the surface of the structure. A-DNA, like the B-form, has two grooves, but with A-DNA the major groove is even deeper, and the minor groove is shallower and broader (Figure 2.19).

In Chapter 9, we will learn that the expression of the biological information contained within a DNA molecule is mediated by DNA-binding proteins, which attach to the double helix and regulate the activity of the genes contained within it. To carry out their function, each DNA-binding protein must

FIGURE 2.18 The structures of *anti*- and *syn*-adenosine. The two conformations differ in the orientation of the base relative to the sugar component of the nucleotide. Rotation around the β-*N*-glycosidic bond converts one form into the other. The three other nucleotides also have *anti* and *syn* conformations.

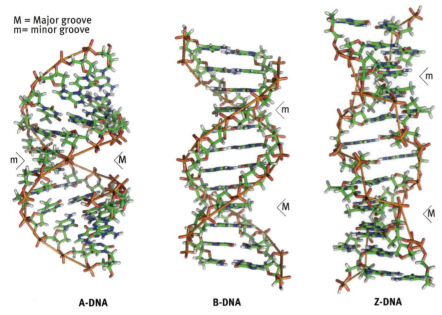

FIGURE 2.19 Comparison between the A-, B- and Z-DNA versions of the double helix. The major and minor groves of each molecule are indicated by 'M' and 'm', respectively. (Courtesy of Richard Wheeler, published under GFDL 1.2.)

TABLE 2.1 FEATURES OF THE DIFFERENT CONFORMATIONS OF THE DNA DOUBLE HELIX

Feature	B-form	A-form	Z-DNA
Type of helix	Right-handed	Right-handed	Left-handed
Number of base pairs per turn	10	11	12
Distance between base pairs (nm)	0.34	0.23	0.38
Distance per complete turn (nm)	3.4	2.5	4.6
Diameter of helix (nm)	2.37	2.55	1.84
Base orientation	*anti*	*anti*	Mixture

attach at a specific position, near to the gene whose activity it must influence. This can be achieved, with at least some degree of accuracy, by the protein reaching down into a groove, within which the DNA sequence can be 'read' without the helix being opened up by breaking the base pairs. This is possible with the B-form, but easier with A-DNA because the bases are more exposed in the minor groove. Some DNA-binding proteins induce the formation of a short stretch of A-DNA when they attach to a DNA molecule.

A third type, **Z-DNA**, is more strikingly different. In this structure, the helix is left-handed, not right-handed as it is with the A- and B-forms, and the sugar–phosphate backbone adopts an irregular zig-zag conformation (see Figure 2.19). Z-DNA is more tightly wound with 12 bp per turn and a diameter of only 1.84 nm (Table 2.1). It is known to occur in regions of a double helix that contain repeats of the motif GC (i.e. the sequence of each strand is ..GCGCGCGC..). In these regions, each G nucleotide has the *syn* conformation of the base relative to the sugar and each C has the *anti* conformation.

2.3 THE MOLECULAR EXPLANATION OF THE BIOLOGICAL ROLE OF DNA

To fulfill its role as the genetic material, a DNA molecule must provide the genes that it contains with the ability to act as units of biological information and units of inheritance. We can now ask ourselves how these requirements are met by the structure of the double helix.

Biological information is contained in the nucleotide sequence of a DNA molecule

DNA is able to act as a store of biological information because of its polymeric structure and because there are four different nucleotides. The order of nucleotides in a DNA molecule – the **DNA sequence** – is, in essence, a language made up of the four letters A, C, G and T. The biological information contained in genes is written in this language, which we call the **genetic code**. The language is read by the process called **gene expression**.

There are no chemical restrictions on the order in which the nucleotides can join together in a DNA molecule. At any point, the nucleotide could be A, C, G or T. This means that a polynucleotide of just ten nucleotides in length could have any one of $4^{10} = 1,048,576$ different sequences (Figure 2.20). The average length of a gene is about 1,000 nucleotides. This length of DNA can exist as $4^{1,000}$ different sequences, which we will look as simply a very big number, greater than the supposed number of atoms in the observable universe, which is a paltry 10^{80}. Bear in mind that we can also have genes 999 and 1,001 nucleotides in length, as well as many other lengths, each of these providing an immense number of possible DNA sequence variations. The early geneticists were mystified by the need for genes to be able to exist in many different forms in order to account for all of the genes present in the myriad of species alive today and which have lived in the past. An answer was so difficult to imagine that some biologists wondered if genes really were physical particles inside cells. Perhaps they were just abstract entities whose invention by geneticists made it possible to explain how biological characteristics are passed from parents to offspring. Now that we understand the structure of DNA the immense variability required by the genetic material is no puzzle at all.

```
AGCTAAGGGT
ACCTAAGGGT
AACTAAGGGT
ATCTAAGGGT
AGGTAAGGGT
AGTTAAGGGT
ACGTAAGGGT
ACGAAAGGGT
AGCTTCGGGT
AGCTGGGGGT
AGCTAAGGGA
TCAATTTTAA
```

FIGURE 2.20 Twelve of the 1,048,576 different possible sequences for a DNA molecule ten nucleotides in length. The orange nucleotides are the ones that are different from the equivalent nucleotide in the topmost sequence.

Complementary base pairing enables DNA molecules to replicate

In order to act as units of inheritance, genes must be able to replicate. Copies of the parent's genes must be placed in the fertilized egg cell so that this cell receives the information it needs to develop into a new living organism that displays a mixture of the biological characteristics of its parents. Genes must also be replicated every time a cell divides, so that each of the two daughter cells can be given a complete copy of the biological information possessed by the parent cell.

Before the structure of the double helix was known, the ability of genes to make copies of themselves was an even greater mystery than their ability to exist in an almost infinite number of different forms. No chemical capable of replication had ever been found in the natural world. The discovery of the double helix provided a clear and obvious solution to the problem. The key lies with the complementary base pairing that links nucleotides that are adjacent to one another in the two strands of the double helix.

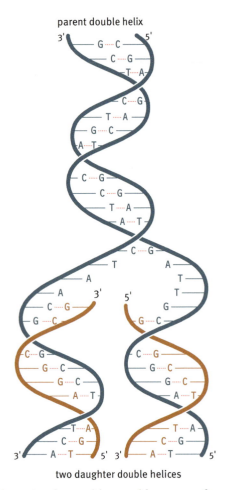

FIGURE 2.21 Complementary base pairing provides a means for making two copies of a double helix. The polynucleotides in the parent double helix are shown in blue, and the newly synthesized strands are in orange.

The rules are that A can only base pair with T and G can only base pair with C. This means that if the two polynucleotides in a DNA molecule are separated, then two perfect copies of the parent double helix can be made simply by using the sequences of these pre-existing strands to dictate the sequences of the new strands (Figure 2.21).

In living cells, DNA molecules are copied by enzymes called **DNA polymerases**. A DNA polymerase builds up a new polynucleotide by adding nucleotides one by one to the 3′ end of the growing strand, using the base-pairing rules to identify which of the four nucleotides should be added at each position (Figure 2.22). This is called **template-dependent DNA synthesis** and it ensures that the double helix that is made is a precise copy of the double helix from which the original polynucleotide was obtained. The structure of DNA therefore explains how genes are able to replicate and hence act as units of inheritance.

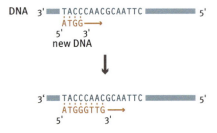

FIGURE 2.22 Template-dependent DNA synthesis. The new DNA, shown in orange, is extended by adding nucleotides one by one to its 3′ end.

THE MOLECULAR EXPLANATION OF THE BIOLOGICAL ROLE OF DNA

KEY CONCEPTS

- Genes were originally thought to be made of protein. Research carried out by Avery and colleagues, and by Hershey and Chase, showed that this hypothesis was incorrect and genes are made of DNA.

- DNA is a polymer in which the individual units are called nucleotides. Each nucleotide comprises the pentose sugar 2′-deoxyribose, a nitrogenous base and a phosphate group.

- There are four different bases. Two are single-ring pyrimidines called cytosine and thymine, and two are double-ring purines called adenine and guanine. This means that there are four different nucleotides, usually referred to as A, C, G and T, the abbreviations of their full chemical names.

- Nucleotides can be linked together in any order to form polynucleotides.

- In living cells, DNA is a double helix. The two polynucleotides of the double helix are held together by hydrogen bonds, in such a way that A can only base pair with T, and G can only pair with C.

- There are variations of the double helix structure, called A-, B- and Z-DNA. The B-form is the commonest type in the cell.

- The sequence of nucleotides in a DNA molecule is a language made up of the four letters A, C, G and T. The biological information contained in a gene is written in this language.

- If the two polynucleotides in a DNA molecule are separated, then two identical copies of the parent double helix can be made by using the sequences of the pre-existing polynucleotides to dictate the sequences of the new polynucleotides. This explains how genes can act as units of inheritance.

QUESTIONS AND PROBLEMS

Key terms

Write short definitions of the following terms:

β-N-glycosidic bond
3′–OH terminus
5′–P terminus
A-form
adenine
amino acid
antiparallel
B-form
bacteria
bacteriophage
base pair
base pairing
base ratio
capsule
chromosome
complementary
cytosine
deoxyribonuclease

DNA
DNA polymerase
DNA sequence
double helix
gene expression
genetic code
genetic material
genetics
guanine
hydrogen bond
major groove
marker
minor groove
molecular approach
nucleoside
nucleotide
pentose
phage

phosphodiester bond
photo 51
polymer
polynucleotide
protease
protein
purine
pyrimidine
radiolabeling
ribonuclease
RNA
serotype
template-dependent DNA synthesis
thymine
Z-DNA

Self-study questions

2.1 Why did biologists originally think that protein is the genetic material?

2.2 Outline the two experiments carried out in the 1940s and 1950s that indicated that genes are made of DNA.

2.3 To what extent is the statement "Genes are made of DNA" consistent with the results of the Avery and Hershey–Chase experiments?

2.4 Draw the structure of a nucleotide.

2.5 What are the complete chemical names of the four nucleotides found in DNA molecules?

2.6 Draw a fully annotated diagram of the structure of a short DNA polynucleotide containing each of the four nucleotides.

2.7 Explain why the two ends of a polynucleotide are chemically distinct.

2.8 Outline how X-ray diffraction analysis was used to help solve the structure of DNA.

2.9 What was the importance of the base ratios in understanding the structure of the double helix?

2.10 What are the important features of the double helix structure?

2.11 What is meant by complementary base pairing, and why is it important?

2.12 If the sequence of one polynucleotide of a DNA double helix is 5′–ATAGCAATGCAA–3′, what is the sequence of the complementary polynucleotide?

2.13 Thirty percent of the nucleotides in the DNA of the locust are As. What are the percentage values for (a) T, (b) G+C, (c) G, and (d) C?

2.14 DNA from the fungus *Neurospora crassa* has an AT content of 46%. What is the GC content?

2.15 What are the main differences between the A- and B-forms of DNA?

2.16 Describe how Z-DNA differs from the A- and B-forms.

2.17 Explain how DNA provides the variability needed by the genetic material.

2.18 Draw a diagram to illustrate the process called template-dependent DNA synthesis.

Discussion topics

2.19 Is the statement "Genes are made of DNA" universally correct?

2.20 An A–T base pair is held together by two hydrogen bonds and a G–C base pair by three hydrogen bonds. In which parts of a DNA molecule might you expect to find AT-rich sequences?

2.21 Discuss why the double helix gained immediate universal acceptance as the correct structure for DNA.

2.22 Human DNA has a GC content of 40.3%. In other words, 40.3% of the nucleotides are either G or C. The GC contents for different organisms vary over a wide range. The DNA of the malaria parasite, *Plasmodium falciparum*, has a GC content of just 19.0%, whereas that of the bacterium *Streptomyces griseolus* is 72.4%. Speculate on the reasons why the GC contents for different species should be so different.

2.23 Explore the reasons why, in the early 20th century, some biologists thought that genes were abstract entities invented by geneticists to explain how biological characteristics are passed from parents to offspring.

2.24 The scheme for DNA replication shown in Figure 2.21 is the same as that proposed by Watson and Crick immediately after their discovery of the double helix structure. Many biologists thought that this process would be impossible in a living cell, especially for the circular DNA molecules present in many bacteria. Why was this?

2.25 A DNA polymerase builds up a new polynucleotide by adding nucleotides one by one to the 3′ end of the growing strand. Enzymes that make DNA in the opposite direction, by adding nucleotides to the 5′ end, are unknown. This fact complicates the process by which a double-stranded DNA molecule is replicated. Explain.

The answers to, or hints on how to answer, the self-study questions and discussion topic questions can be downloaded from the book's product page at this link: www.routledge.com/9781032743530

FURTHER READING

Hershey AD & Chase M (1952) Independent functions of viral protein and nucleic acid in growth of bacteriophage. *Journal of General Physiology* 36, 39–56. *One of the original papers that showed that genes are made of DNA.*

Maddox B (2002) *Rosalind Franklin: The Dark Lady of DNA*. London: HarperCollins. *A biography of one of the key people involved in discovery of the double helix structure, who sadly died just a few years later.*

McCarty M (1985) *The Transforming Principle: Discovering that Genes are Made of DNA*. London: Norton. *Personal account by one of the scientists who worked with Avery.*

Olby R (1974) *The Path to the Double Helix*. London: Macmillan. *A scholarly account of the research that led to the discovery of the double helix.*

Ravichandran S, Subramani VK & Kim KK (2019) Z-DNA in the genome: from structure to disease. *Biophysical Reviews* 11, 383–387.

Rich A & Zhang S (2003) Z-DNA: the long road to biological function. *Nature Reviews Genetics* 4, 566–572.

Watson JD (1968) *The Double Helix*. London: Atheneum. *The most important discovery of twentieth-century biology, written as a soap opera.*

Watson JD & Crick FHC (1953) Molecular structure of nucleic acids: a structure for deoxyribose nucleic acid. *Nature* 171, 737–738. *The scientific report of the discovery of the double helix structure.*

CHAPTER 3

Genes

In Chapter 2, we studied the structure of DNA and asked how this structure enables genes to play their two roles in living cells, as units of biological information and as units of inheritance. We learnt that biological information is encoded in the nucleotide sequence of a DNA molecule and that complementary base pairing enables DNA molecules to be replicated so that copies can be passed from parents to offspring. Now we must focus more closely on the genes themselves. Knowing that biological information can be encoded in a nucleotide sequence is only the first step in understanding how that information is used by the cell. To take the next step forward, we must examine the units of biological information that are contained in individual genes.

The first question that we will address in this chapter is the most fundamental of all. What is the nature of the information contained in individual genes? We know that it is in the form of a nucleotide sequence, but we need to probe deeper and see exactly what that means. This will introduce us to **gene expression**, the process by which the information contained in a gene is utilized by the cell.

In this chapter, we will also ask why the members of a species are subtly different from one another even though they all possess the same set of genes. Why, for example, do some people have blue eyes when others have brown? In answering this question, we will discover that small alterations in the nucleotide sequence can inactivate a gene or alter the precise nature of the information that it contains.

Finally, we will discover that some genes are present in multiple copies and that studying these gene families helps us understand how genes evolve.

3.1 THE NATURE OF THE INFORMATION CONTAINED IN GENES

We know that the information contained in a gene is encoded in its nucleotide sequence. But what is the information for? Before we can begin to answer this question, we must take a closer look at the molecular structures of individual genes.

Genes are segments of DNA molecules

There are many more genes in a cell than there are DNA molecules. Humans, for example, have over 45,000 genes but only 25 different DNA molecules. A single DNA molecule therefore carries a large number of genes. In humans, the number ranges from 37 for the shortest of the 25 DNA molecules to over 4,300 for the longest. A gene is therefore a segment of a DNA molecule.

The shortest genes are less than 100 base pairs (bp) in length, and the longest ones are over 2,400,000 bp. To simplify these larger numbers, we use the terms **kilobase pair** (kb) for 1,000 bp and **megabase pair** or 'Mb' for 1,000,000 bp. According to this notation, the longest genes are over 2,400 kb or 2.4 Mb in length. The genes are separated from one another by **intergenic DNA** (Figure 3.1). Again, there is enormous variation in the actual distances between genes.

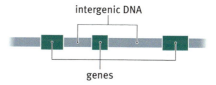

FIGURE 3.1 Genes are separated from one another by intergenic DNA.

DOI: 10.1201/9781003473862-4

Describing the range of gene lengths as between 100 bp and 2.4 Mb is rather misleading when our focus is on the information contents of individual genes. This length difference implies that the longest genes contain 24,000 times as much information as the shortest ones, which is not the case. This is because the lengths of many genes are inflated by the presence within them of long tracts of DNA that do not contain any biological information (Figure 3.2). In a **discontinuous gene** (also called a split or mosaic gene), the sections containing biological information are called **exons** and the intervening sequences are referred to as **introns**.

In most discontinuous genes, the introns are much longer than the exons. In many of the longest genes, the introns added together make up over 90% of the length of the gene. These genes therefore contain much less biological information than suggested by their overall lengths. When just the exons are considered, the longest 'gene' that has so far been discovered is 103 kb, and the average length of 'genes' in all organisms is about 1.2 kb. The average unit of biological information can therefore be encoded in a nucleotide sequence just 1,200 letters in length.

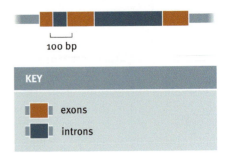

FIGURE 3.2 Exons and introns in a discontinuous gene.

Genes contain instructions for making RNA and protein molecules

Genes contain biological information but on their own they are unable to release that information to the cell. Its utilization requires the coordinated activity of enzymes and other proteins, which participate in a series of events referred to as gene expression.

Gene expression is conventionally looked on as a two-stage process (Figure 3.3). All genes undergo the first stage of gene expression, which is called **transcription** and which results in the synthesis of an RNA molecule. Transcription is a simple copying reaction. RNA, like DNA, is a polynucleotide, the only chemical differences being that in RNA the sugar is ribose rather than 5′-deoxyribose, and the base thymine is replaced by uracil (U) which, like thymine, base pairs with adenine (Figure 3.4). During transcription of a gene, one strand of the DNA double helix acts as a template for the synthesis of an RNA molecule whose nucleotide sequence is determined by the base-pairing rules by the DNA sequence (Figure 3.5).

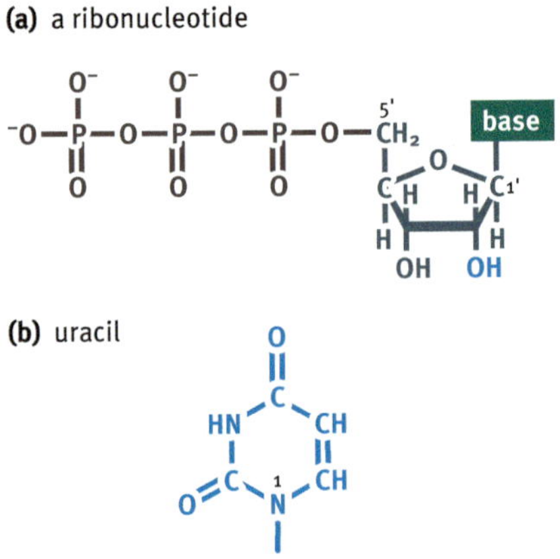

For some genes, the RNA transcript is itself the end product of gene expression. For others, the transcript is a short-lived message that directs the second stage of gene expression, called **translation**. During translation, the RNA molecule (called a **messenger** or **mRNA**) directs the synthesis of a protein. A protein is another type of polymeric molecule but quite different from DNA and RNA. In a protein, the monomers are called amino acids and there are twenty different ones, each with its own specific chemical properties. When a protein is made by translation, its amino acid sequence is determined by the nucleotide sequence of the mRNA. Each triplet of adjacent ribonucleotides specifies a single amino acid of the protein, the identity of the amino acid corresponding to each triplet being set by the **genetic code** (Figure 3.6).

Protein synthesis is the key to the expression of biological information

For all genes, the endpoint of gene expression is synthesis of either an RNA molecule or a protein. How can this simple process enable the information contained in the genes to specify the biological characteristics of a living organism?

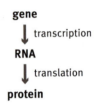

FIGURE 3.3 Gene expression as a two-stage process.

(a) a ribonucleotide

(b) uracil

FIGURE 3.4 The chemical differences between DNA and RNA. (a) RNA contains ribonucleotides, in which the sugar is ribose rather than 2′-deoxyribose. The difference is that a hydroxyl group rather than a hydrogen atom is attached to the 2′-carbon. (b) RNA contains the pyrimidine called uracil instead of thymine.

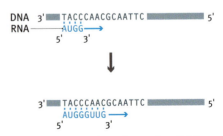

FIGURE 3.5 Template-dependent RNA synthesis. The RNA molecule, shown in blue, is extended by adding nucleotides one by one to its 3′-end. The process is very similar to template-dependent DNA synthesis, which is illustrated in Figure 2.22.

TABLE 3.1 THE FUNCTIONAL DIVERSITY OF PROTEINS	
Type of Protein	Examples
Structural proteins	Collagen, keratin
Motor proteins	Myosin, dynein
Catalytic proteins (enzymes)	Hexokinase, DNA polymerase
Transport proteins	Hemoglobin, serum albumin
Storage proteins	Ovalbumin, ferritin
Protective proteins	Immunoglobulins, thrombin
Regulatory proteins	Insulin, somatostatin, somatotropin

FIGURE 3.6 During translation, each triplet of nucleotides in the RNA specifies a different amino acid. The identity of each amino acid is set by the genetic code. The amino acids indicated here by their three-letter abbreviations are methionine, glycine, cysteine, valine and lysine.

FIGURE 3.7 Part of a molecule of collagen, a structural protein. A collagen molecule is made up of three protein polymers wrapped around each other in a triple helix. This helical arrangement gives the protein a high tensile strength. (After D. S. Goodsell, *Our Molecular Nature*. New York: Springer-Verlag, 1996. With permission from Springer Nature.)

The answer lies with proteins. Proteins with different amino acid sequences can have quite different chemical properties and this enables them to play a variety of roles (**Table 3.1**). Some amino acid sequences, for example, result in fibrous proteins that give rigidity to the framework of an organism. Collagen is an example of one of these **structural proteins** (**Figure 3.7**). Collagen is found in the bones, tendons and cartilage of vertebrates, making up about 20% of the dry weight of the skeleton. Without their collagen component, bones would be fragile and unable to support the body mass. A second example is keratin, present in hair and feathers, and also a major component of the exoskeleton of arthropods such as crabs. In contrast, **motor proteins** have amino acid sequences that give them flexibility. They are able to change their shape enabling organisms to move around. The muscle protein myosin is a motor protein, as is dynein in cilia and flagella (**Figure 3.8**).

Other types of protein have quite different functions. **Enzymes** are proteins whose amino acid sequences enable them to catalyze the multitude of cellular reactions that bring about the release and storage of energy and the synthesis of new compounds (**Figure 3.9**). Other proteins have transport functions and carry compounds around the body, the most important in mammals being hemoglobin, which carries oxygen in the bloodstream (**Figure 3.10**), and serum albumin, which transports fatty acids. Some proteins help to store molecules for future use by the organism, such as ovalbumin, which stores amino acids in egg white, and ferritin, which stores iron in the liver. Another vast range of proteins have protective functions and guard against infectious agents and injury. Examples in mammals are immunoglobulins and other antibodies, which form complexes with foreign proteins (**Figure 3.11**), and thrombin and other components of the blood clotting mechanism.

There are also **regulatory proteins** that control cellular activities. These include well-known hormones such as insulin, which regulates glucose

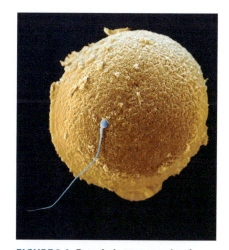

FIGURE 3.8 Dynein is an example of a motor protein. This micrograph shows a human sperm attempting to fertilize a female egg cell. The sperm cell, shown in blue, has a long flagellum, which contains multiple copies of the motor protein called dynein. By changing their shape, these dynein proteins generate a series of waves that move along the flagellum from its base to its tip. This wavelike motion propels the sperm forward, enabling it to swim in search of the female cell. (Courtesy of Eye of Science/Science Photo Library.)

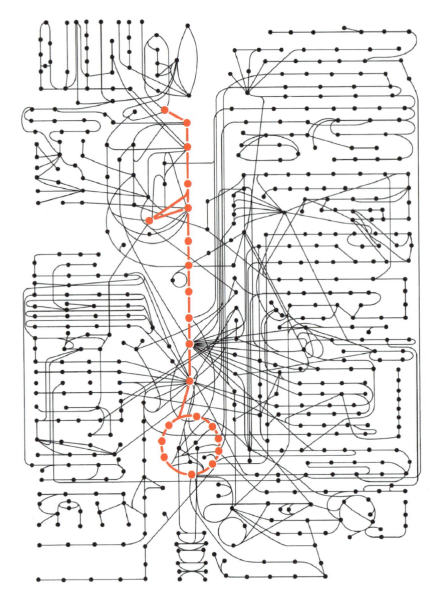

FIGURE 3.9 Enzymes are catalytic proteins. Outline of the metabolic pathways of a typical animal cell. Each dot represents a different biochemical compound. The lines indicate the steps in the network, each of these steps resulting in the conversion of one compound into another. There are approximately 500 steps, each catalyzed by a different enzyme. The glycolysis pathway and the citric acid cycle, which provide energy for metabolism, are shown in red. The pathways shown in black provide substrates for energy generation or assemble molecules such as amino acids and nucleotides from smaller precursors. (Adapted from B. Alberts et al., *Essential Cell Biology*, 3rd ed. New York: Garland Science, 2010.)

metabolism in vertebrates, and the two growth hormones somatostatin and somatotropin. Hormones are extracellular proteins. Although made inside a cell, they are released so they can travel around the body and convey their regulatory messages to other cells (Figure 3.12). Sometimes the message is that certain genes must be switched on and others switched off. Proteins are therefore not just the products of gene expression, some of them also control the way in which biological information is released to the cell.

Proteins can therefore provide organisms with their structure, their ability to move and their ability to carry out biochemical reactions. Proteins carry compounds around the body, store those that are not immediately

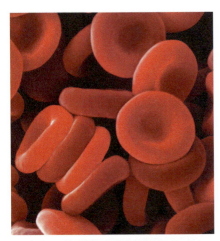

FIGURE 3.10 Hemoglobin is an example of a transport protein. Red blood cells contain large amounts of hemoglobin and hence are able to carry oxygen from one part of the body to another via the bloodstream. (Courtesy of Power and Syred/Science Photo Library.)

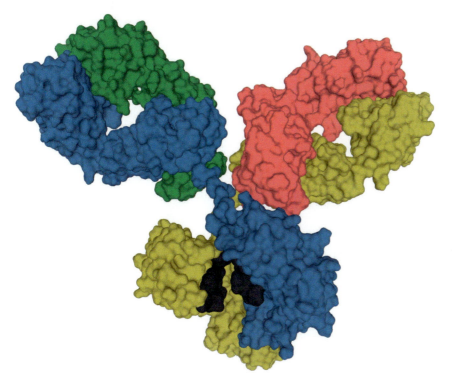

FIGURE 3.11 **An immunoglobulin, an example of a protective protein.** Immunoglobulins recognize foreign proteins and form part of the human body's defense against invading viruses, bacteria and other infectious agents. (Courtesy of Tokenzero, from Wikimedia Commons, published under CC BY-SA 4.0 license.)

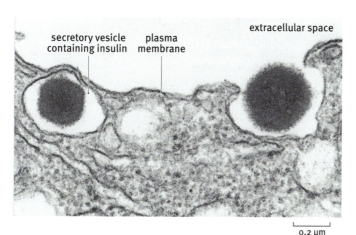

FIGURE 3.12 **Hormones are examples of regulatory proteins.** Electron micrograph showing insulin being secreted by β-cells in the human pancreas. (Courtesy of Lelio Orci, University of Geneva.)

needed, and protect the organism from disease. Proteins also control cellular activities including the gene expression processes that determine which proteins are synthesized. The units of biological information contained in genes give rise to this multiplicity of function by encoding the amino acid sequences of all the proteins that a cell is able to make. In this way, the genes are able to specify the biological characteristics of a living organism.

The RNA molecules that are not translated into protein are also important

Only a small fraction of a cell's RNA, usually no more than 4% of the total, comprises messenger RNA. The vast bulk is made up of **noncoding RNA**, molecules that are not translated into protein. An alternative name is **functional RNA**, which emphasizes that the noncoding RNAs still have essential roles within the cell. Clearly, we must know what the noncoding RNA does in order to fully understand the nature of the biological information contained in the genes.

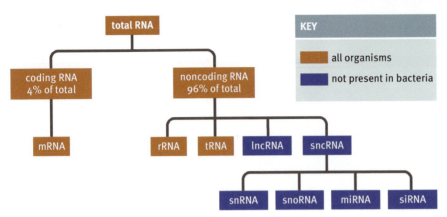

FIGURE 3.13 The various types of RNA that are found in living cells.

The most abundant type of noncoding RNA is **ribosomal** or **rRNA**. There are only four different rRNA molecules in humans, but together they can make up over 80% of the total RNA in an actively dividing cell. These molecules are components of ribosomes, the structures on which protein synthesis takes place. A second important type of noncoding RNA is **transfer** or **tRNA**. These are small molecules that are also involved in protein synthesis, carrying amino acids to the ribosome and ensuring these are linked together in the order specified by the nucleotide sequence of the mRNA that is being translated. Most organisms have genes for between 30 and 50 different tRNAs.

Ribosomal and transfer RNAs are present in all organisms. The other noncoding RNA types are more limited in their distribution and, in particular, are absent in bacteria (Figure 3.13). These RNAs are usually divided into two groups, the **short noncoding RNAs** (**sncRNAs**), comprising RNAs less than 200 nucleotides in length, and the **long noncoding RNAs** (**lncRNAs**), made up of molecules longer than 200 nucleotides. The sncRNAs include **small nuclear RNA** (**snRNA**), which is found in the nuclei. There are 15–20 different ones and most are involved in **RNA splicing**, the process by which copies of introns are removed from mRNA molecules and the exons, the parts containing biological information, joined together. Splicing is an essential part of the expression pathway for a discontinuous gene because the exons have to be linked before the mRNA can be translated into a protein (Figure 3.14). There are also **small nucleolar RNAs** (**snoRNAs**) found in the nucleoli, the parts of the nucleus in which rRNA is transcribed. Extra chemical groups, such as methyl groups, must be added to rRNA molecules before they can be assembled into ribosomes, and snoRNAs aid this process.

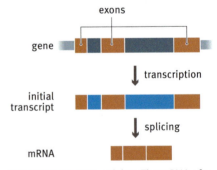

FIGURE 3.14 RNA spicing. The mRNA of a discontinuous gene must be spliced before it can be translated into protein.

This list was thought to be complete until quite recently. Now we know that there are other categories of noncoding RNA whose existence was not suspected until the 1990s. These include the **microRNAs** (**miRNAs**) and **small interfering RNAs** (**siRNAs**). Both are involved in the control of gene expression and the more we find out about them the more important they become. They cause certain mRNAs to be 'silenced' so they cannot be translated, possibly by attaching to these mRNAs by base pairing and causing them to be degraded. We will learn more about this process, called **RNA interference**, in Chapter 5.

The functions of the lncRNAs are less well understood. Humans have over 18,000 lncRNA genes. Most of these are located in intergenic DNA, but some are found within the introns of protein-coding genes (Figure 3.15). A few actually overlap with protein-coding exons, often being transcribed from the other strand of the double helix and hence giving rise to an

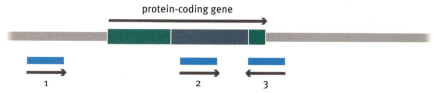

FIGURE 3.15 **Long noncoding RNA genes.** In example 3, the lncRNA gene overlaps with an exon of a protein-coding gene, but is transcribed from the other strand of the DNA. The lncRNA gene will therefore give rise to an antisense version of the mRNA of the exon from the protein-coding gene.

antisense RNA version of the mRNA. Most lncRNAs are synthesized only in certain tissues or at certain developmental stages. Their individual copy numbers are usually very low, with perhaps as few as two or three copies of a particular lncRNA per cell. It seems certain that at least some lncRNAs must play a role in the cell, but so far functions have been assigned to only a few of these molecules.

3.2 VARIATIONS IN THE INFORMATION CONTENT OF INDIVIDUAL GENES

One of the fundamental features of life is that the members of a species are not all identical to one another. It is obvious to us that this is the case with humans but it is equally true for all other species.

How is it possible for individuals to have their own distinctive biological characteristics when everybody has the same set of genes? Part of the distinctiveness might be due to environmental factors, dietary differences and suchlike, but this is only a small part of the answer. Biological characteristics are variable because the genes that code for these characteristics are themselves variable. Our next task is to understand the molecular basis of this variability.

Genetic variants are called alleles

Geneticists use the word **allele** to describe the variants of a biological characteristic. A good example is the round and wrinkled peas that Mendel studied. Here, the biological characteristic is 'pea shape' and there are two alleles, 'round' and 'wrinkled' (**Figure 3.16a**).

We know that the round and wrinkled alleles are specified by two variants of a gene involved in starch synthesis. The main biochemical difference between round and wrinkled peas is that the latter has a higher sucrose content and, as a consequence, absorbs more water while it is developing in the pod. When the pea reaches maturity some of this water is lost, causing the pea to collapse and become wrinkled. Round peas, which have less sucrose but more starch, do not take up so much water when they are developing and so become less dehydrated when they mature. As a result, they stay round (**Figure 3.16b**).

The key difference between wrinkled and round peas therefore lies in the relative amounts of sucrose and starch that they contain. This, in turn, depends on the activities of a series of enzymes that in the 'normal' (round) pea convert the bulk of the sucrose into starch (**Figure 3.16c**). Each of these enzymes is, of course, coded by a gene. One of these genes is called *SBE1* and codes for 'starch branching enzyme type 1'. In wrinkled peas, this enzyme is absent. This means that less sucrose is converted into starch and the pea collapses in the pod. Why is the enzyme absent?

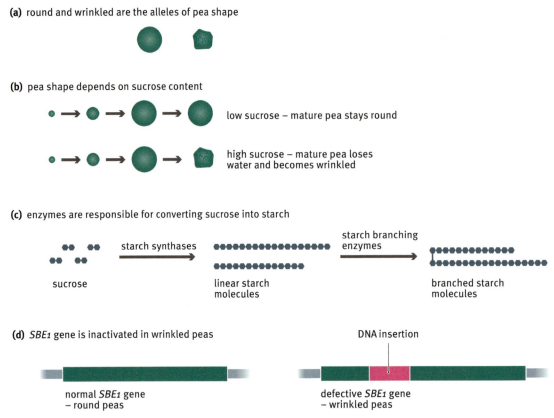

FIGURE 3.16 The genetic basis of pea shape, as studied by Mendel. (a) The two alleles of pea shape are 'round' and 'wrinkled'. (b) The main biochemical difference between round and wrinkled peas is that wrinkled peas have higher sucrose content. (c) The enzymes responsible for converting sucrose into starch include the starch branching enzymes, one of which is coded by the *SBE1* gene. (d) In wrinkled peas, this enzyme is absent because the *SBE1* gene has become inactivated by the insertion of an extra piece of DNA.

This is because in wrinkled peas, an extra piece of DNA has been inserted into the *SBE1* gene (Figure 3.16d). This extra piece of DNA interrupts the nucleotide sequence of the gene and prevents it from being expressed correctly. As a result, no starch branching enzyme type 1 is made.

We therefore understand the molecular basis of the round and wrinkled alleles of the pea shape characteristic. One allele, for round peas, is specified by the correct, functional copy of the gene, and the second allele, for wrinkled peas, arises when that gene is nonfunctional. Strictly speaking, the alleles are the alternative forms of the biological characteristic, but nowadays we also refer to the two versions of the gene as 'alleles'.

There may be many variants of the same gene

Some genes have a whole range of different variant forms. To explore this added complexity, we will look at the human *CFTR* gene, which codes for the cystic fibrosis transmembrane regulator protein. This is a membrane-bound protein involved in the transport of chloride ions into and out of human cells. As its name indicates, if this protein does not function correctly then the disorder called cystic fibrosis might develop, characterized by inflammation and mucus accumulation in the lungs.

The *CFTR* gene is 250 kb in length but most of that length is taken up by 26 introns (Figure 3.17a). If these are ignored, then the coding exons make up 4,440 bp, specifying a protein that is 1,480 amino acids in length – remember that each amino acid in a protein is coded by a triplet of three adjacent nucleotides. In the most common variant of *CFTR*, three nucleotides are deleted from positions 1514, 1515 and 1516 in the

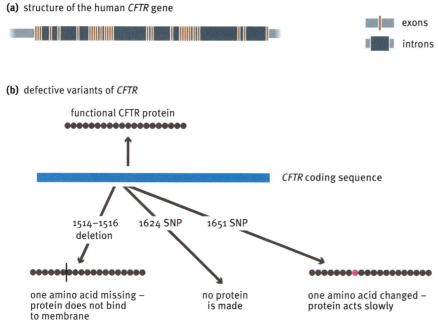

FIGURE 3.17 **Variants of the human CFTR gene.** (a) The structure of the *CFTR* gene, showing the positions of the exons and introns. There are 27 exons and 26 introns, one intron between each pair of adjacent exons. Some of the introns are very short and are not clearly visible when the gene is drawn at this scale. (b) The coding sequence of the *CFTR* gene (the gene minus its introns) is indicated by the blue line. The normal gene gives rise to the functional CFTR protein, as shown above the coding sequence. The products of the three most common variants of the gene are illustrated below the coding sequence.

coding sequence (Figure 3.17b). As a result, the protein that is synthesized has an amino acid missing. This small change prevents the protein from attaching to the cell membrane, which means that the protein cannot carry out its function. Cystic fibrosis might result.

Now things become a little more complicated. There is another variant of the *CFTR* gene in which the nucleotide at position 1624 is changed from a G to T. This is called a **single-nucleotide polymorphism** or **SNP**. The SNP causes the mRNA to be degraded so no cystic fibrosis transmembrane regulator protein is made. Once again the protein function is absent and cystic fibrosis might result.

Another *CFTR* variant has an SNP at position 1651, which does not affect the length of the protein that is made, but changes one of the amino acids. This alters the kinetic properties of the protein so it now transports chloride ions at only 4% of the rate displayed by the normal protein. Again, this can lead to cystic fibrosis. Although the protein is still able to transport chloride ions it does so at such a low rate that the physiological function of the protein is lost. In total, over 2,500 variants of the *CFTR* gene that give rise to a dysfunctional cystic fibrosis transmembrane regulator protein are known.

There are also variants that do not affect the ability of the gene to give rise to a functional protein. The nucleotide sequence changes present in some of these variants do not affect the amino acid sequence of the protein, and in others, they change it in a way that has no impact on the protein's function.

With a gene-protein system as complex as this, the use of 'allele' to describe the alternative forms of the biological characteristic *and* the underlying gene variants can make things seem more complicated than they are. Does the *CFTR* gene have two alleles ('unaffected' and 'cystic fibrosis') or does it have 2,500, the number of gene variants that can give rise to the 'cystic fibrosis' allele? Do we include the gene variants that do not affect the function of the protein, in which case there are many more than 2,500 alleles? To avoid these contradictions, geneticists increasingly use the word **haplotype** to describe a sequence variant of a gene. According to this nomenclature, the *CFTR* gene has more than 2,500 haplotypes, some giving rise to the 'unaffected' allele and some to the 'cystic fibrosis' allele.

3.3 FAMILIES OF GENES

Since the earliest days of DNA sequencing, it has been known that **multigene families** – groups of genes of identical or similar sequences – are common in many organisms. The members of these gene families therefore encode identical or similar units of biological information. Why do these families exist?

Multiple gene copies enable large amounts of RNA to be synthesized rapidly

Families of identical genes are generally assumed to specify RNA molecules or proteins that are needed in large quantities by the cell, continually or at certain times, such as when the cell is dividing. Gene expression is not an instantaneous process and there is a limit to the rate at which the product of an individual gene can be synthesized.

The identities of **simple multigene families**, ones in which all the genes are the same, support this idea. In most organisms, there are multiple copies of the rRNA genes. The bacterium *Escherichia coli*, for example, has seven copies of each of its three rRNA genes. Experiments have suggested that five copies of each rRNA gene are needed to maintain the number of ribosomes needed when the bacterium is dividing at its maximum rate in an unchanging environment. Under these conditions, there may be as many as 70,000 ribosomes in the cell. All seven gene copies are needed when the bacterium is adapting to a new environment, presumably because adaptation requires the synthesis of many new proteins that were not needed under the pre-existing conditions.

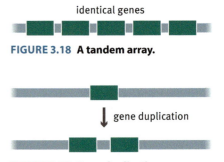

FIGURE 3.18 A tandem array.

Higher organisms also have multiple copies of their rRNA genes. In humans, there are about 280 copies of three of the rRNA genes and over 2,000 copies of the fourth one. These copies are grouped into large clusters where many genes are arranged head-to-tail in a **tandem array** (Figure 3.18). This arrangement probably does not aid the rate at which the genes can be transcribed, but is more a reflection of the process that led to the multiple copies. Each cluster is thought to have evolved by gene duplication, a single gene being duplicated into two copies placed side by side (Figure 3.19).

FIGURE 3.19 Gene duplication. Duplication of the original gene results in two identical copies placed side by side.

In some organisms, even these large numbers of gene copies are apparently unable to meet the demand for ribosomes at certain times. In some amphibians, the 450 or so copies of the rRNA genes in the chromosomes are increased to up to 16,000 copies by **gene amplification**. This involves replication of the rRNA genes into multiple DNA copies, which subsequently exist as independent molecules not attached to the chromosomes (Figure 3.20). Gene amplification typically occurs in oocytes that are actively developing into mature egg cells.

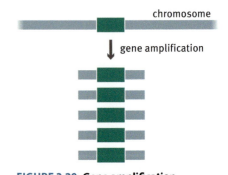

FIGURE 3.20 Gene amplification. The new gene copies resulting from amplification exist as independent molecules, rather than being part of the chromosome.

In some multigene families, the genes are active at different stages of development

In **complex multigene families**, the individual gene members, although having similar nucleotide sequences, are sufficiently different to code for proteins with distinctive properties. These proteins are therefore able to

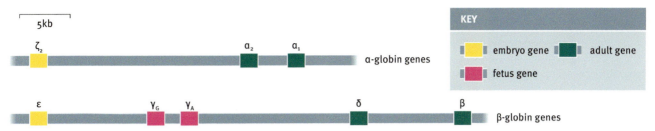

FIGURE 3.21 The human globin gene families. The genes in each family are active at different stages in human development. The α_1 and α_2 genes are the only ones that give rise to identical proteins.

play slightly different roles. The flexibility provided by these families of genes is sometimes exploited during the development of the organism.

The mammalian globin genes are one of the best examples of this type of multigene family. The globins are the blood proteins that combine to make hemoglobin, each molecule of hemoglobin being made up of two α-type and two β-type globin polypeptides. Both of these globin types, α and β, exist as a family of related molecules differing from one another at just a few amino acid positions.

In humans, there are two α-like globins, α and ζ_2, and five β-like globins, β, δ, γ_A, γ_G and ε. These proteins are coded by two multigene families (**Figure 3.21**). The genes were among the first to be sequenced, back in the late 1970s. The sequence data showed that the genes in each family are similar to one another, but by no means identical. In fact, the nucleotide sequences of the two most different genes in the β-type cluster, coding for the β- and ε-globins, display only 79.1% identity. Although this is similar enough for both genes to specify a β-type globin, it is sufficiently different for the proteins to have distinctive biochemical properties. Similar variations are seen in the α-cluster.

Why are the members of the globin gene families so different from one another? The answer was revealed when the expression patterns of the individual genes were studied. It was discovered that the genes are active at different stages in human development. For example, in the β-type cluster gene ε is expressed in the early embryo, γ_A and γ_G (whose protein products differ by just one amino acid) in the fetus, and δ and β in the adult (see Figure 3.21). The different biochemical properties of the resulting globin proteins are thought to reflect slight changes in the physiological role that hemoglobin plays during the course of human development.

The globin families reveal how genes evolve

Multigene families are thought to arise by gene duplication (see Figure 3.19). In a simple multigene family, the sequences of the gene copies stay the same after duplication, so all the members are identical. Exactly how this occurs is not completely understood, because our expectation is that sequences of the individual members of a family should gradually change over time. This is because DNA replication is not an entirely error-free process and because some chemicals and physical agents, such as ultraviolet radiation, can cause DNA sequence alterations by the process called **mutation**. A complex multigene family, in which the members have different but related sequences, is therefore the normal outcome of a series of gene duplications.

Mutations and replication errors accumulate over time, so two members of a gene family that have very similar sequences are likely to have arisen from a duplication that occurred more recently than that giving rise to two genes with less similar sequences (**Figure 3.22**). The pattern of gene

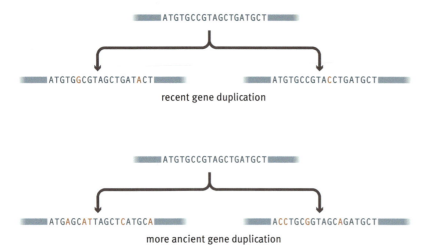

FIGURE 3.22 **Two genes resulting from a recent duplication are likely to be more similar to one another than two genes from a more ancient duplication.** The two new genes resulting from the recent gene duplication differ from their parent gene at two and one nucleotide positions, respectively. These nucleotides are shown in orange. In contrast, the genes resulting from the more ancient duplication differ from their parent at five and four nucleotide positions, respectively.

duplications that gave rise to the genes we see today can therefore be deduced. When we apply this logic to the β-globin family, then we infer that γ_A and γ_G are the result of the most recent duplication, as these are the two most similar genes in the cluster, and that β and δ arose from the preceding duplication (Figure 3.23a).

We can take the analysis a little bit further. In Chapter 20, we will learn how it is possible to work out the rate at which nucleotide sequence changes accumulate in genes. This rate can be converted into a **molecular clock** that enables dates to be assigned to gene duplication events. When we apply the molecular clock to the β-globin genes, we discover that the initial gene duplication that gave rise to the first two members of

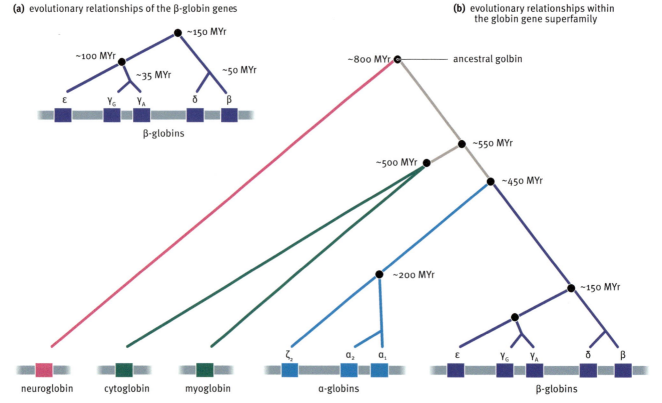

FIGURE 3.23 **The evolutionary relationships between various human globin genes.** (a) The relationships within the β-globin family. (b) The relationships within the globin gene superfamily. The relationships between the genes are estimated from the degree of nucleotide similarity between pairs of genes, with the rate at which mutations accumulate in human genes used as a molecular clock to infer the dates at which gene duplications occurred. MYr, million years ago.

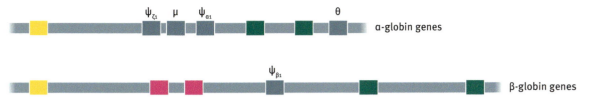

FIGURE 3.24 Pseudogenes in the human globin gene clusters. The θ pseudogene is transcribed into RNA but its protein product is inactive. The μ pseudogene is transcribed but its protein product has never been detected. None of the other pseudogenes is transcribed.

this family occurred approximately 150 million years ago. The times of the subsequent duplications can also be dated.

If we compare not just the β-globins but the members of the α-family as well, then we can make further deductions. The first duplication giving rise to the α-globins took place about 200 million years ago (**Figure 3.23b**). The very first α- and β-globin genes appeared as a result of a duplication 450 million years ago. Even further back there were duplications that gave rise to genes whose protein products are now synthesized in tissues other than blood. These include neuroglobin from the brain, myoglobin from muscles, and cytoglobin which is present in a number of tissues. Each of these proteins is involved in oxygen binding in one way or another, similar to the function of the α- and β-globins in blood cells.

The globin gene clusters also tell us something else about the evolution of genes. When we examine the α and β families, we see several genes whose nucleotide sequences have changed to such an extent that they have lost their function altogether. There are four such genes in the human α family, called $\Psi_{\zeta 1}$, μ, $\Psi_{\alpha 1}$ and θ, and one, $\Psi_{\beta 1}$, in the β cluster (**Figure 3.24**). These are **pseudogenes**, a type of evolutionary relic. They are related to the other genes in their clusters and arose from them by duplication, but since then they have acquired sequence changes that have resulted in their loss of function. Interestingly, the δ gene of the β cluster, which is an active gene in humans, is a pseudogene in some other mammals, including mice and rabbits. The gene clusters have therefore followed different evolutionary pathways in the lineages leading to these other species.

KEY CONCEPTS

- A gene is a segment of a DNA molecule. The shortest genes are less than 100 bp in length, and the longest ones are over 2,400,000 bp.
- Some genes are discontinuous. The biological information that they contain is split into sections called exons, separated by intervening sequences called introns.
- Utilization of the biological information contained in a gene requires the coordinated activity of enzymes and other proteins, in a series of events referred to as gene expression.
- All genes undergo the first stage of gene expression which is called transcription and which results in the synthesis of an RNA molecule.
- For some genes, the RNA transcript is the end product of gene expression. For others, the transcript is a short-lived message that directs the second stage of gene expression, called translation, during which the RNA molecule directs the synthesis of a protein.
- Protein synthesis is the key to the expression of biological information. Proteins of different types have many different functions. By coding for

different proteins, genes are able to specify the biological characteristics of a living organism.
- Many biological characteristics are variable because the genes that code for these characteristics are themselves variable. These variants are called alleles or haplotypes.
- Groups of genes of identical or similar sequences are common in many organisms. These are called multigene families.
- In a simple multigene family, all the genes are identical. Many of these families specify RNA molecules or proteins that are needed in large quantities by the cell.
- In a complex multigene family, the genes have similar sequences and code for related but slightly different proteins. Often these genes are active at different developmental stages.
- Multigene families evolve by gene duplication. The series of gene duplications giving rise to a complex multigene family can be deduced by making comparisons between the nucleotide sequences of the individual genes.

QUESTIONS AND PROBLEMS

Key terms

Write short definitions of the following terms:

allele
antisense RNA
complex multigene family
discontinuous gene
enzyme
exon
functional RNA
gene amplification
gene expression
genetic code
haplotype
intergenic DNA
intron

kilobase pair
long noncoding RNA
megabase pair
messenger RNA
microRNA
molecular clock
motor protein
multigene family
mutation
noncoding RNA
pseudogene
regulatory protein
ribosomal RNA

RNA interference
RNA splicing
short noncoding RNA
simple multigene family
single-nucleotide polymorphism
small interfering RNA
small nuclear RNA
small nucleolar RNA
structural protein
tandem array
transcription
transfer RNA
translation

Self-study questions

3.1 What is a gene?

3.2 Distinguish between the terms base pair, kilobase pair, and megabase pair. How many base pairs would there be in a gigabase pair?

3.3 Why does a human gene that is 2.4 Mb in length not contain 24,000 times the amount of information present in a gene 100 bp in length?

3.4 What is the difference between an exon and an intron? If a gene has six introns, then how many exons would it have?

3.5 Outline the process by which the information contained in a gene is used to direct the synthesis of a protein.

3.6 Write down the nucleotide sequences of the messenger RNA molecules that would be transcribed from these two genes:

5′–TACCCATGATTTCCCGCGATATATATGCCTGCAATC–3′

5′–TACCCTTGGTTTCGCGCGATGTATATACCAGGAATC–3′

3.7 Give examples of the various roles that proteins play in living organisms.

3.8 What is the difference between coding and noncoding RNA?

3.9 List the types of noncoding RNA found in human cells and briefly describe the functions of each one.

3.10 Write a description of the molecular basis for the pea shape characteristic studied by Mendel.

3.11 Give three examples of variants of the *CFTR* gene that give rise to cystic fibrosis.

3.12 Distinguish between the terms 'allele' and 'haplotype'.

3.13 What is the difference between a simple and a complex multigene family?

3.14 Give one example of a simple multigene family in which the individual genes are arranged in a tandem array. What is the biological purpose of this arrangement?

3.15 Give an example of gene amplification.

3.16 Draw a diagram showing the organization of the human α- and β-globin gene families.

3.17 Why are the members of the globin gene families so different from one another?

3.18 What do comparisons between the DNA sequences of the human globin genes tell us about the process by which these gene families evolved?

3.19 What is the special feature of the gene called $\Psi_{\beta 1}$ in the β-globin cluster?

Discussion topics

3.20 Discontinuous genes are common in higher organisms but virtually absent in bacteria. Discuss the possible reasons for this.

3.21 Eukaryotic cells expend a lot of energy in removing introns from the transcripts of discontinuous genes. Yet introns seem to play no role of their own. Discuss.

3.22 What would be the implications for theories of evolution if it were discovered that information can flow from proteins to RNA and thence to DNA?

3.23 There appears to be no biological reason why a DNA polynucleotide could not be directly translated into protein, without the intermediary role played by mRNA. What advantages do cells gain from the existence of mRNA?

3.24 Scientists studying the origins of life have proposed that RNA was the first type of nucleic acid to evolve in the primordial soup, with DNA appearing later. How does the existence of mRNA fit in with this theory?

3.25 Human cells are able to make at least 1,000 different types of microRNA, and these molecules play a central role in controlling the expression of human genes. As there are so many microRNAs and they are so important, why was their existence not suspected until the 1990s?

3.26 Expression of different versions of the α- and β-globin genes at different developmental stages is thought to reflect changes in the physiological role played by hemoglobin during human development. What are these different physiological roles? Do you think this theory can be defended?

The answers to, or hints on how to answer, the self-study questions and discussion topic questions can be downloaded from the book's product page at this link: www.routledge.com/9781032743530

FURTHER READING

Brown DD & Dawid IB (1968) Specific gene amplification in oocytes: oocyte nuclei contain extrachromosomal replicas of genes for ribosomal RNA. *Science* 160, 272–280.

Catania F & Lynch M (2008) Where do introns come from? *PLoS Biology* 6, e283.

Cheetham SW, Faulkner GJ & Dinger ME (2020) Overcoming challenges and dogmas to understand the functions of pseudogenes. *Nature Reviews Genetics* 21, 191–201. *Explores the possibility that some pseudogenes have functions.*

Crick FHC (1970) Central Dogma of molecular biology. *Nature* 227, 561–563. *Summarizes the process of gene expression from DNA to protein.*

Efstratiadis A, Posakony JW, Maniatis T et al. (1980) The structure and evolution of the human beta-globin gene family. *Cell* 21, 653–668.

Fincham JRS (1990) Mendel – now down to the molecular level. *Nature* 343, 208–209. *Describes the molecular basis for round and wrinkled peas.*

Fritsch EF, Lawn RM & Maniatis T (1980) Molecular cloning and characterization of the human α-like globin gene cluster. *Cell* 19, 959–972.

Kumar S (2005) Molecular clocks: four decades of evolution. *Nature Reviews Genetics* 6, 654–662. *Describes in detail the basis of the use of the molecular clock to date events in the past.*

Kung JTY, Colognori D & Lee JT (2013) Long noncoding RNAs: past, present, and future. *Genetics* 193, 651–669.

Veit G, Avramescu RG, Chiang AN et al. (2016) From CFTR biology toward combinatorial pharmacology: expanded classification of cystic fibrosis mutations. *Molecular Biology of the Cell* 27, 424–433.

Transcription of DNA to RNA

In Chapter 3, we learnt that the biological information contained in a gene is encoded in its nucleotide sequence and that this information is utilized by the cell through the process called gene expression. The end product of gene expression is either a noncoding RNA or a protein that has a specific function in the cell. Through the synthesis of these end products, the genes are able to specify the biological characteristics of the organism.

The essential link between the biological information contained in the genes and the characteristics of the living organism is therefore made by the gene expression pathway. We must now turn our attention to this pathway and examine exactly how DNA is transcribed into RNA and mRNA translated into protein. As we are taking the modern molecular approach to genetics, we must acquire much more than a simple overview of this process and instead must look closely at the events that occur at each step in gene expression. An important component of genetics research in the 21st century is concerned with describing those events in greater detail, and in particular in understanding how the expression of individual genes is controlled in healthy cells, and how errors in those control processes might lead to diseases such as cancer.

The details of the gene expression pathway and how it is controlled will take up the next six chapters, beginning in this chapter with the first step, the transcription of DNA to RNA. First, however, we must look at gene expression in outline and identify the key features of the pathway in different types of organism.

4.1 GENE EXPRESSION IN DIFFERENT TYPES OF ORGANISM

A major challenge for a student of genetics is learning the variations in the gene expression pathway that exist in different types of organism. In some organisms, the pathway is much less complicated than in others. When studying genetics for the first time, there is a natural temptation to concentrate on the simplest example of gene expression as this is, obviously, the easiest to remember. The problem is that this means that substantial parts of the pathway operating in humans are ignored. Genetics is not just about humans, but humans are such important organisms, at least to us, that any serious course in genetics must have a strong human element. In this book, we will not shy away from the human complexities. To put these in context, we begin by comparing the gene expression pathways for different types of organism.

There are three major groups of organism

Traditionally, biologists have divided organisms into two groups, called **prokaryotes** and **eukaryotes.** Prokaryotes and eukaryotes are distinguished by their fundamentally different cellular organizations (Figure 4.1). Prokaryotes, which include bacteria, lack an extensive cellular architecture and their DNA is not enclosed in a distinct structure. In contrast, the typical eukaryotic cell is usually larger and more complex, with a membrane-bound nucleus containing the chromosomes and with other distinctive membranous organelles such as mitochondria, vesicles and Golgi bodies.

Most prokaryotes are unicellular though in some species individual cells can associate together to form larger structures, such as the chains of cells formed by *Anabaena* (Figure 4.2). Eukaryotes can be unicellular

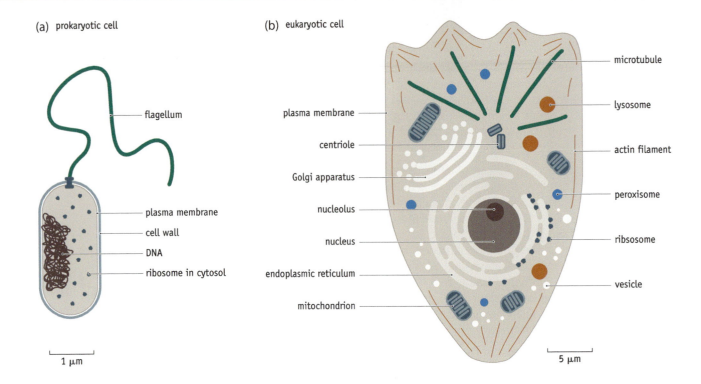

FIGURE 4.1 The different cellular organizations of prokaryotes and eukaryotes. (a) A typical bacterial cell. (b) An example of an animal cell. (Adapted from B. Alberts et al., *Essential Cell Biology*, 3rd ed. New York: Garland Science, 2010.)

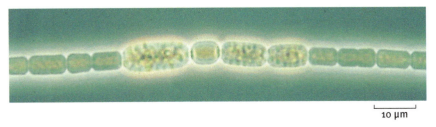

FIGURE 4.2 Part of a chain of cells formed by the photosynthetic bacterium *Anabaena cylindrica*. (Courtesy of David Adams, University of Leeds.)

or multicellular and comprise all the macroscopic forms of life such as plants, animals and fungi (Figure 4.3). Humans are therefore eukaryotes.

Until 1977, it was thought that all prokaryotes were more or less the same. It was recognized that the group included a great diversity of organisms but the differences were thought to be variations on a theme. This assumption has now been overturned and it is accepted that the prokaryotes include two distinct groups of organism, the **bacteria** and the **archaea**. The bacteria comprise most of the prokaryotes that are familiar to us as people and scientists, such as the pathogens *Mycobacterium tuberculosis*, which causes tuberculosis in humans, *Vibrio cholerae*, responsible for cholera, and *Escherichia coli*, some strains of which can cause food poisoning but which is more familiar to geneticists as a bacterium that is frequently studied in the laboratory (Figure 4.4).

The archaea are less well known, partly because they are not widespread in nature. They include methanogens that live at the bottom of lakes and other bodies of water and release methane, which they synthesize as a metabolic by-product. Other types of archaea are confined to extreme environments, such as hot springs where the temperature can be 60°C or higher, brine pools and high-salt lakes such as the Dead Sea, and acidic streams emerging from old mine workings. All of these environments are hostile to most other forms of life (Figure 4.5).

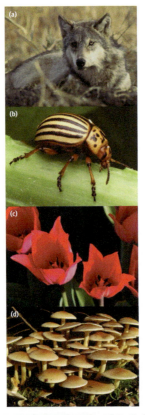

FIGURE 4.3 Examples of multicellular eukaryotic species. (a) A gray wolf, *Canis lupus*. (b) The Colorado potato beetle, *Leptinotarsa decemlineata*. (c) Tulips. (d) *Strophariaceae*, a type of fungus. (a, courtesy of U.S. Fish and Wildlife Service, John & Karen Hollingsworth; b, courtesy of U.S. Department of Agriculture, Scott Bauer.)

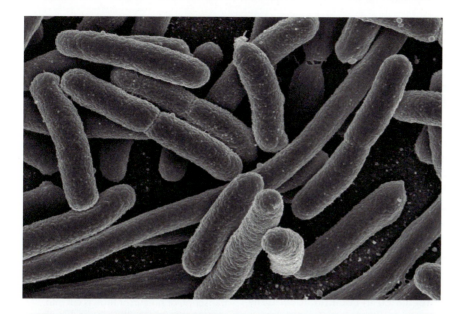

FIGURE 4.4 Micrograph of *Escherichia coli*. This species of bacteria is frequently studied in the laboratory. (Courtesy Rocky Mountain Laboratories, NIAID, NIH.)

FIGURE 4.5 An example of an extreme environment, likely to harbor members of the archaea. The picture shows an acidic mine drainage in Spain. (Courtesy Carol Stoker, NASA Ames Research Center.)

We now look on the eukaryotes, bacteria and archaea as the three main types of living organism, with each group displaying its own distinctive genetic features. Of the three groups, the archaea are by far the least well studied, and at present the gene expression pathway operating in these organisms is not understood in detail. We must therefore focus our attention on the differences in gene expression in bacteria and eukaryotes.

Gene expression is more than simply 'DNA makes RNA makes protein'

In its simplest form, gene expression can be looked on as a two-step process, conveniently described as 'DNA makes RNA makes protein'. The first step, transcription, results in the synthesis of mRNAs and noncoding RNAs, and the second step, translation, uses the mRNAs to direct the synthesis of proteins (Figure 4.6).

For bacteria, 'DNA makes RNA makes protein' is a reasonably accurate description of the process as a whole. Its one weakness is that this description

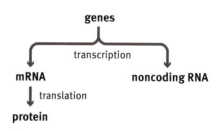

FIGURE 4.6 Gene expression described as a two-step process. The first step is the synthesis of mRNA and noncoding RNA by transcription, and the second step is the translation of the mRNA into protein.

implies that the noncoding RNAs that are made during the first step are immediately able to take up their functions in the cell. In reality, the primary transcripts of these noncoding RNAs have to be cut into segments to release the functional molecules, and some also have additional chemical groups added (Figure 4.7). These are called **processing** events and they are an important part of the gene expression pathway for bacterial noncoding RNAs.

In eukaryotes, gene expression is much more than 'DNA makes RNA makes protein' (Figure 4.8). In eukaryotes, we place much greater emphasis on the start of the process, when a set of proteins, including the enzyme that carries out transcription, is assembled on the DNA adjacent to the gene that is going to be expressed. This preliminary step is more complicated in eukaryotes compared with bacteria, largely because eukaryotes have more sophisticated mechanisms for controlling the expression of individual genes, and many of these control processes operate by regulating the assembly of this **transcription initiation complex**. The same is true for the initiation of translation, which is a second important control point in eukaryotic gene expression and hence is more complex than the equivalent event in bacteria. This is not to suggest that gene expression is not regulated in bacteria. It certainly is, and both transcription and translation are controlled, but the regulatory processes in bacteria have not evolved the degrees of complexity and sophistication that we see in eukaryotes. This will become clear when we examine these regulatory mechanisms, in both types of organism, in Chapter 9.

A second difference is that processing events are more extensive in eukaryotes (Figure 4.9). In addition to the noncoding RNAs, eukaryotic mRNAs must be processed to remove introns from the transcripts of discontinuous genes. This is essential because the exons – the coding segments – must be linked together before the mRNA can be translated. In eukaryotes, many proteins are also processed, some by cutting into segments and some by **post-translational chemical modification**. These modifications include the addition of chemical groups ranging in size from methyl residues (–CH$_3$) to large side chains made up of 5–10 sugar units. As with noncoding RNA processing, these protein processing events are an integral part of gene expression as the end product (in this case, the protein) is not functional until after it has been processed.

In addition to the synthesis of RNA and protein, we must also consider the degradation or **turnover** of these molecules. Turnover is not simply a way of getting rid of unwanted molecules. It plays an active role in controlling gene expression by influencing the amount of an RNA or a protein that is present in the cell at any particular time.

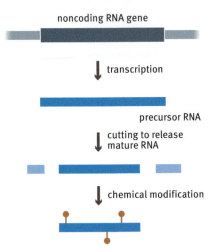

FIGURE 4.7 Processing of the primary transcripts of noncoding RNA genes. Following transcription, the primary transcript, or precursor RNA, is cut into segments to release the mature, functional RNAs. Some of these RNAs may also be modified by the addition of extra chemical groups.

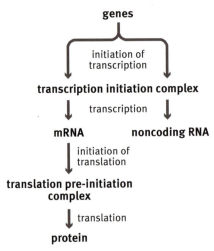

FIGURE 4.8 The more complex series of events involved in gene expression in eukaryotes.

4.2 ENZYMES FOR MAKING RNA

An enzyme that transcribes DNA into RNA is called a **DNA-dependent RNA polymerase**. It is permissible to shorten this to **RNA polymerase**, as the context in which the name is used means that there is rarely confusion with the **RNA-dependent RNA polymerases** that are involved in the replication of some virus RNAs. The RNA polymerase is the central component of the transcription process, and we must therefore familiarize ourselves with the properties of these enzymes.

The RNA polymerase of *Escherichia coli* comprises six subunits

Like many of the proteins involved in gene expression, an RNA polymerase has to perform several tasks, as will be seen when the events occurring during transcription are described later in this chapter. It is not surprising,

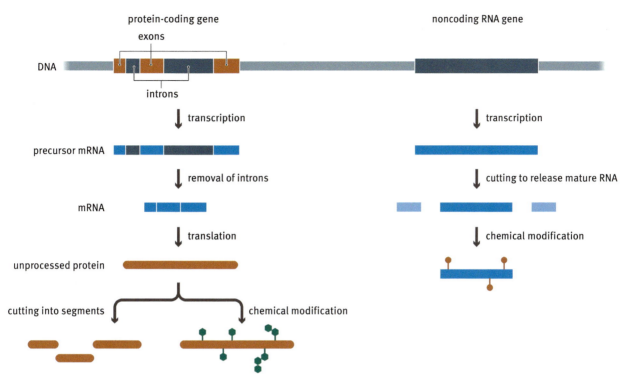

FIGURE 4.9 RNA and protein processing in eukaryotes. On the right, we see the same processing events for noncoding RNA as occur in bacteria. On the left, the processing of eukaryotic mRNAs and proteins is illustrated. If the gene is discontinuous, then the initial transcript is a precursor mRNA whose introns must be removed to give the mature mRNA. After translation of this mRNA, the initial protein product may be cut into segments and may undergo chemical modification. Processing of mRNA is common in eukaryotes, but virtually unknown in bacteria because very few bacterial genes contain introns. Protein processing is also common in eukaryotes but much less frequent in bacteria.

therefore, that the *E. coli* RNA polymerase is a large protein made up of several different polypeptide subunits.

Each *E. coli* cell contains about 7,000 RNA polymerase molecules, between 2,000 and 5,000 of which may be actively involved in transcription at any one time. The structure of the enzyme is described as $\alpha_2\beta\beta'\sigma\omega$, meaning that each molecule is made up of two α subunits plus one each of β, the related β′, σ and ω (**Figure 4.10**). This version of the enzyme is called the **holoenzyme** and has a molecular mass of 460 kDa. It is distinct from a second form, the **core enzyme** (molecular mass 390 kDa), which lacks the σ subunit and is just $\alpha_2\beta\beta'\omega$. The two versions of the enzyme have different roles during transcription, as will be described below.

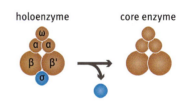

FIGURE 4.10 The *E. coli* RNA polymerase holoenzyme and core enzyme.

Eukaryotes possess more complex RNA polymerases

All *E. coli* genes are transcribed by the same type of RNA polymerase enzyme. In contrast, transcription of eukaryotic nuclear genes requires three different RNA polymerases. These are called **RNA polymerase I**, **RNA polymerase II** and **RNA polymerase III**. The eukaryotic enzymes are larger than the *E. coli* version, each being made up of eight to twelve subunits and having a total molecular mass in excess of 500 kDa. Structurally, these polymerases are quite similar to one another, the three largest subunits being closely related and some of the smaller ones being shared by more than one enzyme. The three largest subunits appear to be equivalent to the σ, β and β′ subunits of the *E. coli* enzyme. This has been deduced by comparing the amino acid sequences of the subunits from *E. coli* and yeast (which is a unicellular eukaryote) and by examining their exact roles during transcription.

Although the three eukaryotic RNA polymerases have related structures, their functions are quite distinct. Each works on a different set of genes, with no interchangeability (**Table 4.1**). Most research attention has been directed at

TABLE 4.1 FUNCTIONS OF THE THREE EUKARYOTIC RNA POLYMERASES

Polymerase	Genes Transcribed
RNA polymerase I	28S, 5.8S and 18S rRNA genes
RNA polymerase II	Protein-coding genes, genes for Sm family of snRNAs, some snoRNAs, siRNA, miRNA, lncRNA
RNA polymerase III	Genes for 5S rRNA, tRNAs, Lsm family of snRNAs, some snoRNAs

RNA polymerase II, as this is the one that transcribes genes that code for proteins and hence synthesizes mRNA. But this is not its only role as RNA polymerase II also transcribes genes specifying various types of noncoding RNA, including one family of snRNAs, some snoRNAs, siRNA, miRNA and lncRNA.

The other two types of eukaryotic RNA polymerase only synthesize noncoding RNA. RNA polymerase I transcribes a set of adjacent genes specifying three of the four rRNA molecules present in eukaryotic ribosomes. The fourth of these rRNAs is synthesized by RNA polymerase III, which also transcribes genes for tRNAs, snoRNAs and a second family of snRNAs.

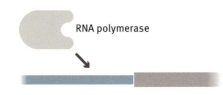

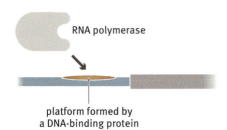

4.3 RECOGNITION SEQUENCES FOR TRANSCRIPTION INITIATION

It is essential that transcription occurs at the correct positions on a DNA molecule. RNA polymerase enzymes must transcribe genes rather than random pieces of DNA and must begin transcription near the start of a gene rather than in the middle. This means that the initial binding of an RNA polymerase to a DNA molecule must occur at a specific position, just in front ('**upstream**') of the gene to be transcribed. Positions where transcription should begin are therefore marked by target sequences that are recognized either by the RNA polymerase itself or by a DNA-binding protein which, once attached to the DNA, forms a platform to which the RNA polymerase binds (Figure 4.11).

FIGURE 4.11 Two ways in which RNA polymerases bind to their promoters. (a) Direct attachment of the RNA polymerase to the DNA molecule, as occurs in bacteria. (b) Indirect attachment of the RNA polymerase via a platform formed by a DNA-binding protein, as occurs in eukaryotes.

Bacterial RNA polymerases bind to promoter sequences

In bacteria, the target sequence for RNA polymerase attachment is called the **promoter**. A promoter is a short nucleotide sequence that is recognized by the bacterial RNA polymerase as a point at which it should bind to DNA in order to begin transcription. Promoters occur just upstream of genes, and nowhere else.

The sequences that make up the *E. coli* promoter were first identified by comparing the regions upstream of over 100 genes. It was assumed that promoter sequences would be very similar for all genes and so should be recognizable when the upstream regions were compared. These analyses showed that the *E. coli* promoter is made up of two distinct components, the –35 box and the –10 box (Figure 4.12). The latter is also called the **Pribnow box**, after the scientist who first discovered it. The names refer to the locations of the boxes on the DNA molecule relative to the position at which transcription starts. The nucleotide at this point is labeled '+1' and is anything between 20 and 600 nucleotides upstream of the start of the coding region of the gene. The spacing between the two boxes is important because it places the two motifs on the same face of the double helix, facilitating their interaction with the DNA-binding component of the RNA polymerase.

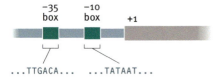

FIGURE 4.12 The two components of an *E. coli* promoter. The position at which transcription starts is labeled '+1'. The two sequences are the consensus sequences of the –35 box and –10 box.

The sequence of the –35 box is 5′–TTGACA–3′ and that of the –10 box is 5′–TATAAT–3′. These are **consensus sequences** and so describe the 'average'

TABLE 4.2 SEQUENCES OF *E. COLI* PROMOTERS

Promoter	Sequence	
	−35 Box	−10 Box
Consensus	5′–TTGACA–3′	5′–TATAAT–3′
Lactose promoter[a]	5′–TTTACA–3′	5′–TATGTT–3′
Tryptophan promoter[a]	5′–TTGACA–3′	5′–TTAACT–3′

[a] These are the promoters for the lactose and tryptophan operons, which we will study in Chapter 9.

of all promoter sequences in *E. coli*. The actual sequences upstream of any particular gene might be slightly different (**Table 4.2**). These variations, together with less well-defined sequence features around the transcription start site and the following 50 or so nucleotides, affect the efficiency of the promoter. Efficiency is defined as the number of productive initiations that are promoted per second, a productive initiation being one that results in the RNA polymerase leaving the promoter and beginning the synthesis of a full-length transcript. The exact way in which the sequence of the promoter affects initiation is not known, but it is clear that different promoters vary 1,000-fold in their efficiencies. The most efficient promoters (called **strong promoters**) direct 1,000 times as many productive initiations as the weakest promoters. We refer to these as differences in the **basal rate** of transcript initiation.

Eukaryotic promoters are more complex

In eukaryotes, the term 'promoter' is used to describe all the sequences that are important in the initiation of transcription of a gene. For some genes, these sequences can be numerous and diverse in their functions. They include the **core promoter**, which is the site at which the initiation complex containing the RNA polymerase is assembled, as well as one or more short sequences found at positions upstream of the core promoter (**Figure 4.13**). Assembly of the initiation complex on the core promoter can usually occur in the absence of the upstream elements, but only in an inefficient way.

Each of the three eukaryotic RNA polymerases recognizes a different type of promoter sequence. Indeed, it is the difference between the promoters that defines which genes are transcribed by which polymerase.

The promoters for RNA polymerase II, the enzyme that transcribes protein-coding genes, are the most complicated, some stretching for several kb upstream of the transcription start site (**Figure 4.14**). The RNA

FIGURE 4.13 The components of a typical promoter for a eukaryotic RNA polymerase. The core promoter is the attachment point for the complex of proteins containing the RNA polymerase. Most core promoters are preceded by a series of upstream promoter elements that play an ancillary role in the initiation of transcription.

FIGURE 4.14 A typical promoter for a gene transcribed by RNA polymerase II. The core promoter is made up of two segments, the TATA box and Inr sequence. The consensus sequences of these two segments are shown. The five boxes shown in light green are regulatory sequences. These will be described in Chapter 9.

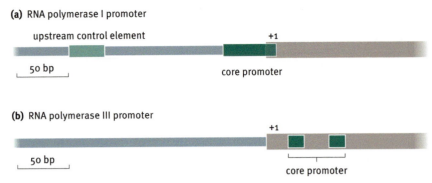

FIGURE 4.15 Promoters for genes transcribed by RNA polymerases I and III. (a) A typical promoter for RNA polymerase I consists of a core promoter that spans the point where transcription will begin and a single upstream control element. (b) RNA polymerase III promoters are variable. In this example, the core promoter is made up of two segments located within the gene.

polymerase II core promoter consists of two main segments. The first of these is the –25 or **TATA box**, which has the consensus sequence 5'–TATAWAAR–3'. In this sequence, 'W' indicates either A or T, which are equally likely to occur at this position, and 'R' is a purine, A or G. The second part of the promoter is the **initiator (Inr) sequence**, which is located around nucleotide +1. In mammals, the consensus of the Inr sequence is 5'–YYANWYY–3', where 'Y' is C or T and 'N' is any of the four nucleotides. Some genes transcribed by RNA polymerase II have only one of these two components of the core promoter, and some, surprisingly, have neither. The latter are called 'null' genes. They are still transcribed, although the start position for transcription is more variable than for a gene with a TATA and/or Inr sequence. In addition to the components of the core promoter, genes transcribed by RNA polymerase II also have various upstream sequence elements to which regulatory proteins bind. These will be described when we study the control of gene expression in Chapter 9.

The promoters for RNA polymerases I and III are not so complicated (Figure 4.15). RNA polymerase I promoters consist of a core promoter region spanning the transcription start point, between nucleotides –45 and +20, and a sequence called an **upstream control element** about 100 bp further upstream. RNA polymerase III promoters are variable, falling into at least three categories. Two of these categories are unusual in that the important sequences are located within the genes whose transcription they promote. Usually, these sequences span 50–100 bp and comprise two segments whose sequences are similar in all promoters of a particular type, separated by a variable region.

Some eukaryotic genes have more than one promoter

In eukaryotes, an additional level of complexity is seen with some protein-coding genes that have **alternative promoters**. This means that transcription of the gene can begin at two or more different sites, giving rise to mRNAs of different lengths.

An example is provided by the human dystrophin gene, which has been extensively studied because defects in this gene result in a genetic disease called Duchenne muscular dystrophy. The dystrophin gene is one of the largest known human genes, stretching over 2.4 Mb and containing 78 introns. It has at least seven alternative promoters (Figure 4.16). These promoters are active in different parts of the body, such as the brain, muscle and retina, enabling different versions of the dystrophin protein to be made in these various tissues. Presumably, the biochemical properties of these variants are matched to the needs of the cells in which they are synthesized.

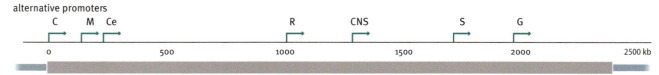

FIGURE 4.16 Alternative promoters. The positions of seven alternative promoters for the human dystrophin gene are shown. Abbreviations indicate the tissue within which each promoter is active: C, cortical tissue; M, muscle; Ce, cerebellum; R, retinal tissue (and also brain and cardiac tissue); CNS, central nervous system (and also kidney); S, Schwann cells; G, general (most tissues other than muscle).

In addition to the tissue-specific pattern of expression directed by the alternative promoters of the dystrophin gene, alternative promoters are also used to generate related versions of a protein at different stages in development and to enable a single gene to direct the synthesis of two or more proteins at the same time in a single tissue. The last point indicates that although usually referred to as *alternative* promoters these are, more correctly, *multiple* promoters as more than one may be active at a particular time. Indeed, this may be the normal situation for many genes. For example, over 10,500 promoters have been shown to be active in human fibroblast cells, but these promoters drive the expression of fewer than 8,000 genes. A substantial number of genes in these cells are therefore being expressed from two or more promoters.

4.4 INITIATION OF TRANSCRIPTION IN BACTERIA AND EUKARYOTES

It is useful to divide the transcription process into three phases, referred to as initiation, elongation and termination. Of these, initiation is looked on as the most important phase because it is a key control point in the gene expression pathway. Regulation of transcript initiation often determines whether or not a gene is active in a particular cell at a particular time.

In a general sense, the initiation of transcription operates along the same lines with each of the four types of RNA polymerase. The bacterial polymerase and the three eukaryotic enzymes all begin by attaching, directly or via accessory proteins, to their promoter or core promoter sequences (Figure 4.17). Next, this **closed promoter complex** is converted into an **open promoter complex** by breakage of a limited number of base pairs around the transcription initiation site. Finally, the RNA polymerase moves away from the promoter. This last step is more complicated than it might appear because some attempts by the polymerase to leave the promoter region are unsuccessful and lead to truncated transcripts that are degraded soon after they are synthesized. True completion of the initiation stage of transcription therefore occurs when the RNA polymerase has made a stable attachment to the DNA and has begun to synthesize a full-length transcript.

Although the scheme shown in Figure 4.17 is correct in outline for all four polymerases, the details are different for each one. We will begin with the more straightforward events occurring in *E. coli* and other bacteria and then move on to the ramifications of initiation in eukaryotes.

The σ subunit recognizes the bacterial promoter

In *E. coli*, a direct contact is formed between the promoter and RNA polymerase. The sequence specificity of the polymerase resides in its σ subunit. The 'core enzyme', which lacks this component, can only make loose and nonspecific attachments to DNA.

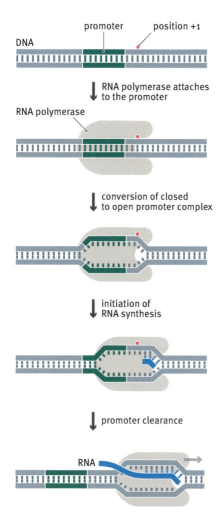

FIGURE 4.17 Generalized scheme for the events occurring during initiation of transcription. The diagram shows the initiation of transcription by a bacterial RNA polymerase, but equivalent events take place with each of the three eukaryotic enzymes. Attachment of the RNA polymerase to the promoter (shown in green) is followed by the breakage of base pairs adjacent to the transcription initiation site (indicated by the red dot) to give the open promoter complex. RNA synthesis then begins. Initiation of transcription is completed by 'promoter clearance', which occurs when the RNA polymerase has made a stable attachment to the DNA and has begun to synthesize a full-length transcript.

Studies of *E. coli* promoters have shown that the ability of the RNA polymerase to bind can be affected if changes are made to the –35 box, whereas changes to the –10 box affect the conversion of the closed promoter complex into the open form. These results have led to the model for *E. coli* initiation shown in Figure 4.18, where recognition of the promoter occurs by an interaction between the σ subunit and the –35 box. This interaction forms a closed promoter complex in which the RNA polymerase spans some 80 bp of DNA, from upstream of the –35 box to downstream of the –10 box. The closed promoter complex is converted to the open form by the combined action of the β' and σ subunits, which break the base pairs within the –10 box. The model is consistent with the fact that the –10 boxes of different promoters are comprised mainly or entirely of A–T base pairs, which are linked by just two hydrogen bonds, and are therefore weaker than G–C pairs, which have three hydrogen bonds (see Figure 2.17).

Opening up of the helix requires that contacts be made between the polymerase and the nontranscribed strand of the gene, again with the σ subunit playing a central role. However, the σ subunit is not all-important because it usually (but not always) dissociates soon after initiation is complete, converting the holoenzyme to the core enzyme which carries out the elongation phase of transcription. Initially, the core enzyme covers some 60 bp of the DNA, but soon after the start of elongation, the polymerase undergoes a second conformational change, reducing its footprint to just 30–40 bp.

Formation of the RNA polymerase II initiation complex

How does the initiation of transcription in *E. coli* compare with the equivalent processes in eukaryotes? RNA polymerase II will show us that eukaryotic initiation involves more proteins and has added complexities.

As illustrated in Figure 4.11, an important difference between initiation of transcription in *E. coli* and eukaryotes is that eukaryotic polymerases do not directly recognize their core promoter sequences. For genes transcribed by RNA polymerase II, the initial contact is made by the protein called **transcription factor IID**, or **TFIID** for short. This is another multisubunit complex, made up of the **TATA-binding protein** (**TBP**) which, as its name implies, is the part that actually recognizes the core promoter, and at least 12 **TBP-associated factors** or **TAFs**. Structural studies of TBP show that it has a saddle-like shape that wraps partially around the double helix, forming a platform onto which the remainder of the initiation complex can be assembled (Figure 4.19). The TAFs assist in the attachment of TBP to the TATA box and, in conjunction with other proteins called **TAF- and initiator-dependent cofactors** (**TICs**), possibly also participate in recognition of the Inr sequence, especially at those promoters that lack a TATA box.

The next step results in the attachment of RNA polymerase II to the complex (Figure 4.20). This involves two more transcription factors. These are TFIIB, which attaches to the DNA-bound TBP and forms the structure that RNA polymerase II recognizes, and TFIIF, which binds to the RNA polymerase and prevents it from attaching to the DNA at the wrong place. Finally, two more transcription factors, TFIIE and TFIIH, are recruited to complete the initiation complex. TFIIH is particularly important as it is a **helicase**, a type of enzyme that is able to break the base pairs that hold the two strands of the double helix together. This activity is needed to convert the closed promoter complex into the open form (see Figure 4.17).

The initiation complex must be activated before it will begin to transcribe the DNA to which it is attached. Activation involves the addition of phosphate groups to the largest subunit of RNA polymerase II, specifically to a series of amino acids within the part of this protein referred to as the **C-terminal**

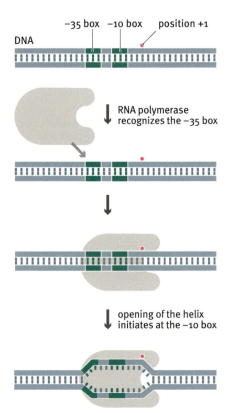

FIGURE 4.18 Initiation of transcription in *E. coli*. The RNA polymerase recognizes the –35 box. The breakage of base pairs that converts the closed promoter complex into the open complex is thought to occur initially within the –10 box.

FIGURE 4.19 The role of the TATA-binding protein (TBP) in initiation of transcription in eukaryotes. TBP is a dimer of two identical subunits, shown here in brown. The protein attaches to the core promoter, forming a platform onto which the initiation complex can be assembled. The DNA is shown in silver – in this view we are looking directly along the double helix. (Courtesy Song Tan, Pennsylvania State University.)

domain (**CTD**). Once these phosphates have been attached, the polymerase is able to leave the initiation complex and begin synthesizing RNA.

After the departure of the polymerase, at least some of the transcription factors detach from the core promoter, but TFIID, TFIIA and TFIIH remain, enabling re-initiation to occur without the need to rebuild the entire assembly from the beginning. Re-initiation is therefore a more rapid process than primary initiation, which means that once a gene is switched on, transcripts can be initiated from its promoter with relative ease until such a time as a new set of signals switches the gene off.

Initiation of transcription by RNA polymerases I and III

We know rather less about the way in which transcription is initiated at RNA polymerase I and III promoters, but it is clear that in some respects the events are similar to those seen with RNA polymerase II. One of the most striking similarities is that TBP, first identified as the key promoter-locating component of the RNA polymerase II initiation complex, is also involved in the initiation of transcription by the two other eukaryotic RNA polymerases.

The RNA polymerase I initiation complex is made up of the polymerase and four additional multisubunit proteins, one of which contains TBP. These proteins locate the core promoter and the upstream control element and hence direct RNA polymerase I to its correct attachment point (**Figure 4.21a**). Originally, it was thought that the initiation complex was built up in a stepwise fashion, but recent research suggests that RNA polymerase I binds to its four protein partners before locating the promoter, the entire assembly attaching to the DNA in a single step.

TBP is also a subunit of the transcription factor called TFIIIB, which is involved in the assembly of RNA polymerase III initiation complexes. The variability displayed by the promoters for this polymerase (see Figure 4.15) is reflected in the processes used for assembly of the initiation complexes, a different set of transcription factors being required for each category of promoter. TFIIIB provides the common link by making the initial contact with the core component of each promoter and, via its TBP subunit, providing the platform onto which RNA polymerase III is attached (**Figure 4.21b**).

4.5 THE ELONGATION PHASE OF TRANSCRIPTION

Once successful initiation has been achieved, the RNA polymerase begins to synthesize the transcript. Synthesis of an RNA molecule involves the polymerization of ribonucleotide subunits and can be summarized as:

$$n(NTP) \rightarrow RNA \text{ of length } n \text{ nucleotides} + n-1 (PPi)$$

The reaction is shown in detail in **Figure 4.22**. During each nucleotide addition, the β- and γ-phosphates are removed from the incoming nucleotide, and the hydroxyl group is removed from the 3'-carbon of the nucleotide at the end of the chain. This results in a loss of a pyrophosphate (PPi) molecule for each bond formed. In transcription, the chemical reaction is modulated by the presence of the DNA template, which directs the order in which the individual ribonucleotides are polymerized into RNA, A base pairing with T or U, and G base pairing with C (**Figure 4.23**). The RNA transcript is therefore built up in a step-by-step fashion in the 5' → 3 direction, with new ribonucleotides added to the free 3' end of the existing polymer. Remember that in order to base pair, complementary polynucleotides must be antiparallel. This means that the transcribed strand of the gene must be read in the 3' → 5' direction.

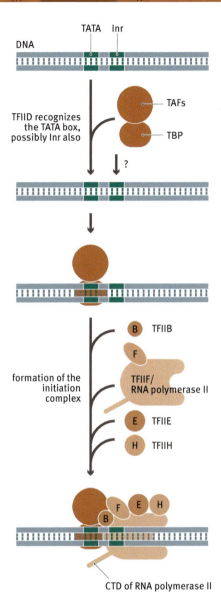

FIGURE 4.20 Initiation of transcription by RNA polymerase II. The first step is the attachment of transcription factor IID (TFIID) to the core promoter. TFIID is made up of the TATA-binding protein (TBP) and the TBP-associated factors (TAFs). TFIID recognizes the TATA box and attaches to the DNA at that point. Additional proteins might also enable the Inr sequence to be recognized as an attachment site. In the second step, the initiation complex is completed by attachment of the RNA polymerase. This step involves at least four additional transcription factors TFIIB, E, F, and H. Transcription begins when the initiation complex is activated by the attachment of phosphate groups to the C-terminal domain (CTD) of the RNA polymerase.

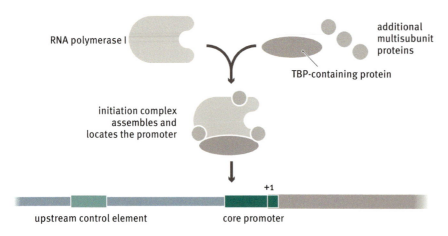

(a) initiation of transcription by RNA polymerase I

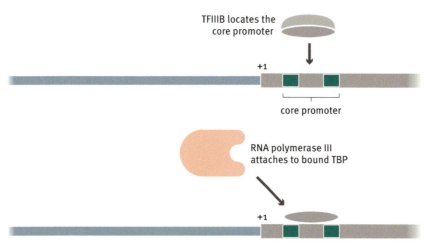

(b) initiation of transcription by RNA polymerase III

FIGURE 4.21 Initiation of transcription by RNA polymerases I and III. (a) RNA polymerase I forms an initiation complex with four multisubunit proteins, one of which contains TBP. This initiation complex then attaches to the RNA polymerase I core promoter. (b) Initiation by RNA polymerase III begins with the attachment of the TBP component of TFIIIB to the core promoter. The RNA polymerase then attaches to the bound TBP.

Bacterial transcripts are synthesized by the RNA polymerase core enzyme

During the elongation stage of transcription, the bacterial RNA polymerase is in its core enzyme form, denoted as $\alpha_2\beta\beta'\omega$. The σ subunit that has the key role in initiation has usually left the complex at this stage. The RNA polymerase covers about 30–40 bp of the template DNA, including the **transcription bubble** of 12–14 bp, within which the growing transcript is held to the transcribed strand of the DNA by approximately eight RNA–DNA base pairs (Figure 4.24).

The RNA polymerase has to keep a tight grip on both the DNA template and the RNA that it is making in order to prevent the transcription complex from falling apart before the end of the gene is reached. However, this grip must not be so tight as to prevent the polymerase from moving along the DNA. Structural studies of actively transcribing RNA polymerase enzymes have shown that the DNA molecule lies between the β and β' subunits, within a trough on the enclosed surface of β' (Figure 4.25).

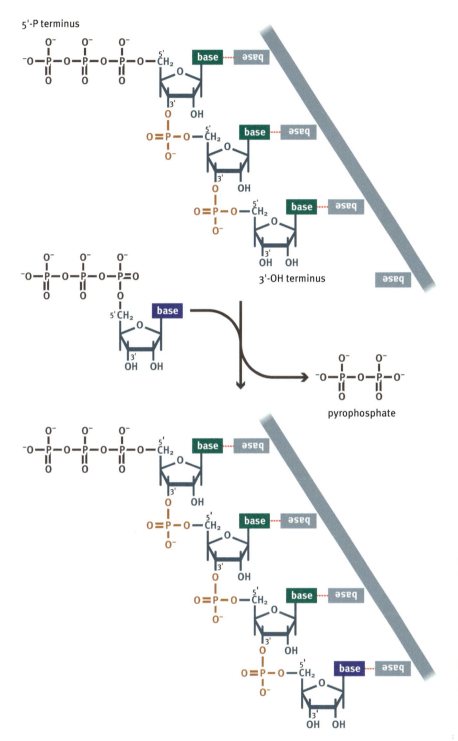

FIGURE 4.22 The polymerization reaction that results in the synthesis of an RNA polynucleotide.

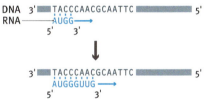

FIGURE 4.23 Template-dependent RNA synthesis. The RNA molecule, shown in blue, is extended by adding nucleotides one by one to its 3′ end.

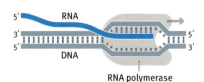

FIGURE 4.24 The *E. coli* transcription elongation complex. The arrow indicates the direction in which the polymerase moves along the DNA. In reality, the polymerase is larger than drawn here and covers 30–40 bp of the template DNA.

The active site for RNA synthesis also lies between these two subunits, with the nontranscribed strand of DNA looping away at this point. The RNA transcript extrudes from the polymerase via a channel formed partly by the β and partly by the β′ subunit.

The polymerase can synthesize RNA at a rate of several hundred nucleotides per minute. The average *E. coli* gene, which is just a few thousand nucleotides in length, can therefore be transcribed in a few minutes. However, the rate of RNA synthesis is not constant. Instead, periods of rapid elongation are interspersed with brief pauses, rarely lasting longer

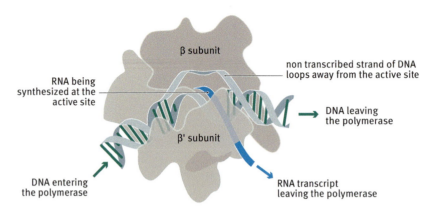

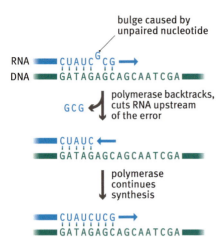

FIGURE 4.25 Synthesis of RNA within a bacterial RNA polymerase. The β and β′ subunits of the RNA polymerase are depicted in gray, the double helix is colored green, and the RNA transcript is blue. The positions at which the DNA enters and leaves the complex are marked, as is the exit point for the RNA transcript. Note how the nontranscribed strand of the DNA molecule loops away from the active site at which RNA is being synthesized.

FIGURE 4.26 Backtracking enables an RNA polymerase to correct a transcription error.

than 6 milliseconds. A pause might be accompanied by the polymerase moving in reverse (**backtracking**) along the template. Backtracking is stimulated by a bulge in the DNA–RNA duplex that occurs at a position that is not base paired (Figure 4.26). The polymerase undergoes a slight structural rearrangement, which enables it to slide back along the template and cut the RNA upstream of the error. The polymerase then reverts to its synthesis structure and moves forward again to retranscribe the previously faulty segment. RNA synthesis is therefore extremely accurate, usually with less than one error per 10^4–10^5 nucleotide additions. These few errors can be tolerated because RNAs are usually multicopy and transcripts that contain errors form only a small proportion of the overall pool.

All eukaryotic mRNAs have a modified 5′ end

The fundamental aspects of transcript elongation are the same in bacteria and eukaryotes. The one major distinction concerns the precise structure of the 5′ end of a eukaryotic mRNA. There is usually a triphosphate group at the 5′ end of a polynucleotide (see Figure 2.12). This applies both to DNA polynucleotides and to the RNA molecules that are synthesized by transcription. As a simple notation, we can describe the 5′ end as pppNpN..., where N is the sugar-base component of the nucleotide and p represents a phosphate group.

The mRNAs made by RNA polymerase II are exceptions to this rule. With these molecules, the 5′ terminus has a more complex chemical structure described as 7–MeGpppNpN..., where 7–MeG is the nucleotide carrying the modified base 7-methylguanosine. This is referred to as the **cap structure** and it is put in place soon after the RNA polymerase leaves the promoter, before the mRNA reaches 30 nucleotides in length.

The first step in 'capping' is the addition of the extra guanosine to the extreme 5′ end of the RNA. Rather than occurring by normal RNA polymerization, capping involves a reaction between the 5′ triphosphate of the terminal nucleotide and the triphosphate of a GTP nucleotide. The γ-phosphate of the terminal nucleotide (the outermost phosphate) is removed, as are the β and γ phosphates of the GTP, resulting in a 5′–5′ bond (Figure 4.27a). The reaction is carried out by the enzyme **guanylyl transferase**.

The second step of the capping reaction converts the new terminal guanosine into 7-methylguanosine by attachment of a methyl group to nitrogen number 7 of the purine ring. This modification is catalyzed by **guanine methyltransferase**. The two capping enzymes make attachments with the RNA polymerase and it is possible that they are intrinsic

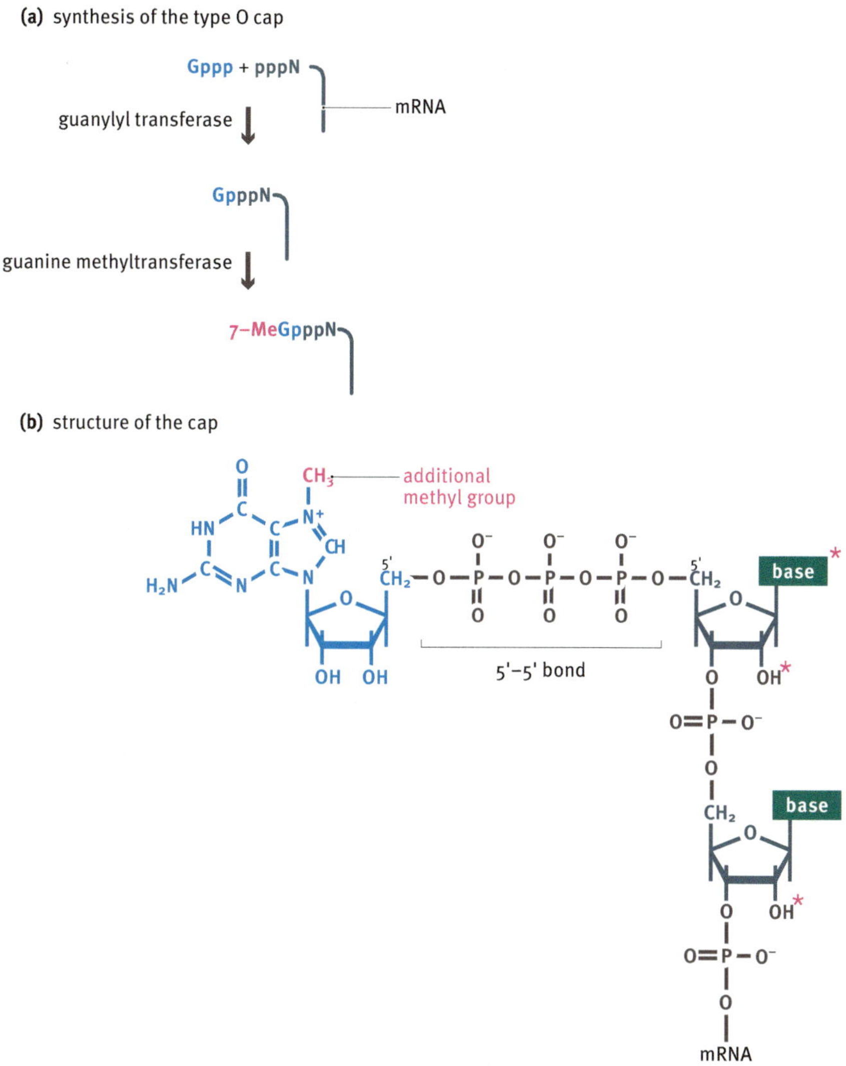

FIGURE 4.27 The cap structure at the 5′ end of a eukaryotic mRNA. (a) Synthesis of the type 0 cap. The nucleotide at the 5′ end of the mRNA is depicted as 'pppN' to indicate that it has a terminal triphosphate. The GTP molecule, in blue, is depicted as 'Gppp', again indicating that it has a triphosphate group. The type 0 cap is synthesized in two steps. First, the GTP is attached to the terminal nucleotide, forming a 5′–5′ triphosphate linkage comprising one phosphate from the GTP and two from the terminal nucleotide of the mRNA. In the second step, a methyl group is added to position 7 of the guanine base. (b) The detailed structure of the type 0 cap. The red stars indicate the positions where additional methyl groups might be added to produce the type 1 and type 2 cap structures.

components of the transcription complex during the early stages of RNA synthesis.

The 7-methylguanosine structure is called a **type 0 cap** and is the commonest form in unicellular eukaryotes such as yeast. In higher eukaryotes, including humans, additional modifications occur (**Figure 4.27b**). The commonest of these affects what is now the second nucleotide in the transcript, where the hydrogen of the 2′–OH group is replaced with a methyl group. This results in a **type 1 cap**. If this second nucleotide is an adenosine, then a methyl group might also be added to the amino group attached to carbon number 6 of the purine. Finally, another 2′–OH methylation might occur at the third nucleotide position, resulting in a **type 2 cap**.

All mRNAs synthesized by RNA polymerase II are capped in one way or another. The cap may be important for the export of transcripts from the nucleus, but its best-defined role is in the translation of mRNAs (Section 8.2).

Some eukaryotic mRNAs take hours to synthesize

The longest bacterial genes are only a few kb in length and can be transcribed in a matter of minutes by the bacterial RNA polymerase, which has a polymerization rate of several hundred nucleotides per minute. In contrast, RNA polymerase II can take hours to transcribe a single gene, even though it can work at up to 2,000 nucleotides per minute. This is because the presence of multiple introns in many eukaryotic genes means that considerable lengths of DNA must be copied. For example, the transcript of the human dystrophin gene is 2,400 kb in length and takes about 20 hours to synthesize.

The extreme length of eukaryotic genes places demands on the stability of the transcription complex. RNA polymerase II on its own is not able to meet these demands. When the purified enzyme is studied in the test tube, its polymerization rate is less than 300 nucleotides per minute and the transcripts that it makes are relatively short. In the nucleus, the transcription complex is stabilized by **elongation factors**, proteins that associate with the polymerase after it has moved away from the promoter. The importance of the elongation factors is shown by the effects of mutations that disrupt the activity of one or other of these proteins. Inactivation of the elongation factor called CSB, for example, results in Cockayne syndrome, a disorder characterized by developmental defects such as intellectual disability. Disruption of a second factor, ELL, causes acute myeloid leukemia.

4.6 TERMINATION OF TRANSCRIPTION

Current thinking views transcription as a discontinuous process, with the polymerase pausing regularly and 'making a choice' between continuing elongation by adding more ribonucleotides to the transcript or terminating by dissociating from the template. Which choice is selected depends on which alternative is more favorable in thermodynamic terms. This means that, in order for termination to occur, the polymerase has to reach a position on the template where dissociation is more favorable than continued RNA synthesis.

Hairpin loops in the RNA are involved in the termination of transcription in bacteria

Bacteria use two distinct strategies for transcription termination, but a common feature of both is that the termination site is marked by an **inverted repeat** in the DNA sequence. An inverted repeat is a segment of DNA or RNA in which a sequence is followed by its reverse complement (Figure 4.28). If the two halves of the inverted repeat are separated by a few intervening nucleotides, as is the case at a termination site, then intrastrand base pairing between the two sequence components can form a hairpin loop in a single-stranded polynucleotide, in this case, the RNA transcript.

At about half the positions in *E. coli* at which transcription terminates the inverted repeat is followed by a run of deoxyadenosine nucleotides in the nontranscribed strand (Figure 4.29). These are called **intrinsic terminators** and the hairpin loops that they specify are relatively stable, more so than the DNA–RNA pairing that normally occurs within the transcription bubble. This means that the formation of the hairpin loop is favored, reducing the number of contacts between the template DNA and transcript and weakening the overall DNA–RNA interaction. The interaction is

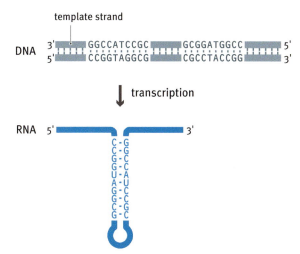

FIGURE 4.28 An inverted repeat in a double-stranded DNA molecule. Following transcription, the inverted repeat gives rise to a hairpin loop in a single-stranded RNA molecule.

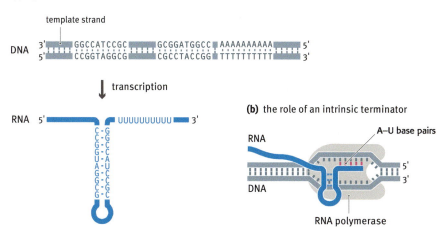

FIGURE 4.29 Termination at an intrinsic terminator. (a) Synthesis of an intrinsic terminator involves the transcription of an inverted repeat sequence followed by a series of A nucleotides. (b) An intrinsic terminator is thought to weaken the DNA–RNA interaction because the presence of the hairpin loop reduces the number of contacts between the template DNA and the transcript, and those contacts that are made are via relatively weak A–U base pairs.

further weakened when the run of As in the DNA is transcribed, because the resulting A–U base pairs have only two hydrogen bonds each, compared with three for each G–C pair. The net result is that detachment of the transcript is favored over continued elongation.

Structural studies suggest that it is also possible that the RNA hairpin makes contact with a flap structure on the outer surface of the RNA polymerase β subunit, adjacent to the exit point of the channel through which the RNA emerges from the complex (**Figure 4.30**). Movement of the flap could affect the positioning of amino acids within the active site, possibly leading to breakage of the DNA–RNA base pairs and termination of transcription.

The second type of bacterial termination signal is **Rho dependent**. These signals usually include an inverted repeat as seen at intrinsic terminators, although the hairpin that is formed is less stable and there is no run of As in the transcribed strand of the DNA. Termination requires the activity of a protein called Rho, which attaches to the transcript and moves along the RNA toward the polymerase. If the polymerase continues to synthesize

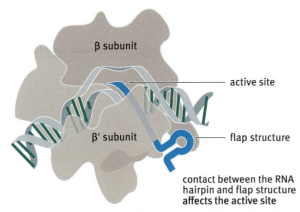

FIGURE 4.30 A model for termination of transcription in bacteria. A flap structure on the surface of the RNA polymerase could mediate termination by responding to the presence of the hairpin loop in the RNA that is leaving the polymerase. Movement of the flap might affect the positioning of amino acids at the active site of the polymerase.

RNA, then it keeps ahead of the pursuing Rho, but at the termination signal the polymerase stalls (Figure 4.31). Exactly why has not been explained, but presumably the hairpin loop that forms in the RNA is responsible in some way. The result is that Rho is able to catch up. Rho is a helicase, which means that it actively breaks base pairs, in this case between the DNA and transcript, resulting in the termination of transcription.

Eukaryotes use diverse mechanisms for the termination of transcription

The three eukaryotic RNA polymerases each use different mechanisms for transcript termination. With RNA polymerase I, termination involves one or more DNA-binding proteins that attach to the DNA close to the point at which transcription terminates. Exactly how the bound proteins cause termination is not known, but a model in which the polymerase becomes stalled because of the presence of the proteins has been proposed (Figure 4.32). Even less is known about RNA polymerase III termination. A run of adenosines in the template is implicated but the process does not involve a hairpin loop, and so it is not analogous to the intrinsic termination system in bacteria.

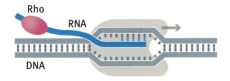
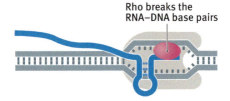

FIGURE 4.31 Rho-dependent termination. Rho follows the RNA polymerase along the transcript. When the polymerase stalls at a hairpin, Rho catches up and breaks the RNA–DNA base pairs, releasing the transcript.

With RNA polymerase II, termination is accompanied by the addition of a series of adenosine nucleotides, called the **poly(A) tail**, to the 3′ end of the mRNA that is being synthesized (Figure 4.33). These adenosines are not transcribed from the DNA – there is no equivalent run of deoxythymidines in the transcribed DNA strand – but instead are added by a DNA-*in*dependent RNA polymerase called **poly(A) polymerase**. This polymerase does not act at the extreme 3′ end of the transcript, but at an internal site that is cleaved to create a new 3′ end to which the poly(A) tail is added. In mammals, this **polyadenylation site** is located between 10 and 30 nucleotides downstream of a signal sequence, almost always 5′–AAUAAA–3′ (Figure 4.34). Most genes have more than one polyadenylation site, which means that termination can occur at different positions, resulting in mRNAs with identical coding properties but with distinctive 3′ ends. The different polyadenylation sites appear to be used in different tissues, suggesting that **alternative polyadenylation** could be an important mechanism for establishing tissue-specific patterns of gene expression.

The signal sequence is the binding site for a multisubunit protein called the **cleavage and polyadenylation specificity factor** (**CPSF**). CPSF binds to the signal sequence as soon as it is transcribed, initiating the

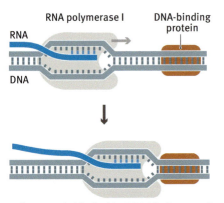

FIGURE 4.32 A possible scheme for termination of transcription by RNA polymerase I. In this model, the polymerase becomes stalled because its progress is blocked by the presence of the DNA-binding protein.

TERMINATION OF TRANSCRIPTION

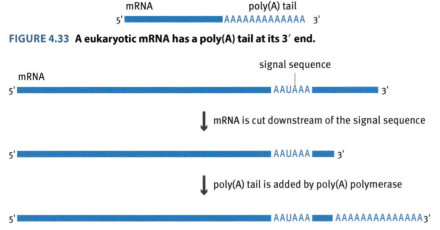

FIGURE 4.33 A eukaryotic mRNA has a poly(A) tail at its 3′ end.

FIGURE 4.34 Polyadenylation of a eukaryotic mRNA. The mRNA is cut at a position between 10 and 30 nucleotides downstream of a signal sequence, and the poly(A) tail is added at this polyadenylation site.

polyadenylation reaction. A third protein, the **cleavage stimulation factor** (**CstF**) makes contact with the polymerase, causing it to detach from the template DNA. As a result, transcription stops soon after the poly(A) signal sequence has been transcribed.

KEY CONCEPTS

- There are three main types of living organism. These are eukaryotes, bacteria and archaea.
- Gene expression is sometimes described as 'DNA makes RNA makes protein', but in reality is more complicated than this, especially in eukaryotes. Additional steps include the processing of the initial RNAs and proteins into their functional forms and the turnover of these molecules.
- The enzymes that carry out transcription are called DNA-dependent RNA polymerases. Bacteria have just one type of RNA polymerase, but eukaryotes have three, each transcribing a different set of genes. Protein-coding genes in eukaryotes are transcribed by RNA polymerase II.
- Transcription must initiate at the correct positions on a DNA molecule, so that genes are transcribed rather than random pieces of DNA. Positions where transcription should begin are marked by sequences called promoters, which are recognized either by the RNA polymerase itself or by a DNA-binding protein that forms a platform to which the RNA polymerase binds.
- The promoters for RNA polymerase II comprise a core element to which the RNA polymerase attaches and various upstream sequences that are involved in the regulation of transcription initiation.
- Initiation of transcription involves opening up a short stretch of the DNA to form an open promoter complex, followed by the start of RNA synthesis and movement of the RNA polymerase away from the promoter region.
- An RNA polymerase does not synthesize its transcript at a constant rate. Instead, synthesis is discontinuous, with periods of rapid elongation interspersed by brief pauses of 6 milliseconds or less.

- All transcripts that are made by RNA polymerase II have a chemical modification called the cap structure at the 5′ end.
- In order for termination to occur, the RNA polymerase has to reach a position on the template where dissociation is more favorable than continued RNA synthesis.
- Termination of transcription by RNA polymerase II is accompanied by attachment of a poly(A) tail to the 3′ end of the transcript.

QUESTIONS AND PROBLEMS

Key terms

Write short definitions of the following terms:

alternative polyadenylation	guanylyl transferase	RNA polymerase I
alternative promoter	helicase	RNA polymerase II
archaea	holoenzyme	RNA polymerase III
backtracking	initiator sequence	strong promoter
bacteria	intrinsic terminator	TAF- and initiator-dependent cofactors
basal rate	inverted repeat	
C-terminal domain	open promoter complex	TATA-binding protein
cap structure	poly(A) polymerase	TATA box
cleavage and polyadenylation specificity factor	poly(A) tail	TBP-associated factor
	polyadenylation site	transcription bubble
cleavage stimulation factor	post-translational chemical modification	transcription factor IID
closed promoter complex		transcription initiation complex
consensus sequence	Pribnow box	turnover
core enzyme	processing	type 0 cap
core promoter	prokaryote	type 1 cap
DNA-dependent RNA polymerase	promoter	type 2 cap
	Rho-dependent terminator	upstream
elongation factor	RNA-dependent RNA polymerase	upstream control element
eukaryote		
guanine methyltransferase	RNA polymerase	

Self-study questions

4.1 Describe the distinguishing features of prokaryotic and eukaryotic cells. What are the names given to the two distinct groups of prokaryotes, and what are the differences between them?

4.2 Why is 'DNA makes RNA makes protein' not a good description of gene expression in eukaryotes?

4.3 What is the difference between a DNA- and an RNA-dependent RNA polymerase? Which type is involved in transcription?

4.4 Distinguish between the core and holoenzyme versions of the E. coli RNA polymerase. What are the roles of the two versions in transcription?

4.5 What are the roles of the three eukaryotic RNA polymerases?

4.6 Define the term 'promoter'. Draw annotated diagrams to illustrate the structures of the promoters for the three eukaryotic RNA polymerases and for the E. coli enzyme.

4.7 Propose a suitable consensus for the following set of sequences:

```
ATAGACATA
ATAGCCATT
ATACACTTA
AGAGAGAAT
ATAGACATA
TTTGACATT
ATAAATATA
```

4.8 What is the link between the sequence of an E. coli promoter and the basal rate of transcription for the gene transcribed from that promoter?

4.9 Give an example of a human gene that is under the control of alternative promoters.

4.10 Explain the roles of the two components of the *E. coli* promoter during initiation of transcription. Be sure to make clear the difference between the closed and open versions of the promoter–RNA polymerase complex.

4.11 What is the role of the σ subunit during promoter recognition in *E. coli*?

4.12 Write an essay on assembly of the RNA polymerase II initiation complex. As part of your essay, compile a table giving the names of the main proteins or groups of proteins involved in assembly of this complex, along with a summary of the role of each one.

4.13 What is the importance of the C-terminal domain of RNA polymerase II?

4.14 How does the TATA-binding protein provide a link between the initiation processes of all three eukaryotic RNA polymerases?

4.15 Draw a detailed diagram of the chemical reaction involved in the synthesis of RNA from individual nucleotides.

4.16 Outline the important features of the elongation phase of transcription in *E. coli*.

4.17 Describe the series of events that result in the capping of a eukaryotic mRNA.

4.18 What are the roles of the elongation factors during eukaryotic transcription?

4.19 Explain what is meant by an 'inverted repeat'.

4.20 Which of the following RNA molecules would be able to form a hairpin structure?

5′–ACGUUUGGCAGUCCAAACU–3′

5′–AGCUAGCUACAGCUAGCUUUGUGAGC-UAGCUUGUG–3′

5′–GGCUAGGUCGAAACGACCUAGCC–3′

4.21 Distinguish between the events that are thought to occur during transcription termination at intrinsic and Rho-dependent terminators.

4.22 Outline our current knowledge regarding the termination of transcription by RNA polymerases I and III.

4.23 Draw a series of diagrams to illustrate how a eukaryotic mRNA becomes polyadenylated.

Discussion topics

4.24 Post-translational chemical modification can alter the function of a protein. Does this contravene the dogma that biological information is encoded in genes?

4.25 Some species of bacteria possess more than one σ subunit. For instance, *Bacillus subtilis* (a bacterium able to produce spores) has at least four. What might be the role of these additional σ subunits?

4.26 With some types of virus, transcription of the host's genes ceases shortly after infection. All the cell's RNA polymerase enzymes start transcribing the virus genes instead. Suggest events that might underlie this phenomenon.

4.27 Construct a hypothesis to explain why eukaryotes have three RNA polymerases. Can your hypothesis be tested?

4.28 Speculate on the reasons why the TATA-binding protein is involved in the initiation of transcription by each of the three eukaryotic RNA polymerases.

4.29 An operon is a group of bacterial genes that are adjacent to one another and are transcribed together as a single mRNA, from a promoter located upstream of the first gene in the cluster. Often the genes in an operon code for proteins that work together in a single biochemical pathway, such as the synthesis of an amino acid. What might be the advantage of having genes grouped into operons?

The answers to, or hints on how to answer, the self-study questions and discussion topic questions can be downloaded from the book's product page at this link: www.routledge.com/9781032743530

FURTHER READING

Aibara S, Schilbach S & Cramer P (2021) Structures of mammalian RNA polymerase II pre-initiation complexes. *Nature* 594, 124–128.

Bujord H (1980) The interaction of *E. coli* RNA polymerase with promoters. *Trends in Biochemical Sciences* 5, 274–278.

Burgess RR (2021) What is in the black box? The discovery of the sigma factor and the subunit structure of *E. coli* RNA polymerase. *Journal of Biological Chemistry* 297, 101310.

Conaway RC & Conaway JW (2019) The hunt for RNA polymerase II elongation factors: a historical perspective. *Nature Structural and Molecular Biology* 26, 771–776.

Green MR (2000) TBP-associated factors (TAFIIs): multiple, selective transcriptional mediators in common complexes. *Trends in Biochemical Sciences* 25, 59–63.

Hao Z, Svetlov V & Nudler E (2021) Rho-dependent transcription termination: a revisionist view. *Transcription* 12, 171–181.

Kadonaga JT (2012) Perspectives on the RNA polymerase II core promoter. *Wiley Interdisciplinary Reviews in Developmental Biology* 1, 40–51.

Khatter H, Vorländer MK & Müller CW (2017) RNA polymerase I and III: similar yet unique. *Current Opinion in Structural Biology* 47, 88–94.

Klug A (2001) A marvellous machine for making messages. *Science* 292, 1844–1846. *Description of the bacterial RNA polymerase.*

Ramanathan A, Robb GB & Chan SH (2016) mRNA capping: biological functions and applications. *Nucleic Acids Research* 44, 7511–7526.

Russell J & Zomerdijk JCBM (2005) RNA-polymerase-I-directed rDNA transcription, life and works. *Trends in Biochemical Sciences* 30, 87–96.

Shi Y & Manley JL (2015) The end of the message: multiple protein-RNA interactions define the mRNA polyadenylation site. *Genes and Development* 29, 889–897.

Tian B & Manley J (2017) Alternative polyadenylation of mRNA precursors. *Nature Reviews Molecular Cell Biology* 18, 18–30.

Tora L & Timmers HT (2010) The TATA box regulates TATA-binding protein (TBP) dynamics in vivo. *Trends in Biochemical Sciences* 35, 309–314. *Details of the recognition of the TATA box by TBP.*

Toulokhonov I, Artsimovitch I & Landick R (2001) Allosteric control of RNA polymerase by a site that contacts nascent RNA hairpins. *Science* 292, 730–733. *A model for termination of transcription in bacteria involving the flap structure on the outer surface of the RNA polymerase.*

Travers AA & Burgess RR (1969) Cyclic re-use of the RNA polymerase sigma factor. *Nature* 222, 537–540. *The first demonstration of the role of the σ subunit.*

Types of RNA Molecule: Messenger RNA

CHAPTER 5

Transcription, as described in Chapter 4, results in the synthesis of RNA molecules whose nucleotide sequences are set, by the base-pairing rules, by the sequences of the genes from which they are copied. These RNA molecules can be grouped into two categories, messenger RNA (mRNA) and noncoding RNA. Messenger RNA is transcribed from protein-coding genes and so undergoes the second stage of gene expression, translation. Messenger RNA therefore acts as the intermediate between a gene and its protein product. Noncoding RNA includes a variety of transcripts that do not code for proteins but instead play roles in the cell as RNA molecules. Now we must look at these two types of transcript in more detail, mRNA in this chapter and noncoding RNA in the next chapter.

Geneticists now use the word **transcriptome** to refer to the RNA content of a cell. This is a useful term that helps us to move away from thinking about the expression of individual genes and instead consider the coordinated activity of groups of genes. In particular, by studying transcriptomes rather than individual RNAs, researchers have been able to make important discoveries about the way in which gene expression patterns change when a cell becomes cancerous. In this chapter, our emphasis will therefore be on transcriptomes rather than on individual mRNAs. We will begin by studying the processes that determine the mRNA content of a transcriptome, which involve more than simply the synthesis of new mRNAs by transcription. The degradation of existing mRNAs is equally important. This is an area of research that has moved rapidly forward over the past few years, thanks to the discovery in eukaryotes of new types of noncoding RNA that participate in the breakdown of mRNAs that are no longer needed.

The processes that influence the mRNA content of a eukaryotic transcriptome also include **splicing**, the mechanism by which introns are removed from the transcripts of discontinuous genes and the exons 'spliced' together. Splicing was initially looked on as a necessary but not particularly exciting process, but that view changed with the discovery that many genes have **alternative splicing** pathways. These pathways enable a single **primary transcript** to give rise to different mRNAs containing different combinations of exons and, of course, specifying different proteins (Figure 5.1). The particular splicing pathways that are operational in a cell greatly influence the composition of that cell's transcriptome. We must therefore look at splicing in some detail.

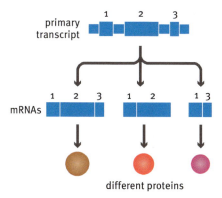

FIGURE 5.1 Alternative splicing. Alternative splicing enables a single gene to specify more than one protein. In this example, a gene has three exons that can be spliced together in three different combinations. On the left, the mRNA contains all three exons. In the middle, exon 3 is omitted from the mRNA, and on the right, exon 2 is omitted. Each of these mRNAs will give rise to a different protein.

DOI: 10.1201/9781003473862-6

Finally, the composition of a transcriptome can be influenced by processes that alter the chemical structure of one or more nucleotides in a mRNA. This will be the final topic that we examine in this chapter.

5.1 THE mRNA COMPONENT OF A TRANSCRIPTOME

Although making up less than 4% of the total RNA in most cells, mRNA is the most significant component of a transcriptome because it specifies the proteins that the cell is able to synthesize. The mRNA content of a transcriptome therefore provides the link between the genes present in a cell and the proteins that the cell makes, which in turn determines the cell's biochemical capabilities.

The mRNA component of a transcriptome is small but complex

Even in the simplest organisms such as bacteria and yeast, many genes are active at any one time. Transcriptomes can therefore contain copies of hundreds, if not thousands, of different mRNAs. In humans, between 10,000 and 15,000 genes are expressed in a single tissue, with the cerebellum and testes being the most complex in this regard.

These gene numbers do not, however, reveal the true complexity of the mRNA content of a transcriptome. In Chapter 4, we learnt that a protein-coding gene could have two or more alternative promoters, giving rise to mRNAs of different lengths. We also noted that eukaryotic mRNAs can have alternative polyadenylation sites which, again, could result in a single gene being transcribed into mRNAs of different lengths. Add to this the possibility that the primary transcript undergoes alternative splicing, and it becomes possible for a single gene to give rise to many different mRNAs (**Figure 5.2**).

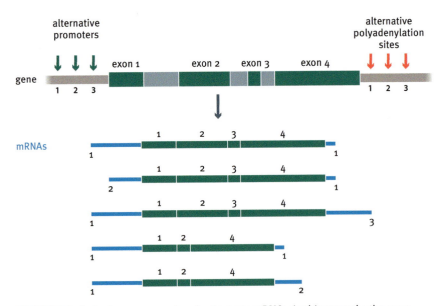

FIGURE 5.2 A single gene can give rise to many mRNAs. In this example, the gene has four exons, three alternative promoters, and three polyadenylation sites. Five of the transcripts that could be obtained from the gene are shown. If every combination of promoter and polyadenylation sites is allowed, and every possible combination of exons is used, then 135 different transcripts could be synthesized from this gene. This assumes that exons can be omitted from the mRNA, but that the order of exons cannot be changed (e.g. 1-3-4 is possible but not 1-4-3). This assumption is in accord with what we know about alternative splicing.

Although the identities of the different mRNAs in a transcriptome can be cataloged with some degree of accuracy, using the method called **RNA-seq** (Chapter 18), measuring the copy numbers of individual mRNAs is much more difficult. The available evidence suggests that the 10,000–15,000 genes that are active in a typical human cell give rise to a transcriptome comprising approximately 200,000 mRNA molecules in total. A transcriptome of this size would contain on average between 13 and 20 mRNAs derived from each expressed gene. Undoubtedly, there is considerable variation around this mean. In human cerebral cells, for example, the **CREB transcription factor**, which controls the expression of genes involved in cell growth and division, has been estimated to have an mRNA copy number of fewer than 25. However, a repressor protein that is involved in the regulation of the same set of genes has an mRNA copy number of up to 240. Much higher copy numbers are also displayed by particular mRNAs in cells with specialized biochemistries. Human reticulocytes are an example. These are immature red blood cells, which are released from the bone marrow into the blood, where they develop into erythrocytes. Reticulocytes therefore synthesize large quantities of hemoglobin, and over 70% of the mRNAs in their transcriptomes are derived from the α_1-, α_2- and β-globin genes (see Figure 3.21).

Transcriptome studies are important in cancer research

Studies of the mRNA contents of transcriptomes have been responsible for many advances in genetics over the last 20 years. Perhaps, the most important of these have been the new insights it has given into the changes in gene expression that occur during cancer.

Cancer is a group of diseases characterized by uncontrolled cell division. Cancers are often described according to the part of the body they affect, and we are all familiar with terms such as 'lung cancer' or 'colon cancer'. This classification hides the fact that a single tissue can be affected by different types of cancer, and understanding as early as possible which one is present in a patient is one of the keys to treatment and possible recovery. Transcriptome studies can diagnose a cancer at an early stage, possibly before there are clear morphological indicators, enabling treatment to start early. The initial breakthrough in this area came in 1999, with studies of different types of leukemia, a cancer of the blood cells. It was shown that the transcriptomes of acute lymphoblastic leukemia cells are different from those of acute myeloid leukemia cells. Twenty-seven lymphoblastic and eleven myeloid cancers were studied and, although all the transcriptomes were slightly different, the distinctions between the two types of cancer were sufficient for unambiguous identifications to be made.

Transcriptome studies can also provide an accurate diagnosis of cancers that are difficult to distinguish by other means. This has been important with non-Hodgkin lymphoma, another cancer that affects white blood cells. The commonest version of this disease is called diffuse large B-cell lymphoma, and for many years it was thought that all cancers of this type were the same. Transcriptome studies changed this view and showed that B-cell lymphoma can be divided into two distinct subtypes. The distinctions between the transcriptomes of the two subtypes enable each one to be related to a different class of B cells, stimulating and directing the search for specific treatments that are tailored to each lymphoma.

Transcriptome analysis is also providing new information on breast cancer. In the early 2000s, this approach enabled breast cancer to be divided into five subtypes, called luminal A, luminal B, HER2-positive, basal-like and normal-like. Each of these subtypes is associated with different patterns of gene expression as revealed by analysis of their transcriptomes. In effect, they are distinct diseases that affect the same tissue.

A critical factor in the potential recovery of a patient from cancer is the avoidance of **metastasis**, which results in cells from one cancer spreading to other places in the body and initiating new tumors (Figure 5.3). Detailed examination of breast cancer transcriptomes has identified groups of genes whose expression profiles enable the risk of metastasis to be predicted at an early stage of the disease. This was an unexpected discovery, as it had been thought that the cellular switches that initiated metastasis were not activated until a much later step in the progression of the cancer. Other transcriptome profiles have been identified that help to predict the long-term prognosis for recovering breast cancer patients, enabling women more likely to have a relapse to be identified at an early stage.

Transcriptome studies have also provided new information on the genetic basis of metastasis. In particular, the risk of metastasis appears to be associated with the presence in tumor transcriptomes of mRNAs more normally found in tissues that are recovering from trauma such as wounding. These mRNAs have been found not just in breast cancer transcriptomes but also in transcriptomes from the more fatal types of stomach and lung cancer. This observation has contributed to the hypothesis that tumors are 'wounds that do not heal', which has stimulated new avenues of research into possible therapies.

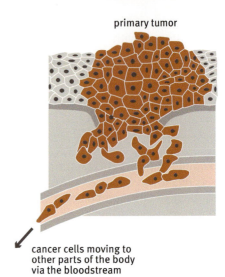

FIGURE 5.3 Metastasis. Cancerous cells from a primary tumor are shown entering the bloodstream. They can therefore move to other parts of the body where they may initiate secondary tumors. (Adapted from B. Alberts et al., *Essential Cell Biology*, 3rd Edition. New York: Garland Science, 2010. With permission from W.W. Norton.)

The mRNA composition of a transcriptome can change over time

The differences between the transcriptomes of normal tissues and cancerous ones alert us to the fact that transcriptomes can change over time. These changes do not occur simply as a result of disease: they are an essential part of the normal functioning of a healthy cell. If transcriptomes were static then cells would not be able to respond to changes in their environment or to signals from other cells, nor would they be able to differentiate into specialized types during the development of an organism. We must therefore examine how the composition of a transcriptome can change.

The most obvious way to change the mRNA composition of a transcriptome is to switch on a new protein-coding gene. Transcription of a gene that previously was silent will add a new type of mRNA to the transcriptome. This is such an obvious way of changing the mRNA content of a cell that we might imagine that it is the key to the dynamism of transcriptomes. In fact, the ability to remove mRNAs from the transcriptome is equally important. If mRNAs were stable, never being degraded or only being turned over very slowly, then switching off the transcription of a gene would have no immediate effect on the synthesis of the protein coded by that gene (Figure 5.4). The gene's mRNA would persist in the transcriptome even though the gene was no longer being transcribed, and repeated rounds of translation of that mRNA would maintain the levels of the protein coded by the gene. Turnover therefore enables control processes that act on transcription to achieve both up- and down-regulation of individual mRNAs.

The rate of degradation of an mRNA can be estimated by determining its **half-life** in the cell. These estimates show that there are considerable variations between and within organisms. Most bacterial mRNAs have a half-life of only a few minutes and so are turned over very quickly. This reflects the rapid changes in protein synthesis patterns that are needed

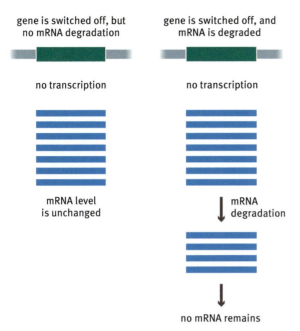

FIGURE 5.4 The influence of mRNA degradation on the composition of a transcriptome. On the left, a gene has been switched off but its mRNA has not been degraded. The mRNA therefore remains in the transcriptome, and the protein coded by the gene will still be synthesized. On the right, we see the importance of mRNA degradation. Now the mRNAs of the inactive gene are removed from the transcriptome, so the synthesis of the protein stops, reflecting the fact that the gene has been switched off.

by an actively growing bacterium with a generation time of 20 minutes or so. Eukaryotic mRNAs are longer lived, with half-lives of, on average, 10–20 minutes for yeast and several hours for mammals.

Within individual cells, the variations are almost equally striking. Some yeast mRNAs have half-lives of only 1 minute whereas for others the figure is more like 35 minutes. But there are a few exceptions and examples of long-lived mRNAs are known. For instance, globin mRNA in human reticulocytes is almost fully stable. In these cells, regulation of globin gene expression is not important because the maximum rate of globin synthesis is required virtually all the time. Other long-lived mRNAs are being discovered as different types of specialized cell are studied.

There are various pathways for nonspecific mRNA turnover

The short half-lives of most RNAs indicate that bacteria and eukaryotes possess biochemical pathways for mRNA turnover. Until recently, all of these appeared to be nonspecific with regard to which mRNAs they targeted. One of the most important developments in genetics over the past 20 years has been the identification in eukaryotes of pathways that target individual mRNAs. Through their ability to remove particular transcripts from a transcriptome, these pathways form a previously unrecognized mechanism for the control of gene expression. First, we will look at the apparently nonspecific pathways and then focus on the newly discovered ones that target individual mRNAs.

Studies of mutant bacteria whose mRNAs have extended half-lives have identified a range of **ribonucleases** that are thought to be involved in mRNA degradation. Most of these enzymes are **exonucleases** that degrade mRNAs by removing nucleotides one by one from the 3′ end of the molecule, but some are **endonucleases** that are able to cut internal phosphodiester bonds (Figure 5.5). The exonucleases are unable to

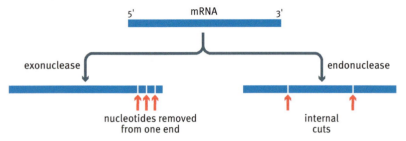

FIGURE 5.5 Exonucleases and endonucleases. An exonuclease degrades an mRNA by removing nucleotides one by one from the end of the molecule. An endonuclease makes cuts within an mRNA. There are also exonucleases and endonucleases that act on other types of RNA and on DNA.

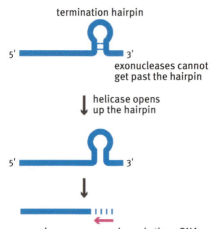

FIGURE 5.6 Degradation of mRNAs in bacteria. The exonucleases responsible for mRNA degradation remove nucleotides one by one from the 3' end of the molecule. Most bacterial mRNAs have a hairpin structure near their 3' end, which prevents the progress of the exonuclease. This hairpin is opened up by a helicase, enabling the exonuclease to progress past this point.

progress past the hairpin structure, involved in the termination of transcription, that is present near the 3' end of most bacterial mRNAs. To solve this problem, the **degradosome**, the multiprotein structure that carries out nonspecific mRNA degradation, also includes a helicase. The helicase opens up the hairpin structure and hence aids the progression of the exonuclease along the mRNA (Figure 5.6).

In eukaryotes, nonspecific mRNA turnover is the role of the **exosome**. This structure comprises a ring of six ribonucleases, with three RNA-binding proteins attached to the top of the ring. Other ribonucleases associate with the exosome in a transient manner. It is thought that RNAs to be degraded are initially captured by the binding proteins and then threaded through the channel in the middle of the ring, where they are exposed to the ribonuclease activities of the ring proteins (Figure 5.7).

Exosomes are present in both the cytoplasm and the nucleus of a eukaryotic cell. The main role of the nuclear exosomes appears to be the rapid turnover of aberrant RNAs that have not been transcribed or processed correctly and which are therefore not released into the cytoplasm. Aberrant mRNAs are detected by a process called **mRNA surveillance**. The 'surveillance' mechanism involves a complex of proteins that scans the mRNA for the nucleotide sequences that mark the position where translation of the mRNA should stop. These are called **termination codons** and we will look at them more closely when we study the genetic code in Chapter 7. The termination codon should normally be near the 3' end of the mRNA, within the final exon if the transcript is from a discontinuous gene (Figure 5.8). If the surveillance proteins find a termination codon in the wrong place then the cap structure is removed from the 5' end of the mRNA, and the molecule is degraded from this end by an exonuclease.

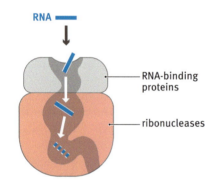

FIGURE 5.7 RNA degradation in a eukaryotic exosome. The RNA is initially captured by the RNA-binding proteins at the top of the exosome and then passed into the channel within the ring of ribonucleases. Within the channel, the RNA is degraded by a combination of exonuclease and endonuclease activities.

Eukaryotes are also able to degrade mRNAs that are no longer being translated into protein. This system is called **deadenylation-dependent decapping**. The first step is the removal of the poly(A) tail, but the cap structure is also removed and the mRNA is degraded from its 5' end (Figure 5.9). Whether or not a particular mRNA is degraded is probably determined by the ability of the decapping enzyme to gain access to the cap structure, which in turn depends on whether or not other proteins are bound to it. Proteins are bound when an mRNA is being translated, so these 'active' mRNAs are protected from decapping and degradation. When an mRNA becomes inactive, the binding proteins detach from the cap and the decapping enzyme can gain access. This pathway might therefore be responsible for removing from the transcriptome mRNAs that are no longer being translated.

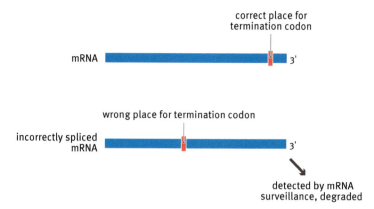

FIGURE 5.8 The basis of mRNA surveillance. If a transcript has been correctly synthesized then its termination codon will be near the 3' end of the mRNA. The surveillance proteins are able to identify if the termination codon is in the wrong place, as might happen if the mRNA has been incorrectly spliced. These mRNAs are degraded.

Individual mRNAs in eukaryotes are degraded by the Dicer protein

Now, we turn our attention to the processes that result in the specific degradation of individual mRNAs. These pathways were discovered by a roundabout route. The genes of some viruses are made of double-stranded RNA, which replicates via a double-stranded RNA intermediate. For several years, it has been known that many eukaryotes possess RNA degradation mechanisms that protect their cells from attack by these viruses. The process is called **RNA silencing** or **RNA interference** and involves a ribonuclease called **Dicer**, which cuts the viral RNA into **short interfering RNAs (siRNAs)** of 21–28 nucleotides in length (Figure 5.10a).

The action of Dicer inactivates the virus genes, but what if some or all of these genes have already been transcribed? If this has occurred then the harmful effects of the virus will already have been initiated and RNA silencing would appear to have failed in its attempt to protect the cell from damage. Remarkably, there is a second stage of the silencing process that is directed specifically at the viral mRNAs. The siRNAs produced by cleavage of the viral RNA are separated into individual strands. One strand of each siRNA then base pairs to any viral mRNAs that are present in the cell. The double-stranded regions that are formed are target sites for the assembly of the **RNA-induced silencing complex (RISC)**, which cleaves and hence silences the mRNAs (Figure 5.10b).

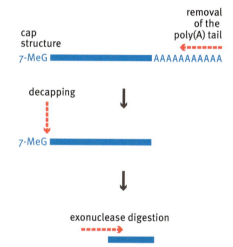

FIGURE 5.9 The deadenylation-dependent decapping pathway for degradation of an mRNA. The poly(A) tail and the cap structure are removed and the mRNA is degraded from its 5' end.

A link between the silencing of viral RNAs and specific degradation of endogenous mRNAs was made when it was discovered that many eukaryotes have more than one type of Dicer protein. The fruit fly, *Drosophila melanogaster*, for example, has two Dicer enzymes. It turns out that the second type of Dicer in *Drosophila* works not with viral RNAs, but with endogenous molecules called **foldback RNAs**, which are coded by the fruit fly DNA and synthesized by RNA polymerase II. The name 'foldback' is given to these RNAs because they can form intrastrand base pairs giving rise to a hairpin structure (Figure 5.11). This hairpin can be cut by Dicer, releasing short double-stranded molecules, each about 21 bp, called **microRNAs**. Each microRNA is complementary to part of a cellular mRNA and hence base pairs with this target, stimulating the assembly of an RISC. Attachment of the microRNA therefore leads to cleavage and silencing of the mRNA.

Human cells are able to make up to 2,300 different microRNAs. These are able to silence the mRNAs of over 6,000 genes, possibly because different mRNAs have identical target sequences, or possibly because a precise match between microRNA and mRNA is not needed for induction of the

RISC. If a few **mismatches** are allowed – positions within the hybrid where base pairs do not form (Figure 5.12) – then the range of sequences that can be recognized by a single microRNA would be greatly increased.

Current thinking is that microRNAs silence sets of mRNAs that are no longer needed when a cell's environment changes or when the cell enters a new phase of a differentiation pathway. The specific degradation of these unwanted mRNAs, combined with the addition of new mRNAs by switching on new genes, enables the composition of the transcriptome to be altered to meet the new demands being placed on the cell.

5.2 REMOVAL OF INTRONS FROM EUKARYOTIC mRNAs

In bacteria, the mRNA molecules that are translated are direct copies of the genes from which they are transcribed. In eukaryotes, the situation is different as many mRNAs are coded by discontinuous genes, ones that contain introns (see Figure 3.2). Transcription produces a faithful copy of the gene, so if the gene contains introns then the initial transcript includes copies of these. The introns must be excised and the exons joined together in the correct order before the transcript can function as a mature mRNA. This process is called splicing.

Splicing occurs in the nucleus. This is evident when the RNA fractions present in the nucleus and cytoplasm are compared. The nucleus can be divided into two regions, the **nucleolus**, in which ribosomal RNA genes are transcribed, and the **nucleoplasm**, where other genes, including those for mRNA, are transcribed. The nucleoplasmic RNA fraction is called **heterogeneous nuclear RNA** or **hnRNA,** the name indicating that it is made up of a complex mixture of RNA molecules, many over 20 kb in length. The mRNA in the cytoplasm is also heterogeneous, but its average length is only 2 kb. If the mRNA in the cytoplasm is derived from hnRNA, then the primary transcripts must be shortened before the mRNA leaves the nucleus (Figure 5.13).

Initially, it was thought that hnRNA was converted to mRNA simply by trimming the ends of the molecules. The existence of introns as noncoding segments within genes was not suspected until 1977 when DNA sequencing was first used on a large scale. The discovery of introns prompted a great deal of research aimed at describing the splicing process and, although this research continues, a fairly complete picture was available by 2000. Another layer of complexity was then added with the realization that many genes have alternative splicing pathways, enabling a single gene to code for more than one protein. Research since 2000 has focused on mapping the alternative pathways for individual genes, and attempting to understand how the choice is made between them.

We begin by examining how introns are removed from transcripts and will then look at what is currently known about alternative splicing.

Intron–exon boundaries are marked by special sequences

When introns were first discovered, it was immediately realized that there must be a very precise mechanism for recognizing the boundary between an exon and an intron and ensuring that the RNA is cut exactly at this position. An error of one nucleotide in either direction would make it impossible for the mRNA to be translated into its correct protein product (Figure 5.14). Research was therefore directed at understanding how splice sites are recognized.

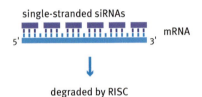

(a) cleavage of a virus RNA by Dicer

(b) silencing of virus mRNAs

FIGURE 5.10 RNA silencing.
(a) The double-stranded RNA molecule of an invading virus is inactivated by the action of the Dicer nuclease. (b) The single-stranded versions of the short interfering RNAs produced by Dicer base pair with viral mRNAs, producing structures that are degraded by the RNA-induced silencing complex (RISC).

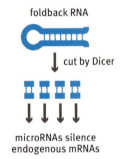

FIGURE 5.11 Synthesis of microRNAs.
The foldback hairpin is cut by Dicer, releasing the microRNA

As more and more genes were sequenced, it became clear that in the vast majority of cases, the first two nucleotides of the intron sequence are 5′–GU–3′ and the last two 5′–AG–3′ (**Figure 5.15**). As intron sequences started to accumulate in the databases, it was further realized that the GU–AG motifs are merely parts of longer consensus sequences that span the 5′ and 3′ splice sites. These consensus sequences vary in different types of eukaryote. In vertebrates, they can be described as

5′ splice site 5′–AG↓GUAAGU–3′

3′ splice site 5′–PyPyPyPyPyPyNCAG↓–3′

In these designations, 'Py' is one of the two pyrimidine nucleotides (U or C), 'N' is any nucleotide, and the arrow indicates the exon–intron boundary. The consensus sequence for the 3′ splice site is therefore six pyrimidines in a row, followed by any nucleotide, followed by a C and then the AG. Remember that these are consensus sequences. The actual sequences might be slightly different, although the GU and AG components are almost always present.

The sequences around these two splice sites were originally thought to be the only regions of nucleotide similarity. It was then discovered that in the yeast *Saccharomyces cerevisiae* there is another conserved sequence, consensus 5′–UACUAAC–3′, located between 18 and 140 nucleotides upstream of the 3′ splice site. Other lower eukaryotes, such as fungi, have a similar sequence, but the motif is not present in vertebrate introns. Vertebrates do, however, have a **polypyrimidine tract** in the same general region; this tract has no clear consensus sequence but, as its name suggests, is usually made up of mainly pyrimidine nucleotides (see Figure 5.15). Although located at similar positions, the polypyrimidine tract and the 5′–UACUAAC–3′ sequence are not functionally equivalent.

The splicing pathway

Having understood how the splice sites could be recognized, attention became focused on the series of events that result in the excision of an intron and the linking together of the two adjacent exons.

In outline, the splicing pathway involves three steps (**Figure 5.16**). The first step is cleavage at the 5′ splice site which, because it is the first to be cut, is also called the **donor site**. The resulting free 5′ end is then attached to an internal site within the intron to form a lariat structure. In yeast, the branch site is the last A in the UACUAAC sequence. In vertebrates, the branch site is also an A, but as vertebrate introns do not have UACUAAC boxes it is not clear how this A is selected. The lariat is formed by creating a phosphodiester bond between the 5′ carbon of the first nucleotide of the intron (the G of the GU motif) and the 2′ carbon of the internal adenosine. The 3′ splice site – the **acceptor site** – is then cleaved and the two exons

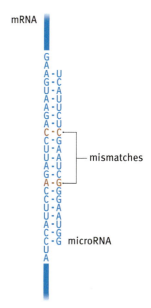

FIGURE 5.12 Mismatches in the hybrid formed between a microRNA and an mRNA.

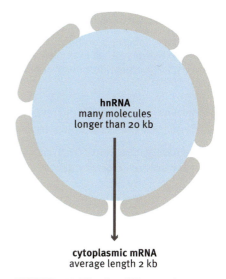

FIGURE 5.13 The size difference between hnRNA and mRNA. The size comparison indicates that mRNA molecules are shortened before they leave the nucleus.

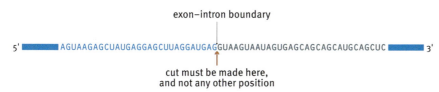

FIGURE 5.14 The importance of precision when making a cut at an exon–intron boundary. The cut must be made at precisely the position shown. An error in either direction would result in a change in the sequence of the mature mRNA. An mRNA that has been incorrectly spliced in this way would no longer be able to direct the synthesis of its protein product.

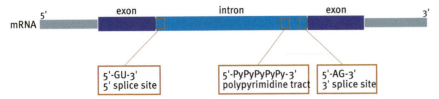

FIGURE 5.15 Conserved sequences in vertebrate introns. Py, pyrimidine nucleotide (U or C).

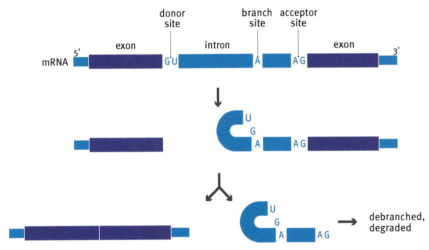

FIGURE 5.16 Splicing in outline. The first step is cleavage at the donor site. The resulting free 5′ end is then attached to an internal branch site within the intron to form a lariat structure. The acceptor site is then cleaved and the two exons are joined together.

are joined together. The intron is released as the lariat structure, which is subsequently converted back to linear RNA and degraded.

The splicing reaction involves a set of RNA–protein complexes called **small nuclear ribonucleoproteins** (**snRNPs**). Each of these contains several proteins and one or two of the noncoding snRNAs. There are a number of different snRNAs in vertebrate nuclei, not all of them involved in splicing, the most abundant being the ones called U1, U2, U3, U4, U5 and U6. The 'U' stands for 'uracil-rich'. They are quite short, varying in size from 106 nucleotides for U6 to 217 nucleotides for U3.

A splicing reaction initiates when U1-snRNP binds to a donor site (Figure 5.17). This attachment is made, at least in part, by RNA–RNA base pairing, as U1-snRNA contains a sequence that is complementary to the consensus for a donor site. At the same time, the accessory proteins SF1, U2AF1 and U2AF2 make protein–RNA contacts with the internal adenosine that acts as the branch site, the polypyrimidine tract and possibly the acceptor site. U2-snRNP then attaches to the branch site, and the accessory proteins detach. The U1- and U2-snRNPs have an affinity for each other, and this draws the donor site toward the branch site. The other snRNPs involved in splicing (the U4-, U5- and U6-snRNPs) attach themselves to the intron at this stage. Their arrival results in additional interactions that bring the acceptor site close to the donor site and the branch site. The U1- and U4-snRNPs then leave the complex, giving rise to the structure called the **spliceosome**. All three key positions in the intron (the donor, acceptor and branch sites) are now in proximity, and the cutting and joining reactions that excise the intron and ligate the exons can take place. These reactions are catalyzed by the U2- and U6-snRNPs. Once the reactions are completed, the spliceosome dissociates into the spliced mRNA and the intron lariat, the latter still attached to the U2- U5- and U6-snRNPs.

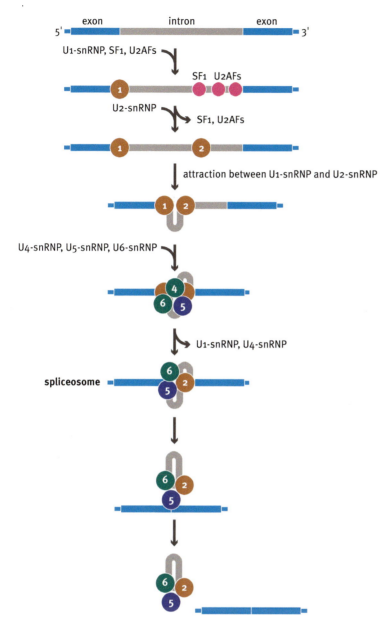

FIGURE 5.17 The roles of the snRNPs and associated proteins during splicing. There are several unanswered questions about the series of events occurring during splicing and it is unlikely that the scheme shown here is entirely accurate.

Enhancer and silencer sequences specify alternative splicing pathways

Alternative splicing is now looked on as a major component of the gene expression process in eukaryotes. In humans, 75% of all protein-coding genes undergo alternative splicing, giving rise to an average of four different spliced mRNAs, or **isoforms**, per gene. Some transcripts undergo such extensive alternative splicing that hundreds or even thousands of different mRNAs are generated from a single gene. An extreme example is the *Drosophila* gene called *Dscam*, which has 95 variable exons that are spliced into 38,016 distinct mRNAs, with up to 50 different mRNAs per cell. The DSCAM protein is a cell-surface receptor that contributes to the ability of nerve cells to adhere to one another. The diversity of *Dscam* mRNAs is such that each nerve cell could synthesize a different repertoire of DSCAM proteins, possibly as a means of establishing individual cell identities.

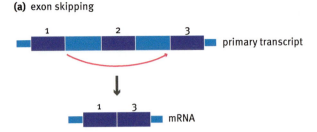

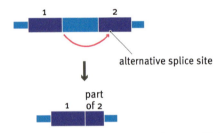

FIGURE 5.18 Alternative splicing events. (a) Exon skipping. (b) Alternative exons. In this example, the combination 1-2-3 would not be permissible. (c) Alternative splice site selection.

The commonest type of alternative splicing event is **exon skipping**, which results in one or more exons being left out of the final mRNA (Figure 5.18a). A variant of exon skipping is **alternative exons**. This occurs when the mRNA contains either of a pair of exons, but not both at the same time (Figure 5.18b). Slightly different is **alternative splice site selection**, in which the usual donor or acceptor site is ignored and a second site is used in its place (Figure 5.18c).

The primary transcript of a protein-coding gene might therefore be able to follow a large number of different splicing pathways, each resulting in a different mRNA. How is this process controlled so that the appropriate spliced mRNAs are synthesized in the correct cells at the correct time? Some geneticists believe that there must be a **splicing code** of some description, whereby regulatory proteins interact with the primary transcript and dictate which splicing pathway is followed (Figure 5.19).

At present, our knowledge of the splicing code is rudimentary, and as such there are still important gaps in our understanding of how alternative splicing is controlled. We do know that short sequences called **splicing**

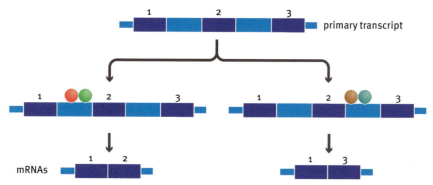

FIGURE 5.19 A splicing code. According to this hypothesis, regulatory proteins interact with the primary transcript and dictate which splicing pathway is followed. In this example, different combinations of regulatory proteins determine if exon 2 or exon 3 is retained in the mRNA.

enhancers and **splicing silencers** are key players in determining which splice sites are used and hence which exons are retained in the mRNA. These enhancers and silencers act as binding sites for proteins that, when attached to the transcript, activate or repress the selection of adjacent splice sites.

Splicing enhancers are binding sites for **SR proteins**. These proteins attach to RNA polymerase II during the initiation of transcription. They are then deposited at splicing enhancer sequences as soon as these are transcribed. When attached to a splicing enhancer, SR proteins help establish the connection between U1-snRNP and the U2AF proteins, at the very beginning of the splicing process. This is one of the critical stages that identifies which sites will be linked.

Less is known about the mode of action of splicing silencers. They act as binding sites for protein–RNA complexes called **heterogeneous nuclear ribonucleoproteins** (**hnRNPs**), but exactly how these influence splicing is not known. They might simply block access to one or more of the splice sites, branch site and/or polypyrimidine site, interfering with the assembly of the spliceosome and therefore preventing splicing.

5.3 CHEMICAL MODIFICATION OF mRNAs

In addition to the cutting and joining events that constitute splicing, some mRNAs are also processed by chemical modification. This occurs when an enzyme attaches to the mRNA and alters the structure of one of the nucleotides (Figure 5.20). In some cases, the structural alteration changes the base-pairing properties of a nucleotide (e.g. converts a C into a U). This is called **RNA editing**. It is important because it has the capacity to alter the amino acid sequence of the protein coded by an mRNA. Other chemical modifications do not affect base pairing but are still important as they might influence the way in which the mRNA is translated.

There are various types of RNA editing

RNA editing involves a chemical modification that alters the base-pairing properties of a nucleotide and hence changes the sequence of the mRNA. The best-studied example of RNA editing occurs with the human mRNA for apolipoprotein B. There are two versions of this protein, called apolipoprotein B48, which is synthesized by intestinal cells, and apolipoprotein B100, which is about twice the size of B48 and is made in the liver.

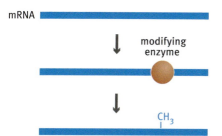

FIGURE 5.20 Chemical modification of an mRNA. A modifying enzyme attaches to the mRNA and, in the example shown, attaches a methyl (–CH$_3$) group to one of the nucleotides.

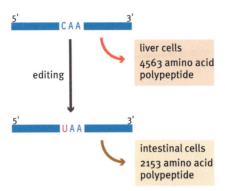

FIGURE 5.21 Editing of the human apolipoprotein B mRNA. Editing of the C at position 6666 in the mRNA converts this nucleotide into a U, creating a signal for termination of translation. A shortened form of apolipoprotein B is therefore synthesized in intestinal cells.

Both proteins are involved in the transport of lipids around the body, but their exact roles are different. The B48 protein forms part of the transport structure called a chylomicron, whereas B100 combines with other proteins to form a complex called a very low density lipoprotein. At first, it was thought that the B100 and B48 versions were coded by different genes but now we know that the same gene specifies both types, the intestinal B48 protein arising by RNA editing. What happens is that in intestinal cells a C within the mRNA is deaminated, by removal of the $-NH_2$ group attached to carbon number 4, converting it into uracil (Figure 5.21). This results in the triplet CAA changing into UAA, which is a signal for the termination of mRNA translation. The consequence is that the protein coded by the edited intestinal mRNA is shorter than that coded by the unedited mRNA present in the liver.

The enzyme that carries out the C→U transformation is called a cytidine deaminase. It locates the correct position on the mRNA by recognizing the nucleotide sequence on either side of the editing site, specificity being ensured because this 21 bp recognition sequence does not occur elsewhere in the apolipoprotein B mRNA. Equally importantly, the sequence appears to be absent in other human mRNAs, so the deaminase does not edit the wrong mRNA by mistake.

Although not common, RNA editing occurs in a number of different organisms and includes a variety of different nucleotide changes (**Table 5.1**). Some editing events have a significant impact on the organism. In humans, for example, editing is partly responsible for the generation of antibody diversity and has also been implicated in the development of some cancers.

One particularly interesting type of editing is the deamination of adenosine to inosine, which is carried out by the **adenosine deaminases acting on RNA** (**ADAR**) enzymes. These enzymes edit their target mRNAs at a limited number of positions specified by segments of the primary transcript that take up a left-handed double-stranded structure, formed by base pairing between the modification site and sequences from adjacent introns.

TABLE 5.1 EXAMPLES OF RNA EDITING IN MAMMALS			
Tissue	Target RNA	Change	Comments
Intestine	Apolipoprotein B	C to U	Creates a signal for termination of translation
Muscle	α-galactosidase	U to A	Converts a phenylalanine amino acid into a tyrosine
B lymphocytes	Immunoglobulin	Various	Contributes to the generation of antibody diversity
Brain	Glutamate receptor	A to inosine	Multiple positions resulting in various changes in the protein

This type of selective editing occurs during the processing of the mRNAs for mammalian glutamate receptors, proteins that play important roles in the transmission of signals between nerve cells.

ADARs also carry out nonselective editing, in which over 50% of the adenosines in the RNA become converted to inosines. This is called **hyperediting** and was initially observed with double-stranded viral RNAs which adopt base-paired structures that act as substrates for ADAR. Hyperediting may have an antiviral effect, but in some cases the editing changes the nature of a viral disease rather than preventing it. With measles, hyperediting might contribute to immune evasion, resulting in a persistent measles infection, as opposed to the more usual transient version of the disease.

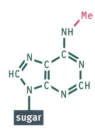

FIGURE 5.22 **The structure of 6-methyladenosine.** The difference between 6-methyadenosine and adenosine is shown in orange.

Chemical modifications that do not affect the sequence of an mRNA

RNA editing can be detected simply by comparing the sequence of an mRNA with that of the gene from which it is transcribed. If the two sequences have nucleotide differences, then editing has occurred. Detecting chemical modifications that do not alter the base-pairing properties of a nucleotide is much more difficult because the mRNA still has the same sequence as the gene. For this reason, it is only recently that geneticists have become aware that this type of chemical modification occurs in eukaryotic mRNA.

The commonest type of modification is the conversion of adenosine into 6-methyladenosine (m^6A; Figure 5.22). Up to ten m^6A modifications are made to every mammalian mRNA, the majority of these close to the 3′ end of the transcript. The methylation is carried out by an enzyme called a **m^6A writer** at target sites with the consensus sequence 5′–DRACH–3′, where D is A, G or U, R is A or G, and H is A, C or U. Such sites are quite common in most mRNAs but only a few are methylated. Why and how particular ones are chosen is not yet known. An m^6A modification can be removed by an 'eraser' enzyme. Between writing and erasing, a variety of 'reader' proteins are able to attach to the modification and influence factors such as the rate at which the mRNA is translated and its stability.

In addition to m^6A, six other types of chemical modification have been identified in eukaryotic mRNAs. Four of these involve methylations, to give m^1A, m^5C, m^7G and an adenosine with a methylation at the 2′ position of the ribose. The other two modifications result in ac^4C (where 'ac' is an acetyl group) and pseudouridine, which is obtained from uridine by rearrangement of the atoms in the pyrimidine base (see Figure 6.11b). So far, specific writers for these other types of modification have not been identified. This has led some researchers to suggest that, unlike m^6A, these modifications have no function. In Section 6.3, we will learn that chemical modification is also important in the processing of rRNA and tRNA. The hypothesis is that the enzymes that modify target sites in rRNA and tRNA also act on mRNA segments that, by chance, have similar sequences or base-paired structures. Some of these mRNA modifications could therefore be 'noise' accompanying the modification of tRNA and rRNA and have no function in the coding transcripts.

KEY CONCEPTS

- The RNA content of the cell is referred to as the transcriptome. The important features of the mRNA component of a transcriptome are the identities of the mRNAs that it contains and their relative amounts.

- Transcriptomes contain copies of hundreds or thousands of different mRNAs. Usually, each mRNA makes up only a small fraction of the transcriptome. Exceptions are those cells that have highly specialized biochemistries, which are reflected by transcriptomes in which one or a few mRNAs predominate.
- The composition of a transcriptome can change over time. Transcription of a gene that previously was silent will add a new type of mRNA to the transcriptome. Turnover of mRNAs means that when a gene is switched off its mRNA will disappear from the transcriptome.
- Most of the processes for mRNA degradation appear to be nonspecific with regard to which mRNAs they target. An exception is the silencing of specific eukaryotic mRNAs by a degradation pathway mediated by microRNAs.
- Transcription produces a faithful copy of a gene, so if the gene contains introns then the initial transcript includes copies of these. These introns must be excised and the exons joined together in the correct order before the transcript can function as a mature mRNA.
- Introns are removed from mRNAs by the spliceosome, which includes a set of RNA–protein complexes called small nuclear ribonucleoproteins.
- Many genes have two or more alternative splicing pathways, enabling the initial transcript to be processed into related but different mRNAs, directing the synthesis of related but different proteins.
- There may be a splicing code that specifies which splicing pathway is followed by a transcript. Sequences called splicing enhancers and silencers might form a part of this code.
- Some mRNAs are processed by chemical modification. RNA editing occurs when a structural alteration changes the base-pairing properties of a nucleotide. Other chemical modifications do not affect base pairing but are still important as they might influence the way in which the mRNA is translated.

QUESTIONS AND PROBLEMS

Key terms

Write short definitions of the following terms:

acceptor site
adenosine deaminase acting on RNA
alternative exons
alternative splice site selection
alternative splicing
CREB transcription factor
cryptic splice site
deadenylation-dependent decapping
degradosome
Dicer
donor site
endonuclease
exon skipping
exonuclease
exosome
foldback RNA
half-life
heterogeneous nuclear RNA
heterogeneous nuclear ribonucleoprotein
hyperediting
isoform
m^6A writer
metastasis
microRNA
mismatch
mRNA surveillance
nucleolus
nucleoplasm
polypyrimidine tract
primary transcript
ribonuclease
RNA editing
RNA-induced silencing complex
RNA interference
RNA-seq
RNA silencing
short interfering RNA
small nuclear ribonucleoprotein
splicing code
splicing enhancer
splicing silencer
spliceosome
splicing
SR protein
termination codon
transcriptome

Self-study questions

5.1 Define the term 'transcriptome'. Typically, what fraction of a transcriptome comprises mRNA?

5.2 Describe the variations in copy number that are seen with different mRNAs.

5.3 Write a short essay on the applications of transcriptome analysis in studies of cancer.

5.4 Why is mRNA degradation an essential part of the processes that enable transcriptomes to change over time?

5.5 Describe the differences in half-life that are seen when bacterial and eukaryotic mRNAs are compared. What variations in half-life are seen in individual cells and organisms?

5.6 What is the difference between an exonuclease and an endonuclease?

5.7 Why is an exonuclease that degrades RNA in the 3′ → 5′ direction unable to degrade a bacterial mRNA? How is this problem solved?

5.8 Explain how termination codons are involved in the eukaryotic RNA degradation process called mRNA surveillance.

5.9 Describe the deadenylation-dependent decapping pathway for the degradation of eukaryotic RNA.

5.10 What is Dicer, and what does it do?

5.11 What is the relationship between a foldback RNA and a microRNA?

5.12 Why is it necessary to remove introns from an mRNA before translation occurs?

5.13 Explain why geneticists were able to conclude that splicing occurs in the nucleus.

5.14 Describe the key sequence features of intron–exon boundaries. What additional sequences within an intron are important in the splicing pathway?

5.15 Give a detailed description of the series of events involved in the RNA splicing pathway. Your answer should explain why the donor and acceptor sites are given those names, and should also make a clear distinction between the roles of different snRNPs.

5.16 Give an example of alternative splicing.

5.17 What processes are thought to ensure that the correct splice sites are selected during splicing?

5.18 Explain how RNA editing leads to the synthesis of two different versions of human apolipoprotein B.

5.19 What is the function of the enzymes called ADARs?

5.20 Describe how an m^6A modification is carried out. What are the functions of m^6A modification of mRNA?

5.21 Draw the structures of the modified nucleotides called m^1A, m^5C, m^7G and ac^4C. (Hint: use Figure 2.11 as a guide).

Discussion topics

5.22 How close are we to understanding how the mRNA of a single gene is targeted for degradation in bacteria and eukaryotes?

5.23 How might the length of the poly(A) tail influence the half-life of a eukaryotic mRNA?

5.24 In addition to the GU–AG introns described in this chapter, some eukaryotic genes include AU–AC introns. Obtain some information on these AU–AC introns and discuss the similarities they display with the GU–AG versions and how they have been used in studies of the splicing pathway.

5.25 Introns are common in higher organisms but virtually absent in bacteria. Discuss the possible reasons for this.

5.26 Discuss the questions raised by the discovery of RNA editing.

The answers to, or hints on how to answer, the self-study questions and discussion topic questions can be downloaded from the book's product page at this link: www.routledge.com/9781032743530

FURTHER READING

Blencowe BJ (2000) Exonic splicing enhancers: mechanism of action, diversity and role in human genetic diseases. *Trends in Biochemical Sciences* 25, 106–110.

Cheang MCU, van de Rijn M & Nielsen TO (2008) Gene expression profiling of breast cancer. *Annual Review of Pathology: Mechanisms of Disease* 3, 67–97. *Applications of transcriptome analysis in cancer studies.*

Cho KH (2017) The structure and function of the gram-positive bacterial RNA degradosome. *Frontiers in Microbiology* 8, 154.

Coller J & Parker R (2004) Eukaryotic mRNA decapping. *Annual Review of Biochemistry* 73, 861–890. *Decapping and RNA degradation.*

Graveley BR (2005) Mutually exclusive splicing of insect *Dscam* pre-mRNA directed by competing intronic RNA secondary structures. *Cell* 123, 65–73.

Mello CC & Conte D (2004) Revealing the world of RNA interference. *Nature* 431, 338–342.

Padgett RA, Grabowski PJ, Konarska MM & Sharp PA (1985) Splicing messenger RNA precursors: branch sites and lariat RNAs. *Trends in Biochemical Science* 10, 154–157. *A good summary of the basic details of RNA splicing.*

Pratt AJ & MacRae IJ (2009) The RNA-induced silencing complex: a versatile gene-silencing machine. *Journal of Biological Chemistry* 284, 17897–17901.

Ruegg MA (2005) Structural and functional diversity generated by alternative mRNA splicing. *Trends in Biochemical Sciences* 30, 515–521.

Smith HC & Sowden MP (1996) Base-modification mRNA editing through deamination: the good, the bad and the unregulated. *Trends in Genetics* 12, 418–424. *Apolipoprotein B and similar editing systems.*

Valcárcel J & Green MR (1996) The SR protein family: pleiotropic functions in pre-mRNA splicing. *Trends in Biochemical Sciences* 21, 296–301.

Wahl MC, Will CL & Lührmann R (2009) The spliceosome: design principles of a dynamic RNP machine. *Cell* 136, 701–718. *Review of RNA splicing with focus on the role of the spliceosome.*

Wilkinson ME, Charenton C & Nagai, K. (2020) RNA splicing by the spliceosome. *Annual Review of Biochemistry* 89, 359–388.

Zaccara S, Ries RJ & Jaffrey SR (2019) Reading, writing and erasing mRNA methylation. *Nature Reviews Molecular Cell Biology* 20, 608–624. *A review on chemical modifications that do not alter the sequence of a mRNA.*

Zinder JC & Lima CD (2017) Targeting RNA for processing or destruction by the eukaryotic RNA exosome and its cofactors. *Genes and Development* 31, 88–100.

Types of RNA Molecule: Ribosomal and Transfer RNA

CHAPTER 6

Only a small fraction of a cell's transcriptome, usually no more than 4% of the total, comprises messenger RNA. The vast bulk is made up of **noncoding RNA**, molecules that are not translated into protein but instead play their roles in the cell as RNA. A typical bacterium contains 0.05–0.10 pg of noncoding RNA, making up about 6% of its total weight. A mammalian cell, being much larger, contains more noncoding RNA, 20–30 pg in all, but this represents only 1% of the cell as a whole.

We have already met three types of noncoding RNA – siRNA, microRNA and snRNA – in Chapter 5 where we looked at the roles of these molecules in mRNA degradation and RNA splicing. In this chapter, our attention will be on two additional types of noncoding RNA, called ribosomal and transfer RNA, both of which play central roles in the translation of mRNAs into protein. We will study those roles in detail in Chapters 7 and 8. First, we must become familiar with the structures of rRNA and tRNA molecules, and we must understand how the initial transcripts of rRNA and tRNA genes are converted into functional versions of these RNAs.

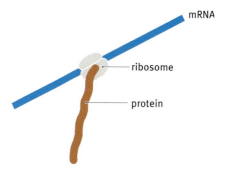

FIGURE 6.1 A ribosome attaches to an mRNA molecule and translates it into a protein.

6.1 RIBOSOMAL RNA

Ribosomal RNA molecules are components of **ribosomes**, which are large multimolecular structures that act as factories for protein synthesis. During this second stage in gene expression, ribosomes attach to mRNA molecules and migrate along them, synthesizing proteins as they go (Figure 6.1).

Ribosomes are made up of rRNA molecules and proteins and are extremely numerous in most cells. An actively growing bacterium may contain over 20,000 ribosomes, comprising about 80% of the total cell RNA and 10% of the total protein. Originally, ribosomes were looked on as passive partners in protein synthesis, merely the structures on which translation occurs. This view has changed over the years and ribosomes are now considered to play an active role, with one of the rRNAs catalyzing the synthesis of bonds between amino acids, the central chemical reaction occurring during translation.

Our understanding of ribosome structure has gradually developed over the last 50 years as more and more powerful techniques have been applied to the problem. We will follow through this research to see what it has revealed.

Ribosomes and their components were first studied by density gradient centrifugation

Originally called 'microsomes', ribosomes were first observed in the early decades of the 20th century as tiny particles almost beyond the resolving power of light microscopy. In the 1940s and 1950s, the first electron micrographs showed that bacterial ribosomes are oval shaped, with dimensions of 29 × 21 nm. Eukaryotic ribosomes are slightly larger, varying a little in size depending on species but averaging about 32 × 22 nm (Figure 6.2). In the mid-1950s, the discovery that ribosomes are the sites of protein synthesis stimulated attempts to define the structures of these particles in greater detail.

The initial progress in understanding the detailed structure of the ribosome was made by analyzing the particles by **density gradient centrifugation** (Figure 6.3). In this procedure, a cell extract is not centrifuged in a normal

DOI: 10.1201/9781003473862-7

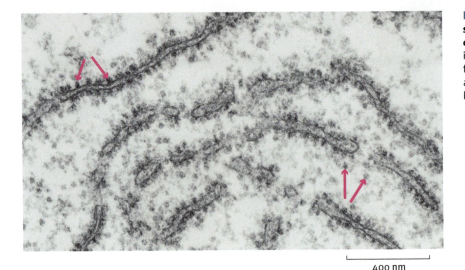

FIGURE 6.2 Electron micrograph showing ribosomes inside an animal cell. The ribosomes appear as black dots, indicated by the red arrows. Some occur free in the cytoplasm, and others are attached to membranes. (Courtesy of the late Professor George Palade.)

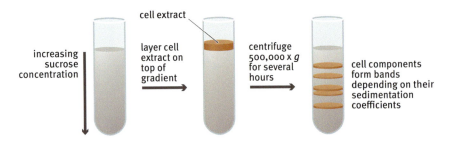

FIGURE 6.3 Density gradient centrifugation. A sucrose solution is layered into a centrifuge tube in such a way that a density gradient is formed, the solution being more concentrated toward the bottom of the tube. The cell extract is placed on the top of the gradient and the tube is centrifuged at a very high speed. The rate of migration of a cell component through the gradient depends on its sedimentation coefficient.

aqueous solution. Instead, a sucrose solution is layered into the tube in such a way that a density gradient is formed, the solution more concentrated and hence denser toward the bottom of the tube. The cell extract is placed on the top of the gradient and the tube is centrifuged at a very high speed, at least 500,000 × g for several hours. Under these conditions, the rate of migration of a cell component through the gradient depends on its **sedimentation coefficient**. The sedimentation coefficient is expressed as an S or Svedberg value, named after the Swedish chemist The Svedberg, who built the first ultracentrifuge in the early 1920s. The S value is dependent on several factors, notably molecular mass and shape.

The centrifugation studies showed that eukaryotic ribosomes have a sedimentation coefficient of 80S, and bacterial ones, reflecting their smaller size, are 70S (**Figure 6.4**). Each type of ribosome is made up of two subunits. In eukaryotes, these subunits are 60S and 40S, and in bacteria, they are 50S and 30S. There are no mistakes here! Sedimentation coefficients are not additive because they depend on shape as well as mass, so it is perfectly acceptable for the intact ribosome to have an S value less than the sum of its two subunits.

In eukaryotes, the large subunit contains three rRNA molecules, called the 28S, 5.8S and 5S rRNAs. The bacterial large subunit has just two rRNAs, 23S and 5S. This is because in bacteria the equivalent of the eukaryotic 5.8S rRNA is contained within the 23S molecule. The small subunit contains just a single rRNA in both eukaryotes and bacteria, 18S in eukaryotes and 16S in bacteria. Both subunits contain a variety of **ribosomal proteins**, the numbers detailed in Figure 6.4. The ribosomal proteins of the small subunit are called S1, S2, etc. and those of the large subunit are L1, L2, etc. There is just one of each protein per ribosome, except for L7 and L12, which are present as dimers.

RIBOSOMAL RNA

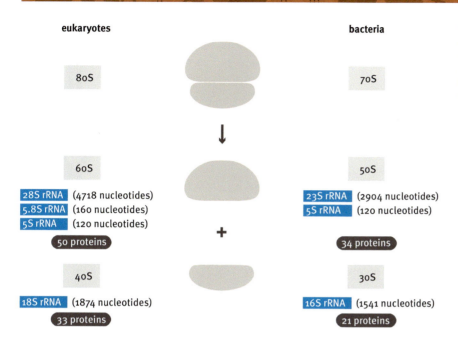

FIGURE 6.4 **The composition of eukaryotic and bacterial ribosomes.** The details refer to a 'typical' eukaryotic ribosome and the *Escherichia coli* ribosome. Variations between different species mainly concern the number of ribosomal proteins.

Understanding the fine structure of the ribosome

The traditional view of ribosome structure is that the rRNA molecules act as a scaffolding on to which the proteins, which provide the functional activity of the ribosome, are attached. To fulfill this role, the rRNA molecules must be able to take up a stable three-dimensional structure. This is achieved by inter- and intramolecular base pairs, with different rRNAs of a subunit base pairing in an ordered fashion with each other, and also more importantly with different parts of themselves.

A two-dimensional representation of the base-paired structure of the *Escherichia coli* 16S rRNA, the component of the small subunit of the bacterial ribosome, is shown in Figure 6.5. This model has been developed by a combination of computational and experimental approaches. First, the nucleotide sequence of the *E. coli* 16S rRNA was examined using computer programs designed to identify regions that could form base pairs with one another. Comparisons were also made between the *E. coli* sequence and the slightly different sequences of the equivalent rRNAs of other bacteria, as it would be expected that the regions of base pairing in the *E. coli* molecule would also be present in other bacteria. The computer predictions were then compared with the results of experiments in which the base-paired rRNA was treated with ribonucleases that act on single-stranded RNA but have no effect on base-paired regions. The double-stranded regions remaining after treatment with these ribonucleases were examined to see if they corresponded with those predicted by the computer model (Figure 6.6).

Additional experiments have located those regions of the 16S rRNA to which the protein components of the *E. coli* small subunit attach. One important technique here is **nuclease protection**, which identifies regions of an rRNA protected from attack by a ribonuclease enzyme. A complex between an rRNA and an individual ribosomal protein is treated with a nonspecific ribonuclease, one that digests both single- and double-stranded RNA. If no protein is present, then the rRNA is completely degraded into nucleotides as every phosphodiester bond in the molecule is open to attack (Figure 6.7a). When the protein is bound, this does not happen as the ribonuclease cannot gain access to all the phosphodiester bonds because some are shielded ('protected') by the bound protein. After

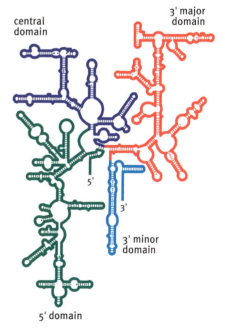

FIGURE 6.5 **The base-paired structure of the *Escherichia coli* 16S rRNA.** The rRNA is a single continuous molecule, but the base-paired structure is looked on as comprising four domains, as shown by the different colors. In addition to the standard base pairs (G–C, A–U), nonstandard base pairs (e.g. G–U) can also form in double-stranded regions of an RNA molecule. These nonstandard pairs are shown as dots.

digestion, the ribonuclease is inactivated and the bound protein removed, revealing two or more intact segments of rRNA whose sequences indicate the parts of the rRNA to which the protein was attached (Figure 6.7b).

Electron microscopy and X-ray crystallography have revealed the three-dimensional structure of the ribosome

The base-paired structures illustrated in Figure 6.5 are two-dimensional representations that tell us little about the real three-dimensional structure of the ribosome. For many years progress in moving from two- to three-dimensional representation of ribosomes was slow, but over the last 20 years there have been remarkable advances due to the application of modern electron microscopy and X-ray diffraction techniques.

Ribosomes are so small that they are close to the resolution limit of the electron microscope, and in the early days of this technique, the best that could be achieved were approximate three-dimensional reconstructions built up by analyzing the fuzzy ribosome images that were obtained. As electron microscopy gradually became more sophisticated, the overall structure of the ribosome could be resolved in greater detail, and the development of innovations such as **immunoelectron microscopy** provided important new information. Before examination by immunoelectron microscopy, ribosomes are first mixed with antibodies that bind specifically to individual ribosomal proteins (Figure 6.8). The antibodies, which are themselves proteins, are labeled with an electron-dense material such as gold particles so that they stand out strongly in the resulting images, enabling the positions of the ribosomal proteins to be located on the ribosome surface.

The most exciting insights into ribosome structure have resulted from X-ray diffraction analysis. With an object as large as a ribosome, the diffraction patterns that are obtained are very complex and the data analysis is a huge task, but modern techniques have proved to be up to the challenge. Structures have now been deduced for entire ribosomes, including ones attached to mRNA and tRNAs (Figure 6.9), greatly increasing our understanding of how protein synthesis works, as we will see in Chapter 8.

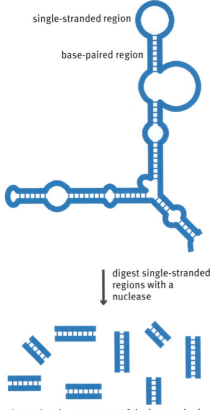

FIGURE 6.6 Using a single-strand specific ribonuclease to identify the base-paired regions in an rRNA molecule.

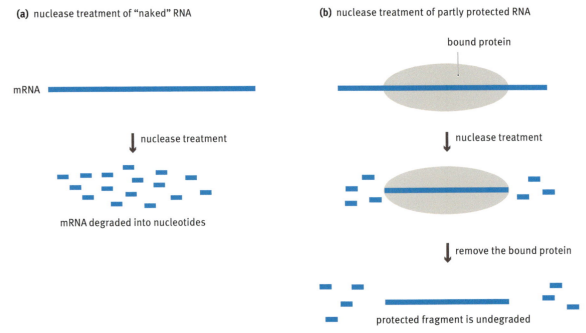

FIGURE 6.7 The nuclease protection technique for identifying protein binding sites on an RNA molecule. (a) If no protein is present, the nuclease can attack every phosphodiester bond, and the rRNA is completely degraded into nucleotides. (b) When a protein is bound to the RNA, the ribonuclease cannot gain access to all the phosphodiester bonds, because some are protected by the protein.

6.2 TRANSFER RNA

Transfer RNA molecules are also involved in protein synthesis but the part they play is completely different from that of rRNA. They form the link between the mRNA that is being translated and the protein that is being synthesized. We will learn exactly how tRNA carries out its function in Chapters 7 and 8 when we deal with the genetic code and the mechanism of protein synthesis. First, in this chapter, we will look at the structures of tRNA molecules.

All tRNAs have a similar structure

Transfer RNA molecules are relatively small, most between 74 and 95 nucleotides in length. Each organism synthesizes a number of different tRNAs, each in multiple copies. However, virtually every tRNA molecule in every organism can be folded into a similar base-paired structure referred to as the **cloverleaf** (Figure 6.10).

The cloverleaf structure has five components. The first is the **acceptor arm**, which is formed by seven base pairs between the 5′ and 3′ ends of the molecule. During protein synthesis, an amino acid is attached to the 3′ end of the acceptor arm of the tRNA, to the adenosine of the invariant CCA terminal sequence.

The second part of the cloverleaf structure is the **D arm**, named after the modified nucleotide dihydrouridine (Figure 6.11), which is always present in this structure. Next is the **anticodon arm**. 'Anticodon' is an annoying jargon term, but its meaning will become clear in Chapter 7 when we look at the role of this part of the tRNA in protein synthesis.

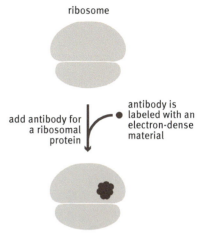

FIGURE 6.8 Identifying the position of a protein on a ribosome by immunoelectron microscopy.

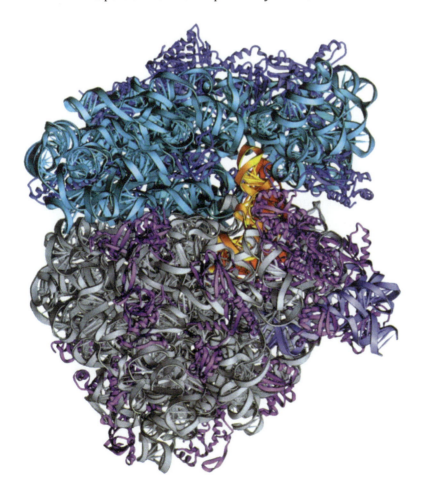

FIGURE 6.9 The bacterial ribosome. The picture shows the ribosome of the bacterium *Thermus thermophilus*. The small subunit is at the top, with the 16S rRNA in light blue and the small subunit ribosomal proteins in dark blue. The large subunit rRNAs are in gray and the proteins are in purple. The gold area is the A site – the point at which tRNAs enter the ribosome during protein synthesis. (From M. B. Mathews & T. Pe'ery, *Trends Biochem. Sci.* 26:585–587, 2001. With permission from Elsevier.)

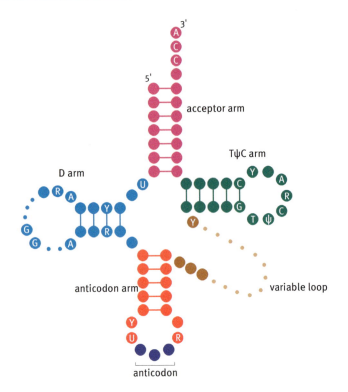

FIGURE 6.10 The cloverleaf structure of a tRNA. The tRNA is drawn in the conventional cloverleaf structure, with the different components labeled. Invariant nucleotides (A, C, G, T, U, Ψ, where Ψ = pseudouridine) and semi-invariant nucleotides (R, purine; Y, pyrimidine) are indicated. Optional nucleotides not present in all tRNAs are shown as smaller dots.

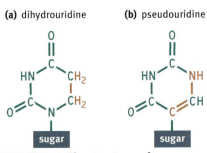

FIGURE 6.11 The structures of two modified nucleotides found in tRNAs. Dihydrouridine and pseudouridine are both modified versions of uridine, the standard U nucleotide found in RNA. The differences from the structure of uridine are indicated in orange.

Following the anticodon arm is the **extra, optional or variable loop**, which may be a loop of just three to five nucleotides or a much larger hairpin structure of 13–21 nucleotides with up to five base pairs in the stem. Types of tRNA with the smaller loop are called Class I and make up 75% of all tRNAs. Those with the larger loop are Class II. The final part of the tRNA is the **TΨC arm**, named after the sequence thymidine–pseudouridine–cytosine, which is always present. Pseudouridine is another modified nucleotide (see Figure 6.11).

The cloverleaf structure can be formed by virtually all tRNAs. In addition to having this common secondary structure, different tRNAs also display a certain amount of nucleotide sequence conservation. Some positions are invariant, and in all tRNAs are occupied by the same nucleotide. Others are semi-invariant and always contain the same type of nucleotide, either one of the pyrimidine nucleotides or one of the purine nucleotides. The invariant and semi-invariant positions are shown in Figure 6.10.

You should keep in mind that although the cloverleaf is a convenient way to draw the structure of a tRNA, it is only a representation and, in the cell, tRNAs have a different three-dimensional structure. This structure has been determined by X-ray diffraction analysis and is shown in Figure 6.12. The base pairs in the arms of the cloverleaf are still present in the three-dimensional structure, but several additional base pairs form between nucleotides in the D and TΨC loops, which appear widely separated in the cloverleaf. This folds the molecule into a compact L-shaped conformation. Many of the nucleotides involved in these extra base pairs are the invariant or semi-invariant ones that are the same in different tRNAs. Note that the three-dimensional conformation places the acceptor arm and anticodon loop at opposite ends of the molecule.

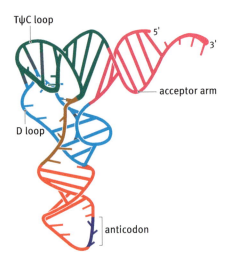

FIGURE 6.12 The three-dimensional structure of a tRNA. Additional base pairs, shown in black and mainly between the D and TΨC loops, fold the cloverleaf structure into this L-shaped configuration. The color scheme is the same as in Figure 6.10.

6.3 PROCESSING OF PRECURSOR rRNA AND tRNA MOLECULES

The rRNAs and tRNAs that play active roles in protein synthesis are quite different from the RNA molecules that are initially transcribed from the rRNA and tRNA genes. These initial transcripts are precursor molecules, longer than the mature RNAs and often containing more than one rRNA and/or tRNA sequence (Figure 6.13). These precursors must therefore be cut into smaller pieces to release the rRNAs and tRNAs. Most rRNAs and tRNAs also contain unusual nucleotides that are not present in the initial transcripts. We have already discovered that two unusual nucleotides – dihydrouridine and pseudouridine – are present in important regions of the tRNA cloverleaf structure (see Figure 6.11), and there are others in tRNA molecules and some in rRNAs also. These unusual nucleotides are produced by chemical modification of the initial transcripts.

All of these processing reactions must be carried out with precision. The cuts must be made at exactly the right positions in the precursor molecules, because if they are not then the rRNA and tRNA molecules that are released will have extra sequences, or may lack segments, and might not be able to carry out their functions in the cell. The chemical modifications must be equally precise. The reasons for many of these modifications are unknown, but we assume they have important roles because each copy of a particular type of tRNA or rRNA is modified at exactly the same nucleotide positions.

We must therefore examine the processing of precursor rRNA and tRNA molecules, not just the events themselves but also the mechanisms that enable the necessary precision to be achieved.

Ribosomal RNAs are transcribed as long precursor molecules

Each ribosome contains one copy of each of the different rRNA molecules, three rRNAs for the bacterial ribosome or four for the eukaryotic version (see Figure 6.4). The most efficient system would be for the cell to synthesize equal numbers of each of these molecules. Of course, the cell could make different amounts of each one but this would be wasteful because some copies would be left over when the least abundant rRNA was all used up.

Synthesis of equal numbers of different rRNA molecules can be achieved by transcribing a set of rRNA molecules as a single unit. The product of transcription, the primary transcript, is therefore a long RNA precursor, the **pre-rRNA**, containing each rRNA separated by short spacers. In bacteria, all three rRNAs are transcribed into a single pre-rRNA with a sedimentation coefficient of 30S, containing the mature molecules in the order 16S–23S–5S (Figure 6.14a). Cutting events are therefore needed to release the mature rRNAs. The same is true for eukaryotic rRNA, with the exception that only the 28S, 18S and 5.8S genes are transcribed together (Figure 6.14b). The 5S genes occur elsewhere on the eukaryotic chromosomes and are transcribed independently of the main unit.

A variety of ribonucleases are involved in cutting the pre-rRNA molecules so that the mature rRNAs are released. Most of these are specific for double-stranded RNA. They cut the pre-rRNA by digesting short segments of

FIGURE 6.13 The initial transcripts of rRNA and tRNA genes are precursor molecules. These pre-RNAs must be cut into smaller pieces to release the mature RNAs.

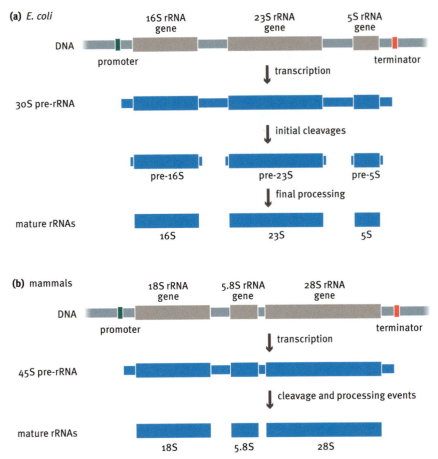

FIGURE 6.14 Processing of pre-rRNA in *E. coli* and mammals.

double-stranded RNA formed by base pairing between different parts of the precursor (Figure 6.15). The base pairing, which of course is determined by the sequence of the pre-rRNA, therefore ensures that these cuts are made at the correct positions.

As well as needing equal numbers of each rRNA, most cells need lots of them. An actively growing *E. coli* cell contains 20,000 ribosomes and divides once every 20 minutes or so. Therefore, every 20 minutes it needs to synthesize 20,000 new ribosomes, an entire complement for one of the

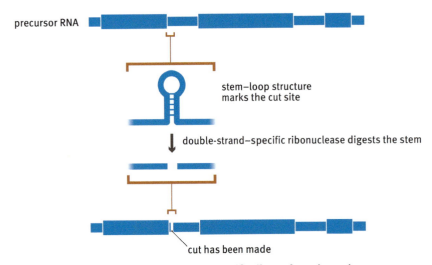

FIGURE 6.15 The role of a double-strand specific ribonuclease in cutting precursor rRNA molecules.

FIGURE 6.16 A tandem array of mammalian rRNA transcription units. Each green box is a transcription unit comprising one set of 18S, 5.8S and 28S rRNA genes.

two daughter cells. This necessitates a considerable amount of rRNA transcription, to such an extent that a single transcription unit would not be able to meet the demand. In fact, the E. coli chromosome contains seven copies of the rRNA transcription unit. In eukaryotes, there can be an even greater demand for rRNA synthesis and 50–5,000 identical copies of the rRNA transcription unit are present depending on species. In eukaryotes, these units are usually arranged into multigene families, with large numbers of copies following one after the other, separated by nontranscribed spacers (Figure 6.16).

Under certain circumstances, the demand for rRNA synthesis is so great that the transcriptional abilities of the genes are outstripped, even when they are in multiple copies. In some eukaryotic cells (for example, amphibian oocytes), an additional strategy known as **gene amplification** may be called upon. This involves replication of rRNA genes into multiple DNA copies which subsequently exist as independent molecules not attached to the chromosomes (Figure 6.17). Transcription of the amplified copies then produces additional rRNA molecules. Gene amplification is not restricted to rRNA genes and occurs with a few other genes whose transcription is required at a greatly enhanced rate in certain situations.

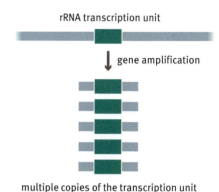

FIGURE 6.17 Production of multiple copies of rRNA genes by amplification. The amplified DNA copies exist as independent molecules and are not attached to any of the chromosomes.

Transfer RNAs are also cut out of longer transcription units

With tRNAs, unlike rRNA, there is no obvious reason why the cell would need equal amounts of the different types that it can make. Nonetheless, in both bacteria and eukaryotes, tRNAs are transcribed initially as precursor-tRNA, which is subsequently processed to release the mature molecules. In E. coli, there are several separate tRNA transcription units, some containing just one tRNA gene and some with as many as seven different tRNA genes in a cluster. In some bacteria, tRNA genes also occur as infiltrators in the rRNA transcription units. This is the case with E. coli, which has either one or two tRNA genes between the 16S and 23S genes in each of its seven rRNA transcription units.

All pre-tRNAs are processed in a similar though not identical way (Figure 6.18). The tRNA sequence within the precursor molecule adopts its base-paired cloverleaf structure and two additional hairpin structures form, one on either side of the tRNA. Processing begins with a cut by ribonuclease E or F forming a new 3′ end just upstream of one of the hairpins. Ribonuclease D, which is an exonuclease, trims seven nucleotides from this new 3′ end and then pauses while ribonuclease P makes a cut at the start of the cloverleaf, forming the 5′ end of the mature tRNA. Ribonuclease D then removes two more nucleotides, creating the 3′end of the mature molecule.

All mature tRNAs must end with the trinucleotide 5′–CCA–3′. With some tRNAs, the terminal CCA is present in the pre-RNA and is not removed by ribonuclease D, but with some other pre-tRNAs this sequence is absent or is removed by the processing ribonucleases. If the CCA is absent or removed during processing, it has to be added by one or more template-independent RNA polymerases such as **tRNA nucleotidyltransferase**.

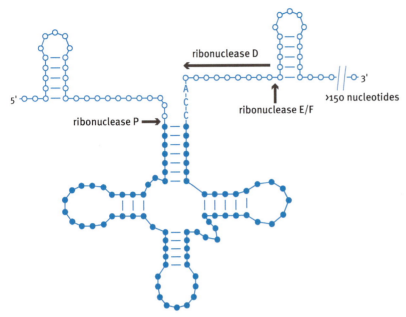

FIGURE 6.18 **Processing of a pre-tRNA in *E. coli*.** The nucleotides of the mature tRNA are shown as closed circles, and those of the remainder of the precursor molecule as open circles. The tRNA has adopted its base-paired structure, and two additional hairpins have formed in the pre-tRNA, one on either side of the cloverleaf. Ribonuclease E or F makes a cut adjacent to one of these hairpins, forming a new 3' end. Ribonuclease D then removes seven nucleotides from this new 3' end. Ribonuclease P makes a cut at the start of the cloverleaf, forming the 5' end of the mature tRNA. Ribonuclease D then removes two more nucleotides, creating the 3' end of the mature molecule. In this example, the trinucleotide 5'–CCA–3' is present in the pre-tRNA and is not removed from the 3' end of the mature tRNA. If this sequence is absent or removed during processing, then it is added by tRNA nucleotidyltransferase.

Transfer RNAs display a diverse range of chemical modifications

In addition to being cut out of their pre-RNAs, tRNAs are also processed by chemical modification. This involves the conversion of certain nucleotides into unusual forms by alteration of their chemical structures. The reasons for many of these modifications are unknown, although roles have been assigned to some specific cases. Some modifications in the D and TΨC arms are recognized by the enzymes that attach an amino acid to the tRNA. These modifications therefore provide some of the specificity that ensures that the correct amino acid is attached to the correct tRNA. Some modifications in the anticodon arm are critical for the correct reading of the genetic code, as we will see in Chapter 7.

A number of different types of chemical modification are known in tRNAs (Figure 6.19). One of the commonest is methylation, which involves the addition of one or more methyl groups ($-CH_3$) to the base or sugar component of the nucleotide. Examples include the conversion of guanosine to 7-methylguanosine. Some modifications involve the removal of groups from the original nucleotide, such as deamination, which is the removal of an amino group ($-NH_2$), as in the conversion of adenosine into inosine. Replacement of an oxygen atom with sulfur gives a thio-substituted nucleotide, an example being 4-thiouridine, obtained by sulfur substitution of uridine.

More complex modifications include base rearrangements, which result in the positions of atoms in the purine or pyrimidine ring becoming changed, as in the conversion of uridine to pseudouridine. Another type of rearrangement is double-bond saturation, which involves the conversion of a

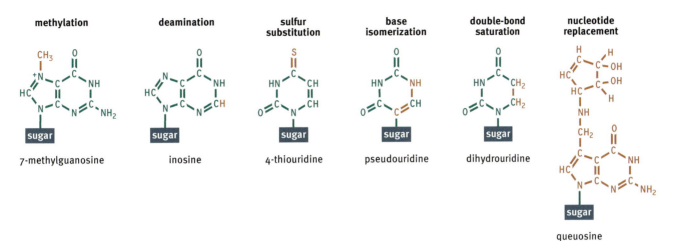

FIGURE 6.19 Examples of chemical modifications occurring with nucleotides in tRNAs. The differences between these modified nucleotides and the standard ones from which they are derived are shown in orange.

double bond in the base to a single bond. An example is the modification of uridine to give dihydrouridine. A few modifications involve the replacement of an entire base with a different, more complex one. The nucleotide called queosine is produced in this way from guanosine.

Over 50 types of chemical modification have been discovered so far in tRNAs. The enzymes that carry out these modifications are thought to recognize particular nucleotide sequences or base-paired structures in the tRNA, or possibly a combination of both, and so modify only the appropriate nucleotides.

Ribosomal RNAs are also modified, but less extensively

Chemical modification is not simply a feature of tRNAs. Ribosomal RNAs are also modified, though not to such a great extent.

In eukaryotic rRNAs, the two commonest types of modification are 2′-O-methylation, in which the hydrogen of the –OH group attached to the 2′-carbon is replaced by a methyl group (Figure 6.20), and the conversion of uridine to pseudouridine. Modification by 2′-O-methylation occurs at just over 100 positions in each set of mature human rRNAs, in other words at about one in every 70 nucleotides. These modified positions are, to a certain extent, the same in different species, and some similarities in modification patterns are even seen when bacteria and eukaryotes are compared, although bacterial rRNAs are less heavily modified than eukaryotic ones. Functions for the modifications have not been identified, although most occur within those parts of rRNAs thought to be most critical to the activity of these molecules in ribosomes. Modified nucleotides might, for example, be involved in rRNA-catalyzed reactions such as the synthesis of peptide bonds.

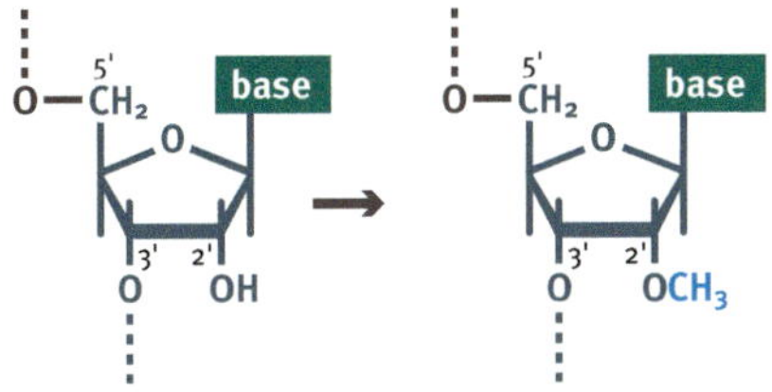

FIGURE 6.20 2′-O-methylation of a nucleotide.

In bacteria, rRNAs are modified by enzymes that directly recognize the sequence and/or structure of the regions of RNA that contain the nucleotides to be modified. Often two or more nucleotides in the same region are modified at once. In eukaryotes, a more complex machinery exists to ensure that the modifications are made at the correct positions. The nucleolus (the part of the eukaryotic nucleus where rRNA transcription and processing occur) contains short RNAs, between 70 and 100 nucleotides in length, called **small nucleolar RNAs (snoRNAs)**, that are involved in the modification process. By base pairing to the relevant region, snoRNAs pinpoint positions at which the pre-rRNA must be methylated. The base pairing involves only a few nucleotides, not the entire length of the snoRNA, but these nucleotides are always located immediately upstream of a sequence called the D box, which is present in all snoRNAs (Figure 6.21). The base pair involving the nucleotide that will be modified is five positions away from the D box.

The first snoRNAs to be discovered were all involved in methylation. It was then realized that a different family of snoRNAs carries out the same guiding role in conversion of uridines to pseudouridines. These snoRNAs do not have D boxes but still have conserved motifs that can be recognized by the modifying enzyme, and each is able to form a specific base-paired interaction with its target site, indicating the nucleotide to be modified.

There is a different snoRNA for each modified position in a pre-rRNA, except possibly for a few sites that are close enough together to be dealt with by a single snoRNA. This means that there must be a few hundred snoRNAs per cell. At one time, this seemed unlikely because very few snoRNA genes could be located. Now it appears that only a fraction of all the snoRNAs are transcribed from standard genes, most being specified by sequences within the introns of other genes and released by cutting up the intron after splicing (Figure 6.22).

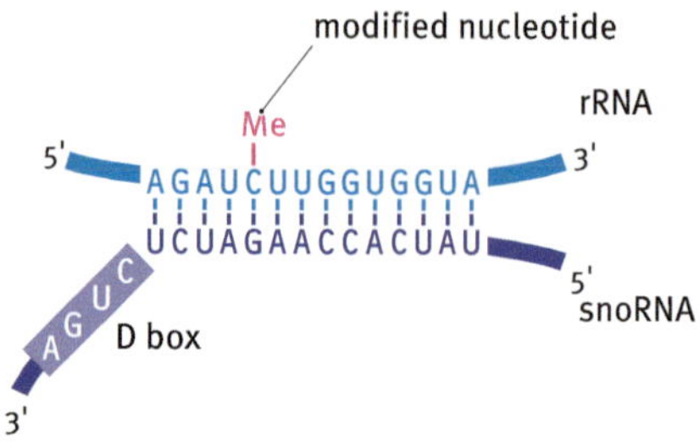

FIGURE 6.21 The role of a snoRNA in methylation of a specific nucleotide in a rRNA molecule. The D box of the snoRNA is highlighted. Modification always occurs at the base pair five positions away from the D box. Note that the interaction between the rRNA and the snoRNA involves an unusual G–U base pair, which is permissible between RNA polynucleotides.

6.4 REMOVAL OF INTRONS FROM PRE-rRNAs AND PRE-tRNAs

Introns are not only found in genes that are transcribed into mRNA – they are also found in some eukaryotic rRNA and tRNA genes. These introns are not the same as those found in mRNA. They lack the characteristic consensus sequences, such as GU and AG at the splice sites, and they are not associated with spliceosomes. The mechanisms by which they are spliced are therefore different from the process that we studied in Chapter 5.

Splicing of pre-rRNAs and pre-tRNAs appears simply to be a part of the series of processing events needed to produce the mature molecules. There is no equivalent of alternative splicing with rRNA and tRNA transcripts. But these introns are still interesting, especially the ones in rRNAs, as their discovery forced scientists to reappraise one of the fundamental notions of biology.

Some rRNA introns are enzymes

The fundamental notion that was overturned by rRNA introns was the assumption that only proteins can have enzymatic activity and hence catalyze biochemical reactions. The introns present in some pre-rRNAs are enzymes. The biochemical reaction that they catalyze is their own splicing.

Introns are not common in rRNA genes but are found in certain protozoa, notably the ciliate *Tetrahymena*. These introns are able to fold up by intramolecular base pairing into a complex structure in which the two splice sites are brought close together (Figure 6.23). The intron is then cut out and the two exons joined together in the complete absence of any protein molecules. The intron catalyzes the reaction by acting as an RNA enzyme or **ribozyme**. The *Tetrahymena* self-splicing intron was the first known example of a ribozyme and caused quite a stir when it was discovered in 1982, as many biochemists at that time were unwilling to believe that RNA could have enzymatic activity.

The self-splicing intron is a member of the 'Group I' class of introns. Group I introns are also found in the rRNA genes of other protozoa and are also

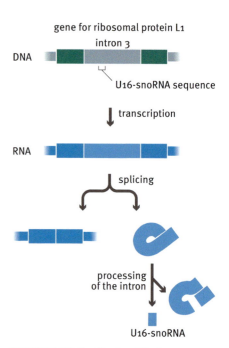

FIGURE 6.22 Synthesis of the human U16-snoRNA. Many snoRNAs are specified by sequences within the introns of other genes. This example shows synthesis of the human U16-snoRNA by processing of the intron spliced from the mRNA for ribosomal protein L1.

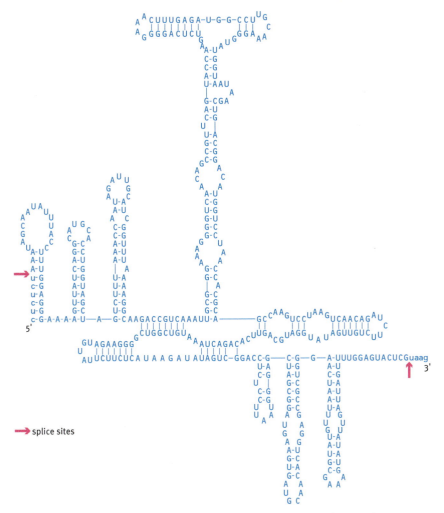

FIGURE 6.23 The base-paired structure of the *Tetrahymena* rRNA intron. The sequence of the intron is shown in capital letters, with the exons in lowercase. Additional interactions fold the intron into a three-dimensional structure that brings the two splice sites (indicated by the arrows) close together.

present in the Myxomycetes, a type of slime mold. All introns of the Group I type can take up a similar base-paired structure, and most of those that have been studied in detail are able to self-splice in a manner similar to the *Tetrahymena* intron.

Some eukaryotic pre-tRNAs contain introns

Transfer RNA introns are relatively common in lower eukaryotes but less frequent in vertebrates – introns are present in only 6% of all human tRNA genes. Introns in eukaryotic pre-tRNAs are 14–60 nucleotides in length and are usually found at the same position in the transcript, within the anticodon arm. The intron sequence is variable, but includes a short region complementary to part of the anticodon arm. Base pairing between the complementary sequences forms a short stem between two loops in the unspliced pre-tRNA (Figure 6.24).

Unlike the mRNA and Group I introns, splicing of pre-tRNA introns involves an endonuclease. This enzyme contains four nonidentical subunits, one of which uses the structure of the base-paired intron as a guide to identify the correct positions at which the RNA should be cut. The upstream and downstream cuts are then made by two of the other enzyme subunits.

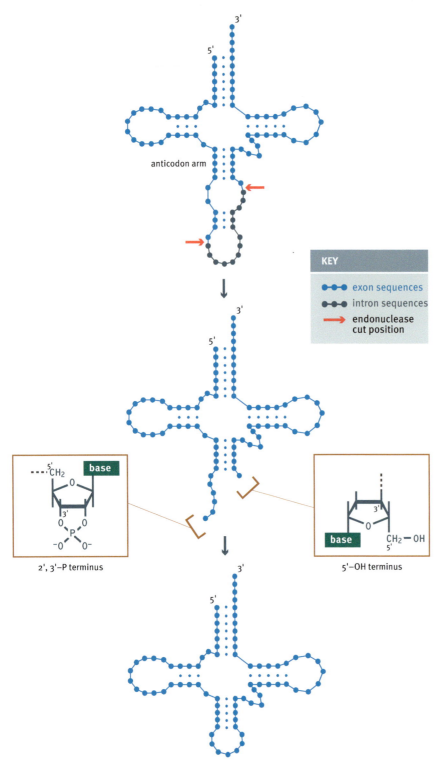

FIGURE 6.24 Removal of an intron from a pre-tRNA. In the top drawing, the tRNA has folded into its cloverleaf structure, but with an extra stem-loop, formed by the intron, in the anticodon arm. The nucleotides of the mature tRNA are shown in blue and those of the intron in black. Cuts are made by endonucleases at the exon–intron boundaries, indicated by red arrows. The unusual structures that are left at the cut sites are shown in the boxes. These are converted to standard 3′–OH and 5′–P termini, which are ligated to one another to complete the splicing process.

Cleavage leaves an unusual cyclic phosphate structure attached to the 3′ end of the upstream exon and a hydroxyl group at the 5′ end of the downstream exon. Before the ends can be joined, the cyclic phosphate must be converted to a 3′–OH end, and the 5′–OH terminus to 5′–P.

The two ends are held in proximity by the natural base pairing adopted by the tRNA sequence and are ligated together. The enzymatic activities needed to convert the ends and join them together are all provided by a single protein.

KEY CONCEPTS

- Most of the RNA in a cell is noncoding. Noncoding RNAs are molecules that are not translated into protein but instead play their roles in the cell as RNA.

- Ribosomal RNA is the most abundant type of noncoding RNA. Ribosomal RNA molecules are components of ribosomes, which are large multi-molecular structures that act as factories for protein synthesis.

- Transfer RNA molecules are also involved in protein synthesis but the part they play is completely different from that of rRNA. Transfer RNAs form the link between the mRNA that is being translated and the protein that is being synthesized.

- Almost all tRNAs can adopt a base-paired two-dimensional structure called the cloverleaf. In the cell, the cloverleaf folds into a compact L-shaped conformation.

- The rRNAs and tRNAs that play active roles in protein synthesis are initially transcribed as precursor molecules. The precursors are processed by cutting events, which might include RNA splicing, and chemical modification of certain nucleotides.

- The precursor rRNA molecules contain copies of three types of rRNA. This ensures that equal amounts of each rRNA are synthesized.

- Chemical modification of tRNA and bacterial rRNA is carried out by enzymes that recognize the sequence and/or structure of the regions of RNA that contain the nucleotides to be modified. Modification of eukaryotic rRNA is aided by snoRNAs, which bind to the rRNA and guide the modifying enzyme to the correct nucleotide.

- Some rRNAs contain introns that are self-splicing. These were the first types of ribozyme to be discovered and forced a reappraisal of the notion that enzymes are always made of protein.

QUESTIONS AND PROBLEMS

Key terms

Write short definitions of the following terms:

acceptor arm
anticodon arm
cloverleaf
D arm
density gradient centrifugation
extra, optional, or variable loop
gene amplification
immunoelectron microscopy
noncoding RNA
nuclease protection
pre-rRNA
ribosomal protein
ribosome
ribozyme
sedimentation coefficient
small nucleolar RNA
tRNA
nucleotidyltransferase
TΨC arm

Self-study questions

6.1 What is the definition of noncoding RNA?

6.2 Explain how density gradient centrifugation is used to measure the sizes of cell components. How does the structure of a cell component influence its sedimentation coefficient?

6.3 What are the sizes of typical bacterial and eukaryotic ribosomes?

6.4 Construct two tables giving details of the RNA and protein components of bacterial and eukaryotic ribosomes.

6.5 Describe how methods that make use of nucleases have been used to study the structure of the ribosome.

6.6 An RNA molecule has the sequence:

5'–AUUAGCAUGUAAAUGCUAUCUGGCAAC–3'

After digestion with a single-strand-specific ribonuclease, the following base-paired molecule is obtained:

```
5'-AGCAU-3'
   |||||
3'-UCGUA-5'
```

What was the base-paired structure of the original molecule?

6.7 A second RNA molecule has the following sequence:

5'–CUCCUCAAAUGAUUCAGUGCAUCUAACC CUAUUUAAACGCACUAGCUCAUGAGAGG AG–3'

After digestion with the same enzyme, the following are obtained:

```
5'-UAA-3'   5'-AUGA-3'   5'-AGUGC-3'
   |||         ||||         |||||
3'-AUU-5'   3'-UACU-5'   3'-UCACG-5'

5'-CUCCUC-3'
   ||||||
3'-GAGGAG-5'
```

Draw the structure of the original molecule.

6.8 Describe the methods that have been used and are being used to study the three-dimensional structure of the ribosome. What are the applications and limitations of each technique?

6.9 Draw and annotate the cloverleaf structure of tRNA.

6.10 To what extent is the cloverleaf a true representation of the actual structure of tRNA?

6.11 Here is the sequence of the yeast alanine tRNA:

5'–GGGCGUGUGGCGUAGUCGGUAG CGCGCUCCCUUIGCIUGGGAGAGGUCU CCGGUUCGAUUCCGGACUCGUC CACCA–3'

Make a fully annotated drawing of a possible cloverleaf structure for this tRNA.

6.12 Give one possible reason why pre-rRNAs contain copies of more than one gene.

6.13 Describe the cutting events involved in processing of pre-rRNA.

6.14 Why do cells require large amounts of rRNA? How are these demands met?

6.15 Describe how a tRNA molecule is cut out of its primary transcript.

6.16 List six types of chemical modification that occur with nucleotides in tRNA. In each case, draw the structure of an example of a nucleotide resulting from the modification.

6.17 What are the possible roles of the chemical modifications found in tRNAs?

6.18 What types of chemical modification are found in eukaryotic rRNA?

6.19 Describe how rRNAs are chemically modified in bacteria.

6.20 How does chemical modification of rRNA in eukaryotes differ from the process occurring in bacteria?

6.21 Why is the *Tetrahymena* rRNA intron referred to as 'self-splicing'?

6.22 Describe how introns are removed from pre-tRNA molecules.

Discussion Topics

6.23 To what extent have studies of ribosome structure been of value in understanding the detailed process by which proteins are synthesized?

6.24 Ribosomal and transfer RNA molecules are relatively long-lived in the cell. In contrast, mRNAs are subject to quite rapid turnover rates. Discuss the possible reasons and consequences.

6.25 In eukaryotes, the 28S, 18S and 5.8S genes are transcribed together, but the 5S genes are transcribed separately. What implications does this have for the hypothesis that the most efficient system is for the cell to synthesize equal numbers of each of its rRNAs?

6.26 What are the possible reasons for the observation that virtually all tRNA molecules in all organisms adopt a similar base-paired structure?

6.27 Discuss the reasons why tRNA and rRNA molecules are chemically modified.

6.28 The existence of ribozymes is looked on as evidence that RNA evolved before proteins and therefore at one time, during the earliest stages of evolution, all enzymes were made of RNA. Assuming that this hypothesis is correct, explain why some ribozymes persist to the present day.

The answers to, or hints on how to answer, the self-study questions and discussion topic questions can be downloaded from the book's product page at this link: www.routledge.com/9781032743530

FURTHER READING

Björk GR, Ericson JU, Gustafsson C et al. (1987) Transfer RNA modification. *Annual Review of Biochemistry* 56, 263–287. *Information on modified nucleotides in tRNAs.*

Clark BFC (2001) The crystallization and structural determination of tRNA. *Trends in Biochemical Sciences* 26, 511–514. *Determination of the three-dimensional structure of a tRNA.*

Hedberg A & Johansen SD (2103) Nuclear group I introns in self-splicing and beyond. *Mobile DNA* 4, 17.

Holley RW, Apgar J, Everett GA et al. (1965) Structure of a ribonucleic acid. *Science* 147, 1462–1465. *The first complete sequence of a tRNA and discovery of the cloverleaf structure.*

Hopper AK, Pai DA & Engelke DR (2010) Cellular dynamics of tRNAs and their genes. *FEBS Letters* 584, 310–317. *Review of tRNA processing including splicing.*

Jobe A, Liu Z, Gutierrez-Vargas C & Frank J (2019) New insights into ribosome structure and function. *Cold Spring Harbor Perspectives in Biology* 11, a032615.

Kruger K, Grabowski PJ, Zaug AJ et al. (1982) Self-splicing RNA: autoexcision and autocyclization of the ribosomal RNA intervening sequence of Tetrahymena. *Cell* 31, 147–157. *The first indication that the Tetrahymena intron is a ribozyme.*

Steitz TA & Moore PB (2003) RNA, the first macromolecular catalyst: the ribosome is a ribozyme. *Trends in Biochemical Sciences* 28, 411–418. *The discovery that rRNAs act as enzymes during protein synthesis.*

Tollervey D (1996) Small nucleolar RNAs guide ribosomal RNA methylation. *Science* 273, 1056–1057.

Venema J & Tollervey D (1999) Ribosome synthesis in *Saccharomyces cerevisiae*. *Annual Review of Genetics* 33, 261–311. *Extensive details on rRNA processing.*

Wilson DN & Cate JHD (2012) The structure and function of the eukaryotic ribosome. *Cold Spring Harbor Perspectives in Biology* 4, a011536.

Yusupov MM, Yusupova GZ, Baucom A et al. (2001) Crystal structure of the ribosome at 5.5Å resolution. *Science* 292, 883–896. *An early and important X-ray crystallography study of ribosome structure.*

CHAPTER 7

The Genetic Code

The three major types of RNA molecule that are produced by transcription – messenger, ribosomal and transfer RNA – work together to synthesize proteins by the process called translation. The key feature of translation is that the sequence of amino acids in the protein being synthesized is specified by the sequence of nucleotides in the mRNA molecule that is being translated (Figure 7.1).

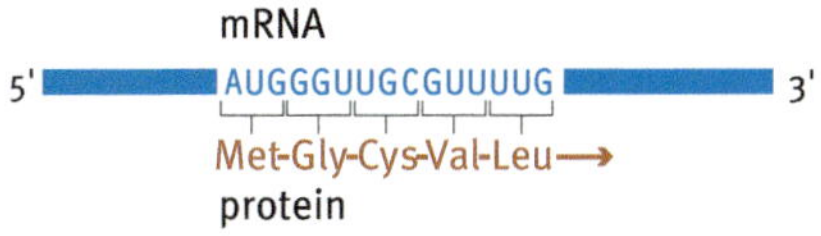

We will study the mechanism by which proteins are synthesized in Chapter 8. Before doing that we must cover two underlying aspects of translation, both of which are vital to our understanding of the molecular basis to genetics. First, we must look more closely at proteins. We established in Chapter 3 that proteins with different amino acid sequences can have quite different chemical properties and that this enables them to play a variety of roles in the cell. The link between a protein's amino acid sequence and its function is therefore a vital part of the process by which the biological information contained in a gene is expressed and made use of by the cell. As such, we must understand the nature of this link. The first part of this chapter is therefore devoted to a short overview of protein structure.

The second underlying aspect of translation that we must examine is the genetic code. The genetic code is the set of rules that determines which sequence of nucleotides specifies which sequence of amino acids. It is therefore central to the translation of mRNA into protein and demands our detailed attention. Not only must we be aware of the features of the genetic code, we must also understand how the rules of the genetic code are enforced during the translation of an individual mRNA. That enforcement is the role of tRNA, whose function in protein synthesis we will look at in the last part of this chapter.

FIGURE 7.1 During translation, each triplet of nucleotides in the RNA specifies a particular amino acid. The identity of each amino acid is set by the genetic code. The amino acids indicated here by their three-letter abbreviations are methionine, glycine, cysteine, valine and leucine.

7.1 PROTEIN STRUCTURE

A protein, like a DNA molecule, is a linear unbranched polymer. In proteins, the monomeric subunits are called **amino acids** and the resulting polymers, or **polypeptides**, are rarely more than 2,000 units in length.

As with DNA, the key features of protein structure were determined in the first half of the 20th century. This phase of protein biochemistry culminated in the 1940s and early 1950s with the elucidation of the major conformations or **secondary structures** taken up by polypeptides. In recent years, interest has focused on how these secondary structures combine to produce the complex three-dimensional shapes of proteins.

Amino acids are linked by peptide bonds

Twenty different amino acids are found in protein molecules (Table 7.1). Each has the general structure shown in Figure 7.2, comprising a central α-carbon atom to which four groups are attached. These are a hydrogen atom, a carboxyl group (–COO⁻), an amino group (–NH₃⁺), and an **R group**, which is different for each amino acid (Figure 7.3).

The R groups vary considerably in chemical complexity. For glycine, the R group is simply a hydrogen atom, whereas for tyrosine, phenylalanine and tryptophan, the R groups are complex aromatic side chains. The majority of R groups are uncharged, though two amino acids have negatively charged R groups (aspartic acid and glutamic acid) and three have

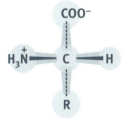

FIGURE 7.2 The general structure of amino acid. The four bonds have a tetrahedral arrangement around the central carbon, so the amino group and hydrogen atom are, in effect, above the plane of the paper, and the carboxyl and R groups are below it.

DOI: 10.1201/9781003473862-8

TABLE 7.1 THE 20 AMINO ACIDS FOUND IN PROTEINS

Amino Acid	Abbreviation	
	Three Letter	One Letter
Alanine	Ala	A
Arginine	Arg	R
Asparagine	Asn	N
Aspartic acid	Asp	D
Cysteine	Cys	C
Glutamic acid	Glu	E
Glutamine	Gln	Q
Glycine	Gly	G
Histidine	His	H
Isoleucine	Ile	I
Leucine	Leu	L
Lysine	Lys	K
Methionine	Met	M
Phenylalanine	Phe	F
Proline	Pro	P
Serine	Ser	S
Threonine	Thr	T
Tryptophan	Trp	W
Tyrosine	Tyr	Y
Valine	Val	V

positively charged R groups (lysine, arginine and histidine). Some R groups are **polar** e.g. (serine and threonine) and prefer to be in an aqueous environment, others are **nonpolar** (e.g. alanine, valine and leucine) and lack an affinity for water. These differences mean that although all amino acids are closely related, each has its own specific chemical properties.

The 20 amino acids shown in Figure 7.3 are the ones that are conventionally looked on as being specified by the genetic code. They are therefore the amino acids that are linked together when mRNA molecules are translated into proteins. However, these 20 amino acids do not on their own represent the limit of the chemical diversity of proteins. At least two additional amino acids – selenocysteine and pyrrolysine (Figure 7.4) – can be inserted into a polypeptide chain during protein synthesis, their insertion directed by a modified reading of the genetic code. Also, after a protein has been synthesized, some amino acids might be modified by the addition of new chemical groups, for example, by phosphorylation, or by attachment of large side chains made up of sugar units.

A polypeptide contains amino acids linked together by **peptide bonds**, formed by condensation between the carboxyl group of one amino acid and the amino group of a second amino acid (Figure 7.5a). The structure of a tripeptide, comprising three amino acids, is shown in Figure 7.5b.

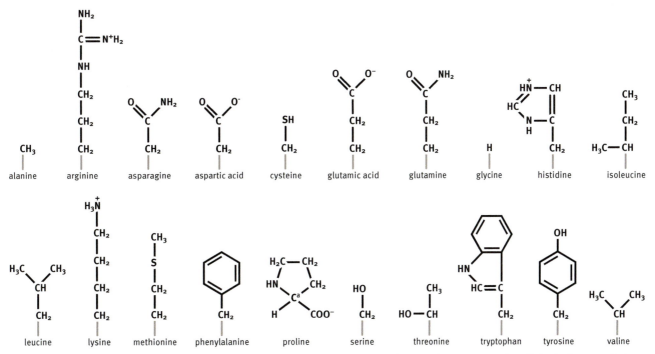

FIGURE 7.3 The structures of the amino acid R groups. These 20 amino acids are the ones that are conventionally looked upon as being specified by the genetic code. Note that the entire structure of proline is shown, not just its R group. The central α carbon is labeled. The entire structure is shown because the R group of proline forms a bond not just with the α carbon but also with the amino group attached to this carbon.

Note that, as with a polynucleotide, the two ends of the polypeptide are chemically distinct. One has a free amino group and is called the **amino**, **NH₂–** or **N terminus**, and the other has a free carboxyl group and is called the **carboxyl**, **COOH–** or **C terminus**. The direction of the polypeptide can therefore be expressed as either N→C (left to right in Figure 7.5b) or C→N (right to left in Figure 7.5b).

There are four levels of protein structure

Proteins are traditionally looked on as having four distinct levels of structure. These levels are hierarchical, the protein being built up stage by stage, with each level of structure depending on the one below it.

The first of these structural levels, called the **primary structure**, is the linear sequence of amino acids. The next level, the **secondary structure**, refers to the different conformations that can be taken up by the polypeptide. The two main types of secondary structure are the **α helix** and **β sheet** (Figure 7.6). These are stabilized mainly by hydrogen bonds that form between different amino acids in the polypeptide. Usually, different regions of a polypeptide take up different secondary structures, so that the whole is made up of a number of α helices and β sheets, together with less organized regions.

The **tertiary structure** results from folding the secondary structural components of the polypeptide into a three-dimensional configuration (Figure 7.7). The tertiary structure is stabilized by various chemical forces, notably hydrogen bonding between individual amino acids, electrostatic interactions between the R groups of charged amino acids, and hydrophobic forces. The hydrophobic forces dictate that amino acids with nonpolar side groups must be shielded from water by embedding within the internal regions of the protein. There may also be covalent linkages

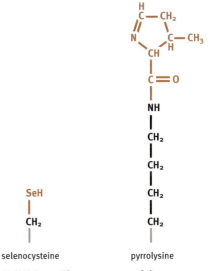

FIGURE 7.4 The structures of the R groups of selenocysteine and pyrrolysine. The parts shown in orange indicate the differences between these amino acids and cysteine and lysine, respectively.

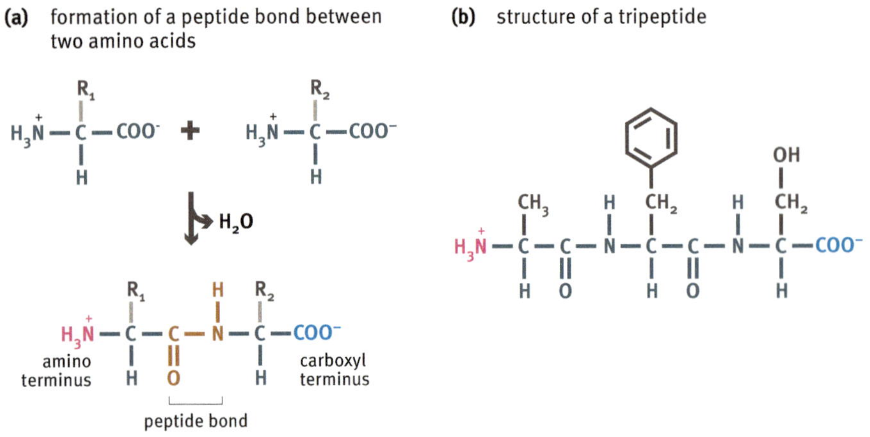

(a) formation of a peptide bond between two amino acids

(b) structure of a tripeptide

FIGURE 7.5 Amino acids are linked together by peptide bonds. (a) The chemical reaction that results in two amino acids becoming linked together by a peptide bond. (b) The structure of a tripeptide with the sequence alanine-phenylalanine-serine.

called **disulfide bridges** between cysteine amino acids at various places in the polypeptide (Figure 7.8).

Finally, the **quaternary structure** involves the association of two or more polypeptides, each folded into its tertiary structure, into a multisubunit protein. Not all proteins form quaternary structures, but it is a feature of many proteins with complex functions, including many involved in gene expression. Some quaternary structures are held together by disulfide bridges between the different polypeptides, resulting in stable multisubunit proteins that cannot easily be broken down into the component parts. Other quaternary structures comprise looser associations of subunits stabilized by hydrogen bonding and hydrophobic effects, which means that these proteins can revert to their component polypeptides, or change their subunit composition, according to the functional requirements. The quaternary structure may involve several molecules of the same polypeptide or may comprise different polypeptides. An example of the latter is the bacterial RNA polymerase, whose subunit composition is described as $\alpha_2\beta\beta'\sigma\omega$ (see Figure 4.10). In some cases, the quaternary structure is built up from a very large number of polypeptide subunits, to give a complex array. The best examples are the protein coats of viruses, such as that of tobacco mosaic virus, which is made up of 2,130 identical protein subunits (Figure 7.9).

The amino acid sequence is the key to protein structure

Each of the higher levels of protein structure – secondary, tertiary and quaternary – is specified by the primary structure, the amino acid sequence itself. This is most clearly understood at the secondary level, where it is recognized that certain amino acids, because of the chemical and physical properties of their R groups, stimulate the formation of an α helix, whereas others promote the formation of a β sheet. Conversely, certain amino acids more frequently occur outside regular structures and may act to determine the end point of a helix or sheet. These factors are now so well understood that rules to predict the secondary structures taken up by amino acid sequences have been developed.

Although less well characterized, it is nonetheless clear that the tertiary and quaternary structures of a protein also depend on the amino acid sequence. The interactions between individual amino acids at these levels are so complex that predictive rules, although attempted, are still unreliable. However, it has been established for some years that if a protein is **denatured**, for instance, by mild heat treatment or adding a chemical

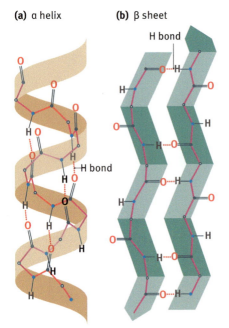

FIGURE 7.6 The two main secondary structural units found in proteins. The polypeptide chains are shown in outline. Hydrogen bonds are shown by dotted lines. The R groups have been omitted for clarity. The β-sheet conformation that is shown is antiparallel, the two chains running in opposite directions. Parallel β sheets also occur.

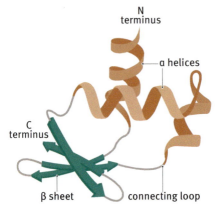

FIGURE 7.7 The tertiary structure of a protein. This imaginary protein structure comprises three α helices, shown as coils, and a four-stranded β sheet, indicated by the arrows.

denaturant such as urea, so that it loses its higher levels of structure and takes up a nonorganized conformation, it still retains the innate ability upon **renaturation** (by cooling down again, for example) to reform spontaneously the correct tertiary structure (Figure 7.10). Once the tertiary structure has formed, subunit assembly into a multimeric protein again occurs spontaneously. This shows that the instructions for the tertiary and quaternary structures must reside in the amino acid sequence.

Amino acid sequence also determines protein function

The amino acid sequence not only determines the secondary, tertiary and quaternary structures of a protein. Through these, it also specifies the protein's function. This is because a protein, in order to perform its function, must interact with other molecules. The precise nature of these interactions is set by the overall shape of the protein and the distribution of chemical groups on its surface.

Understanding the interactions of proteins with other molecules (including other proteins) is a complex area of modern biochemistry, but the basic principle that amino acid sequence determines function is easily illustrated. Let us consider one of the many proteins that must attach themselves to a DNA molecule in order to perform their function. These DNA-binding proteins form a large and diverse group that includes, for instance, RNA polymerase and the repressors and activators that regulate the initiation of transcription. An example is Cro, a regulatory protein that controls the expression of a number of genes of the bacteriophage called λ.

Cro is an example of the **helix-turn-helix** family of DNA-binding proteins, the name indicating that the binding motif is made up of two α helices separated by a turn (Figure 7.11). The latter is not a random conformation but a specific structure, referred to as a **β turn**, made up of four amino acids, the second of which is usually glycine. This turn, in conjunction with the first α helix, positions the second α helix on the surface of the protein in an orientation that enables it to fit inside the major groove of a DNA molecule. This second α helix is therefore called the **recognition helix**, because it makes the vital contacts with the DNA.

The active form of the Cro protein is a dimer. The two recognition helices, one from each polypeptide in the dimer, are exactly 3.4 nm apart and therefore fit into two adjacent sections of the major groove of a DNA molecule (Figure 7.12). Hence, the shape of the protein is critical to its function. If either recognition helix were absent or if they were orientated incorrectly on the surface of the protein, then Cro would not be able to bind to DNA. But this is only part of the story. Cro does not attach randomly to the λ DNA molecule. Like the vast majority of regulatory proteins, it binds to specific sequences adjacent to the genes whose expression it controls. This requires that precise contacts be made between chemical groups in the recognition helices of the Cro protein and the bases in the DNA sequence to which Cro binds. If the helices have the wrong chemical groups, or if the groups are positioned incorrectly, then the binding sequence will not be recognized.

The function of Cro therefore depends on its amino acid sequence in three ways. First, the recognition structure is present because the amino acid sequence of that particular part of the Cro polypeptide promotes the formation of an α helix. Second, the amino acid sequence of the protein as a whole specifies a quaternary structure in which the two recognition helices of the dimeric protein are orientated in precisely the correct way. Finally, the amino acid sequence of each recognition helix provides the

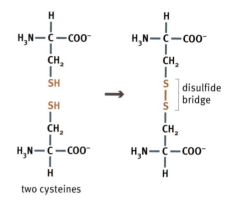

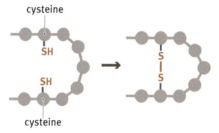

FIGURE 7.8 Disulfide bridges. The upper drawing shows the chemical structure of a disulfide bridge. Below is the effect that the formation of a disulfide bridge can have on the structure of a polypeptide.

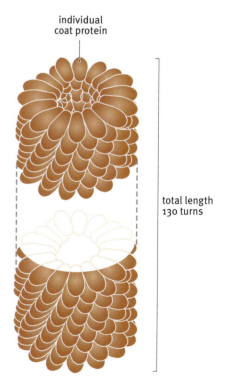

FIGURE 7.9 The tobacco mosaic virus coat. The virus coat is made up of 2,130 identical protein subunits.

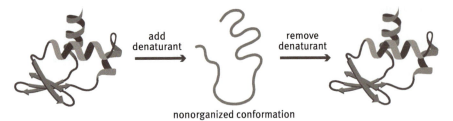

FIGURE 7.10 A denatured protein can regain its tertiary structure after the removal of the denaturant.

particular combination of R groups that enables the protein to recognize its specific binding sequence. There is therefore a precise link between the amino acid sequence of the Cro polypeptide and the function of the dimeric Cro protein.

7.2 THE GENETIC CODE

Now that we have a clear grasp of the link between a protein's amino acid sequence and its function, we can turn our attention to the rules that govern the order in which amino acids are joined together during translation of an mRNA. These rules are called the genetic code.

Understanding the genetic code was the main pre-occupation of geneticists in the years immediately after it was realized that genes are made of DNA and that biological information is contained in the nucleotide sequences of DNA molecules. The culmination of this work came in 1966 when the genetic code operating in the bacterium *Escherichia coli* was completely solved. Since then the code itself has been studied less intensively, but important new discoveries have still been made, including variations of the code that operate in some species, including humans. We will first review the key features of the genetic code and then examine the coding variations that are known.

There is a colinear relationship between a gene and its protein

When research into the genetic code began in the 1950s, it was more or less assumed that there is a colinear relationship between a gene and its protein. By 'colinear' we mean that the order of nucleotides in the gene correlates directly with the order of amino acids in the corresponding protein (Figure 7.13). This is clearly the most straightforward way in which genes could code for proteins.

Assumptions are dangerous in science and must always be tested by experiments. The easiest way to test the hypothesis that genes and proteins are colinear would be to alter the nucleotide sequence of a gene at a specific point and see if the resulting change in the amino acid sequence of the corresponding protein occurs at the same relative position or elsewhere (Figure 7.14). This type of experiment was first carried out with the *E. coli* gene coding for one of the two subunits – subunit A – of the enzyme called tryptophan synthetase. As its name implies, this enzyme catalyzes the final step in the biochemical pathway that results in the synthesis of tryptophan. The experiments confirmed that a change in the subunit A nucleotide sequence gives rise to an amino acid alteration at the equivalent position in the subunit A protein. The results showed that the subunit A gene is colinear with the subunit A protein, with the amino terminus of the protein corresponding to the 5′ end of the gene.

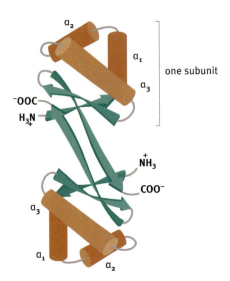

FIGURE 7.11 The Cro protein. The protein is a dimer. Each subunit has three α helices. The helix-turn-helix structure comprises the helices labeled α_2 and α_3. The recognition helix, which makes the important contact with the DNA, is α_3.

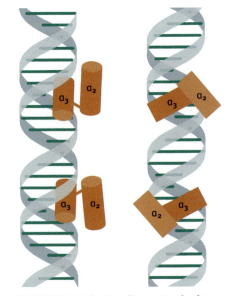

FIGURE 7.12 The Cro dimer attached to DNA. The two recognition helices of a Cro dimer fit perfectly into two adjacent sections of the major groove of a DNA double helix. Only the recognition helices are shown, the rest of the dimeric protein being left out for clarity. The two drawings show two different views of the helix, rotated 90°.

Establishing that genes and proteins are colinear was an important step forward in understanding the genetic code. As such, it is fortunate that these experiments were done with bacteria and not with a eukaryote. The presence of introns means that a discontinuous eukaryotic gene is not, strictly speaking, colinear with its protein. The relationship is only *linear*, because two nucleotide changes on either side of an intron, which could be several kb apart, will result in amino acid alterations that are much closer in the corresponding protein (Figure 7.15). The relationship is even more complicated with genes that have alternative splicing pathways. Assumptions are indeed dangerous in science.

Each codeword is a triplet of nucleotides

A second fundamental question about the genetic code is the size of the codeword, or **codon**, the group of nucleotides that code for a single amino acid. Codons cannot be just single nucleotides (A, T, G or C) because that would allow only four different codewords when 20 are required, one for each of the 20 amino acids found in proteins. Similarly, a doublet code (codons such as AT, TA, TT and GC) seems unlikely as this would contain only $4^2 = 16$ different codons. However, the next stage up, a triplet code (codons AAA, AAT, TAT, GCA, etc.) would be feasible as this would yield $4^3 = 64$ codewords, which would be more than enough.

As with colinearity, the hypothesis that codons are triplets of nucleotides was first tested by experiments with *E. coli*. These experiments made use of a group of chemicals called the **acridine dyes**, of which **proflavin** is an example. Acridine dyes cause base pair deletions or additions in double-stranded DNA molecules (Figure 7.16). The rationale was as follows. Some proteins contain segments where the amino acid sequence can be changed without altering the function of the protein. What if a series of insertions and/or deletions are introduced into the region of a gene coding for one of these tolerant segments of a protein? If the code is triplet then a single insertion or deletion would give rise to a nonfunctional protein, because all the codewords downstream of the insertion/deletion would be altered, including those in the nontolerant segment following the tolerant region (Figure 7.17). Two insertions or deletions (although not one of each) would have the same effect. However, three insertions or deletions would restore the correct **reading frame** in the nontolerant region and would be predicted to have no effect on the function of the protein.

An elegant experiment of this kind was first carried out successfully with a gene from the *E. coli* bacteriophage called T4. This work established the triplet nature of the code and, although assumptions are still dangerous, this particular one holds for all organisms.

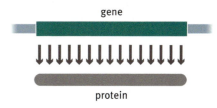

FIGURE 7.13 The colinear relationship between a gene and its protein. The order of nucleotides in the gene correlates directly with the order of amino acids in the protein.

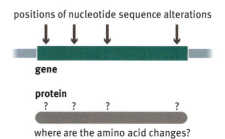

FIGURE 7.14 An experimental strategy for testing whether a gene is colinear with its protein.

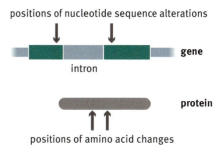

FIGURE 7.15 A gene that contains an intron is not colinear with its protein.

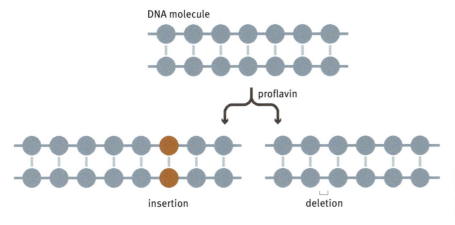

FIGURE 7.16 The effect of proflavin on DNA. An acridine dye such as proflavin can cause an insertion or deletion in a double-stranded DNA molecule.

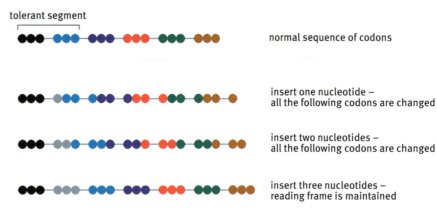

FIGURE 7.17 An experiment to test if codons are triplets of nucleotides. A series of codons are shown, each in a different color. Insertion of a single nucleotide changes all the following codons, as does an insertion of two nucleotides. Insertion of three nucleotides restores the reading frame.

Elucidation of the genetic code

Now we have established the principles of (co)linearity and triplet codons, we can focus on the central question of which codons specify which amino acids. The work that led to the elucidation of the genetic code was carried out during 1961–1966 and at that time was the most exciting and talked about research in all of biology.

Successful assignment of codons to amino acids was made possible by two innovations. The first was the discovery of **polynucleotide phosphorylase**, an enzyme that degrades RNA in the cell, but which, in the test tube, will catalyze the reverse reaction and synthesize RNA. This reaction does not require a DNA template and is unrelated to transcription. It enables artificial RNA molecules of known or predicted sequences to be made. The rationale was that if protein synthesis could be directed by an RNA molecule of known nucleotide sequence, then it would be possible to assign individual codons by looking at the amino acid sequence of the protein that was synthesized. To do this, a cell extract able to synthesize proteins is needed. A **cell-free protein synthesizing system** must contain all the cellular components necessary for protein synthesis, including ribosomes, tRNAs, amino acids and suchlike. It must, however, lack endogenous mRNA so that protein synthesis occurs only when the artificial message is added.

The first codon to be assigned a meaning was 5′–UUU–3′, which specifies phenylalanine. This was worked out by showing that poly(U), an RNA molecule that contains just U nucleotides (an example of a homopolymer), directs the synthesis of polyphenylalanine in the cell-free system (Figure 7.18). Equivalent experiments enabled 5′–AAA–3′ to be assigned to lysine and 5′–CCC–3′ to proline. For unexplained reasons, poly(G) gave no protein product.

The next step was to construct heteropolymers, artificial RNA molecules containing more than just one nucleotide. This can be achieved by polymerizing mixtures of nucleotides with polynucleotide phosphorylase. The problem is that the random nature of polymerization means that the actual sequence of the resulting RNA molecule is not known. For example, a random heteropolymer of A and C contains eight different codons – AAA, AAC, ACA, CAA, ACC, CCA, CAC and CCC – but these can occur in any order in the RNA molecule. When used in the cell-free system, this poly(A,C) heteropolymer directed synthesis of a protein that contained six amino acids – proline, histidine, threonine, asparagine, glutamine and lysine.

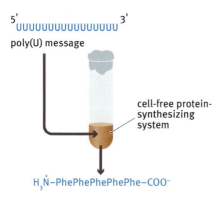

FIGURE 7.18 Cell-free translation of poly(U) gives polyphenylalanine.

random heteropolymer comprising 5C:1A

probability of **C** being included in a triplet = 5/6
probability of **A** being included in a triplet = 1/6

possible triplets	probability		
CCC	$(5/6)^3$	=	57.9%
CCA, CAC, ACC	$(1/6)(5/6)^2$	=	11.6%
CAA, ACA, AAC	$(1/6)^2(5/6)$	=	2.3%
AAA	$(1/6)^3$	=	0.4%

results of cell-free protein synthesis

amino acid	amount in polypeptide	interpretation
proline	69%	CCC + one of CCA, CAC, ACC
threonine	14%	one of CCA, CAC, ACC + one of CAA, ACA, AAC
histidine	12%	one of CCA, CAC, ACC
asparagine	2%	one of CAA, ACA, AAC for each
glutamine	2%	
lysine	1%	AAA

FIGURE 7.19 Analysis of the results of a typical experiment with a random heteropolymer.

Clearly, these six amino acids are coded by the eight possible codons, but which is coded by which? An answer can be obtained, at least partially, by using different amounts of each nucleotide in the reaction mixture for RNA synthesis. The frequencies of each of the possible codons in the heteropolymers can then be worked out and compared with the amounts of each amino acid in the resulting polypeptides (Figure 7.19). The codon allocations provided by this method are not definite but are statistically probable and can be cross-checked by changing the composition of the heteropolymer.

Although homopolymers and random heteropolymers allowed most of the genetic code to be worked out, unambiguous identification of each codon required three additional types of experiment. The first of these made use of ordered heteropolymers, which are made by polymerizing not mononucleotides, but known dinucleotides such as AC. This would give poly(AC), whose sequence is ACACACAC, and which contains two codons, ACA and CAC. Trinucleotides such as UGU could also be polymerized, giving the sequence UGUUGUUGU, with the codons UGU, GUU and UUG. The proteins produced by these messages allowed the meaning of several difficult codons to be determined.

The next breakthrough was the development of a modification of the cell-free protein synthesizing system called the **triplet binding assay**. It was discovered that purified ribosomes can attach to an mRNA molecule of only three nucleotides, and having done so will then bind the amino acid that is specified by that particular codon. As triplets of known sequence could be synthesized in the laboratory, it was possible with this binding assay to check the previously assigned codons and to allocate virtually all of the remaining ones.

The genetic code is degenerate and includes punctuation codons

With 64 possible codons and just 20 amino acids, one expectation might be that the genetic code is **degenerate**, which in this context means that some amino acids are specified by more than one codon. This is indeed the case (Figure 7.20). All amino acids except methionine and tryptophan have more than one codon, so that all the possible triplets have a

UUU Phe UUC Phe UUA Leu UUG Leu	UCU UCC Ser UCA UCG	UAU Tyr UAC Tyr UAA stop UAG stop	UGU Cys UGC Cys UGA stop UGG Trp
CUU CUC Leu CUA CUG	CCU CCC Pro CCA CCG	CAU His CAC His CAA Gln CAG Gln	CGU CGC Arg CGA CGG
AUU AUC Ile AUA AUG Met	ACU ACC Thr ACA ACG	AAU Asn AAC Asn AAA Lys AAG Lys	AGU Ser AGC Ser AGA Arg AGG Arg
GUU GUC Val GUA GUG	GCU GCC Ala GCA GCG	GAU Asp GAC Asp GAA Glu GAG Glu	GGU GGC Gly GGA GGG

FIGURE 7.20 The genetic code.

meaning, despite there being 64 triplets and only 20 amino acids. Most synonymous codons are grouped into families. For example, 5′–GGA–3′, 5′–GGU–3′, 5′–GGG–3′ and 5′–GGC–3′ all code for glycine. This similarity between synonymous codons is relevant to the way the code is deciphered during protein synthesis, as we will see when we examine the role of tRNA later in this chapter.

The genetic code also contains **punctuation codons**, special ones that indicate the start and end of the nucleotide sequence that must be translated into protein (Figure 7.21). The triplet 5′–AUG–3′ occurs at the start of most genes and marks the position where translation should begin. This triplet is therefore the **initiation codon** and because it codes for methionine most newly synthesized polypeptides have this amino acid at the amino terminus, though the methionine may subsequently be removed after the protein has been made. Note that 5′–AUG–3′ is the only codon for methionine, so AUGs that are not initiation codons may be found in the internal region of a gene. With a few genes, a different triplet such as 5′–GUG–3′ or 5′–UUG–3′ is used as the initiation codon.

Three triplets, 5′–UAA–3′, 5′–UAG–3′ and 5′–UGA–3′, do not code for amino acids and instead act as **termination codons**. One of these always occurs at the end of a gene at the point where translation must stop.

The genetic code is not universal

It was originally thought that the genetic code must be the same in all organisms. The argument was that, once established, it would be impossible for the code to change because giving a new meaning to any single codon would result in widespread disruption of the amino acid sequences of an organism's proteins.

This reasoning seems sound, so it is surprising that, in reality, the genetic code is not universal. The code shown in Figure 7.20 holds for the vast majority of genes in the vast majority of organisms, but deviations are widespread. In particular, genes present in human mitochondrial DNA use a nonstandard code (Table 7.2). Mitochondrial DNA is a small circular molecule (Section 20.4), located inside the energy-generating organelles called mitochondria. Mitochondrial DNA contains 13 protein-coding

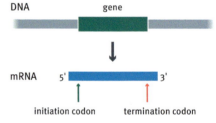

FIGURE 7.21 The positions of the punctuation codons in an mRNA.

TABLE 7.2 NONSTANDARD CODONS IN HUMAN MITOCHONDRIAL DNA

Codon	Should Code for	Actually Codes for
UGA	Stop	Tryptophan
AGA, AGG	Arginine	Stop
AUA	Isoleucine	Methionine

genes, several of which have the sequence 5′–UGA–3′, which normally codes for termination, at internal positions where protein synthesis is not expected to stop. Comparison with the amino acid sequences of the proteins coded by these genes showed that 5′–UGA–3′ is a tryptophan codon in human mitochondria and that this is just one of four code deviations in this particular genetic system. Mitochondrial genes in other organisms also display code deviations, although at least one of these – the use of 5′–CGG–3′ as a tryptophan codon in plant mitochondria – is probably corrected by RNA editing before translation occurs.

Nonstandard codes are also known in the nuclear genes of lower eukaryotes. Often, a modification is restricted to just a small group of organisms, and frequently, it involves the reassignment of the termination codons (**Table 7.3**). Modifications are less common among bacteria.

A second type of code variation is **context-dependent codon reassignment**, which occurs when the protein to be synthesized contains either selenocysteine or pyrrolysine. Proteins containing pyrrolysine are rare and are probably only present in some archaea and a very small number of bacteria, but proteins containing selenocysteine are widespread in many organisms (**Table 7.4**). One example is the enzyme glutathione peroxidase, which helps protect the cells of humans and other mammals against

TABLE 7.3 EXAMPLES OF NONSTANDARD CODONS IN NUCLEAR GENOMES

Organism	Codon	Should Code for	Actually Codes for
Several protozoa	UAA, UAG	Stop	Glutamine
Candida (yeast)	CUG	Leucine	Serine
Euplotes (ciliated protozoan)	UGA	Stop	Cysteine

TABLE 7.4 EXAMPLES OF PROTEINS THAT CONTAIN SELENOCYSTEINE

Organism	Protein
Mammals	Glutathione peroxidase
	Thioredoxin reductase
	Iodothyronine deiodinase
Bacteria	Formate dehydrogenase
	Glycine reductase
	Proline reductase
Archaea	Hydrogenase
	Formyl-methanofuran dehydrogenase

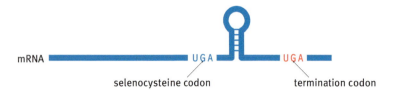

FIGURE 7.22 Distinction between a termination codon and one for selenocysteine. A 5′–UGA–3′ codon that specifies selenocysteine is distinguished from a termination codon by the presence of a hairpin in the mRNA. This drawing shows a bacterial mRNA, in which the hairpin is adjacent to the selenocysteine codon.

oxidative damage. Selenocysteine is coded by 5′–UGA–3′, which therefore has a dual meaning because it is still used as a termination codon in the organisms concerned. A 5′–UGA–3′ codon that specifies selenocysteine is distinguished from true termination codons by the presence of a hairpin structure in the mRNA, positioned just downstream of the selenocysteine codon in prokaryotes and in the 3′ untranslated region (i.e. the part of the mRNA after the termination codon) in eukaryotes (Figure 7.22). Recognition of the selenocysteine codon requires interaction between the hairpin and a special protein that is involved in translation of these mRNAs. A similar system probably operates with pyrrolysine, which is specified by a second termination codon, 5′–UAG–3′.

7.3 THE ROLE OF tRNAS IN PROTEIN SYNTHESIS

Transfer RNAs perform the key role of ensuring that the rules laid down by the genetic code are followed when an mRNA is translated into a protein. To do this, tRNAs form a physical link between the mRNA and the protein that is being synthesized, binding to both the mRNA and the growing protein (Figure 7.23). To understand the role of tRNAs, we must therefore examine **aminoacylation**, the process by which the correct amino acid is attached to each tRNA, and **codon–anticodon recognition**, the interaction between tRNA and mRNA.

Aminoacyl-tRNA synthetases attach amino acids to tRNAs

Bacteria contain 30–45 different tRNAs and eukaryotes have up to 50. As only 20 amino acids are designated by the genetic code, this means that all organisms have at least some **isoaccepting tRNAs**, different tRNAs that are specific for the same amino acid. The terminology used when describing tRNAs is to indicate the amino acid specificity by a superscript suffix, with the numbers 1, 2, etc., distinguishing different isoacceptors. According to this notation, two tRNAs specific for glycine would be written as tRNA^{Gly1} and tRNA^{Gly2}.

Each tRNA molecule forms a covalent linkage with its specific amino acid by a process called aminoacylation or **charging**. The amino acid becomes attached to the end of the acceptor arm of the tRNA cloverleaf. The linkage forms between the carboxyl group of the amino acid and the 2′–OH or 3′–OH group of the terminal nucleotide of the tRNA (Figure 7.24). Remember that this terminal nucleotide is always A, because all tRNAs have the sequence 5′–CCA–3′ at their 3′ ends (see Figure 6.10).

Aminoacylation is catalyzed by a group of enzymes called the **aminoacyl-tRNA synthetases**. In most cells, there is a single aminoacyl-tRNA synthetase for each amino acid, meaning that one enzyme can charge each member of a series of isoaccepting tRNAs. Although aminoacyl-tRNA synthetases are a fairly heterogeneous group of enzymes, each catalyzes the same reaction (Figure 7.25). The energy required to attach the amino acid to the tRNA is provided by the cleavage of ATP to adenosine monophosphate (AMP) and pyrophosphate. The reaction takes place in two distinct

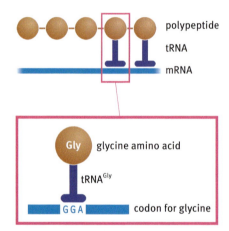

FIGURE 7.23 The role of tRNA in translation. The top drawing shows two tRNAs forming an attachment between the polypeptide and the mRNA. The lower drawing shows how the amino acid that is attached to the tRNA is the one specified by the codon that the tRNA recognizes.

steps, the first resulting in an activated amino acid intermediate in which the carboxyl group has formed a link with AMP. This intermediate remains bound to the enzyme until the AMP is replaced in the second stage of the reaction by the tRNA molecule, producing aminoacyl-tRNA and free AMP.

Aminoacylation must be carried out accurately. The correct amino acid must be attached to the correct tRNA if the rules of the genetic code are to be followed during protein synthesis. To achieve the necessary degree of accuracy, each aminoacyl-tRNA synthetase forms an extensive interaction with its tRNA, with contacts made to the acceptor arm and anticodon loop as well as to individual nucleotides in the D and TΨC arms. These interactions enable the enzyme to distinguish the specific sequence features of different tRNAs and hence to recognize the correct one. The interaction between enzyme and amino acid is, of necessity, less extensive, amino acids being much smaller than tRNAs, and presents a greater problem with regard to specificity because several pairs of amino acids are structurally similar. Errors do therefore occur, at a very low rate for most amino acids but possibly as frequently as one aminoacylation in 80 for difficult pairs such as isoleucine and valine. Most of these errors are corrected by the aminoacyl-tRNA synthetase before the charged tRNA is released.

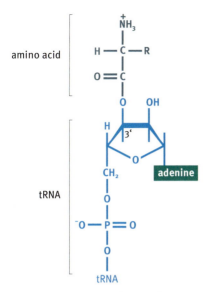

FIGURE 7.24 Aminoacylation of a tRNA. The result of aminoacylation by a class II aminoacyl-tRNA synthetase is shown, the amino acid being attached via its –COOH group to the 3′–OH of the terminal nucleotide of the tRNA. A class I aminoacyl-tRNA synthetase attaches the amino acid to the 2′–OH group.

Unusual types of aminoacylation

In most organisms, aminoacylation is carried out by the process just described, but a few unusual events have been documented. These include a number of instances where the aminoacyl-tRNA synthetase attaches the incorrect amino acid to a tRNA, this amino acid subsequently being transformed into the correct one by a second, separate chemical reaction. This was first discovered in the bacterium *Bacillus megaterium*, which lacks an aminoacyl-tRNA synthetase capable of attaching glutamine to tRNA^{gln}. Instead, the glutamic acid tRNA synthetase attaches a glutamic acid to tRNA^{gln}, this 'mistake' being corrected by a second enzyme (an aminotransferase), which replaces the –O$^-$ group of the glutamic acid with an –NH$_2$ group, converting it into glutamine (Figure 7.26a). The same process is used by various other bacteria, although not *E. coli*.

In the example given above, the amino acid resulting from the correction process is one of the 20 specified by the genetic code. There are also cases where chemical modification results in an unusual amino acid. One such chemical modification results in the conversion of methionine to *N*-formylmethionine (Figure 7.26b). This produces a special aminoacyl-tRNA which, as we will see in Chapter 8, is used during initiation of bacterial protein synthesis.

Chemical modification is also responsible for the synthesis of tRNAs aminoacylated with selenocysteine. Context-dependent reading of 5′–UGA–3′ triplets as selenocysteine codons involves a special tRNA^{SeCys}, but there is no aminoacyl-tRNA synthetase that is able to attach selenocysteine to this tRNA. Instead, the tRNA is aminoacylated with a serine by the seryl-tRNA synthetase and then modified by replacement of the –OH group of the serine with an –SeH to give selenocysteine (Figure 7.26c).

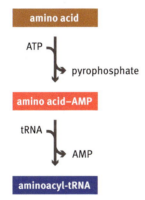

FIGURE 7.25 The two-step reaction catalyzed by an aminoacyl-tRNA synthetase. The first step results in an activated amino acid intermediate whose carboxyl group has formed a link with AMP. In the second stage of the reaction, the AMP is replaced by the tRNA molecule, producing aminoacyl-tRNA and free AMP.

The mRNA sequence is read by base pairing between the codon and anticodon

Aminoacylation represents the first level of specificity displayed by a tRNA. Once the correct amino acid has been attached to the acceptor arm of the tRNA, the aminoacylated molecule must complete the link between

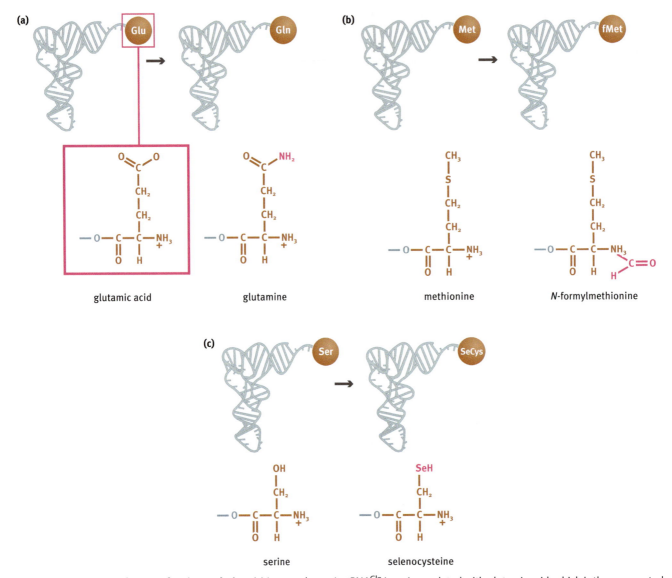

FIGURE 7.26 Unusual types of aminoacylation. (a) In some bacteria, tRNA^{Gln} is aminoacylated with glutamic acid, which is then converted to glutamine by an aminotransferase enzyme. (b) The special tRNA used in initiation of translation in bacteria is aminoacylated with methionine, which is then converted to N-formylmethionine. (c) tRNA^{SeCys} in various organisms is initially aminoacylated with serine.

mRNA and protein by recognizing and attaching to the correct codon, one coding for the amino acid that it carries.

Codon recognition is a function of the anticodon loop of the tRNA, specifically of the triplet of nucleotides called the **anticodon** (see Figure 6.10). This triplet is complementary to the codon and can therefore attach to it by base pairing (**Figure 7.27**). The specificity of the genetic code is ensured because the anticodon present on a particular tRNA is one that is complementary to a codon for the amino acid with which the tRNA is charged.

From what has been said so far it might be imagined that there are 61 different types of tRNA molecule in each cell, one for each of the codons that specify an amino acid. In fact, it has been known since the early 1960s that there are substantially fewer than 61 different tRNA molecules, usually between 30 and 50 depending on the organism. The explanation for this is provided by the **wobble hypothesis**. This hypothesis is based on the fact that the anticodon, being contained in a loop, is not a perfectly linear trinucleotide. The short double helix formed by base pairing between the codon and anticodon does not therefore have the precise configuration

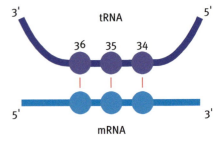

FIGURE 7.27 The interaction between a codon and its anticodon. The anticodon is complementary to the codon and can therefore attach to it by base pairing. The numbers refer to the positions of the nucleotides within the tRNA sequence, position 1 being the nucleotide at the extreme 5′ end.

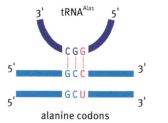
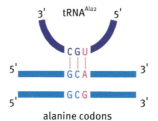

FIGURE 7.28 Wobble involving a G–U base pair. In this example, wobble enables the four-codon family for alanine to be decoded by just two tRNAs. The wobble position is highlighted in magenta.

of a standard RNA helix. Instead, its dimensions are slightly altered. As a result, nonstandard base pairs can form at the 'wobble position', between the third nucleotide of the codon and the first nucleotide of the anticodon.

Because of the wobble pairing, a single anticodon may be able to base pair with more than one codon. This means that a single tRNA might decode more than one member of a codon family. However, the base-pairing rules do not become totally flexible at the wobble position, and only a few types of unusual base pairs are allowed. G–U base pairs are a common example. By allowing G to pair with U as well C, an anticodon with the sequence 3'–∇∇G–5' can base pair with both 5'–∆∆C–3' and 5'–∆∆U–3' (Figure 7.28). Similarly, the anticodon 3–∇∇U–5' can base pair with both 5'–∆∆A–3' and 5'–∆∆G–3'. The consequence is that the four members of a codon family (e.g. 5'–GCN–3', all coding for alanine) can be decoded by just two tRNAs.

A second type of wobble involves inosine, one of the modified nucleotides present in tRNA. Inosine can base pair with A, C and U. The triplet 3'–UAI–5' is sometimes used as the anticodon in a tRNA^{Ile} molecule because it pairs with 5'–AUA–3', 5'–AUC–3' and 5'–AUU–3' (Figure 7.29). These triplets form the three-codon family for isoleucine in the standard genetic code.

Wobble reduces the number of tRNAs needed in a cell by enabling one tRNA to read two or possibly three codons. Hence, bacteria can decode their mRNAs with as few as 30 tRNAs. Eukaryotes also make use of wobble but in a restricted way. Humans, who in this regard are fairly typical of higher eukaryotes, have 48 tRNAs. Of these, 16 are predicted to use wobble to decode two codons each, with the remaining 32 being specific for just a single triplet. Although reducing the number of tRNAs that are needed, wobble does not violate the rules of the genetic code, and the protein that is made during translation is always synthesized strictly in accordance with the nucleotide sequence of the relevant mRNA.

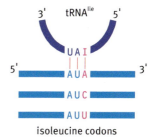

FIGURE 7.29 Wobble involving inosine. The wobble position is highlighted in magenta.

KEY CONCEPTS

- A protein is a linear unbranched polymer. The monomeric subunits are called amino acids and the resulting polymers, or polypeptides, are rarely more than 2,000 units in length.

- Twenty different amino acids are found in protein molecules. Their R groups vary considerably in chemical complexity. These differences mean that although all amino acids are closely related, each has its own specific chemical properties.

- Proteins have four distinct levels of structure. These levels are hierarchical, the protein being built up stage by stage, with each level of structure depending on the one below it. Each of the higher levels of protein structure is specified by the amino acid sequence.

- Through the higher levels of structure, the amino acid sequence also specifies the function of a protein. This is because a protein, in order to perform its function, must interact with other molecules, and the precise nature of these interactions is set by the overall shape of the protein and the distribution of chemical groups on its surface.
- The genetic code contains the rules that govern the order in which amino acids are joined together during the translation of an mRNA.
- The codewords of the genetic code are three-letter triplets called codons. Most amino acids are specified by more than one codon. There are also initiation and termination codons.
- The genetic code was elucidated during the 1960s by experiments involving artificial RNA molecules of known or predicted sequences that were translated in a cell-free protein synthesizing system.
- The genetic code is not universal.
- Transfer RNAs ensure that the rules laid down by the genetic code are followed when an mRNA is translated into a protein.
- The correct amino acid must be attached to each tRNA. This is ensured by the specificity of the aminoacyl-tRNA synthetase enzymes.
- The aminoacylated tRNA must recognize and attach to the correct codon, one coding for the amino acid that it carries. Codon recognition is a function of the anticodon of the tRNA. The anticodon is complementary to the codon and can therefore attach to it by base pairing.
- Wobble reduces the number of tRNAs needed in a cell by enabling one tRNA to read two or possibly three codons. Wobble does not violate the rules of the genetic code, and the protein that is made during translation is always synthesized strictly in accordance with the nucleotide sequence of the relevant mRNA.

QUESTIONS AND PROBLEMS

Key terms

Write short definitions of the following terms:

α helix
β sheet
β turn
acridine dye
amino acid
amino, NH_2^- or N terminus
aminoacyl-tRNA synthetase
aminoacylation
anticodon
carboxyl, COOH– or C terminus
cell-free protein synthesizing system

charging
codon
codon–anticodon recognition
context-dependent codon reassignment
degenerate
denaturation
disulfide bridge
helix-turn-helix
initiation codon
isoaccepting tRNAs
nonpolar

peptide bond
polar
polynucleotide phosphorylase
polypeptide
primary structure
proflavin
punctuation codon
quaternary structure
R group
reading frame
recognition helix
renaturation

Self-study questions

7.1 Draw the structure of an amino acid and indicate which parts of the molecule are the same for each amino acid and which parts are variable.

7.2 How are amino acids linked together to form a polypeptide?

7.3 Describe the four levels of protein structure. Your answer should include an indication of the types of chemical interaction that are important at each structural level.

7.4 What experimental evidence supports the contention that the instructions for the higher levels of protein structure reside in the amino acid sequence? Using Cro as an example, explain how the function of a protein is specified by its amino acid sequence.

7.5 Explain what is meant by colinearity between gene and protein. How was colinearity first demonstrated for genes and proteins in *E. coli*?

7.6 What considerations suggested that each codon would comprise three nucleotides? How was this fact proven experimentally?

7.7 Calculate the number of codons that would be possible if each codon contained four nucleotides.

7.8 Explain what is meant by degeneracy with respect to the genetic code.

7.9 List the key features of the genetic code.

7.10 A messenger RNA contains the following sequence:

5′–AUGUUAGCUGAUCCGGAAAUGAUGUUA UAUAUAAUAUAUGCCCAAUAG–3′

What would be the amino acid sequence of the protein specified by this mRNA?

7.11 Distinguish between the way homopolymers and heteropolymers were used in the experiments that elucidated the genetic code.

7.12 List the codons that would be contained in a random heteropolymer comprising A and G nucleotides. What amino acids would a polypeptide synthesized from this heteropolymer contain?

7.13 Explain why the triplet binding assay was important in studies of the genetic code.

7.14 Give examples of nonstandard genetic codes found in eukaryotes.

7.15 Using examples, explain what is meant by context-dependent codon reassignment.

7.16 Explain how tRNA molecules ensure that protein synthesis follows the rules laid down by the genetic code.

7.17 Outline the terminology used to distinguish between isoaccepting tRNAs.

7.18 Describe how an amino acid is attached to a tRNA molecule.

7.19 Give three examples where the incorrect amino acid is attached to a tRNA, this amino acid subsequently being transformed into the correct one.

7.20 Explain why fewer than 61 tRNAs are sufficient to decode the entire genetic code.

7.21 Draw a series of diagrams to illustrate the codon–anticodon interactions that occur during wobble involving G–U base pairs and inosine.

Discussion topics

7.22 The 20 amino acids shown in Figure 7.3 are not the only amino acids found in living cells. Devise a hypothesis to explain why these 20 amino acids are the only ones that are specified by the genetic code. Can your hypothesis be tested?

7.23 Discuss the reasons why polypeptides can take up a variety of structures whereas polynucleotides cannot.

7.24 During the 1950s, it was suggested that adjacent codons may overlap, so that ACAUG might contain two codons: ACA and AUG. Design an experiment to test this proposition.

7.25 How can the genetic code change if an alteration in a codon assignment is likely to cause an alteration to every protein in the cell?

7.26 Most organisms display a distinct codon bias in their genes. For example, leucine is specified by six codons in the genetic code (TTA, TTG, CTT, CTC, CTA and CTG), but in human genes leucine is most frequently coded by CTG and is only rarely specified by TTA or CTA. Similarly, of the four valine codons, human genes use GTG four times more frequently than GTA. It has been suggested that a gene that contains a relatively high number of unfavored codons might be expressed at a relatively slow rate. Explain the thinking behind this hypothesis and discuss its ramifications.

7.27 In human mitochondria, protein synthesis requires only 22 different tRNAs. What implications does this have for the rules governing codon–anticodon interactions in this system?

7.28 Discuss the connection between the wobble hypothesis and the degeneracy of the genetic code.

The answers to, or hints on how to answer, the self-study questions and discussion topic questions can be downloaded from the book's product page at this link: www.routledge.com/9781032743530

FURTHER READING

Agris PF, Eruysal ER, Narendran A et al. (2018) Celebrating wobble decoding: half a century and still much is new. *RNA Biology* 15, 537–553.

Branden CI & Tooze J (1998) *Introduction to Protein Structure*, 2nd ed. Abingdon: Garland Science.

Brennan RG & Matthews BW (1989) Structural basis of DNA-protein recognition. *Trends in Biochemical Sciences* 14, 286–290. *The binding of the Cro repressor to DNA.*

Crick FHC, Barrett FRSL, Brenner S & Watts-Tobin RJ (1961) General nature of the genetic code for proteins. *Nature* 192, 1227–1232. *The experimental proof of the triplet nature of the code.*

Hall BD (1979) Mitochondria spring surprises. *Nature* 282, 129–130. *Reviews the first reports of unusual genetic codes in mitochondrial genes.*

Koonin EV & Novozhilov AS (2009) Origin and evolution of the genetic code: the universal enigma. *IUBMB Life* 61, 99–111.

Nirenberg MW & Leder P (1964) RNA codewords and protein synthesis. *Science* 145, 1399–1407. *Results of the triplet binding assay.*

Nirenberg MW & Matthaei JH (1961) The dependence of cell-free protein synthesis in *E. coli* upon naturally occurring or synthetic polyribonucleic acids. *Proceedings of the National Academy of Sciences USA* 47, 1588–1602. *An early publication on elucidation of the genetic code.*

Low SC & Berry MJ (1996) Knowing when not to stop: selenocysteine incorporation in eukaryotes. *Trends in Biochemical Sciences* 21, 203–208. *The questions raised by the discovery that UGA can code for both stop and selenocysteine.*

Rubio Gomez MA & Ibba M. (2020) Aminoacyl-tRNA synthetases. *RNA* 26, 910–936.

Saier MH (2019) Understanding the genetic code. *Journal of Bacteriology* 201, e00091–19. *A discussion of the early evolution of the code.*

Yanofsky C, Carlton BC, Guest JR et al. (1964) On the colinearity of gene structure and protein structure. *Proceedings of the National Academy of Sciences USA* 51, 266–272. *The first demonstration of colinearity.*

Protein Synthesis

CHAPTER 8

We now reach the stage of the gene expression pathway where the information contained in the nucleotide sequence of an mRNA is translated into a protein. We have already learnt that this process follows the rules laid down by the genetic code, with tRNAs enforcing these rules. Now we must examine the mechanism by which all this happens. This will require us to study the role of ribosomes in protein synthesis.

It is often forgotten that translation is only the first stage in protein synthesis. Before becoming functional in the cell, the linear amino acid sequence that results from translation must adopt its correct secondary, tertiary and, possibly, quaternary structures. In some cases, it must also undergo processing, possibly by chemical modification or, less frequently, by removal of some segments of the polypeptide chain (Figure 8.1). These post-translational events are inherent steps in the final stage of gene expression, because until they have been carried out the biological information contained in the gene has not been made fully available to the cell. We will therefore spend some time becoming familiar with the most important of these post-translational processes.

The collection of proteins in a cell is called the **proteome**. Within the proteome, the identity and relative abundance of individual proteins represent a balance between the synthesis of new proteins and the degradation of existing ones (Figure 8.2). Exactly as with mRNAs in the transcriptome, degradation is needed so that individual proteins can be down- as well as up-regulated by changing their rate of synthesis. The terminal step in the expression pathway for a gene is therefore the process by which its protein, when no longer needed, is removed from the proteome. We will conclude this chapter with a brief overview of protein degradation.

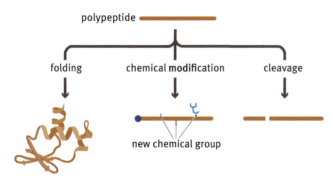

FIGURE 8.1 A summary of protein processing events. All proteins must be folded in order to become functional, and some are also processed by chemical modification and/or cleavage of the polypeptide into segments.

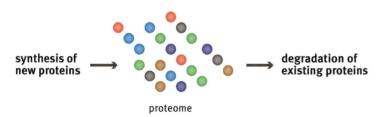

FIGURE 8.2 Synthesis and degradation of the components of a proteome. The composition of a proteome reflects the balance between synthesis of new proteins and degradation of existing ones. In this drawing, dots of different colors represent different proteins.

DOI: 10.1201/9781003473862-9

8.1 THE ROLE OF RIBOSOMES IN PROTEIN SYNTHESIS

Messenger RNAs are translated within the structures called ribosomes. We will follow through the series of events involved in initiation of translation, elongation of the polypeptide chain and termination of translation. These events are similar in bacteria and eukaryotes, though the details are different, most strikingly during the initiation phase.

Initiation in bacteria requires an internal ribosome binding site

The main difference between initiation of translation in bacteria and eukaryotes is that in bacteria the translation initiation complex is built up directly at the initiation codon, the point at which protein synthesis will begin, whereas eukaryotes use a more indirect process for locating the initiation point.

When not actively participating in protein synthesis, ribosomes separate into their subunits, which remain in the cytoplasm waiting to be used for a new round of translation. In bacteria, the process initiates when a small subunit attaches to the **ribosome binding site** (also called the **Shine–Dalgarno sequence**). This is a short sequence, consensus 5′–AGGAGGU–3′ in *E. coli* (**Table 8.1**), located about 3–10 nucleotides upstream of the initiation codon, the point at which translation will begin (**Figure 8.3**). The ribosome binding site is complementary to a region at the 3′ end of the 16S rRNA, the one present in the small subunit, and it is thought that base pairing between the two is involved in the attachment of the small subunit to the mRNA.

Attachment to the ribosome binding site positions the small subunit of the ribosome over the initiation codon (**Figure 8.4**). As described in Chapter 7, this codon is usually 5′–AUG–3′, which codes for methionine, although 5′–GUG–3′ and 5′–UUG–3′ are sometimes used. All three codons can be recognized by the same initiator tRNA, the last two by wobble. The initiator tRNA joins the small subunit of the ribosome along with a molecule of GTP, which will be used as an energy source at the start of the elongation phase. This initiator tRNA is the one that was aminoacylated with methionine and subsequently modified by conversion of the methionine to *N*-formylmethionine (see Figure 7.26). The modification attaches a formyl group (–COH) to the

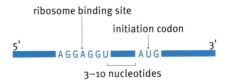

FIGURE 8.3 The *E. coli* ribosome binding site. In *E. coli*, the ribosome binding site is located between 3 and 10 nucleotides upstream of the initiation codon.

TABLE 8.1 EXAMPLES OF RIBOSOME BINDING SITE SEQUENCES IN *E. COLI*			
Gene	Codes for	Ribosome Binding Sequence	Nucleotides from the Initiation Codon
E. coli consensus	–	5′–AGGAGGU–3′	3–10
Lactose operon	Lactose utilization enzymes	5′–AGGA–3′	7
galE	Hexose-1-phosphate uridyltransferase	5′–GGAG–3′	6
rplJ	Ribosomal protein L10	5′–AGGAG–3′	8

amino group, which means that only the carboxyl of the initiator methionine is free to participate in peptide bond formation. This ensures that polypeptide synthesis can take place only in the N → C direction.

The formyl group remains attached until translation has proceeded into the elongation phase but it is then removed from the growing polypeptide, either on its own or along with the rest of the initial methionine. Note that the tRNA$_i^{Met}$ is only able to decode the initiation codon. It cannot enter the complete ribosome during the elongation phase of translation. During elongation, the internal 5′–AUG–3′ codons are recognized by a different tRNA^{Met} carrying an unmodified methionine.

In addition to the small subunit of the ribosome and the *N*-formylmethionine tRNA, initiation also requires three proteins, called **initiation factors**. These are not permanent components of the ribosome, but attach at the appropriate times in order to perform their functions (**Table 8.2**). The roles of two of these initiation factors are fairly well understood. The one called IF-2 directs the initiator tRNA to its correct position on the small subunit of the ribosome, and IF-3 prevents the large subunit from joining the complex until it is needed, which is important as the complete ribosome is unable to initiate translation. The function of IF-1 is less clear. It attaches to the complex toward the end of the initiation phase and appears to coordinate the transition from initiation to elongation, but exactly how it does this is not known. Possibly, it causes a change in the conformation of the small subunit, enabling the large subunit to attach, or possibly it prevents the second aminoacyl-tRNA from entering the complex until it is needed.

Initiation in eukaryotes is mediated by the cap structure and poly(A) tail

Only a small number of eukaryotic mRNAs have internal ribosome binding sites. Instead, with most mRNAs, the small subunit of the ribosome makes its initial attachment at the 5′ end of the molecule and then **scans** along the sequence until it locates the initiation codon.

The process works as follows. The first step involves assembly of the **pre-initiation complex**, the principal components of which are the small subunit of the ribosome and the initiator tRNA, the latter associated with the molecule of GTP that will be used at the start of the elongation phase (**Figure 8.5a**). As in bacteria, the initiator tRNA is distinct from the normal tRNA^{Met} that recognizes internal 5′–AUG–3′ codons but, unlike bacteria, it is aminoacylated with normal methionine, not the formylated version.

After assembly, the pre-initiation complex attaches to the cap structure at the extreme 5′ end of the mRNA. The **initiation complex**, as it is now called, then scans along the molecule to find the initiation codon

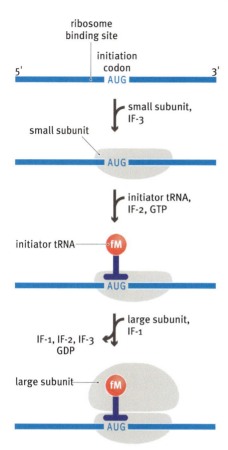

FIGURE 8.4 Initiation of translation in *Escherichia coli*. The process begins when the small subunit of the ribosome attaches to its binding site on the mRNA. The small subunit is accompanied by IF-3, which prevents the large subunit from attaching until the appropriate time. The initiator tRNA is now brought to the complex by IF-2, along with a molecule of GTP. The initiation process is completed by attachment of the large subunit. IF-1 is also involved at this stage but its role is unclear. Attachment of the large subunit requires energy, obtained by conversion of the bound GTP to GDP, and results in the release of the initiation factors. fM, *N*-formylmethionine.

TABLE 8.2 FUNCTIONS OF THE INITIATION FACTORS IN BACTERIA	
Initiation Factor	Function
IF-1	Unclear. Might cause conformational changes that prepare the small subunit for attachment to the large subunit or might prevent premature entry of the second aminoacyl-tRNA
IF-2	Directs the initiator tRNA to its correct position in the initiation complex
IF-3	Prevents premature reassociation of the large and small subunits of the ribosome

(Figure 8.5b). The leader regions of eukaryotic mRNAs can be several tens, or even hundreds, of nucleotides in length and often contain several 5′–AUG–3′ triplets that the complex must ignore before it reaches the true initiation codon. This discrimination is possible because the initiation codon, but not the upstream 5′–AUG–3′ triplets, is contained in a short consensus sequence, 5′–ACCAUGG–3′, referred to as the **Kozak consensus**.

There are two further aspects of eukaryotic translation initiation that we must consider. The first is the involvement of the poly(A) tail of the mRNA, which somehow promotes the binding of the pre-initiation complex to the cap structure. We know that the poly(A) tail is involved at this stage because the length of the tail influences the extent of translation initiation that occurs with a particular mRNA, and removal of the tail is one of the steps that leads to inactivation of an mRNA whose translation product is no longer needed. But exactly how the poly(A) tail associates with the pre-initiation complex is not yet known.

The second issue is the role of the plethora of initiation factors possessed by eukaryotes, at least 12 at the last count (**Table 8.3**). Five of these – eIF-1, eIF-1A, eIF-2, eIF-2B and eIF-3 – are components of the pre-initiation complex. A further three – eIF-4A, eIF-4E and eIF-4G – form a structure called

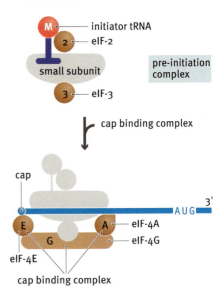

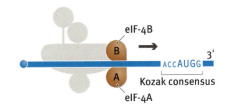

FIGURE 8.5 Initiation of translation in eukaryotes. (a) Assembly of the pre-initiation complex and its attachment to the mRNA. The pre-initiation complex comprises the small subunit of the ribosome, the initiator tRNA and various initiation factors including eIF-2 and eIF-3. The pre-initiation complex attaches to the cap structure at the 5′ end of the tRNA, assisted by the cap binding complex, which is made up of eIF-4A, eIF-4E and eIF-4G. (b) The initiation complex scans along the mRNA until it reaches the initiation codon, which is located within the Kozak consensus sequence. Two of the initiation factors, eIF-4A and eIF-4B, are involved in the scanning process. M, methionine.

TABLE 8.3 FUNCTIONS OF INITIATION FACTORS IN EUKARYOTES

Initiation Factor	Function
eIF-1	Component of the pre-initiation complex
eIF-1A	Component of the pre-initiation complex
eIF-2	Binds to the initiator tRNA within the pre-initiation complex; phosphorylation of eIF-2 results in a global repression of translation
eIF-2B	Component of the pre-initiation complex
eIF-3	Component of the pre-initiation complex; makes direct contact with eIF-4G and so forms the link with the cap binding complex
eIF-4A	Component of the cap binding complex; a helicase that aids scanning by breaking intramolecular base pairs in the mRNA
eIF-4B	Aids scanning, possibly by acting as a helicase that breaks intramolecular base pairs in the mRNA
eIF-4E	Component of the cap binding complex, possibly the component that makes direct contact with the cap structure at the 5′ end of the mRNA
eIF-4F	The cap binding complex, comprising eIF-4A, eIF-4E, and eIF-4G, which makes the primary contact with the cap structure at the 5′ end of the mRNA
eIF-4G	Component of the cap binding complex; forms a bridge between the cap binding complex and eIF-3 in the pre-initiation complex; in at least some organisms, eIF-4G also forms an association with the poly(A) tail
eIF-4H	In mammals, aids scanning in a manner similar to eIF-4B
eIF-5	Aids release of the other initiation factors at the completion of initiation
eIF-6	Associated with the large subunit; prevents large subunits from attaching to small subunits in the cytoplasm

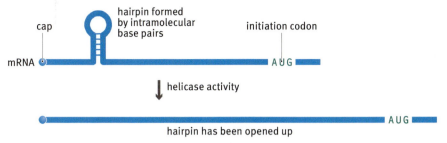

FIGURE 8.6 **The role of helicases in eukaryotic translation.** The helicase activities of eIF-4A and possibly eIF-4B enable hairpin structures resulting from intramolecular base pairing to be broken in the leader region of the mRNA.

the **cap binding complex** (unhelpfully also called eIF-4F), which makes the initial contact with the 5′ end of the mRNA and mediates subsequent attachment of the pre-initiation complex. Two factors, eIF-4A and eIF-4B, are involved in the scanning process. Both of these have a helicase activity, which enables them to break intramolecular base pairs that sometimes form in the mRNA leader region (Figure 8.6). Finally, eIF-5 aids the release of all the other initiation factors at the end of the initiation phase, and eIF-6 plays the same role as IF-3 in bacteria, preventing the large subunit of the ribosome from joining the initiation complex until it is needed.

Translation of a few eukaryotic mRNAs initiates without scanning

The scanning system for initiation of translation does not apply to every eukaryotic mRNA. This was first recognized with the picornaviruses, a group of viruses that includes the poliovirus and the human rhinovirus, the latter responsible for the common cold. These viruses are not themselves eukaryotes, but their genes are expressed within the eukaryotic cells that they infect. They make use of at least some of the endogenous noncoding RNAs and proteins, the ones made by the cell, which are subverted by the viruses to their own ends.

Picornavirus mRNAs are not capped but instead have an **internal ribosome entry site** (**IRES**) which is similar in function to the ribosome binding site of bacteria, although the sequences of IRESs and their positions relative to the initiation codon are more variable than the bacterial versions. The presence of IRESs on their mRNAs means that picornaviruses can block protein synthesis in the host cell by inactivating the cap binding complex, without affecting translation of their own mRNAs, although this is not a normal part of the infection strategy of all picornaviruses.

Remarkably, no virus proteins are required for recognition of an IRES by a host ribosome. In other words, the normal eukaryotic cell possesses proteins and/or other factors that enable it to initiate translation by the IRES method. Because of their variability, IRESs are difficult to identify, but it is becoming clear that a few cellular mRNAs possess them and that these are translated, at least under some circumstances, via their IRES rather than by scanning. These cellular mRNAs include a few that are translated when the cell is put under stress, for example, by exposure to heat, irradiation or low oxygen conditions. Under these circumstances, cap-dependent translation is globally suppressed by inactivation of eIF-2, one of the initiation factors present in the pre-initiation complex. The presence of IRESs on the 'survival' mRNAs therefore enables these to undergo preferential translation at the time when their products are needed.

Elongation of the polypeptide begins with formation of the first peptide bond

The initiation phase of translation is completed when the large subunit of the ribosome attaches to the small subunit, forming a complete ribosome positioned over the initiation codon. Attachment of the large subunit requires energy, which is generated by hydrolysis of the GTP that was bound during initiation (see Figure 8.4), and results in release of the initiation factors. We then move into the elongation phase, during which the protein is synthesized step by step, individual amino acids being attached to the carboxyl terminus of the growing polypeptide. Elongation is similar in bacteria and eukaryotes, so we can deal with the two types of organism together, pointing out the important distinctions as we come to them.

Bringing together the two ribosome subunits creates two sites at which aminoacyl-tRNAs can bind. The first of these, the **P** or **peptidyl site**, is already occupied by the initiator tRNA$_i^{Met}$, charged with N-formylmethionine or methionine, and base paired with the initiation codon. The second site, the **A** or **aminoacyl site**, covers the second codon in the open reading frame (**Figure 8.7**). The structures revealed by X-ray crystallography show that these sites are located in the cavity between the large and small subunits of the ribosome, the codon–anticodon interaction being associated with the small subunit and the aminoacyl end of the tRNA with the large subunit (**Figure 8.8**).

To begin elongation, the A site becomes filled with the appropriate aminoacyl-tRNA, which in *E. coli* is brought into position by the **elongation factor** EF-1A. This factor helps to ensure that translation is accurate as it only allows a tRNA that carries its correct amino acid to enter the ribosome, with mischarged tRNAs being rejected at this point. EF-1A also binds a molecule of GTP, which will be hydrolyzed to release energy when the peptide bond is formed. In eukaryotes, the equivalent factor is called eEF-1 (**Table 8.4**).

Once in the A site, the anticodon of the tRNA forms base pairs with the next codon of the mRNA. Again, it is essential that mistakes are not made. The tRNA and mRNA must fit exactly within the A site, and this can only happen if the codon and anticodon are perfectly matched, with each of the three base pairs formed correctly. If there is a mispair at any position in the codon–anticodon interaction then the tRNA is rejected.

When the correct aminoacyl-tRNA has entered the A site, a peptide bond is formed between the two amino acids. The bond is formed by the enzyme called **peptidyl transferase**. In both bacteria and eukaryotes, peptidyl transferase is an example of a ribozyme – an RNA enzyme. The catalytic

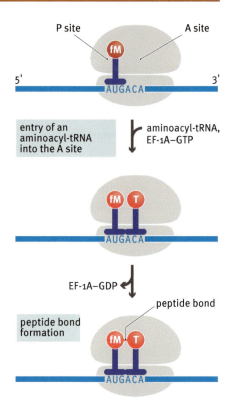

FIGURE 8.7 The first steps in the elongation phase of protein synthesis in *E. coli*. The ribosome contains two sites at which aminoacyl-tRNAs can bind. To begin with, the P site is occupied by the initiator tRNA$_i^{Met}$, and the A site is empty. Elongation begins with the entry of the second aminoacyl-tRNA into the A site, accompanied by elongation factor EF-1A and a molecule of GTP. A peptide bond is now formed between the two amino acids. This requires energy released by conversion of the GTP molecule to GDP. fM, *N*-formylmethionine; T, threonine.

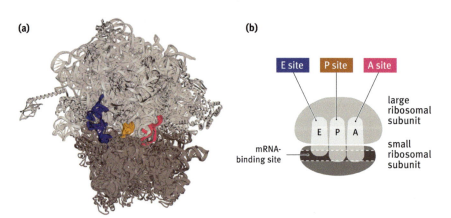

FIGURE 8.8 The structure of a bacterial ribosome during the elongation phase of translation. (a) Detailed structure of the ribosome with the small subunit in dark brown and the large subunit in light brown. The locations of the P and A sites are shown in orange and red, respectively. The E or exit site, which the tRNA passes through as it leaves the ribosome after its amino acid has been attached to the growing polypeptide, is shown in blue. (b) Schematic diagram illustrating the positions of the E, P and A sites in the ribosome. The mRNA is associated with the small subunit and the aminoacyl end of each tRNA with the large subunit. (Adapted from M. M. Yusupov et al., *Science* 292: 883–896, 2001. With permission from AAAS.)

TABLE 8.4 FUNCTIONS OF TRANSLATION ELONGATION FACTORS IN BACTERIA AND EUKARYOTES

Elongation Factor	Function
Bacteria	
EF-1A	Directs the next tRNA to its correct position in the ribosome
EF-1B	Regenerates EF-1A after the latter has yielded the energy contained in its attached GTP molecule
EF-2	Mediates translocation
Eukaryotes	
eEF-1	Complex of four subunits (eEF-1a, eEF-1b, eEF-1d and eEF-1g); directs the next tRNA to its correct position in the ribosome
eEF-2	Mediates translocation

activity for peptide bond formation is provided not by a protein but by the largest of the rRNAs present in the large subunit of the ribosome. Peptide bond formation is energy dependent and requires hydrolysis of the GTP attached to EF-1A (eEF-1 in eukaryotes). This inactivates EF-1A, which is ejected from the ribosome and regenerated by EF-1B. A eukaryotic equivalent of EF-1B has not been identified, and it is possible that one of the subunits of eEF-1 has regenerative activity.

Elongation continues until a termination codon is reached

Formation of the first peptide bond results in a dipeptide corresponding to the first two codons of the open reading frame. The attachment between the first amino acid and its tRNA is broken at this stage, leaving the dipeptide attached to the tRNA located in the A site (Figure 8.9).

The next step is **translocation**, during which the ribosome moves three nucleotides along the mRNA, so a new codon enters the A site. This moves the dipeptide-tRNA to the P site, which in turn displaces the deacylated tRNA. In eukaryotes, the deacylated tRNA is simply ejected from the ribosome, but in bacteria the deacylated tRNA departs via a third position, the **E** or **exit site**. This site was originally looked on as a simple exit point from the ribosome, but it is now known to have an important role in ensuring that translocation moves the ribosome along the mRNA by precisely three nucleotides, thereby ensuring that the ribosome keeps to the correct reading frame.

Translocation requires hydrolysis of a molecule of GTP and is mediated by EF-2 in bacteria and by eEF-2 in eukaryotes. Electron microscopy of ribosomes at different intermediate stages in translocation shows that, in order to move along the mRNA, the ribosome adopts a less compact structure, with the two subunits rotating slightly in opposite directions. This opens up the space between them and enables the ribosome to slide along the mRNA. Translocation results in the A site becoming vacant, allowing a new aminoacyl-tRNA to enter.

The elongation cycle is now repeated and continues until a termination codon is reached. After several cycles of elongation, the start of the mRNA molecule is no longer associated with the ribosome, and a second ribosome can attach and begin to synthesize another copy of the protein. The end result is a **polysome**, an mRNA that is being translated by several ribosomes at once. Polysomes have been seen in electron microscopic images of both prokaryotic and eukaryotic cells (Figure 8.10).

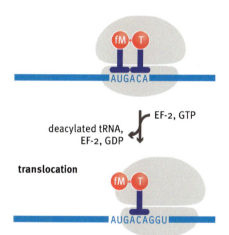

FIGURE 8.9 The first cycle of translocation during protein synthesis in *E. coli*. The ribosome moves three nucleotides along the mRNA, ejecting the deacylated initiator tRNA from the P site, which now becomes occupied by the second tRNA attached to the first two amino acids of the polypeptide. Translocation requires hydrolysis of a molecule of GTP and is mediated by EF-2. fM, *N*-formylmethionine; T, threonine.

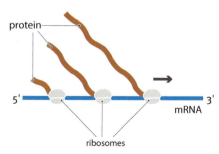

FIGURE 8.10 Polysomes. Three ribosomes are attached to a single mRNA. The arrow shows the direction in which the ribosomes are moving along the mRNA.

TABLE 8.5 RELEASE AND RIBOSOME RECYCLING FACTORS IN BACTERIA AND EUKARYOTES

Factor	Function
Bacteria	
RF-1	Recognizes the termination codons 5′–UAA–3′ and 5′–UAG–3′
RF-2	Recognizes 5′–UAA–3′ and 5′–UGA–3′
RF-3	Stimulates dissociation of RF-1 and RF-2 from the ribosome after termination
RRF	Ribosome release factor, responsible for separation of the ribosome subunits after translation has terminated
Eukaryotes	
eRF-1	Recognizes the termination codon; causes the ribosome subunits to disassociate after termination of translation
eRF-3	Stimulates dissociation of eRF-1 from the ribosome after termination

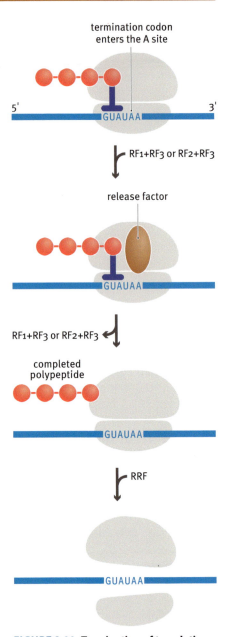

FIGURE 8.11 Termination of translation in *E. coli*. In the top drawing, the ribosome has reached a termination codon. Either RF-1 or RF-2, along with RF-3, enters the A site and releases the completed polypeptide. The ribosome is then separated into its two subunits by the ribosome recycling factor (RRF). V, valine.

Termination requires special release factors

Protein synthesis ends when one of the three termination codons enters the A site. There are no tRNA molecules with anticodons able to base pair with any of the termination codons, and instead a protein **release factor** enters the A site in order to terminate translation (Figure 8.11).

Bacteria have three release factors (Table 8.5). RF-1 recognizes the termination codons 5′–UAA–3′ and 5′–UAG–3′, and RF-2 recognizes 5′–UAA–3′ and 5′–UGA–3′. RF-3 stimulates release of RF1 and RF2 from the ribosome after termination, in a reaction requiring energy from the hydrolysis of GTP.

The bacterial release factors terminate translation but they do not appear to be responsible for disassociation of the ribosomal subunits. This is the function of an additional protein called **ribosome recycling factor (RRF)**, which probably enters the P or A site and 'unlocks' the ribosome (see Figure 8.11). Disassociation requires energy, which is released from GTP by EF-2, one of the elongation factors, and also requires the initiation factor IF-3 to prevent the subunits from attaching together again. The disassociated ribosome subunits enter the cytoplasmic pool, where they remain until used again in another round of translation.

Eukaryotes have just two release factors. The first of these, eRF-1, recognizes the termination codon and also acts as the ribosome recycling factor. The second, eRF-3, is responsible for release of eRF-1 from the ribosome after termination is complete.

8.2 POST-TRANSLATIONAL PROCESSING OF PROTEINS

If we were not taking the modern molecular approach to genetics, then we might be tempted to end our study of the gene expression pathway at this point. Our argument would be that 'DNA makes RNA makes protein' and that we have reached the stage where a protein has been made, so clearly that is the end of the story. But it is not. Translation results in the synthesis of a polypeptide, but this polypeptide is simply a linear chain of amino acids. It is inactive and as such is not the final product of the gene expression pathway. Utilization of the biological information contained in a gene requires the synthesis of an *active* protein, one able to perform its function in the cell. We must therefore study the events that convert the newly synthesized polypeptide into a functional protein. Only by studying these post-translational processing events will we acquire a complete understanding of how a gene is able to perform its role as a unit of biological information.

The most important type of post-translational processing is protein folding. All polypeptides have to be folded into their correct secondary and tertiary structures before becoming active in the cell. Some proteins also undergo chemical modification, by attachment of new chemical groups, and some also are processed by cutting events carried out by enzymes called **proteases**. These cutting events may remove segments from one or both ends of the polypeptide, resulting in a shortened form of the protein, or they may cut the polypeptide into a number of different segments, all or some of which are active.

Often these different types of processing occur together, a polypeptide being modified and cut at the same time that it is folded. If this is the case, then the chemical modifications and/or cuts might be needed in order for the polypeptide to take up its correct three-dimensional conformation. Alternatively, a chemical modification or a cutting event might occur after the protein has been folded, possibly as part of a regulatory mechanism that converts a folded but inactive protein into an active form.

Some proteins fold spontaneously in the test tube

In Chapter 7, we examined the four levels of protein structure and learnt that all of the information that a polypeptide needs in order to fold into its correct three-dimensional structure is contained within its amino acid sequence. This inviolate link between the amino acid sequence of a protein and its tertiary structure is one of the keys to gene expression, because it completes the pathway that begins with the biological information contained within a gene and ends with a protein function.

The experiments that showed that the amino acid sequence of a protein specifies its tertiary structure were first carried out with ribonuclease. Ribonuclease is a small protein, just 124 amino acids in length, containing four disulfide bridges and with a tertiary structure that is made up predominantly of β sheet, with very little α helix (**Figure 8.12**). Studies of its folding

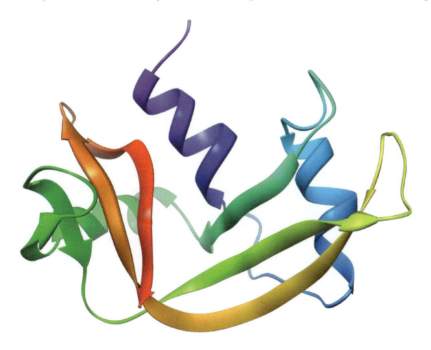

FIGURE 8.12 The tertiary structure of ribonuclease A. The colors indicate each of the various secondary structural components of the protein.

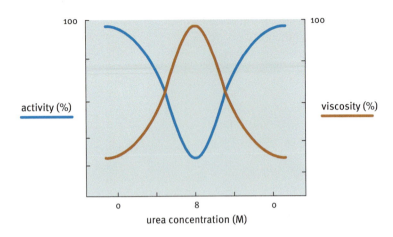

FIGURE 8.13 **Denaturation and spontaneous renaturation of ribonuclease.** As the urea concentration increases to 8 M, the protein becomes denatured by unfolding. Its activity decreases and the viscosity of the solution increases. When the urea is removed by dialysis, this small protein re-adopts its folded conformation. The activity of the protein increases back to the original level and the viscosity of the solution decreases.

can be carried out with purified ribonuclease that has been resuspended in buffer. The addition of urea, a compound that disrupts hydrogen bonding, results in a decrease in the activity of the enzyme, measured by testing its ability to cut RNA. At the same time, there is an increase in the viscosity of the solution (**Figure 8.13**). Both observations indicate that the protein has been denatured by unfolding into an unstructured polypeptide chain. This unfolded protein is no longer able to act as a ribonuclease, hence the loss of enzymatic activity, and the chains will have a tendency to get tangled up, explaining the increase in viscosity.

What if the urea is removed from the solution? Now the viscosity decreases and the enzyme activity reappears. The conclusion is that the protein refolds spontaneously when the denaturant is removed. The same result is obtained when the urea treatment is combined with the addition of a reducing agent to break the disulfide bonds. The activity is still regained on renaturation. This shows that the disulfide bonds are not critical to the protein's ability to refold, and they merely stabilize the tertiary structure once it has been adopted.

A more detailed study of the spontaneous folding of ribonuclease and other small proteins has suggested that the secondary structural motifs along the polypeptide chain form within a few milliseconds of the denaturant being removed. At this stage, the protein adopts a compact, but not folded, organization, with its hydrophobic groups on the inside, shielded from water (**Figure 8.14**). During the next few seconds or minutes, the secondary structural motifs interact with one another and the tertiary structure gradually takes shape, often via a series of intermediate conformations that make up the protein's **folding pathway**. There may be more than one possible pathway that a protein can follow to reach its correctly folded structure, and the pathways might also have side branches into which the protein can be diverted, leading to an incorrect structure. If the incorrect structure is sufficiently unstable, then partial or complete unfolding may occur, allowing the protein a second opportunity to pursue a correct folding pathway (**Figure 8.15**).

Inside cells, protein folding is aided by molecular chaperones

The experiments described above confirm that the amino acid sequence of a protein specifies its tertiary structure, but they do not help us understand how protein folding occurs in the cell. Folding in the cell is probably not an entirely spontaneous process, at least not for all proteins, especially larger ones with structures more complex than that of ribonuclease. When studied in the test tube, these proteins tend to form insoluble aggregates when the denaturant is removed. This is probably because they collapse into interlocked networks when they attempt to protect their hydrophobic

linear
polypeptide
↓
secondary structure
motifs form
↓
protein adopts compact
but unfolded structure
↓
correct tertiary structure
gradually emerges

FIGURE 8.14 **A typical protein folding pathway.**

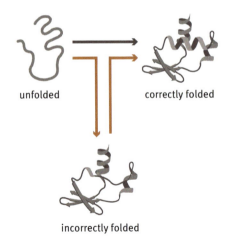

FIGURE 8.15 **An incorrectly folded protein might be able to refold into its correct conformation.** The black arrow represents the correct folding pathway, leading from the unfolded protein on the left to the active protein on the right. The orange arrow leads to an incorrectly folded conformation, but this conformation is unstable and the protein is able to unfold partially, return to its correct folding pathway and, eventually, reach its active conformation.

groups from water at the start of their folding pathway. Some also get stuck in nonproductive side branches of their folding pathways, taking on an intermediate form that is incorrectly folded but which is too stable to unfold. Experimentally, correct folding can be achieved by diluting the protein, but this is not an option that the cell can take to prevent its unfolded proteins from aggregating or following an incorrect pathway.

The difficulties that larger proteins experience when folding in the test tube indicate that within cells there must be mechanisms that help proteins adopt their correct structures. We now know that the necessary assistance is provided by proteins called **molecular chaperones**. The existence of molecular chaperones does not compromise the underlying principle that the information for folding is carried by a protein's amino acid sequence. Molecular chaperones do not specify the tertiary structure of a protein, and they merely help the protein find that correct structure.

There are two types of molecular chaperones. The first are the **Hsp70 chaperones**, which bind to hydrophobic regions of proteins, including proteins that are still being translated (Figure 8.16). The latter is an important point because if a protein begins to fold before it has been fully synthesized, when only part of the polypeptide is available, then there might be an increased possibility of incorrect branches of the folding pathway being followed. Hsp70 chaperones are thought to hold the protein in an open conformation and somehow to modulate the association between those parts of the polypeptide which form interactions in the folded protein. Exactly how this is achieved is not understood but it involves repeated binding and attachment of the Hsp70 protein, each cycle of which requires energy provided by the hydrolysis of ATP. In addition to protein folding, the Hsp70 chaperones are also involved in other processes that require shielding of hydrophobic regions in proteins, such as transport through membranes, association of proteins into multisubunit complexes and disaggregation of proteins that have been damaged by heat stress.

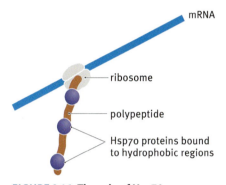

FIGURE 8.16 The role of Hsp70 chaperones. Hsp70 chaperones bind to hydrophobic regions in unfolded polypeptides, including those that are still being translated, and hold the protein in an open conformation to aid its folding.

The second group of molecular chaperones are called the **chaperonins**. These work in a quite different way. The best-studied examples are the bacterial GroEL/GroES complex and the eukaryotic chaperonin called TRiC. The GroEL/GroES complex is a multisubunit structure that looks like a hollowed-out bullet with a central cavity (Figure 8.17). A single unfolded protein enters the cavity and emerges folded. The mechanism for this is not known but it is postulated that GroEL/GroES acts as a cage that prevents the unfolded protein from aggregating with other proteins and that the inside surface of the cavity changes from hydrophobic to hydrophilic in such a way as to promote the burial of hydrophobic amino

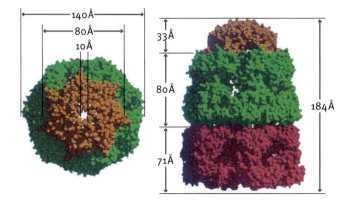

FIGURE 8.17 The GroEL/GroES chaperonin. On the left is the view from the top, and on the right is the view from the side. The GroES part of the structure is made up of seven identical protein subunits and is shown in gold. The GroEL components consist of 14 identical proteins arranged into two rings (shown in red and green), each containing seven subunits. The main entrance into the central cavity is through the bottom of the structure shown on the right. 1 Å is equal to 0.1 nm. (Courtesy of Zhaohui Xu, University of Michigan.)

acids within the protein. This is not the only hypothesis. Another possibility is that the cavity unfolds proteins that have folded incorrectly, passing these unfolded proteins back to the cytoplasm so they can have a second attempt at adopting their correct tertiary structure.

Although chaperonins are present in both bacteria and eukaryotes, it is thought that protein folding depends mainly on the action of the Hsp70 proteins. Only 5% of bacterial proteins are folded exclusively by GroEL/GroES, with another 5% folded by a combination of Hsp70 proteins and GroEL/GroES. The remaining 90% are folded by Hsp70 proteins without any involvement of the chaperonin. In eukaryotes, only 10% of all proteins, mostly proteins involved in the cell cycle and cytoskeleton, are folded by TRiC.

Some proteins are chemically modified

The standard genetic code specifies 20 different amino acids, and two others – selenocysteine and pyrrolysine – can be inserted during translation by context-dependent reassignment of 5′–UGA–3′ and 5′–UAG–3′ codons. This repertoire is increased dramatically by post-translational chemical modification of proteins, which results in a vast array of different amino acid types.

Because of chemical modification, the amino acid sequence of a mature, functioning protein might be different from that of the polypeptide coded by the gene. There is still, however, an inviolate link between the gene and the modified protein, because the chemical modifications do not occur at random. Instead, they are carried out in a highly specific manner, with the same amino acids being modified in the same way in every copy of the protein. In essence, the biological information that specifies the chemical modifications resides not in the gene for the protein that is being modified, but in the genes coding for the enzymes that carry out the modifications (**Figure 8.18**). The structures of these modifying enzymes, as coded by their gene sequences, determine their abilities to make specific chemical modifications at the correct positions on their target proteins.

The simplest types of chemical modification occur in all organisms and involve the addition of a small chemical group (e.g. an acetyl, methyl or phosphate group; **Table 8.6**) to an amino acid side chain or to the amino or carboxyl groups of the terminal amino acids in a polypeptide. Over 150 different modified amino acids have been documented in different proteins. Some proteins undergo an array of different modifications, an example being mammalian histone H3, which can be modified by

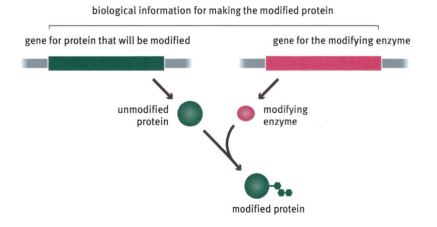

FIGURE 8.18 Biological information and chemical modification of proteins. The biological information needed to synthesize a chemically modified protein resides in both the gene for that protein and the gene for the enzyme that carries out the modification.

TABLE 8.6 EXAMPLES OF POST-TRANSLATIONAL CHEMICAL MODIFICATIONS

Modification	Amino Acids that are Modified	Examples
Addition of small chemical groups		
Acetylation	Lysine	Histones
Methylation	Lysine	Histones
Phosphorylation	Serine, threonine, tyrosine	Some proteins involved in signal transduction
Hydroxylation	Proline, lysine	Collagen
N-formylation	N-terminal glycine	Melittin
Addition of sugar side chains		
O-linked glycosylation	Serine, threonine	Many membrane proteins and secreted proteins
N-linked glycosylation	Asparagine	Many membrane proteins and secreted proteins
Addition of lipid side chains		
Acylation	Serine, threonine, cysteine	Many membrane proteins
N-myristoylation	N-terminal glycine	Some protein kinases involved in signal transduction
Addition of biotin		
Biotinylation	Lysine	Various carboxylase enzymes

FIGURE 8.19 Post-translational modification of human histone H3. The first 29 amino acids of the protein are shown, along with the modifications that can occur in this segment. Ac, acetylation; Me, methylation; P, phosphorylation.

acetylation, methylation and phosphorylation at a number of positions along its polypeptide chain (**Figure 8.19**). Histones are components of nucleosomes, the protein structures around which DNA is wound in eukaryotic chromosomes. The way in which nucleosomes interact with one another is determined by the nature of the chemical modifications on their histone proteins. Chemical modification is therefore a means of modifying a protein's activity, possibly to change its function in a subtle but important way, as with histone modification, or possibly to activate a protein whose function is only needed at particular times.

A more complex type of modification, found predominantly in eukaryotes, is **glycosylation**, the attachment of large carbohydrate side chains to polypeptides. There are two general types of glycosylation (**Figure 8.20**). **O-linked glycosylation** is the attachment of a sugar side chain via the hydroxyl group of a serine or threonine amino acid, and **N-linked glycosylation** involves attachment through the amino group on the side chain of asparagine.

In some eukaryotes, glycosylation can result in attachment to the protein of large structures comprising branched networks of 10–20 sugar units of various types. These side chains help to target proteins to particular sites in cells and to increase the stability of proteins circulating in the bloodstream. Another type of large-scale modification involves attachment of long-chain lipids, often to serine or cysteine amino acids. This process is called **acylation** and occurs with many proteins that become associated with membranes.

Some proteins are processed by proteolytic cleavage

Proteolytic cleavage of proteins is relatively common in eukaryotes but less frequent in bacteria. It has two functions (Figure 8.21). The more frequent of these is to remove short pieces from the N- and/or C-terminal regions of polypeptides, leaving a single shortened molecule that folds into the active protein. Less commonly, proteolytic cleavage is used to cut a **polyprotein** into segments, all or some of which are active proteins. As with chemical modification, these cleavage events are carried out in a specific manner and, in many cases, are used to activate a protein or, less frequently, to change its function.

The first type of proteolytic processing – removal of N- or C-terminal segments – is common with secreted polypeptides whose biochemical activities might be deleterious to the cell producing the protein. An example is provided by melittin, the most abundant protein in bee venom and the one responsible for causing cell lysis after injection of the bee sting into the person or animal being stung. Melittin lyses cells in bees as well as animals and so must initially be synthesized as an inactive precursor. This precursor, promelittin, has 22 additional amino acids at its N terminus. The presequence is removed by an extracellular protease that cuts it at 11 positions, releasing the active venom protein. The protease does not cleave within the active sequence because its mode of action is to release dipeptides with the sequence X–Y, where X is alanine, aspartic acid or glutamic acid, and Y is alanine or proline, motifs that do not occur in the active sequence (Figure 8.22).

A similar type of processing occurs with insulin, the protein made in the islets of Langerhans in the vertebrate pancreas and responsible for controlling blood sugar levels. Insulin is synthesized as preproinsulin, which is 110 amino acids in length (Figure 8.23). The processing pathway involves the removal of the first 24 amino acids to give proinsulin, followed by two additional cuts that excise a central segment. This leaves two active parts of the protein, the A and B chains, which link together by the formation of two disulfide bridges to form mature insulin. The first segment to be removed, the 24 amino acids from the N terminus, is a **signal peptide**,

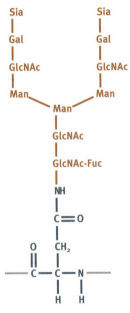

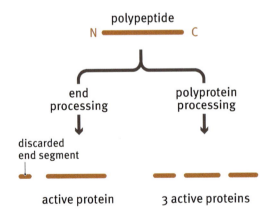

FIGURE 8.21 Two different types of protein processing by proteolytic cleavage. On the left, the protein is processed by removal of the N-terminal segment. C-terminal processing also occurs with some proteins. On the right, a polyprotein is processed to give three different proteins. Not all proteins undergo proteolytic cleavage.

FIGURE 8.20 Glycosylation. (a) O-linked glycosylation. The structure shown is found in a number of glycoproteins. It is drawn here attached to a serine amino acid but it can also be linked to a threonine. (b) N-linked glycosylation usually results in larger sugar structures than are seen with O-linked glycosylation. The drawing shows a typical example of a complex structure attached to an asparagine amino acid. Fuc, fucose; Gal, galactose; GalNAc, N-acetylgalactosamine; GlcNAc, N-acetylglucosamine; Man, mannose; Sia, sialic acid.

FIGURE 8.22 Processing of promelittin, the bee-sting venom. The arrows indicate the cut sites.

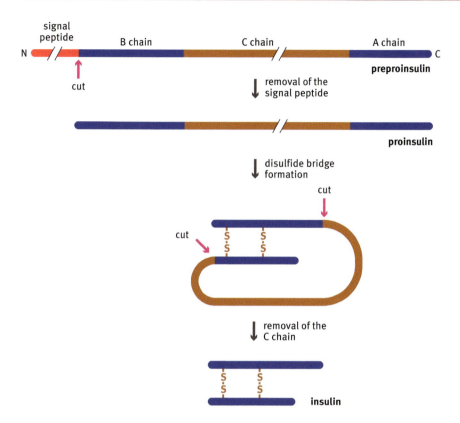

FIGURE 8.23 **Processing of preproinsulin.** The first 24 amino acids are removed to give proinsulin, which then forms two disulfide bridges. Two additional cuts remove the central segment, to form mature insulin.

a highly hydrophobic stretch of amino acids that attaches the precursor protein to a membrane prior to transport across that membrane and out of the cell. Signal peptides are commonly found on proteins that bind to and/or cross membranes, in both eukaryotes and prokaryotes.

Polyproteins are long polypeptides that contain a series of mature proteins linked together in a head-to-tail fashion. Cleavage of the polyprotein releases the individual proteins, which may have very different functions from one another. Polyproteins are not uncommon in eukaryotes. For example, the polyprotein called proopiomelanocortin, made in the pituitary gland, contains at least ten different peptide hormones. These are released by proteolytic cleavage of the polyprotein (Figure 8.24), but not all can be produced at once because of overlaps between individual peptide sequences. Instead, the exact cleavage pattern is different in different cells.

8.3 PROTEIN DEGRADATION

The protein synthesis and processing events that we have studied so far in this chapter result in new active proteins that take their place in the cell's proteome. These proteins either replace existing ones that have

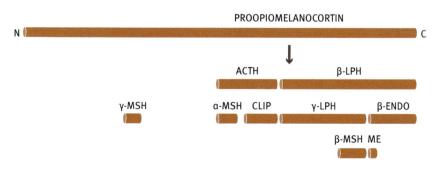

FIGURE 8.24 **Processing of the proopiomelanocortin polyprotein.** Two additional peptides are not shown. One of these is an intermediate in the processing events leading to γ-MSH, and the function of the other is unknown. Note that although met-enkephalin can theoretically be obtained by processing of proopiomelanocortin, as shown here, most met-enkephalin made by humans is probably obtained from a different peptide hormone precursor called proenkephalin. ACTH, adrenocorticotropic hormone; CLIP, corticotropin-like intermediate lobe protein; ENDO, endorphin; LPH, lipotropin; ME, met-encephalin; MSH, melanotropin.

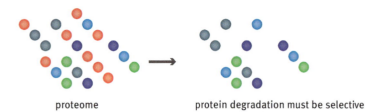

proteome → protein degradation must be selective

FIGURE 8.25 Protein degradation must be selective. Only those proteins whose functions are no longer needed are removed from a proteome.

reached the end of their working lives or provide new protein functions in response to the changing requirements of the cell. However, the proteome cannot simply accumulate new protein functions over time. It must also lose those proteins whose functions are no longer needed (Figure 8.25). This removal must be highly selective so that only the correct proteins are degraded and must also be rapid in order to account for the abrupt changes that occur under certain conditions.

Degradation of eukaryotic proteins involves ubiquitin and the proteasome

For many years, protein degradation was an unfashionable subject and it was not until the 1990s that real progress was made in understanding how protein turnover is linked with changes in cellular activity. Even now, our knowledge centers largely on descriptions of general protein breakdown pathways and less on regulation of those pathways and the mechanisms used to target specific proteins. There appear to be a number of different types of breakdown pathway whose interconnectivities have not yet been traced. This is particularly true in bacteria, which seem to have a range of proteases that work together in the controlled degradation of proteins. In eukaryotes, most breakdown involves a single system, involving **ubiquitin** and the **proteasome**.

A link between ubiquitin and protein degradation was first established when it was shown that this abundant 76-amino-acid protein is involved in energy-dependent proteolysis reactions in rabbit cells. Subsequent research identified a three-step process that attaches ubiquitin molecules, singly or in chains, to lysine amino acids in proteins that subsequently get broken down (Figure 8.26). A ubiquitin molecule is initially attached to an activator protein, then transferred to a conjugating enzyme, and finally transferred to the target protein by a ubiquitin ligase enzyme. Most species possess multiple versions of the conjugating enzymes and many types of ubiquitin ligase: humans, for example, have over 50 conjugating enzymes and several hundred ligases. Different pairs of conjugating enzyme and ligase are thought to have specificity for different proteins, with individual pairs activated by signals from within or outside of the cell. In response to these signals, particular proteins are degraded, altering the composition of the proteome.

The second component of the ubiquitin-dependent degradation pathway is the proteasome, the structure within which ubiquitinated proteins are broken down. In eukaryotes, the proteasome is a large, multisubunit structure comprising a hollow cylinder and two 'caps' (Figure 8.27). The entrance into the cavity within the proteasome is narrow, and a protein must be unfolded and its ubiquitin tags removed before it can enter. These steps require energy, obtained from hydrolysis of ATP, which is catalyzed by proteins present in the cap. After unfolding, the protein can enter the proteasome, within which it is cleaved into short peptides 4–10 amino acids in length. These are released back into the cytoplasm where they are broken down into individual amino acids, which can be reutilized in protein synthesis.

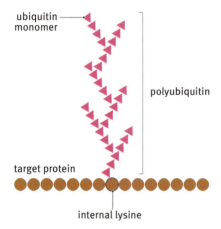

FIGURE 8.26 Ubiquitin attachment target proteins for degradation. To act as a label for degradation, chains of ubiquitin molecules are linked to lysine amino acids in the target protein.

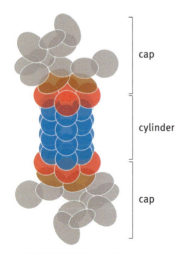

FIGURE 8.27 The eukaryotic proteasome. The protein components of the two caps are shown in grey, orange and red and those forming the cylinder in blue.

KEY CONCEPTS

- The collection of proteins in a cell is called the proteome. Within the proteome, the identity and relative abundance of individual proteins represent a balance between the synthesis of new proteins and the degradation of the existing ones.

- During protein synthesis, the information contained in the nucleotide sequence of an mRNA is translated into the amino acid sequence of a protein. This process follows the rules laid down by the genetic code, with tRNAs enforcing these rules.

- In bacteria, the translation initiation complex is built up directly at the initiation codon, the point within the mRNA at which protein synthesis will begin. In eukaryotes, the equivalent complex makes its initial attachment at the 5′ end of the mRNA and then scans along the sequence until it locates the initiation codon.

- During the elongation phase of translation, the protein is synthesized step by step, with individual amino acids being attached to the carboxyl terminus of the growing polypeptide.

- Protein synthesis ends when the ribosome reaches one of the three termination codons. There are no tRNA molecules with anticodons able to base pair with any of the termination codons, and instead a protein release factor enters the ribosome in order to terminate translation.

- Utilization of the biological information contained in a gene requires synthesis of an active protein, one able to perform its function in the cell. The initial products of translation are not active proteins. Protein processing is therefore an integral component of the gene expression pathway.

- All proteins must be folded into their correct three-dimensional structures before becoming active in the cell.

- Some proteins also undergo chemical modification, by attachment of new chemical groups. Chemical modification might activate an inactive protein or change the function of an already active protein.

- Some proteins are also processed by cutting events that may remove segments from one or both ends of the polypeptide, resulting in a shortened form of the protein, or may cut the polypeptide into a number of different segments, all or some of which are active.

- Pathways for protein breakdown have been identified but little is known about how those pathways are regulated. In eukaryotes, protein degradation involves attachment of ubiquitin molecules followed by breakdown in the proteasome.

QUESTIONS AND PROBLEMS

Key terms

Write short definitions of the following terms:

acylation
aminoacyl site
cap binding complex
chaperonin

elongation factor
exit site
folding pathway
glycosylation

Hsp70 chaperone
initiation complex
initiation factor
internal ribosome entry site

Kozak consensus
molecular chaperone
N-linked glycosylation
O-linked glycosylation
peptidyl site
peptidyl transferase
polyprotein
polysome
pre-initiation complex
protease
proteasome
proteome
release factor
ribosome binding site
ribosome recycling factor
scanning
Shine–Dalgarno sequence
signal peptide
translocation
ubiquitin

Self-study questions

8.1 What is the proteome?

8.2 Outline the events involved in the formation of the initiation complex during protein synthesis in *E. coli*.

8.3 What is the importance of *N*-formylmethionine during initiation of protein synthesis in *E. coli*?

8.4 Describe the role of the initiation factors during protein synthesis in *E. coli*.

8.5 Distinguish between the eukaryotic structures called the pre-initiation complex and the initiation complex.

8.6 Explain how the initiation complex is able to ignore certain AUG codons when it scans along a eukaryotic mRNA, and how it identifies the correct initiation codon.

8.7 What is thought to be the role of the poly(A) tail during initiation of protein synthesis in eukaryotes?

8.8 Briefly describe the roles of initiation factors in protein synthesis in eukaryotes.

8.9 How are eukaryotic mRNAs that lack a cap structure translated?

8.10 Give a detailed description of the elongation phase of translation in bacteria and eukaryotes.

8.11 What are the roles of the P, A and E sites during protein synthesis in *E. coli*? Which of these sites is absent in a eukaryotic ribosome?

8.12 Outline the roles of the three release factors and the ribosome recycling factor during termination of translation in *E. coli*. Which proteins play the equivalent roles during termination in eukaryotes?

8.13 Describe the experiments that showed that a small protein such as ribonuclease can fold spontaneously in the test tube. Why are larger proteins unable to fold spontaneously?

8.14 Distinguish between the activities of Hsp70 chaperones and chaperonins in protein folding.

8.15 Construct a table that lists, with examples, the various types of post-translational chemical modification that occur with different proteins.

8.16 What is glycosylation? Your answer should distinguish between O- and N-linked processes.

8.17 Outline the role of proteolytic processing in the synthesis of insulin.

8.18 Give an example of proteolytic processing of a polyprotein.

8.19 Describe the processes thought to be responsible for protein degradation in eukaryotes.

Discussion topics

8.20 Speculate on the reasons why the poly(A) tail of a eukaryotic mRNA is involved in initiation of translation of the mRNA.

8.21 Bacterial and some eukaryotic mRNAs have internal recognition sequences to which the small subunit of the ribosome attaches in order to initiate protein synthesis. If this relatively straightforward process is suitable for some mRNAs, then why should initiation of most eukaryotic mRNAs have to involve an elaborate scanning mechanism?

8.22 Discuss the extent to which the ribosome is an active or a passive partner in protein synthesis.

8.23 Devise an experiment to demonstrate that peptidyl transferase is a ribozyme.

8.24 To what extent are protein folding studies that are conducted in the test tube good models for protein folding in living cells?

 The answers to, or hints on how to answer, the self-study questions and discussion topic questions can be downloaded from the book's product page at this link: www.routledge.com/9781032743530

FURTHER READING

Caskey CT (1980) Peptide chain termination. *Trends in Biochemical Sciences* 5, 234–237.

Clark B (1980) The elongation step of protein biosynthesis. *Trends in Biochemical Sciences* 5, 207–210.

Daggett V & Fersht AR (2003) Is there a unifying mechanism for protein folding? *Trends in Biochemical Sciences* 28, 18–25.

Drickamer K & Taylor ME (1998) Evolving views of protein glycosylation. *Trends in Biochemical Sciences* 23, 321–324.

Hellen CUT (2018) Translation termination and ribosome recycling in eukaryotes. *Cold Spring Harbor Perspectives in Biology* 10, a032656.

Hellen CUT & Sarnow P (2001) Internal ribosome entry sites in eukaryotic mRNA molecules. *Genes and Development* 15, 1593–1612.

Hinnebusch AG & Lorsch JR. (2012) The mechanism of eukaryotic translation initiation: new insights and challenges. *Cold Spring Harbor Perspectives in Biology* 4, a011544.

Hunt T (1980) The initiation of protein synthesis. *Trends in Biochemical Sciences* 5, 178–181.

Jackson RJ, Hellen CUT & Pestova TV (2010) The mechanism of eukaryotic translation initiation and principles of its regulation. *Nature Reviews Molecular Cell Biology* 11, 113–127.

Kapp LD & Lorsch JR (2004) The molecular mechanics of eukaryotic translation. *Annual Review of Biochemistry* 73, 657–704.

Rosenzweig R, Nillegoda NB, Mayer MP et al. (2019) The Hsp70 chaperone network. *Nature Reviews Molecular Cell Biology* 20, 665–680.

Schmeing TM & Ramakrishnan V (2009) What recent ribosome structures have revealed about the mechanism of translation. *Nature* 461, 1234–1242.

Steitz TA & Moore PB (2003) RNA, the first macromolecular catalyst: the ribosome is a ribozyme. *Trends in Biochemical Sciences* 28, 411–418. *Describes the evidence that led to the discovery that peptidyl transferase is a ribozyme.*

Varshavsky A (1997) The ubiquitin system. *Trends in Biochemical Sciences* 22, 383–387. *Protein degradation.*

Voges D, Zwickl P & Baumeister W (1999) The 26S proteasome: a molecular machine designed for controlled proteolysis. *Annual Review of Biochemistry* 68, 1015–1068. *More on protein degradation.*

Xu B, Liu L & Song G. (2022) Functions and regulation of translation elongation factors. *Frontiers in Molecular Biosciences* 8, 816398.

Yébenes H, Mesa P, Muñoz IG et al. (2011) Chaperonins: two rings for folding. *Trends in Biochemical Sciences* 36, 424–432.

CHAPTER 9

Control of Gene Expression

In the previous five chapters, we followed the gene expression pathway from the initiation of transcription of a gene through to the synthesis, processing and eventual degradation of its protein product. Now that we understand how the biological information contained in a gene is utilized by the cell, we must look back over the process and examine how gene expression is controlled.

We have already acknowledged that the rate at which a gene is expressed can change and that many genes might be completely switched off at a particular time. This prompted us to include mRNA and protein degradation within our study of gene expression, as we realized that degradation was needed in order for the composition of the transcriptome and proteome to respond to changes in the rate of expression of individual genes. We have also looked at specific mechanisms for altering the expression pathway of a gene, for example, through the use of alternative promoters, by changing the way in which an mRNA is spliced or by RNA editing. But so far we have only scratched the surface of gene regulation.

First, we must ask why it is necessary to control gene expression. In answering this question, we will begin to appreciate that the underlying principles of gene regulation are the same in all organisms, but that the control processes are more sophisticated in eukaryotes compared with bacteria. This reflects the greater need that multicellular organisms have for gene regulation during differentiation and development.

Next, we will attempt to distinguish between those regulatory events that determine whether a gene is switched on or off and those that affect the rate at which the RNA or protein product is synthesized once a gene is switched on. This will quickly lead us to focus on how the initiation of transcription is regulated in bacteria and eukaryotes, as this is looked on as the key step at which the cell controls which genes are active at any one time. We will examine transcriptional regulation in some detail and discover once again that, despite the process being more complex in eukaryotes, the underlying principles are the same in all types of organisms.

Finally, we will survey the most important of the control mechanisms that target steps in gene expression downstream from transcription initiation, asking how these mechanisms work and what influence they have on the synthesis of the RNA or protein product.

Throughout the chapter, our attention will be specifically on the regulation of individual genes. This is a narrow focus but one that is necessary in order to deal with a complex topic in an accessible way. In Chapter 19, we will place gene regulation in its broader context by examining the genetic basis of differentiation and development.

9.1 THE IMPORTANCE OF GENE REGULATION

The entire complement of genes in a single cell represents a staggering amount of biological information. Some of this information is needed by the cell at all times. For example, most cells continually synthesize ribosomes and so have a continuous requirement for transcription of the rRNA and ribosomal protein genes. Similarly, genes coding for enzymes such as RNA polymerase or those involved in the basic metabolic pathways are active in virtually all cells all of the time. Genes that are active all the time are called **housekeeping genes**, reflecting the role in the cell of the biological information they carry.

Other genes have more specialized roles and their biological information is needed by the cell only under certain circumstances. All organisms are therefore able to regulate expression of their genes, so that those genes whose RNA or protein products are not needed at a particular time are switched off. This is a straightforward concept but it has extensive ramifications. To illustrate this point, we will briefly consider some examples of gene regulation in bacteria and eukaryotes.

Gene regulation enables bacteria to respond to changes in their environment

Even the simplest bacterium is able to control the expression of its genes. By doing this, it is able to respond to changes in its environment.

The response of bacteria to environmental changes was originally studied with cultures of *Escherichia coli* grown in the laboratory. A culture medium must provide a balanced mixture of nutrients, including compounds that can be metabolized to release the energy that the bacteria need to power their cellular processes. Usually, this energy is supplied in the form of sugar. Most bacteria are not particularly fussy about what type of sugar is provided – *E. coli* can use glucose, maltose, lactose, galactose, arabinose or any of several others. A different set of enzymes is needed to release the energy from each of these compounds, but *E. coli* has genes coding for all of these enzymes.

Exactly which enzymes are required by an individual bacterium depends on which sugars are present in the medium. Each bacterium could continuously express all its sugar utilizing genes and so have molecules of each enzyme available all the time, but this would waste energy, something all organisms try to avoid. Why synthesize enzymes for the utilization of a sugar that is not available? Instead, *E. coli* expresses only those genes coding for the enzymes it needs to metabolize the sugars that are present. The genes for all the other enzymes are inactive, and switched off as their gene products are not required (Figure 9.1).

What if the growth medium is changed by adding a new sugar, for example, lactose instead of glucose? Now, *E. coli* quickly switches on expression of the genes whose products it needs and switches off the ones that have become redundant. So by regulating expression of its genes, *E. coli* is able to respond quickly to changes in the growth medium without wasting

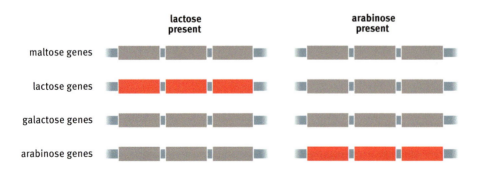
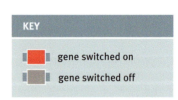

FIGURE 9.1 *E. coli* **only expresses the genes coding for those enzymes that it needs to metabolize the sugars present in its environment.** If the only sugar present is lactose, then the genes needed for lactose utilization are switched on, and those for utilization of other sugars are switched off. Conversely, when only arabinose is present, the arabinose genes are switched on and the others are switched off.

energy by maintaining in an active state a gene whose product is of no immediate use.

Gene regulation in eukaryotes must be responsive to more sophisticated demands

At the most basic level, gene regulation achieves the same thing in both bacteria and eukaryotes. It enables a cell to change its biochemical capabilities. The difference is that with eukaryotes we see a greater degree of sophistication with respect to both the signals that influence gene expression and the impact that gene regulation has on the organism.

Eukaryotic cells, just like bacteria, can alter their gene expression patterns in response to changes in their environment. For example, the yeast *Saccharomyces cerevisiae* regulates its genes for sugar utilization enzymes in a manner analogous to *E. coli*, and plant cells switch on genes for photosynthetic proteins in response to light (Figure 9.2a). In a multicellular organism, individual cells, or groups of cells, also respond to stimuli that originate from within the organism. Hormones, growth factors and other regulatory molecules are produced by one type of cell and cause changes in gene expression in other cells (Figure 9.2b). In this way, the activities of groups of cells can be coordinated in a manner that is beneficial to the organism as a whole.

In addition to these transient changes in gene expression, most eukaryotes (as well as some bacteria) are able to switch genes on and off at different stages of development. An example is provided by the human globin genes, different members of the α and β gene families being expressed in

(a) plant genes respond to light

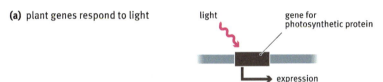

(b) hormones and other regulatory molecules control gene expression

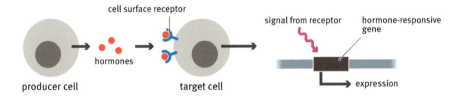

(c) β-globin genes are developmentally regulated

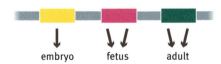

(d) specialized cells express different genes

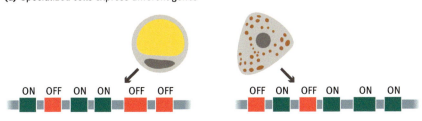

FIGURE 9.2 Examples of gene regulation in multicellular eukaryotes. (a) Some genes respond to changes in the environment. This example shows a plant gene being activated by exposure to light. (b) Hormones, growth factors and other regulatory molecules are produced by one type of cell and cause changes in gene expression in other cells. (c) Some genes are subject to developmental regulation and are expressed at different stages of the life cycle. (d) Different sets of genes are expressed in different types of specialized cell.

the embryo, fetus and adult (see Figure 3.21). The distinctive biochemical properties of the resulting globin proteins are matched to the specific physiological roles that hemoglobin plays at these different phases in human development. Developmental regulation therefore enables the biochemistry of an organism to be altered in response to the particular requirements of each stage of its life cycle (Figure 9.2c).

Gene regulation in multicellular organisms is also the key to cell specialization. In humans, there are over 400 specialized cell types, each with a different morphology and biochemistry. The distinctions between these cell types are the result of differences in gene expression patterns (Figure 9.2d). A liver cell is very different from a muscle cell because it expresses a different set of genes. These gene expression patterns are permanent – a liver cell can never become a muscle cell – and are laid down very early in development.

The outcomes of gene regulation are therefore more sophisticated in eukaryotes, especially multicellular ones like humans, than they are in bacteria. Now, we must move on to consider the various ways in which expression of a gene can be controlled. Are these more sophisticated in eukaryotes also?

The underlying principles of gene regulation are the same in all organisms

Although gene expression responds to a wider range of regulatory signals in eukaryotes, and the control processes have more sophisticated outcomes, the underlying principles of gene regulation are the same in all organisms. The difference is in the detail.

Control of gene expression is, in essence, control over the amount of gene product present in the cell. This amount is a balance between the rate of synthesis (how many molecules of the gene product are made per unit time) and the degradation rate (how many molecules are broken down per unit time) (Figure 9.3). The result of this balance is a different steady-state concentration for each gene product in the cell. If either the synthesis rate or the degradation rate changes, then the steady-state concentration also changes. Although we have already acknowledged the importance of mRNA and protein degradation in the gene expression pathway, geneticists at present believe that the critical variations in the steady state of a gene product arise because of changes in its rate of synthesis. This is true for both proteins and noncoding RNA.

How can the synthesis rate of a gene product be regulated? The answer is by exerting control over any one, or a combination, of the various steps in the gene expression pathway (Figure 9.4). The possibilities include, first of all, transcription. If the number of transcripts synthesized per unit time changes, then the amount of the gene product that is synthesized also changes. Increase the transcription rate and there will be more gene product, decrease the transcription rate and there will be less gene product. Moving along the pathway, control could also be exerted over

steady-state concentration of gene product

FIGURE 9.3 Controlling the amount of gene product in a cell. The amount of a gene product that is present in a cell is a balance between the rate of synthesis and the rate of degradation.

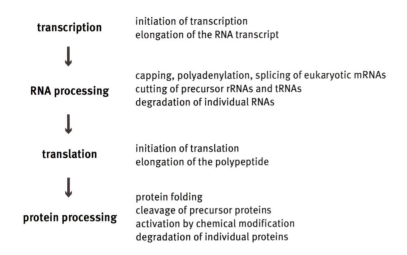

FIGURE 9.4 **Potential control points during gene expression.** The steps in the expression pathway for a eukaryotic gene are shown on the left. Points at which control could be exerted over synthesis of the gene product are listed on the right.

RNA processing events. These include not just splicing of mRNA but also capping and polyadenylation, both of which are prerequisites for translation, as well as the various processing events involved in the synthesis of functional noncoding RNAs. Slow down one or more of these processing events and product synthesis will fall. Finally, regulation could be exerted during protein synthesis, by controlling either the initial attachment of ribosomes to an mRNA or the rate at which individual ribosomes translocate along an mRNA while making a protein, or by regulating one or more of the various post-translational processing events.

It is becoming clear that organisms use many, possibly all, of the above strategies for controlling expression of their genes. In this regard, eukaryotes do display greater complexity than bacteria simply because the gene expression pathway in eukaryotes has more steps at which regulation could be exerted. Control over mRNA processing, for example, is possible only in eukaryotes because mRNA is not processed in bacteria. But when we consider gene regulation, we must be careful to distinguish between the details and the underlying principles. The principle of gene regulation is the need to control synthesis of RNA and protein products by switching genes on and off and to modulate the rate of expression of those genes that are on. At this primary level, bacteria and eukaryotes are the same because in all organisms the key events that determine if a gene is on or off take place during the initiation of transcription. Later events in the gene expression pathway are able to change the rate of expression of a gene that is switched on and, in eukaryotes, control the synthesis of alternative products of a single gene (Figure 9.5). But initiation of transcription is the primary control point, and it is this event that we must

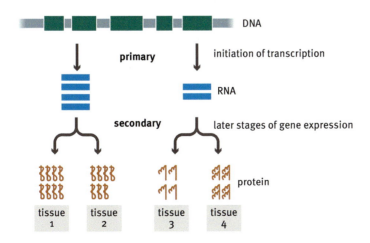

FIGURE 9.5 **Initiation of transcription is the primary control point.** The primary events that determine if a gene is switched on or off take place during the initiation of transcription. Secondary events that occur later in the gene expression pathway influence the amount of gene product that is made, as illustrated by tissues 1 and 2, or its identity, as in tissues 3 and 4.

study in detail in order to understand how gene expression is regulated in bacteria and eukaryotes.

9.2 REGULATION OF TRANSCRIPTION INITIATION IN BACTERIA

The foundation to our understanding of how bacteria regulate transcription of their genes was laid by François Jacob and Jacques Monod of the Institut Pasteur in Paris, in a paper, published in 1961, that is now considered a classic example of experimental analysis and deductive reasoning. Jacob and Monod based their propositions on an intricate genetic analysis of lactose utilization by *E. coli* and described an elegant regulatory system that has subsequently been confirmed in virtually every detail. We will use the lactose system as our central example of how gene regulation works in bacteria.

Four genes are involved in lactose utilization by *E. coli*

Lactose is a disaccharide sugar composed of a single glucose unit attached to a single galactose unit (**Figure 9.6a**). In order to utilize lactose as an energy source, an *E. coli* cell must first transport lactose molecules from the extracellular environment into the cell and then split the molecules, by hydrolysis, into glucose and galactose (**Figure 9.6b**). These reactions are catalyzed by three enzymes, each of which has no function other than in lactose utilization. The enzymes are lactose permease, which transports lactose into the cell, β-galactosidase, which is responsible for the splitting reaction, and β-galactoside transacetylase, whose precise role in the process is unknown.

In the absence of lactose, only a small number of molecules of each enzyme are present in an *E. coli* cell, probably less than five of each. When the bacterium encounters lactose, enzyme synthesis is rapidly induced and within a few minutes levels of up to 5,000 molecules of each enzyme per cell are reached. Induction of the three enzymes is coordinated, meaning that each is induced at the same time and to the same extent. This

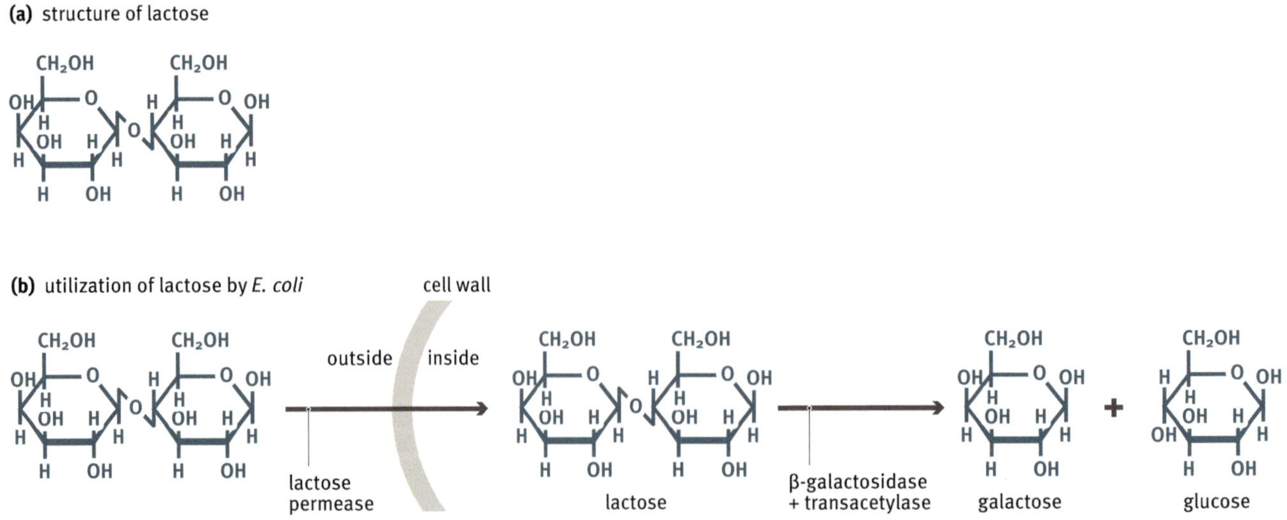

FIGURE 9.6 Lactose and its utilization as an energy source by *E. coli*. (a) The lactose molecule consists of a glucose unit attached to a galactose unit. (b) In order to utilize lactose as an energy source, the *E. coli* cell must import lactose molecules from the extracellular environment and then split the molecules into their glucose and galactose components.

provides a clue to the arrangement of the relevant genes on the *E. coli* DNA molecule.

The three enzymes involved in lactose utilization are coded by genes called *lacZ* (β-galactosidase), *lacY* (permease) and *lacA* (transacetylase). These names follow the convention for genes in *E. coli*, as well as most other organisms, comprising a three-letter abbreviation indicating the function of the gene followed by one or more letters and/or numbers to distinguish between genes of related function. The three genes lie in a cluster with only a very short distance between the end of one gene and the start of the next (**Figure 9.7a**). There is just one promoter sequence, immediately upstream of *lacZ* – there are no promoters in the small gaps between each pair of genes – and the three genes are transcribed together as a single mRNA molecule. This type of organization is called an **operon**.

Jacob and Monod used genetic analysis techniques to identify *lacZ*, *lacY* and *lacA* and to determine their relative positions in the *E. coli* DNA molecule. They also discovered an additional gene, which they designated *lacI*. This gene lies just upstream of the lactose gene cluster but is not itself a part of the operon, because it is transcribed from its own promoter and has its own terminator (**Figure 9.7b**).

The gene product of *lacI* is intimately involved in lactose utilization but is not an enzyme directly required for the uptake or hydrolysis of the sugar. Instead, the *lacI* product regulates the expression of the other three genes. This is evident from the effect of mutations on *lacI*. If *lacI* is inactivated, the lactose operon becomes switched on continuously, even in the absence of lactose. The terminology used by Jacob and Monod is still important today. We refer to *lacZ*, *lacY* and *lacA* as **structural genes** because their products contribute to the 'biochemical structure' of the cell. In contrast, *lacI* is called a **regulatory gene** because its function is to control the expression of the other genes.

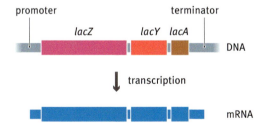

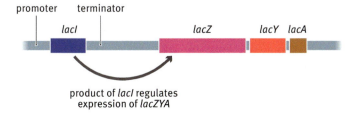

FIGURE 9.7 The lactose operon and the position of the regulatory gene *lacI*. (a) The three genes of the lactose operon are transcribed as a single unit. (b) The fourth gene involved in lactose utilization, *lacI*, has its own promoter and terminator. This gene controls expression of the lactose operon.

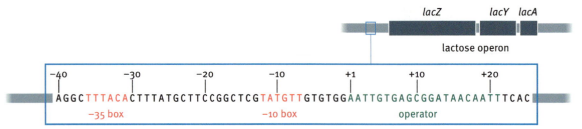

FIGURE 9.8 The DNA sequence immediately upstream of the *lacZ* gene of *E. coli*. The sequence shows the position of the operator relative to the –35 and –10 components of the promoter.

The regulatory gene codes for a repressor protein

The gene product of *lacI* is a protein that Jacob and Monod called the lactose **repressor**. This protein is able to attach to the *E. coli* DNA molecule at a site between the promoter for the lactose operon and the start of *lacZ*, the first gene in the cluster (Figure 9.8). The attachment site is called the **operator** and was also located by Jacob and Monod by genetic means. The operator, in fact, overlaps the promoter so that when the repressor is bound, access to the promoter is blocked so that RNA polymerase cannot attach to the DNA and transcription of the lactose operon cannot occur (Figure 9.9). This is what happens if lactose is not available to the cell. To summarize, *if there is no lactose, transcription of the lactose operon does not occur because the promoter is blocked by the repressor.*

Transcription is induced by allolactose. Allolactose is an isomer of lactose and is synthesized as an byproduct during the splitting of lactose into glucose and galactose. When the bacterium encounters a new supply of lactose, it takes up a few molecules and splits them. It is able to do this because of the small number of lactose utilization enzymes that are always present. The allolactose that is formed as a byproduct then binds to the repressor, causing a change in the conformation of the protein in such a way that the repressor is no longer able to attach to the operator. The repressor–allolactose complex dissociates from the DNA molecule, enabling RNA polymerase to locate the promoter and begin transcription of the operon (Figure 9.10). This results in the synthesis of the much larger numbers of enzyme molecules needed to take up and metabolize the rest of the lactose. Allolactose therefore acts as the **inducer** of the operon. *If lactose is present, transcription of the lactose operon occurs as the repressor–inducer complex does not bind to the operator.*

Eventually, the lactose utilization enzymes exhaust the available supply of lactose. The binding of allolactose to the repressor is an equilibrium event (Figure 9.11), so when the free lactose concentration decreases the number of repressor–allolactose complexes also decreases and free repressor molecules start to predominate. These free repressors have regained their original conformation and so can attach once again to the operator. *When the lactose supply is used up, the lactose operon is switched off as the repressor reattaches to the operator.*

Glucose also regulates the lactose operon

The presence or absence of lactose is not the only factor that influences expression of the lactose operon. If a bacterium has a sufficient source of glucose (one of the breakdown products of lactose) for its energy needs, then it does not metabolize lactose even if lactose is also available in the environment. Only when all of the glucose has been used up will the cell start to make use of the lactose (Figure 9.12).

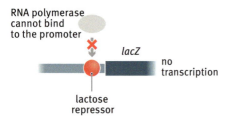

FIGURE 9.9 The role of the lactose repressor. The RNA polymerase cannot gain access to the promoter when the lactose repressor is bound to the operator.

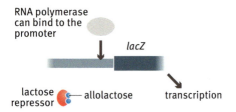

FIGURE 9.10 The repressor–allolactose complex does not bind to the operator. In the presence of allolactose, RNA polymerase is able to gain access to the promoter.

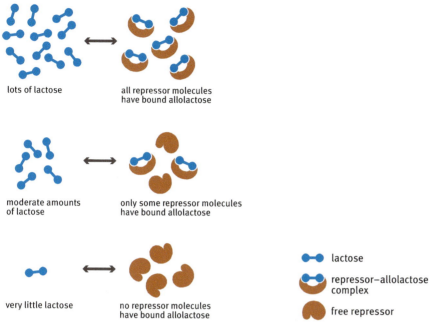

FIGURE 9.11 Repressor–inducer binding is an equilibrium event.

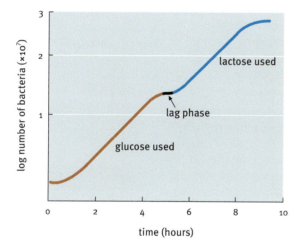

FIGURE 9.12 **A typical diauxic growth curve.** The growth curve illustrates the events that occur when *E. coli* is grown in a medium containing a mixture of glucose and lactose. During the first few hours, the bacteria divide exponentially, using the glucose as its energy source. When the glucose is used up, there is a brief lag period while the lactose genes are switched on before the bacteria return to exponential growth, now using up the lactose. Note that the *y*-axis is a logarithmic scale, and therefore the exponential phases of bacterial growth appear as straight lines on the graph.

This phenomenon, called **diauxie**, was discovered by Monod in 1941. When the details of the lactose operon were worked out some 20 years later, it became clear that the diauxie between glucose and lactose must involve a mechanism whereby the presence of glucose can override the inductive effect that lactose usually has on its operon. In the presence of both lactose and glucose, the lactose operon remains switched off, even though some of the lactose in the mixture is converted into allolactose which binds to the lactose repressor, the situation that under normal circumstances results in the operon being transcribed (Figure 9.13).

The explanation for the diauxic response is that glucose prevents expression of the lactose operon, as well as other sugar utilization operons, through an indirect influence on the **catabolite activator protein**. This protein binds to a recognition sequence at various sites in the *E. coli* DNA and activates transcription initiation at downstream promoters, probably by interacting with the σ subunit of the RNA polymerase. Productive initiation of transcription at these promoters is dependent on the presence of

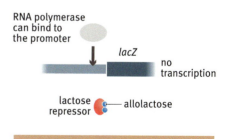

FIGURE 9.13 **Glucose overrides the lactose repressor.** If lactose is present, then the repressor detaches from the operator and the lactose operon should be transcribed, but it remains silent if glucose is also present.

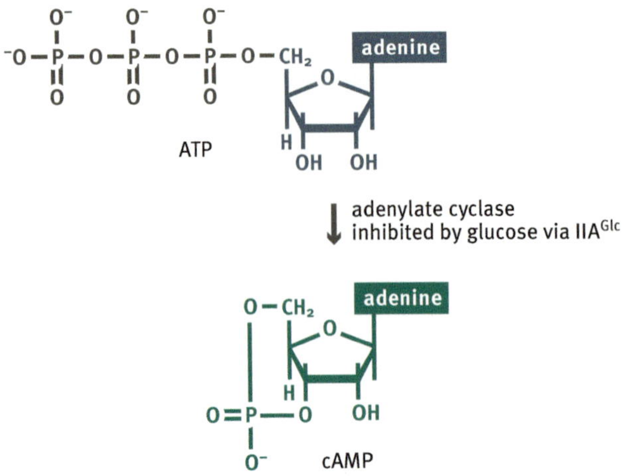

FIGURE 9.14 Glucose controls the level of cAMP in the cell. Glucose inhibits the activity of adenylate cyclase, the enzyme that synthesizes cAMP from ATP. Inhibition is mediated by IIA^{Glc}, a component of the multiprotein complex that transports sugars into the bacterium.

bound catabolite activator protein. If the protein is absent, then the genes controlled by the promoter are not transcribed.

Glucose does not itself interact with the catabolite activator protein. Instead, glucose controls the level in the cell of the modified nucleotide **cyclic AMP** (**cAMP**; Figure 9.14). It does this by inhibiting the activity of **adenylate cyclase**, the enzyme that synthesizes cAMP from ATP. Inhibition is mediated by a protein called IIA^{Glc}, a component of a multiprotein complex that transports sugars into the bacterium. When glucose is being transported into the cell, IIA^{Glc} becomes modified by removal of phosphate groups that are usually attached to its surface. The modified version of IIA^{Glc} inhibits adenylate cyclase activity. This means that if glucose levels are high, the cAMP content of the cell is low.

The catabolite activator protein can bind to its target sites only in the presence of cAMP, so when glucose is present the protein remains detached and the operons it controls are switched off. In the specific case of diauxie involving glucose plus lactose, the indirect effect of glucose on the catabolite activator protein means that the lactose operon remains inactivated, even though the lactose repressor is not bound. The glucose in the medium is therefore used up first. When the glucose is gone, the cAMP level rises and the catabolite activator protein binds to its target sites, including the site upstream of the lactose operon, and transcription of the lactose genes is activated (Figure 9.15).

Operons are common in prokaryotes

There are over 200 operons in the *E. coli* DNA molecule, each containing two or more genes. Operons are also present in most other bacteria and are common in archaea.

Operons can be broadly divided into two categories. The first type are called **inducible operons** and are typified by the lactose operon. An inducible operon codes for a set of enzymes involved in a metabolic pathway, one that results in the breakdown of a substrate in the same way that the lactose enzymes break down lactose. The presence of the substrate induces expression of the genes. Other examples of inducible operons are the galactose operon and the arabinose operon.

The second type are called **repressible operons**. These code for enzymes involved in a biosynthetic pathway, one that results in the synthesis of a product. The product controls expression of the operon. An example is the tryptophan operon, which is made up of five genes involved in the synthesis

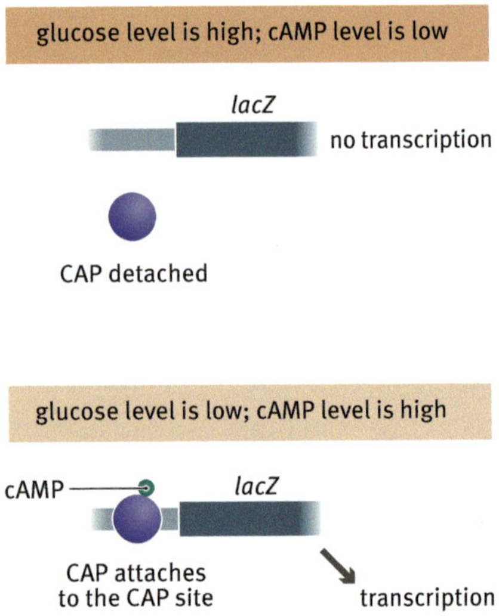

FIGURE 9.15 The catabolite activator protein can attach to its DNA-binding site only in the presence of cAMP. If glucose is present, the cAMP level is low, so CAP does not bind to the DNA and does not activate the RNA polymerase. Once the glucose has been used up, the cAMP level rises, allowing CAP to bind to the DNA and activate transcription of the lactose operon.

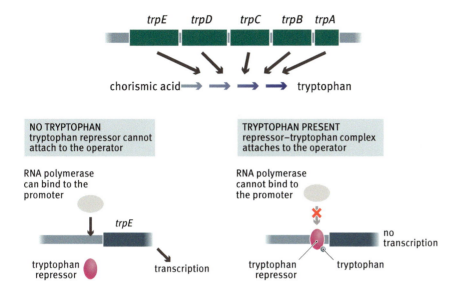

FIGURE 9.16 The tryptophan operon. This operon contains five genes coding for enzymes involved in the multistep biochemical pathway that converts chorismic acid into the amino acid tryptophan.

FIGURE 9.17 Regulation of the tryptophan operon of *E. coli*. When tryptophan is present, and so does not need to be synthesized, the operon is switched off because the repressor–tryptophan complex binds to the operator. In the absence of tryptophan, the repressor cannot attach to the operator, and the operon is expressed.

of the amino acid tryptophan (Figure 9.16). Expression of the operon is controlled by the tryptophan repressor, which attaches to the tryptophan operator and prevents transcription. In this case though, the repressor on its own cannot attach to the operator. Repression of the operon occurs only when the tryptophan repressor binds tryptophan (Figure 9.17). This is of course completely logical as tryptophan is the product of the biochemical pathway controlled by the operon. If tryptophan is absent, then the enzymes for its synthesis are needed, and the operon must be transcribed. Tryptophan is called the **co-repressor**. A second example of a repressible operon is the phenylalanine operon.

9.3 REGULATION OF TRANSCRIPTION INITIATION IN EUKARYOTES

Now, we turn our attention to the regulation of transcription initiation in eukaryotes. We have just learnt that in bacteria the initiation of transcription is influenced by DNA-binding proteins, such as the lactose repressor and the catabolite activator protein, which bind to specific sequences located near the attachment site for the RNA polymerase. This is also the basis for transcriptional control in eukaryotes. We will look first at the regulatory sequences.

RNA polymerase II promoters are controlled by a variety of regulatory sequences

RNA polymerase II provides the best illustration of transcription regulation in eukaryotes. Promoters recognized by this polymerase are subject to the most complex control patterns, reflecting the need to ensure that each protein-coding gene is expressed at precisely the level that is appropriate for the prevailing conditions inside and outside of the cell.

We have already examined the promoters for genes transcribed by RNA polymerase II and seen that, as well as the TATA box and Inr sequence that make up the core promoter, these genes are preceded by a variety of regulatory sequences located in the several kb immediately upstream of the transcription start site (see Figure 4.14). A single gene might be under the control of a number of different regulatory sequences, reflecting the wide range of internal and external signals that a cell must be able

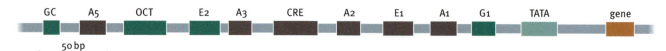

FIGURE 9.18 Regulatory sequences upstream of the human insulin gene. The A5, A3, CRE, A2, E1 and A1 sequences, shown in brown, are cell-specific modules, which ensure that the insulin gene is active only in the β-cells of the pancreas. The GC, OCT, E2 and G1 modules, shown in dark green, control the rate of transcription of the gene when it is switched on. GC, GC box; OCT, octamer sequence.

to respond to in order to perform its function in a multicellular organism. The human insulin gene, for example, has at least ten regulatory sequences (Figure 9.18) even though this gene is not subject to a particularly sophisticated control regime.

As in bacteria, the regulatory sequences that control expression of a eukaryotic gene are the binding sites for proteins that influence the activity of the RNA polymerase. There is, however, one important difference. The bacterial RNA polymerase has a strong affinity for its promoter and the basal rate of initiation is relatively high for all but the weakest promoters. With most eukaryotic genes, the reverse is true. The RNA polymerase II initiation complex does not assemble efficiently and the basal rate of transcript initiation is therefore very low. In order to achieve effective initiation, formation of the complex must be activated by additional proteins. This means that most of the proteins that bind to eukaryotic regulatory sequences are activators of transcription, and repressors are relatively rare.

Some regulatory sequences and their binding proteins mediate short-term changes in gene expression, enabling the genes they control to be switched on temporarily when their products are needed. An example in humans is the **heat shock module**, which is recognized by the HSP70 protein. HSP70 is thought to detect cellular damage caused by stresses such as heat shock, switching on transcription of genes whose products help repair the damage and protect the cell from further stress. There are also cell-specific modules, whose binding proteins ensure that genes are switched on only in those cell types in which their gene products are needed. The insulin gene has a number of cell-specific modules as insulin is only made in the β-cells of the pancreas, the gene being switched off in all other tissues. Other regulatory sequences mediate the expression of genes that are active at specific developmental stages.

A final group of regulatory sequences have a rather different function. These do not respond to signals from inside or outside of the cell but instead control the basal rate of transcription of the gene to which they are attached. The proteins that bind to these sequences therefore ensure that when the gene is switched on, transcription takes place at an appropriate rate. The human insulin gene has four sequences of this type, including copies of the **GC box** and **octamer sequence**, both of which are found upstream of many protein-coding genes in eukaryotes. Another common element, though not present in the insulin control region, is the **CAAT box**.

In addition to the regulatory sequences in the region immediately upstream of a gene, the same and other sequences are also contained in **enhancers**. These are longer DNA regions, 200–300 bp in length, that are located some distance away from their target genes (Figure 9.19). Often a single enhancer controls expression of more than one gene.

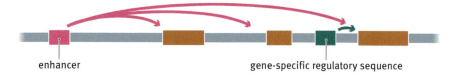

FIGURE 9.19 An enhancer is located some distance away from its target gene. The regulatory sequence shown in green is located upstream of a gene and influences transcription initiation only at that single gene. In contrast, the enhancer influences transcription of all three genes.

The RNA polymerase II initiation complex is activated via a mediator protein

How can a DNA-binding protein activate transcription of a gene? A physical contact of some kind must be made between the activator protein and the transcription initiation complex. The separation between a regulatory sequence and the transcription start site is not a problem because DNA is flexible and so can bend to enable contact to be made between a bound regulatory protein and the initiation complex. But this does not help us to understand how the regulatory protein influences the activity of the RNA polymerase. How this happens at RNA polymerase II promoters was a mystery for several years, with apparently conflicting evidence coming from work with different organisms. The solution to the problem was found when a large multisubunit protein called the **mediator** was identified.

The yeast mediator comprises 21 subunits that form a structure with distinct head, middle and tail domains. The tail forms a physical contact with an activator protein attached to its regulatory sequence, and the middle and head sections interact with the initiation complex (Figure 9.20). Therefore, rather than the activator protein associating directly with the initiation complex, the association is indirect with the activation signal being transmitted by the mediator.

Mediators in higher eukaryotes are larger than the yeast protein, with 30 or more subunits making up the human version. One feature of the mammalian mediator is that its subunit composition is variable, raising the possibility that there are several versions, each one responding to a different, although possibly overlapping, set of activators.

How does the mediator affect transcription? The initiation complex is unable to begin transcription until a set of phosphate groups have been added to the C-terminal region of the largest subunit of RNA polymerase II. Initially, it was thought that one of the subunits of the mediator acted as a **kinase** enzyme and was responsible for phosphorylating the polymerase, but more recent evidence has cast doubt on this hypothesis. The key observation was that yeast cells that lack the kinase subunit of the mediator still have normal levels of polymerase phosphorylation. Now, it seems more likely that the transcription factor TFIIH is responsible for adding the phosphates. The latest research suggests that the mediator is involved in the earlier stages of assembly of the initiation complex and influences both attachment of the TATA-binding protein, TBP, to the promoter and recruitment of the RNA polymerase into the complex.

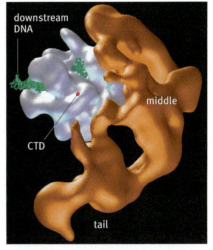

FIGURE 9.20 Interaction between the yeast mediator and the RNA polymerase II initiation complex. The mediator is shown in brown. Its middle and tail sections are visible, and its head component is on the other side of the structure. The middle and head parts of the mediator make contact with the initiation complex, shown in white, which in turn is attached to the DNA, shown in green. The position of the C-terminal domain (CTD) of the RNA polymerase is indicated. (From J. A. Davis et al., *Mol. Cell* 10:409–415, 2002. With permission from Elsevier.)

Signals from outside of the cell must be transmitted to the nucleus in order to influence gene expression

We now understand how regulatory sequences and the proteins that bind to them activate transcription by RNA polymerase II. The final question is how signals that originate from outside of the cell influence the attachment of a protein to its DNA-binding site. We must bear in mind that, in eukaryotes, transcription occurs in the nucleus. A signal from outside of the cell must therefore be transmitted from the cell surface, through the cytoplasm, and into the nucleus (Figure 9.21). There are several ways in which this can occur. The simplest method is for the signaling compound, which might be a hormone or growth factor, to pass through the cell membrane and enter the cell. This is what happens with the **steroid hormones**, which coordinate a range of physiological activities in the cells of higher eukaryotes. They include the sex hormones (estrogens

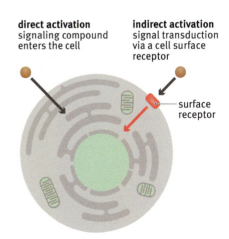

FIGURE 9.21 Two ways in which an extracellular signaling compound can influence events occurring within a cell.

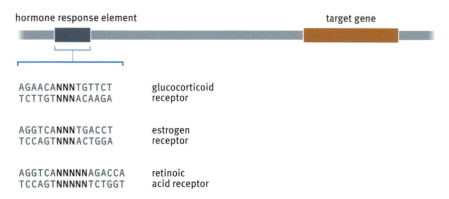

FIGURE 9.22 Hormone response elements. The sequences of three elements are shown. Each element has two parts, which often are exact repeats, possibly in opposite orientations. This is the case with the elements for the glucocorticoid and estrogen receptors. N, any nucleotide.

for female sex development, androgens for male sex development) and the glucocorticoid and mineralocorticoid hormones. Once inside the cell, each steroid hormone binds to its own **steroid receptor** protein, which is usually located in the cytoplasm. After binding, the activated receptor migrates into the nucleus, where it attaches to a **hormone response element** upstream of a target gene (Figure 9.22). Response elements for each receptor are located upstream of 50–100 genes in humans, often within enhancers, so a single steroid hormone can induce a large-scale change in the biochemical properties of the cell.

Steroid hormones are hydrophobic molecules and so can easily penetrate the cell membrane. This is not the case for many other signaling compounds. Instead, these must bind to a **cell surface receptor**, a protein that spans the cell membrane and is able to transmit the signal into the cell. Attachment of the signaling compound to the part of the receptor on the outside of cell results in a conformational change. Often this change is dimerization of the receptor, two subunits combining to form a single structure. This is possible because the liquid nature of the cell membrane allows a limited amount of lateral movement by membrane proteins, enabling the two subunits of a receptor to associate and disassociate in response to the presence or absence of the extracellular signal (Figure 9.23).

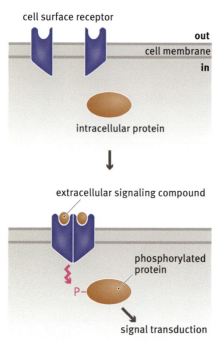

FIGURE 9.23 The role of a cell surface receptor. Binding of the extracellular signaling compound to the outer surface of the receptor protein causes a conformational change, often involving dimerization, that results in activation of an intracellular protein, for example, by phosphorylation. 'P' indicates a phosphate group, PO_3^{2-}.

The change in the structure of the receptor induces a biochemical event within the cell, such as attachment of phosphate groups to a cytoplasmic protein. This protein might be a transcription activator which, once phosphorylated, moves into the nucleus and binds to its regulatory sequences. Many cytokines – signaling proteins that control cell growth and division – control gene expression in this way. Their cell surface receptors respond to cytokine binding by phosphorylating transcription activators called **STATs**, which then move to the nucleus and switch on their target genes (Figure 9.24). Many genes can be activated by STATs but the overall response is modulated by other proteins that interact with different STATs and influence which genes are switched on under a particular set of circumstances. Complexity is entirely expected because the cellular processes that STATs mediate include growth and division, which are themselves complex. We therefore anticipate that changes in these processes will require extensive remodeling of the proteome and hence large-scale alterations in gene activity.

The direct phosphorylation of a transcription activator by a cell surface receptor is a straightforward but rather uncommon way of transmitting a signal from the cell surface to nucleus. With the more prevalent forms of **signal transduction**, the cell surface receptor is just the first component of a more complicated multistep pathway. An example is provided by the **MAP** (mitogen-activated protein) **kinase system** (Figure 9.25). This

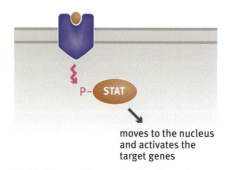

FIGURE 9.24 The mode of action of a STAT. A STAT is a transcription activator which, once phosphorylated, moves to the nucleus in order to switch on its target genes.

pathway responds to many extracellular signals, including mitogens, compounds with similar effects to cytokines but which specifically stimulate cell division. Binding of the signaling compound results in dimerization of the mitogen receptor, which activates an internal signaling protein called Raf. This protein initiates a cascade of phosphorylation reactions. It phosphorylates Mek, which, in turn, phosphorylates the MAP kinase. The activated MAP kinase now moves into the nucleus where it switches on, again by phosphorylation, a series of transcription activators. As with STATs, exactly which genes are activated by this pathway depends on interactions with other proteins that transduce signals from other extracellular signaling compounds.

9.4 OTHER STRATEGIES FOR REGULATING GENE EXPRESSION

We have looked closely at how gene expression is controlled by DNA-binding proteins that attach to regulatory sequences and repress or activate initiation of transcription. We have focused on this aspect of gene regulation because geneticists believe that initiation of transcription is the key step that determines if a gene is switched on or off. But we must not be misled into thinking that this is the only important type of gene regulation. As discussed above, virtually every step in the gene expression pathway is subject to some form of regulation (see Figure 9.4). These other steps might not be primary control points that switch genes on and off, but they have equally important influences on the rate of gene product synthesis.

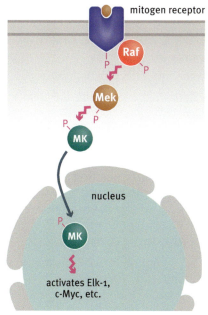

FIGURE 9.25 The MAP kinase signal transduction system. The signaling compound causes the mitogen receptor to dimerize, which results in phosphorylation of the internal protein called Raf. Activation of Raf initiates a cascade that leads via Mek to the MAP kinase. The activated MAP kinase moves into the nucleus, where it switches on various transcription activators, such as Elk-1 and c-Myc. MK, MAP kinase.

The problem for the student of genetics attempting to understand gene regulation for the first time is that a broad range of control strategies have evolved in different organisms. It is easy to become swamped in a mass of detail. To avoid this difficulty, we will not attempt a comprehensive survey of all of the control mechanisms that are known. Instead, we will study just three examples that together illustrate the most important of these other ways in which gene expression can be regulated.

Modification of the bacterial RNA polymerase enables different sets of genes to be expressed

The first of these other types of gene regulation also involves transcription initiation, but the control strategy is quite different from any that we have looked at so far. It is known only in bacteria, and it enables sets of genes, not just those present in a single operon, to be switched on or off at the same time. It is used by some bacteria as a means of responding rapidly to emergencies such as heat shock and by others as the basis of cellular differentiation.

The key to this type of control is the σ subunit of the bacterial RNA polymerase. Although there is just one type of bacterial RNA polymerase, some species possess more than one version of the σ subunit and so can assemble a series of RNA polymerase enzymes with slightly different properties. Because the σ subunit is the part of the polymerase that has the sequence-specific DNA-binding capability, polymerases with different σ subunits will recognize different promoter sequences.

In *E. coli*, the standard σ subunit, which recognizes the consensus promoter sequence shown in Figure 4.12, is called σ^{70} (its molecular mass is approximately 70 kDa). *E. coli* also has a second σ subunit, σ^{32}, which is made when the bacterium is exposed to a heat shock. This subunit recognizes a different promoter sequence (Figure 9.26), which is found upstream of genes coding for special proteins that help the bacterium withstand heat stress. The bacterium is therefore able to switch on a whole range of new genes by making one simple alteration to the structure of its RNA polymerase.

Alternative σ subunits are also known in other bacteria. For example, *Klebsiella pneumoniae* uses the system to control expression of genes involved in nitrogen fixation, this time with the σ^{54} subunit. The most sophisticated example of this control strategy is provided by *Bacillus* species, which use a whole range of different σ subunits to switch on and off groups of genes during the changeover from normal growth to formation of spores. *Bacillus* is one of several genera of bacteria that produce spores in response to unfavorable environmental conditions (Figure 9.27). These spores are highly resistant to physical and chemical abuse and can survive for decades or even centuries.

The standard *Bacillus subtilis* σ subunits are called σ^A and σ^H. These subunits are synthesized in nonsporulating cells and enable the RNA polymerase to recognize promoters for all the genes it needs to transcribe in order to maintain normal growth and cell division. When sporulation begins, the cell divides into two compartments, one of which will become the spore and the second of which will become the mother cell that eventually dies when the spore is released (Figure 9.28). The two compartments must therefore follow separate differentiation pathways. This is brought about by replacement of the standard σ subunits by two new subunits, σ^F in the prespore and σ^E in the mother cell. Each of these recognizes its own promoter sequence, which are attached to genes specific for prespore or mother cell development. Later in the sporulation process, another change occurs when the σ^F and σ^E subunits are replaced by σ^G and σ^K, respectively. This second change switches on the genes needed in the later stages of spore and mother cell formation. The succession of σ subunits therefore brings about the time-dependent changes in gene activity that underlie differentiation of the original bacterium into a spore.

Transcription termination signals are sometimes ignored

Now, we move further into the gene expression pathway and look at regulatory events occurring after the initiation of transcription. Two mechanisms for controlling the synthesis of RNA after transcription has been initiated are known in bacteria.

The first of these is called **antitermination**. This occurs when the RNA polymerase ignores a termination signal and continues elongating its transcript until a second signal is reached. Antitermination provides a mechanism whereby one or more of the genes at the end of an operon can be switched off or on by the polymerase recognizing or not recognizing a termination signal located upstream of those genes.

(a) an *E. coli* heat shock gene

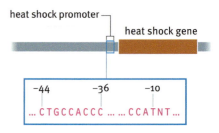

(b) recognition by the σ^{32} subunit

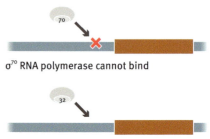

σ^{70} RNA polymerase cannot bind

σ^{32} RNA polymerase binds to the heat shock promoter

FIGURE 9.26 Recognition of an *E. coli* heat shock gene by the σ^{32} subunit. (a) The sequence of the heat shock promoter is different from that of the normal *E. coli* promoter. (b) The heat shock promoter is not recognized by the normal *E. coli* RNA polymerase containing the σ^{70} subunit, but is recognized by the σ^{32} RNA polymerase that is active during heat shock. N, any nucleotide.

green endospores inside bacterial cells

FIGURE 9.27 Endospores. (Courtesy of CNX OpenStax, published under CC BY 4.0 license.)

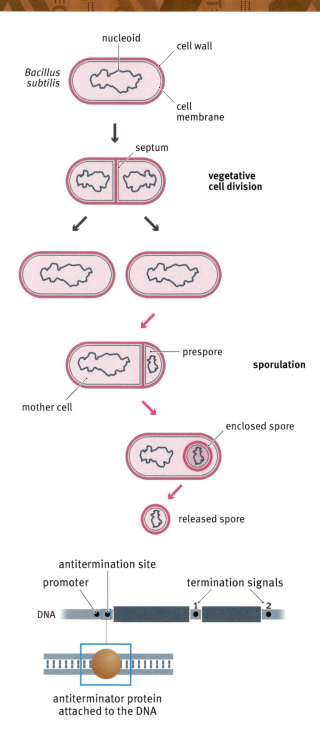

FIGURE 9.28 Sporulation in *Bacillus subtilis*. The top part of the diagram shows the normal vegetative mode of cell division, involving formation of a septum across the center of the bacterium and resulting in two identical daughter cells. The lower part of the diagram shows sporulation, in which the septum forms near one end of the cell, leading to a mother cell and prespore of different sizes. Eventually, the mother cell completely engulfs the prespore. At the end of the process, the mature resistant spore is released.

FIGURE 9.29 Antitermination. The antiterminator protein attaches to the DNA at the antitermination site and transfers to the RNA polymerase as it moves past, subsequently enabling the polymerase to continue transcription through termination signal number 1, so the second of the pair of genes in this operon is transcribed.

Antitermination is controlled by an **antiterminator protein**, which attaches to the DNA near the beginning of the operon and then transfers to the RNA polymerase as it moves past on its way to the first termination signal (Figure 9.29). The presence of the antiterminator protein causes the enzyme to ignore the termination signal, so transcription continues into the downstream region. Exactly how this works is not known, but presumably the antiterminator protein is able to stabilize an intrinsic terminator and/or prevent stalling of the polymerase at a Rho-dependent one.

Although the mechanics of antitermination are unclear, the impact that it can have on gene expression has been described in detail. It is especially important during the infection cycle of bacteriophage λ. This is one of the various viruses that infect *E. coli*. As bacteriophages have relatively few genes – λ has just 48 – they were used in the early days of molecular

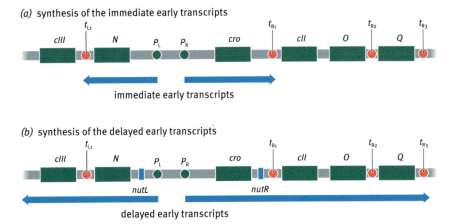

FIGURE 9.30 **Antitermination during the infection cycle of bacteriophage λ.** (a) Transcription from the promoters p_L and p_R initially results in synthesis of two 'immediate early' mRNAs, these terminating at positions t_{L1} and t_{R1}. (b) The mRNA transcribed from p_L to t_{L1} codes for the N protein, which attaches to the antitermination sites *nutL* and *nutR*. Now the RNA polymerase continues transcription downstream of t_{L1} and t_{R1}. Transcription from p_R also ignores terminator t_{R2} and continues until t_{R3} is reached.

genetics as model systems with which to study the basic principles of gene expression. Studies of bacteriophage λ were particularly instructive with regard to gene regulation.

Successful progress of λ through its infection cycle requires that different sets of bacteriophage genes be switched on and off in the correct order. Immediately after entering an *E. coli* cell, transcription of the λ DNA is initiated by the bacterial RNA polymerase attaching to two promoters, p_L and p_R. Transcription from these promoters results in the synthesis of two 'immediate early' mRNAs, these terminating at positions t_{L1} and t_{R1} (Figure 9.30a). The mRNA transcribed from p_R to t_{R2} codes for the N protein, which is an antiterminator. The N protein attaches to the λ DNA at sites *nutL* and *nutR* and transfers to the RNA polymerase as it passes. Now the RNA polymerase ignores the t_{L1} and t_{R1} terminators and continues transcription downstream of these points. The resulting mRNAs encode the 'delayed early' proteins (Figure 9.30b). Antitermination controlled by the N protein therefore ensures that the immediate early and delayed early proteins are synthesized in the correct order during the λ infection cycle. One of the delayed early proteins, Q, is a second antiterminator that controls the switch to the later stages of the infection cycle.

Attenuation is a second control strategy targeting transcription termination

The second regulatory process used by bacteria to control termination of transcription is called **attenuation**. This occurs primarily with operons that code for enzymes involved in amino acid biosynthesis, although a few other examples are also known.

The basis for attenuation is the coupling between transcription and translation that occurs in bacterial cells, ribosomes attaching to and beginning to translate an mRNA while that molecule is still being synthesized by an RNA polymerase (Figure 9.31). The tryptophan operon of *E. coli* illustrates

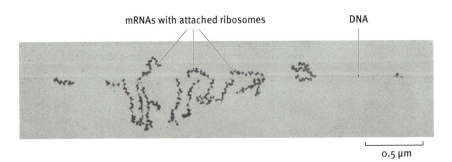

FIGURE 9.31 **In bacteria, transcription and translation are often coupled.** This electron micrograph shows a part of the *E. coli* DNA molecule. Several mRNAs are being transcribed from the DNA. Each of these mRNAs has ribosomes attached to it – seen as small dark dots. The mRNAs are therefore being translated even though they have not yet been completely transcribed. (From O. L. Miller, B. A. Hamkalo & C. A. Thomas Jr, *Science* 169: 392–395, 1970. With permission from AAAS.)

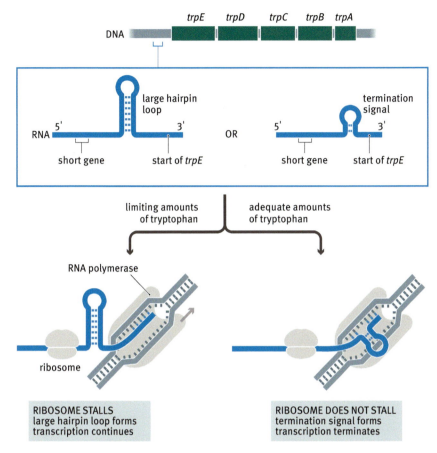

FIGURE 9.32 Attenuation at the tryptophan operon. The two hairpins that can form in the region between the start of the transcript and the beginning of *trpE* are shown in the box. The smaller of these hairpins acts as a termination signal. The lower part of the diagram shows the different events that occur in the absence or presence of tryptophan. If the amount of tryptophan is limiting, then the ribosome stalls when it reaches the short gene. This enables the polymerase to move ahead so the larger hairpin forms and transcription continues. If the amount of tryptophan is adequate, then the ribosome does not stall but instead keeps up with the polymerase. Now the termination signal forms.

how attenuation works. In this operon, two hairpins can form in the region between the start of the transcript and the beginning of *trpE*, the first of the five genes in the operon (Figure 9.32). The smaller of these hairpins acts as a termination signal, but the larger one, which is closer to the start of the transcript, is more stable. The larger structure overlaps with the termination hairpin, so only one can form at any one time.

Which hairpin forms depends on the relative positioning between the RNA polymerase and a ribosome which attaches to the 5′ end of the transcript as soon as it is synthesized. If the ribosome is unable to keep up with the polymerase, then the larger hairpin forms and transcription continues. However, if the ribosome keeps pace with the RNA polymerase, then it disrupts the larger hairpin by attaching to the RNA that makes up part of the stem of this structure. When this happens, the termination hairpin is able to form and transcription stops.

The critical issue is therefore whether or not the ribosome keeps pace with the polymerase. It could fall behind because just upstream of the termination signal is a short gene. This gene codes for a very small protein, comprising just 14 amino acids, two of which are tryptophan. If the amount of tryptophan available to the bacterium is low, then the ribosome is held up while translating this gene, because it has to wait for the two molecules of tryptophan that it needs. The ribosome therefore falls behind, enabling the polymerase to move ahead so the larger hairpin forms and transcription continues. Because the resulting transcript contains copies of the genes coding for the biosynthesis of tryptophan, its continued elongation addresses the deficiency of this amino acid in the cell. When the amount of tryptophan in the cell reaches a satisfactory level, the attenuation system prevents further transcription of the tryptophan operon, because now the ribosome does not fall behind while making the small

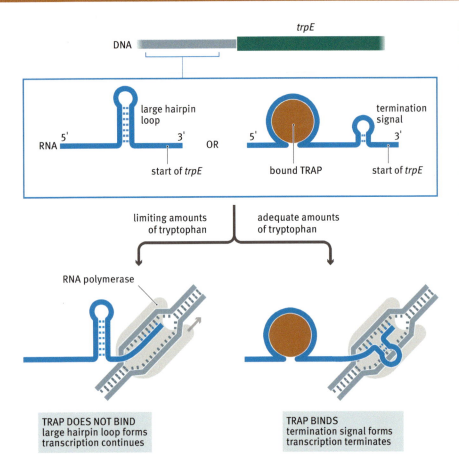

FIGURE 9.33 Regulation of the tryptophan operon of *Bacillus subtilis*. The box shows the two hairpins that can form in the region between the start of the transcript and the beginning of *trpE*. The smaller of the hairpins, which acts as a termination signal, can form only if the protein called TRAP attaches to the leader region of the transcript. The lower part of the diagram shows the different events that occur in the absence or presence of tryptophan. When the amount of tryptophan is limiting, TRAP does not bind so the larger hairpin forms and transcription continues. When the amount of tryptophan is adequate, TRAP attaches to the mRNA, so that the termination signal forms.

protein. Instead, it keeps pace with the polymerase, allowing the termination signal to form.

The *E. coli* tryptophan operon is controlled not only by attenuation but also by a repressor. It is thought that repression provides the basic on–off switch and attenuation modulates the precise level of gene expression that occurs. Other *E. coli* operons, such as those for biosynthesis of histidine, leucine and threonine, are controlled solely by attenuation.

Interestingly, in some bacteria, including *Bacillus subtilis*, the tryptophan operon is one of those that does not have a repressor system and so is regulated entirely by attenuation. In these bacteria, attenuation is mediated not by the speed at which the ribosome tracks along the mRNA, but by an RNA-binding protein called **trp RNA-binding attenuation protein** (**TRAP**). In the presence of tryptophan, this protein attaches to the mRNA in the region equivalent to the short gene of the *E. coli* transcript (Figure 9.33). Attachment of TRAP leads to formation of the termination signal and cessation of transcription.

Bacteria and eukaryotes are both able to regulate the initiation of translation

We are now well aware that the initiation of transcription is an important control point at which expression of individual genes can be regulated. The same is true, though to a lesser extent, for initiation of translation.

The best-understood example involves the operons for the ribosomal protein genes of *E. coli* (Figure 9.34). The leader region of the mRNA transcribed from each operon contains a sequence that acts as a binding site for one of the proteins coded by the operon. When this protein is synthesized, it can either attach to its position on the ribosomal RNA or

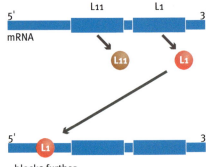

FIGURE 9.34 Regulation of ribosomal protein synthesis in bacteria. The L11 operon of *Escherichia coli* is transcribed into an mRNA carrying copies of the genes for the L11 and L1 ribosomal proteins. When the L1 binding sites on the available 23S rRNA molecules have been filled, L1 binds to the leader region of the mRNA, blocking further initiation of translation.

bind to the leader region of the mRNA. The rRNA attachment is favored and occurs if there are free rRNAs in the cell. Once all the free rRNAs have been assembled into ribosomes, the ribosomal protein binds to its mRNA, blocking translation initiation and hence switching off further synthesis of the ribosomal proteins coded by that particular mRNA. Similar events involving other mRNAs ensure that synthesis of each ribosomal protein is coordinated with the amount of free rRNA in the cell.

Eukaryotes also have mechanisms for control of translation initiation. In mammals, an interesting example is provided by the mRNA for ferritin, an iron-storage protein. In the absence of iron, ferritin is not needed. Its synthesis is inhibited by proteins that bind to sequences called **iron-response elements** located in the leader region of the ferritin mRNA (Figure 9.35). The bound proteins block the ribosome as it attempts to scan along the mRNA in search of the initiation codon. When iron is present, the binding proteins detach and the mRNA is translated. The mRNA for a related protein – the transferrin receptor involved in the uptake of iron by the cell – also has iron-response elements. In this case, detachment of the binding proteins in the presence of iron results not in translation of the mRNA but in its degradation. This is logical because when iron is present in the cell, there is less requirement for transferrin receptor activity because there is less need to import iron from the outside.

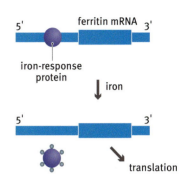

FIGURE 9.35 Regulation of ferritin protein synthesis in mammals. In the absence of iron, the iron-response protein attaches to the ferritin mRNA and prevents its translation. When iron is present, it binds to the iron-response protein, which detaches from the mRNA, enabling ferritin to be synthesized.

Initiation of translation of some bacterial mRNAs can also be regulated by short RNAs that attach to recognition sequences within the mRNAs. This RNA-mediated mechanism can result in both activation and inhibition of translation of the target mRNAs. One such regulatory RNA in *E. coli* is called OxyS, which is 109 nucleotides in length and regulates translation of 40 or so mRNAs. Synthesis of the OxyS RNA is activated by hydrogen peroxide and other reactive oxygen compounds, which can cause oxidative damage to the cell. Once synthesized, OxyS switches on translation of some mRNAs whose products help protect the bacterium from oxidative damage and switches off translation of other mRNAs whose products would be deleterious under these circumstances. It is not yet clear how the same RNA can have this differential effect, activating translation of some mRNAs and inhibiting translation of others. Inhibition might occur because the regulatory RNA binds to its target mRNA and blocks access to the ribosome binding site. Activation of translation probably involves a more subtle mechanism. In some cases, the target mRNA is able to form a hairpin structure, with the ribosome binding site present in the stem region and hence inaccessible to the components of the initiation complex. Attachment of the regulatory RNA disrupts the hairpin, exposing the ribosome binding site so that translation initiation can now occur.

KEY CONCEPTS

- Some genes are expressed all the time, but all organisms are able to regulate expression of at least some of their genes. Gene regulation enables bacteria to respond to changes in their environment and for cells in multicellular organisms to respond to signals from other cells. Gene regulation also underlies differentiation and development.

- Control of gene expression is control over the amount of gene product present in the cell. This amount is a balance between the rate of synthesis (how many molecules of the gene product are made per unit time) and the degradation rate (how many molecules are broken down per unit time).

- The rate of synthesis of a gene product can be regulated by exerting control over any one, or a combination, of the various steps in the gene expression pathway.
- In all organisms, the key events that determine if a gene is on or off take place during the initiation of transcription. Later events in the gene expression pathway are able to change the rate of expression of a gene that is switched on and, in eukaryotes, control the synthesis of alternative products of a single gene.
- The lactose operon of *E. coli* illustrates the basic principles of control over initiation of transcription. Initiation of transcription is influenced by DNA-binding proteins, such as the lactose repressor and the catabolite activator protein, which bind to specific sequences located near the attachment site for the RNA polymerase and influence the ability of the RNA polymerase to access the promoter and initiate RNA synthesis.
- Regulatory sequences and proteins that bind to those sequences also underlie control of transcription initiation in eukaryotes. One difference is that the basal rate of transcription by a eukaryotic RNA polymerase is low, so most eukaryotic regulatory proteins are activators of transcription.
- In eukaryotes, signals from outside of the cell must be transmitted from the cell surface, through the cytoplasm, and into the nucleus in order to influence gene expression.
- Later steps in the gene expression pathway, subsequent to initiation of transcription, are also subject to control. Examples are termination of transcription and initiation of translation.

QUESTIONS AND PROBLEMS

Key terms

Write short definitions of the following terms:

adenylate cyclase
antitermination
antiterminator protein
attenuation
CAAT box
catabolite activator protein
cell surface receptor
co-repressor
cyclic AMP
diauxie
enhancer
GC box
heat shock module
hormone response element
housekeeping genes
inducer
inducible operon
iron-response element
kinase
MAP kinase system
mediator
octamer sequence
operator
operon
regulatory gene
repressible operon
repressor
signal transduction
STAT
steroid hormone
steroid receptor
structural gene
trp RNA-binding attenuation protein

Self-study questions

9.1 Using examples, explain why gene regulation is important in bacteria and eukaryotes.

9.2 To what extent are the underlying principles of gene regulation the same in all organisms?

9.3 Distinguish between the terms 'structural' and 'regulatory' genes, using lactose utilization by *E. coli* as an example.

9.4 Describe how the operator and repressor interact to control expression of the lactose operon. Why is this called an inducible operon?

9.5 In *E. coli*, the galactose operon comprises three structural genes, called *galK*, *galT* and *galE*. The proteins coded by these genes are involved in metabolism of the sugar galactose. There is also a regulatory gene, *galR*, which codes for a repressor protein that binds to an operator upstream of the *gal* operon. Galactose acts as an inducer. Describe the events that you would expect to occur (a) in the absence of galactose and (b) in the presence of galactose.

9.6 A mutation is an alteration in nucleotide sequence that may alter the function of a gene, promoter or regulatory site. Deduce and explain the effects of each of the following mutations on expression of the *lac* operon:

 a. A mutation in the operator, so that the repressor is no longer able to bind

 b. A mutation in *lacI*, so that the repressor is no longer synthesized

 c. A mutation in the promoter, so that RNA polymerase no longer binds

 d. A mutation in *lacI*, so that the repressor no longer binds allolactose

 e. A mutation in *lacY* that causes transcription to terminate in the middle of the gene

 f. A mutation, location unknown, that prevents the *lac* mRNA from being degraded

9.7 What is meant by the term 'diauxie'?

9.8 Describe how glucose influences expression of the lactose operon.

9.9 Distinguish between an inducible and a repressible operon.

9.10 Give examples of regulatory modules found upstream of genes in eukaryotes.

9.11 What are the special features of an enhancer?

9.12 Outline the process by which steroid hormones regulate gene activity in higher eukaryotes.

9.13 Outline the role that STATs play in regulation of gene expression.

9.14 What is meant by signal transduction? Give one example of a signal transduction pathway.

9.15 Describe the differences between the yeast and mammalian mediators, and explain how the mediator is thought to influence the initiation of transcription.

9.16 Give two examples to illustrate the regulation of bacterial gene expression by alternative σ subunits.

9.17 Describe the process called antitermination and explain how it is involved in expression of genes in bacteriophage λ.

9.18 What is attenuation? Your answer should distinguish between the processes occurring in *E. coli* and in *B. subtilis*.

9.19 Giving examples, describe how the initiation of translation is controlled in bacteria and eukaryotes.

Discussion topics

9.20 To what extent are the underlying principles of gene regulation *not* the same in all organisms?

9.21 Why is allolactose, rather than lactose, the inducer of the lactose operon?

9.22 To what extent is *E. coli* a good model for the regulation of transcription initiation in eukaryotes? Justify your opinion by providing specific examples of how extrapolations from *E. coli* have been helpful and/or unhelpful in the development of our understanding of equivalent events in eukaryotes.

9.23 Operons are very convenient systems for achieving coordinated regulation of expression of related genes. Operons are common in bacteria, yet they are absent in eukaryotes. Discuss.

9.24 The tryptophan operon of *E. coli* is regulated both by a repressor and by attenuation. Other operons coding for amino acid biosynthetic enzymes are controlled only by attenuation. Discuss.

The answers to, or hints on how to answer, the self-study questions and discussion topic questions can be downloaded from the book's product page at this link: www.routledge.com/9781032743530

FURTHER READING

Cargnello M & Roux PP (2011) Activation and function of the MAPKs and their substrates, the MAPK-activated protein kinases. *Microbiology and Molecular Biology Reviews* 75, 50–83. *MAP kinase systems.*

Henkin TM (1996) Control of transcription termination in prokaryotes. *Annual Review of Genetics* 30, 35–57. *A detailed account of antitermination and attenuation.*

Hershey JWB, Sonenberg N & Mathews MB (2012) Principles of translational control: an overview. *Cold Spring Harbor Perspectives in Biology* 4, a011528. *Details of the ways in which translation can act as a control point in eukaryotic gene expression.*

Horvath CM (2000) STAT proteins and transcriptional responses to extracellular signals. *Trends in Biochemical Sciences* 25, 496–502.

Jacob F & Monod J (1961) Genetic regulatory mechanisms in the synthesis of proteins. *Journal of Molecular Biology* 3, 318–389. *The original proposal of the operon theory for control of bacterial gene expression.*

Kadonaga JT (2004) Regulation of RNA polymerase II transcription by sequence-specific DNA binding factors. *Cell* 116, 247–257.

Karin M & Hunter T (1995) Transcriptional control by protein phosphorylation: signal transmission from the cell surface to the nucleus. *Current Biology* 5, 747–757.

Koo BM, Rhodius VA, Campbell EA et al. (2009) Dissection of recognition determinants of *Escherichia coli* σ32 suggest a composite −10 region with an 'extended −10' motif and a core −10 element. *Molecular Microbiology* 72, 815–829. *The E. coli heat shock promoter.*

Losick RL & Sonenshein AL (2001) Turning gene regulation on its head. *Science* 293, 2018–2019. *Describes the attenuation systems at the tryptophan operons of E. coli and B. subtilis.*

Naville M & Gautheret D (2009) Transcription attenuation in bacteria: themes and variations. *Briefings in Functional Genomics and Proteomics* 8, 482–492.

Richter WF, Nayak S, Iwasa J et al. (2022) The Mediator complex as a master regulator of transcription by RNA polymerase II. *Nature Reviews Molecular Cell Biology* 23, 732–749.

Riley EP, Schwarz C, Derman AI et al. (2021) Milestones in *Bacillus subtilis* sporulation research. *Microbial Cell* 8, 1–16. *Describes how different σ subunits are involved in sporulation.*

Rojo F (1999) Repression of transcription initiation in bacteria. *Journal of Bacteriology* 181, 2987–2991.

Tsai M-J & O'Malley BW (1994) Molecular mechanisms of action of steroid/thyroid receptor superfamily members. *Annual Review of Biochemistry* 63, 451–486. *Control of gene expression by steroid hormones.*

Zubay G, Schwartz D & Beckwith J (1970) Mechanisms of activation of catabolite-sensitive genes: a positive control system. *Proceedings of the National Academy of Sciences USA* 66, 104–110. *An early description of the CAP system.*

CHAPTER 10

DNA Replication

We have learnt how the process of gene expression converts the biological information contained in an organism's genes into a set of RNA and protein molecules that specify the biochemical capabilities of the cell in which the genes reside. In order to continue performing this function, copies of the genes must be made every time that the cell divides. This means that the entire DNA content of the cell must be replicated. In this chapter, we will examine how this is carried out. First, we will look at the overall pattern of replication and ask how a single DNA double helix gives rise to two identical daughter helices. We know that complementary base pairing enables polynucleotides to be copied but there are several possible ways of organizing the replication process, and we must discover which of these is correct. Then, we must study the biochemistry and enzymology of DNA replication. What proteins are involved and how are the new polynucleotides synthesized?

10.1 THE OVERALL PATTERN OF DNA REPLICATION

One of the reasons why the discovery of the double helix structure was an important breakthrough in genetics is because it immediately suggested a way in which DNA molecules could be copied, and hence solved the great mystery of how genes are able to replicate. Complementary base pairing is the key because this enables each strand of the double helix to act as a template for synthesis of its partner (Figure 10.1). But there are various ways in which template-dependent DNA synthesis could bring about replication of a double helix. We must therefore look at the possible replication processes and discover which actually operates in living cells.

DNA replicates semiconservatively, but this causes topological problems

There are three different ways in which complementary base pairing could be used to make a copy of a DNA double helix (Figure 10.2). The first of these is **semiconservative replication**, in which each daughter molecule contains one polynucleotide derived from the original molecule and one newly synthesized strand. The second possibility is **conservative replication**. In this process, one daughter molecule contains both parent polynucleotides and the other daughter contains both newly synthesized strands. The third option is **dispersive replication**, in which each strand of each daughter molecule is composed partly of the original polynucleotide and partly of the newly synthesized polynucleotide.

Semiconservative replication appears to be the most straightforward scheme and was the one favored by Watson and Crick when they discovered the double helix structure in 1953. The conservative model seems less likely because it would require that the two newly synthesized strands detach from their templates and reassociate with one another, which would appear to be a pointless extra step in the replication pathway (Figure 10.3a). Dispersive replication would require repeated template switching, which again would be a seemingly unnecessary elaboration (Figure 10.3b). Based on these considerations, it may seem surprising to learn that during the 1950s, after the double helix structure had been published, many geneticists thought that the semiconservative process was the *least* likely of the three possibilities.

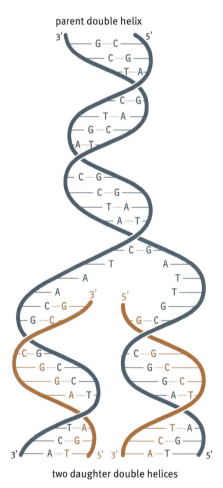

FIGURE 10.1 Complementary base pairing enables a double helix to be replicated. The polynucleotides of the parent double helix are shown in blue, and the newly synthesized strands are in orange.

DOI: 10.1201/9781003473862-11

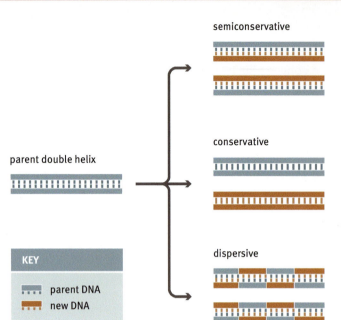

FIGURE 10.2 **Three possible schemes for DNA replication.**

What was the perceived problem with semiconservative replication? The answer lies with the structure of the double helix. The double helix is **plectonemic**, meaning that the two strands cannot be separated without unwinding. Therefore, during replication, it would be necessary for the parent double helix to rotate so that its strands can be unwound prior to each being copied. With one turn occurring for every 10 bp of the double helix, complete replication of the DNA molecule in human chromosome 1, which is 249 Mb in length, would require 25 million rotations of the chromosomal DNA. It is difficult to imagine how this could occur within the constrained volume of the nucleus, but the unwinding of a linear chromosomal DNA molecule is not physically impossible. In contrast, a circular double-stranded molecule, such as those found inside bacteria, would not be able to rotate in the required manner and so, apparently, could not be replicated by the semiconservative process.

The debate continued until 1958, when an experiment was carried out that distinguished between the three modes of replication and confirmed that semiconservative replication actually operates. This experiment, carried out at the California Institute of Technology by Matthew Meselson and Franklin Stahl, is the starting point for our study of DNA replication.

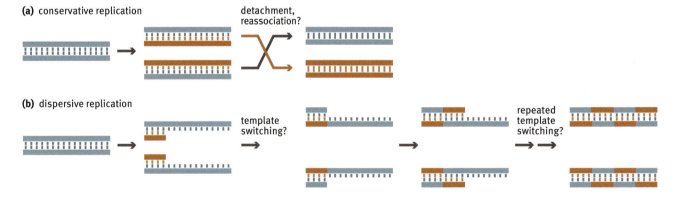

FIGURE 10.3 **Complications with conservative and dispersive replication.** (a) Conservative replication requires that the two newly synthesized strands detach from their templates and reassociate with one another. (b) Dispersive replication requires repeated template switching.

The Meselson–Stahl experiment proved that replication is semiconservative

Like many advances in biology, the Meselson–Stahl experiment depended on the use of chemical isotopes, of nitrogen in this case. Nitrogen exists in several isotopic forms. In addition to ^{14}N, which predominates in the environment and has an atomic weight of 14.008, there are a number of other isotopes that occur in much smaller amounts. These include ^{15}N, which because of its greater atomic weight is called 'heavy nitrogen'. If *Escherichia coli* cells are provided with heavy nitrogen in the form of $^{15}NH_4Cl$, then the bacteria incorporate the isotope into new DNA molecules that they synthesize.

DNA molecules containing only heavy nitrogen can be separated from DNA molecules containing only the normal isotope by density gradient centrifugation. We met this procedure in Chapter 6, where we learned how centrifugation through a sucrose density gradient is used to measure the sedimentation coefficients of ribosomes and other cell components. In a second type of density gradient centrifugation, a solution such as 6 M cesium chloride is used, which is substantially denser than the sucrose solution used to measure sedimentation coefficients. The starting CsCl solution is uniform, the gradient being established during centrifugation. Cellular components migrate all the way down through the centrifuge tube, but molecules such as DNA and proteins do not reach the bottom. Instead, each one comes to rest at a position where its own **buoyant density** equals the cesium chloride density. Meselson and Stahl used this version of density gradient centrifugation because ^{15}N-DNA has a greater buoyant density than ^{14}N-DNA and therefore forms a band at a lower position in the gradient (**Figure 10.4**).

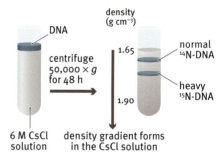

FIGURE 10.4 The use of density gradient centrifugation to separate ^{14}N- and ^{15}N-DNA molecules.

Meselson and Stahl grew a culture of *E. coli* in the presence of $^{15}NH_4Cl$ so that the DNA molecules in the cells became labeled with heavy nitrogen. The cells were then transferred to normal medium, containing $^{14}NH_4Cl$, and allowed to undergo one round of cell division. This takes roughly 20 minutes for *E. coli*, during which time each DNA molecule replicates just once. Some cells were then taken from the culture, their DNA purified, and a sample analyzed by density gradient centrifugation. The result was a single band of DNA at a position corresponding to a buoyant density intermediate between the values expected for ^{14}N-DNA and ^{15}N-DNA (**Figure 10.5**). This shows that after one round of replication each DNA double helix contained roughly equal amounts of ^{14}N-polynucleotide and ^{15}N-polynucleotide. The culture was then left for another 20 minutes so the bacteria could undergo a second round of cell division. Again, cells were removed and their DNA molecules were analyzed in a density gradient. Two bands appeared, one representing the same hybrid molecules as before and the second corresponding to double helices made entirely of ^{14}N-DNA.

How do we interpret the results of the Meselson–Stahl experiment? If we reexamine the three schemes for DNA replication, we see that the banding pattern obtained after a single round of cell division enables conservative replication to be discounted (**Figure 10.6**). This scheme predicts that after one round of replication there will be two different types of double helix, one containing just ^{14}N and the other containing just ^{15}N. As just one band is seen, we can rule out conservative replication.

To distinguish between semiconservative and dispersive replication, we have to look at the results obtained after a second round of cell division. Now the density gradient revealed two bands of DNA, the first corresponding to a hybrid composed of equal parts of newly synthesized and old

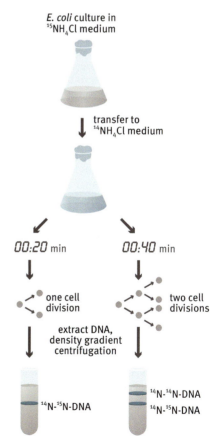

FIGURE 10.5 The results of the Meselson–Stahl experiment. On the left is the banding pattern seen in the density gradient after a single round of DNA replication, and on the right is the pattern after two rounds of replication.

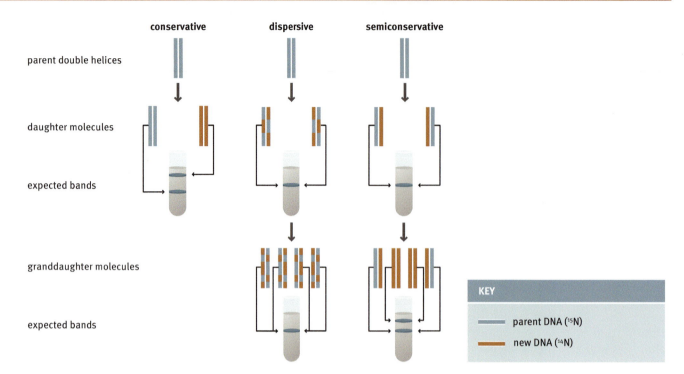

FIGURE 10.6 Interpretation of the Meselson–Stahl experiment. The banding patterns predicted by each of the three possible replication schemes are shown. The single ^{14}N-^{15}N-DNA band that was seen after one round of DNA replication allows conservative replication to be discounted, as this scheme would result in one band of ^{14}N-DNA and a second band of ^{15}N-DNA. The single hybrid band does not, however, allow dispersive and semiconservative replication to be distinguished. Under either of these schemes, the daughter molecules will be made up of equal amounts of ^{14}N- and ^{15}N-DNA. We therefore move on to the results after two rounds of DNA replication. The prediction for semiconservative replication is that now the ^{14}N-^{15}N-DNA band will be accompanied by a new band made up entirely of ^{14}N-DNA. This is the result that was obtained. It is incompatible with dispersive replication, as this mode predicts that after two rounds of replication all the molecules will still be hybrids.

DNA, and the second corresponding to molecules made up entirely of new DNA. This result is in agreement with the semiconservative mode of replication, because according to this scheme there are now some granddaughter molecules that are made up entirely of ^{14}N-polynucleotides. In contrast, the dispersive mode can be discounted because that method would still produce only hybrid molecules and, in fact, would continue to do so for a very large number of cell generations. We can therefore conclude that DNA replication is semiconservative.

DNA topoisomerases solve the topological problem

The Meselson–Stahl experiment showed that DNA replication is semiconservative, but did not provide a solution to the topological problem. Geneticists were still faced with the apparent impossibility of replicating a circular DNA molecule by this method. The solution to the topological problem was not found until the 1970s when the activities of the enzymes called **DNA topoisomerases** were characterized. DNA topoisomerases are enzymes that separate the two strands in a DNA molecule without actually rotating the double helix. They achieve this feat by causing transient breakages in the polynucleotide backbone.

There are two different types of DNA topoisomerase. **Type I topoisomerases** introduce a break in one polynucleotide and pass the second polynucleotide through the gap that is formed (Figure 10.7). The two ends

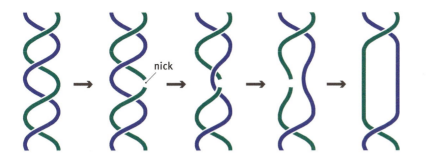

FIGURE 10.7 **The mode of action of a type I DNA topoisomerase.** A type I topoisomerase introduces a break in one polynucleotide and passes the second polynucleotide through the gap that is formed.

of the broken strand are then religated. This mode of action results in the linking number (the number of times one strand crosses the other in a circular molecule) being changed by one. **Type II topoisomerases**, on the other hand, break both strands of the double helix, creating a 'gate' through which a second segment of the helix is passed (Figure 10.8). This changes the linking number by two. Despite their different mechanisms, both type I and type II topoisomerases achieve the same end result. They enable the helix to be 'unzipped', with the two strands pulled apart sideways without the molecule having to rotate (Figure 10.9).

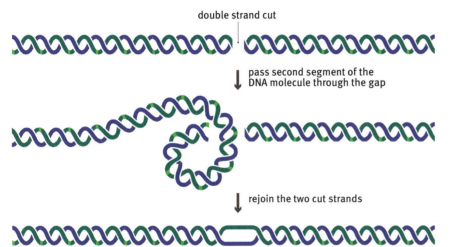

FIGURE 10.8 **The mode of action of a type II DNA topoisomerase.** A type II topoisomerase breaks both strands of the double helix, creating a gap through which a second segment of the helix is passed.

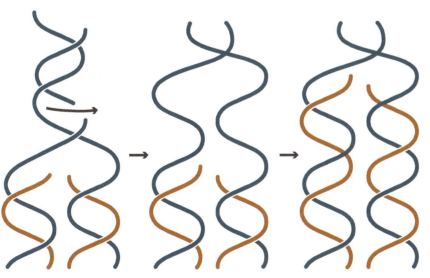

FIGURE 10.9 **Unzipping the double helix.** During replication, the double helix is 'unzipped' as a result of the action of DNA topoisomerases. The replication fork is therefore able to proceed along the molecule without the helix having to rotate.

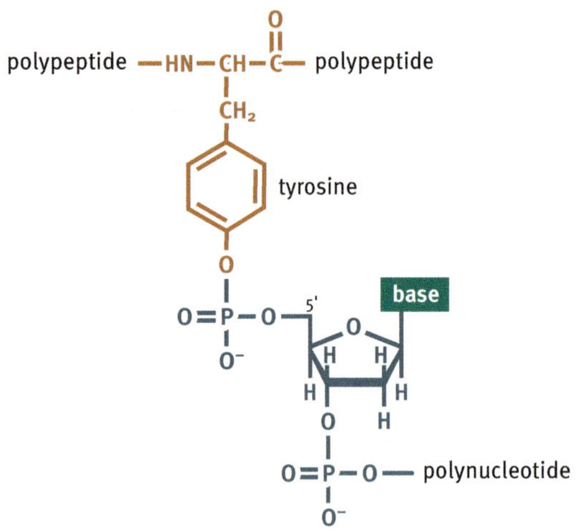

FIGURE 10.10 The linkage between a topoisomerase and a polynucleotide. The linkage is formed between a tyrosine present in the topoisomerase and a phosphate group at one of the ends of the cut polynucleotide. The diagram shows a type IA or IIA topoisomerase, which forms an attachment with the 5′ end of the polynucleotide. Type IB and IIB topoisomerases attach to the 3′ end of the cut polynucleotide.

Breaking one or both DNA strands might appear to be a drastic solution to the topological problem, leading to the possibility that the topoisomerase might occasionally fail to rejoin a strand and hence inadvertently break a DNA molecule into two sections. This possibility is reduced by the mode of action of these enzymes. One end of each cut polynucleotide becomes covalently attached to a tyrosine amino acid at the active site of the enzyme, ensuring that these ends are held tightly in place while the free ends are being manipulated. Type I and II topoisomerases are subdivided according to the precise chemical structure of the polynucleotide-tyrosine linkage. With IA and IIA enzymes, the link involves a phosphate group attached to the free 5′ end of the cut polynucleotide, and with IB and IIB enzymes the linkage is via a 3′ phosphate group (Figure 10.10). The A and B topoisomerases probably evolved separately. Both types are present in eukaryotes but IB and IIB enzymes are very uncommon in bacteria.

Replication is not the only activity that is complicated by the topology of the double helix, and it is becoming increasingly clear that DNA topoisomerases have equally important roles during transcription, and other processes that can result in over- or underwinding of DNA. Most topoisomerases are only able to relax DNA, but some bacterial type II enzymes can carry out the reverse reaction and introduce extra turns into DNA molecules. An example is the enzyme called **DNA gyrase**.

Variations on the semiconservative theme

No exceptions to the semiconservative mode of DNA replication are known but there are several variations on this basic theme. DNA copying via a **replication fork**, as shown in Figure 10.1, is the predominant system, being used by chromosomal DNA molecules in eukaryotes and by the circular DNA molecules of bacteria.

Some smaller circular molecules, such as those present in animal mitochondria, use a slightly different process called **displacement replication**. In these molecules, the point at which replication begins is marked by a **D-loop**, a region of approximately 500 bp where the double helix is disrupted by an RNA molecule that is base paired to one of the DNA strands (Figure 10.11). This RNA molecule acts as the starting point for synthesis of one of the daughter polynucleotides. This polynucleotide is synthesized by continuous copying of one strand of the helix, the second strand being displaced and subsequently copied after synthesis of the first daughter molecule has been completed.

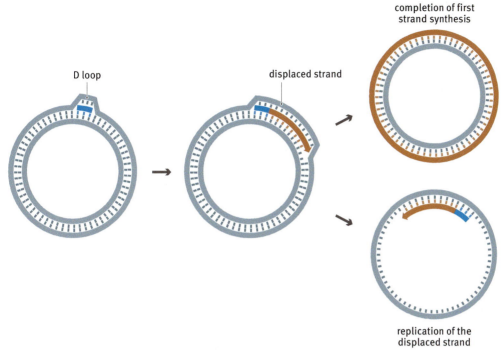

FIGURE 10.11 Displacement replication. The D loop contains a short RNA molecule (shown in blue) that acts as the starting point for synthesis of the first daughter polynucleotide. This polynucleotide is synthesized by continuous copying of one strand of the helix to give the double-stranded molecule at the top right. The displaced strand (bottom right) is then copied by attachment of a second RNA molecule, which acts as the starting point for synthesis of the second daughter polynucleotide.

The advantage of displacement replication as performed by animal mitochondrial DNA is not clear. In contrast, the special type of displacement process called **rolling-circle replication** is an efficient mechanism for the rapid synthesis of multiple copies of a circular DNA molecule. Rolling-circle replication, which is used by λ and various other bacteriophages, initiates at a nick which is made in one of the parent polynucleotides. The free 3′ end that results is extended, displacing the 5′ end of the polynucleotide. Continued DNA synthesis 'rolls off' a complete copy of the molecule, and further synthesis eventually results in a series of identical molecules linked head to tail (Figure 10.12). These molecules are single stranded

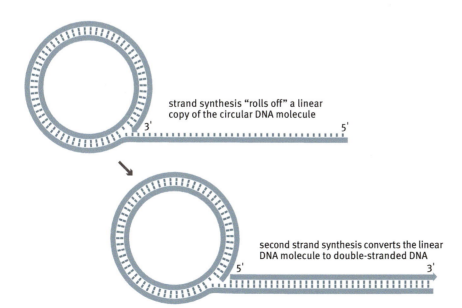

FIGURE 10.12 Rolling-circle replication. The top drawing shows a single strand that has been 'rolled off' of the circular molecule by DNA synthesis at its 3′ end. The rolled-off strand is then converted to double-stranded DNA by complementary strand synthesis.

and linear, but can easily be converted to double-stranded circular molecules by complementary strand synthesis, followed by cleavage at the junction points and circularization of the resulting segments.

10.2 DNA POLYMERASES

Now that we have established that DNA replication is semiconservative, we can turn our attention to the details of the copying process. The central players in this process are the enzymes that synthesize the new daughter strands of DNA. An enzyme able to build up a new DNA polynucleotide using an existing DNA strand as a template is called a **DNA-dependent DNA polymerase**. It is acceptable to shorten the name simply to 'DNA polymerase', but it is important not to get confused with the different, but equally important class of enzymes called **RNA-dependent DNA polymerases**, which are involved in replication of the RNA molecules of certain viruses (Chapter 12).

DNA polymerases synthesize DNA but can also degrade it

The chemical reaction catalyzed by a DNA polymerase is very similar to that of RNA polymerase except of course that the new polynucleotide that is assembled is built up of deoxyribonucleotide subunits rather than ribonucleotides (Figure 10.13). The sequence of the new polynucleotide is dependent on the sequence of the template and is determined by complementary base pairing and, as with RNA synthesis, DNA polymerization can occur only in the 5'→3' direction (Figure 10.14).

Although the main function of a DNA polymerase is DNA synthesis, most of these enzymes also have at least one type of exonuclease activity, meaning that they can remove nucleotides one by one from the end of a DNA polynucleotide. Because the two ends of a polynucleotide are chemically distinct, there are two equally distinct types of exonuclease activity. These are the **3'→5' exonuclease** activity which enables nucleotides to be removed from the 3' end of a polynucleotide, and the **5'→3' exonuclease** activity which removes nucleotides from the 5' end. Some polymerases have both activities, others have just one or neither of them (Table 10.1).

It may seem odd that DNA polymerases can degrade polynucleotides as well as synthesize them, but when we consider the details of the exonuclease activities the natural logic becomes apparent. The 3'→5' exonuclease activity, for example, enables a template-dependent DNA polymerase to remove nucleotides from the 3' end of a strand that it has just synthesized. This is looked on as a **proofreading** activity because it allows the polymerase to correct errors by removing nucleotides that have been inserted incorrectly (Figure 10.15a).

The 5'→3' exonuclease activity is less common but equally important for the functioning of those polymerases that possess it. It is typically a feature of a polymerase whose role in DNA replication requires that it must be able to remove at least part of a polynucleotide that is already attached to the template strand that the polymerase is copying (Figure 10.15b). As we will see later, there is an essential requirement for this activity during DNA replication in bacteria.

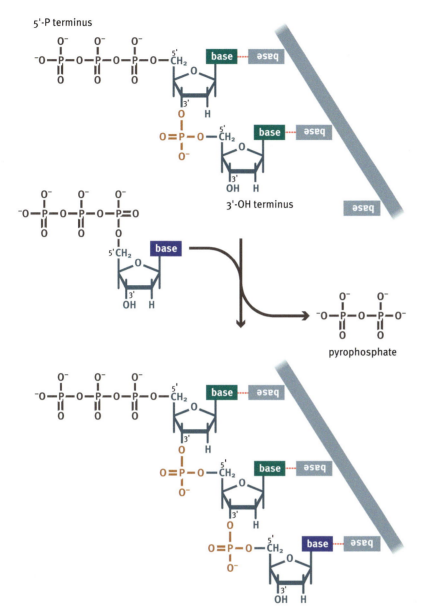

FIGURE 10.13 Template-dependent synthesis of DNA.

TABLE 10.1 EXONUCLEASE ACTIVITIES OF THOSE DNA POLYMERASES INVOLVED IN DNA REPLICATION		
Enzyme	Exonuclease Activities	
	3′ → 5′	5′ → 3′
Bacterial DNA polymerases		
DNA polymerase I	Yes	Yes
DNA polymerase III	Yes	No
Eukaryotic DNA polymerases		
DNA polymerase α	No	No
DNA polymerase γ	Yes	No
DNA polymerase δ	Yes	No
DNA polymerase ε	Yes	No

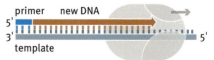

FIGURE 10.14 The action of a DNA polymerase. A DNA polymerase synthesizes DNA in the 5′ → 3′ direction, with the sequence of the new polynucleotide determined by complementary base pairing. Note that a DNA polymerase cannot initiate DNA synthesis unless there is already a short double-stranded region to act as a primer.

Bacteria and eukaryotes possess several types of DNA polymerase

The search for DNA polymerases began in the mid-1950s, as soon as it was realized that DNA synthesis was the key to replication of genes. It was thought that bacteria would probably have just a single DNA polymerase, and when the enzyme now called **DNA polymerase I** was identified in 1957 there was a widespread assumption that this was the main replicating enzyme. The discovery that almost complete inactivation of the *E. coli polA* gene, which codes for DNA polymerase I, had no effect on the ability of the cells to replicate their DNA, therefore came as something of a surprise. A second enzyme, **DNA polymerase II**, was then isolated, but this was shown to be involved mainly in the repair of damaged DNA rather than DNA replication. It was not until 1972 that the main replicating polymerase of *E. coli*, **DNA polymerase III**, was eventually discovered.

DNA polymerases I and II are single polypeptides but DNA polymerase III, befitting its role as the main replicating enzyme, is multisubunit. The core enzyme comprises three main subunits, called α, ε and θ. The polymerase activity is specified by the α subunit and the 3′→5′ exonuclease activity of the enzyme resides in ε. The function of the θ subunit is not clear. It appears to enhance the proofreading function of the ε subunit in some way, though it may have a purely structural role in bringing together the other two core subunits and in assembling the various accessory subunits. There are seven of these, the most important being β, which acts as a 'sliding clamp' holding the polymerase complex tightly to the DNA.

Eukaryotes have at least 15 DNA polymerases, which in mammals are distinguished by Greek suffices (α, β, γ, δ, etc.), an unfortunate choice of nomenclature as it tempts confusion with the identically named subunits of *E. coli* DNA polymerase III. The main replicating enzymes are **DNA polymerase δ** and **DNA polymerase ε**, which work in conjunction with an accessory protein called the **proliferating-cell nuclear antigen (PCNA)**. PCNA is the functional equivalent of the β subunit of *E. coli* DNA polymerase III and holds the enzyme tightly to the DNA that is being copied. **DNA polymerase α** also has an important function in DNA replication, and **DNA polymerase γ** is responsible for replicating the DNA molecules in mitochondria. The other 11 eukaryotic DNA polymerases are involved in the repair of damaged DNA.

The limitations of DNA polymerase activity cause problems during replication

Although DNA polymerases have evolved to replicate DNA, these enzymes have two features that complicate the way in which replication occurs in the cell.

The first difficulty is caused by the limitation that DNA polymerases can synthesize polynucleotides only in the 5′→3′ direction, which means that the template strands must be read in the 3′→5′ direction (see Figure 10.14). For one strand of the parent double helix, called the **leading strand**, this is not a problem because the new polynucleotide can be synthesized continuously (Figure 10.16). However, the second or **lagging strand** cannot be copied in a continuous fashion because this would necessitate 3′→5′ DNA synthesis. Instead, the lagging strand has to be replicated in sections. A portion of the parent helix is disassociated and a short stretch of the lagging strand is replicated, a bit more of the helix is disassociated and another segment of the lagging strand is replicated and so on. At first, this process was just a hypothesis but the isolation in 1969 of **Okazaki**

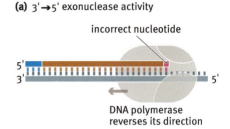
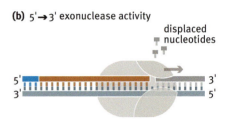

FIGURE 10.15 The exonuclease activities possessed by some DNA polymerases. (a) The 3′ → 5′ exonuclease activity enables a DNA polymerase to remove nucleotides from the 3′ end of a strand that it has just synthesized. (b) The 5′ → 3′ exonuclease activity enables a DNA polymerase to remove part of a polynucleotide that is already attached to the template strand that it is copying.

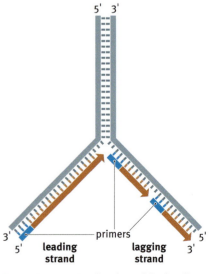

FIGURE 10.16 Replication of the leading and lagging strands of a DNA molecule.

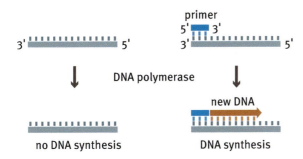

FIGURE 10.17 A primer is needed to initiate DNA synthesis by a DNA polymerase.

fragments, short pieces of single-stranded DNA associated with DNA replication, confirmed that the suggestion is correct.

The second difficulty is that, unlike an RNA polymerase, DNA polymerases cannot initiate DNA synthesis unless there is already a short double-stranded region to act as a primer (Figure 10.17). How can this occur during DNA replication? The answer is that the very first few nucleotides attached to either the leading or the lagging strand are not deoxyribonucleotides but ribonucleotides that are put in place by an RNA polymerase enzyme.

In bacteria, primers are synthesized by **primase**, a special RNA polymerase unrelated to the transcribing enzyme, with each primer 10–12 nucleotides in length and most starting with the sequence 5′–AG–3′. Once the primer has been completed, strand synthesis is continued by DNA polymerase III (Figure 10.18a).

In eukaryotes, the situation is slightly more complex because the primase is tightly bound to DNA polymerase α and cooperates with this enzyme in the synthesis of the first few nucleotides of a new polynucleotide. This eukaryotic primase synthesizes an RNA primer of 7–12 nucleotides and then hands over to DNA polymerase α, which extends the RNA primer by adding about 20 nucleotides of DNA. This DNA stretch often has a few ribonucleotides mixed in, but it is not clear if these are incorporated by DNA polymerase α or by intermittent activity of the primase. After completion of the RNA–DNA primer, DNA synthesis is continued by DNA polymerase ε on the leading strand and DNA polymerase δ on the lagging strand (Figure 10.18b).

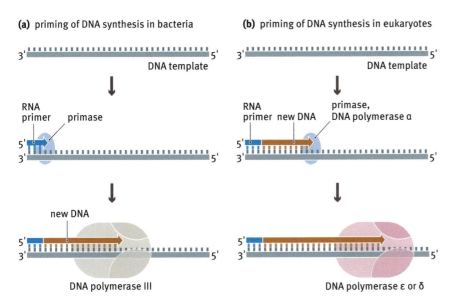

FIGURE 10.18 **Priming of DNA synthesis in bacteria and eukaryotes.** (a) In bacteria, the primer is synthesized by an RNA polymerase called primase. Strand synthesis is then continued by DNA polymerase III. (b) In eukaryotes, the primase is bound to DNA polymerase α and works with this enzyme in the synthesis of the first 30 or so nucleotides. DNA synthesis is then continued by DNA polymerase δ or DNA polymerase ε.

Priming needs to occur just once on the leading strand, because once primed, the leading-strand copy is synthesized continuously until replication is completed. On the lagging strand, priming is a repeated process that must occur every time a new Okazaki fragment is initiated. In *E. coli*, which makes Okazaki fragments of 1,000–2,000 nucleotides in length, approximately 4,000 priming events are needed every time the cell's DNA is replicated. In eukaryotes, the Okazaki fragments are much shorter, perhaps less than 200 nucleotides in length, and priming is a highly repetitive event.

10.3 DNA REPLICATION IN BACTERIA

There are many similarities between DNA replication in bacteria and eukaryotes, and the easiest way to deal with the topic is first to consider the less complex and better-understood process occurring in bacteria and then to move on to the special features of replication in eukaryotes.

As with many processes in genetics, we conventionally look on DNA replication as being made up of three phases – initiation, elongation and termination. Initiation of replication involves recognition of the position(s) on a DNA molecule where replication will begin. Elongation concerns the events occurring at the replication fork, where the parent polynucleotides are copied. Termination, which in general is only vaguely understood, occurs when the parent molecule has been completely replicated. We will consider each of these stages of replication in turn.

E. coli has a single origin of replication

When a DNA molecule is being replicated, only a limited region is ever in a non-base-paired form. The breakage in base pairing starts at a distinct position called the **origin of replication** (Figure 10.19) and gradually progresses along the molecule, possibly in both directions, with synthesis of the new polynucleotides occurring as the double helix unzips.

The circular *E. coli* DNA molecule, which is typical of most bacteria, has just one origin of replication. This origin spans approximately 245 bp of DNA and comprises an AT-rich **DNA unwinding element** (**DUE**) and 11 binding sites for a protein called DnaA (Figure 10.20). Once attached to their binding sites, the DnaA proteins associate with one another to form a right-handed filament structure with the double-stranded DNA attached to its outer surface. This creates torsional stress that opens up, or 'melts', the double helix within the AT-rich DUE (Figure 10.21).

Melting of the helix initiates a series of events that construct a nascent replication fork at either end of the open region. The first step is the attachment of a **prepriming complex** at each of these two positions. Each prepriming complex initially comprises six copies of the DnaB protein and six copies of DnaC. DnaC has a transitory role and is released from the complex soon after it is formed, its function probably being simply to aid the attachment of DnaB. The latter is a **helicase**, an enzyme that can break base pairs.

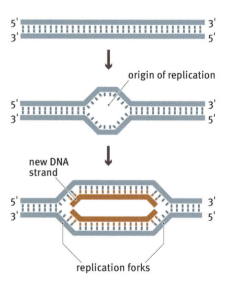

FIGURE 10.19 An origin of replication.

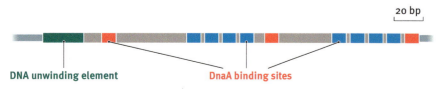

FIGURE 10.20 The *Escherichia coli* origin of replication. The *E. coli* origin of replication is called *oriC* and is approximately 245 bp in length. It contains a DNA unwinding element (DUE) and eleven binding sites for the DnaA protein. These sites have varying binding affinities, shown here as high affinity (red boxes) and low affinity (blue boxes). The high-affinity sites at the extreme left and right of the array have the sequence 5′–TGTGGATAA–3′ and the central high-affinity site has the sequence 5′–TGTGATAA–3′. The low-affinity sites are similar lengths but have more variable sequences.

DnaB begins to increase the size of the open region by increasing the lengths of single-stranded DNA that are exposed. These single-stranded regions are naturally 'sticky' and the two separated polynucleotides would immediately reform base pairs after the enzyme has passed, if allowed to. The single strands are also highly susceptible to nuclease attack and are likely to be degraded if not protected in some way. To avoid these unwanted outcomes, **single-strand binding proteins** (**SSBs**) attach to the polynucleotides and prevent them from reassociating or being degraded (Figure 10.22).

As soon as DnaB has begun to increase the size of the open region, the enzymes involved in the elongation phase of DNA replication are able to attach. The replication forks now start to progress away from the origin and DNA copying begins.

The elongation phase of bacterial DNA replication

The first enzyme involved in DNA copying is recruited soon after the DnaB helicase has bound to the origin. This enzyme is the primase, the special type of RNA polymerase that synthesizes the primers needed before DNA polymerase III can begin to make DNA. Addition of the primase converts the prepriming complex into the structure called the **primosome**, and replication of the leading strand is initiated.

After 1,000–2,000 nucleotides of the leading strand have been replicated, the first round of discontinuous strand synthesis on the lagging strand can begin. The primase, which is still associated with the DnaB helicase in the primosome, makes an RNA primer which is then extended by DNA polymerase III (Figure 10.23). This means that there are now two DNA polymerase core enzymes attached to the double helix, one copying the leading strand and one copying the lagging strand.

There is also a single **γ complex**, which is made up of five of the accessory subunits of DNA polymerase III (called γ, δ, δ', χ and ψ). The main function of the γ complex (sometimes called the 'clamp loader') is to interact with the β subunit (the 'sliding clamp') of each polymerase and hence control the attachment and removal of the enzymes from the template DNA. This function is required primarily during lagging-strand replication when the polymerase has to attach and detach repeatedly at the start and end of each Okazaki fragment.

Some models of the replication complex place the two polymerase enzymes in opposite orientations to reflect the different directions in which DNA synthesis occurs, toward the replication fork on the leading strand and away from it on the lagging strand. It is more likely, however, that the pair of enzymes face the same direction and the lagging strand forms a loop. This would enable DNA synthesis on the two strands to proceed in parallel as the complex moves forward, keeping pace with the progress of the replication fork (Figure 10.24). The combination of the two DNA polymerase III enzymes, the γ complex and the primosome, migrating along the parent DNA and carrying out most of the replicative functions, is called the **replisome**.

After the replisome has passed, the replication process must be completed by joining up the individual Okazaki fragments. This is not a trivial event because one member of each pair of adjacent Okazaki fragments still has its RNA primer attached at the point where ligation should take place (Figure 10.25). The primer cannot be removed by DNA polymerase III, because this enzyme lacks the required 5'→3' exonuclease activity (see Table 10.1). At this point, DNA polymerase III releases the lagging

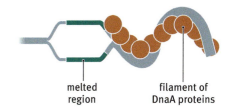

FIGURE 10.21 Melting the *E. coli* origin of replication. Bound DnaA proteins associate with one another to form a right-handed filament structure. This creates torsional stress that opens up the double helix within the DUE.

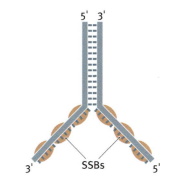

FIGURE 10.22 The role of single-strand binding proteins (SSBs) during DNA replication. SSBs attach to the unpaired polynucleotides produced by helicase action and prevent the strands from base pairing with one another or being degraded by nucleases.

strand and its place is taken by DNA polymerase I, which does have a 5'→3' exonuclease and so removes the primer. Usually, it also removes the start of the DNA component of the Okazaki fragment, extending the 3' end of the adjacent fragment into the region of the template that is exposed. The two Okazaki fragments now abut, with the terminal regions of both composed entirely of DNA. All that remains is for the missing phosphodiester bond to be put in place by a **DNA ligase**, linking the two fragments and completing replication of this region of the lagging strand.

Replication of the *E. coli* DNA molecule terminates within a defined region

Replication forks proceed along linear DNA molecules, or around circular ones, generally unimpeded except when a region that is being transcribed is encountered. DNA synthesis occurs at approximately five times the rate of RNA synthesis, so the replication complex can easily overtake an RNA polymerase, but this probably does not happen. Instead, it is thought that the replication fork pauses behind the RNA polymerase, proceeding only when the transcript has been completed. Eventually, the replication fork reaches the end of a linear molecule or, with a circular one, meets a second replication fork moving in the opposite direction. What happens next is one of the less well-understood aspects of DNA replication.

Bacterial DNA molecules are replicated bidirectionally from a single point (Figure 10.26), which means that the two replication forks should meet at a position diametrically opposite the origin of replication on the DNA molecule. However, if one fork is delayed, possibly because it has to

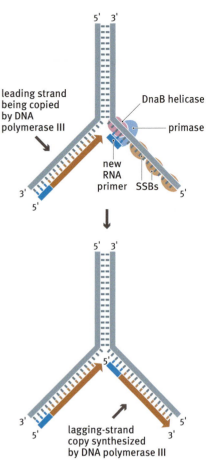

FIGURE 10.23 Priming and synthesis of the lagging-strand copy during DNA replication in *E. coli*. In the top drawing, replication of the leading strand has begun, and the primase has synthesized a primer on the lagging strand. In the lower drawing, this primer has been extended to begin replication of the lagging strand. Note that this is just the first round of discontinuous strand synthesis on the lagging strand. As the replication fork progresses along the parent molecule, additional rounds of lagging-strand replication are needed.

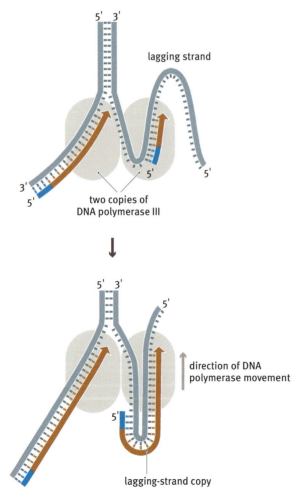

FIGURE 10.24 A model for parallel replication of the leading and lagging strands. There are two copies of DNA polymerase III, one for each strand. It is thought that the lagging strand loops through its copy of the polymerase, in the manner shown. This would enable the leading and lagging strands to be replicated in parallel as the two polymerase enzymes move in the same direction along the parent DNA molecule.

replicate extensive regions where transcription is occurring, then it might be possible for the other fork to overshoot the halfway point and continue replication on the 'other side' of the molecule (Figure 10.27). It is not immediately apparent why this should be undesirable, the daughter molecules presumably being unaffected, but it is not allowed to happen because of the presence of **terminator sequences**. Ten of these have been identified in the *E. coli* DNA molecule (Figure 10.28), each one acting as the recognition site for a sequence-specific DNA-binding protein called the **terminator utilization substance** (**Tus**).

The mode of action of Tus is quite unusual. When bound to a terminator sequence, a Tus protein allows a replication fork to pass if the fork is moving in one direction, but blocks progress if the fork is moving in the opposite direction. The directionality is set by the orientation of the Tus protein on the double helix. When approaching from one direction, the DnaB helicase, which is responsible for progression of the replication fork, encounters an impenetrable wall of β-strands on one side of the Tus protein (Figure 10.29). The replication fork cannot get through this wall and so stops. But coming from the other direction, DnaB meets a less rigid part of the Tus structure and hence is able to displace the protein, enabling the replication fork to continue its progress along the DNA.

The orientation of the termination sequences, and hence of the bound Tus proteins, on the *E. coli* DNA molecule is such that both replication forks become trapped within a relatively short region (see Figure 10.28). This ensures that termination always occurs at or near the same position. Exactly what happens when the two replication forks meet is unknown, but meeting is followed by disassembly of the replisomes, either spontaneously or in a controlled fashion. The result is two interlinked daughter molecules, which are separated by topoisomerase IV.

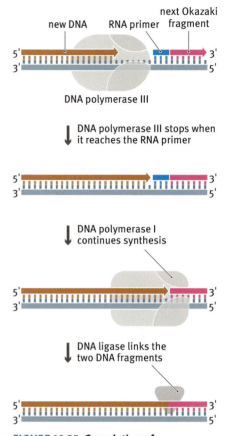

FIGURE 10.25 Completion of discontinuous stand synthesis in *E. coli*. DNA polymerase III detaches when it reaches the RNA primer of the next Okazaki fragment. DNA polymerase I removes the RNA primer and replaces it with DNA. The final phosphodiester bond is synthesized by DNA ligase.

10.4 DNA REPLICATION IN EUKARYOTES

As mentioned above, in outline DNA replication is similar in both bacteria and eukaryotes. Eukaryotic replication does, however, have its own special features. We will now take a look at these.

Eukaryotic DNA has multiple replication origins

Most circular bacterial DNA molecules have a single origin of replication. Eukaryotic chromosomes, in contrast, have multiple origins, 400 in the yeast *Saccharomyces cerevisiae* and thousands in humans. Yeast and other simple eukaryotes have discrete origins of replication, similar in some ways to that of *E. coli*. In *S. cerevisiae*, an origin is about 200 bp in length and comprises four segments (Figure 10.30). Two of these make up the **origin recognition sequence**, a stretch of some 40 bp in total that is the binding site for the **origin recognition complex** (**ORC**), a set of six proteins that attach to the origin. The ORC proteins have been described as yeast versions of the *E. coli* DnaA proteins, but this interpretation is probably not strictly correct because ORCs remain attached to yeast origins even when replication is not being initiated. Rather than being genuine initiator proteins, ORCs are involved in the regulation of DNA replication, acting as mediators between replication origins and the regulatory signals that coordinate the initiation of DNA replication with division of the cell. The real equivalents of DnaA probably bind to a different segment of a yeast origin.

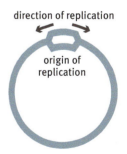

FIGURE 10.26 Bidirectional replication of a circular bacterial DNA molecule.

Attempts to identify replication origins in more complex eukaryotes such as humans have been less successful. **Initiation regions** (parts of the chromosomal DNA where replication initiates) have been identified in mammals, suggesting that there are specific regions in mammalian chromosomes where replication begins, but some researchers are doubtful whether these regions contain replication origins equivalent to those in yeast and bacteria. One alternative hypothesis is that replication does not initiate at specific sequences within the DNA, but instead at positions where the helix is less tightly wound than normal.

The eukaryotic replication fork: variations on the bacterial theme

As in bacteria, the progress of a eukaryotic replication fork is maintained by helicase activity, although which of the several eukaryotic helicases that have been identified are primarily responsible for DNA melting during replication has not been established. The separated polynucleotides are prevented from reattaching by SSBs, the main one of these in eukaryotes being **replication protein A** (**RPA**).

We begin to encounter unique features of the eukaryotic replication process when we examine the method used to prime DNA synthesis. As described above, the eukaryotic DNA polymerase α cooperates with the primase enzyme to put in place the RNA–DNA primers at the start of the leading-strand copy and at the beginning of each Okazaki fragment. DNA polymerase α is not capable of lengthy DNA synthesis, presumably because it lacks the stabilizing effect of a sliding clamp equivalent to the β subunit of *E. coli* DNA polymerase III or the PCNA accessory protein that aids the eukaryotic DNA polymerases δ and ε. This means that although DNA polymerase α can extend the initial RNA primer with about 20 nucleotides of DNA, it must then be replaced by the main replicative enzyme, DNA polymerase ε on the leading strand or DNA polymerase δ on the lagging strand (see Figure 10.18b).

As in *E. coli*, completion of lagging-strand synthesis requires removal of the RNA primer from each Okazaki fragment. There appears to be no eukaryotic DNA polymerase with the 5′→3′ exonuclease activity needed for this purpose and the process is therefore very different to that described

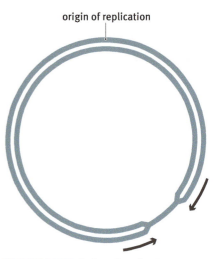

FIGURE 10.27 A situation that is not allowed to occur during replication of the circular *E. coli* DNA molecule. One of the replication forks has proceeded some distance past the halfway point.

FIGURE 10.28 The positions of the ten terminator sequences on the *E. coli* DNA molecule. The arrowheads indicate the direction in which each terminator sequence can be passed by a replication fork.

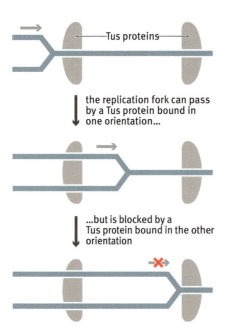

FIGURE 10.29 The role of Tus proteins during termination of replication in *E. coli*. Bound Tus proteins allow a replication fork to pass when the fork approaches from one direction but not when it approaches from the other direction.

for bacterial cells. The central player is the **flap endonuclease** (**FEN1**), which associates with DNA polymerase δ at the 3′ end of an Okazaki fragment. A helicase breaks the base pairs holding the primer to the template strand, enabling the primer to be pushed aside by DNA polymerase δ as it extends the adjacent Okazaki fragment into the region thus exposed (Figure 10.31). The flap that results can be cut off by FEN1, which has an unusual endonuclease activity that enables it to cut the phosphodiester bond at the branch point that is formed when the 5′ end of a polynucleotide is displaced from its template.

It is thought that both the RNA primer and all of the DNA originally synthesized by DNA polymerase α are removed. This is necessary because DNA polymerase α has no 3′→5′ proofreading activity (see Table 10.1) and therefore synthesizes DNA in a relatively error-prone manner. Removal of this region as part of the flap cleaved by FEN1, followed by resynthesis by DNA polymerase δ (which does have a proofreading activity and so makes a highly accurate copy of the template) would prevent these errors from becoming permanent features of the daughter double helix.

Termination of replication in eukaryotes

As in bacteria, a eukaryotic replication fork progresses along the DNA until it meets a second replication fork moving in the opposite direction. Unlike bacteria, there appears to be little or no control over the positions at which replication forks meet. Fork blocking proteins that act in a similar manner to Tus are known in eukaryotes, but there are only a small number of terminator sequences, called **replication fork barriers**. One of these barriers prevents a replication fork from having a head-on collision with an RNA polymerase that is transcribing a set of rRNA genes. The positioning of this barrier ensures that these DNA segments are always replicated by a fork that follows the RNA polymerase (Figure 10.32). A head-on collision is thought to be undesirable because it can inactivate the helicase component of the replisome, causing the replication fork to stall and possibly become destabilized. Head-on collisions must occur at other positions where genes are being transcribed but there are no replication fork barriers, but presumably the high level of transcription that occurs in the rRNA regions presents a particular danger to the replication process.

In eukaryotes, the interlinked daughter molecules that result when two replication forks meet are separated by the type II topoisomerase called Top2. The replisomes are then disassembled. The first step is removal of the helicase, followed by the other replisome components, which enter the ubiquitin-directed degradation pathway (Section 8.3). Degradation of the replisome components, rather than their reuse, is to be expected as they will not be needed until the next round of DNA replication, which even in rapidly dividing eukaryotic cells will occur only after a delay of some 24 hours.

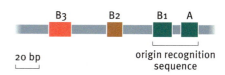

FIGURE 10.30 The structure of a yeast origin of replication.

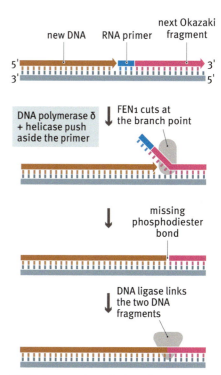

FIGURE 10.31 Completion of discontinuous strand synthesis in eukaryotes. A helicase helps DNA polymerase δ to push aside the primer of the next Okazaki fragment. The resulting flap is then cut off by FEN1, and the final phosphodiester bond synthesized by DNA ligase.

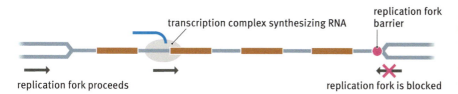

FIGURE 10.32 The role of the replication fork barrier adjacent to a rRNA region in eukaryotic DNA. Multiple copies of the rRNA genes are located in tandem arrays (see Figure 6.16). In this diagram, the genes are transcribed in the left-to-right direction. The presence of the replication fork barrier ensures that the replication fork progressing from the right is blocked so that it does not meet the transcription complex head-on. The fork progressing from the left is unimpeded and follows the transcription complex through the region containing the rRNA genes.

DNA replication in eukaryotes must be coordinated with the cell cycle

The final aspect of eukaryotic DNA replication that we must consider is the way in which the process is coordinated with the **cell cycle**. This term refers to the series of four stages that each cell passes through between one division and the next (Figure 10.33). The first of these stages is a gap period called the **G1 phase**. This is generally the longest phase of the cell cycle and for some cells may last for weeks or even months. The second stage is **S phase** or 'synthesis' phase. This is the period when the DNA molecules are replicated. Despite the presence of multiple origins on each DNA molecule, S phase can take 6–8 hours to complete. There then follows a second gap period called the **G2 phase**. The cell is now committed to division so G2 rarely lasts longer than 3–4 hours. This leads into **M phase**, when cell division takes place.

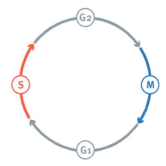

FIGURE 10.33 The cell cycle. The lengths of the individual phases vary in different cells. Abbreviations: G1 and G2, gap phases; M, mitosis (cell division); S, synthesis phase.

The S and M phases must be coordinated so that each DNA molecule is completely replicated, but replicated only once, before division occurs. To ensure this happens, there are a series of **cell cycle checkpoints**, progression past a checkpoint committing the cell to the next phase of the cycle. The most important of these is the **G1-S checkpoint**, because only when this point has been passed is the cell able to replicate its DNA. If the cell suffers trauma or the DNA has been damaged in some way, then the cell cycle can be arrested at the G1-S checkpoint while the damage is repaired. Later in the cell cycle, the **G2-M checkpoint** ensures that the cell is ready to enter division, the central requirement being that its DNA has been replicated correctly, every part replicated once and no parts replicated multiple times.

Recent research has led to a model that defines **origin licensing** as the basis to the preparation of the cell for passage through the G1-S checkpoint. Origin licensing involves construction on a replication origin of a set of proteins, called the **pre-replication complex (pre-RC)**. Each origin is already marked by the six proteins that make up the ORC, which is assembled onto multiple origins soon after completion of the previous cell division. To become licensed, an origin must initially recruit two additional proteins, Cdc9 and Cdt1. Origin licensing enables the cell to pass through the G1-S checkpoint, but the pre-RCs are at this stage inactive and unable to initiate DNA replication. To become active, each pre-RC must recruit additional proteins, including the helicase and other components of the replisome. DNA synthesis can now begin.

Identification of the components of the pre-RC takes us some distance toward understanding how DNA replication is initiated, but still leaves open the question of how replication is coordinated with other events in the cell cycle. Cell cycle control is a complex process mediated largely by **cyclin-dependent kinase (CDK)** proteins, which phosphorylate and activate enzymes and other proteins that have specific functions during the cell cycle. The activities of these CDKs change throughout the cell cycle, being lowest at the beginning of the G1 phase, and increasing rapidly during S phase. The CDK level is thought to influence the activities of other proteins whose functions are needed at different stages, thereby ensuring that the cell cycle progresses in an orderly manner. CDKs are themselves regulated by **cyclins**, proteins that vary in abundance at different stages of the cell cycle (Figure 10.34). CDKs are also regulated by inhibitory proteins such as geminin, which accumulates during the S, G2 and M phases and prevents the recopying of segments of DNA that have already been replicated.

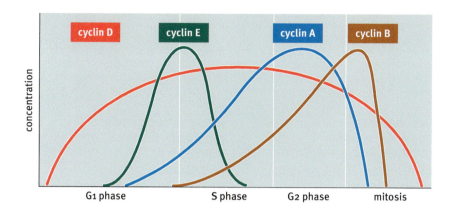

FIGURE 10.34 Variations in abundance of mammalian cyclins during the cell cycle. The abundance of each of the four cyclins varies during the cell cycle, resulting in stage-specific changes in cyclin-dependent kinase (CDK) activity.

KEY CONCEPTS

- The discovery of the double helix structure suggested a way in which DNA molecules could be copied. Experiments carried out soon after the discovery confirmed that DNA replication occurs in a semiconservative manner.

- Semiconservative replication presents a topological problem as the two strands of the double helix cannot be separated without unwinding. The problem is solved by the action of DNA topoisomerases.

- An enzyme able to build up a new DNA polynucleotide using an existing DNA strand as a template is called a DNA-dependent DNA polymerase. Most DNA polymerases have exonuclease as well as polymerase activity.

- DNA polymerases can synthesize polynucleotides only in the 5' → 3' direction. This means that one strand of the parent double helix has to be replicated in sections.

- DNA polymerases cannot initiate DNA synthesis unless there is already a short double-stranded region to act as a primer. To solve this problem, the first few nucleotides that are attached are ribonucleotides that are put in place by an RNA polymerase enzyme.

- Most bacterial DNA molecules have a single origin or replication, but eukaryotic DNA molecules have many origins.

- Progress of a replication fork along a DNA molecule requires the coordinated activities of several different enzymes and proteins.

- In bacteria, replication forks traveling in opposite directions around a circular DNA molecule become trapped in a region defined by terminator sequences. These sequences prevent one or other fork from progressing too far around the DNA.

- In eukaryotes, there are only a small number of replication fork barriers and therefore less control over the positions where replication forks meet. One of the barriers prevents a replication fork from having a head-on collision with an RNA polymerase that is transcribing a set of rRNA genes.

- Origin licensing and other events controlled by cyclins and cyclin-dependent kinases ensure that DNA replication is coordinated with the cell cycle.

QUESTIONS AND PROBLEMS

Key terms

Write short definitions of the following terms:

γ complex
3′ → 5′ exonuclease
5′ → 3′ exonuclease
buoyant density
cell cycle
cell cycle checkpoint
conservative replication
cyclin
cyclin-dependent kinase
D loop
dispersive replication
displacement replication
DNA gyrase
DNA ligase
DNA polymerase I
DNA polymerase II
DNA polymerase III
DNA polymerase α
DNA polymerase γ
DNA polymerase δ

DNA polymerase ε
DNA topoisomerase
DNA unwinding element
DNA-dependent DNA polymerase
flap endonuclease
G1 phase
G1-S checkpoint
G2 phase
G2-M checkpoint
helicase
initiation region
lagging strand
leading strand
M phase
Okazaki fragment
origin licensing
origin of replication
origin recognition complex
origin recognition sequence
plectonemic

pre-replication complex
prepriming complex
primase
primosome
proliferating-cell nuclear antigen
proofreading
replication fork
replication fork barrier
replication protein A
replisome
RNA-dependent DNA polymerase
rolling-circle replication
S phase
semiconservative replication
single-strand binding protein
terminator sequence
terminator utilization substance
type I topoisomerase
type II topoisomerase

Self-study questions

10.1 Distinguish between the three conceivable ways by which the double helix could be replicated.

10.2 Describe the experiment that showed that DNA replicates in a semiconservative manner.

10.3 Draw the density gradient banding patterns that Meselson and Stahl would have obtained after two rounds of replication if DNA replication had turned out to be (a) conservative or (b) dispersive.

10.4 What is the topological problem in DNA replication, and how was this problem solved by the discovery of DNA topoisomerases?

10.5 Distinguish between the mode of action of type I and type II DNA topoisomerases.

10.6 With the aid of diagrams, indicate how displacement replication and rolling-circle replication differ from DNA copying via a replication fork.

10.7 What are the roles of the exonuclease activities possessed by some DNA polymerases?

10.8 Distinguish between the roles of DNA polymerase I and DNA polymerase III during DNA replication in *E. coli*, and between DNA polymerase α, DNA polymerase δ and DNA polymerase ε in eukaryotes.

10.9 What impact does the inability of DNA polymerases to synthesize DNA in the 3′→5′ direction have on DNA replication?

10.10 Explain why DNA replication must be primed and describe how the priming problem is solved by *E. coli* and by eukaryotes.

10.11 Give a detailed description of the structure of the *E. coli* origin of replication and outline the role of each component of the origin in the initiation of replication.

10.12 What are the roles of helicases and single-strand binding proteins during DNA replication?

10.13 Give a detailed description of the events occurring at the replication fork in *E. coli*. Your answer should make a clear distinction between the complexes referred to as the primosome and the replisome.

10.14 How are Okazaki fragments joined together in *E. coli*?

10.15 Why does replication of the *E. coli* DNA molecule always terminate in a defined region?

10.16 Describe the features of a yeast replication origin.

10.17 In what ways do events at the eukaryotic replication fork differ from those occurring in *E. coli*?

10.18 How are Okazaki fragments joined together in eukaryotes?

10.19 Outline how replication terminates in eukaryotes.

10.20 Explain what is meant by the 'cell cycle' and distinguish between the different stages of the cell cycle.

10.21 Outline our current knowledge regarding the way in which DNA replication is coordinated with the cell cycle.

Discussion topics

10.22 Would it be possible to replicate the DNA molecules present in living cells if DNA topoisomerases did not exist?

10.23 Why is inactivation of the *E. coli polA* gene, coding for DNA polymerase I, not lethal?

10.24 What might be the role of DNA polymerase enzymes in the repair of damaged DNA?

10.25 Construct a hypothesis to explain why all DNA polymerases require a primer in order to initiate the synthesis of a new polynucleotide. Can your hypothesis be tested?

10.26 Biologists have been aware for many years that repeated rounds of DNA replication ought to result in the gradual shortening of a linear double-stranded molecule. Explain why this should be the case.

The answers to, or hints on how to answer, the self-study questions and discussion topic questions can be downloaded from the book's product page at this link: www.routledge.com/9781032743530

FURTHER READING

Berguis BA, Raducanu V-S, Elshenawy MM et al. (2018) What is all this fuss about Tus? Comparison of recent findings from biophysical and biochemical experiments. *Critical Reviews in Biochemistry and Molecular Biology* 53, 49–63. *Describes hypotheses regarding the mode of action of Tus proteins.*

Dewar JM & Walter JC (2017) Mechanisms of DNA replication termination. *Nature Reviews Molecular Cell Biology* 18, 507–516. *Describes the topological problems that arise when replication forks meet, and how these problems are solved in order to separate the daughter molecules.*

Dueva R & Iliakis G (2020) Replication protein A: a multisubunit protein with roles in DNA replication, repair and beyond. *NAR Cancer* 2, zcaa022.

Ekundayo B & Bleichert F (2019) Origins of DNA replication. *PLoS Genetics* 15, e1008320.

Finger LD, Atack JM, Tsutakawa S et al. (2012) The wonders of flap endonucleases: structure, function, mechanism and regulation. *Subcellular Biochemistry* 62, 301–326.

Hübscher U, Nasheuer H-P & Syväoja JE (2000) Eukaryotic DNA polymerases: a growing family. *Trends in Biochemical Sciences* 25, 143–147.

Johnson A & O'Donnell M (2005) Cellular DNA replicases: components and dynamics at the replication fork. *Annual Review of Biochemistry* 74, 283–315. *Details of replication in bacteria and eukaryotes.*

Kornberg A (1960) Biologic synthesis of deoxyribonucleic acid. *Science* 131, 1503–1508. *A description of DNA polymerase I.*

Kornberg A (1984) DNA replication. *Trends in Biochemical Sciences* 9, 122–124. *An early description of events at the replication fork.*

Leonard AC & Méchali M (2013) DNA replication origins. *Cold Spring Harbor Perspectives in Biology* 5, a010116.

Li H & Stillman B (2012) The origin recognition complex: a biochemical and structural view. *Subcellular Biochemistry* 62, 37–58.

McKie SJ, Neuman KC & Maxwell A (2021) DNA topoisomerases: advances in understanding of cellular roles and multi-protein complexes via structure-function analysis. *Bioessays* 43, e2000286.

Meselson M & Stahl F (1958) The replication of DNA in *Escherichia coli*. *Proceedings of the National Academy of Sciences USA* 44, 671–682. *The Meselson-Stahl experiment.*

Okazaki T & Okazaki R (1969) Mechanisms of DNA chain growth. *Proceedings of the National Academy of Sciences USA* 64, 1242–1248. *The discovery of Okazaki fragments.*

Pomerantz RT & O'Donnell M (2007) Replisome mechanics: insights into a twin polymerase machine. *Trends in Microbiology* 15, 156–164. *Details of how DNA is synthesized at the replication fork.*

Stillman B (1996) Cell cycle control of DNA replication. *Science* 274, 1659–1664.

Vos SM, Tretter EM, Schmidt BH et al. (2011) All tangled up: how cells direct, manage and exploit topoisomerase function. *Nature Reviews Molecular Cell Biology* 12, 827–841.

CHAPTER 11

Mutation and DNA Repair

DNA sequences are not static. If they were then every organism would be exactly the same as every other organism. Individual variations, and indeed individual species, would not exist. The variations that we see among the members of a population, and the differences between species, arise because DNA sequences can change over time. These changes occur as a result of the cumulative effects of small-scale sequence alterations caused by **mutation**. Mutations can arise in two ways. The first is from an error in DNA replication. Although DNA replication is virtually error-free, mistakes do occasionally occur. An error in replication results in a daughter molecule that contains a **mismatch**, a position at which a base pair does not form because the nucleotides opposite each other in the double helix are not complementary (Figure 11.1a). When the

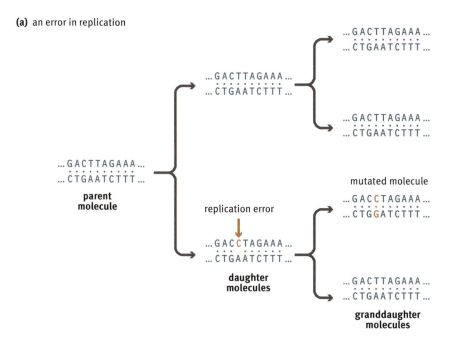

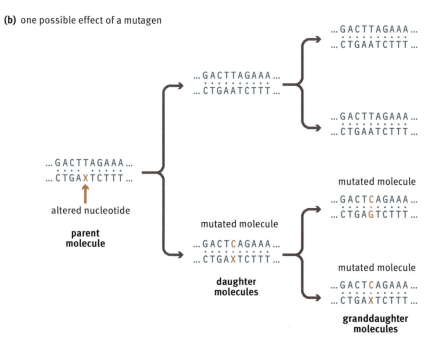

FIGURE 11.1 Examples of mutations. (a) An error in replication leads to a mismatch in one of the daughter double helices. When the mismatched molecule is itself replicated, it gives one double helix with the correct sequence and one with a mutated sequence. (b) A mutagen has altered the structure of an A in the lower strand of the parent molecule, giving nucleotide X, which does not base pair with T in the other strand. When the parent molecule is replicated, X base pairs with C, giving a mutated daughter molecule. When this daughter molecule is replicated, both granddaughters inherit the mutation.

DOI: 10.1201/9781003473862-12

mismatched daughter molecule is itself replicated, the two granddaughter molecules that are produced are not identical. One has the correct nucleotide sequence, but the second contains a mutation.

Mutations can also be caused by a chemical or physical **mutagen**. Various chemicals, as well as heat and some types of radiation, react with DNA molecules and change the structure of one or more nucleotides within the double helix. Changing the structure of a nucleotide may affect its base-pairing properties, resulting in a mutation when the DNA molecule is replicated (Figure 11.1b).

As we will see, mutations can be harmless, or they may have a serious, even lethal, effect on the organism. Because they can be harmful, all organisms have **DNA repair** mechanisms which correct the vast majority of the mutations that occur in their DNA molecules. Despite these measures, a few mutations slip through. A mutation that has not been repaired might be inherited by one of that organism's progeny. It might then become a new and permanent feature of the genetic diversity of the species.

In this chapter, we will look first at the processes that give rise to mutations and the mechanisms that are used to repair mutated DNA molecules. We will then examine the various possible effects that a mutation can have on a cell or organism.

11.1 THE CAUSES OF MUTATIONS

A mutation is a change in the nucleotide sequence of a DNA molecule. The simplest type of sequence change is a **point mutation**, in which one nucleotide is replaced by another (Figure 11.2). Point mutations are also called, simple mutations or single-site mutations. A point mutation is classified as a **transition** if it is a purine to purine (A ↔ G) or a pyrimidine to pyrimidine (T ↔ C) change, or a **transversion** if the alteration is purine to pyrimidine or vice versa (A or G ↔ T or C).

Other types of mutations include **insertions** and **deletions**, which are the addition or removal of anything from one base pair up to quite extensive pieces of DNA, and **inversions**, which involve the excision of a portion of the double helix followed by its reinsertion at the same position but in the reverse orientation.

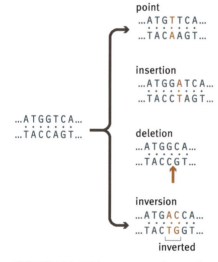

FIGURE 11.2 Different types of mutation.

Some mutations arise from errors in DNA replication that escape the proofreading functions of the replicating enzymes, and others are due to the action of mutagens that occur naturally in the environment and cause structural changes in DNA molecules. We begin this chapter by examining the molecular basis of these two types of events.

Errors in replication are a source of point mutations

During DNA replication, the sequence of the new strand of DNA that is being made is determined by complementary base pairing with the template polynucleotide (Figure 11.3). When considered purely as a chemical reaction, complementary base pairing is not particularly accurate. Nobody has yet devised a way of carrying out the template-dependent synthesis of DNA without the aid of enzymes, but if the process could be carried out simply as a chemical reaction in a test tube then the resulting polynucleotide would probably have point mutations at 5–10 positions out of every hundred. This represents an error rate of 5%–10%, which would be completely unacceptable during DNA replication in a living cell.

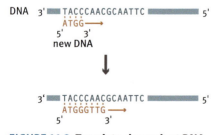

FIGURE 11.3 Template-dependent DNA replication.

The accuracy of DNA replication in the cell is increased by a checking process that is operated by the DNA polymerase enzyme and reduces the possibility of an incorrect nucleotide being incorporated into the growing strand (Figure 11.4). This selection process acts at different stages during the polymerization reaction. The identity of the nucleotide is checked when it is first bound to DNA polymerase and checked again when the nucleotide is moved to the active site of the enzyme. At either of these stages, the enzyme is able to reject the nucleotide if it recognizes that it is not the right one.

Should an incorrect nucleotide evade this surveillance system and become attached to the 3' end of the polynucleotide that is being synthesized, then all is not lost. Most DNA polymerase enzymes possess a 3'→5' exonuclease activity and so are able to go back over their work and remove a stretch of a newly synthesized polynucleotide that contains one or more incorrect pairings. This is the proofreading activity that we met when we studied the properties of DNA polymerases (see Figure 10.15a).

Despite these precautions, some errors do creep through, each one causing a mismatch in the daughter molecule and a permanent point mutation in one of the granddaughter molecules, as shown in Figure 11.1. But not all of these point mutations can be blamed on the polymerase enzymes. Sometimes an error occurs even though the enzyme adds the 'correct' nucleotide, the one that base pairs with the template. This is because each nucleotide base can exist as either of two alternative structural forms called **tautomers**. For example, thymine exists as two tautomers, the *keto* and *enol* forms, with individual molecules occasionally undergoing a shift from one tautomer to the other (Figure 11.5). The equilibrium is biased very much toward the *keto* form but every now and then the *enol* version of thymine occurs in the template DNA at the precise time that the replication fork is moving past. This will lead to an 'error', because *enol*-thymine base pairs with G rather than A. The same problem can occur with adenine, the rare *imino* tautomer of this base preferentially forming a pair with C, and with guanine, *enol*-guanine pairing with thymine. After replication, the rare tautomer will inevitably revert to its more common form, leading to a mismatch in the daughter double helix.

Replication errors can also lead to insertion and deletion mutations

Not all errors in replication are point mutations. Aberrant replication can also result in small numbers of extra nucleotides being inserted into the polynucleotide being synthesized, or some nucleotides in the template not being copied. An insertion or deletion that occurs within the coding region of a gene might result in a frameshift mutation, which changes the reading frame used for the translation of the protein specified by the gene (Figure 11.6). There is a tendency to use 'frameshift' to describe all insertions and deletions, but this is inaccurate because inserting or deleting three nucleotides, or multiples of three, simply adds or removes codons or parts of adjacent codons without affecting the reading frame. Also, of course, many insertions/deletions occur outside of genes, within the intergenic regions of a DNA molecule.

Insertion and deletion mutations are particularly prevalent when the template DNA contains short repeated sequences. This is because repeated sequences can induce **replication slippage**, in which the template strand and its copy shift their relative positions so that part of the template is either copied twice or missed out. The result is that the new

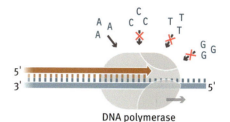

FIGURE 11.4 Ensuring the accuracy of DNA replication. During DNA replication, the DNA polymerase selects the correct nucleotide to insert at each position.

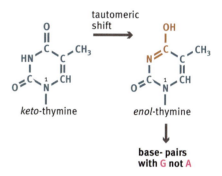

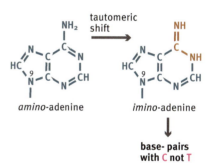

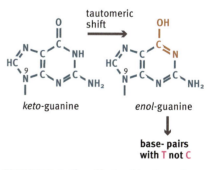

FIGURE 11.5 The effects of tautomerism on base pairing. In each of these three examples, the two tautomeric forms of the base have different pairing properties. Cytosine also has *amino* and *imino* tautomers but both pair with guanine.

polynucleotide has a larger or smaller number, respectively, of the repeat units (Figure 11.7).

Replication slippage is probably responsible for the **trinucleotide repeat expansion diseases** that have been discovered in humans in recent years. Each of these neurodegenerative diseases is caused by a relatively short series of trinucleotide repeats becoming elongated to two or more times its normal length. For example, the human *HTT* gene contains the sequence 5′–CAG–3′ repeated between 6 and 29 times in tandem, coding for a series of glutamines in the protein product. In Huntington's disease, this repeat expands to a copy number of 38–180, increasing the length of the polyglutamine tract and resulting in a dysfunctional protein (Figure 11.8).

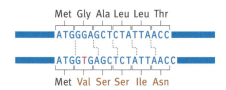

FIGURE 11.6 **An insertion mutation.** Insertion of a nucleotide changes the reading frame used for translation of the protein coded by a gene.

Mutagens are one type of environmental agent that causes damage to cells

Many chemicals that occur naturally in the environment have mutagenic properties and these have been supplemented in recent years with other chemical mutagens that result from human industrial activity. Physical agents such as radiation are also mutagenic. Most organisms are exposed to greater or lesser amounts of these various mutagens, and their DNA molecules suffer damage as a result.

Mutagens cause mutations in three different ways (Figure 11.9). Some act as **base analogs** and are mistakenly used as substrates when new DNA is synthesized at the replication fork. Others react directly with DNA,

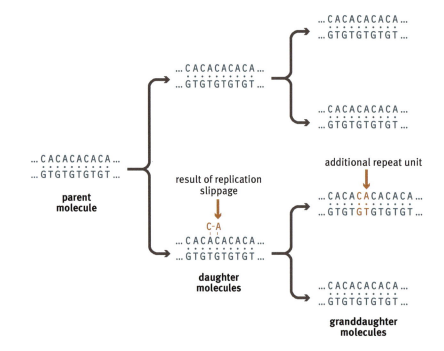

FIGURE 11.7 **Replication slippage.** The diagram shows the replication of a short repeat sequence. Slippage has occurred during replication of the parent molecule, inserting an additional repeat unit into the newly synthesized polynucleotide of one of the daughter molecules. When this daughter molecule replicates, it gives a granddaughter molecule whose repeat sequence is one unit longer than that of the original parent.

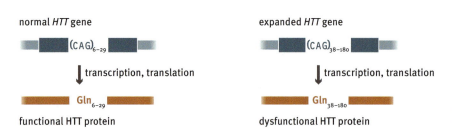

FIGURE 11.8 **The genetic basis of Huntington's disease.**

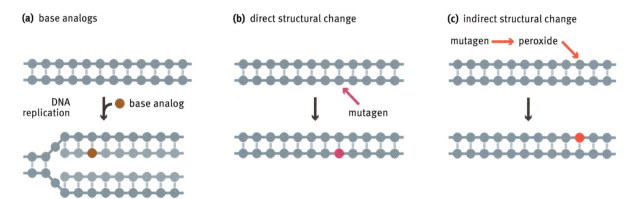

FIGURE 11.9 Three ways in which mutagens can cause mutations. (a) A base analog can be mistakenly used as a substrate during DNA replication. (b) Some mutagens react directly with DNA, causing a structural change that might be miscopied during DNA replication. (c) Other mutagens act indirectly on DNA, by causing the cell to synthesize chemicals that have a direct mutagenic effect.

causing structural changes that lead to miscopying of the template strand when the DNA is replicated. These structural changes are diverse, as we will see when we look at individual mutagens. And finally, some mutagens act indirectly on DNA. They do not themselves affect DNA structure, but instead cause the cell to synthesize chemicals such as peroxides that have a direct mutagenic effect.

It is important to understand that a mutagen, defined as 'a chemical or physical agent that causes mutations', is distinct from other types of environmental agent that cause damage to cells in ways other than by causing mutations. These include **carcinogens**, which cause cancer (the neoplastic transformation of cells), **oncogens** which cause tumor formation and **teratogens** which cause developmental abnormalities. There are overlaps between these categories (for example, some mutagens are also carcinogens) but each type of agent has a distinct biological effect. The definition of 'mutagen' also makes a distinction between true mutagens and other agents that damage DNA without causing mutations, such as **clastogens**, which cause breaks in DNA molecules. This type of damage might block replication and cause the cell to die, but it is not a mutation in the strict sense of the term and the causative agents are therefore not mutagens.

There are many types of chemical mutagen

The range of chemical mutagens is so vast that it is difficult to devise an all-embracing classification. We will therefore restrict our study to the most important types (**Table 11.1**).

First, there are base analogs. These are purine and pyrimidine bases that are similar enough to the standard bases to be incorporated into

TABLE 11.1 IMPORTANT TYPES OF CHEMICAL MUTAGEN	
Type	Examples
Base analogs	5-Bromouracil, 2-aminopurine
Deaminating agents	Nitrous acid, sodium bisulfite
Alkylating agents	Ethylmethane sulfonate, dimethylnitrosamine, methyl halides
Intercalating	Ethidium bromide

nucleotides when these are synthesized by the cell. The resulting unusual nucleotides can then be used as substrates for DNA synthesis during DNA replication. For example, **5-bromouracil** (**5-bU**; Figure 11.10a) has the same base-pairing properties as thymine, and nucleotides containing 5-bU can be added to the daughter polynucleotide at positions opposite As in the template. The mutagenic effect arises because the equilibrium between the two tautomers of 5-bU is shifted more toward the rarer *enol* form than is the case with thymine. This means that during the next round of replication, there is a relatively high chance of the polymerase encountering *enol*-5bU which, like *enol*-thymine, pairs with G rather than A (Figure 11.10b). This results in a point mutation (Figure 11.10c). **2-Aminopurine** acts in a similar way. It is an analog of adenine with an *amino*-tautomer that pairs with thymine and an *imino*-tautomer that pairs with cytosine. The *imino* form is more common than *imino*-adenine and induces a greater number of T-to-C transitions during DNA replication.

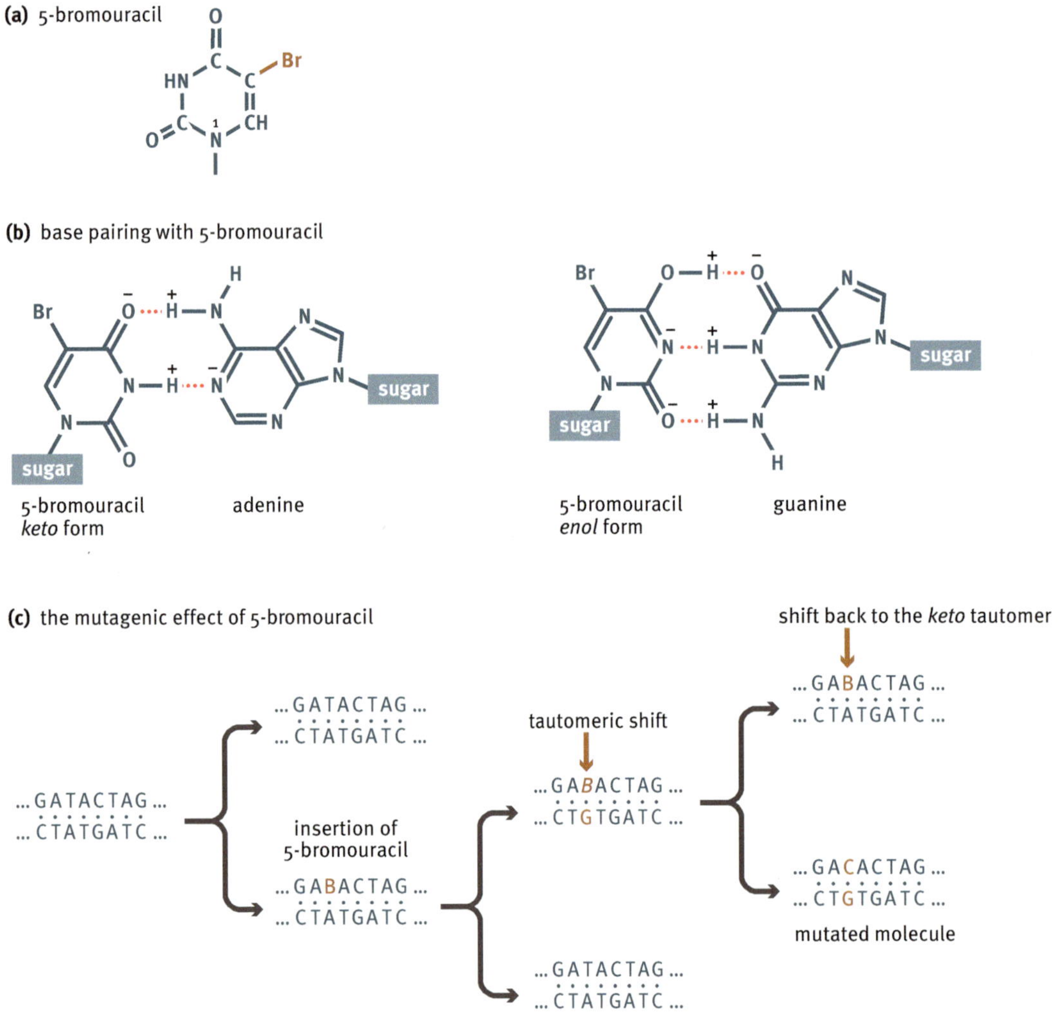

FIGURE 11.10 5-Bromouracil and its mutagenic effect. (a) 5-Bromouracil is a modified form of thymine. (b) The *keto* form of 5-bromouracil base pairs with adenine, and the *enol* form with guanine. (c) Three cycles of DNA replication are shown. During the first cycle, the *keto* form of 5-bromouracil acts as a base analog and replaces a thymine nucleotide in one of the daughter molecules. A tautomeric shift then converts the inserted 5-bromouracil into the *enol* tautomer. This results in a base-pairing change during the second round of replication. The third round of replication converts this error into a point mutation.

Deaminating agents also cause point mutations. A certain amount of base deamination (removal of an amino group) occurs spontaneously in cellular DNA molecules, with the rate being increased by chemicals such as nitrous acid, which deaminates adenine, cytosine and guanine, and sodium bisulfite, which acts only on cytosine. Thymine, of course, has no amino group and so cannot be deaminated. Deamination of adenine gives hypoxanthine (Figure 11.11), which pairs with C rather than T, and deamination of cytosine gives uracil, which pairs with A rather than G. Deamination of these two bases therefore results in point mutations when the template strand is copied. Deamination of guanine is bad for the cell because the resulting base, xanthine, blocks replication when it appears in the template polynucleotide, but this is not a mutagenic effect according to our definition of mutation.

Alkylating agents are a third type of mutagen that can give rise to point mutations. Chemicals such as **ethylmethane sulfonate (EMS)** and dimethylnitrosamine add alkyl groups to nucleotides in DNA molecules, as do methylating agents such as methyl halides which are present in the atmosphere, and the products of nitrite metabolism. The effect of alkylation depends on the position at which the nucleotide is modified and the type of alkyl group that is added. Methylations, for example, often result in modified nucleotides with altered base-pairing properties and so lead to point mutations. Other alkylations block replication by forming crosslinks between the two strands of a DNA molecule or by adding large alkyl groups that prevent the progress of the replisome.

The final type of chemical mutagen that we will look at are the **intercalating agents**. The best-known mutagen of this type is **ethidium bromide**, which is sometimes used as a stain for DNA because it fluoresces when exposed to ultraviolet radiation. Ethidium bromide and other intercalating agents are flat molecules that can slip between base pairs in the double helix, slightly unwinding the helix and hence increasing the distance between adjacent base pairs (Figure 11.12). It has been assumed that this is likely to lead specifically to an insertion or deletion, but exposure to an intercalating agent can also lead to other types of mutations.

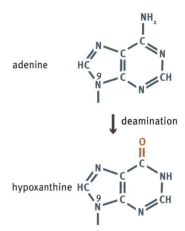

FIGURE 11.11 Hypoxanthine is a deaminated version of adenine. The nucleoside formed by hypoxanthine is called inosine.

There are also several types of physical mutagen

It is not only the chemical composition of the environment that can cause mutations in DNA molecules. There is also a variety of physical mutagens. An example is ultraviolet radiation, which induces dimerization of adjacent pyrimidine bases, especially if these are both thymines. This results in a **cyclobutyl dimer** (Figure 11.13a). Other pyrimidine combinations also form dimers, the order of frequency being 5′–CT–3′ > 5′–TC–3′ > 5′–CC–3′. Purine dimers are much less common. UV-induced dimerization usually results in a deletion mutation when the modified strand is copied. Another type of UV-induced **photoproduct** is the **(6–4) lesion** in which carbons number 4 and 6 of adjacent pyrimidines become covalently linked (Figure 11.13b).

Ionizing radiation is also mutagenic, having various effects on DNA depending on the type of radiation and its intensity. Point, insertion and/or deletion mutations might arise, as well as more severe forms of DNA damage that prevent replication. Some types of ionizing radiation act directly on DNA, and others act indirectly by stimulating the formation of reactive molecules such as peroxides in the cell.

Heat is mutagenic. Heat stimulates the water-induced cleavage of the β-N-glycosidic bond that attaches the base to the sugar component of the nucleotide (Figure 11.14a). This occurs more frequently with purines than

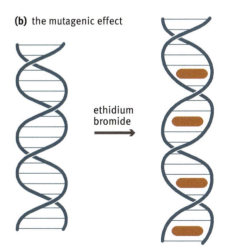

FIGURE 11.12 The mutagenic effect of ethidium bromide. (a) Ethidium bromide is a flat plate-like molecule that is able to slot in between the base pairs of the double helix. (b) Ethidium bromide molecules, viewed sideways on, are shown intercalated into a double helix, increasing the distance between adjacent base pairs.

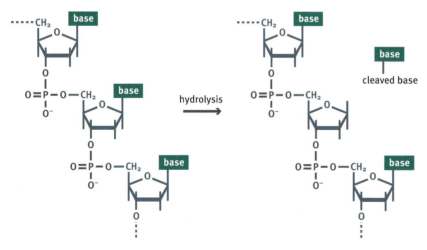

FIGURE 11.13 **Photoproducts induced by UV irradiation.** A segment of a polynucleotide containing two adjacent thymine bases is shown. (a) A thymine dimer contains two UV-induced covalent bonds, one linking the carbons at position 6 and the other linking the carbons at position 5. (b) The (6–4) lesion involves the formation of a UV-induced covalent bond between carbons 4 and 6 of the adjacent nucleotides. In (a) and (b), the UV-induced covalent bonds are shown as dotted lines.

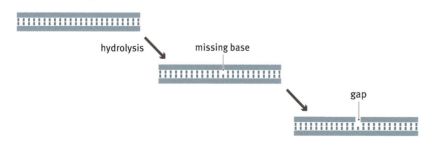

FIGURE 11.14 **The mutagenic effect of heat.** (a) Heat induces hydrolysis of β-N-glycosidic bonds, resulting in a baseless site in a polynucleotide. (b) Schematic representation of the effect of heat-induced hydrolysis on a double-stranded DNA molecule. The baseless site is unstable and degrades, leaving a gap in one strand.

with pyrimidines and results in an **AP** (apurinic/apyrimidinic) or **baseless site**. The sugar–phosphate that is left is unstable and rapidly degrades, leaving a gap if the DNA molecule is double stranded (Figure 11.14b). This reaction is not normally mutagenic because cells have effective systems for repairing gaps, which is reassuring when one considers that 10,000 AP sites are generated in each human cell per day.

11.2 DNA REPAIR

In view of the thousands of damage events that DNA molecules suffer every day, coupled with the errors that occur when DNA replicates, it is essential that cells possess efficient repair systems. Without these repair systems, it would only be a few hours before key genes became inactivated by DNA damage. Similarly, cell lineages would accumulate replication errors at such a rate that their DNA molecules would become dysfunctional after a few thousand cell divisions.

Most cells possess four different types of DNA repair system (Figure 11.15). The first of these are the **direct repair systems** which, as the name suggests, act directly on damaged nucleotides, converting each one back to its original structure. The second type is **excision repair**. This involves excision of a segment of the polynucleotide containing a damaged site, followed by resynthesis of the correct nucleotide sequence by a DNA polymerase. The segment removed may be just one nucleotide in length or much longer. A special type of excision repair is **mismatch repair**, which corrects errors in replication, again by excising a stretch of single-stranded DNA containing the offending nucleotide and then repairing the resulting gap. Finally, **break repair** is used to mend single- and double-strand breaks.

The repair processes are not particularly complicated, but we need to spend some time examining how each one works and what they achieve.

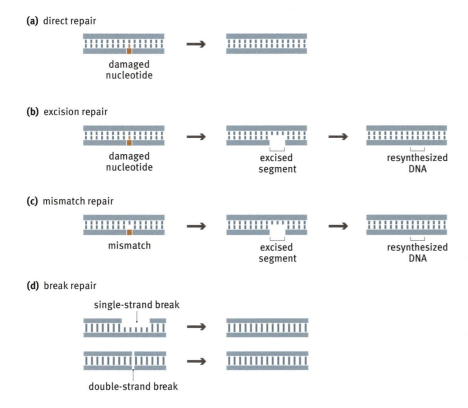

FIGURE 11.15 Four different types of DNA repair system. (a) Direct repair converts a damaged nucleotide back to its original structure. (b) Excision repair involves the removal of a segment of the polynucleotide containing a damaged nucleotide, followed by the resynthesis of the correct DNA sequence. (c) Mismatch repair is similar to excision repair, but is initiated by the detection of a base-pairing error. (d) Break repair mends single- and double-strand breaks.

Direct repair systems fill in nicks and correct some types of nucleotide modification

Most of the types of DNA damage that are caused by chemical or physical mutagens can only be repaired by excision of the damaged nucleotide followed by resynthesis of a new stretch of DNA, as shown in Figure 11.15b. Only a few types of damage can be repaired directly without removal of the nucleotide. These include some of the products of alkylation, which are directly reversible by enzymes that transfer the alkyl group from the nucleotide to their own polypeptide chains. Enzymes capable of doing this are present in many different organisms, and include the **Ada enzyme** of *E. coli*, which is involved in an <u>ada</u>ptive process that this bacterium is able to activate in response to DNA damage. Ada removes alkyl groups attached to the oxygen atoms at positions 4 and 6 of thymine and guanine, respectively, and can also repair phosphodiester bonds that have become methylated. Other alkylation repair enzymes have more restricted specificities, an example being human **MGMT** (O^6-methylguanine–DNA methyltransferase) which, as its name suggests, only removes alkyl groups from position 6 of guanine.

Cyclobutyl dimers resulting from UV damage can also be repaired directly, by a light-dependent system called **photoreactivation**. In *E. coli*, the process involves the enzyme called **DNA photolyase**. When stimulated by light with a wavelength between 300 and 500 nm, this enzyme binds to cyclobutyl dimers and converts them back to the original nucleotides. Photoreactivation is a widespread but not universal type of repair. It is known in many but not all bacteria and also in quite a few eukaryotes, including some vertebrates, but is absent in humans and other placental mammals. A similar type of photoreactivation involves the **(6–4) photoproduct photolyase** and results in the repair of (6–4) lesions. Neither *E. coli* nor humans have this enzyme but it is possessed by a variety of other organisms.

Many types of damaged nucleotide can be repaired by base excision

Base excision is the least complex of the various repair systems that involve the removal of one or more damaged nucleotides followed by the resynthesis of DNA to span the resulting gap. It is used to repair many modified nucleotides whose bases have suffered relatively minor damage resulting from, for example, exposure to alkylating agents or ionizing radiation.

The process is initiated by a **DNA glycosylase** that cleaves the β-*N*-glycosidic bond between a damaged base and the sugar component of the nucleotide (Figure 11.16a). Each DNA glycosylase has a limited specificity, the specificities of the glycosylases possessed by a cell determining the range of damaged nucleotides that can be repaired in this way. Most organisms are able to deal with deaminated bases such as uracil (deaminated cytosine) and hypoxanthine (deaminated adenine), oxidation products such as 5-hydroxycytosine and thymine glycol, and methylated bases such as 3-methyladenine, 7-methylguanine and 2-methylcytosine.

A DNA glycosylase removes a damaged base by 'flipping' the structure to a position outside of the helix and then detaching it from the polynucleotide. This creates an AP or baseless site, which is converted into a single-nucleotide gap in the second step of the repair pathway (Figure 11.16b). This step can be carried out in a variety of ways. The standard method makes use of an **AP endonuclease**, which cuts the phosphodiester bond on the 5′ side of the AP site. Some AP endonucleases can also remove the

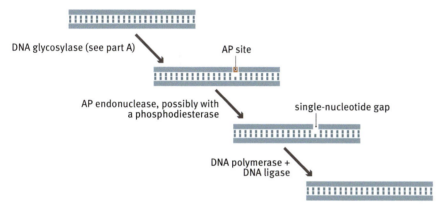

FIGURE 11.16 Base excision repair. (a) Excision of a damaged nucleotide by a DNA glycosylase. (b) Schematic representation of the base excision repair pathway.

sugar from the AP site, this being all that remains of the damaged nucleotide, but others lack this ability and so work in conjunction with a separate **phosphodiesterase** that cuts out the sugar. An alternative pathway for converting the AP site into a gap utilizes the endonuclease activity possessed by some DNA glycosylases, which can make a cut at the 3′ side of the AP site, probably at the same time that the damaged base is removed, followed again by removal of the sugar by a phosphodiesterase.

The single-nucleotide gap is filled by a DNA polymerase, using base pairing with the undamaged base in the other strand of the DNA molecule to ensure that the correct nucleotide is inserted. In *E. coli*, the gap is filled by DNA polymerase I and in mammals by DNA polymerase β. After gap filling, the final phosphodiester bond is put in place by a DNA ligase.

Nucleotide excision repair is used to correct more extensive types of damage

Nucleotide excision repair is a second type of excision repair system, able to deal with more extreme forms of damage such as intrastrand crosslinks and bases that have become modified by attachment of large chemical groups. It is also able to correct cyclobutyl dimers by a **dark repair** process, providing those organisms that do not have the photoreactivation system (such as humans) with a means of repairing this type of damage.

In nucleotide excision repair, a segment of single-stranded DNA containing the damaged nucleotide(s) is excised and replaced with new DNA.

The process is therefore similar to base excision repair except that it is not preceded by selective base removal, and a longer stretch of polynucleotide is cut out. The best-studied example of nucleotide excision repair is the **short patch** process of *E. coli*, so-called because the region of polynucleotide that is excised and subsequently 'patched' is relatively short, usually 12 nucleotides in length.

Short patch repair is initiated by a multienzyme complex called the **UvrABC endonuclease**. In the first stage of the process, a trimer comprising two UvrA proteins and one copy of UvrB attaches to the DNA at the damaged site (Figure 11.17). How the site is recognized is not known but the broad specificity of the process indicates that individual types of damage are not directly detected and that the complex must search for a more general attribute of DNA damage such as distortion of the double helix. UvrA may be the part of the complex most involved in damage location because it dissociates once the site has been found and plays no further part in the repair process.

Departure of UvrA allows UvrC to bind, forming a UvrBC dimer that cuts the polynucleotide on either side of the damaged site. The first cut is made by UvrB at the fifth phosphodiester bond downstream of the damaged nucleotide, and the second cut is made by UvrC at the eighth phosphodiester bond upstream. This results in the 12 nucleotide excision, although there is some variability, especially in the position of the UvrB cut site.

The excised segment is then removed, usually as an intact piece of DNA, by DNA helicase II, which detaches the segment by breaking the base pairs holding it to the second strand. UvrC also detaches at this stage, but UvrB remains in place and bridges the gap produced by the excision. The bound UvrB is thought to prevent the single-stranded region that has been exposed from base pairing with itself. Alternatively, the role of UvrB could be to prevent this strand from becoming damaged or possibly to direct the DNA polymerase to the site that needs to be repaired. As in base excision repair, the gap is filled by DNA polymerase I and the last phosphodiester bond is synthesized by DNA ligase.

E. coli also has a **long patch** nucleotide excision repair system that involves Uvr proteins but differs in that the piece of DNA that is excised can be anything up to 2 kb in length. Long patch repair has been less well studied and the process is not understood in detail, but it is presumed to work on more extensive forms of damage, possibly regions where groups of nucleotides, rather than just individual ones, have become modified. Eukaryotes have just one type of nucleotide excision repair pathway, resulting in the replacement of only 24–32 nucleotides of DNA, and seemingly unrelated to either of the excision pathways in bacteria.

Mismatch repair corrects errors in replication

Each of the repair systems that we have looked at so far – direct, base excision and nucleotide excision repair – recognize and act upon DNA damage caused by mutagens. This means that they search for abnormal chemical structures such as modified nucleotides and cyclobutyl dimers. They cannot, however, correct mismatches resulting from errors in replication because the mismatched nucleotide is not abnormal in any way; it is simply an A, C, G or T that has been inserted at the wrong position. As these nucleotides look exactly like any other nucleotide, the mismatch repair system that corrects replication errors has to detect not the mismatched nucleotide itself but the absence of base pairing between the parent and daughter strands. Once it has found a mismatch, the repair system excises

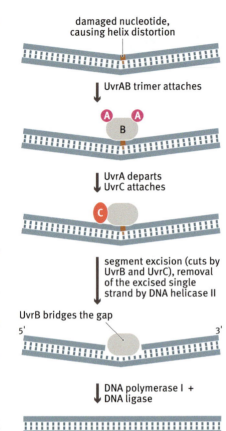

FIGURE 11.17 Shortpatch nucleotide excision repair in *E. coli*. The process begins with recognition of the damaged site by the UvrAB trimer. UvrA leaves the complex and UvrC joins, causing a cut to be made on either side of the damaged site. UvrB bridges the gap while it is repaired by DNA polymerase I and DNA ligase. In the top part of the drawing, the damaged nucleotide is shown distorting the helix because this is thought to be one of the recognition signals for the UvrAB trimer.

part of the daughter polynucleotide and fills in the gap, in a manner similar to base and nucleotide excision repair.

There is one difficulty. The repair must be made in the daughter polynucleotide because it is in this newly synthesized strand that the error has occurred. The parent polynucleotide, on the other hand, has the correct sequence. How does the repair process know which strand is which? In *E. coli*, the answer is that the parent strand is methylated and so can be distinguished from the newly synthesized polynucleotide, which is not methylated.

E. coli DNA is methylated because of the activities of two enzymes. These are the **DNA adenine methylase** (**Dam**), which converts adenines to 6-methyladenines in the sequence 5′–GATC–3′, and the **DNA cytosine methylase** (**Dcm**), which converts cytosines to 5-methylcytosines in 5′–CCAGG–3′ and 5′–CCTGG–3′. These methylations are not mutagenic: the modified nucleotides have the same base-pairing properties as the unmodified versions. There is a delay between DNA replication and methylation of the daughter strand, and it is during this window of opportunity that the repair system scans the DNA for mismatches and makes the required corrections in the unmethylated, daughter strand (Figure 11.18).

E. coli has at least three mismatch repair systems, called 'long patch', 'short patch' and 'very short patch', the names indicating the relative lengths of the excised and resynthesized segments. The long patch system replaces up to a kb or more of DNA and requires the MutH, MutL and MutS proteins, as well as the DNA helicase II that we met during nucleotide excision repair. MutS recognizes the mismatch and MutH distinguishes the two strands by binding to unmethylated 5′–GATC–3′ sequences (Figure 11.19). After binding, MutH cuts the phosphodiester

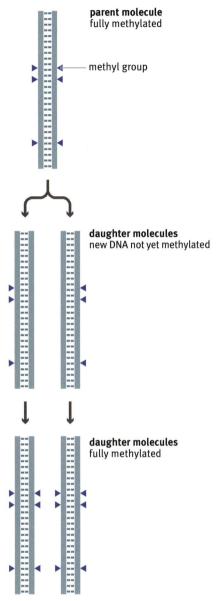

FIGURE 11.18 **Absence of methylation enables the daughter strands to be identified after DNA replication in *E. coli*.** Methylation of newly synthesized DNA in *E. coli* does not occur immediately after replication, providing a window of opportunity for the mismatch repair proteins to recognize the daughter strands and correct replication errors.

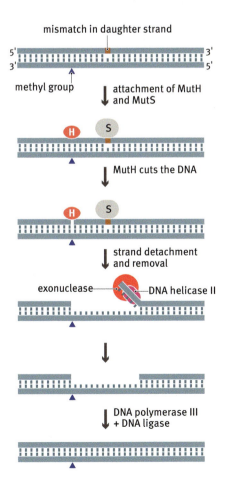

FIGURE 11.19 **Longpatch mismatch repair in *E. coli*.** The process begins with the attachment of MutS, which recognizes the mismatch, and MutH, which distinguishes the methylated and unmethylated polynucleotides. MutH cuts the unmethylated polynucleotide at the methylation sequence, and DNA helicase II detaches the single strand around the mismatch. The detached single strand is degraded by an exonuclease and the gap is filled by DNA polymerase III and DNA ligase.

bond immediately upstream of the G in the methylation sequence and DNA helicase II detaches the single strand from this point. There does not appear to be an enzyme that cuts the strand downstream of the mismatch. Instead, the detached single-stranded region is degraded by an exonuclease that follows the helicase and continues beyond the mismatch site. The gap is then filled in by DNA polymerase III and DNA ligase.

The mismatch repair processes of eukaryotes probably work in a similar way, although methylation might not be the method used to distinguish between the parent and daughter polynucleotides. Methylation has been implicated in mismatch repair in mammalian cells, but the DNA of some eukaryotes, including fruit flies and yeast, is not extensively methylated. It is therefore thought that in these organisms a different method must be used to identify the daughter strand. Possibilities include an association between the repair enzymes and the replisome, so that repair is coupled with DNA synthesis, or the use of single-strand binding proteins to mark the parent strand.

DNA breaks can also be repaired

Finally, we must study how breaks in DNA molecules are repaired. A single-stranded break in a double-stranded DNA molecule, such as is produced by some types of oxidative damage, does not present the cell with a critical problem as the double helix retains its overall intactness. In humans, the exposed strand is coated with PARP1 proteins, which protects it from damage. The break can then be filled in by DNA polymerase and ligase enzymes (Figure 11.20).

A double-stranded break is more serious because this converts the original double helix into two separate fragments, which have to be brought back together again in order for the break to be repaired. The two broken ends must be protected in some way because the loss of nucleotides would result in a deletion mutation appearing at the repaired break point. The repair processes must also ensure that the correct ends are joined. If there are two broken DNA molecules in a nucleus, then the correct pairs must be brought together so that the original structures are restored. Experimental studies indicate that achieving this outcome is difficult and if two molecules are broken then misrepair resulting in hybrid structures occurs relatively frequently. Even if only one molecule is broken, there is still a possibility that a natural end could be confused as a break and an incorrect repair made.

Double-strand breaks are generated by exposure to ionizing radiation and some chemical mutagens, and breakages can also occur during DNA replication. These breaks can be repaired by the **nonhomologous end-joining** (**NHEJ**) process, which involves a pair of proteins, called the **Ku70–Ku80 heterodimer**, which forms a loop structure that encloses each cut end of a DNA molecule (Figure 11.21). It is thought that the heterodimers have an affinity for one another, which means that the two broken ends of the DNA molecule are brought into proximity and can be joined back together by a DNA ligase.

NHEJ was originally thought to be restricted to eukaryotes, but searches of the protein databases have uncovered bacterial versions of the Ku proteins. Experimental studies have indicated that these act in conjunction with bacterial ligases in a simplified type of double-strand break repair.

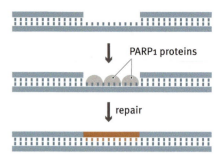

FIGURE 11.20 Single-strand break repair.

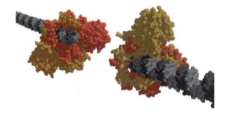

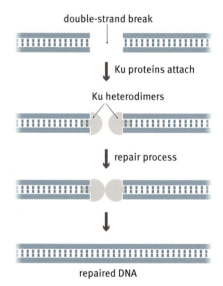

FIGURE 11.21 Repair of a double-strand break by nonhomologous end joining. (a) A model showing two Ku70–Ku80 heterodimers bound to the ends of a broken DNA molecule. The Ku70 subunit is colored red and Ku80 is yellow. DNA is depicted with one dark gray and one light gray strand. (b) The repair process. It is thought that an affinity between the two heterodimers draws the two ends together so the break can be repaired by a DNA ligase. (a, courtesy of Jonathan Goldberg, Howard Hughes Medical Institute.)

In an emergency, damaged nucleotides can be bypassed during DNA replication

If part of a DNA molecule has suffered extensive damage, then it is conceivable that the repair processes will be overwhelmed. The cell then faces a stark choice between dying or attempting to replicate the damaged region even though this replication may be error-prone and result in mutated daughter molecules. When faced with this choice, E. coli cells invariably take the second option, by inducing one of several emergency procedures for bypassing sites of major damage.

The best studied of these bypass processes occurs as part of the **SOS response**. This system enables an E. coli cell to replicate its DNA even though the template polynucleotides contain AP sites and/or photoproducts that would normally block, or at least delay, the replication complex. Bypass of these sites requires the construction of a **mutasome**, comprising DNA polymerase V (also called the UmuD′$_2$C complex) and several copies of a single-stranded DNA-binding protein called RecA. In this bypass system, RecA coats the damaged strands, enabling DNA polymerase V to carry out its own, error-prone, DNA synthesis until the damaged region has been passed (Figure 11.22). The main replicating enzyme, DNA polymerase III, then takes over once again.

The SOS response is primarily looked as the last best chance that the bacterium has to replicate its DNA and hence survive under adverse conditions. However, the price of survival is an increased mutation rate because the mutasome does not repair damage, it simply allows a damaged region of a polynucleotide to be replicated. When it encounters a damaged position, DNA polymerase V selects a nucleotide more or less at random, although with some preference for placing an A opposite an AP site.

Defects in DNA repair underlie human disorders, including cancers

The importance of DNA repair is emphasized by the number and severity of inherited human disorders that have been linked with defects in one of the repair processes. One of the best characterized of these is xeroderma pigmentosum, which results from a mutation in any one of several genes for proteins involved in nucleotide excision repair. Nucleotide excision is the only way in which human cells can repair cyclobutyl dimers and other photoproducts, so it is no surprise that the symptoms of xeroderma pigmentosum include hypersensitivity to UV radiation, patients suffering more mutations than normal on exposure to sunlight, and often suffering from skin cancer.

Defects in nucleotide excision repair are also linked with breast and ovarian cancers, and Cockayne syndrome, a complex disorder manifested by growth and neurological disorders. Ataxia telangiectasia, the symptoms of which include sensitivity to ionizing radiation, results from defects in the *ATX* gene, which is involved in the damage detection process. Other disorders that are associated with a breakdown in DNA repair are Bloom's and Werner's syndromes, which are caused by inactivation of a DNA helicase that may have a role in NHEJ, the cancer-susceptibility syndrome called HNPCC (hereditary nonpolyposis colorectal cancer), which results from a defect in mismatch repair, and some types of spinocerebellar ataxia, which arise from defects in the pathway used to repair single-strand breaks.

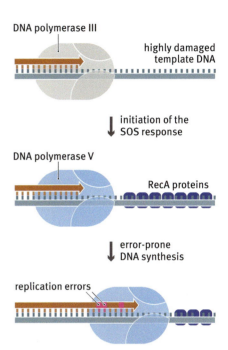

FIGURE 11.22 The SOS response of *E. coli*. During DNA replication, a highly damaged region can be bypassed by the SOS response. DNA polymerase III, the main replicating enzyme, detaches and is replaced by the mutasome, a combination of DNA polymerase V and the RecA protein. RecA coats the damaged polynucleotides, enabling DNA polymerase V to carry out error-prone DNA synthesis through the damaged region. DNA polymerase III then takes over again.

11.3 THE EFFECTS OF MUTATIONS ON GENES, CELLS AND ORGANISMS

Now that we understand how mutations arise in DNA molecules and appreciate that many of them are repaired before they can be propagated by DNA replication, we must consider the effect of those few mutations that slip through. We will think about their effects at three different levels: on genes, on cells and on organisms.

Mutations have various effects on the biological information contained in a gene

Many mutations are **silent** because they occur in intergenic regions. They result in changes in the nucleotide sequence but have no effect on the biological information contained in a DNA molecule. But what are the likely effects if a mutation occurs in the coding region of a gene? To answer this question, we will invent a short imaginary gene (Figure 11.23) which, although coding for a protein of only six amino acids in length, will show us most of the possible consequences.

FIGURE 11.23 A short imaginary gene.

The first possibility is that a point mutation takes place at the third nucleotide position of a codon and changes the codon but, owing to the degeneracy of the genetic code, not the amino acid. In the example shown in Figure 11.24, nucleotide 15 is changed from A to G. This changes the fifth codon of our gene from 5′-TTA-3′, coding for leucine, to 5′-TTG-3′, which also codes for leucine. This is called a **synonymous mutation**, and as it has no effect on the amino acid sequence of the gene product it is silent as far as the cell is concerned.

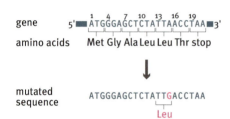

FIGURE 11.24 A silent mutation within a gene. Nucleotide 15 has been changed from A to G, but because of the degeneracy of the genetic code, the amino acid sequence of the protein remains the same.

A **nonsynonymous mutation**, on the other hand, is a point change that does change the amino acid sequence of the protein coded by the gene. Most point changes at the first or second nucleotide positions of a codon result in nonsynonymous mutations, as do a few third-position changes. In our gene, changing nucleotide 4 from G to A produces an arginine codon instead of a glycine codon (Figure 11.25). Similarly, changing nucleotide 15 from A to T specifies phenylalanine rather than leucine. A nonsynonymous mutation gives rise to a protein with a single amino acid change. Whether or not this changes the activity of the protein depends on the precise role of the mutated amino acid. Most proteins can tolerate some changes in their amino acid sequence, although a nonsynonymous mutation that alters an amino acid essential for structure or function will inactivate the protein.

Other mutations are much more likely to result in a nonfunctioning protein. A **nonsense mutation** (Figure 11.26) is a point mutation that changes a codon specifying an amino acid into a termination codon. The result is a truncated gene that codes for a protein that has lost a segment

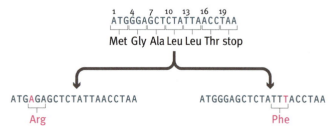

FIGURE 11.25 Two nonsynonymous mutations. Changing nucleotide 4 from G to A produces an arginine codon instead of a glycine codon, and changing nucleotide 15 from A to T gives phenylalanine rather than leucine.

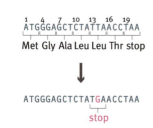

FIGURE 11.26 A nonsense mutation. Changing nucleotide 14 from T to G converts the leucine codon into a termination codon.

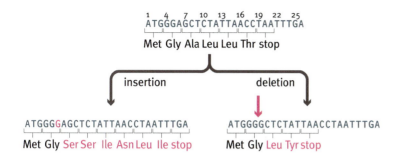

FIGURE 11.27 **Two frameshift mutations.** The frameshift mutation on the left is caused by inserting a G between nucleotides 5 and 6. The one on the right involves deleting nucleotide 6.

at its carboxyl terminus. In many cases, although not always, this segment will include amino acids essential for the protein's activity. A **frameshift mutation** (Figure 11.27) is the usual consequence of an insertion or deletion event. This is because the addition or removal of any number of base pairs that is not a multiple of three causes the ribosome to read a completely new set of codons downstream from the mutation. Finally, a **readthrough mutation** (Figure 11.28) converts a termination codon into one specifying an amino acid. This results in a readthrough of the stop signal so the protein is extended by an additional series of amino acids at its carboxyl terminus. Most proteins can tolerate short extensions without an effect on function, but longer extensions might interfere with the folding of the protein and so result in reduced activity.

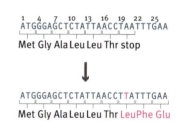

FIGURE 11.28 **A readthrough mutation.** Changing nucleotide 20 from A to T converts the termination codon into one specifying leucine.

Mutations have various effects on multicellular organisms

In order to affect the activity of a cell, a nucleotide sequence alteration must give rise to a protein that is unable to fulfill its function. In many cases, the cell will be unable to tolerate the loss of function and will die. Such mutations are therefore called **lethal mutations**. However, not all mutations that produce a change in protein activity are so drastic in their effect. Some mutations inactivate proteins that are not essential to the cell, and others result in proteins with reduced or modified activities. What effects would these nonlethal mutations have on a multicellular organism?

We must distinguish between mutations that occur in somatic cells, the ones that make up the bulk of the organism but are not passed on during reproduction and those in sex cells. Because somatic cells do not pass copies of their DNA molecules to the next generation, a somatic cell mutation is important only for the organism in which it occurs. It has no potential evolutionary impact. In fact, most somatic cell mutations have no significant effect, even if they result in cell death, because there are many other identical cells in the same tissue and the loss of one cell is immaterial. An exception is when a mutation causes a somatic cell to malfunction in a way that is harmful to the organism, for instance, by inducing tumor formation or other cancerous activity.

Mutations in sex cells are more important because they can be transmitted to members of the next generation and will then be present in all the cells of any individual who inherits the mutation. Most mutations, including all silent ones and many in coding regions, will not affect the organism in any significant way. Those that do have an effect can be divided into two categories, **loss of function** and **gain of function**.

Loss of function is the normal result of a mutation that reduces or abolishes a protein activity. Because most eukaryotes are diploid, meaning that they have two copies of each DNA molecule and hence two copies of each gene, a loss-of-function mutation might not be critical, as the second copy of the gene is still active and compensates for the loss. There

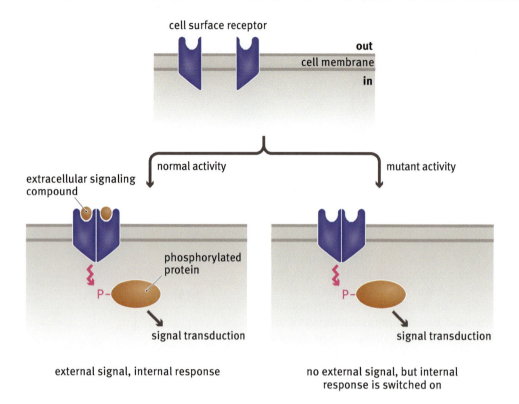

FIGURE 11.29 One way in which a gain of function can arise. The top drawing shows a cell surface receptor. The normal activity is shown on the left. The presence of the external signaling compound causes the receptor to dimerize and activate an internal protein. This protein initiates a signal transduction pathway that switches on a particular set of genes whose products are needed only when the external stimulus is present. On the right is the mutant activity. Now the receptor protein dimerizes even though there is no external signaling compound. The genes targeted by the signal transduction pathway are therefore switched on all the time, and the mutant cell gains the function specified by these genes.

are some exceptions, one example being **haploinsufficiency**, where the organism is unable to tolerate the approximately 50% reduction in protein activity resulting from the loss of one of the gene copies.

Gain-of-function mutations are much less common. The mutation must be one that confers an abnormal activity on a protein. Examples are known with proteins that act as cell surface receptors. These proteins respond to external stimuli by sending into the cell signals that switch genes on or off as appropriate. A mutation could result in the receptor protein sending signals into the cell even though the external stimulus is absent (Figure 11.29). The cell therefore acts as if it is continuously being affected by the stimulus, resulting in a gain of function. Gain-of-function mutations cannot be compensated by the presence of the second, unmutated gene copy in a diploid organism.

A second mutation may reverse the phenotypic effect of an earlier mutation

There are various ways in which the effect of a mutation can be reversed by a second mutation, the simplest being when the second mutation restores the original nucleotide sequence of the DNA molecule. A point mutation can be reversed by a second point mutation, an insertion event by a subsequent deletion and so on. These events are called **back mutations** and are not very likely unless the site at which the original mutation occurred has some natural predisposition toward mutation.

Many mutations can also be corrected by a **second-site reversion**. This occurs when a second mutation reverses the effect of the first mutation, but does not return the DNA sequence to its precise unmutated form. To illustrate one of the ways in which this might occur, we will return to the made-up gene that we used earlier to illustrate some of the effects of mutations. One of the nonsynonymous mutations shown in Figure 11.25 involved changing nucleotide 15 from A to T, changing a leucine codon (5'–TTA–3') into a codon for phenylalanine (5'–TTT–3'). If we now introduce a second mutation, changing nucleotide 13 from T to C, the codon becomes 5'–CTT–3', which codes for leucine once more (Figure 11.30). The nonsynonymous mutation has been corrected and the protein coded by the gene returned to normal even though the nucleotide sequence remains different from the original.

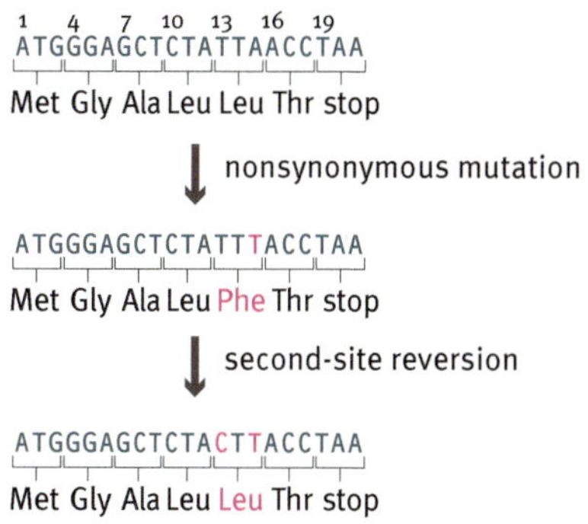

A second-site reversion occurs within the same gene as the original mutation. The final way of reversing the effect of a mutation is by **suppression**, in which the second mutation occurs in a different gene. There are several examples of this phenomenon but the most common is the reversal of a nonsense mutation by a suppressive mutation in the anticodon of a tRNA. For instance, the anticodon for tRNA^{trp} is 5'–CCA–3', to decode the tryptophan codon 5'–UGG–3'. If the first nucleotide of the anticodon in the tRNA is changed by mutation from C to U, then the anticodon becomes 5'–UCA–3' and now decodes the termination codon 5'–UGA–3' (Figure 11.31a). This termination codon is therefore read as a tryptophan codon during protein synthesis. The shortened protein that we produced by a nonsense mutation in Figure 11.26 would now be restored to full length (Figure 11.31b). Note, however, that the amino acid inserted into the protein when the suppressed termination codon is read is not the one found at the corresponding position in the unmutated version. As with a nonsynonymous mutation, such a small alteration in the amino acid sequence can probably be tolerated by the protein, so the activity is restored.

FIGURE 11.30 An example of second-site reversion. The first mutation changes nucleotide 15 from A to T, converting a leucine codon into one for phenylalanine. The second mutation changes nucleotide 13 from T to C, causing a reversion back to a leucine codon, without restoring the original nucleotide sequence.

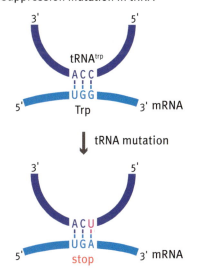

(a) suppression mutation in tRNA^{trp}

(b) suppression of a nonsense mutation

FIGURE 11.31 Suppression. (a) If the first nucleotide (reading in the 5' → 3' direction) of the tRNA^{trp} anticodon is changed from C to U, then the new anticodon will be able to base pair with the termination codon 5'–UGA–3'. (b) A nonsense mutation in our imaginary gene has converted a leucine codon into a termination codon. This mutation could be suppressed by the mutant tRNA^{trp}, which will read the termination codon as one specifying tryptophan.

KEY CONCEPTS

- A mutation is a change in the nucleotide sequence of a DNA molecule. Mutations can arise as errors in replication or from the effects of chemical or physical mutagens.

- All organisms have DNA repair mechanisms, which correct the vast majority of the mutations that occur in their DNA molecules.

- Most cells possess four different types of DNA repair system. Direct repair systems act directly on damaged nucleotides, converting each one back to its original structure. Excision repair involves the excision of a segment of the polynucleotide containing a damaged site, followed by the resynthesis of the correct nucleotide sequence. Mismatch repair corrects errors of replication. Break repair is used to mend single- and double-strand breaks.

- In bacteria, mismatches resulting from errors in replication can be repaired correctly because the parent strand, which contains the correct sequence, is methylated. Some eukaryotes use a different method to distinguish the parent and daughter strands.

- The importance of DNA repair is emphasized by the number and severity of inherited human diseases that have been linked with defects in one of the repair processes.

- Mutations can have various effects on the biological information contained in a gene. Some have no effect. At the other extreme, some mutations result in the loss of a protein function.

- Some mutations alter the protein product in such a way that a gain of function results. The mutated cell possesses a biochemical activity that is absent from the unmutated version.

- The effects of a mutation can be reversed by back mutation, second-site reversion or suppression.

QUESTIONS AND PROBLEMS

Key terms

Write short definitions of the following terms:

(6–4) lesion
(6–4) photoproduct photolyase
Ada enzyme
alkylating agent
2-aminopurine
AP endonuclease
AP site
back mutation
base analog
base excision
baseless site
break repair
5-bromouracil
carcinogen
clastogen
cyclobutyl dimer
dark repair
deaminating agent
deletion
direct repair
DNA adenine methylase
DNA cytosine methylase
DNA glycosylase
DNA photolyase
DNA repair
ethidium bromide
ethylmethane sulfonate
excision repair
frameshift mutation
gain of function
haploinsufficiency
insertion
intercalating agent
inversion
Ku70–Ku80 heterodimer
lethal mutation
long patch repair
loss of function
MGMT
mismatch
mismatch repair
mutagen
mutasome
mutation
nonhomologous end joining

nonsense mutation
nonsynonymous mutation
nucleotide excision
oncogen
phosphodiesterase
photoproduct
photoreactivation
point mutation
readthrough mutation
replication slippage
second-site reversion
short patch repair
silent mutation
SOS response
suppression
synonymous mutation
tautomer
teratogen
transition
transversion
trinucleotide repeat expansion disease
UvrABC endonuclease

Self-study questions

11.1 What is the difference between a transition and a transversion mutation?

11.2 Distinguish between the following terms: (a) point mutation, (b) insertion mutation, (c) deletion mutation and (d) inversion mutation.

11.3 Describe how tautomerism of nucleotide bases can lead to errors occurring during the replication process.

11.4 What is replication slippage, and how can it lead to an error in replication?

11.5 What is the link between replication slippage and a trinucleotide repeat expansion disease?

11.6 Describe the difference between a mutagen and a clastogen.

11.7 Using examples, outline how base analogs result in mutations.

11.8 Distinguish between the mutagenic effects of deaminating agents, alkylating agents and intercalating agents.

11.9 What is the difference between a cyclobutyl dimer and a (6–4) lesion? How does each type of structure arise?

11.10 Why is heat mutagenic?

11.11 Distinguish between the direct, excision and mismatch systems for DNA repair.

11.12 Give examples that illustrate the direct repair of DNA damage in bacteria and eukaryotes.

11.13 Describe the base excision repair process, paying particular attention to the mode of action of DNA glycosylases.

11.14 Explain how nucleotide excision repair results in the correction of mutations in *E. coli*.

11.15 Describe the importance of DNA methylation in the mismatch repair system of *E. coli*.

11.16 Outline how double-strand breaks are repaired by nonhomologous end joining.

11.17 What is a mutasome, and why is the error-prone replication that it directs sometimes of value to an *E. coli* cell?

11.18 Give examples of human diseases that have been linked to defects in DNA repair.

11.19 Explain why nonsense and frameshift mutations are more likely to result in an altered protein function than nonsynonymous mutations.

11.20 Which of the following types of mutation could be caused by a point change in a nucleotide sequence? (a) Nonsynonymous, (b) nonsense, (c) frameshift, (d) back, (e) second-site reversion and (f) suppression. Which could be caused by an insertion?

11.21 Here is the nucleotide sequence of a short gene:

5′–ATGGGTCGTACGACCGGTAGTTACTGGTTCAGTTAA–3′

Write out the amino acid sequence of the protein coded by this gene. Now introduce the following mutations and describe their effects on the protein: (a) a silent mutation, (b) a nonsynonymous mutation, (c) a nonsense mutation and (d) a frameshift mutation.

11.22 Give one example of the way in which the effects of a mutation can be reversed by suppression.

Discussion topics

11.23 Explain why a purine-to-purine or pyrimidine-to-pyrimidine point mutation is called a transition, whereas a purine-to-pyrimidine (or vice versa) change is called a transversion.

11.24 What would be the anticipated ratio of transitions to transversions in a large number of mutations?

11.25 In some eukaryotes, the mismatch repair process is able to recognize the daughter strand of a double helix even though the two strands lack distinctive methylation patterns. Propose a mechanism by which the daughter strand can be recognized in the absence of methylation. How would you test your hypothesis?

11.26 Explain how point mutations could result in each of the following codon changes:

   a. a glycine codon to an alanine codon
   b. a tryptophan codon to a termination codon
   c. a cysteine codon to an arginine codon
   d. a serine codon to an isoleucine codon
   e. a serine codon (either AGT or AGC) to a threonine codon
   f. the previous mutation, back to serine but without recreating the original nucleotide sequence.

11.27 The example of suppression described in the text ought to result in all UGA termination codons being read through. Clearly, this would result in a lot of elongated polypeptides. Would you expect this to be lethal? If not, why not? If so, then what might be the precise nature of this type of suppressive mutation?

11.28 Propose a scheme for a suppressive mutation that does not involve changing a tRNA gene.

The answers to, or hints on how to answer, the self-study questions and discussion topic questions can be downloaded from the book's product page at this link: www.routledge.com/9781032743530

FURTHER READING

Bertrand C, Thibessard A, Bruand C et al. (2019) Bacterial NHEJ: a never ending story. *Molecular Microbiology* 111, 1139–1151.

Caldecott KW (2020) Mammalian base excision repair: dancing in the moonlight. *DNA Repair* 93, 102921.

Chang HHY, Pannunzio NR, Adachi N et al. (2017) Non-homologous DNA end joining and alternative pathways to double-strand break repair. *Nature Reviews Molecular Cell Biology* 18, 495–506.

Chatterjee N & Walker GC (2017) Mechanisms of DNA damage, repair, and mutagenesis. *Environmental and Molecular Mutagenesis* 58, 235–263.

Hearst JE (1995) The structure of photolyase: using photon energy for DNA repair. *Science* 268, 1858–1859.

Hsieh P & Zhang Y (2017) The Devil is in the detail for DNA mismatch repair. *Proceedings of the National Academy of Sciences USA* 114, 3552–3554.

Hung K-F, Sidorova JM, Nghiem P et al. (2020) The 6–4 photoproduct is the trigger of UV-induce replication blockage and ATR activation. *Proceedings of the National Academy of Sciences USA* 117, 12806–12816.

Kamileri I, Karakasilioti I & Garinis GA (2012) Nucleotide excision repair: new tricks with old bricks. *Trends in Genetics* 28, 566–573.

Krokan HE & Bjørås M (2013) Base excision repair. *Cold Spring Harbor Perspectives in Biology* 5, a012583.

Kunkel TA & Bebenek K (2000) DNA replication fidelity. *Annual Review of Biochemistry* 69, 497–529. *Covers the processes that ensure that the minimum number of errors are made during DNA replication.*

Kusajabe M, Onishi Y, Tada H et al. (2019) Mechanism and regulation of DNA damage recognition in nucleotide excision repair. *Genes and Environment* 41, 2.

Lieber MR (2010) The mechanism of double-strand break repair by the nonhomologous DNA end-joining pathway. *Annual Review of Biochemistry* 79, 181–211.

O'Driscoll M (2012) Diseases associated with defective responses to DNA damage. *Cold Spring Harbor Perspectives in Biology* 4, a012773.

Paulson H (2019) Repeat expansion diseases. *Handbook of Clinical Neurology* 147, 105–123.

Sutton MD, Smith BT, Godoy VG & Walker GC (2000) The SOS response: recent insights into *umuDC*-dependent mutagenesis and DNA damage tolerance. *Annual Review of Genetics* 34, 479–497.

Yi C & He C (2013) DNA repair by reversal of DNA damage. *Cold Spring Harbor Perspectives in Biology* 5, a012575.

GENES AS UNITS OF INHERITANCE

12 Inheritance of Genes During Virus Infection Cycles

13 Inheritance of Genes in Bacteria

14 Inheritance of Genes During Eukaryotic Cell Division

15 Inheritance of Genes During Eukaryotic Sexual Reproduction

16 Inheritance of Genes in Populations

CHAPTER 12

Inheritance of Genes During Virus Infection Cycles

We have completed our study of genes as units of biological information. We have learnt that biological information is coded within the nucleotide sequence of a gene and that this information is made available to the cell by the multistep process called gene expression. We have followed that process from the initial transcription of a gene through to the eventual synthesis of its noncoding RNA or protein product, and we have examined how those products are degraded when they are no longer needed by the cell. We discovered that various strategies are used to control gene expression, either by determining if a gene is switched on or off or by influencing the rate at which the product is synthesized once a gene is switched on. Finally, we studied the process by which DNA molecules replicate and we surveyed the ways in which mutations occur and are repaired.

We must now focus on the second role of genes, as units of inheritance. This means that we must understand how a copy of the **genome** is passed to each of the progeny when an organism reproduces. The genome is the entire collection of DNA molecules, containing all the genes and hence all the biological information possessed by an organism. We will approach this question in stages, starting with the simplest forms of life – the viruses – and gradually progressing to the more complex issues involving genome inheritance in eukaryotes.

Viruses are so simple in biological terms that we have to ask ourselves if they can really be thought of as living organisms. Doubts arise partly because viruses are constructed differently from all other forms of life – viruses are not cells – and partly because of the nature of the virus life cycle. Viruses are obligate parasites of the most extreme kind. They reproduce only within a host cell, and in order to replicate and express their genes they must subvert at least part of the host's genetic machinery to their own ends. Some viruses possess genes coding for their own DNA polymerase and RNA polymerase enzymes, but many depend on the host enzymes for DNA replication and transcription. All viruses make use of the host's ribosomes and translation apparatus for the synthesis of polypeptides that make up the protein coats of their progeny. This means that virus genes must be matched to the host genetic system. Viruses are therefore quite specific for particular organisms, and individual types cannot infect a broad spectrum of species.

There is a multitude of different types of virus and planning a strategy for studying them can be quite difficult. The best approach is to follow history. Geneticists in the 1930s chose **bacteriophages** – the viruses that infect bacteria – as model organisms with which to study virus infection cycles. This was because it is relatively easy to grow cultures of bacteria that are infected with bacteriophages. In later years, attention turned to viruses that infect eukaryotes, including the **retroviruses**, the group to which the human immunodeficiency viruses that cause AIDS belong.

We will follow this same approach. First, we will become familiar with the basic features of virus life cycles by examining the bacteriophages. Then, we will move on to eukaryotic viruses with our focus largely on the retroviruses, not because these are the only important eukaryotic viruses, but because they display features that we do not encounter among the bacteriophages. Our study of retroviruses will then lead us to the very edge of life.

DOI: 10.1201/9781003473862-14

12.1 BACTERIOPHAGES

Bacteriophages are common but frequently overlooked members of the natural environment. From time to time, research has been directed at the possibility that they might be used to control or treat bacterial infections. As the antibiotic age draws to a close, bacteriophages may once again be looked on as potential benefactors of the human race. We are mainly interested in how genes are inherited during bacteriophage infection cycles, but before studying this topic we must become familiar with the bacteriophages themselves.

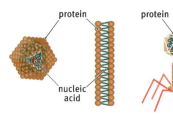

FIGURE 12.1 The three types of capsid structure commonly displayed by bacteriophages.

Bacteriophages have diverse structures and equally diverse genomes

Bacteriophages are constructed primarily from two components, protein and nucleic acid. The protein forms a coat or **capsid** within which the nucleic acid genome is contained.

There are three basic capsid structures (Figure 12.1). The first of these is **icosahedral**, in which the individual protein subunits (**protomers**) are arranged into a three-dimensional geometric structure that surrounds the nucleic acid. Examples of icosahedral bacteriophages are MS2, which infects *Escherichia coli*, and PM2, which infects *Pseudomonas aeruginosa*. The second type has **filamentous** capsids, in which the protomers are arranged in a helix producing a rod-shaped structure. The *E. coli* bacteriophage called M13 is an example. The **head-and-tail** bacteriophages combine the features of the other two types. Their capsid is made up of an icosahedral head, containing the nucleic acid, and a filamentous tail that facilitates entry of the nucleic acid into the host cell. They may also have other structures, such as the 'legs' possessed by the *E. coli* bacteriophage T4.

The term 'nucleic acid' has to be used when referring to bacteriophage genomes because in some cases these molecules are made of RNA. Viruses are the one form of 'life' in which the genetic material is not always DNA. Bacteriophages and other viruses also break another rule. Their genomes, whether of DNA or RNA, can be single stranded as well as double stranded.

A whole range of different genome structures is known among the bacteriophages, as summarized in Table 12.1. With most types of bacteriophage, there is a single DNA or RNA molecule that comprises the entire genome. A few RNA bacteriophages have **segmented genomes**, meaning that their genes are carried by a number of different RNA molecules. The sizes of bacteriophage genomes vary enormously, from about 1.6 kb for the smallest bacteriophages to over 150 kb for large ones such as T2, T4 and T6.

Bacteriophage genomes, being relatively small, were among the first to be studied comprehensively by the rapid and efficient DNA sequencing methods that were developed in the late 1970s. Gene numbers vary from just three in the case of MS2 to over 200 for the more complex head-and-tail bacteriophages (see Table 12.1). The smaller bacteriophage genomes of course contain relatively few genes, but these can be organized in a very complex manner. Phage φX174, for example, manages to pack into its genome 'extra' biological information as several of its genes overlap (Figure 12.2). These **overlapping genes** share nucleotide sequences (gene *B*, for example, is contained entirely within gene *A*), but code for different gene products as the transcripts are translated from different

TABLE 12.1 FEATURES OF SOME TYPICAL BACTERIOPHAGES AND THEIR GENOMES

Phage	Host	Capsid Structure	Genome Structure	Genome Size (kb)	Number of Genes
λ	Enterobacteria	Head-and-tail	Double-stranded linear DNA	48.5	73
M13	Enterobacteria	Filamentous	Single-stranded circular DNA	6.4	10
MS2	Enterobacteria	Icosahedral	Single-stranded linear DNA	3.6	4
φ6	*Pseudomonas*	Icosahedral	Double-stranded segmented linear RNA	2.9, 4.0, 6.4	13
φX174	Enterobacteria	Icosahedral	Single-stranded circular DNA	5.4	11
PM2	*Pseudoalteromonas*	Icosahedral	Double-stranded circular DNA	10.0	22
SP01	*Bacillus*	Head-and-tail	Double-stranded linear DNA	133	204
T4	Enterobacteria	Head-and-tail	Double-stranded linear DNA	169	278
T7	Enterobacteria	Head-and-tail	Double-stranded linear DNA	39.9	60

The genome structure is that in the phage capsid. Some genomes exist in different forms within the host cell.

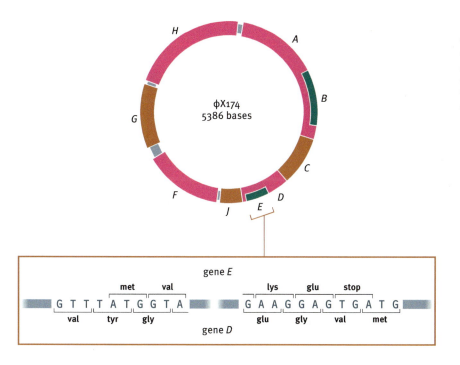

FIGURE 12.2 The φX174 genome contains overlapping genes. The expanded region shows the DNA sequence at the start and end of the overlap between genes *E* and *D*, showing that the reading frames used by these two genes are different.

start positions and in different reading frames. Overlapping genes are not uncommon in viruses.

The larger bacteriophage genomes contain more genes, reflecting the more complex capsid structures of these bacteriophages and a dependence on a greater number of bacteriophage-encoded enzymes during the infection cycle. The T4 genome, for example, includes some 50 genes involved solely in the construction of the bacteriophage capsid. Despite

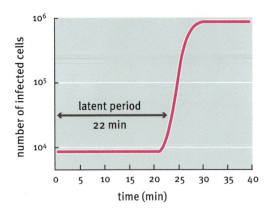

FIGURE 12.3 **The time course of the T4 infection cycle.** There is no change in the number of infected cells during the first 22 minutes of infection. Then, the number of infected cells starts to increase, showing that lysis of the original hosts is occurring and that the new bacteriophages that are being produced are infecting other cells.

their complexity, even these large bacteriophages still require at least some host-encoded proteins and RNAs in order to carry through their infection cycles.

The lytic infection cycle of bacteriophage T4

Bacteriophages are classified into two groups according to their life cycle. The two types are called **lytic** and **lysogenic**. The fundamental difference between these groups is that a lytic bacteriophage kills its host bacterium very soon after the initial infection, usually within 30 minutes, whereas a lysogenic bacteriophage can remain quiescent within its host for a substantial period of time, even throughout numerous generations of the host cell. These two life cycles are typified by two *E. coli* bacteriophages, the lytic (or **virulent**) T4 and the lysogenic (or **temperate**) λ.

The T4 infection cycle can be followed by adding T4 bacteriophages to a culture of *E. coli*, waiting 3 minutes for the bacteriophages to attach to the bacteria, and then measuring the number of infected cells over a period of 40 minutes (Figure 12.3). The first interesting feature of the infection cycle is that there is no change in the number of infected cells during the first 22 minutes of infection. This **latent period** is the time needed for the bacteriophages to reproduce within their hosts. After 22 minutes, the number of infected cells starts to increase, showing that lysis of the original hosts is occurring and that the new bacteriophages that have been produced are now infecting other cells in the culture. This is called the **one-step growth curve**.

The molecular events occurring during the one-step growth curve are as follows. The initial event is the attachment of the bacteriophage to a receptor protein on the outside of the bacterium (Figure 12.4). Different bacteriophages have different receptors. For T4, the receptor is a protein called OmpC. OmpC is a type of outer membrane protein called a porin, which forms a channel through the membrane and facilitates the uptake of nutrients. After attachment, the bacteriophage injects its DNA genome into the cell through its tail structure.

The latent period now begins. The name is, in fact, a misnomer. Although to the outside observer, nothing very much seems to happen, inside the cell there is frenzied activity directed at the synthesis of new bacteriophage particles. Immediately after entry of the bacteriophage DNA, the synthesis of host DNA, RNA and protein stops, and transcription of the bacteriophage genome begins. Within 5 minutes, the bacterial DNA molecule has been broken down and the resulting nucleotides are being utilized in replication of the T4 genome. After 12 minutes, new bacteriophage capsid proteins start to appear and the first complete bacteriophage

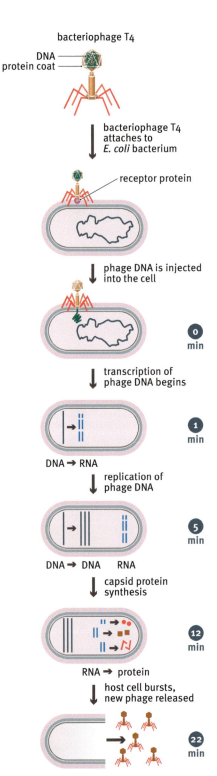

FIGURE 12.4 **Events occurring during the T4 infection cycle.**

particles are assembled. Finally, at the end of the latent period, the cell bursts and the new bacteriophages are released. A typical infection cycle produces 200–300 T4 bacteriophages per cell, all of which can go on to infect other bacteria.

The lytic infection cycle is regulated by the expression of early and late genes

With all bacteriophages, but especially those with larger genomes, the question arises as to how gene expression is regulated in order to ensure that the correct activities occur at the right time during the infection cycle. With most bacteriophages, genome replication precedes the synthesis of capsid proteins. Similarly, the synthesis of lysozyme, the enzyme that causes the bacterium to burst, must be delayed until the very end of the infection cycle. Individual bacteriophage genes must therefore be expressed at different times in order for the infection cycle to proceed correctly.

One of the simplest strategies for regulating a lytic infection cycle is displayed by ϕX174. This bacteriophage exerts no control over the transcription of its genes. All 11 genes are transcribed by the host RNA polymerase as soon as the bacteriophage DNA enters the cell. Genome replication and capsid synthesis occur more or less at the same time, but lysozyme synthesis is delayed as the mRNA for this enzyme is translated slowly.

With most other bacteriophages, there are distinct phases of gene expression. Traditionally, two groups of genes are recognized. These are **early genes**, whose products are needed during the early stages of infection, and **late genes**, which remain inactive until toward the end of the cycle. There may, however, be other divisions within these groups, some bacteriophages having 'very early' genes, for example. A number of strategies are employed by bacteriophages to ensure that these groups of genes are expressed in the correct order. Many of these schemes utilize a **cascade system**, meaning that the appearance in the cell of the translation products from one set of genes switches on the transcription of the next set of genes (Figure 12.5).

With T4, for example, the very first genes to be expressed are transcribed by the *E. coli* RNA polymerase from a few standard *E. coli* promoter sequences present on the bacteriophage genome. These very early gene products include proteins that modify the σ subunit of the host RNA polymerase so it no longer recognizes *E. coli* promoters, thereby switching off host gene expression. Instead, the RNA polymerase now specifically transcribes a second set of bacteriophage genes (Figure 12.6). One of these genes specifies a new σ subunit, σ^{55}, which replaces the host's σ^{70}

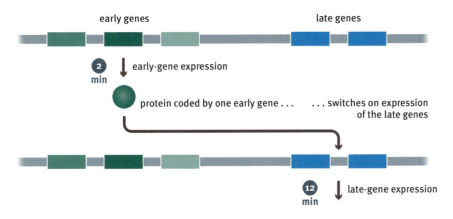

FIGURE 12.5 A cascade system for ensuring sequential gene expression. The cascade ensures the correct timing of early- and late-gene expression during a bacteriophage infection cycle. The proteins coded by the early genes include one that, after it has been synthesized, switches on expression of the late genes.

(a) first set of T4 genes are transcribed by the *E. coli* RNA polymerase

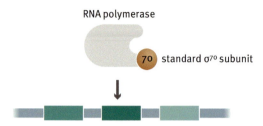

(b) modification of the σ70 subunit leads to transcription of a second set of T4 genes

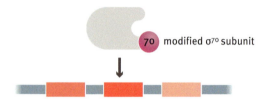

(c) replacement of σ70 with σ55 leads to transcription of a third set of T4 genes

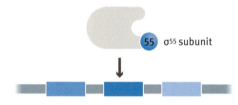

FIGURE 12.6 Sequential expression of T4 genes. Modifications to the *E. coli* RNA polymerase ensure that the correct bacteriophage genes are expressed at the correct time.

version, so the RNA polymerase now transcribes a third set of bacteriophage genes. The individual groups of genes are therefore expressed in the correct order, the products of one set switching on the expression of the next set.

The lysogenic infection cycle of bacteriophage λ

Bacteriophage λ, like most temperate bacteriophages, can follow a lytic infection cycle but is more usually associated with the alternative lysogenic cycle. The distinction is that during a lysogenic cycle, the bacteriophage genome becomes integrated into the host DNA.

Integration occurs immediately after entry of the bacteriophage DNA into the cell, and results in a quiescent form of the bacteriophage, called the **prophage** (Figure 12.7). The integrated prophage can be retained in the host DNA molecule for many cell generations, being replicated along with the bacterial genome and passed with it to the daughter cells. However, the switch to the lytic mode of infection occurs if the prophage is **induced** by any one of several chemical or physical stimuli. Each of these appears to be linked to DNA damage and possibly therefore signals the imminent death of the host by natural causes.

In response to these stimuli, the bacteriophage genome is excised from the host DNA and converted into a circular molecule. The λ genome then replicates by the rolling-circle mechanism. This produces a series of **concatemers** (linear genomes linked end to end) that are cleaved by an endonuclease (coded by λ gene *A*) at a 12-bp recognition sequence called the **cos** site. The linear genomes that result are packaged into λ bacteriophage particles and these new bacteriophages are released from the cell. The genomes recircularize immediately after injection into a new host.

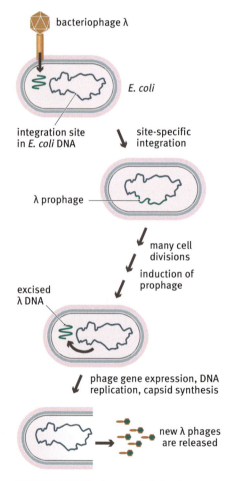

FIGURE 12.7 The lysogenic infection cycle of bacteriophage λ. Integration of λ DNA occurs by a site-specific mechanism, so the quiescent prophage is always located at the same position in the *E. coli* genome. The prophage can be retained in the host DNA molecule for many cell divisions. Following induction, the λ genome is excised from the host DNA and directs synthesis of new bacteriophage particles.

Lysogeny adds an additional level of complexity to the bacteriophage life cycle and ensures that the bacteriophage is able to adopt the particular infection strategy best suited to the prevailing conditions.

Many genes are involved in establishment and maintenance of lysogeny

Lysogeny adds an additional layer of complexity to the pattern of bacteriophage gene expression that occurs during the infection cycle. In particular, it raises three questions. How does the bacteriophage 'decide' whether to follow the lytic or lysogenic cycle, how is lysogeny maintained and how is the prophage induced to break lysogeny? A considerable amount is known about gene expression during λ infection, so much so that fairly complete answers can be given for each of these three questions. What follows is a summary.

We will deal with the second question – how is lysogeny maintained – before numbers 1 and 3. The first step in the lytic infection cycle is the expression of the early λ genes. These are transcribed from two promoters, p_L and p_R, located on either side of a regulatory gene called cI (**Figure 12.8**). During lysogeny, p_L and p_R are switched off because the cI gene product, which is a repressor protein, is bound to the operators O_L and O_R. As a result, the early genes are not expressed and the bacteriophage cannot enter the lytic cycle. Lysogeny is maintained for numerous cell divisions because the cI gene is continuously expressed, albeit at a low level, so that the amount of cI repressor present in the cell is always enough to keep p_L and p_R switched off. This continued expression of cI occurs because the cI repressor, when bound to O_R, not only blocks transcription from p_R, but also stimulates transcription from the promoter for the cI gene. The dual role of the cI repressor is therefore the key to lysogeny.

How does the bacteriophage 'decide' whether to follow the lytic or lysogenic cycle? This depends on the outcome of a race between the cI and cro proteins. When a λ DNA molecule enters an *E. coli* cell, the host's RNA polymerase enzymes attach to the various promoters on the molecule and start transcribing the λ genes. Once the cI gene is expressed, the cI repressor binds to O_L and O_R and blocks expression of the early genes, preventing entry into the lytic cycle and enabling lysogeny to be established. But lysogeny is not always the outcome of a λ infection. This is because a second gene, cro, also codes for a repressor, but in this case one that prevents transcription of cI (**Figure 12.9**). Both the cI and cro genes are expressed immediately after the λ DNA molecule enters the cell. If the cI repressor is synthesized more quickly than the Cro repressor, then early gene expression is blocked and lysogeny follows. However, if the Cro repressor wins the race, it blocks expression of the cI gene before

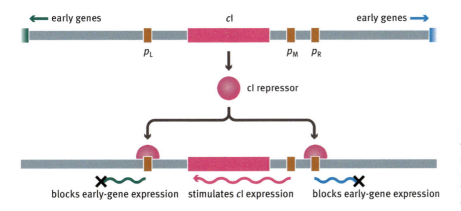

FIGURE 12.8 The role of the cI repressor. The cI repressor maintains lysogeny by blocking transcription from the early gene promoters p_L and p_R. It also stimulates transcription from p_M, the promoter for the cI gene.

FIGURE 12.9 **The role of the Cro repressor.** The Cro repressor prevents lysogeny by blocking transcription of the cI gene.

enough cI repressor has been synthesized to switch the early genes off. As a result, the bacteriophage enters the lytic infection cycle. The decision between lysis and lysogeny therefore depends on which of the two gene products, cI and Cro, accumulates the quickest. The decision is influenced by the products of other λ genes, which are able to assess the physiological state of the host cell, and hence ensure that the appropriate choice is made between lysogeny and immediate lysis.

Finally, how is lysogeny ended? This requires the inactivation of the cI repressor. Lysogeny is maintained as long as the cI repressor is bound to the operators O_L and O_R. The prophage will therefore be induced if the levels of active cI repressor decline below a certain point. This may happen by chance, leading to spontaneous induction, or may occur in response to physical or chemical stimuli. These stimuli activate a general protective mechanism in *E. coli*, the **SOS response**. Part of the SOS response is the expression of an *E. coli* gene, *recA*, coding for the RecA protein. RecA inactivates the cI repressor by cutting it in half (Figure 12.10). This switches on the expression of the early genes, enabling the bacteriophage to enter the lytic cycle. Inactivation of the cI repressor also means that transcription of the *cI* gene is no longer stimulated, avoiding the possibility of lysogeny being re-established through the synthesis of more cI repressor. Inactivation of the cI repressor therefore leads to induction of the prophage.

There are some unusual bacteriophage life cycles

Although lysis and lysogeny are the two most typical bacteriophage life cycles, they are not the only ones. One or two other bacteriophages display unusual infection cycles that are neither truly lytic nor truly lysogenic. An example is provided by a third *E. coli* bacteriophage, M13, which is unusual in that its genome is a *single*-stranded, circular DNA molecule, 6,400 nucleotides in length.

After injection into the bacterium, the M13 genome is replicated by synthesis of the complementary strand, producing a double-stranded, circular DNA molecule (Figure 12.11). This molecule then undergoes further replication until there are over 100 copies of it in the cell. At this stage, the infection cycle takes on characteristics of both lytic and lysogenic bacteriophages. As with lytic bacteriophages, M13 coat proteins are synthesized, and new bacteriophage particles are assembled and released from the

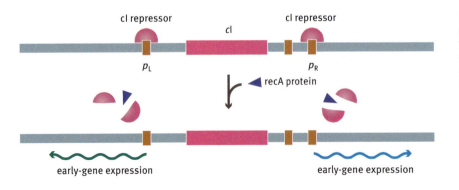

FIGURE 12.10 **The role of the RecA protein.** As part of the SOS response, the RecA protein cleaves the cI repressor, enabling the early genes to be expressed and the bacteriophage to enter the lytic infection cycle.

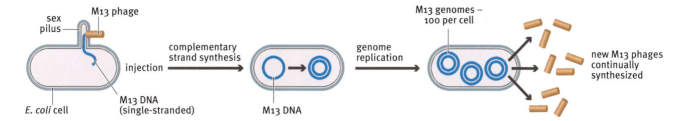

cell. However, as with lysogenic bacteriophages, cell bursting does not occur and the infected bacteria continue to grow and divide. Copies of the bacteriophage genome are passed to daughter bacteria during cell division, and M13 assembly and release continues. The M13 infection cycle is therefore partly lytic and partly lysogenic.

FIGURE 12.11 The unusual infection cycle of bacteriophage M13. The bacteriophage attaches to a cell-surface structure called the sex pilus in order to inject its DNA into the host cell.

12.2 VIRUSES OF EUKARYOTES

All eukaryotes act as hosts for viruses of one kind or another. Indeed, most eukaryotes are susceptible to infection by a broad range of virus types. Just think of the number of viral diseases that humans can catch. Because of the medical relevance, the viruses of humans and animals have received the most research attention, with plant viruses, capable of destroying crops, a rather distant second.

In many respects, eukaryotic viruses resemble bacteriophages, but the fact that their hosts are eukaryotes rather than bacteria forces some differences upon them. Their genes have to be expressed within eukaryotic cells and so must resemble eukaryotic genes. They therefore need the complex upstream sequences required to activate transcription by RNA polymerase II, and they may contain introns.

Eukaryotic viruses have diverse structures and infection strategies

The capsids of eukaryotic viruses are either icosahedral or filamentous – the head-and-tail structure is unique to bacteriophages. A second distinct feature of eukaryotic viruses, especially those with animal hosts, is that the capsid may be surrounded by a lipid membrane, forming an additional component to the virus structure (Figure 12.12). This membrane is derived from the host when the new virus particle leaves the cell and may subsequently be modified by the insertion of virus-specific proteins.

Virus genomes display a great variety of structures (Table 12.2). They may be DNA or RNA, single- or double-stranded (or partly double-stranded with single-stranded regions), linear or circular, segmented or nonsegmented. For reasons that no one has ever understood, the vast majority of plant viruses have RNA genomes. Genome sizes cover approximately the same range as seen with bacteriophages, although the largest viral genomes (e.g. vaccinia virus at 195 kb) are rather larger than the largest bacteriophage genomes.

Most eukaryotic viruses follow only the lytic infection cycle, but few take over the host cell's genetic machinery in the way that a bacteriophage does. Many viruses coexist with their host cells for long periods, possibly years. The host cell functions cease only toward the end of the infection cycle, when the virus progeny that have been stored in the cell are released (Figure 12.13a). Other viruses have infection cycles similar to M13 in

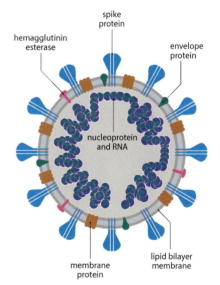

FIGURE 12.12 The structure of the Covid-19 virus. Covid-19 is caused by the severe acute respiratory syndrome coronavirus 2 (SARS-CoV-2), which displays many of the typical features of a eukaryotic virus. The capsid is surrounded by a lipid membrane to which additional viral proteins are attached. The proteins form the 'spikes' on the outer surface of the virus. These include the spike protein which mediates entry of the virus into a host cell.

TABLE 12.2 FEATURES OF SOME TYPICAL EUKARYOTIC VIRUSES AND THEIR GENOMES		
Virus	**Host**	**Genome Structure**
Adenovirus	Vertebrates	Double-stranded linear DNA
Anellovirus	Vertebrates	Single-stranded circular DNA
Coronavirus	Mammals, birds	Single-stranded linear RNA
Hepatovirus A	Vertebrates	Single-stranded linear RNA
Influenza A virus	Mammals, birds	Single-stranded segmented linear RNA
Parvovirus	Mammals	Single-stranded linear DNA
Poliovirus	Humans	Single-stranded linear RNA
Reovirus	Wide host range	Double-stranded segmented linear RNA
Retroviruses	Mammals, birds	Single-stranded linear RNA
Tobacco mosaic virus	Plants	Single-stranded linear RNA
Vaccinia virus	Mammals	Double-stranded linear DNA

The genome structure is that in the phage capsid. Some genomes exist in different forms within the host cell

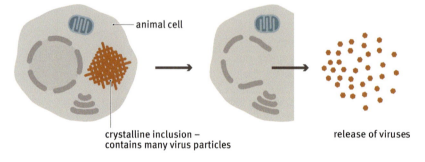

(a) release of stored viruses

animal cell

crystalline inclusion – contains many virus particles

release of viruses

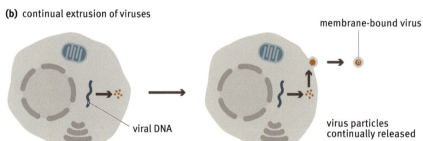

(b) continual extrusion of viruses

viral DNA

membrane-bound virus

virus particles continually released

FIGURE 12.13 Infection strategies for eukaryotic viruses. The progeny of a eukaryotic virus might be (a) stored in the host cell prior to release or (b) continually extruded from the infected cell. In (a), the stored viruses form a crystalline inclusion inside the cell. In (b), the virus particles become coated with a small part of the outer cell membrane as they are released from the cell.

E. coli, continually synthesizing new virus particles that are extruded from the cell (Figure 12.13b).

Viral retroelements integrate into the host cell DNA

Many eukaryotic viruses can set up long-term infections without integrating into the host cell DNA, but this does not mean that there are no eukaryotic equivalents to lysogenic bacteriophages. A number of DNA and RNA viruses are able to integrate into the genomes of their hosts, sometimes with drastic effects on the host cell. The **viral retroelements** are important examples of integrative eukaryotic viruses. These viruses are also interesting because their replication pathways include a novel step in which an RNA version of the genome is converted into DNA.

There are two kinds of viral retroelement. These are the **retroviruses**, whose capsids contain the RNA version of the genome, and the **pararetroviruses**, whose encapsidated genome is made of DNA. The ability of viral retroelements to convert RNA into DNA requires the enzyme called **reverse transcriptase**, which is capable of making a DNA copy of an RNA template (Figure 12.14).

The typical retroviral genome is a single-stranded RNA molecule, 7–12 kb in length. After entry into the cell, the genome is copied into double-stranded DNA by a few molecules of reverse transcriptase that the virus carries in its capsid. The double-stranded version of the genome then integrates into the host DNA (Figure 12.15).

Integration of the viral genome into the host DNA is a prerequisite for the expression of the retrovirus genes. There are three of these, called *gag*, *pol* and *env* (Figure 12.16). Each gene codes for a polyprotein that is cleaved after translation into two or more functional gene products. These products include the virus coat proteins (from *env*) and the reverse transcriptase (from *pol*). The protein products combine with full-length RNA transcripts of the retroviral genome to produce new virus particles.

The causative agents of HIV/AIDS (human immunodeficiency virus infection and acquired immune deficiency syndrome) were shown to be retroviruses in 1983–1984. The first AIDS virus to be isolated is called human immunodeficiency virus 1, or HIV-1, and is responsible for the most prevalent and pathogenic form of AIDS. A related virus, HIV-2, is less widespread and causes a milder form of the disease. The HIVs attack certain types of lymphocyte in the bloodstream, thereby depressing the immune response of the host. These lymphocytes carry on their surfaces multiple copies of a protein called CD4, which acts as a receptor for the virus. An HIV particle binds to a CD4 protein and then enters the lymphocyte after fusion between its lipid envelope and the cell membrane.

Some retroviruses cause cancer

The human immunodeficiency viruses are not the only retroviruses capable of causing diseases. Several retroviruses can induce **cell transformation**, possibly leading to cancer. Cell transformation involves changes in cell morphology and physiology. In cell cultures, transformation results in a loss of control over growth, so that transformed cells grow as a disorganized mass, rather than as a monolayer (Figure 12.17). In whole

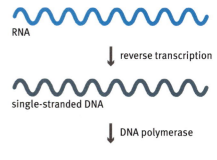

FIGURE 12.14 Reverse transcription of an RNA template into DNA. The single-stranded DNA resulting from reverse transcription can be copied into double-stranded DNA by a DNA-dependent DNA polymerase.

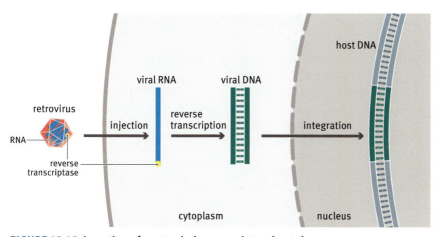

FIGURE 12.15 Insertion of a retroviral genome into a host chromosome.

FIGURE 12.16 A retrovirus genome. LTR, long terminal repeat.

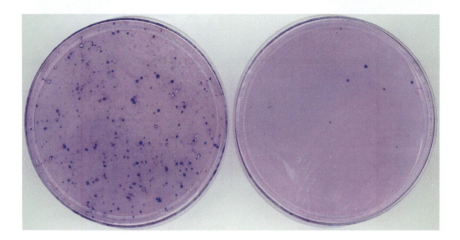

FIGURE 12.17 Transformation of cultured avian cells. In the Petri dish on the right, normal avian fibroblasts are growing in a monolayer. On the left, the cells have been transformed. Piled-up cells forming microtumors can be seen, showing that some of the processes that normally control cell growth have been disrupted. (Courtesy of Klaus Bister and Markus Hartl, University of Innsbruck.)

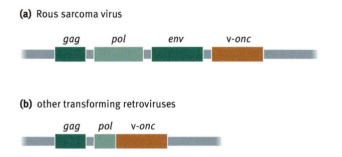

FIGURE 12.18 The genomes of transforming retroviruses. (a) Rous sarcoma virus and (b) other transforming retroviruses. The gene labeled v-*onc* is the cellular gene that has been captured by the virus.

animals, cell transformation is thought to underlie the development of tumors.

There appears to be two distinct ways in which retroviruses can cause cell transformation. With some retroviruses, such as the leukemia viruses, cell transformation is a natural consequence of infection, although it may be induced only after a long latent period during which the integrated provirus lies quiescent within the host genome. Other retroviruses cause cell transformation because of abnormalities in their genome structures. These transforming viruses carry cellular genes that they have captured by some undefined process. With at least one transforming retrovirus (Rous sarcoma virus), this cellular gene is in addition to the standard retroviral genes (Figure 12.18a). With others, the cellular gene replaces part of the retroviral gene complement (Figure 12.18b). In the latter case, the retrovirus may be **defective**, meaning that it is unable to replicate and produce new viruses, as it has lost genes coding for vital replication enzymes and/or capsid proteins. These defective retroviruses are not always inactive as they can make use of proteins provided by other retroviruses in the same cell (Figure 12.19).

The ability of a transforming virus to cause cell transformation lies with the nature of the cellular gene that has been captured. Often this captured gene (called a v-*onc*, with 'onc' standing for **oncogene**) codes for a protein involved in cell proliferation. The normal cellular version of the gene is subject to strict regulation and is expressed only in limited quantities when needed. It is thought that expression of the v-*onc* follows a different, less controlled pattern, either because of changes in the gene structure or because of the influence of promoters within the retrovirus. One result of this altered expression pattern could be a loss of control over cell division, leading to the transformed state.

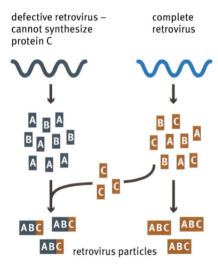

FIGURE 12.19 Replication of a defective retrovirus. A defective retrovirus may be able to give rise to infective virus particles if it shares the cell with a nondefective retrovirus. The latter acts as a 'helper', providing the proteins that the defective virus is unable to synthesize.

12.3 TOWARD AND BEYOND THE EDGE OF LIFE

Viruses occupy the boundary between the living and nonliving worlds. At the edge of this boundary, and arguably beyond it, lie a variety of genetic elements that might or might not be classified as viruses. We will complete our survey of virus infection cycles by examining the most important of these enigmatic entities.

There are cellular versions of viral retroelements

Many eukaryotic genomes contain multiple copies of sequences, called **RNA transposons**, that are able to move from place to place in the genome by a process that involves an RNA intermediate. **Retrotransposition**, as it is called, begins with the synthesis of an RNA copy of the sequence by the normal process of transcription (Figure 12.20). The transcript is then copied into double-stranded DNA, which initially exists as an independent molecule outside of the genome. Finally, the DNA copy of the transposon integrates into the genome, possibly back into the same chromosome occupied by the original unit, or possibly into a different chromosome.

If we compare the mechanism for retrotransposition with that for replication of a viral retroelement, as shown in Figure 12.15, then we see that the two processes are very similar. The one significant difference is that the RNA molecule that initiates retrotransposition is transcribed from an endogenous sequence, whereas the one that initiates replication of a viral retroelement comes from outside of the cell. This close similarity alerts us to the possibility that a relationship exists between RNA transposons and viral retroelements.

RNA transposons can be broadly classified into two types, those that have **long terminal repeats** (**LTRs**) and those that do not. LTRs are also possessed by viral retroelements (see Figure 12.16). It is now clear that these viruses and the endogenous LTR retrotransposons are members of the same superfamily of elements.

The first LTR retrotransposon to be discovered was the *Ty* sequence of yeast, which is 6.3 kb in length and has a copy number of about 50 in most *Saccharomyces cerevisiae* genomes. There are several types of *Ty* element in yeast genomes. The most abundant of these, *Ty1*, is similar to the *copia* retrotransposon of the fruit fly. These elements are therefore now called the *Ty1/copia* family. If we compare the structure of a *Ty1/copia* retrotransposon with that of a viral retroelement, then we see clear family relationships (Figure 12.21a and b). Each *Ty1/copia* element contains two genes, called *TyA* and *TyB* in yeast, which are similar to the *gag* and *pol* genes of a viral retroelement. In particular, *TyB* codes for a polyprotein that includes the reverse transcriptase that plays the central role in the transposition of a *Ty1/copia* element. Note, however, that the *Ty1/copia* retrotransposon lacks an equivalent of the viral *env* gene, the one that codes for the viral coat proteins. This means that *Ty1/copia* retrotransposons cannot form infectious virus particles and therefore cannot escape from their host cell. They do, however, form virus-like particles (VLPs) consisting of the RNA and DNA copies of the retrotransposon attached to core proteins derived from the TyA polyprotein (Figure 12.22). In contrast, the members of a second family of LTR retrotransposons, called *Ty3/gypsy* (again after the yeast and fruit fly versions), do have an equivalent of the *env* gene (Figure 12.21c) and at least some of these can form infectious viruses. Although classed as endogenous retrotransposons, these infectious versions should be looked upon as viral retroelements.

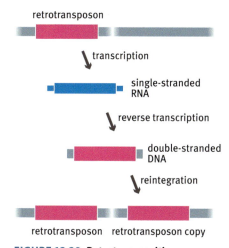

FIGURE 12.20 Retrotransposition. An RNA copy of the retrotransposon is converted into double-stranded DNA, which reintegrates into the genome.

(a) Viral retroelement
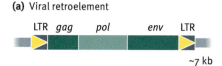

(b) *Ty1/copia* retroelement
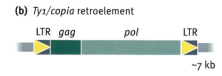

(c) *Ty3/gypsy* retroelement
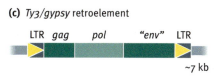

FIGURE 12.21 Genome structures for LTR retroelements.

LTR retrotransposons make up substantial parts of many eukaryotic genomes and are particularly abundant in larger plant genomes, especially those of grasses such as maize. They also make up an important component of invertebrate and some vertebrate genomes. However, in the genomes of humans and other mammals, all the LTR elements appear to be decayed viral retroelements rather than true retrotransposons. These sequences are called **endogenous retroviruses (ERVs)** and they make up 9% of the human genome. Human ERVs are 6–11 kb in length and have copies of the *gag*, *pol* and *env* genes. Although most contain mutations that inactivate one or more of these genes, a few members of the HERV-K group have functional sequences. By comparing the positions of the HERV-K elements in the genomes of different individuals, it has been inferred that at least some of these are active transposons, unlike the majority of human ERVs which are inactive sequences that are not capable of additional proliferation.

Satellite RNAs, virusoids, viroids and prions are all probably beyond the edge of life

When we move to very edge of life, we meet a variety of infectious molecules related to, but different from, viruses. The **satellite RNAs** or **virusoids** are examples. These are RNA molecules, some 320–400 nucleotides in length, each containing a single or very small number of genes. Satellite RNAs and virusoids cannot construct their own capsids and instead move from cell to cell within the capsids of helper viruses. The distinction between the two groups is that a satellite virus shares the capsid with the genome of the helper virus whereas a virusoid RNA molecule becomes encapsidated on its own. They are generally looked on as parasites of their helper viruses, although there appear to be a few cases where the helper cannot replicate without the satellite RNA or virusoid, suggesting that at least some of the relationships are symbiotic.

Satellite RNAs and virusoids are both found predominantly in plants, as is a more extreme group called the **viroids.** These are RNA molecules of 240–475 nucleotides that contain no genes. Viroids never become encapsidated, spreading from cell to cell as naked RNA. They include some economically important pathogens, such as the citrus exocortis viroid that reduces the growth of citrus fruit trees.

Finally, there are the **prions**. These are infectious, disease-causing particles that contain no nucleic acid. Prions are responsible for scrapie in sheep and goats and their transmission to cattle has led to the new disease called BSE – bovine spongiform encephalopathy. Whether their further transmission to humans causes a variant form of Creutzfeldt–Jacob disease (CJD) is controversial but accepted by many biologists. At first, prions were thought to be viruses but it is now clear that they are made solely of protein.

The normal version of the prion protein, called PrP^C, is coded by a mammalian nuclear gene and synthesized in the brain, although its function is unknown. PrP^C is easily digested by proteases whereas the infectious version, PrP^{SC} has a more highly β-sheeted structure that is resistant to proteases and forms fibrillar aggregates that are seen in infected tissues. Once inside a cell, PrP^{SC} molecules are able to convert newly synthesized PrP^C proteins into the infectious form, by a mechanism that is not yet understood, resulting in the disease state. Transfer of one or more of these PrP^{SC} proteins to a new animal results in the accumulation of new PrP^{SC} proteins in the brain of that animal, transmitting the disease (Figure 12.23). Infectious proteins with similar properties are known in lower eukaryotes,

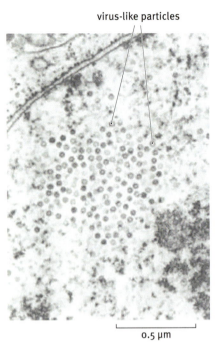

FIGURE 12.22 Electron micrograph showing virus-like particles. These virus-like particles of the *copia* retrotransposon are present in the nucleus of a *Drosophila melanogaster* cell. (From K. Yoshioka et al., *EMBO J.* 9:535–541, 1990. With permission from Macmillan Publishers Ltd.)

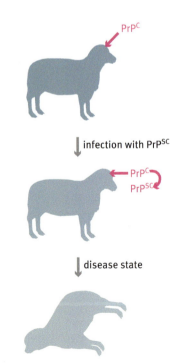

FIGURE 12.23 The mode of action of a prion.

examples being the Ure3 and Psi+ prions of *Saccharomyces cerevisiae*. It is clear, however, that prions are *gene products* rather than genetic material and despite their infectious properties, which led to the initial confusion regarding their status, they are unrelated to viruses or to subviral particles such as viroids and virusoids.

KEY CONCEPTS

- Viruses are the simplest form of life on the planet. They are so simple in biological terms that it is doubtful if they can really be thought of as living organisms.

- Bacteriophages are classified into two groups according to their life cycle. The two types are called lytic and lysogenic. The fundamental difference between these groups is that a lytic bacteriophage kills its host bacterium very soon after the initial infection, whereas a lysogenic bacteriophage can remain quiescent throughout numerous generations of the host cell.

- Expression of bacteriophage genes is regulated so that the correct genes are expressed at the appropriate times in the infection cycle.

- Most eukaryotic viruses follow only the lytic infection cycle, but few take over the host cell's genetic machinery in the way that a bacteriophage does. Many viruses coexist with their host cells for long periods, possibly years.

- Viral retroelements are examples of eukaryotic viruses whose genomes become integrated into the host DNA. Their replication pathways include a novel step in which an RNA version of the genome is converted into DNA.

- There are two kinds of viral retroelement. These are the retroviruses, whose capsids contain the RNA version of the genome, and the pararetroviruses, whose encapsidated genome is made of DNA.

- The causative agents of HIV/AIDS are retroviruses, as are some of the viruses that cause cell transformation, which may lead to cancer.

- Many eukaryotic genomes contain multiple copies of RNA transposons, which are able to move from place to place in the genome by a process that involves an RNA intermediate. Some of these transposons are closely related to viral retroelements.

- Satellite RNAs, virusoids and viroids are subviral particles that are probably beyond the edge of life.

- Prions are infectious, disease-causing particles that contain no nucleic acid. They are made of protein and so are gene products rather than genes.

QUESTIONS AND PROBLEMS

Key terms

Write short definitions of the following terms:

bacteriophage
capsid
cascade system
cell transformation
concatemer
cos site

CHAPTER 12: Inheritance of Genes During Virus Infection Cycles

defective retrovirus
early gene
endogenous retrovirus
filamentous bacteriophage
genome
head-and-tail
icosahedral
induction of prophage
late gene
latent period
long terminal repeat

lysogenic infection cycle
lytic infection cycle
oncogene
one-step growth curve
overlapping genes
pararetrovirus
prion
prophage
protomer
retrotransposition
retrovirus

reverse transcriptase
RNA transposon
satellite RNA
segmented genome
SOS response
temperate bacteriophage
viral retroelement
viroid
virulent bacteriophage
virusoid

Self-study questions

12.1 Describe the different capsid structures seen with bacteriophages.

12.2 Describe the different types of bacteriophage genome.

12.3 What are overlapping genes, as found in some viral genomes?

12.4 List the key differences between the infection cycles of lytic and lysogenic bacteriophages.

12.5 Draw and annotate a graph that illustrates the one-step growth curve. Explain how this graph relates to events occurring during infection of *E. coli* with T4 bacteriophage.

12.6 Describe what is meant by the terms 'early genes' and 'late genes'. Outline a typical strategy by which a bacteriophage ensures that its genes are expressed at the correct time.

12.7 Describe the process by which a bacteriophage λ prophage is excised from the *E. coli* genome and is packaged into new bacteriophage particles.

12.8 Explain how bacteriophage λ makes the decision to enter lysogeny and how the state is maintained.

12.9 Outline the link between the *E. coli* SOS response and induction of a λ prophage.

12.10 Explain in what respects the M13 infection cycle is lytic and in what respects it is lysogenic.

12.11 Describe the ways in which the structures of eukaryotic viruses are similar to and distinct from the structures of bacteriophages.

12.12 Distinguish between a retrovirus and a pararetrovirus.

12.13 Describe the infection cycle of a retrovirus. Your answer should explain the roles of the three genes in the retrovirus genome.

12.14 What is the distinctive feature of a transforming retrovirus, and how does this relate to its ability to cause cell transformation?

12.15 Explain why an oncogene can be harmless when present in the human genome but result in cancer when carried by a retrovirus.

12.16 Using examples, describe the relationships between RNA transposons and viral retroelements.

12.17 Describe the differences between the *Ty1/copia* and *Ty3/gypsy* families of LTR retrotransposons.

12.18 What are the features of the LTR elements present in the human genome?

12.19 Describe the key features of, and differences between, virusoids and viroids.

12.20 What is a prion, and how does it cause disease?

Discussion topics

12.21 To what extent can viruses be considered a form of life?

12.22 Bacteriophages with small genomes (for example, φX174) are able to replicate very successfully in their hosts. Why, then, should other bacteriophages, such as T4, have large and complicated genomes?

12.23 Bacteriophages were extensively used in the middle decades of the 20th century as model systems with which to study genes. Many of the basic principles of gene structure and regulation were discovered through this work. Discuss the advantages, and disadvantages, of bacteriophages for this type of research.

12.24 Devise a research project aimed at discovering whether HIV-1 was originally an animal virus that jumped into the human population.

12.25 Why do LTR retroelements have long terminal repeats?

12.26 At what point in the series virus-virusoid-viroid-prion do we reach the stage where we are no longer studying genetics?

The answers to, or hints on how to answer, the self-study questions and discussion topic questions can be downloaded from the book's product page at this link: www.routledge.com/9781032743530

FURTHER READING

Casjens SR & Hendrix RW (2015) Bacteriophage lambda: early pioneer and still relevant. *Virology* 479–480, 310–330. *Details of the bacteriophage λ genome and the genetic basis to the lytic and lysogenic infection cycles.*

Curcio MJ, Lutz S & Lesage P (2015) The Ty1 LTR-retrotransposon of budding yeast, *Saccharomyces cerevisiae. Microbiology Spectrum* 3, 1–35.

Dimmock NJ, Easton AJ & Leppard KN (2016) *An Introduction to Modern Virology*, 7th ed. Oxford: Blackwell Scientific Publishers. *The best general text on viruses.*

Flores R, Gas M-E, Molina-Serrano D et al. (2009) Viroid replication: rolling-circles, enzymes and ribozymes. *Viruses* 1, 317–334.

Grandi N & Tramontano E (2018) Human endogenous retroviruses are ancient acquired elements still shaping innate immune responses. *Frontiers in Immunology* 12, 169.

Lesbats P, Engelman AN & Cherepanov P (2016) Retroviral DNA integration. *Chemical Reviews* 116, 12730–12757.

Louten J (2016) Virus structure and classification. *Essential Human Virology*. 2016, 19–29.

Maeda N, Fan H & Yoshikai Y (2008) Oncogenesis by retroviruses: old and new paradigms. *Reviews in Medical Virology* 18, 387–405.

Marvin DA (1998) Filamentous phage structure, infection and assembly. *Current Opinion in Structural Biology* 8, 150–158. *Bacteriophage M13.*

Mastrianni JA (2010) The genetics of prion diseases. *Genetics in Medicine* 12, 187–195.

Miller ES, Kutter E, Mosig G et al. (2003) Bacteriophage T4 genome. *Microbiology and Molecular Biology Reviews* 67, 86–156.

Palukaitis P (2016) Satellite RNAs and satellite viruses. *Molecular Plant-Microbe Interactions* 29, 181–186.

Song SU, Gerasimova T, Kurkulos M et al. (1994) An env-like protein encoded by a *Drosophila* retroelement: evidence that *gypsy* is an infectious retrovirus. *Genes and Development* 8, 2046–2057.

Terry C & Wadsworth JDF (2019) Recent advances in understanding mammalian prion structure: a mini review. *Frontiers in Molecular Neuroscience* 2, 169.

Inheritance of Genes in Bacteria

CHAPTER 13

We will now examine the structures of bacterial genomes and ask how these are replicated and how complete copies are passed to the daughter cells when a bacterium divides. In doing this, we will make three important discoveries that will challenge our understanding of genomes and inheritance. First, we will learn that some bacteria possess a variety of DNA molecules, and defining exactly which ones constitute the organism's genome can be difficult. Second, we will discover that genome size and gene content vary among the members of a single bacterial species, so much so that two bacteria of the same species can have quite different genomes. Third, we will see that linear descent – the repeated passage of the genome from parent to daughter cells – is not the only way in which bacteria obtain genes. Some of the genes in bacterial genomes have been acquired from other bacteria, possibly other species, by **horizontal** or **lateral gene transfer** (Figure 13.1).

13.1 BACTERIAL GENOMES

Most of the early research in bacterial genetics was carried out with *Escherichia coli*. This is a common bacterium that normally lives in the lower intestine of humans and other mammals (Figure 13.2). Most strains are harmless but a few types have genes that code for proteins that are toxic to the host. These pathogenic types are often responsible for outbreaks of food poisoning. *E. coli* was first isolated in 1885. The ease with which it can be grown on a solid agar medium or in a liquid broth culture led to the species gradually being adopted as a model bacterium for all types of microbiology research, including bacterial genetics.

How typical is *E. coli* of bacteria in general? In the early days of bacterial genetics, there tended to be an assumption that most if not all bacteria would prove to have very similar genomes and therefore that discoveries made with *E. coli* would also apply to other species. We now know that this view was incorrect and that in some respects *E. coli* is quite an unusual bacterium.

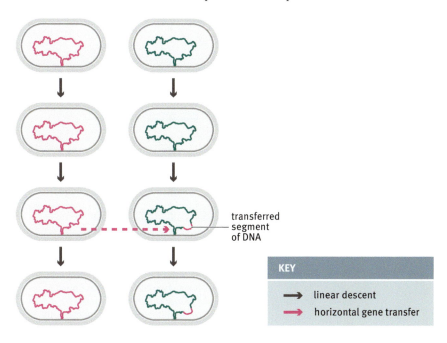

FIGURE 13.1 Inheritance of genes by linear descent and by horizontal gene transfer.

Bacterial genomes vary greatly in size and some are linear DNA molecules

One feature of the *E. coli* genome that is 'typical' is its size. At 4,642 kb, *E. coli* is in the midrange for bacterial genomes. Most bacterial genomes are less than 5 Mb, although the variation among those genomes that have been completely sequenced is from 100 kb to over 25 Mb (**Table 13.1**).

In most bacterial genomes, the genes make up 85%–90% of the DNA sequence. This means that genome size is proportional to gene number. Gene numbers therefore vary over an extensive range, with these numbers reflecting the nature of the ecological niches within which different species of prokaryote live. Many of the smallest genomes belong to species that are obligate parasites. *Nasuia deltocephalinicola*, for example, is an endosymbiont of leafhoppers, living inside specialized structures within the insect's abdomen. The bacteria receive various nutrients from the insect, which means that *Nasuia* is able to dispense with many of the enzymes needed by free-living bacteria for the synthesis of metabolites and for energy generation. As a consequence, the *Nasuia* genome is just 112 kb and contains only 137 protein-coding genes, the majority of these involved in essential functions such as DNA replication, transcription and translation.

At the other end of the scale, the largest genomes tend to belong to free-living species that are found in the soil, an environment that provides a broad range of physical and biological conditions. The 14.8 Mb genome of *Sorangium cellulosum* contains 10,400 protein-coding genes, including some specifying enzymes that enable this bacterium to break down cellulose into sugars, and others coding for enzymes that synthesize antibacterial and antifungal compounds that help it to compete in the complex soil ecosystem. There are also genes for proteins involved in cell-to-cell communication, enabling the bacteria to migrate together in swarms and to associate into a multicellular fruiting body that produces resistant spores.

The *E. coli* genome is a single, circular DNA molecule, often referred to as the **bacterial chromosome**. The vast majority of bacterial genomes that have been studied comprise a single circular chromosome, but an increasing number of linear versions are being found. The first of these, for *Borrelia burgdorferi*, the organism that causes Lyme disease, was described in 1989, and during the following years, similar discoveries were made for *Streptomyces coelicolor* and *Agrobacterium tumefaciens*.

Linear molecules have free ends that must be distinguishable from DNA breaks, so that the DNA repair processes do not attempt to 'mend' them. In *Borrelia* and *Agrobacterium*, the real chromosome ends are distinguishable because a covalent linkage is formed between the 5′ and 3′ ends of the polynucleotides in the DNA double helix, and in *Streptomyces* the ends appear to be capped with special binding proteins (**Figure 13.3**).

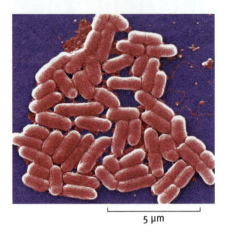

FIGURE 13.2 Scanning electron micrograph of *Escherichia coli* cells. (Courtesy of Janice Carr, Centers for Disease Control and Prevention.)

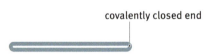

FIGURE 13.3 Linear bacterial chromosomes. The ends of a linear chromosome must be distinguished from a DNA break, so they are not recognized by DNA repair enzymes. (a) In *Borrelia* and *Agrobacterium*, the ends of the polynucleotides are joined together. (b) In *Streptomyces*, the ends are protected by proteins.

TABLE 13.1 GENOME SIZES FOR VARIOUS BACTERIA

Species	Size of Genome (Mb)
Nasuia deltocephalinicola NAS-ALF	0.12
Mycoplasma genitalium G37	0.58
Mycobacterium tuberculosis H37Rv	4.41
Escherichia coli K12	4.64
Sorangium cellulosum So0157–2	14.78

Plasmids are independent DNA molecules within a bacterial cell

In addition to its genome, the cytoplasm of a bacterial cell might contain other DNA molecules called **plasmids** (Figure 13.4). Most plasmids are circular, though some linear ones are known. Their sizes vary from approximately 1 kb for the smallest to over 250 kb for the larger ones. Some plasmids are restricted to just a few related species and are found in no other bacteria, but others have a broad host range and can exist in numerous species, although possibly displaying a preference for a few kinds of bacteria in which they are more common.

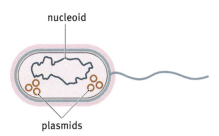

FIGURE 13.4 Plasmids are small, usually circular DNA molecules that are found inside some bacterial cells.

All plasmids possess at least one DNA sequence that can act as an origin of replication, so they are able to multiply in the cell independently of the bacterial chromosome (Figure 13.5a). The smaller plasmids make use of the cell's own DNA replicative enzymes to make copies of themselves, whereas some of the larger ones carry genes that code for special enzymes that are specific for plasmid replication. A few types of plasmid are also able to replicate by inserting themselves into the chromosome (Figure 13.5b). These integrative plasmids or **episomes** may be stably maintained in this form through numerous cell divisions, but always at some stage exist as independent elements.

Plasmids almost always carry one or more genes, and often these genes are responsible for a useful characteristic. Different types of plasmid can therefore be recognized, depending on the genes that they carry and the characteristics that those genes confer on the host bacterium. According to this classification, there are five main types of plasmid (Table 13.2). The most common type is **resistance** or **R plasmids**. These carry genes conferring on the host bacterium resistance to one or more antibacterial agents, such as chloramphenicol, ampicillin or mercury. R plasmids are very important in clinical microbiology because their spread through natural populations can have profound consequences for the treatment of bacterial infections.

Some species of bacteria, including *E. coli*, often have copies of a second type of plasmid called a **fertility** or **F plasmid**. These are able to

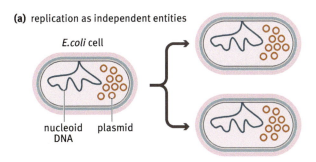
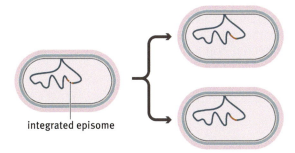

FIGURE 13.5 Two mechanisms for plasmid replication. (a) Some plasmids replicate as independent entities. (b) Episomes are able to insert into the bacterial chromosome.

TABLE 13.2 FEATURES OF TYPICAL PLASMIDS		
Type of Plasmid	Gene Functions	Examples
Resistance	Resistance to antibacterial agents	Rbk of *Escherichia coli* and other bacteria
Fertility	Conjugation and DNA transfer between bacteria	F of *E. coli*
Killer	Synthesis of toxins that kill other bacteria	Col of *E. coli*, for colicin production
Degradative	Enzymes for metabolism of unusual molecules	TOL of *Pseudomonas putida*, for toluene metabolism
Virulence	Pathogenicity	Ti of *Agrobacterium tumefaciens*, conferring the ability to cause crown gall disease in dicotyledonous plants

direct **conjugation** between different bacteria, a process that enables two bacteria to join together so that plasmids, and possibly parts of the main genome, can be passed from one cell to another. Other plasmids carry genes that code for toxic proteins, ones not directed against the host but which kill other bacteria. We assume that bacteria carrying these plasmids gain an advantage in the competition for scarce resources. Plasmids of this type are not common in bacterial species as a whole, but the **Col plasmids**, which code for toxins called **colicins**, are an example occurring in *E. coli*. In species other than *E. coli*, we also find **degradative plasmids** and **virulence plasmids**. Degradative plasmids allow the host bacterium to metabolize unusual molecules such as toluene and salicylic acid. Several examples occur in the *Pseudomonas* genus of bacteria, such as the TOL plasmid of *Pseudomonas putida*. Virulence plasmids confer pathogenicity on the host bacterium. The best-known example is the Ti plasmid of *Agrobacterium tumefaciens*, which induces crown gall disease in dicotyledonous plants.

Each type of plasmid has its own characteristic **copy number**, this being the number of copies of the plasmid that are present in a single bacterial cell. Some plasmids, especially the larger ones, are **stringent** and have a low copy number of perhaps just one or two per cell. Others, called **relaxed** plasmids, are present in multiple copies of 50 or more. There can also be more than one type of plasmid in a single cell – cells of *E. coli* have been known to contain up to seven different plasmids at once. To be able to coexist in the same cell, different plasmids must be **compatible**. If two plasmids are incompatible, then one or the other will be quite rapidly lost from the cell. Different types of plasmid can therefore be assigned to different **incompatibility groups** on the basis of whether or not they can coexist. Plasmids from a single incompatibility group (which cannot coexist) are often related to one another in various ways.

Some bacteria have multipartite genomes

An *E. coli* cell contains a single copy of its circular chromosome and possibly one or more types of plasmid. Although the plasmids might provide the bacterium with a useful characteristic, such as resistance to an antibiotic, the plasmids are not necessary for the survival of the cell. The genes they contain provide optional extras rather than essential components of the cell's biochemistry. This simple interpretation of the role of plasmids becomes rather more complicated when we examine other bacteria which have multipartite genomes – genomes that are divided into two or more DNA molecules.

TABLE 13.3 EXAMPLES OF GENOME ORGANIZATION IN BACTERIA		
Species	DNA Molecules	Size (Mb)
Escherichia coli K12	One circular molecule	4.642
Vibrio cholerae El Tor N16961	Two circular molecules	2.937, 1.094
Deinococcus radiodurans R1	Four circular molecules	2.649, 0.412, 0.177, 0.046

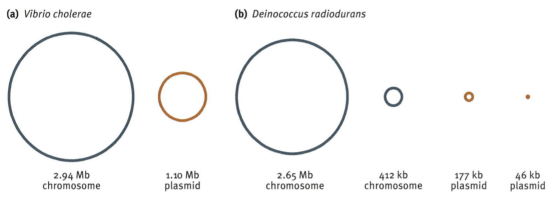

FIGURE 13.6 **Multipartite bacterial genomes.**

Vibrio cholerae, the pathogenic bacterium that causes cholera, is a case in point (Table 13.3). Members of this species have two circular DNA molecules, one of 2.94 Mb and the other of 1.09 Mb, with 72% of the organism's 3,720 genes on the larger of these molecules (Figure 13.6a). Both molecules carry essential genes, although most of those specifying the central cellular activities, such as gene expression and energy generation, as well as the genes that confer pathogenicity, are located on the larger molecule. It would appear obvious that these two DNA molecules together constitute the *Vibrio* genome. The complications arise when the smaller of the two molecules is examined more closely. This molecule contains an **integron**, a set of genes and other DNA sequences that enable plasmids to capture genes from bacteriophages and other plasmids. It therefore appears that the smaller molecule is a plasmid rather than a chromosome, even though some of the genes it carries are essential to the bacterium.

Deinococcus radiodurans, whose genome is of particular interest because it contains many genes that help this bacterium resist the harmful effects of radiation, is constructed on similar lines. Its essential genes are distributed among two circular DNA molecules, considered to be genuine chromosomes, and two plasmids (Figure 13.6b).

The complications posed by bacteria such as *Vibrio* and *Deinococcus* have prompted microbial geneticists to invent a new term – **chromid** – to describe a plasmid that carries essential genes. This means that we now distinguish between three, rather than just two, types of DNA molecule that might be found in a bacterium (Figure 13.7). First, there are one or more bacterial chromosomes, carrying essential genes. Second, genuine plasmids, whose genes are nonessential to the bacterium. Third, chromids, which resemble plasmids but which carry genes that the bacterium needs to survive. According to this nomenclature, *Vibrio cholerae* has one chromosome and one chromid, and *Deinococcus radiodurans* has two chromosomes and two chromids.

Genome sizes and gene numbers vary within individual species

It has always been difficult to define exactly what constitutes a species in the bacterial world. The early taxonomists such as Linnaeus described species in morphological terms, all members of one species having the same or very similar structural features. This form of classification was first applied to microorganisms in the 1880s by Robert Koch and others, who used staining and biochemical tests to distinguish between bacterial species. However, it was recognized that many of the resulting species were made up of a variety of types with quite different properties. An example is provided by *E. coli*, which includes strains with distinctive pathogenic characteristics, ranging from harmless to lethal.

A century later, when DNA sequencing was first applied to bacterial genomes, it became clear that different strains of a single species can have very different genome sequences and often have individual sets of strain-specific genes. This was first shown by a comparison between two strains of *Helicobacter pylori*, which causes gastric ulcers and other diseases of the human digestive tract. The two strains were isolated in the United Kingdom and the United States and had genomes of 1.67 and 1.64 Mb, respectively. The larger genome contained 1,552 genes and the smaller one 1,495 genes, with 1,406 of these genes being present in both strains.

A much more extreme distinction between strains was revealed when the genome sequence of the common laboratory strain of *E. coli*, K12, was compared with that of one of the most pathogenic strains, O157:H7. The lengths of the two genomes are significantly different – 4.64 Mb for K12 and 5.59 Mb for O157:H7 – with the pathogenic strain containing over 1,300 genes not present in *E. coli* K12. Many of these 1,300 genes code for toxins and other proteins that are involved in the pathogenic properties of O157:H7. But it is not simply a case of O157:H7 containing extra genes that make it pathogenic. K12 also has over 500 genes that are absent from O157:H7. The situation, therefore, is that *E. coli* O157:H7 and *E. coli* K12 each has a set of strain-specific genes, which make up approximately 25% and 12% of their gene catalogs, respectively.

The differences in genome sizes and gene contents that occur within a bacterial species have led to the **pan-genome concept**. According to this concept, the genome of a species is divided into two components (Figure 13.8). The first component is the **core genome**, which contains the set of genes possessed by all members of the species. The second component is the **accessory genome**, which is the entire collection of additional genes present in different strains and isolates of that species.

The core genome can therefore be looked on as specifying the basic biochemical and cellular activities that define a particular species, whereas the accessory genome describes the complete biological capability of the species as a whole, components of which are expressed by individual strains. In some species, the pan-genome is only slightly larger than the core genome. An example is *Bacillus anthracis*, whose pan-genome has been estimated to contain 2,978 genes, of which 2,893 form the core. In other species, the difference is much greater. The *E. coli* core genome is thought to contain as few as 800 genes, even though as many as 23,000 accessory genes have been identified in different strains of this species. It has been suggested that the relatively small number of accessory genes possessed by a species with a small pan-genome reflects a more limited ecological range compared to a species with a large pan-genome, whose vast array of accessory genes presumably enables the species to colonize a larger variety of ecological niches.

chromosome – located in nucleoid, carries essential genes

chromid – uses plasmid partitioning system, carries essential genes

plasmid – uses plasmid partitioning system, carries nonessential genes

FIGURE 13.7 The relationship between chromosomes, chromids and plasmids. The plasmid partitioning system is the process used to distribute plasmids to daughter cells when a bacterium divides (Section 13.2).

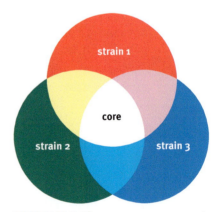

FIGURE 13.8 The pan-genome concept. The gene contents of the genomes of three strains of a bacterium are depicted. Each gene set is represented by a circle, the overlap between the three circles (shown in white) being the core genome. The accessory genome comprises those genes that lie outside of the core set. These genes can be further subdivided into those which are present in just one genome (shown in red, dark blue and green), and genes shared by two genomes (yellow, pink and light blue).

13.2 INHERITANCE OF CHROMOSOMES AND PLASMIDS DURING BACTERIAL CELL DIVISION

When a bacterium divides, each daughter cell must receive a copy of the chromosomal DNA molecule, along with a copy of any chromids that carry genes needed for the bacterium's survival. Inheritance of plasmids carrying nonessential genes is not such an absolute requirement, but it is clear that during the vast majority of cell divisions, the plasmids present in the parent cell are also passed on to the daughters.

The bacterial chromosome resides in the **nucleoid**, a structure that takes up about one-third of the volume of the cell (Figure 13.9). The nucleoid is surrounded by a peripheral region that is usually referred to simply as the cytoplasm. In order to understand how a bacterial chromosome is inherited during cell division, we must first examine the physical structure of the nucleoid.

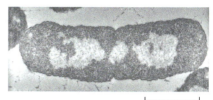

FIGURE 13.9 The *Escherichia coli* nucleoid. This transmission electron micrograph shows the cross section of a dividing *E. coli* cell. The nucleoid is the lightly staining area in the center of the cell. (Courtesy of Conrad Woldringh, University of Amsterdam.)

The *E. coli* nucleoid contains supercoiled DNA attached to a protein core

Most of what we know about nucleoid structure relates specifically to *E. coli*, but it appears that the nucleoids of at least the majority of other bacterial species are organized in a similar manner.

The critical feature of the nucleoid is the need to fit a relatively long DNA molecule into a very small space. The circular *E. coli* chromosome has a contour length (i.e. circumference) of approximately 1.6 mm. This molecule has to be squeezed into a cell that is about 1 µm by 2 µm. This is not impossible as a DNA molecule is very thin and so does not take up much space when it is folded up tightly. This tight folding can be achieved by **supercoiling**. To become supercoiled, a DNA molecule must be circular or, if linear, it must be attached to proteins or other structures that prevent the free rotation of the ends of the molecule. A DNA topoisomerase can then introduce additional turns into the double helix (positive supercoiling) or remove existing turns (negative supercoiling). In each case, the torsional stress that is introduced causes the molecule to wind around itself to form a more compact structure (Figure 13.10). Supercoiling is therefore an ideal way of packaging a circular molecule into a small space.

Evidence that supercoiling is involved in packaging the circular *E. coli* chromosome was first obtained in the 1970s from examination of isolated nucleoids and subsequently confirmed as a feature of DNA in living cells

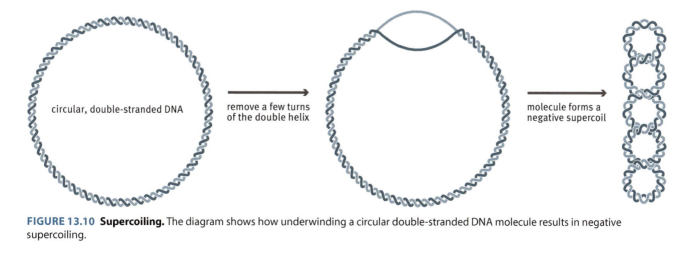

FIGURE 13.10 Supercoiling. The diagram shows how underwinding a circular double-stranded DNA molecule results in negative supercoiling.

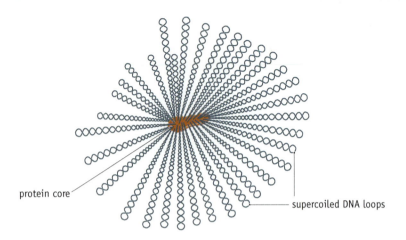

FIGURE 13.11 A model for the structure of the *Escherichia coli* nucleoid.

in 1981. In *E. coli*, the supercoiling is thought to be generated and controlled by the type II topoisomerase called DNA gyrase.

The supercoiling introduced into the *E. coli* chromosome must be organized in such a way that the genes within the molecule are still accessible, so they can be transcribed as and when necessary. It must also be possible to replicate the DNA and separate the daughter molecules without everything getting tangled up. This means that the DNA molecule must be folded up in a very ordered manner. Currently, we believe that the *E. coli* DNA is attached to a protein core from which supercoiled loops, each containing 40–300 kb of DNA, radiate out into the cell (Figure 13.11). The loops are called **chromosomal interaction domains** (**CIDs**). During active growth, there are 31 CIDs, but the number is variable with fewer (and hence longer) CIDs when there are lower amounts of available nutrients. The protein component of the nucleoid is made up of a variety of **nucleoid-associated proteins**, which are presumed to include those that are involved in the packaging of the chromosome. These include the **HU family** of proteins. Each HU protein is a dimer of two subunits, either two HUα, two HUβ, or a heterodimer comprising one of each type of monomer. HU proteins induce bends in the DNA in order to facilitate the formation of the supercoiled loops.

Partitioning systems control the inheritance of bacterial chromosomes and plasmids

Now that we have established that the bacterial chromosome is carefully packaged inside the cell, we can move on to the process by which copies of it are passed to the daughter cells when the bacterium divides.

Partitioning of the bacterial chromosome is closely linked with DNA replication. When the chromosome begins to replicate, the origins of replication of the two daughter DNA molecules move apart from one another, each toward a different end of the cell. The replicated DNA molecules therefore take up the appropriate positions needed for them to become enclosed within the daughter cells that are formed when the bacterium divides into two. In a rich growth medium, when the bacteria are able to divide every 20 minutes or so, the septum begins to form across the middle of the parent cell before chromosome replication has been completed (Figure 13.12). If necessary, the last parts of the replicated DNA molecules can be 'pumped' across the septum by special translocase enzymes.

How is partitioning achieved? In most species that have been studied the process involves a three-component system called **ParABS**. Two of these components are proteins, and the third, *parS*, is a DNA sequence. Single

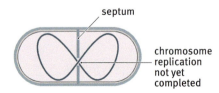

FIGURE 13.12 Chromosome partitioning in rapidly dividing *E. coli* cells. Septum formation begins before the chromosomal DNA has been completely replicated. Should part of either daughter DNA molecule become trapped in the septum then it can be released by a translocase enzyme.

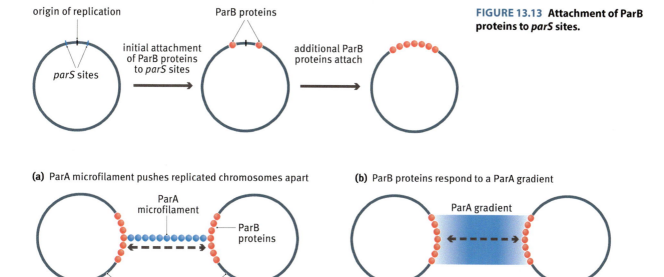

FIGURE 13.13 Attachment of ParB proteins to *parS* sites.

FIGURE 13.14 Two models for chromosome partitioning by the ParABS system. (a) The replicated chromosomes are pushed apart by growth of a ParA microfilament. (b) ParA proteins form a gradient to which the ParB proteins respond.

or multiple copies of *parS* are located on the bacterial chromosome close to the origin of replication. These act as binding sites for ParB proteins which, once the initial attachments are made, recruit additional copies of ParB that cover the region of DNA adjacent to *parS* (Figure 13.13). The second protein, ParA, then associates with ParB-bound DNA. Initially, it was thought that the replicated chromosomes were pushed apart by the growth of a microfilament made up of ParA proteins (Figure 13.14a). This now seems unlikely as careful observations of dividing bacterial cells have failed to detect these microfilaments. Instead, researchers are exploring models in which ParA proteins form a gradient within the cell, to which the bound ParB proteins respond in order to move each of the daughter chromosomes in the appropriate direction (Figure 13.14b).

ParABS is called the Type I partitioning system. It is also used by several low-copy-number plasmids such as the *E. coli* F plasmid. The related type II or ParMRC system is employed by several of the clinically important R plasmids of *E. coli*. Relaxed plasmids, which are present in high copy numbers, are not thought to have an organized partitioning system. Instead, their distribution throughout the parent bacterium ensures that at least some are inherited by each of the daughters.

13.3 TRANSFER OF GENES BETWEEN BACTERIA

So far, we have only considered the way in which bacterial genes, whether on a chromosome, chromid or plasmid, are inherited by linear descent – by daughter cells when a parent cell divides. Bacteria can also acquire genes directly from other bacteria. This is called horizontal or lateral gene transfer.

It has been known since the 1940s that plasmids and occasionally chromosomal genes can move horizontally, but it was thought that the transfer could only take place between members of the same species, or occasionally between closely related species. This assumption was overturned in the 1990s when complete genome sequences were obtained for a variety of bacterial species. Comparisons of these sequences revealed that the same genes are present in quite distantly related species, suggesting that the barriers to horizontal gene transfer are much less rigid than previously thought. It also became clear that although we know a great deal about the processes by which plasmids and chromosomal genes move within species, we know much less about how genes are transferred from one species to another. We will look first at our understanding of these within-species processes and then examine the implications of the much broader examples of horizontal gene transfer revealed by genome sequencing.

Plasmids can be transferred between bacteria by conjugation

One of the major breakthroughs in 20th-century microbiology was made in 1946 when Joshua Lederberg, a 21-year-old graduate student at the University of Yale, discovered that bacteria can exchange genes by a process that was subsequently called conjugation. During the next decade, Lederberg and other microbial geneticists showed that the transfer of genes is unidirectional from donor to recipient cells. The donor cells are referred to as **F⁺** ('F-plus', F for fertility), and the recipient cells are called **F⁻** ('F-minus') cells.

Conjugation involves a tube-like structure, called the **pilus**, which F⁺ cells are able to construct (Figure 13.15). An F⁺ cell forms a connection with an F⁻ cell via its sex pilus, and as the pilus is hollow it has been assumed that DNA transfer occurs through it. An alternative is that the pilus simply brings the F⁺ and F⁻ cells into close contact (Figure 13.16). The contact is possibly so close that the outer membranes on the surfaces of the two bacteria fuse, enabling DNA to be transferred directly from one cell to the other.

The difference between an F⁺ and an F⁻ cell is that the former contains a fertility plasmid. The best studied of these is the F plasmid of *E. coli*. The F plasmid is 99 kb in size and carries a large operon that contains the *tra* (for transfer) genes. These genes code for proteins involved in the synthesis and assembly of the pilus and in the DNA transfer process itself.

Conjugation is always between an F⁺ cell and an F⁻ one. The F⁺ cell, the one that contains an F plasmid, initiates conjugation by making a pilus which subsequently attaches to an F⁻ cell (Figure 13.17). The two cells become drawn together and a copy of the F plasmid is transferred from the F⁺ to the F⁻ bacterium. Replication is by the rolling-circle mechanism, with the parent plasmid remaining in the F⁺ cell and the copy transferring

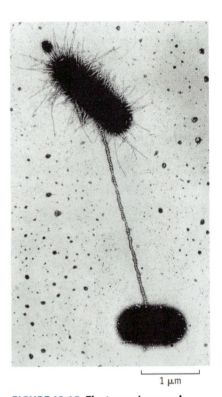

FIGURE 13.15 Electron micrograph showing two bacteria joined together by a pilus. One of the two bacteria also has a number of shorter pili radiating from its surface. (Courtesy of the late Charles Brinton.)

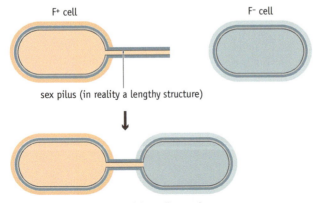

FIGURE 13.16 The role of the sex pilus in bringing two bacteria together.

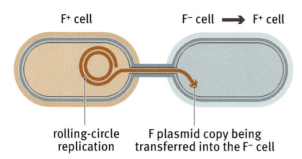

FIGURE 13.17 Transfer of the F plasmid during conjugation between an F⁺ and an F⁻ cell The transfer is shown to occur through the sex pilus, although it may be that the transfer occurs directly across the walls of the two cells.

to the recipient F⁻ cell as it is rolled off of the parent. The F⁻ cell therefore becomes F⁺. Theoretically, this should mean that, over time, all *E. coli* bacteria will become F⁺. This is not the case, so presumably the F plasmid is occasionally lost from an F⁺ bacterium, so there are always F⁻ cells in a population.

A large number of bacterial species are able to conjugate in the same way as *E. coli* and in all of these the process is controlled by a plasmid analogous to the F plasmid. Fertility plasmids of this type are called self-transmissible, which means that they can set up conjugation and mobilize themselves into the recipient cell. Conjugation (i.e. setting up the initial contact) and mobilization (passage of plasmid DNA from donor to recipient) are two distinct characteristics and only self-transmissible plasmids are able to direct both. A few other plasmids can set up the conjugation contact between cells but not mobilize on their own, and some can mobilize, but only if they are coexisting in the cell with a second plasmid that can set up the initial contact (Figure 13.18). Other plasmids are totally nonfertile and can neither conjugate nor mobilize.

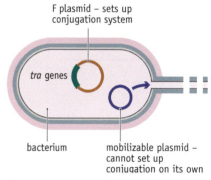

FIGURE 13.18 Mobilization of a nonconjugative plasmid. Some plasmids cannot set up conjugation but can be mobilized if they share a cell with a conjugative plasmid.

Chromosomal genes can also be transferred during conjugation

Bacterial conjugation is important not only because it enables plasmids to be transferred between bacteria. Chromosomal genes can also be passed from donor cell to recipient. How does this come about?

There are several possible ways, the first being simply that a small random piece of the donor cell's genome is transferred along with the F plasmid (Figure 13.19). This probably occurs fairly infrequently.

More important are the gene transfer properties of two special types of donor cells, called **Hfr** and **F′** ('F-prime'). In Hfr cells, the F plasmid has become integrated into the *E. coli* chromosomal DNA. The integrated form of the F plasmid can still direct conjugal transfer, but in this case in addition to transferring itself it also carries into the F⁻ cell a copy of at least some of the *E. coli* DNA molecule to which it is attached (Figure 13.20). It takes approximately 100 minutes for the entire *E. coli* chromosome to be transferred in this way, but conjugation rarely continues for this long. Termination of conjugation interrupts the DNA transfer, so usually only a part of the chromosomal DNA is passed to the recipient cell.

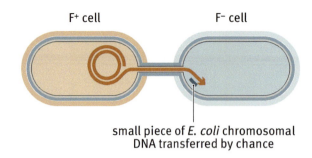

FIGURE 13.19 Transfer of a segment of the bacterial chromosome during conjugation between an F⁺ and an F⁻ cell.

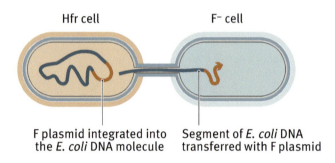

FIGURE 13.20 Transfer of chromosomal DNA during conjugation between an Hfr and an F⁻ cell.

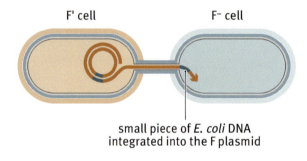

FIGURE 13.21 Transfer of chromosomal DNA during conjugation between an F′ and an F⁻ cell.

F′ cells, the second type of donor cell regularly associated with the transfer of chromosomal genes, occasionally arise from Hfr cells when the integrated F plasmid excises from the genome. Normally, this event results in an F⁺ cell, but sometimes excision of the F plasmid is not entirely accurate, and a small segment of the adjacent bacterial DNA is also snipped out. This leads to an F′ plasmid that carries a segment of the bacterium's genome, possibly including a few genes (Figure 13.21). Conjugation involving an F′ cell always results in the transfer of these plasmid-borne bacterial genes.

Bacterial genes can also be transferred without contact between donor and recipient

In addition to conjugation, bacteria can also acquire segments of chromosomal DNA from other cells by processes that do not involve direct cell-to-cell contact. There are two of these processes, called **transformation** and **transduction**.

During transformation, a bacterium simply takes up DNA that it encounters in its local environment. Some bacteria, such as *Bacillus* and *Haemophilus* species, have efficient mechanisms for DNA uptake. The outer membranes of these species contain proteins that bind DNA and transfer it into the cell

(Figure 13.22). Other species, including *E. coli*, do not have DNA uptake proteins, but some DNA can still penetrate the outer surface of the bacterium and enter the cell. In the laboratory, artificial treatments can improve the transformation efficiency of *E. coli* cells by rendering them more competent for DNA uptake.

The second type of DNA transfer that does not require cell-to-cell contact is transduction. In transduction, the transfer is mediated by a bacteriophage, one of the viruses that infect bacterial cells. We learnt in Chapter 12 that some bacteriophages do not immediately kill the host cell. Instead, the bacteriophage genome becomes integrated into the host bacterium's chromosome (see Figure 12.7). At some stage, the integrated prophage will excise from the host chromosome and replicate, and new bacteriophage particles will be constructed around these replicated genomes. This is accompanied by a breakdown of the bacterial chromosome into small fragments, some of which, by chance, will be about the same size as the bacteriophage genome. By mistake, one of these small chromosome fragments might be packaged into a bacteriophage particle (Figure 13.23). The resulting particle is still infective, as infection is solely a function of the proteins within which the DNA is enclosed. The fragment of bacterial DNA can therefore be transferred to a new cell when that recipient becomes infected with the transducing bacteriophage particle.

Transferred bacterial DNA can become a permanent feature of the recipient's genome

What happens to the DNA that a recipient cell acquires by conjugation, transformation or transduction? If the DNA is an intact plasmid, then often it will be able to replicate in its new host, adding to the plasmid complement of the recipient cell. This is the outcome of most conjugation events and might also arise from transformation, as it is quite possible for plasmid DNA to be taken up from the environment in this way. Horizontal transfer of plasmids is immensely important in microbiology because it is responsible for the spread of antibiotic resistance genes. This has brought about the evolution of strains of pathogenic bacteria that can no longer be controlled with the antibiotics that previously were effective against them. An important example is methicillin-resistant *Staphylococcus aureus* (MRSA). It is sometimes thought that the 'M' in MRSA stands for 'multiple' but it does not. It is bad enough that MRSA has become resistant to the methicillin antibiotics usually used to prevent *S. aureus* from invading wounds in hospital patients. Should it genuinely become 'multiply resistant', then the problems in controlling it will be substantially worse.

What about the chromosomal DNA that is transferred into a recipient cell by transformation, transduction and occasionally conjugation? This DNA might be broken down and used as a source of carbon, nitrogen and phosphorus for the construction of new molecules within the recipient bacterium. But sometimes the DNA can survive in the cell. For this to happen, the DNA must become integrated into the bacterial chromosome or into a chromid. This is possible if the transferred DNA has the same or similar sequence to a segment of the recipient cell's chromosome, as will be the case if it comes from the same species. The segment of transferred DNA might then replace the equivalent stretch of DNA in the chromosome (Figure 13.24a). If this happens, then the recipient bacterium will inherit the genes from the donor. As the genes are the same as the ones that it had before, the replacement might appear to be immaterial. But bacterial genes, just like eukaryotic ones, have variant forms called alleles, each one with a slightly different DNA sequence. Transfer might therefore

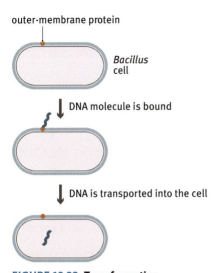

FIGURE 13.22 Transformation. Transformation of *Bacillus* is shown. DNA initially becomes bound to an outer-membrane protein and is then transported into the cell.

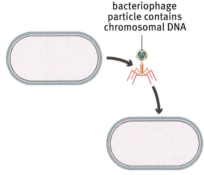

FIGURE 13.23 Transduction. A small chromosome fragment has been packaged into a bacteriophage particle. This DNA will be transferred to a new bacterium when that cell becomes infected with the particle.

(a) transferred DNA has sequence similarity to bacterial chromosome

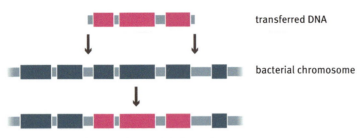

integrated DNA replaces equivalent genes

(b) transferred DNA has no similarity to bacterial chromosome

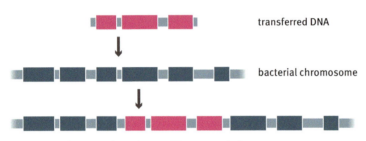

integrated DNA inserted between existing genes

FIGURE 13.24 Integration of transferred DNA into a bacterial chromosome. Two possibilities are shown. (a) The transferred DNA contains genes already present in the bacterial chromosome. Integration replaces those genes so the bacterium might acquire new alleles. (b) The transferred segment has no sequence similarity with the bacterial chromosome, so integration occurs at a random position without the loss of any of the bacterium's genes.

result in the recipient acquiring a different allele for one or more genes. Its biological characteristics might therefore change in a small way.

Genome sequencing projects have told us that the transfer of DNA also occurs between quite different species. Exactly how this happens is not known. We presume that the transfer is by transformation, because unrelated species are unable to conjugate, and transduction can only take place within the host range of the transducing bacteriophage, which is always quite narrow. The incoming DNA is unlikely to have extensive sequence similarity with any part of the recipient's chromosome, so integration is assumed to occur randomly (**Figure 13.24b**). Once it becomes a part of the chromosome, the new DNA, and any genes it contains, will become a permanent part of the recipient's genome. The genes will be passed to daughter cells along with the rest of the genome when the parent cell divides.

Transfer across a species boundary might not happen very frequently, but it occurs often enough for bacteria to acquire, from time to time, new genes from other species. It has been estimated that almost 10% of *E. coli* genes have been acquired by horizontal transfer, and in other species, the number is substantially higher. The plant pathogen *Xylella fastidiosa* appears to have obtained one-fifth of its genes in this way. The genome of the thermophilic bacterium *Thermotoga maritima* has 1,895 genes, a quarter of which appear to have been obtained from other species, including members of the second prokaryotic kingdom, the archaea. These new genes provide the basis for the ability of *T. maritima* to tolerate high temperatures. Similarly, the transfer of approximately one thousand genes from bacteria to an anaerobic ancestor of the Haloarchaea might have enabled this genus to evolve a tolerance to oxygen and adopt an aerobic lifestyle, albeit as extremophiles in brine pools and other high salt environments. The picture that is emerging is one in which prokaryotes living in similar ecological niches exchange genes with one another in order to increase their individual fitness for survival in their particular environment.

TRANSFER OF GENES BETWEEN BACTERIA

KEY CONCEPTS

- Bacterial genomes range in size from 0.1 to 25 Mb. The larger genomes belong to species that occupy complex ecological niches such as soil.
- Most bacterial genomes are circular DNA molecules, but some linear examples are known.
- Plasmids are DNA molecules, usually circular, that lead an independent existence within a bacterium. Plasmids carry genes but these are not essential for the survival of the bacterium. The plasmid genes may, however, confer a useful characteristic such as resistance to an antibiotic.
- Some bacteria have multipartite genomes. In some of these species, it is difficult to distinguish between chromosomal DNA molecules, which carry essential genes, and plasmids. A new term, chromid, has been introduced to describe molecules that have plasmid-like features but which carry essential genes.
- Gene numbers vary within a bacterial species. The core genome is the set of genes possessed by all members of the species. The accessory genome is the additional genes present in different strains and isolates of that species.
- Most of the DNA in a bacterium is contained in the nucleoid. The nucleoid makes up about one-third of the volume of the cell.
- The *E. coli* nucleoid contains the bacterial chromosome attached to a protein core from which supercoiled loops, each containing 40–300 kb of DNA, radiate out into the cell.
- During division of a bacterium, the chromosome and low-copy-number plasmids are partitioned into the daughter cells by proteins that move the replicated molecules apart.
- Conjugation, transformation and transduction are processes that can result in the transfer of DNA between bacteria. Conjugation and transduction can only result in transfer between members of the same or related species. Transformation can, in theory, occur between members of any two species.
- Horizontal gene transfer between members of different species is evident from comparisons between bacterial genome sequences.
- Prokaryotes, both bacteria and archaea, living in similar ecological niches exchange genes with one another in order to increase their individual fitness for survival in their particular environment.

QUESTIONS AND PROBLEMS

Key terms

Write short definitions of the following terms:

accessory genome
bacterial chromosome
chromid
chromosomal interaction domain
Col plasmid
colicin
compatible
conjugation
copy number
core genome
degradative plasmid
episome
F^+
F^-
F'

fertility plasmid
Hfr
HU family
horizontal gene transfer
incompatibility group
integron
nucleoid

nucleoid-associated protein
pan-genome concept
ParABS
partitioning
pilus
plasmid
relaxed

resistance plasmid
stringent
supercoiling
transduction
transformation
virulence plasmid

Self-study questions

13.1 The *E. coli* genome is a single, circular DNA molecule. What other types of genome structure are found among bacteria?

13.2 How is genome size in a bacterium related to gene number?

13.3 What are the typical features of a bacterium with (A) a small number of genes and (B) a large number of genes?

13.4 Explain what a plasmid is, and provide examples of the different types of plasmids that are known.

13.5 Distinguish between a stringent and relaxed plasmid.

13.6 Describe the genome organizations of *Vibrio cholerae* and *Deinococcus radiodurans*.

13.7 What are the distinguishing features of bacterial chromosomes, plasmids and chromids?

13.8 What are the differences between the gene contents of the K12 and O157:H7 strains of *E. coli*?

13.9 Describe the pan-genome concept.

13.10 What is supercoiling, and what effect does it have on a circular DNA molecule?

13.11 What proteins are present in the *E. coli* nucleoid?

13.12 Describe how the bacterial chromosome is partitioned to daughter cells when a bacterium divides.

13.13 How are plasmids partitioned during bacterial cell division?

13.14 Describe how plasmids are transferred between bacterial cells by conjugation.

13.15 Distinguish between F+, F−, Hfr and F′ cells.

13.16 Describe two ways in which genes can be transferred without contact between bacteria.

13.17 How can transferred DNA become incorporated into the genome of the recipient bacterium?

13.18 Outline the evidence for the occurrence of horizontal gene transfer between prokaryotes of different species.

Discussion topics

13.19 Should the traditional view of the bacterial genome as a single, circular DNA molecule be abandoned? If so, what new definition of 'bacterial genome' should be adopted?

13.20 The obligate intracellular parasitic bacterium *Mycoplasma genitalium* has just 519 genes. Why does this organism require so few genes?

13.21 Genetic elements that reproduce within or along with a host genome, but confer no benefit on the host, are called 'selfish'. Discuss this concept, in particular as it applies to plasmids.

13.22 Discuss possible ways in which plasmid copy number could be controlled.

13.23 Plasmid R100 carries a gene for resistance to streptomycin. You have an F+ strain of *E. coli* that contains R100. Design an experiment to test whether R100 is able to mobilize on its own.

13.24 The biological species concept states that a species is a group of interbreeding individuals that are reproductively isolated from other such groups. Does this definition of species hold for prokaryotes? If not, then how might a prokaryotic species be defined?

The answers to, or hints on how to answer, the self-study questions and discussion topic questions can be downloaded from the book's product page at this link: www.routledge.com/9781032743530

FURTHER READING

Alm RA, Ling L-SL, Moir DT et al. (1999) Genomic sequence comparison of two unrelated isolates of the human gastric pathogen *Helicobacter pylori*. *Nature* 397, 176–180.

Arnold BJ, Huang IT & Hanage, WP (2022) Horizontal gene transfer and adaptive evolution in bacteria. *Nature Reviews Microbiology* 20, 206–218.

Chiang YN, Penadés JR & Chen J (2019) Genetic transduction by phages and chromosomal islands: the new and noncanonical. *PLoS Pathogens* 15, e1007878.

Dame RT, Rashid F-ZM & Grainger DC (2020) Chromosome organization in bacteria: mechanistic insights into genome structure and function. *Nature Reviews Genetics* 21, 227–242.

Dillon S & Dorman CJ (2010) Bacterial nucleoid-associated proteins, nucleoid structure and gene expression. *Nature Reviews Microbiology* 8, 185–195.

Harrison PW, Lower RPJ, Kim NKD & Young JP (2010) Introducing the bacterial "chromid": not a chromosome, not a plasmid. *Trends in Microbiology* 18, 141–148.

Heidelberg JF, Eisen JA, Nelson WC et al. (2000) DNA sequence of both chromosomes of the cholera pathogen *Vibrio cholerae*. *Nature* 406, 477–483.

Helinski DR (2022) A brief history of plasmids. *EcoSal Plus* 10, eESP00282021.

Jalal ASB & Le TBK (2020) Bacterial chromosome segregation by the ParABS system. *Open Biology* 10, 200097.

Mell JC & Redfield RJ (2014) Natural competence and the evolution of DNA uptake specificity. *Journal of Bacteriology* 196, 1471–1483.

Perna NT, Plunkett G, Burland V et al. (2001) Genome sequence of enterohaemorrhagic *Escherichia coli* O157:H7. *Nature* 409, 529–533.

Rouli L, Merhej V, Fournier PE et al. (2015) The bacterial pangenome as a new tool for analysing pathogenic bacteria. *New Microbes and New Infections* 7, 72–85.

Virolle C, Goldlust K, Djermoun S et al. (2020) Plasmid transfer by conjugation in Gram-negative bacteria: from the cellular to the community level. *Genes* 11, 1239.

White O, Eisen JA, Heidelberg JF et al. (1999) Genome sequence of the radioresistant bacterium *Deinococcus radiodurans* R1. *Science* 286, 1571–1577.

Willetts N & Skurray R (1980) The conjugation system of F-like plasmids. *Annual Review of Genetics* 14, 47–76.

CHAPTER 14

Inheritance of Genes During Eukaryotic Cell Division

In the previous two chapters, we studied the inheritance of genes in viruses and bacteria. In this chapter, we address the more complex issues relating to gene inheritance in eukaryotes. We will ask how a complete set of genes is passed to the daughter cells when a eukaryotic cell, such as a human cell, divides.

To understand how DNA molecules are correctly distributed during eukaryotic cell division, we must explore some of the areas of research where genetics and cell biology overlap. We must recognize that the problem we are tackling is not the distribution of 'naked' DNA molecules from parent to daughter cells, but the transmission of chromosomes. We must therefore understand what chromosomes are and how DNA is organized within them. We will then be in a position to study the events occurring during cell division and to see how these ensure that the chromosomes are correctly distributed to the daughter cells.

14.1 EUKARYOTIC GENOMES AND CHROMOSOMES

Eukaryotes have two or possibly three distinct genomes. First there is the **nuclear genome**, contained in the chromosomes and comprising the vast bulk of the cell's DNA and almost all of its genes. Often, when geneticists refer to the 'genome' they simply mean the nuclear genome.

Most nuclear genomes are over 10 Mb in length, and the largest are over 100,000 Mb (Figure 14.1). This size range coincides to a certain extent with the complexity of the organism, the simplest eukaryotes such as fungi and protozoans having the smallest genomes.

All eukaryotes also have a **mitochondrial genome**. This is a small, usually circular molecule present in multiple copies in each of the energy-generating organelles called mitochondria. The mitochondrial genome is much smaller than the nuclear genome – the human mitochondrial genome is only 16.57 kb in length and carries just 37 genes. These genes code for proteins and RNAs involved in the biochemical activities that occur inside the mitochondria. We will examine the human mitochondrial genome more closely when we study the human genome in Chapter 20.

In addition to the mitochondrial genome, plants and other photosynthetic eukaryotes have a **chloroplast genome**. Again, this is a small molecule, about 225 kb in most flowering plants and with about 200 genes.

Eukaryotic nuclear genomes are contained in chromosomes

All eukaryotic nuclear genomes are made up of linear DNA molecules, each of these molecules contained in a different chromosome. The number of chromosomes varies between species and appears to be unrelated to the biological features of the organism. Some quite simple eukaryotes have multiple chromosomes, such as the yeast *Saccharomyces cerevisiae* which has 16, while more complex organisms have relatively few. The ant *Myrmecia pilosula* has just one chromosome, and the Indian muntjac deer has only four (Figure 14.2). Nor is chromosome number linked to genome size. Some salamanders have genomes 30 times bigger than the human version but split into half the number of chromosomes.

DOI: 10.1201/9781003473862-16

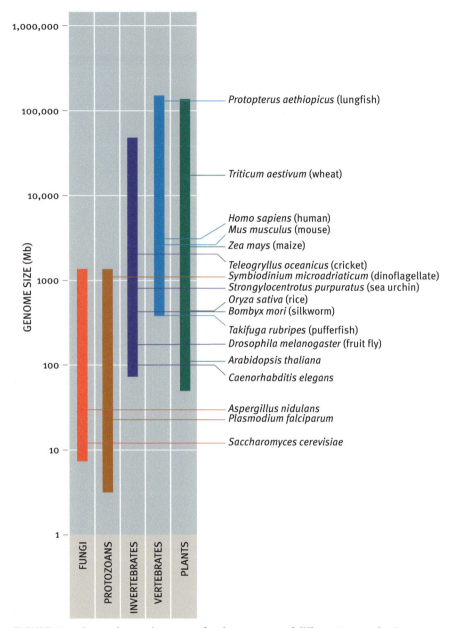

FIGURE 14.1 Approximate size ranges for the genomes of different types of eukaryote.

The human nuclear genome is typical of those of all multicellular animals. It comprises approximately 3,100 Mb of DNA, divided into 24 molecules, the shortest 47 Mb in length and the longest 249 Mb. The 24 chromosomes consist of 22 **autosomes** and the two sex chromosomes, X and Y. The vast majority of human cells are **diploid** and so have two copies of each autosome, plus two sex chromosomes, XX for females or XY for males – 46 chromosomes in all. These are called **somatic cells**, in contrast to **sex cells** or **gametes**, which are **haploid** and have just 23 chromosomes, comprising one of each autosome and one sex chromosome.

Chromosomes contain DNA and proteins

Chromosomes are made up of about one-half protein and one-half DNA. The complex between the two is referred to as **chromatin**. This term was first used by the German cytologist Walther Flemming in 1879 to describe the fibrous material in the nuclei of cells that becomes visible by light microscopy after staining with dyes such as hematoxylin (Figure 14.3).

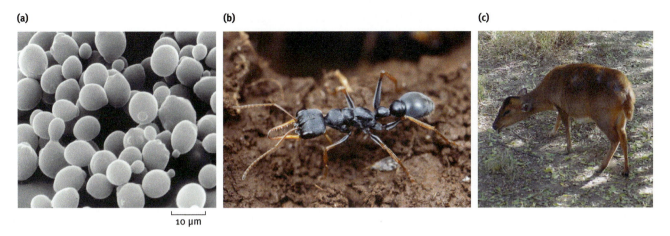

FIGURE 14.2 Three eukaryotes with different numbers of chromosomes. (a) The yeast *Saccharomyces cerevisiae*, which has 16 chromosomes. (b) The ant *Myrmecia pilosula*, which has just one chromosome. (c) The Indian muntjac deer, which has four chromosomes. (a, Courtesy of Eric Schabtach; b, Courtesy of Pavel Krasensky/Shutterstock.com; c, Courtesy of Susan Hoffman, Miami University.)

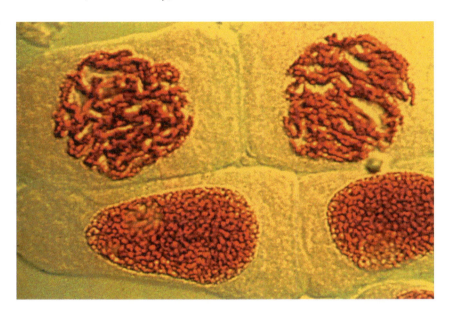

FIGURE 14.3 Light micrograph of cells of the common bluebell. The material is stained so that the DNA content of the nuclei appears red. Microscopic images similar to this one led cytologists in the 19th century to use the term 'chromatin' to describe the material present in the nucleus. 'Chroma' is Greek for color. (Courtesy of G. Giminez-Martin/Science Photo Library.)

A few years later, it became clear that this fibrous material forms discrete structures, which became known as chromosomes.

Some of the protein component of a chromosome is made up of a heterogeneous mixture of molecules involved in DNA replication and gene expression, including DNA and RNA polymerases and regulatory proteins. The remainder, associated most intimately with the DNA, consists of a group of proteins called **histones**.

Histones contain a high proportion of basic amino acids (lysine, histidine and arginine) and are remarkable in that they are very similar in all species. If, for example, the H4 histone proteins of pea and cow are compared, then we find that only two of the 103 amino acids in the polypeptides are different (Figure 14.4). This is a much greater degree of similarity than

```
cow MSGRGKGGKGLGKGGAKRHRKVLRDNIQGITKPAIRRLARRGGVKRISGLIYEETRGVLKVFLENVIRDAVTYTEHAKRKTVTAMDVVYALKRQGRTLYGFGG
pea MSGRGKGGKGLGKGGAKRHRKVLRDNIQGITKPAIRRLARRGGVKRISGLIYEETRGVLKIFLENVIRDAVTYTEHARRKTVTAMDVVYALKRQGRTLYGFGG
    **************************************************************  *****************  ************************
```

FIGURE 14.4 Amino acid sequences of histone H4 from cow and pea. The sequences are written using the one-letter abbreviations for the amino acids (see Table 7.1). Asterisks mark the positions where the same amino acid is present in both sequences.

we expect for proteins with equivalent functions in such widely divergent organisms. The conservation indicates that histones were among the first proteins to evolve (before the common ancestor of peas and cows lived) and that their structures have not changed for hundreds of millions of years. This in turn suggests that histones play a fundamental role within chromosomes, a role that was set out during the earliest stages of evolution of the eukaryotic cell.

Histones are constituents of the nucleosome

The important breakthroughs in understanding the exact nature of the association between DNA and histones were made in the early 1970s by a combination of biochemical analysis and electron microscopy. The biochemical work included **nuclease protection** experiments. In this procedure, a DNA–protein complex is treated with an endonuclease, an enzyme that cuts polynucleotides at internal phosphodiester bonds (Figure 14.5a). The endonuclease has to gain access to the DNA in order to cut it and hence can only attack the phosphodiester bonds that are not masked ('protected') by attachment to a protein.

These experiments were carried out with chromatin that had been gently extracted from nuclei by methods designed to retain as much of the chromosome structure as possible. The sizes of the resulting DNA fragments were not random as might be expected, but instead were all 146 bp in length when human chromatin was used (Figure 14.5b). The conclusion is that chromatin has a uniform structure, in which each protein or protein complex is closely associated with, and hence protects, 146 bp of DNA.

The biochemical results were complemented by electron microscopic observations of chromatin. The images showed linear arrays of spherical structures that appear as beads, attached at regular intervals on a string of DNA. This structure has a width of approximately 10 nm and is called the **10 nm fiber** (Figure 14.6).

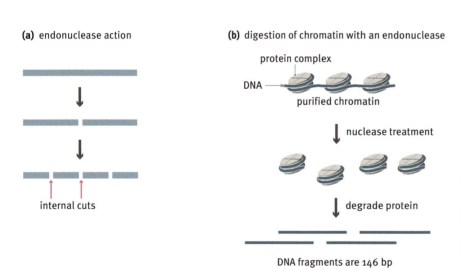

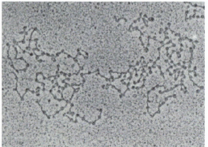

FIGURE 14.5 The nuclease protection technique and its use with purified chromatin. (a) An endonuclease makes cuts at internal phosphodiester bonds within a DNA molecule. The longer the exposure to the nuclease, the greater the number of cuts that are made, until eventually only mononucleotides remain. (b) If proteins are bound to a DNA molecule, then not all phosphodiester bonds can be cut by an endonuclease, as some are protected by the bound protein. When human chromatin is treated with an endonuclease, DNA fragments of 146 bp are obtained.

FIGURE 14.6 The 10 nm fiber. Electron micrograph of the 10 nm fiber, showing protein beads attached at regular intervals along a string of DNA. (From S. Tesoro, I. Ali & A. N. Morozov, *Phys. Biol.* 13: 016004. Doi: 10.1088/1478-3975/13/1/016004. Published under CC BY 3.0.)

The spherical particles are called **nucleosomes** and are made up of equal amounts of each histone except H1. The histones form a barrel-shaped **core octamer** consisting of two molecules each of H2A, H2B, H3 and H4, with the DNA molecule wound twice around each nucleosome (Figure 14.7a). In humans, exactly 146 bp of DNA is associated with the nucleosome and hence protected from endonuclease digestion. The only parts of the DNA molecule available to the endonuclease are the 50–70 bp stretches of **linker DNA** that join the individual nucleosomes to one another.

The one histone not found in the core octamer of the nucleosome is H1. We now know that histone H1 is not a single protein but a group of proteins, all closely related to one another and now collectively called the **linker histones**. In humans, there are eleven of these. Six are present in all somatic cells (H1.1–H1.5 and H1x), and the remaining five are only found in sex cells (H1t, H1T2, H1oo and H1LS1) or differentiated cells (H1.0). A single linker histone is attached to each nucleosome, to form the **chromatosome**. The linker histone is located on the outer surface of the nucleosome, possibly acting as a clamp, preventing the coiled DNA from detaching from the nucleosome (Figure 14.7b).

The 10 nm fiber is present in nondividing cells

Between cell divisions, chromosomes cannot be distinguished as individual structures unless specialized techniques are used. When nondividing nuclei are examined by electron microscopy all that can be seen is a mixture of light and dark staining areas (Figure 14.8). The light areas are called **euchromatin** and the dark areas are called **heterochromatin**. The euchromatin is thought to comprise those parts of the chromosomes that contain genes that are being expressed. The DNA in these regions cannot be too condensed as it must remain accessible to the proteins involved in the transcription process. We believe that it is structured as the 10 nm fiber.

Genes that are not active in a particular cell are located in the heterochromatin regions of the nucleus. The DNA in heterochromatin is more condensed than in euchromatin, with the 10 nm fibers packaged into globular structures that are responsible for the darker staining seen with the electron microscope (Figure 14.9).

Two types of heterochromatin are recognized. The first is **constitutive heterochromatin**, which is a permanent feature of all cells and represents segments of a DNA molecule that contain no genes and so can always be retained in a compact organization. In contrast, **facultative heterochromatin** is not a permanent feature but is seen in some cells some of the time. Facultative heterochromatin is thought to contain genes that are inactive in some cells or at some periods of the cell's life cycle. When these genes are inactive, their DNA regions are compacted into heterochromatin.

Higher levels of DNA packaging are needed when a cell divides

Although chromosomes are present throughout the lifetime of a cell, they are most apparent during cell division when the familiar structures more accurately referred to as **metaphase chromosomes** reveal themselves (Figure 14.10). Metaphase chromosomes are much shorter than the DNA molecules that they contain, which means that a packaging system is needed to fit a DNA molecule into its chromosome.

The problem is immense. The DNA that makes up the human genome has a total length of about 1 m. This is shared between 24 chromosomes, each of which is only a few thousandths of a millimeter in length in its

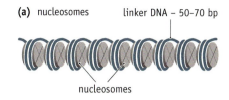

(a) nucleosomes   linker DNA – 50–70 bp

nucleosomes

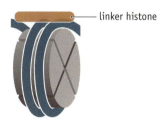

(b) a chromatosome

linker histone

FIGURE 14.7 Nucleosomes and the chromatosome. (a) The model for the 10 nm fiber in which each bead is a barrel-shaped nucleosome with the DNA wound twice around the outside. (b) On the surface of the chromatosome, the linker histone acts as a clamp, preventing the DNA from detaching from the outside of the nucleosome.

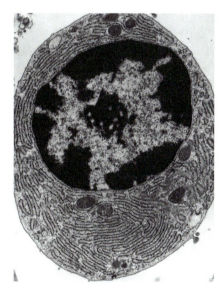

FIGURE 14.8 Euchromatin and heterochromatin. Electron micrograph of a plasma cell from guinea pig. Within the circular nucleus, the light areas are euchromatin and the dark areas are heterochromatin. (Courtesy of Don Fawcett/Science Photo Library.)

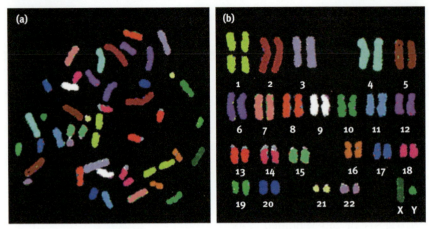

FIGURE 14.10 Human metaphase chromosomes. Each chromosome has been 'painted' by labeling with a different fluorescent marker. (a) The chromosomes as originally obtained from the cell. (b) The same image manipulated so that the two copies of each chromosome are lined up next to one another. Note that the cell was male. (From E. Schröck et al., *Science* 273: 494–497, 1996. With permission from AAAS.)

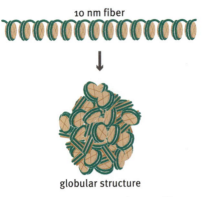

FIGURE 14.9 Packaging of 10 nm fibers into globular chromatin structures.

metaphase form but contains, on average, over 4 cm of DNA. The 10 nm fiber only reduces the length of a DNA molecule by one-sixth. Each molecule will still be a million times longer than its metaphase chromosome. There must be a highly organized packaging system to fit such lengthy DNA molecules into such small structures.

Understanding these higher levels of packaging is proving a challenge. For several years, it was thought that the 10 nm fiber could form a coiled structure called the **30 nm fiber** (Figure 14.11). However, the 30 nm fiber has only been identified in cell extracts and never directly observed in living cells, even with the most sophisticated imaging systems now available. It is now looked on as an artifact that can form in cell extracts, rather than being a significant structure within the nucleus.

Currently, researchers are exploring various models for the possible ways in which the 10 nm fiber can be compacted to the extent needed to form a metaphase chromosome. One suggestion is that higher levels of packaging are achieved by stacking nucleosomes into plate-like structures that can be placed side by side and in multiple layers (Figure 14.12a). An alternative model envisages a protein core from which loops of the 10 nm fiber radiate, the loops condensing by interactions between the nucleosomes to form a disordered but compact structure similar to the globules thought to be present in heterochromatin (Figure 14.12b).

FIGURE 14.11 The 30 nm fiber. The nucleosomes have been left out of the drawing in order to show the way in which the DNA molecule was thought to be coiled within the fiber. The upper drawing shows the view from the side, and the lower drawing is the view along the axis of the fiber. The 30 nm fiber is now thought to be an artifact only seen in cell extracts.

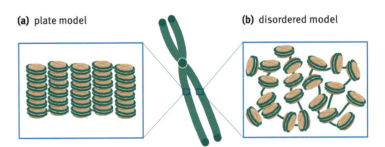

FIGURE 14.12 Two models for the higher order packaging of the 10 nm fiber into a metaphase chromosome. (a) The plate model, in which nucleosomes are stacked into plate-like structures that can be packed side by side and in multiple layers. (b) A disordered model, in which the nucleosomes form a disordered but compact structure.

14.2 THE SPECIAL FEATURES OF METAPHASE CHROMOSOMES

Although metaphase chromosomes are transient structures that are formed only during cell division, they are important because their visual appearance enables individual chromosomes to be distinguished and their important morphological features to be recognized.

Individual metaphase chromosomes have distinct morphologies

The typical appearance of a metaphase chromosome is shown in Figure 14.13. This is in reality two chromosomes joined together because by the stage of the cell cycle when chromosomes condense and become visible by light microscopy, DNA replication has already occurred. A metaphase chromosome therefore contains the two daughter chromosomes linked together at their **centromeres**. The centromere is also the position at which the chromosome attaches to the microtubules that draw the daughters into their respective nuclei during cell division (Figure 14.14).

The position of the centromere is characteristic of a particular chromosome (Figure 14.15). The centromere can be in the middle of the chromosome (**metacentric**), a little off-center (**submetacentric**), toward one end of the chromosome (**acrocentric**), or located very close to one end (**telocentric**). The regions either side of the centromere are called the chromosome arms or **chromatids**.

The differing positions of the centromere mean that individual chromosomes can be recognized. Further distinguishing features are revealed when chromosomes are stained. There are a number of different staining techniques (Table 14.1), each resulting in a banding pattern that is characteristic of a particular chromosome. The set of chromosomes possessed by an organism can therefore be represented as a **karyogram**, in which the banded appearance of each one is depicted. The human karyogram is shown in Figure 14.16.

Centromeres contain repetitive DNA and modified nucleosomes

DNA is present along the entire length of a chromosome including within the centromeres. This centromeric DNA is made up largely of repeat sequences. In humans, these repeats are 171 bp in length and are called **alphoid DNA**. The alphoid DNA repeats have sequence variability and are arranged in head-to-tail arrays. For example, the centromere of the human X chromosome is made up of approximately 18,000 alphoid repeats spanning a region of 3.1 Mb. Either side of the centromere is a **pericentromeric** region of 300–600 kb which, as well as alphoid DNA units, also contains other repeat sequences, including some that are common in other parts of the chromosomes.

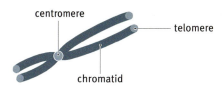

FIGURE 14.13 The typical appearance of a metaphase chromosome. Metaphase chromosomes are formed after DNA replication has taken place, so each one is, in effect, two chromosomes linked together at the centromere. The arms are called the chromatids. A telomere is the extreme end of a chromatid.

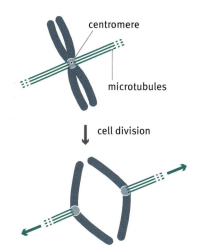

FIGURE 14.14 The role of the centromeres during cell division. During cell division, individual chromosomes are drawn apart by the contraction of microtubules attached to the centromere.

FIGURE 14.15 Different centromere positions.

TABLE 14.1 STAINING TECHNIQUES USED TO PRODUCE CHROMOSOME BANDING PATTERNS

Technique	Procedure	Banding Pattern
G banding	Mild proteolysis followed by staining with Giemsa	Dark bands are AT-rich Pale bands are GC-rich
R banding	Heat denaturation followed by staining with Giemsa	Dark bands are GC-rich Pale bands are AT-rich
Q banding	Staining with quinacrine	Dark bands are AT-rich Pale bands are GC-rich
C banding	Denaturation with barium hydroxide and then staining with Giemsa	Dark bands contain constitutive heterochromatin

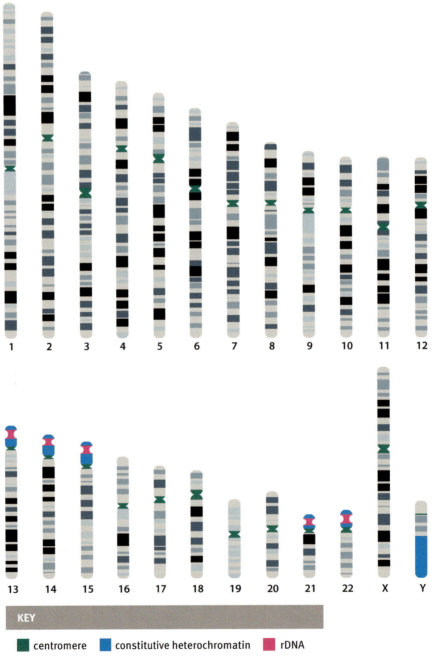

FIGURE 14.16 The human karyogram. The chromosomes are shown with the G-banding pattern obtained after Giemsa staining. Chromosome numbers are given below each structure. 'rDNA' is a region containing a cluster of repeat units for the ribosomal RNA genes.

The centromeres of most eukaryotes also contain nucleosomes, similar to those in other regions of the chromosome, but some of them contain the protein CENP-A instead of histone H3. CENP-A-containing nucleosomes are more compact and structurally rigid than those containing H3. It has been suggested that the arrangement of CENP-A and H3 nucleosomes along the DNA is such that the CENP-A versions are located on the surface of the centromere. Here, they form an outer shell on which the **kinetochores** are constructed. These are the structures that act as the attachment points for the microtubules which draw the divided chromosomes into the daughter nuclei. Kinetochores form on the metaphase chromosomes, each of the daughter chromosomes constructing its own kinetochore on the conjoined centromere. One kinetochore is on one side of the centromere, and one on the other (Figure 14.17).

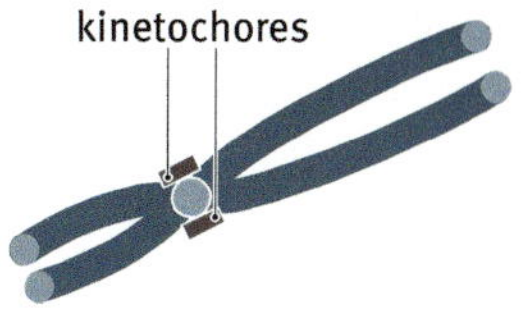

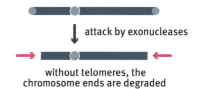

FIGURE 14.17 **The positions of the kinetochores on a metaphase chromosome.**

Telomeres protect chromosome ends

The ends of a chromosome are called the **telomeres**. These are specialized structures with a number of important roles. The first is simply to protect the ends of the DNA molecule contained within the chromosome from attack by exonuclease enzymes, which otherwise might degrade some of the terminal regions (Figure 14.18). The natural ends of the chromosomal DNA also need protection from the cell's DNA repair systems. These include very efficient mechanisms for repairing a broken chromosome, driven by proteins that bind to the ends of the DNA molecule that are exposed at the breakpoint. If these repair proteins confuse a natural chromosome end for a break, then they might join two chromosomes together. The telomeres prevent this from happening.

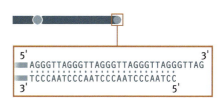

FIGURE 14.18 **One of the roles of the telomeres.** The proteins attached to a telomere prevent exonucleases from degrading the ends of a chromosome.

Like centromeres, the telomeres contain special DNA sequences. Telomeric DNA is made up of hundreds of copies of a short sequence, 5'–TTAGGG–3' in humans (Figure 14.19). On the other strand, the repeat sequence is 5'–CCCTAA–3', which means that one of the strands in the telomeric region is rich in G nucleotides, while the other is rich in Cs. This is true for all eukaryotes, even though the actual repeat sequence varies. The G-rich strand, which contributes the 3' end of the DNA molecule, extends for up to 200 nucleotides beyond the terminus of the C-rich strand, giving a single-stranded overhang.

Nucleosomes are not present in the telomeric regions of chromosomes. Instead, special proteins bind to the repeat sequences. In humans, these include TRF1, TRF2 and POT1. Together with other proteins, they make up a structure called a **shelterin** that protects the telomeres from degradation and accidental repair.

FIGURE 14.19 **The DNA sequence at the end of a human telomere.** The single-strand extension is up to 200 nucleotides in length.

Chromosomes should get progressively shorter during multiple rounds of replication

Telomeres must also overcome a problem posed by DNA replication. In theory, every round of DNA replication should result in a slight decrease in the length of a chromosome, something that does not happen in practice, at least not in all cells.

To understand why chromosomes should gradually get shorter, we must recall one of the basic features of DNA replication. DNA synthesis can occur only in the 5'→3' direction, which means that the lagging strand must be copied discontinuously, as a series of Okazaki fragments (see Figure 10.16). This presents two problems at the ends of a linear DNA molecule such as those found in eukaryotic chromosomes. The first is that the extreme 3' end of the lagging strand might not be copied because the final Okazaki fragment cannot be primed, the natural position for the priming site being beyond the end of the lagging strand (Figure 14.20a). The absence of this Okazaki fragment means that the lagging-strand copy is shorter than it should be. If the copy remains this length, then when it acts as a parent polynucleotide in the next round of replication the resulting daughter molecule will be shorter than its grandparent.

A second problem arises if the primer for the last Okazaki fragment is placed at the extreme 3′ end of the lagging strand. Shortening will still occur, although to a lesser extent, because this terminal RNA primer cannot be converted into DNA by the standard processes for primer removal (Figure 14.20b). This is because the methods for primer removal (see Figure 10.31) require an extension of the 3′ end of an adjacent Okazaki

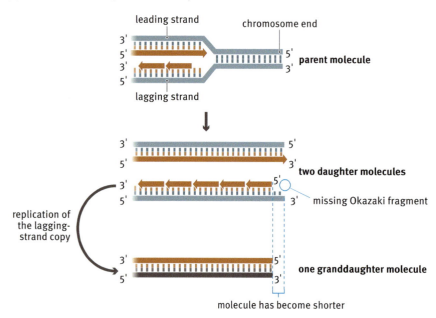

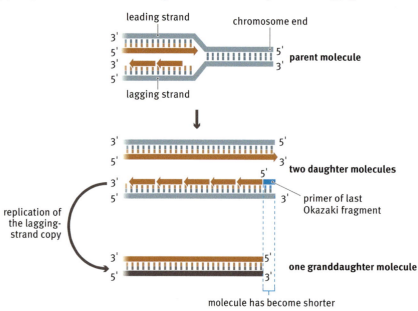

FIGURE 14.20 Two of the reasons why linear DNA molecules could become shorter after DNA replication. In both examples, the parent molecule is replicated in the normal way. A complete copy is made of its leading strand, but in (a) the lagging-strand copy is incomplete because the last Okazaki fragment is not made. This is because primers for Okazaki fragments are synthesized at positions approximately 200 bp apart on the lagging strand. If one Okazaki fragment begins at a position less than 200 bp from the 3′ end of the lagging strand, then there will not be room for another priming site, and the remaining segment of the lagging strand will not be copied. The resulting daughter molecule therefore has a 3′ overhang and, when replicated, gives rise to a granddaughter molecule that is shorter than the original parent. In (b), the final Okazaki fragment can be positioned at the extreme 3′ end of the lagging strand, but its RNA primer cannot be converted into DNA because this would require the extension of another Okazaki fragment positioned beyond the end of the lagging strand. If the primer is lost or is not copied into DNA, then one of the granddaughter molecules will be shorter than the original parent.

fragment, which cannot occur at the very end of the molecule because the necessary Okazaki fragment is absent.

Telomeres are designed to solve these problems in the following way. Most of the telomeric DNA is copied in the normal fashion during DNA replication, but this is not the only way in which it can be synthesized. To compensate for the limitations of the replication process, telomeres can be extended by an independent mechanism catalyzed by the enzyme **telomerase**. This is an unusual enzyme in that it consists of both protein and RNA. In the human enzyme, the RNA component is 450 nucleotides in length and contains near its 5′ end the sequence 5′–CUAACCCUAAC–3′. Note that the central region of this sequence is the reverse complement of the human telomere repeat 5′–TTAGGG–3′. Telomerase RNA can therefore base pair to the single-stranded DNA overhang that is present at the end of the telomere (Figure 14.21). Base pairing provides a template that enables the 3′ end of the DNA to be extended by a few nucleotides. The telomerase RNA then translocates to a new base-pairing position slightly further along the DNA polynucleotide and the molecule is extended by a few more nucleotides. The process can be repeated until the chromosome end has been extended by a sufficient amount.

Telomerase can only add nucleotides to the end of the G-rich strand. It is not clear how the other polynucleotide – the C-rich strand – is extended, but it is presumed that when the G-rich strand is long enough, the primase–DNA polymerase α complex attaches and initiates the synthesis of a new Okazaki fragment (Figure 14.22). This requires the use of a new RNA primer, which explains why the C-rich strand is always shorter than the G-rich one. The important point is that the overall length of the chromosomal DNA has not been reduced.

14.3 PARTITIONING OF CHROMOSOMES DURING EUKARYOTIC CELL DIVISION

Eukaryotic cells, or more precisely the nuclei contained within them, display two distinct types of division. These are called **mitosis** and **meiosis**. Mitosis is more common, being the process by which the diploid nucleus of a somatic cell divides to produce two daughter nuclei, both of which are diploid. Approximately 10^{17} mitoses are needed to produce all the cells required during a human lifetime. Meiosis, on the other hand, occurs only in reproductive cells, and results in a diploid cell giving rise to four haploid gametes, each of which can subsequently fuse with a gamete of the opposite sex during sexual reproduction.

We begin with mitosis, as this is the more straightforward of the two processes.

Partitioning of chromosomes during mitosis

Mitosis is conventionally looked on a comprising five phases (Figure 14.23). The first of these is **prophase**. This is the period when the chromosomes condense and become visible under the light microscope. It is also the period when, in most eukaryotes, the nuclear membrane breaks down and the two **centrosomes** of the cell move to positions on either side of the nuclear region.

Prophase is followed by **prometaphase**. The chromosomes, now fully condensed, begin to attach to the microtubules radiating out from the centrosomes. In **metaphase**, the chromosomes line up in the middle of

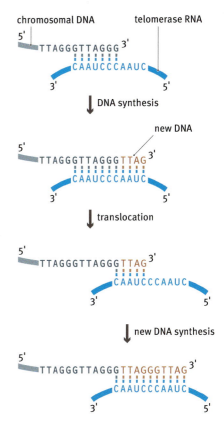

FIGURE 14.21 Extension of the end of a human chromosome by telomerase. The 3′ end of a human chromosomal DNA molecule is shown. The sequence comprises repeats of the human telomere motif 5′–TTAGGG–3′. The telomerase RNA base pairs to the end of the DNA molecule, which is then extended a short distance. The telomerase RNA translocates to a new base-pairing position slightly further along the DNA polynucleotide and the molecule is extended by a few more nucleotides.

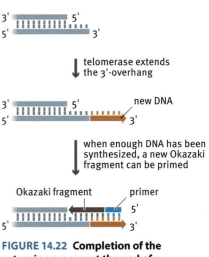

FIGURE 14.22 Completion of the extension process at the end of a chromosome. It is believed that after telomerase has extended the 3′ end of the chromosome, a new Okazaki fragment is primed and synthesized, resulting in a double-stranded end.

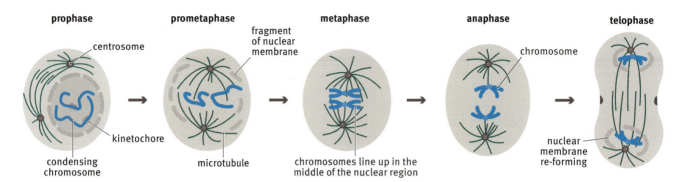

FIGURE 14.23 The five phases of mitosis. The chromosomes begin to condense during prophase. During prometaphase, microtubules attach to the chromosomes. In metaphase, the chromosomes line up in the middle of the nuclear region, and then, during anaphase, each pair of replicated chromosomes separates. Finally, during telophase, the chromosome pairs become fully separated and new nuclear membranes are formed around each set.

the nuclear region, and then, during **anaphase**, each pair of replicated chromosomes separates. One member of each pair is drawn toward one of the centrosomes. During the later part of anaphase, the cell begins to constrict around its middle, prior to division. Finally, during **telophase**, the chromosome pairs become fully separated and new nuclear membranes are formed around each set.

How is the appropriate partitioning of chromosomes assured? The answer lies with the structure of the metaphase chromosome, specifically the positioning of the kinetochores on the surface of the conjoined centromeres. As indicated in Figure 14.17, two kinetochores form, one on each centromere. The two kinetochores are therefore on opposite sides of the chromosome pair.

The kinetochores are the attachment points for the microtubules that radiate out from the centrosomes. The centrosomes are themselves located on the opposite sides of the **mitotic spindle**, the microtubular arrangement that now occupies the region of the cell previously taken up by the nucleus (Figure 14.24). During prometaphase, the microtubules of the spindle begin to attach to the chromosomes, migrating along the chromatids until they locate one or other of the kinetochores. The first microtubule to locate a kinetochore will orientate that kinetochore toward the centrosome to which the microtubule is attached (Figure 14.25). The other kinetochore is now more likely to be caught by a microtubule from the other centrosome.

Eventually, up to 40 microtubules will attach to either kinetochore. The two sets of microtubules begin to pull in opposite directions, the tension that is set up confirming that the correct attachments have been made. If both kinetochores have been captured by microtubules from the same centrosome, then this tension is not generated and mitosis cannot continue.

At the start of anaphase, the microtubules begin to draw the freed daughters toward the centrosomes. The natural outcome is that one of the daughter chromosomes of each pair moves toward one centrosome, and the other toward the other centrosome. The result is that a complete set of chromosomes is assembled at each of the poles of the mitotic spindle. During telophase, each set becomes enclosed within a new nucleus. Division of the cell – **cytokinesis** – occurs across the plane of the mitotic spindle, ensuring that each daughter cell forms around one of the divided nuclei (Figure 14.26). Each daughter cell therefore inherits a complete set of chromosomes and hence a complete copy of the parent cell's genome.

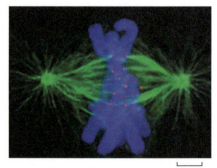

FIGURE 14.24 Fluorescence micrograph of a cell undergoing mitosis. The mitotic spindle is shown in green, with chromosomes in blue and the kinetochores visible as red dots. (From A. Desai, *Curr. Biol.* 10: R508, 200. With permission from Elsevier.)

(a) chromosomes are randomly oriented at the start of prometaphase

(b) the first microtubule orients a kinetochore toward one of the centrosomes

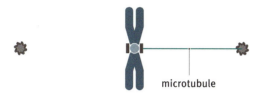

(c) the other kinetochore will now be captured by a microtubule from the other centrosome

(d) during anaphase, daughter chromosomes are pulled in opposite directions

FIGURE 14.25 The attachment of microtubules to kinetochores. The way in which the microtubules attach to the kinetochores ensures that the two daughter chromosomes are drawn toward different centrosomes.

cytokinesis

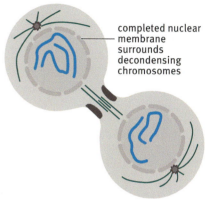

FIGURE 14.26 Cytokinesis.

Meiosis involves two successive nuclear divisions

In order to understand meiosis, we must first take a step back and remind ourselves what it means to be diploid. Each somatic cell has two copies of each chromosome. In humans, this means 46 chromosomes in all, two copies of each of the 22 autosomes, and a pair of sex chromosomes, XX for a female cell and XY for a male. We refer to the two copies of each chromosome as a **homologous** pair or as **homologous chromosomes**.

At the start of nuclear division, whether meiosis or mitosis, every homologous chromosome has itself replicated, the pairs of daughter chromosomes remaining attached at their centromeres. The cell at this point is therefore tetraploid, possessing *four* copies of each chromosome (Figure 14.27).

Meiosis involves two successive cell divisions, with no DNA replication occurring between them. The two divisions are called meiosis I and meiosis II. Meiosis I results in two cells, each containing one copy of each of the replicated homologous chromosomes. The cells resulting from meiosis I are therefore diploid. During the second meiotic division, the attachment that holds each pair of daughter chromosomes together breaks. One daughter chromosome passes into one gamete, and the second daughter into the second gamete. The gametes are therefore haploid.

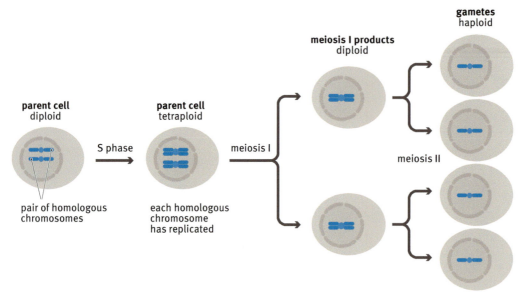

FIGURE 14.27 Outline of the events occurring during meiosis. The changes in ploidy that occur during the process are indicated. During the S phase of the cell cycle, each chromosome is replicated, and the daughters remain attached to one another at their centromere. At this stage, the cell is therefore tetraploid.

During meiosis I, bivalents are formed between homologous chromosomes

As meiosis I begins, the replicated chromosomes condense and migrate to the middle of the nuclear region. It is at this stage that we see the most important difference between meiosis and mitosis. During mitosis, homologous chromosomes remain separate from one another. In meiosis I, however, the pairs of homologous chromosomes are by no means independent. Instead, each chromosome finds its homolog and forms a **bivalent** (Figure 14.28). It is these bivalents, not the independent chromosomes, that line up in the middle of the nucleus.

After formation of the bivalents, meiosis I continues in a similar manner to a mitotic cell division (Figure 14.29). Microtubules radiate out from the centrosomes, attach to the kinetochores and begin to pull in opposite directions. The tension exerted on a bivalent breaks it apart, and the two homologous chromosomes are pulled in opposite directions. A complete set of chromosomes is therefore assembled at each of the poles of the mitotic spindle. Meiosis I is then completed by cytokinesis.

Meiosis II begins almost immediately (Figure 14.30). In each of the cells, the chromosomes again migrate to the middle of the nucleus. Each cell

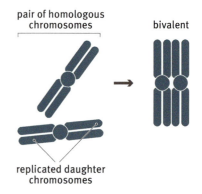

FIGURE 14.28 Formation of a bivalent between a pair of homologous chromosomes.

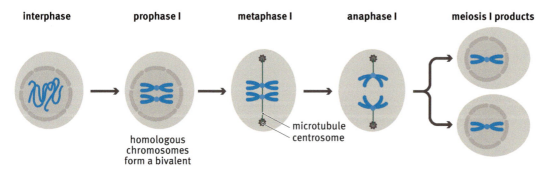

FIGURE 14.29 Meiosis I. The homologous chromosomes form bivalents during prophase I, and microtubules attach to the kinetochores during metaphase I. The chromosomes are then pulled toward opposite centrosomes during anaphase I, with meiosis I being completed by cytokinesis.

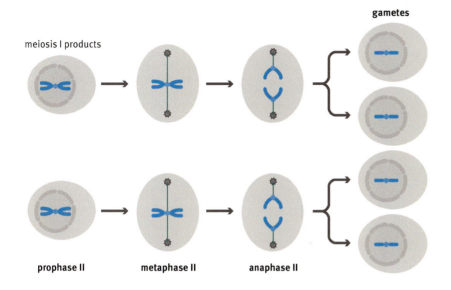

FIGURE 14.30 **Meiosis II.** The events are similar to those occurring during meiosis I, except that each starting cell has only one member of each pair of homologous chromosomes, so no bivalents are formed.

has only one member of each pair of homologous chromosomes, so the chromosomes remain independent of one another. Microtubules attach to the kinetochores and pull the chromosomes apart, separating the two daughters which again move in opposite directions. A second round of cytokinesis creates the four gametes.

Formation of bivalents ensures that siblings are not identical to one another

To a geneticist, the most exciting thing about sexual reproduction is the formation of bivalents. One reason why formation of bivalents is so exciting is that it is the reason why siblings – children who have the same parents – are not exactly the same. This is true for all siblings except identical twins, who derive from a single fertilized egg cell.

The difference between siblings is such an obvious fact of life that it is easy to forget to ask why it should be the case. We should ask this question because we know that mitosis results in identical daughter cells, each of which contains the same set of DNA molecules. If meiosis had the same outcome, then all the gametes produced by an individual would be identical to one another (Figure 14.31). The mother's gametes would all be the same, as would those of the father. Fusion of maternal and paternal gametes would produce identical fertilized egg cells that would develop into identical siblings.

Why are siblings different from one another, each a blend of the parents' biological characteristics, but each with its own individuality? There are two reasons, both of them outcomes of the formation of bivalents. The first results from the separation of the bivalents during the anaphase period of meiosis I. The two homologous chromosomes in a bivalent are not identical. They contain the same set of genes but will possess different alleles of many of those genes (Figure 14.32). When the bivalents break apart, the two chromosomes are pulled toward opposite centrosomes. Which chromosome goes in which direction is entirely random, depending only on the orientation of the bivalent with respect to the two spindle poles.

Imagine a cell with two pairs of homologous chromosomes. Meiosis can result in gametes with any of four different chromosome combinations, depending on the directions in which the members of each homologous pair segregate during anaphase I (Figure 14.33a). Each of these

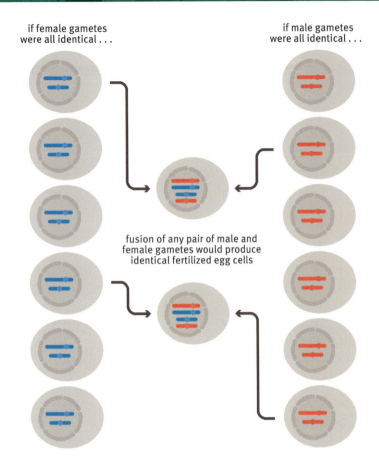

FIGURE 14.31 If meiosis had the same outcome as mitosis, then all the gametes produced by an individual would be identical. Fusion of gametes from female and male partners would always result in identical fertilized egg cells. Siblings developing from these egg cells would be genetically identical.

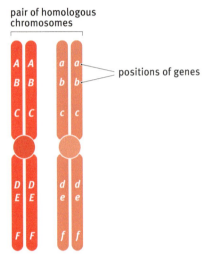

FIGURE 14.32 The pair of homologous chromosomes in a bivalent are not identical. They are not identical as one member of the pair was originally inherited from the father, and the other member from the mother. In this example, *A* and *a*, etc., are different alleles of the same gene.

chromosome combinations corresponds to a gamete with a different set of alleles. For a cell with three chromosomes, there are $2^3 = 8$ possible combinations (Figure 14.33b). For human cells, with 46 chromosomes in the diploid set, there are 2^{23} possible ways in which the pairs of homologous chromosomes can be distributed during meiosis I. That is more than 8 million combinations, each combination giving a gamete with a different set of alleles.

Recombination occurs between homologous chromosomes within a bivalent

Random segregation of chromosomes during anaphase I is not the only factor influencing the variability of the gametes resulting from meiosis. The bivalent contributes to the resulting variability in a second, even more important way.

Within the bivalent, the chromosome arms – the chromatids – can exchange segments of DNA (Figure 14.34). This exchange is called **crossing over** or **recombination**. It was first discovered by a Belgian cell biologist, F.A. Janssens, in 1909. Janssens had made a very close examination of the bivalents formed during meiosis in salamanders. He did not actually see crossovers between the chromatids in a bivalent, as this was not possible with the types of microscope available at that time. But in a flash of insight, he suggested that the close proximity of the chromatids could lead to breakage and transfer of segments between chromosomes. Later, with more sophisticated microscopes, cell biologists were able to see the actual crossover points, or **chiasmata**. The micrograph shown in Figure 14.35 is fairly typical, with three separate chiasmata along the length of a chromosome.

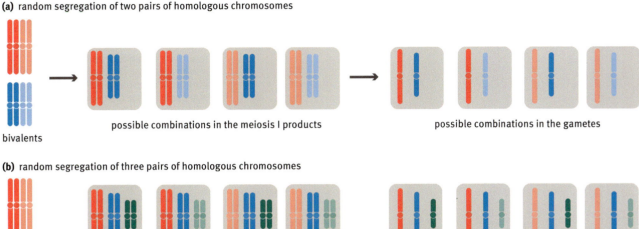
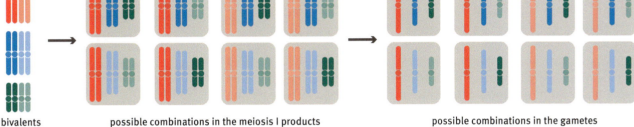

FIGURE 14.33 Meiosis results in gametes with different chromosome combinations. Random segregation of homologous chromosomes during anaphase I means that the gametes resulting from meiosis are not all identical. The individual members of each pair of homologous chromosomes are given dark and light shading. The various combinations that can arise in the diploid meiosis I products and in the gametes are shown.

What impact does crossing over have on the genetic variability of the gametes resulting from meiosis? If the pair of homologous chromosomes in a bivalent exchange a segment of DNA, then any genes present in that segment will also be exchanged. The alleles present in one chromosome will be transferred to the other and vice versa. Furthermore, the two chromosomes in each member of the homologous pair can participate in different exchanges. The outcome could be that after crossing over the four chromosomes in the bivalent are all different from one another (Figure 14.36).

Recombination increases the variability of the gametes resulting from meiosis, by changing the allele combinations within individual chromosomes prior to the random segregation of those chromosomes during anaphase I. Random segregation on its own can give rise to any of 8 million different chromosome combinations in the gametes resulting from human meiosis. The number of possible *allele* combinations in those gametes is infinitely higher than 8 million, because of the randomness of the DNA exchanges that are possible within each bivalent.

Recombination involves a DNA heteroduplex

Homologous recombination is the name given to the type of recombination that occurs during crossing over between homologous chromosomes. It occurs between two double-stranded DNA molecules that have regions where the nucleotide sequences are the same or at least very similar.

Homologous recombination begins when the two double-stranded molecules line up adjacent to one another (Figure 14.37). A double-strand cut is made in one of the molecules, breaking this one into two pieces. One strand in each half of this molecule is then shortened by the removal of a few nucleotides, giving each end a 3' overhang.

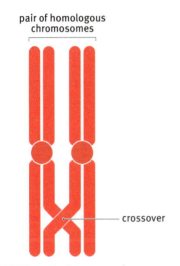

FIGURE 14.34 A crossover between two chromatids in a bivalent.

FIGURE 14.35 Electron micrograph of a bivalent in a grasshopper nucleus. Three crossovers have occurred between this pair of homologous chromosomes. (Courtesy of the late Professor Bernard John.)

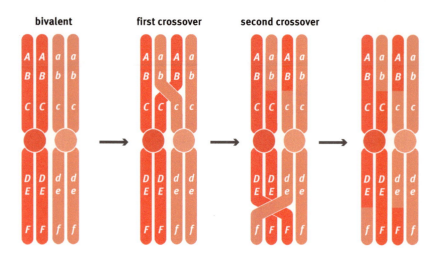

FIGURE 14.36 One possible outcome of two crossovers in a bivalent. At the end of the process, the four chromosomes are all different from one another. Many other results are possible, depending on the number of crossovers, the positions where they occur, and the members of the bivalent that participate in these crossovers.

The partnership between the two chromosomes is set up when one of the 3′ overhangs invades the uncut DNA molecule, displacing one of its strands and forming a **D loop**. Strand invasion is stabilized by base pairing between the transferred segment of polynucleotide and the intact polynucleotide of the recipient molecule. This base pairing is possible because of the sequence similarity between the two molecules. The invading strand is extended by new DNA synthesis, enlarging the D loop, and at the same time, the other broken strand is also extended.

After strand extension, the free polynucleotide ends are joined together. This gives a structure called a **heteroduplex**, in which the two double-stranded molecules are linked together by a pair of **Holliday structures**. These are named after the scientist, Robin Holliday, who proposed the first model for recombination and realized that the process must involve the formation of a heteroduplex. Each Holliday structure is dynamic and can move along the heteroduplex. This **branch migration** results in the exchange of longer segments of DNA (Figure 14.38).

Separation, or **resolution**, of the heteroduplex back into individual double-stranded molecules occurs by cleavage of the Holliday structures. The two-dimensional representation used in Figure 14.37 is rather unhelpful in this regard because it obscures the true topology of the Holliday structures. In particular, it fails to reveal that each Holliday structure can be cleaved in either of two orientations, with very different outcomes. To understand this point, we must depict a Holliday structure in its true three-dimensional configuration, called the **chi form**, as revealed by electron microscopic observation of recombining DNA molecules (Figure 14.39).

Now the two possible cleavages become much clearer. First, the cut can be made left to right across the chi form. This is equivalent to cleavage directly across the Holliday structure as drawn at the top of Figure 14.39. Cleavage of a Holliday structure in this way, followed by the joining of the cut strands, gives two molecules that have exchanged short segments of polynucleotide. As the exchanged strands have similar sequences, the effect on the genetic constitution of each molecule is relatively minor.

Alternatively, the Holliday structure can be resolved by cutting the chi structure in the second of the two possible orientations, up-down in Figure 14.39. This is the type of cleavage that is not readily apparent from the two-dimensional representation of the Holliday structure. But it is vitally important because up-down cleavage results in **reciprocal strand exchange**, double-stranded DNA being transferred between the

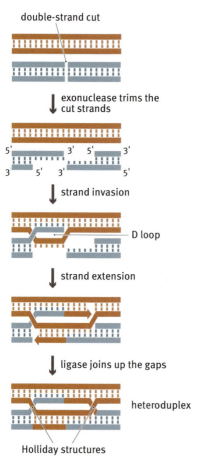

FIGURE 14.37 The initial steps in homologous recombination. This part of the process results in the formation of a heteroduplex with two Holliday structures.

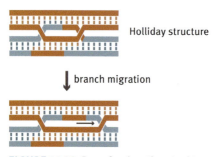

FIGURE 14.38 Branch migration. In this example, one of the two Holliday structures resulting from the recombination process shown in Figure 14.37 migrates in such a way that the heteroduplex is extended.

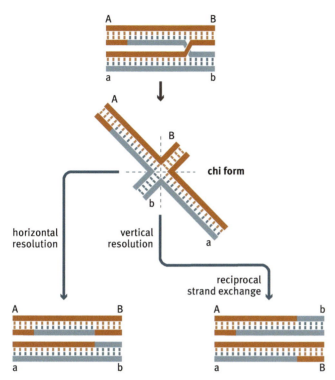

FIGURE 14.39 The two possible ways of resolving a Holliday structure.

two molecules so that the end of one molecule is exchanged for the end of the other molecule. This is the DNA transfer that occurs during crossing over.

The biochemical pathway for homologous recombination

Now that we have explored the way in which the DNA molecules behave during homologous recombination, we can turn our attention to the enzymes and other proteins that are responsible for carrying out the process. In this regard, it is important to recognize that homologous recombination also occurs in bacteria. It is the process by which a segment of DNA transferred into a cell by, for example, conjugation can become integrated into the recipient cell's chromosome (Figure 14.40). Research into the biochemistry of recombination in *E. coli* is well advanced, but much less is known about the equivalent events in eukaryotes such as humans. This is a problem because there appear to be some important

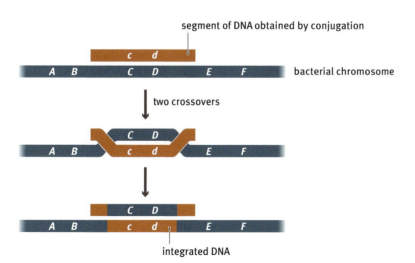

FIGURE 14.40 Homologous recombination enables a segment of DNA to be integrated into the host bacterial genome. The transferred DNA must have the same or similar sequence to a segment of the recipient cell's chromosome, as will be the case if it comes from the same species by, for example, conjugation. The transferred DNA lines up with the equivalent stretch of DNA in the chromosome. Two crossovers can then set up a reciprocal exchange, so that the transferred DNA becomes integrated into the chromosome, replacing the pre-existing sequence. In the example shown, integration results in a change in the alleles present at genes C and D. This is the molecular explanation for the event shown in outline in Figure 13.24a.

differences between the recombination processes in bacteria and eukaryotes, so we cannot be certain that everything that has been learnt from studies of *E. coli* is directly applicable to eukaryotes. With that caveat in mind, we will begin by examining what we know about the enzymes and proteins involved in homologous recombination in bacteria.

At the biochemical level, there are three distinct recombination systems in *E. coli*, each involving a different set of proteins. These are called the RecBCD, RecFOR and RecBFI pathways. The RecBCD pathway is the best studied and illustrates the key points of all three pathways.

The central player in the RecBCD pathway is the **RecBCD enzyme** which, as its name implies, is made up of three different proteins. Two of these – RecB and RecD – are helicases, enzymes capable of breaking the base pairs that hold polynucleotides together. To initiate homologous recombination, one copy of the RecBCD enzyme attaches to the free ends of a chromosome at a double-strand break. Using its helicase activity, RecBCD then progresses along the DNA molecule at a rate of approximately 1 kb per second until it reaches the first copy of the eight-nucleotide consensus sequence 5′–GCTGGTGG–3′. This is called the **chi (crossover hotspot initiator) site**. The sequence occurs on average once every 5 kb in *E. coli* DNA. Translocation of the RecBCD complex along the DNA is accompanied by 3′→5′ degradation of the upper strand, due to the exonuclease activity of RecB. When a chi site is encountered, the exonuclease activity is suppressed. RecB now acts as an endonuclease and cleaves the lower strand. The result is a double-stranded molecule with a 3′ overhang, as is required for the initiation of recombination (Figure 14.41).

The next step is the establishment of the heteroduplex. This stage is mediated by the RecA protein, which forms a protein-coated DNA filament that is able to invade the intact double helix and set up the D-loop (Figure 14.42). An intermediate in the formation of the D-loop is a **triplex** structure, a three-stranded DNA helix in which the invading polynucleotide lies within the major groove of the intact helix and forms hydrogen bonds with the base pairs it encounters.

Branch migration is then catalyzed by the RuvA and RuvB proteins, both of which attach to a Holliday structure. Four copies of RuvA bind directly to the branch point, forming a core to which two RuvB rings, each consisting of eight proteins, attach, one ring to either side (Figure 14.43). The resulting structure might act as a 'molecular motor', rotating the helices in the required manner so that the branch point moves.

Branch migration does not appear to be a random process. Instead, it stops preferentially at the sequence $5'-\frac{A}{T}TT\frac{G}{C}-3'$, where $\frac{A}{T}$ and $\frac{G}{C}$ denote that either of the two nucleotides can be present at the positions indicated. This sequence occurs frequently in the *E. coli* genome, so presumably migration does not halt at the first instance of the motif that is reached. When branch migration has ended, the RuvAB complex detaches and is replaced by two RuvC proteins. RuvC is a nuclease and so can carry out the cleavage that resolves the Holliday structure. The cuts are made between the second T and the $\frac{G}{C}$ components of the recognition sequence. The orientation of the attachment of the RuvC proteins to the Holliday structure determines the direction of the cuts that are made to resolve the structure.

Researchers studying homologous recombination in eukaryotes are gradually identifying proteins with equivalent functions to RecBCD and RuvABC. For example, two yeast proteins called RAD51 and DMC1 are homologs of RecA, and proteins similar to RAD51 and DMC1 are also known in several other eukaryotes including humans. The biggest puzzle

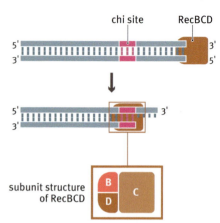

FIGURE 14.41 The role of the RecBCD enzyme in homologous recombination in *E. coli*. Translocation of the RecBCD complex along the DNA is accompanied by 3′→5′ degradation of the upper strand, due to the exonuclease activity of RecB. When a chi site is encountered, the exonuclease activity is suppressed and the RecB endonuclease cleaves the lower strand, giving the 3′ overhang. If RecBCD translocates some distance along the DNA before reaching a chi site, then the 5′ overhang is repeatedly trimmed by the endonuclease, so it is never more than a few tens of bp in length.

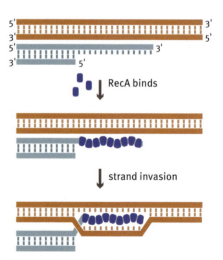

FIGURE 14.42 The role of the RecA protein during homologous recombination in *E. coli*. RecA forms a protein-coated DNA filament that invades the intact double helix and sets up the D-loop.

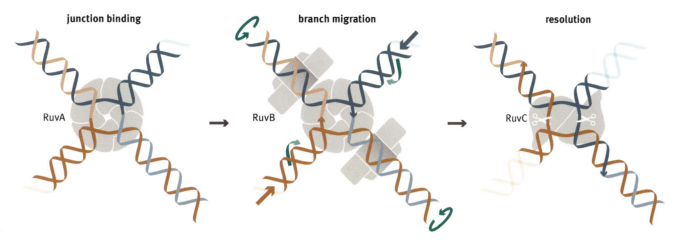

FIGURE 14.43 The role of the Ruv proteins in homologous recombination in *E. coli*. Branch migration is induced by a structure comprising four copies of RuvA bound to the Holliday junction with a RuvB ring on either side. After branch migration, two RuvC proteins bind to the junction, the orientation of their attachment determining the direction of the cuts that resolve the structure. In the central drawing, the orange and blue arrows indicate the direction in which the double helices are being driven through the RuvAB complex, and the green arrows show how the strands rotate during branch migration.

has been the process by which Holliday structures are resolved in eukaryotes. For many years, eukaryotic proteins similar to RuvC of *E. coli* were sought but not found. Other types of nuclease that can cut Holliday structures in the test tube were discovered, but the cleaved ends that they leave cannot be ligated to one another easily. It therefore seems unlikely that they have any role in recombination in living cells. During this search, it was realized that, even in bacteria, it is possible to resolve a Holliday structure without the use of a nuclease enzyme. A combination of helicases and topoisomerases can separate the heteroduplex back into independent chromosomes. It began to look possible that there is no eukaryotic equivalent of RuvC, but in 2008 the puzzle was solved with isolation of the GEN1 protein from human cells. This protein cuts the Holliday structure in almost exactly the same way as RuvC, and, importantly, the cut DNA ends resulting from GEN1 cleavage can be joined together easily.

KEY CONCEPTS

- All eukaryotic genomes are made up of linear DNA molecules, each of these molecules contained in a different chromosome. The number of chromosomes varies between species.
- Chromosomes contain DNA and protein. The histone proteins form nucleosomes, which can be seen in electron micrographs as beads on a string of DNA. This structure is called the 10 nm fiber.
- Chromosomes are only visible in dividing cells. Between cell divisions, the nucleus contains a mixture of light and dark staining areas. The light areas are euchromatin made up mainly of 10 nm fiber. The dark areas are called heterochromatin and are made up of DNA that is more condensed than in euchromatin.
- A packaging system is needed to fit a DNA molecule into its chromosome. One model suggests that the nucleosomes are stacked into plate-like structures.

- A metaphase chromosome contains two replicated daughter chromosomes linked together at the centromere. The centromere is also the position at which the chromosome attaches to the microtubules that draw the daughters into their respective nuclei during cell division.
- The ends of a chromosome are called the telomeres. Telomeres have various roles. Importantly, they prevent the ends of the chromosome from becoming progressively shorter.
- Mitosis ensures the correct partitioning of chromosomes during somatic cell division.
- Meiosis is the type of cell division that gives rise to the sex cells or gametes. Meiosis results in haploid cells because it is, in effect, two successive cell divisions, with no DNA replication occurring between them.
- During meiosis I, the two members of a pair of homologous chromosomes line up alongside one another to form a bivalent. Within the bivalent, crossing over might occur.
- Recombination involves the formation of a heteroduplex, a structure within which two double-stranded DNA molecules are linked together by a pair of crossovers.
- Resolution of a heteroduplex can lead to an exchange of chromosome segments.
- The biochemistry of the recombination pathway in *E. coli* is well understood. The equivalent events in eukaryotes are less well studied, but proteins similar to those involved in bacterial recombination are known.

QUESTIONS AND PROBLEMS

Key terms

Write short definitions of the following terms:

10 nm fiber	D loop	mitosis
30 nm fiber	diploid	mitotic spindle
acrocentric	euchromatin	nuclear genome
alphoid DNA	facultative heterochromatin	nuclease protection
anaphase	gamete	nucleosome
autosome	haploid	pericentromeric
bivalent	heterochromatin	prometaphase
branch migration	heteroduplex	prophase
centromere	histone	RecBCD enzyme
centrosome	Holliday structure	reciprocal strand exchange
chi form	homologous chromosomes	recombination
chi site	homologous recombination	resolution of a heteroduplex
chiasmata	karyogram	sex cell
chloroplast genome	kinetochore	shelterin
chromatid	linker DNA	somatic cell
chromatin	linker histone	submetacentric
chromatosome	meiosis	telocentric
constitutive heterochromatin	metacentric	telomerase
core octamer	metaphase	telomere
crossing over	metaphase chromosome	telophase
cytokinesis	mitochondrial genome	triplex helix

Self-study questions

14.1 When chromosome number is considered, how typical are humans of eukaryotes in general?

14.2 Describe the role of histones in eukaryotic chromosomes.

14.3 Explain the difference between euchromatin and heterochromatin, and between the two different types of heterochromatin.

14.4 Describe the main morphological features of a human metaphase chromosome.

14.5 Draw four diagrams to show the differences between metacentric, submetacentric, acrocentric and telocentric chromosomes.

14.6 What are the special features of the DNA sequences and the nucleosomes present in centromeres?

14.7 Summarize the roles that telomeres play. What are the special features of the DNA present in telomeres?

14.8 Explain why the ends of a chromosomal DNA molecule could become shortened after repeated rounds of DNA replication and show how telomerase prevents this from occurring.

14.9 Draw a series of diagrams to illustrate the main stages of mitosis.

14.10 What roles do kinetochores play during mitosis?

14.11 Describe how the correct partitioning of chromosomes is ensured during mitosis.

14.12 Explain how meiosis results in haploid cells whereas mitosis gives rise to diploid cells.

14.13 Explain what is meant by the term 'homologous chromosomes'. How many pairs of homologous chromosomes are present in a diploid human cell?

14.14 Describe how bivalents are formed and how they separate during meiosis I.

14.15 How was crossing over between homologous chromosomes first discovered?

14.16 What is a heteroduplex, and how is it formed during recombination?

14.17 Draw a series of diagrams that illustrate the events occurring during homologous recombination.

14.18 Describe the different ways in which the chi form of a Holliday structure can be cleaved and explain the effect that each of these cleavages has on the structures of the resulting DNA molecules.

14.19 What is the role of the RecBCD enzyme during homologous recombination in *E. coli*?

14.20 Explain how the triplex intermediate is formed during homologous recombination in *E. coli*.

14.21 Describe the molecular basis of branch migration and cleavage of the Holliday structure in *E. coli*.

14.22 Explain why the molecular mechanism behind the resolution of Holliday structures in eukaryotes was a puzzle, and describe how this puzzle has been solved.

Discussion topics

14.23 Although the size of a eukaryotic genome coincides to a certain extent with the complexity of the organism, the relationship is only approximate. For example, the salamander genome is over 20 times the length of the human genome. Few humans would accept that we are only one-twentieth as complex as a salamander. What factors might be responsible for the large genomes of certain organisms?

14.24 Why are chromosome numbers so different in different organisms, and why does the number not relate to the complexity of the organism?

14.25 Discuss the impact that the presence of nucleosomes is likely to have on the expression of individual genes.

14.26 Crossing over between homologous chromosomes is the basis of methods used to map the positions of genes on eukaryotic chromosomes. How might crossing over be used in this way?

14.27 Some biologists believe that the main function of recombination is in DNA repair. Explain how recombination could be used to repair a damaged DNA molecule.

The answers to, or hints on how to answer, the self-study questions and discussion topic questions can be downloaded from the book's product page at this link: www.routledge.com/9781032743530

FURTHER READING

Cech TR (2004) Beginning to understand the end of the chromosome. *Cell* 116, 273–279. *Reviews all aspects of telomerase.*

Chen Z, Yang H & Pavietich NP (2008) Mechanism of homologous recombination from the RecA-ssDNA/dsDNA structures. *Nature* 453, 489–494. *Details of how the triplex structure is set up by the RecA protein.*

Cutter AR & Hayes JJ (2015) A brief review of nucleosome structure. *FEBS Letters* 589, 2022–2914.

de Lange T (2005) Shelterin: the protein complex that shapes and safeguards human telomeres. *Genes and Development* 19, 2100–2110.

Dillingham MS & Kowalczykowski SC (2008) RecBCD enzyme and the repair of double-stranded DNA breaks. *Microbiology and Molecular Biology Reviews* 72, 642–671. *The mode of action of the RecBCD enzyme during homologous recombination and other events in which it is involved.*

Fierz B (2019) Revealing chromatin organization in metaphase chromosomes. *EMBO Journal* 38, e101699.

Harshman SW, Young NL, Parthun MR & Freitas MA (2013) H1 histones: current perspectives and challenges. *Nucleic Acids Research* 41, 9593–9609.

Mazia D (1961) How cells divide. *Scientific American* 205 (September), 100–120.

Miga KH, Koren S, Rhie A et al. (2020) Telomere-to-telomere assembly of a complete human X chromosome. *Nature* 585, 79–84. *Includes details of the organization of DNA in the centromere.*

Murray AW & Szostak JW (1985) Chromosome segregation in mitosis and meiosis. *Annual Review of Cell Biology* 1, 289–315.

Öztürk MA, Cojocaru V & Wade RC (2018) Toward an ensemble view of chromatosome structure: a paradigm shift from one to many. *Structure* 26, 1050–1057. *Details of the possible location of the linker histone in the chromatosome.*

Persky NS & Lovett ST (2008) Mechanisms of recombination: lessons from *E. coli*. *Critical Reviews in Biochemistry and Molecular Biology* 43, 347–370.

Punatar RS, Martin MJ, Wyatt HDM et al. (2017) Resolution of single and double Holliday junction recombination intermediates by GEN1. *Proceedings of the National Academy of Sciences USA* 114, 443–450.

Pyle AM (2004) Big engine finds small breaks. *Nature* 432, 157–158. *The structure of the RecBCD complex.*

Rafferty JB, Sedelnikova SE, Hargreaves D et al. (1996) Crystal structure of DNA recombination protein RuvA and a model for its binding to the Holliday junction. *Science* 274, 415–421.

Roeder GS (1997) Meiotic chromosomes: it takes two to tango. *Genes and Development* 11, 2600–2621. *Describes the formation of bivalents and exchange of DNA by recombination.*

Shinagawa H & Iwasaki H (1996) Processing the Holliday junction in homologous recombination. *Trends in Biochemical Sciences* 21, 107–111.

Szostak JW, Orr-Weaver TK, Rothstein RJ & Stahl EW (1983) The double-strand-break repair model for recombination. *Cell* 33, 25–35. *The molecular basis of recombination.*

Wyatt HDM & West SC (2014) Holliday junction resolvases. *Cold Spring Harbor Perspectives in Biology* 6, a023192.

Inheritance of Genes During Eukaryotic Sexual Reproduction

CHAPTER 15

From our study of meiosis, we have learnt how chromosomes, and the DNA molecules that they contain, are partitioned during the formation of gametes. Now we must examine the events that occur when a pair of gametes participate in sexual reproduction.

Sexual reproduction begins when two gametes, one from the male parent and one from the female, fuse to give a fertilized egg cell (Figure 15.1). Each gamete is haploid, so the egg cell is diploid. Repeated division of the egg cell and its progeny by mitosis gives rise to the somatic cells in the new organism.

In this chapter, we will ask two questions. First, what happens if the pair of gametes contain different alleles of a particular gene? In Chapter 3, we learnt that alleles specify different variants of a biological characteristic, an example being the pair of alleles responsible for the round and wrinkled variants of pea shape. If an organism inherits two different alleles for a particular gene, then which of the two variants of the biological characteristic does it display?

The second question concerns interactions between groups of genes. Many biological characteristics are specified by groups of genes working together. These include relatively simple characteristics such as the color of a flower as well as more complex ones such as susceptibility to certain diseases in humans. How do the allele combinations that are inherited for these groups of genes influence the biological characteristic?

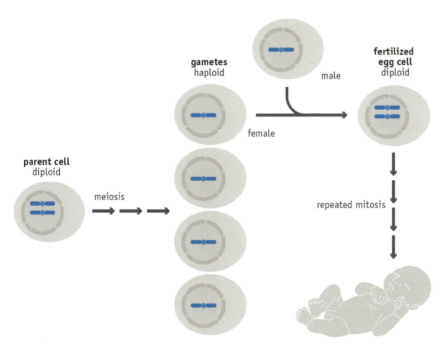

FIGURE 15.1 The cellular events occurring during sexual reproduction.

DOI: 10.1201/9781003473862-17

Answering these questions takes us back to the beginning of genetics, as the issues that we will address include those that were studied by Gregor Mendel and the other early geneticists. This early work is important not just because it established the principles of genetics. It also established the principles of **genetic analysis**. These are the methods, based on examination of the progeny resulting from breeding experiments, that enable interactions between the alleles of the same and different genes to be studied, including information that is difficult or impossible to obtain solely by DNA sequencing and other molecular techniques.

15.1 WHAT MENDEL DISCOVERED

Mendel was not the first biologist to carry out breeding experiments as a means of studying inheritance. In the first half of the 19th century, several biologists set up crosses between different varieties of plant and observed the features of the hybrids that were obtained, using methods very similar to those that Mendel later employed. However, prior to Mendel, the tendency was to consider all of the features of the parents and offspring together, and to interpret the underlying process of inheritance in vague terms such as 'cytoplasmic mixing'. Mendel avoided this problem by studying the inheritance of discrete characteristics, such as the height of a plant and the color of its flowers. This made possible a statistical analysis of the outcome of his breeding experiments, which in turn enabled Mendel to follow the inheritance of individual genes. His work therefore provides the model for the planned breeding experiments that we still use today.

Mendel's experiments were carefully planned

Mendel decided to work with the garden pea, *Pisum sativum* (now more correctly called *Lathyrus oleraceus*; Figure 15.2). Pea plants are easy to grow, and their life cycle is relatively short, so several generations can be obtained in a single growing season. They are therefore ideal experimental organisms for studying the way in which inherited characteristics are passed from parents to offspring.

Two considerations were taken into account by Mendel when choosing the particular characteristics to study with his garden peas. The first and overriding one was that there should be alternative forms to each characteristic. That is to say, the plant could display one form of the characteristic or another, but not both at once. The best-known example of the contrasting pairs of characteristics chosen by Mendel is plant height, which he classified as tall or short.

The second consideration in choosing characteristics to study was a purely experimental one. Mendel anticipated that he would need to determine which characteristic of a pair was displayed by each of several hundred progeny plants in order to be able to distinguish a pattern of inheritance, should one exist. Consequently, his characteristics had to be scored easily. It would be tedious and difficult to score characteristics that were not obvious and needed dissection or chemical analysis before they could be assigned.

FIGURE 15.2 The garden pea, *Pisum sativum*. (From Wikimedia Commons, courtesy of Rasbak, published under CC BY-SA 3.0 license.)

TABLE 15.1	THE PAIRS OF CONTRASTING CHARACTERISTICS STUDIED BY MENDEL
Characteristic	**Contrasting Forms**
Plant height	Tall, short
Pod location	Axial, terminal
Pod shape	Full, constricted
Pod color	Green, yellow
Pea shape	Round, wrinkled
Pea color	Yellow, green
Flower color	Violet, white

In the end, Mendel selected seven pairs of characteristics, each of which had two distinct forms (**Table 15.1**). In addition to plant height, he decided to study the inheritance of three pairs of characteristics that could be scored simply by looking at the pea pods (their position on the plant, their shape and their color), two pairs involving the peas themselves (color and shape) and a single pair of contrasting flower colors.

Mendel's crosses revealed a regular pattern in the inheritance of characteristics

In his first set of experiments, Mendel allowed his pea plants to **self-fertilize**, the female carpels within the flowers being fertilized with pollen from the same plant. Mendel discovered that the progeny resulting from self-fertilization always displayed the characteristics of the parent plant (**Figure 15.3**). So, for example, self-fertilization of a tall plant gave rise to many progeny plants, but all of these were tall. Nowadays, we would say that the plants are **pure-breeding**.

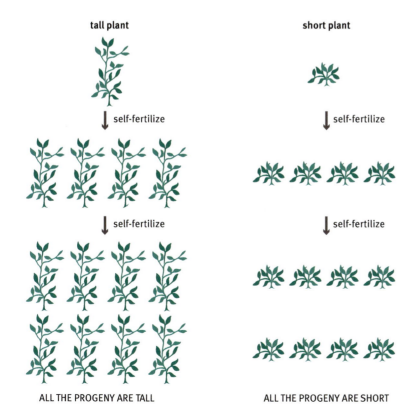

FIGURE 15.3 The characteristics chosen by Mendel were stable. The progeny plants produced from several rounds of self-fertilization displayed only the parental characteristics.

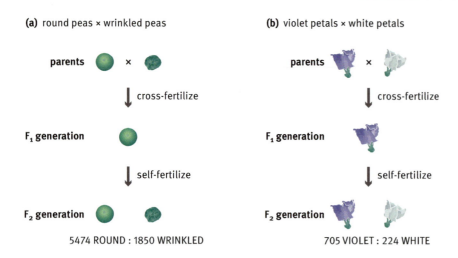

FIGURE 15.4 **Two examples of the monohybrid crosses carried out by Mendel.** (a) Plants with round peas × plants with wrinkled peas. (b) Violet petals × white petals.

Mendel then studied the inheritance of each pair of characteristics by **cross-fertilizing** plants with alternative forms of the same character. A clear and consistent pattern emerged from these seven **monohybrid crosses**. In each case, the first generation (F_1) of plants arising from the cross displayed only one of the parental characteristics. For example, in the cross between plants with round peas and plants with wrinkled peas, all the F_1 plants had round peas and none had wrinkled peas (Figure 15.4a). However, if members of the F_1 generation were self-fertilized, giving rise to a second generation (F_2) of plants, the characteristic that had been absent in the F_1 generation reappeared. In the case of round pea × wrinkled pea, the F_2 generation yielded a total of 5,474 round peas and 1,850 wrinkled peas which, Mendel noted, was a ratio of 2.96:1.

Each of the seven monohybrid crosses produced a similar pattern. Let us take as a second example the cross involving flower color (violet petals × white petals; Figure 15.4b). In this case, all the F_1 plants had violet petals, and in the F_2 generation 705 plants had violet petals and 224 had white petals, a ratio of 3.15:1. The results of each of the monohybrid crosses are summarized in **Table 15.2**. In each cross, only one of the pair of characteristics is displayed in the F_1 generation, but both appear in the F_2 generation. When all of the monohybrid crosses are added together, the F_2 ratio that is obtained works out to 2.98 plants with the F_1 characteristic to every one plant with the second characteristic. Mendel decided that within the limits of experimental error, this was a real ratio of 3:1.

TABLE 15.2	THE RESULTS OF MENDEL'S MONOHYBRID CROSSES		
Cross	**F_1 Generation**	**F_2 Generation**	
		Numbers	Ratio
Tall × short plants	All tall	787 tall, 277 short	2.84:1
Axial × terminal pods	All axial	651 axial, 207 terminal	3.15:1
Full × constricted pods	All full	882 full, 299 constricted	2.95:1
Green × yellow pods	All green	428 green, 152 yellow	2.82:1
Round × wrinkled peas	All round	5,474 round, 1,850 wrinkled	2.96:1
Yellow × green peas	All yellow	6,022 yellow, 2,001 green	3.01:1
Violet × white flowers	All violet	705 violet, 224 white	3.15:1

Mendel's interpretation of the results of the monohybrid crosses

To explain the consistent pattern of inheritance that he had discovered, Mendel proposed that each trait (for example, plant height) is controlled by a **unit factor** and that each unit factor can exist in more than one form, with different forms responsible for different characteristics. According to this interpretation, the unit factor for pea morphology can exist in two alternative forms, one responsible for round peas (we would now designate this version of the unit factor as '*R*') and one responsible for wrinkled peas (designated as '*r*'). Unit factors are exactly the same things that we now call genes and, as we discovered in Chapter 3, the alternative forms of a gene are now called alleles.

Mendel realized that the most important aspect of the monohybrid crosses was that, in each cross, one of the parental characteristics disappeared in the F_1 generation and then reappeared in the F_2 generation. In the cross between plants with round peas and plants with wrinkled peas, for example, the wrinkled characteristic was absent in the F_1 generation, but present again in the F_2 plants. He deduced that this means that the F_1 plants must contain two alleles, one for round peas and one for wrinkled. If this were not the case, then how could both alleles be transmitted to the F_2 plants?

We would therefore describe the **genotype** (the genetic constitution) of the F_1 plants as *Rr*. We would refer to the F_1 plants as **heterozygotes**, indicating that they have two different alleles for the pea morphology gene. A **homozygote**, on the other hand, has two alleles the same (in this case, *RR* or *rr*).

Although the genotype of the F_1 plants is *Rr*, the **phenotype** (the observable characteristic) is round pea. Therefore, the *R* allele must dominate over the *r* allele, so that only the characteristic that it controls – round peas – appears in the F_1 plants. We say that the *R* allele is **dominant** and the *r* allele **recessive**.

Why is there a 3:1 ratio in the F_2 generation?

Having established how one of the parental characteristics could disappear in the F_1 generation and reappear in the F_2 plants, Mendel went on to explain how the 3:1 ratios arise in the F_2 generation. As the F_1 plants have the genotype *Rr*, the F_1 cross can be written as

Rr × *Rr*

This notation is acceptable even though the F_1 cross is a self-fertilization, meaning that in reality it does not involve two individual plants, but instead different parts of the same plant.

We know that the F_1 cross produces some plants with wrinkled peas, whose genotypes must be *rr*. These F_2 plants must have obtained an *r* allele from each F_1 'parent': there is no other way in which they could arise. This suggests that during sexual reproduction the alleles of each parent separate (we use the term **segregate**), producing intermediate structures that contain just one allele. These intermediate structures are the gametes. Mendel correctly postulated that two gametes, one from each parent, fuse to bring together the pair of alleles carried by a member of the next generation (**Figure 15.5**).

How do the events shown in Figure 15.5 result in the 3:1 ratio? The useful construction called the **Punnett square** (**Figure 15.6**) helps us to see that the 3:1 ratio arises naturally, as long as the F_1 gametes are able to

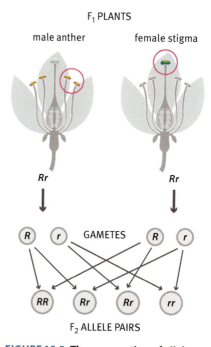

FIGURE 15.5 **The segregation of alleles during the F_1 cross involving round and wrinkled peas.** The genotype of the F_1 plant is *Rr*. The male and female reproductive parts produce gametes, each gamete containing just a single allele, *R* or *r* in this case. Two alleles, one from each 'parent', fuse to produce the pair of alleles carried by an F_2 plant.

segregate in an entirely random fashion. If segregation is random, then the genotypes displayed by the F₂ plants will be 1*RR* : 2*Rr* :1*rr*. As both *RR* and *Rr* plants have round peas, the phenotypic ratio will be three round peas to one wrinkled pea, exactly as Mendel observed.

Crosses involving two pairs of characteristics

Next, Mendel examined whether different pairs of characteristics are inherited together or independently. He investigated this by carrying out a series of **dihybrid crosses**. As an example, we will consider the cross between plants with round, yellow peas (homozygous dominant *RRYY*) and plants with wrinkled, green peas (homozygous recessive *rryy*). This cross and its outcome are illustrated in Figure 15.7. The F₁ generation all have the same genotype (*RrYy*) so all are double heterozygotes and have round, yellow peas.

In the F₂ generation, each possible combination of characteristics is represented. In addition to the two parental combinations (round, yellow and wrinkled, green), the F₂ plants include some with the nonparental combinations (round, green and wrinkled, yellow). These different combinations occur in the ratio:

9 plants with round, yellow peas

3 plants with round, green peas

3 plants with wrinkled, yellow peas

1 plant with wrinkled, green peas

Different pairs of characteristics are therefore not inherited together. But are they inherited completely independently? The Punnett square treatment of the dihybrid cross tells us that the answer is 'yes': the 9:3:3:1 ratio

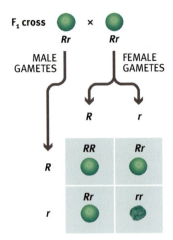

FIGURE 15.6 The Punnett square for the F₁ cross involving the *R* and *r* alleles. The Punnett square enables the possible combinations of alleles to be worked out and shows the numbers of each genotype that result. Note that the 3:1 ratio arises only if the gametes of each parent segregate randomly.

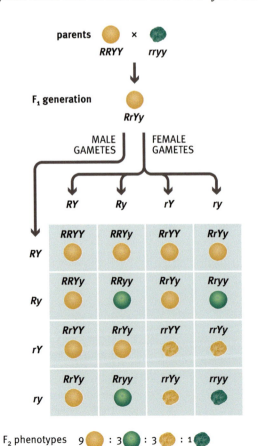

FIGURE 15.7 The dihybrid cross between a plant with round, yellow peas and one with wrinkled, green peas.

arises only if the two pairs of characteristics segregate in an independent fashion.

The dihybrid cross is most easily interpreted if it is looked on simply as two concurrent monohybrid crosses. Taking just the round:wrinkled component of the cross, the expected ratio of 3:1 in the F_2 generation holds true, although it is written above as 12:4. If we now examine just the F_2 plants with round peas, then we would expect three-quarters of the round peas to be yellow and one-quarter to be green, and this is indeed the case – 9 out of 12 yellow and 3 out of 12 green. Now consider the plants with wrinkled peas. Three-quarters of these peas should be yellow and one-quarter green. Again, this is the case – 3 out of 4 yellow and 1 out of 4 green.

Mendel pre-empted the discovery of meiosis, but did he cheat?

Mendel published the results of his experiments in 1865, 20 years before the first description of meiosis and 40 years before the chromosome theory of inheritance was established. Mendel therefore pre-empted almost half a century of cell biology by deducing the behavior of genes during sexual reproduction. He showed that his unit factors occur in pairs, which we now know is because most cells of higher organisms are diploid, meaning that each cell contains two copies of each chromosome and hence two copies of each gene. He also showed that unit factors are passed to offspring via intermediate structures that contain just one allele. This is precisely what happens during meiosis when the double set of chromosomes is reduced by half, producing cells (the gametes) that contain just one copy of each chromosome and hence one copy of each gene (see Figure 15.1). There are no restrictions regarding which of the four gametes resulting from a round of meiosis give rise to a member of the next generation. Each gamete has an equal chance of fusing with a gamete derived from the other parent. The entire process occurs at random, just as postulated by Mendel.

Despite Mendel being correct about most things, there is one serious shortcoming in the conclusions that he drew from his results. His interpretation of the dihybrid crosses was that alleles of different genes segregate independently of one another. For this to happen, the genes must be on different chromosomes. If they are on the same chromosome, then they would be expected to segregate together – what we now call **linkage** (Figure 15.8). Initially, it was assumed that Mendel had been very lucky in that each of the genes he studied was on a different chromosome, and so no pair of genes displayed linkage. We now know that this is not the case. Two of the genes that Mendel studied are on pea chromosome 1 and two or possibly three are on chromosome 4 (Figure 15.9). However, with one exception, the genes are so far apart on these chromosomes that in almost every meiosis a crossover will occur between them. This means that the linkage between them will not be evident in the results of a dihybrid cross, and they will in effect segregate independently as Mendel reported (Figure 15.10).

The exception concerns the genes for plant height and pod shape. These are both on chromosome 4 and are so close together that they will only occasionally be unlinked by a crossover. Mendel did not include the dihybrid cross for plant height and pod shape in the paper he published in 1866 describing his work. Was this because the results of this cross did not agree with his conclusions? In fact, Mendel did describe a *tetrahybrid* cross involving plant height, pod shape, pod color and pea color in a letter he wrote to the Swiss botanist Carl Nägeli the following year. However, it

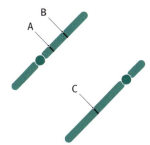

FIGURE 15.8 Genes on the same chromosome should display linkage. Genes A and B are on the same chromosome and so should be inherited together. Mendel's conclusion that alleles of different genes segregate independently should therefore not apply to genes A and B. Gene C is on a different chromosome, so will segregate independently of A and B.

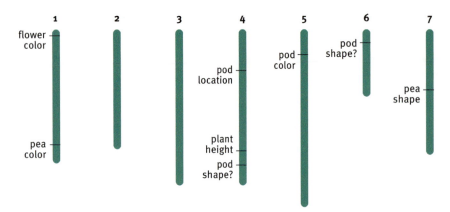

FIGURE 15.9 **The locations of the genes studied by Mendel on the pea chromosomes** Peas have seven chromosomes. Two of the genes that Mendel studied are on chromosome 1, and two or possibly three are on chromosome 4. The doubt concerns the gene for pod shape. There are two of these, one on chromosome 3 and one on chromosome 6, and it is not certain which of these was responsible for the characteristic that Mendel studied.

is not clear if the way he carried out this cross would have enabled him to detect the linkage between plant height and pod shape.

There is also a second more convincing defense of Mendel's integrity as a scientist. Peas have two genes that control pod shape. In addition to the gene on chromosome 4, there is a second pod-shape gene on chromosome 6 (see Figure 15.9). We simply do not know which of these genes was responsible for the different versions of pod shape in the actual plants studied by Mendel. If it was the gene chromosome 6, then there would have been no linkage between pod shape and plant height, and no need to question the accuracy of Mendel's scientific reports.

15.2 RELATIONSHIPS BETWEEN PAIRS OF ALLELES

If an organism inherits two different alleles of a gene, then which of the two variants of the biological characteristic specified by that gene does it display? Mendel began to answer this question by demonstrating that the alternative forms of each of the seven pea characteristics that he studied displayed a simple dominant–recessive relationship. We will examine the molecular basis of this type of relationship, and then look at different types of allele relationships that Mendel did not encounter, but which were discovered by the early geneticists that followed in his footsteps.

Recessive genes are often nonfunctional

In a heterozygote, only one allele, the dominant one, contributes to the phenotype. The recessive allele is, in effect, silent. This is quite easily understood when we remember that genes code for proteins that have particular functions in the cell. Let us envisage a simple system in which a gene can exist as either of two alleles: functional and nonfunctional (Figure 15.11a). We will use *G* to designate the functional allele and *g* for the nonfunctional allele, and we will assume that allele *g* is nonfunctional because it contains a mutation that prevents the protein specified by the gene from being synthesized.

Now let us look at the different possible genotypes and deduce what phenotypes result (Figure 15.11b). First, there is the homozygote, *GG*. An organism with this genotype carries two functional copies of the gene and so is able to synthesize the protein coded by the gene. The characteristic that the protein is responsible for is therefore displayed by the organism. Second, there is the other homozygote, *gg*. Now there are two nonfunctional versions of the gene. The organism cannot synthesize the protein and the characteristic it is responsible for is absent. Finally, there is the

FIGURE 15.10 **Genes on the same chromosome might not display linkage.** Two genes on the same chromosome will not display linkage if a crossover occurs between them during meiosis. In this example, genes A and C are so far apart that a crossover that unlinks them is likely to occur in every meiosis. Genes A and C will therefore appear to be unlinked and their alleles will segregate randomly, as observed by Mendel for each gene pair that he studied.

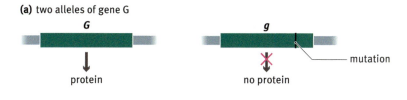
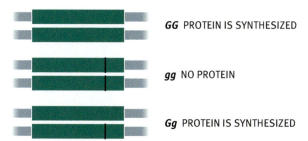

FIGURE 15.11 A molecular explanation for dominance and recessiveness. In this example, allele G is dominant and g is recessive.

heterozygote, *Gg*, with one functional and one nonfunctional copy of the gene. This organism contains a functional gene and so is able to synthesize the protein. The characteristic is therefore displayed. In a cross such as those carried out by Mendel, *G* will be the dominant allele and *g* the recessive allele.

How does this molecular explanation of dominance and recessiveness correlate with the characteristics actually studied by Mendel? We have already answered this question with regard to pea shape. In Section 3.2, we learnt that the difference between a round pea and a wrinkled one is that a round pea has a functional gene for the starch branching enzyme SBE1 and is therefore able to convert the sucrose it produces by photosynthesis into starch (Figure 15.12). The presence of starch reduces water uptake as the pea develops, so the pea stays round when it reaches maturity and dehydrates. The *r* allele that Mendel studied is a mutated version of the *SBE1* gene, so a plant with the genotype *rr* does not produce this enzyme. This means that *rr* peas have a high sucrose content as the sucrose cannot be converted into starch. These peas absorb much more water as they develop, and so dehydrate more when they reach maturity, causing the pea to collapse and become wrinkled. In the heterozygous state, *Rr*, the *R* allele directs the synthesis of the SBE1 enzyme

FIGURE 15.12 The basis of the round and wrinkled pea phenotypes. (a) Round peas have a functional *SBE1* gene, so convert sucrose into starch. This means that the pea stays round when it matures. (b) Wrinkled peas have a nonfunctional *SBE1* gene and convert less sucrose into starch. They lose water when they mature and become wrinkled.

in sufficient amounts to reduce the sucrose level. *Rr* plants therefore have round seeds: *R* is dominant to *r*.

A second example is provided by plant height. The plant height gene specifies an enzyme called gibberellin 3-oxidase, which is responsible for the synthesis of the plant growth hormone gibberellin. The recessive allele has a point mutation that converts a guanine into an adenine in the DNA sequence, changing an alanine codon into one for threonine (Figure 15.13). This alteration reduces the activity of the enzyme, so less gibberellin is synthesized and the plant has a dwarf morphology. The dominant allele lacks the mutation and so specifies fully functional gibberellin 3-oxidase, resulting in sufficient gibberellin synthesis to direct the growth of tall plants.

Some pairs of alleles display incomplete dominance

Mendel's characteristics illustrate the most simple type of relationship between a pair of alleles, where one is dominant and the other is recessive, and the dominant allele contributes to the phenotype of the heterozygote. After the rediscovery of Mendel's work in 1900, the new generation of geneticists performed breeding experiments to study the inheritance of a wide range of characteristics in various species of plants. These experiments revealed that the relationships between pairs of alleles are not always as straightforward as envisaged by Mendel.

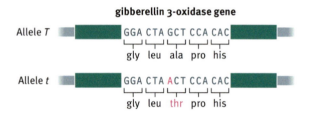

FIGURE 15.13 The plant height alleles studied by Mendel. Plant height is specified by the gibberellin 3-oxidase gene. A mutation that replaces a G with an A in allele *t* converts an alanine in the protein to a threonine. This change results in a decrease in the activity of the gibberellin 3-oxidase enzyme, giving a dwarf phenotype.

(a) the cross between red and white carnations

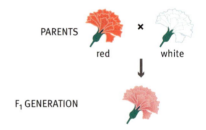

(b) the molecular explanation

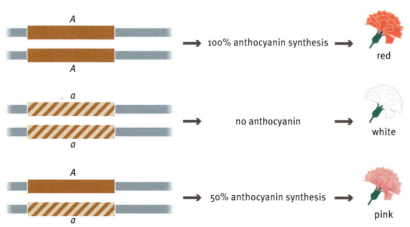

FIGURE 15.14 Incomplete dominance as displayed by flower color in carnations. (a) When carnations with red flowers are crossed with carnations with white flowers, all the F₁ plants have white flowers. (b) The *A* allele specifies a functional gene for the synthesis of the anthocyanin pigment. The *a* allele is nonfunctional. The genotype *AA* therefore produces plenty of pigment and the flowers are red. The other homozygote, *aa*, synthesizes no pigment and the flowers are white. The heterozygote, *Aa*, makes an intermediate amount of anthocyanin, giving pink flowers.

One of the first of these 'post-Mendelian' discoveries was **incomplete dominance**. This is displayed by the flower color gene for various species of plant, including carnations. In a cross between carnations with red flowers and ones with white flowers, the F_1 generation of plants all have pink flowers (Figure 15.14a). In other words, the heterozygous form has a phenotype (pink in this case) that is intermediate between the phenotypes of the two homozygous parents (red and white).

The molecular explanation of this example of incomplete dominance is quite straightforward. Anthocyanin is a type of flavonoid that gives carnation petals their deep red color. It is synthesized from a white precursor by the enzyme anthocyanin synthase, which is coded by the ANS gene. This gene has two alleles: A, corresponding to the functional copy of the gene, and a, the inactive version.

A plant that is homozygous for allele A, and hence has the genotype AA, is able to make anthocyanin synthase and so produces lots of anthocyanin and has red flowers (Figure 15.14b). The other homozygote has the genotype aa and makes no anthocyanin synthase and hence no pigment. Its flowers are white.

If the A and a alleles had a simple dominant–recessive relationship, then a heterozygous plant, Aa, would have red flowers. But in this case, the 50% reduction in overall gene expression that occurs in the heterozygote – because there is only one functional copy of the gene – is insufficient to maintain the amount of anthocyanin synthase in the cells at the required level. Only a limited amount of anthocyanin is made, and the flowers are the rather attractive pink color.

Lethal alleles result in the death of a homozygote

In the two examples that we have studied so far, a homozygote that contains two copies of the nonfunctional allele is unable to synthesize a particular enzyme but otherwise is healthy. The absence of the SBE1 enzyme in pea plants simply means that the peas are wrinkled rather than round, and the absence of the pigment synthesis enzyme in carnations only affects the color of the flowers.

More serious consequences arise if the organism cannot tolerate the loss of gene product that occurs in a homozygote with two nonfunctional alleles. In this case, the nonfunctional allele is said to be **lethal**, as the homozygote cannot survive.

One of the most interesting examples of a lethal allele concerns the golden or *aurea* variety of snapdragon. These plants have pale yellow rather than green leaves. When a cross is performed with two *aurea* plants, a mixture of green and *aurea* plants is obtained, in a 2:1 ratio (Figure 15.15a). From this ratio, we can make two deductions. First, the presence of two phenotypes (green and *aurea*) in the progeny tells us that the *aurea* version of snapdragon is heterozygous for the *aurea* gene. Second, because we see a 2:1 rather than the expected 3:1 ratio in the progeny, we can deduce that one of the homozygotes resulting from the cross is unable to survive.

Once again, the molecular explanation clarifies exactly what is happening. The *aurea* gene codes for a protein involved in chlorophyll synthesis. The plant must make chlorophyll in order to photosynthesize, and if it cannot photosynthesize then it cannot survive. All is well in a plant that has two functional copies of the *aurea* gene. Its genotype is usually denoted $L^G L^G$. It has two functional alleles and so makes plenty of chlorophyll, appearing green in color and able to grow healthily (Figure 15.15b).

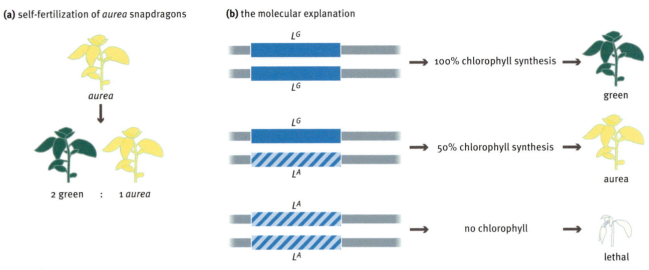

FIGURE 15.15 An example of a lethal allele. (a) When the *aurea* variety of snapdragons are allowed to self-fertilize, green and *aurea* plants are obtained in a 2:1 ratio. (b) The *aurea* gene codes for a protein involved in chlorophyll synthesis. The homozygote, $L^G L^G$, synthesizes plenty of chlorophyll and is green. The heterozygote, $L^G L^A$, makes enough chlorophyll to survive but is a pale yellow color. The recessive homozygote, $L^A L^A$, cannot synthesize chlorophyll. The seeds are able to germinate but the plants quickly die.

In the heterozygote, genotype $L^G L^A$, the single active L^G allele is able to direct synthesis of some chlorophyll, enough for the plant to survive even though its leaves are pale yellow rather than green. However, a homozygote with two inactive copies of the gene is unable to synthesize chlorophyll and cannot photosynthesize. Homozygous $L^A L^A$ seeds are able to germinate but the seedlings are pure white and do not survive for long. So, L^A and L^G display incomplete dominance and L^A is a lethal allele.

The *aurea* gene provides an example of a recessive lethal allele, one which is lethal only in the homozygous form. Dominant lethal alleles are also known, where possession of just one allele causes death. This means that neither the dominant homozygote nor the heterozygote is able to survive (Figure 15.16). At first glance, this appears to present a genetic impossibility. If every fertilized egg cell containing a particular allele is unable to survive, then how is it possible for that allele to propagate within a population? The answer is that all of the dominant lethal alleles that are known specify a **delayed onset** phenotype. This means that their lethality is only expressed late in the life of the organism, possibly after it has reproduced and passed the lethal allele onto its offspring. An example in humans is the neurodegenerative disorder called Huntington's disease, which usually does not cause symptoms until the affected patient is at least 30 years of age.

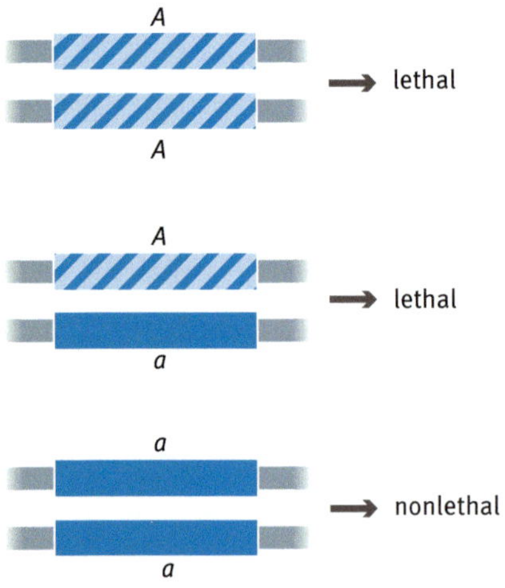

FIGURE 15.16 A dominant lethal allele. Allele *A* is dominant but gives rise to a lethal condition. Only the recessive homozygote, *aa*, is able to survive.

Some alleles are codominant

The final type of allele relationship that we will study is **codominance**. To display codominance, both members of a pair of alleles must be functional. Codominance then occurs when both alleles are active in the heterozygous state.

Normally, to observe codominance, it is necessary to examine the biochemistry of the organism. This is the case with one of the best-understood examples, the MN blood group series in humans. Whether a person has blood group M or N depends on the identity of certain protein molecules in the blood. The M and N proteins are coded by a pair of alleles designated L^M and L^N (Figure 15.17). A person who has the homozygous $L^M L^M$ genotype is able to make just M proteins and so has blood group M. Similarly, the homozygous $L^N L^N$ genotype produces N proteins and gives rise to blood group N. However, in the heterozygote, both alleles are

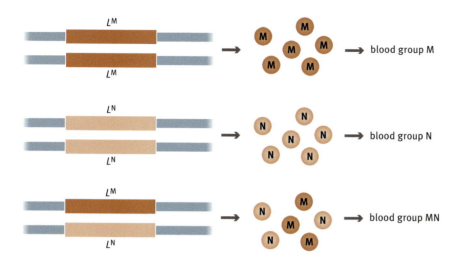

FIGURE 15.17 Codominance of the M and N blood protein genes. The heterozygote $L^M L^N$ is able to make both the M and N proteins and so has blood type MN.

present ($L^M L^N$) and both still direct synthesis of proteins. The blood type is MN. Neither allele is dominant.

A second human blood grouping, ABO, provides a slightly more complicated example. With this grouping, there are three different alleles, I^A, I^B and I^O. Blood protein A is specified by the I^A allele, and protein B by the I^B allele. Allele I^O is nonfunctional and does direct synthesis of any protein. I^A and I^B are codominant, and both are dominant over I^O. These allele relationships underlie the four common human blood groups, A, B, AB and O (Figure 15.18).

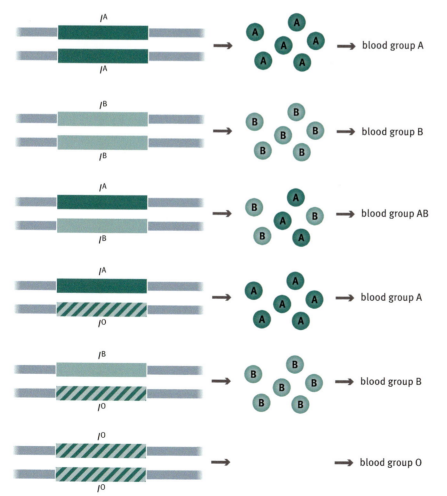

FIGURE 15.18 The ABO blood groups. There are three alleles, I^A, I^B and I^O. I^A and I^B are codominant, and both are dominant over the nonfunctional I^O.

15.3 INTERACTIONS BETWEEN ALLELES OF DIFFERENT GENES

Although many phenotypes can be linked to the activity of just a single gene, there are numerous others that result from interactions between two or more separate genes. Understanding these interactions is crucial to our awareness of how the genome as a whole specifies the biological characteristics of an organism.

The question that we must ask is how the more complex characteristics that are controlled by groups of genes are influenced by the presence of different alleles of those genes. We will begin by looking at relatively simple interactions that involve just two genes working together and then examine the greater complexity presented by **quantitative traits**, which are specified by the combined activities of large numbers of genes.

Functional alleles of interacting genes can have additive effects

Flower pigmentation illustrates various ways in which different genes can interact to produce a single phenotype. Some plant species are able to synthesize several different flower pigments of the same or similar colors. An example is the daylily of the genus *Hemerocallis* (Figure 15.19), which makes two types of anthocyanin pigment, called cyanidin and pelargonidin. Both are red in color. Cyanidin synthase is specified by gene C and pelargondin synthase by gene P. Each gene has a functional allele, *C* or *P*, and a nonfunctional version, *c* or *p*. Each pair of alleles, *C* and *c*, and *P* and *p*, displays incomplete dominance. You will realize that each gene is therefore equivalent to the anthocyanin synthase gene of carnation that we studied earlier.

FIGURE 15.19 A daylily of the genus *Hemerocallis*.

Armed with our knowledge of the phenotypes resulting from different combinations of pigment synthesis alleles in carnations, we can begin to deduce the effect that the two interacting genes, C and P, will have on flower color in daylilies. A double homozygote with two copies of each functional allele, genotype, *CCPP*, will synthesize cyanidin and pelargonidin at their maximum levels and have deep red flowers (Figure 15.20a). At the other end of the scale, the other double homozygote, *ccpp*, will make neither pigment and have pure white flowers.

So far the outcome is the same as if there was just one pigment synthesis gene. The differences arise when we consider the other possibilities (Figure 15.20b). In addition to the double homozygotes, we can have a variety of other genotypes when the two genes are considered together. These are *CcPP*, *CcPp*, *Ccpp*, *CCPp*, *CCpp*, *ccPP* and *ccPp*. What will be the phenotypes of plants with each of these genotypes?

The easiest way to address this problem is to look on each functional allele copy as contributing a unit of pigment (Table 15.3). According to this interpretation, the double homozygotes *CCPP* makes 4 units of pigment, which makes a deep red flower. The other double homozygote *ccpp* makes zero units of pigment and so is white. In between, we have genotypes that have 1, 2 or 3 functional alleles, contributing 1, 2 or 3 units of pigment. This means there will be three intermediate shades of flower color between red and white. These would be light pink, pink and light red, though they would probably be called something more adventurous by the company selling daylilies, perhaps seashell, baby pink and salmon.

Table 15.3 makes two important points. The first is that, as we have seen, alleles of different genes can combine in an additive fashion to determine

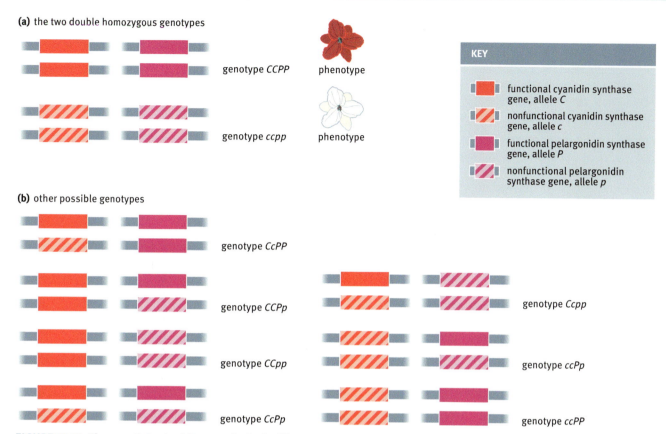

FIGURE 15.20 The genetic basis of flower color in daylilies. (a) The phenotypes of the two double homozygotes can be predicted. (b) There are seven additional genotypes in which one or both genes are heterozygous.

TABLE 15.3	PIGMENT PRODUCTION BY TWO INTERACTING SYNTHESIS GENES	
Number of Pigment Units	Genotypes	Phenotype
4	CCPP	Red
3	CCPp, CcPP	Light red
2	CCpp, ccPP, CcPp	Pink
1	Ccpp, ccPp	Light pink
0	ccpp	White

the nature of a phenotype. The second point is that the number of genotypes is not the same for each of the five possible phenotypes. There is only one way of making 'red' and one way of making 'white', but two possible genotypes each for 'light pink' and 'light red' and three for 'pink'. In a large population of plants, we would therefore expect relatively few to have red or white flowers, and we would expect pink to be the commonest color. We would go further and predict that the numbers of each flower color would form a **normal** or **Gaussian distribution**, the distribution commonly called a bell-shaped curve (Figure 15.21). This is an important observation that we will return to when we study quantitative traits specified by multiple genes.

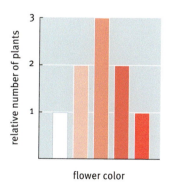

FIGURE 15.21 The expected distribution of flower colors in a large population of daylilies.

Important interactions occur between genes controlling different steps in a biochemical pathway

The example of flower color that we have just considered involves two genes that are responsible for the synthesis of different types of red pigment. Their functional alleles therefore have an additive effect on the phenotype – on the color of the flowers that are produced.

A more puzzling interaction between genes for flower color was discovered by William Bateson and Reginald Punnett in 1905. They crossed two white-flowered sweet peas and obtained an F_1 generation comprised entirely of plants with purple flowers. When these plants were allowed to self-fertilize, they gave rise to an F_2 ratio of 9 purple to 7 white (Figure 15.22a).

At first glance, this cross may seem difficult to understand but it is easy enough if we deal with it step by step. First, the 9:7 ratio is clearly a modified 9:3:3:1, indicating that we are dealing with two genes. Having established that this is a dihybrid cross, we can construct a Punnett square to show the genotypes of the F_2 plants (Figure 15.22b). We now see that the 9:7 ratio could be explained if we assume that the phenotype is purple only if both dominant alleles, A and B, are present. According to this hypothesis, pigment synthesis does not occur if either pair of alleles is homozygous recessive, so seven of the genotypes result in white flowers.

The molecular explanation of these results is that genes A and B code for enzymes that catalyze two steps in the pigment synthesis pathway

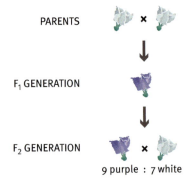

FIGURE 15.22 The cross between sweet peas with white flowers. (a) The cross gave a ratio of 9 purple flowers to 7 white flowers in the F_2 generation. (b) The Punnett square shows how the phenotypes of the F_2 plants arise.

(Figure 15.23). When two genes work together in a single pathway, at least one functional allele of each gene must be present in order for the product to be synthesized. In this example, a functional allele of gene A is needed for the precursor to be converted into the intermediate compound, and a functional allele of gene B is needed to convert the intermediate into the pigment. This means that there are four genotypes that give rise to pigment synthesis: *AABB*, *AABb*, *AaBB* and *AaBb*. Each of the other five possible genotypes – *aaBB*, *aaBb*, *AAbb*, *Aabb*, *aabb* – lack a functional allele of either gene A or gene B and so cannot make the pigment.

This type of relationship is common in biology as it holds for all cases where two or more genes specify enzymes responsible for different steps in a biochemical pathway. We have used pigment synthesis as our example, but it could equally well be a pathway for the synthesis of an amino acid or an essential vitamin. The relationship is sometimes called **complementary gene action**, as a particular combination of alleles at two separate genes is needed in order to produce the phenotype.

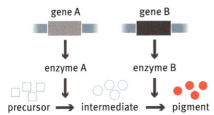

FIGURE 15.23 Complementary gene action. Genes A and B code for enzymes involved in different steps in a biochemical pathway leading to synthesis of a pigment. The gene A product converts the colorless precursor compound into a colorless intermediate. The gene B product converts this intermediate into the pigment.

Epistasis occurs when alleles of one gene mask the effect of a second gene

Although often referred to as complementary gene action, it is more accurate to describe the relationship we have just studied between genes A and B as **epistasis**. Epistasis is where the alleles present at one gene mask or cancel those at a second gene. In the example shown in Figure 15.23, the *aa* genotype masks the genotype for gene B. It makes no difference if this genotype is *BB*, *Bb* or *bb*, the same phenotype – no pigment synthesis – always occurs. Similarly, *bb* masks the genotype at gene A. The two genes are epistatic toward one another.

In the example that we have just considered the masking effect is conferred by a recessive, nonfunctional allele of the epistatic gene. It is also possible for a dominant, functional allele to be epistatic toward a second gene. Continuing our plant pigment theme, an example is provided by fruit color in squash. The orange color of the mesocarp or fleshy part of squashes is due to the presence of carotenoid pigments (Figure 15.24), which also increase the nutritional value of the fruit. The carotenoid pigment is synthesized from a green precursor by an enzyme coded by gene G (Figure 15.25). The functional allele *G* displays dominance over nonfunctional allele *g*, so genotypes *GG* and *Gg* give orange fruits, and *gg* gives green ones.

The epistatic gene is called W. When this gene is present as one or more of its dominant alleles, *WW* or *Ww*, the fruit is white. The molecular explanation for this example of epistasis is that the *W* allele codes for a protein that inhibits the synthesis of the green precursor. As the green compound is not made, the mesocarp is nonpigmented and appears white. The genotype of gene G is immaterial because there is no green precursor for its enzyme to convert into the carotenoid pigment. Gene W therefore displays dominant epistasis toward gene G.

Interactions between multiple genes result in quantitative traits

So far we have only considered the ways in which pairs of genes can interact. As there are thousands of genes in a eukaryotic genome, we might anticipate that a simple pairwise interaction would be a relatively uncommon phenomenon, with multiple gene interactions being much more prevalent. We must therefore examine situations where many genes contribute to a single phenotype.

FIGURE 15.24 A butternut squash showing the orange fleshy mesocarp. (Courtesy of Andy Roberts, published under CC BY 2.0 license.)

An indication of the effect that multiple genes might have was provided by the additive effects of the daylily pigmentation genes that we considered above. We recognized five different levels of pigmentation from white to red, these different degrees of color resulting from interaction between just two genes. The five levels of pigmentation are expected to occur at different frequencies in a random population of plants, the distribution as a whole forming a bell curve (see Figure 15.21).

The bell curve is a characteristic feature of the phenotype resulting from interactions between multiple genes. This is true whether there are three, four or many hundred interacting genes, the only difference being that the curve becomes smoother as more genes are added (Figure 15.26). A continuous Gaussian distribution pattern is precisely what is found if a quantitative trait such as height, which is thought to be specified by many genes, is examined in the human population (Figure 15.27). Many quantitative traits, including height, are also affected by environmental factors such as diet, complicating attempts to work out their genetic basis.

Today, quantitative traits are extensively studied because of their applied importance. For example, the productivity of most crops and farm animals is determined by quantitative traits such as seed size and the meat-to-weight ratio. In medicine, quantitative traits are important because susceptibility to certain diseases depends on interactions between multiple genes. Susceptibility therefore displays a Gaussian distribution in the population as a whole. Typically, those individuals falling at one extreme of the distribution are likely to develop the disease (Figure 15.28).

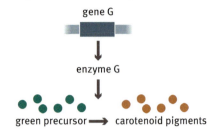
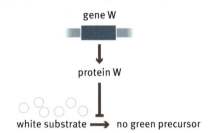

FIGURE 15.25 Carotenoid pigment synthesis in squash. (a) Synthesis of carotenoid pigments by the product of gene G. (b) The epistatic effect of gene W.

15.4 A SINGLE GENE CAN GIVE RISE TO A COMPLEX PHENOTYPE

The link between genotype and phenotype is not always clear-cut and unambiguous. A single gene can sometimes give rise to a complex or variable phenotype. Analysis of such a phenotype often provides valuable information about the function of the gene. To illustrate this point, we will consider **pleiotropy** and **incomplete penetrance**.

Pleiotropic genes affect multiple characteristics

Pleiotropy is the situation that arises when a mutation in a single gene affects more than one trait. Mendel's gene for flower color is an example as it also influences the color pattern of the leaves of the plant. This

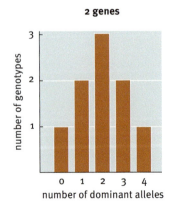
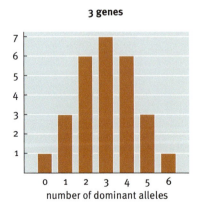
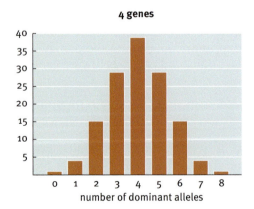

FIGURE 15.26 Bell curves for different numbers of interacting genes. The histograms show the predicted distributions of phenotypes for two, three, and four genes acting in an additive fashion and displaying incomplete dominance.

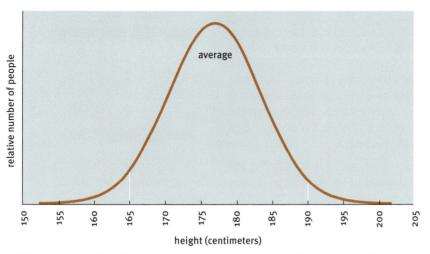

FIGURE 15.27 **The distribution pattern for height among male adults from the United States**.

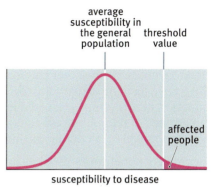

FIGURE 15.28 **Susceptibility to certain diseases is a quantitative trait in humans.** A disease of this type displays a Gaussian distribution when the population as a whole is considered. All individuals will display some susceptibility, but only those whose particular combination of alleles places them to the right of the threshold value are likely to develop the disease.

indicates that the trait interpreted by Mendel as 'flower color' is more correctly 'pigment synthesis'. As a result, the gene affects all the pigmented parts of the pea plant, not just the petals.

Pleiotropy is common in higher organisms, because the complexity of eukaryotic biochemistry means that a genetic defect that leads to the loss of an enzyme activity rarely has a clearcut phenotypic effect. Cystic fibrosis, for example, results from a mutation in a single human gene, coding for a protein involved in chloride ion secretion by epithelial cells. This defect leads to the complex phenotype that we call 'cystic fibrosis', which manifests itself as a variety of symptoms including lung disease, a deficiency in pancreatic enzymes and excessive loss of salt in the sweat (Figure 15.29). In extreme cases, the sheer complexity of a pleiotropic phenotype can obscure the nature of the primary biochemical defect, but once the latter has been identified the pleiotropic effects are a bonus as they enable the function of the gene in different tissues to be assessed.

Not all deleterious alleles have an immediate effect

Alleles that are incompletely penetrant are ones that do not always give rise to their associated phenotype. Typically, incomplete penetrance occurs when a mutation merely provides a predisposition to developing the phenotype, which appears only in response to a later trigger (Figure 15.30). Incomplete penetrance is displayed by a number of human genetic cancers, with the triggering factor possibly linked to the environment or to the person's lifestyle. If the trigger is never encountered during an individual's lifetime, then the person remains healthy until dying of some other cause.

An example is provided by the human disease retinoblastoma, which is a cancer of the eye. Retinoblastoma is inherited, indicating that it is a genetic disease and is controlled by a gene or set of genes. However, the disease is incompletely penetrant and of every 10 individuals that, according to their genotype, should have retinoblastoma, on average only nine actually develop the disease (a penetrance of 90%). This observation provides an important clue to the genetic basis of retinoblastoma. It indicates that the relevant gene or genes do not directly specify the disease, but merely provide vulnerability toward it.

We now know that retinoblastoma is associated with a single gene (*Rb*) that regulates the division of cells in the retina. Inactivation of the gene, for example, by environmental radiation or other mutagenic agents, results

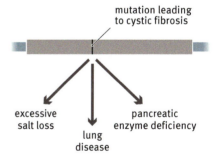

FIGURE 15.29 **A single gene can give rise to a complex, multifactorial phenotype.**

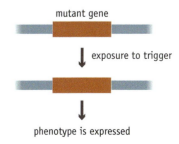

FIGURE 15.30 **Incomplete penetrance.**

in over-proliferation of retinal cells, leading to retinoblastoma. Most individuals are homozygous for *Rb*, so each retinal cell has two copies of the gene. It is very unlikely that inactivating mutations will occur in both genes of a single cell, so homozygotes virtually never develop the disease. Heterozygotes, on the other hand, are more vulnerable to retinoblastoma. They have just one active copy of *Rb* per cell, and so, a single inactivating mutation in any retinal cell could lead to retinoblastoma. The incomplete penetrance of retinoblastoma tells us that there is a 90% chance that such a mutation will occur during the individual's lifetime.

Incomplete penetrance often has an age factor, meaning that the chances of expressing the phenotype increase with the age of the individual. These delayed onset phenotypes are expressed only late in an individual's lifetime, possibly again due to the need for a trigger, but in this case one that results from the accumulation of some factor over time (Figure 15.31). Many of the debilitating human afflictions of old age, such as Huntington's disease which we met earlier in this chapter, fall into this category. As with other types of incomplete penetrance, an individual with a predisposing genotype might never express the phenotype simply because he or she does not accumulate the trigger to a critical level during their lifetime. The difference is that with incomplete penetrance the phenotype may never appear however long the person lives, but with delayed onset it will inevitably appear if the person lives long enough.

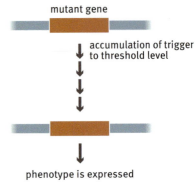

FIGURE 15.31 Delayed onset.

KEY CONCEPTS

- Mendel's work established the principles of genetic analysis, which makes use of data from breeding experiments to study the inheritance of genes.
- Many pairs of alleles display a relationship in which one allele is dominant and the other is recessive. The phenotypic effect of the recessive allele is masked in a heterozygote.
- If a pair of alleles display a simple dominant–recessive relationship, then usually the recessive allele codes for a nonfunctional protein.
- Some pairs of alleles display incomplete dominance. The phenotype of the heterozygote is intermediate between those of the two homozygotes.
- Some alleles are lethal when present in the homozygous condition. A few are lethal when present in a heterozygote. These usually display delayed onset.
- Codominance occurs when both alleles are active in the heterozygous state.
- Many phenotypes can be linked to the activity of just a single gene, but there are numerous others that result from interactions between two or more separate genes. Understanding these interactions is crucial to our awareness of how the genome as a whole specifies the biological characteristics of an organism.
- The alleles of two different genes can combine in an additive fashion to determine the nature of a phenotype such as flower color.
- Sometimes, the alleles present at one gene mask or cancel those at a second gene. This is called epistasis.
- The phenotypes resulting from a quantitative trait, one specified by many genes acting together, display a Gaussian distribution when examined in the population as a whole.

- Pleiotropy arises when a mutation in a single gene affects more than one trait.
- Alleles that are incompletely penetrant do not always give rise to their associated phenotype, which may never appear however long the organism lives. This is distinct from delayed onset, where the phenotype will inevitably appear if the organism lives long enough.

QUESTIONS AND PROBLEMS

Key terms

Write short definitions of the following terms:

codominance	genotype	phenotype
complementary gene action	heterozygote	pleiotropy
cross-fertilization	homozygote	Punnett square
delayed onset	incomplete dominance	pure breeding
dihybrid cross	incomplete penetrance	quantitative trait
dominant	lethal allele	recessive
epistasis	linkage	segregation
Gaussian distribution	monohybrid cross	self-fertilization
genetic analysis	normal distribution	unit factor

Self-study questions

15.1 Distinguish between the terms 'genotype' and 'phenotype'.

15.2 Carefully explain how Mendel was able to deduce that alleles occur in pairs and that they display dominance and recessiveness.

15.3 Describe the molecular basis for the dominant–recessive relationship between the round and wrinkled pea alleles.

15.4 Use a Punnett square to explain why there is a 3:1 ratio of phenotypes in the F_2 generation of a monohybrid cross.

15.5 Green pods are dominant to yellow pods in pea plants. What will be the phenotypes and genotypes of the F_1 and F_2 plants from a green pod (GG) and yellow pod (gg) cross?

15.6 What would be the outcome of crossing an F_1 plant from the cross described in Question 15.5 with the yellow pod parent (this is called a backcross)?

15.7 The genes for pod color and pea shape are on different chromosomes in peas. What would be the ratio of phenotypes in the F_1 and F_2 generations resulting from a cross between a parent plant with green pods and round peas, genotype GGRR, and one with yellow pods and wrinkled peas, ggrr?

15.8 Why did Mendel conclude that different genes segregate independently, and why is this conclusion only partly correct?

15.9 What would be the ratio of phenotypes in the F_1 and F_2 generations resulting from a cross between two parent plants with the genotypes AA and aa, if A and a display incomplete dominance?

15.10 Explain what is meant by the term 'lethal allele'. What is the molecular basis of the lethality of the *aurea* allele of snapdragon?

15.11 What would be the ratio of phenotypes resulting from a cross between a pair of heterozygotes, genotype Aa, if a was a recessive lethal allele?

15.12 What is codominance, and how is it usually detected?

15.13 What can flower pigmentation tell us about interactions between genes?

15.14 Give an example of complementary gene action.

15.15 Describe how fruit color in squash illustrates the principle of epistasis.

15.16 Explain why the phenotypes resulting from a quantitative trait display a Gaussian distribution.

15.17 What is pleiotropy?

15.18 Distinguish between incomplete penetrance and delayed onset.

Discussion topics

15.19 What features are desirable for an organism that is to be used for extensive studies of gene inheritance?

15.20 A cross was set up to study the inheritance of red eyes and white eyes in the fruit fly. The male parent had white eyes and the female parent had red eyes. The F_1 generation resulting from the cross all had red eyes. When the F_1 flies were allowed to interbreed, a F_2 population was obtained that had two red-eyed females to one red-eyed male and one white-eyed male. To gain further data, a second cross was set up between a red-eyed male and white-eyed female. Now the F_1 generation were a 1:1 mix of white-eyed males and red-eyed females, and the F_2 generation comprised a 1:1:1:1 ratio of red-eyed male, white-eyed male, red-eyed female and white-eyed female. Explain these results.

15.21 Most wild mice have black hairs with fine yellow bands, giving them the coat color referred to as agouti, which has evolved to provide the best camouflage in the mouse's natural environment. The black coloration is coded by gene C and the yellow banding by gene A. The following observations have been made:

- Mice with the genotype C_A_ (this notation indicates that each mouse possesses at least one dominant allele for each of the two genes) have black hairs (coded by C) and yellow bands (coded by A) and so have agouti coats.

- Mice that lack the dominant allele of gene A (genotypes C_aa) still have black hairs but this time without the yellow bands. They have black coats.

- Mice with the double recessive genotype, ccaa, have white hairs without yellow bands and their coats are white.

- We would anticipate that mice that lack the dominant allele of gene C but do have at least one dominant allele for gene A (genotypes ccA_) would have white hairs with yellow bands. But, in fact, these mice have pure white hairs exactly the same as the double recessive ones.

Suggest a hypothesis that takes account of each of these observations. How might your hypothesis be tested?

15.22 For several years, it has been possible to screen human DNA for the presence of alleles specifying delayed-onset diseases such as Huntington's. Discuss the ethical issues raised by the discovery that a young healthy individual carries the susceptibility gene for a delayed-onset neurodegenerative disease.

The answers to, or hints on how to answer, the self-study questions and discussion topic questions can be downloaded from the book's product page at this link: www.routledge.com/9781032743530

FURTHER READING

Abbott S & Fairbanks DJ (2106) Experiments on plant hybrids by Gregor Mendel. *Genetics* 204, 407–422. *A translation of Mendel's paper.*

Baur E (1908) Die Aurea-Sippen von Antirrhinum majus. *Zeitschrift für induktive Abstammungs- und Vererbungslehre* 1, 124–125. *The original publication of the lethal aurea allele of snapdragon.*

Cooper DN, Krawczak M, Polychronakos C et al. (2013) Where genotype is not predictive of phenotype: towards an understanding of the molecular basis of reduced penetrance in human inherited disease. *Human Genetics* 132, 1077–1130.

Cordell HJ (2002) Epistasis: what it means, what it doesn't mean, and statistical methods to detect it in humans. *Human Molecular Genetics* 20, 2463–2468.

Fairbanks DJ (2002) *Gregor Mendel: His Life and Legacy*. Buffalo: Prometheus Books.

Fairbanks DJ & Rytting B (2001) Mendelian controversies: a botanical and historical review. *American Journal of Botany* 88, 737–752.

Fincham JRS (1990) Mendel – now down to the molecular level. *Nature* 343, 208–209. *Describes the molecular basis of round and wrinkled peas.*

Gudbjartsson DF, Walter GB, Thorleifsson G et al. (2008) Many sequence variants affecting diversity of adult human height. *Nature Genetics* 40, 609–615.

Mackay TFC & Anholt RRH (2022) Gregor Mendel's legacy in quantitative genetics. *PLoS Biology* 20, e3001692.

Reid JB & Ross JJ (2011) Mendel's genes: toward a full molecular characterization. *Genetics* 189, 3–10.

Inheritance of Genes in Populations

CHAPTER 16

In Chapter 11, we learnt that the DNA molecules present in living cells undergo sequence alterations due to mutation, caused by errors in replication or the effects of chemical and physical mutagens. However, only a very small number of the mutations occurring in a multicellular organism are passed on to the next generation. To be inherited in this way, a mutation must occur in a sex cell, evade the cell's repair enzymes and be passed to a gamete that participates in the formation of a fertilized egg cell (Figure 16.1). Working out how many such mutations are inherited per generation is very difficult, but it is thought that each human being possesses over 100 new mutations that are absent from the somatic cells of their mother and father, and which therefore arose in the sex cells of those two parents prior to the individual's conception. This might sound like a minute degree of change, bearing in mind that the human genome contains 3,100 Mb and only 4% of this comprises coding DNA. Most, if not all, of those 100 unique mutations possessed by an individual will therefore be in intergenic regions, but occasionally one will occur in the coding part of a gene, creating a new allele.

In this chapter, we will explore the possible fates of the new alleles that arise in this way. This means that we must investigate how genes are inherited not by individuals but by populations. Geneticists use the term 'population' to describe a group of organisms whose members are likely to breed with one another. A population could therefore be all the members of a species. Alternatively, a species might be split into two or more

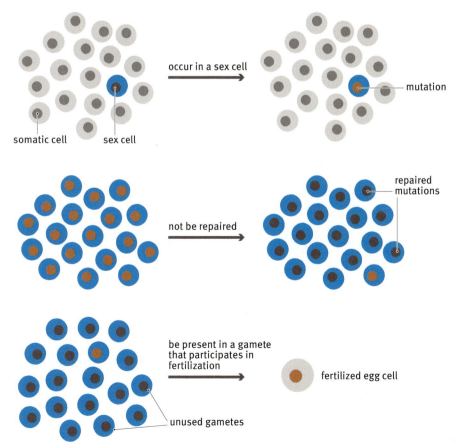

FIGURE 16.1 Only a small number of mutations are passed on to the next generation.

populations with little interbreeding between them. Usually, this is because the populations have become separated from one another by geographical barriers (Figure 16.2).

Within a population, the number of individuals who possess a particular allele can change over time. These changes in allele frequencies are referred to as **microevolution**. They occur as a result of random processes called **genetic drift** and more directed processes called **selection**. We will examine these processes in order to understand how they might affect the fate of a new allele created by mutation.

Populations exist in an environment that is constantly changing. We must therefore examine what happens if there is a sudden reduction in the size of the population, or if a new geographical barrier splits a population into two independent subpopulations. This will lead us to the final topic that we will consider in our study of the gene as a unit of inheritance, the process by which new species arise.

16.1 FACTORS AFFECTING THE FREQUENCIES OF ALLELES IN POPULATIONS

Imagine a gene called A which has two alleles referred to, naturally, as A and a. Imagine also that within a population, 95% of all the alleles for gene A are A, which means that 5% will be a (Figure 16.3a). The frequency of allele A is therefore 0.95 and that for a is 0.05. We usually denote these frequencies as p and q, so in this case $p = 0.95$ and $q = 0.05$. Now imagine that 1,000 years pass and we again determine the frequencies of A and a in the same population. We find that things are different because the value for p is now 0.85 and that of q is 0.15 (Figure 16.3b). The allele frequencies have changed and the population as a whole has undergone microevolution. How has this come about?

Allele frequencies do not change simply as a result of mating

In the early 1900s, when genetics first became a major topic of conversation for scientists as a whole, a common misconception was that recessive alleles should gradually be lost from a population. This notion is most famously (and probably unfairly) credited to Udny Yule, a Scottish statistician, who observed that in humans, abnormally short fingers are dominant to fingers of normal length. The number of people with normal fingers should therefore decrease over time until everybody has short fingers. Is this what geneticists are saying must happen?

Yule's argument was based on the expected phenotypes of the children of parents who were heterozygous for the finger-length gene. If we call this the F gene, then the two alleles will be F for short fingers and f for normal fingers. If the two parents are heterozygotes, then both will have the genotype Ff and both will produce equal numbers of gametes carrying the F and f alleles. A child is therefore equally likely to inherit an F or f allele from either parent and so could have any of the three possible genotypes, FF, Ff or ff. Mendel had shown experimentally with his pea plants that if a sufficient number of crosses occur, then the ratio of genotypes in the next generation will be $1FF : 2Ff : 1ff$. The point that Yule seized on was that when rewritten in terms of the phenotype, this outcome is three short-fingered children to every one child with normal fingers. This is because in the heterozygote the dominant allele, in this case short fingers, determines the phenotype that is observed. If every mating between heterozygotes

(a) all members of a species can form a single population

(b) a species can be split into populations separated by a geographical barrier

FIGURE 16.2 Populations. (a) All the members of a species could be a single population. (b) A species may be split into two or more populations.

(a) population in which $p = 0.95$, $q = 0.05$

(b) population in which $p = 0.85$, $q = 0.15$

FIGURE 16.3 Populations displaying different allele frequencies. Gene A has two alleles, A and a, whose frequencies are p and q, respectively.

results in three children with the dominant phenotype and only one with the recessive phenotype, then surely the number of people displaying the recessive phenotype would decrease over time?

Well, no actually. Yule would only be correct if heterozygotes sought out one another in order to raise families. Within a population, matings occur much more randomly between individuals of different genotypes. Yule's argument was shown to be a fallacy in 1908 by Godfrey Hardy, an English mathematician, and Wilhelm Weinberg, a German doctor. They applied the Mendelian principle that alleles segregate randomly to a population of organisms rather than just a single mating pair. We can reproduce their logic with our imaginary gene A, with allele A present in the population with a frequency of p, and allele a present at frequency q. As we are dealing with a population, we can look on alleles A and a as forming a pool, from which pairs of gametes are taken, at random, to produce a new individual. We can therefore use a Punnett square to calculate the outcome of these random matings (Figure 16.4). This analysis shows us that the frequencies of the AA, Aa and aa genotypes in the population arising from these random matings will be p^2, $2pq$ and q^2, respectively. In other words,

$$p^2 + 2pq + q^2 = 1$$

This is the **Hardy–Weinberg equilibrium**. The important point is that the equilibrium remains the same however many matings occur.

The Hardy–Weinberg equilibrium therefore shows that mating, by itself, has no effect on the allele frequencies present in a population. But in Figure 16.3, we considered a scenario in which allele frequencies *did* change over time, and we have learnt that such changes are considered important by biologists because they underlie microevolution. We must therefore consider the factors that disrupt the Hardy–Weinberg equilibrium for a set of alleles within a population and lead to changes in the frequencies of those alleles over time.

Genetic drift occurs because populations are not of infinite size

A central assumption of the Hardy–Weinberg equilibrium is that all possible matings within a population occur at their predicted statistical frequencies. This assumption is only true if the population is of infinite size, because any finite population will be subject to **stochastic effects** – the deviations from statistical perfection that occur with real events in the real

	pA	qa
pA	AA $p \times p = p^2$	Aa $p \times q$
qa	Aa $p \times q$	aa $q \times q = q^2$

FIGURE 16.4 The expected genotype frequencies for a gene with two alleles. The Punnett square shows the frequencies of genotypes that are expected as a result of random matings for a gene whose alleles, A and a, have frequencies of p and q, respectively. The total genotype frequency is $p^2 + pq + pq + q^2 = 1$. This frequency remains unchanged however many rounds of random mating take place.

world. Statistics predicts the way events are most likely to turn out, but we are all aware that deviations from statistical predictions occur. This is true with a simple process such as spinning a coin and is equally true with a more complex process such as the inheritance of alleles by a population.

Stochastic effects are important in populations because, for each new generation to arise, a finite number of alleles are sampled from the preceding generation. We can refer to the allele frequencies in these samples as pS and qS for alleles A and a, respectively. The allele frequencies in the sample *might* be precisely equal to the allele frequencies in the population as a whole, so that $pS=p$ and $qS=q$, but this is by no means certain. Any deviation from $pS=p$ and $qS=q$ will lead to a change in the allele frequencies of the succeeding generation (Figure 16.5). This process is called **genetic drift**.

No real population can be of infinite size, but larger populations are closer to infinite size than smaller ones. We would therefore expect that samples taken from larger populations will, on average, deviate from $pS=p$ and $qS=q$ less than samples taken from smaller populations (Figure 16.6). If this assumption is true then genetic drift should have a more dramatic effect on the allele frequencies in smaller populations. Computer simulations have shown that this prediction is indeed correct (Figure 16.7). In these simulations, the effects of genetic drift are modeled for an allele that begins with a frequency of $p=0.5$ in populations of different sizes. When the population number, N, is set at 20, the allele frequency fluctuates dramatically, whereas when $N=1,000$ the fluctuation, although still occurring, is less extreme.

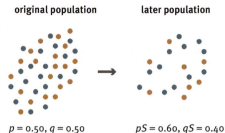

$p = 0.50, q = 0.50$ $pS = 0.60, qS = 0.40$

FIGURE 16.5 A possible effect of sampling on allele frequencies. The sampling that occurs when each new generation arises might affect the allele frequencies in a population. This example considers a single gene with two alleles A and a. The frequencies of A and a in the original population are $p=0.50$ and $q=0.50$, respectively. However, the frequencies in the sample (i.e. the later population) are different, $pS=0.60$ and $qS=0.40$.

Alleles become fixed at a rate determined by the effective population size

The simulations shown in Figure 16.7 illustrate an important outcome of genetic drift. Over time allele frequencies will drift such that at some point one will become lost from the population. The corollary is that the other allele reaches **fixation**, the situation when all members of the population are homozygous for that allele. In Figure 16.7, this occurred after 26 generations in both simulations with a population size of 20. It did not occur

(a) sampling a large population

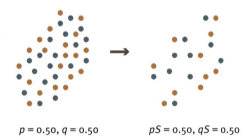

$p = 0.50, q = 0.50$ $pS = 0.50, qS = 0.50$

(b) sampling a small population

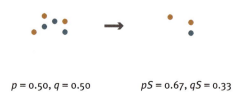

$p = 0.50, q = 0.50$ $pS = 0.67, qS = 0.33$

FIGURE 16.6 Sampling effects are greater with a small population. Again, we consider a single gene with two alleles A and a, whose frequencies are p and q in the original population, and pS and qS in a sample that is half the size of the original population. (a) and (b) illustrate the expectation that a sample taken from a larger population will display a smaller deviation in allele frequencies than a sample taken from a smaller population.

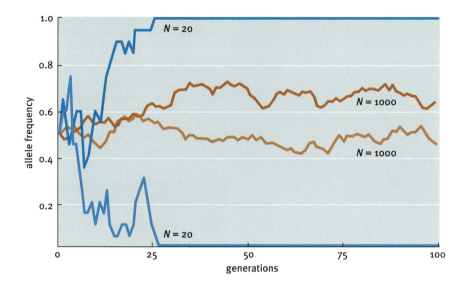

FIGURE 16.7 **Computer simulation of genetic drift.** Four simulations are shown for one member of an allele pair with a starting frequency of $p = 0.5$. Two of the simulations assume a population size, N, of 20 and two assume a population of 1,000. (From M. A. Jobling, M. Hurles & C. Tyler-Smith, *Human Evolutionary Genetics*. New York: Garland Science, 2003.)

at all during the timespan of the simulations when the population size was 1,000, but the statistical prediction is that one member of a pair of alleles will always become fixed as a result of genetic drift. This has been confirmed by the results of simulations that have been allowed to run for a greater number of generations.

On average, therefore, the time to fixation is longer in larger populations. We might imagine that this time to fixation would be *proportional* to the population size, but this is not strictly correct. All of the issues that we have so far considered regarding the inheritance of alleles within a population assume that all members of the population participate in reproduction so that their alleles contribute to the next generation. This is certainly not true of human populations and it is equally untrue of most other animal populations. In any population, there will be some members who have not yet reached reproductive age and others who are too old to contribute their alleles to the next generation (Figure 16.8).

The part of the population that does participate in reproduction is called the **effective population size**, N_e. It is not simply the number of individuals of reproductive age because the figure is influenced by a variety of factors including the relative numbers of males and females, but it is almost always smaller than the total population size and often is substantially smaller.

The more rapid loss and fixation of alleles in smaller populations is one of the reasons why the population size of an endangered species is so critical. We become concerned not simply that the last remaining individuals will die but also that a small population might quickly lose genetic variability. Without variability, the population might not be able to respond to a disease which, in a larger and more variable population, would not have such a devastating effect because some individuals will have a greater

FIGURE 16.8 **Only some members of a population can participate in sexual reproduction.**

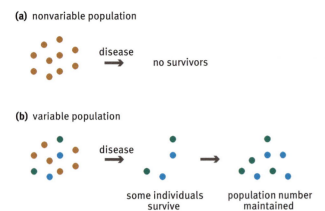

FIGURE 16.9 A possible problem for a population with low genetic variability. Individuals with the orange genotype are susceptible to disease. (a) The population has low variability. When disease strikes, there are no survivors. (b) The population has a greater degree of variability. Disease kills the individuals with the orange genotype, but other individuals survive and are able to re-establish the original population number.

tolerance and hence be able to survive (Figure 16.9). Understanding the effective population size of an endangered species, and its degree of genetic variability, is essential for designing planned breeding programs aimed at rescuing the species from impending extinction (Figure 16.10).

New alleles are continually created by mutation

If genetic drift leads inevitably to the fixation and elimination of alleles, then why do populations display any genetic variability at all? We might expect that, over time, all variation will be lost because all but one of the alleles for any particular gene will become eliminated. The answer is, of course, that new alleles are continually created by mutation.

Initially, a new allele will have a very low frequency in the population as a whole – indeed it will be possessed by just one individual. The chances of it being eliminated by drift are therefore relatively high. Elimination is not, however, inevitable, and might not occur immediately. Computer simulations show that some new alleles increase in frequency to such an extent as a result of drift that eventually they in turn become fixed in the population (Figure 16.11).

The creation of new alleles by mutation, and the subsequent drift of those alleles to fixation or elimination, means that over time the genetic features of a population will change. These molecular events enable a population to undergo microevolution, but in the absence of any other forces the rate of microevolutionary change will be very slow. In real populations, additional factors contribute to the rate of microevolution by bringing about changes in allele frequencies. The most important of these is **natural selection**.

Natural selection favors individuals with high fitness for their environment

Darwin's evolutionary theory holds that natural selection favors individuals with characteristics that provide them with a reproductive advantage over other members of a population. This reproductive advantage can take a number of different forms. There can be an increased likelihood that an individual survives to reproductive age, an increased likelihood that the individual finds a mate, or increased fertility, so that a greater proportion of an individual's matings are productive (Figure 16.12).

The outcome of natural selection is that advantageous traits increase in a population over time, because the individuals possessing those traits, being more successful at reproduction, make a greater contribution to later generations than do those individuals with less advantageous traits

FIGURE 16.10 The black-footed ferret, a successful captive breeding program. The black-footed ferret, *Mustela nigripes*, has been rescued from extinction by a captive breeding program designed to increase the genetic diversity of the population. By 1981, the species had been reduced to 18 individuals in a single wild population near Meeteetse, Wyoming. The breeding program has enabled new populations to be established at various places in western North America, with approximately 370 animals now living in the wild.

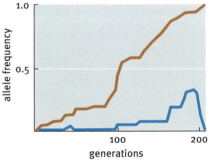

FIGURE 16.11 Possible fates of a new allele. The orange line shows a new allele that eventually becomes fixed in its population. The blue line shows the more common situation where the allele is lost.

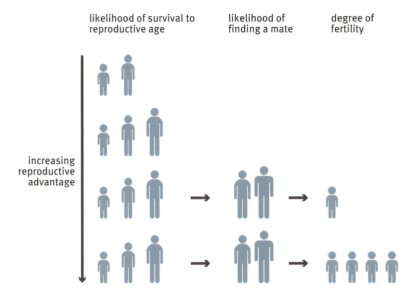

FIGURE 16.12 **Various factors that influence the reproductive advantage of an individual.**

(Figure 16.13). The degree of reproductive advantage possessed by an individual member of a population is called its **fitness**. A key feature of fitness is that it varies according to the environment. An individual who has high reproductive success under one set of conditions might have lower reproductive success if the environment changes.

The importance of the environment in determining the fitness of an individual is illustrated by one of the classic examples of natural selection, melanism of the peppered moth, *Biston betularia* (Figure 16.14). This species is common throughout Britain and until the mid-19th century existed predominantly as a light-colored form, the variety called *typica*. In 1848, a dark-colored variety, *carbonaria*, was first observed near Manchester, which at that time was one of the most industrialized cities in Europe. The dark body and wings of *carbonaria* moths are caused by increased synthesis of the dark pigment melanin in the skin cells.

During the second half of the 19th century, the *carbonaria* version of the peppered moth became more frequent, to the extent that by 1900 over 90% of the moths in the Manchester area had dark bodies and wings

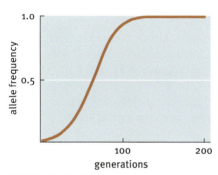

FIGURE 16.13 **The effect of positive natural selection on an allele.** An allele that specifies an advantageous trait can display a relatively rapid increase in frequency due to positive natural selection. Compare this plot with those in Figure 16.11, where the new alleles were not subject to selection.

FIGURE 16.14 **Pale and dark forms of the peppered moth, *Biston betularia*.** (From Wikimedia Commons, courtesy of Siga, published under CC BY-SA 4.0 license.)

(Figure 16.15). This increase is put down to natural selection on the grounds that the dark coloration provides a camouflage when the moth alights on a tree trunk or other structure that has become blackened due to the airborne pollution typical of industrial areas in the 19th century. In contrast, the lighter *typica* version of the peppered moth is more conspicuous against a blackened background and so is more likely to be seen and eaten by a passing bird. The dark-winged moths are therefore more likely to survive until reproductive age, and their progeny will therefore increase in frequency over time. Similar increases in the frequencies of dark versions of other species of moth were recorded in industrial areas elsewhere in Europe and in North America.

We do not know when the *carbonaria* mutation occurred in peppered moths, but we suspect that it was sometime before the Industrial Revolution. In the pre-industrial environment, these dark moths had relatively low fitness and hence were uncommon in the population as a whole. Industrialization, however, resulted in a change in the environment, such that the fitness of the *carbonaria* moths increased and that of the *typica* version decreased. Consequently, the dark moths increased in numbers and became more frequent in the population. In rural parts of Britain, where there was no industrial pollution and hence the environment did not change, the fitness of *carbonaria* moths remained low and their numbers did not increase during the 19th century. During the 20th century, when air pollution became reduced in Britain due to stricter controls on factory emissions, *carbonaria* moths became less frequent again, their fitness declining as the environment changed back to its less blackened, pre-industrial state.

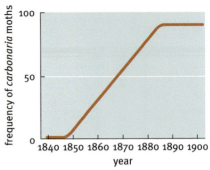

FIGURE 16.15 The impact of natural selection on *carbonaria* moths. The graph shows the change in frequency of the *carbonaria* version of the peppered moth that occurred in the vicinity of Manchester, England, during the nineteenth century.

Natural selection acts on phenotypes but causes changes in the frequencies of the underlying genotypes

Natural selection is a biological rather than genetic process. It acts on the phenotypes of individuals, as it is the phenotype that determines the fitness of the individual for its environment. Natural selection does not act directly on genes, but the changes that it brings about within a population have important effects on allele frequencies. To explore this point, we will look at the genetic changes occurring in peppered moth populations as a result of selection for the *carbonaria* variety.

Crosses between *carbonaria* and *typica* moths show that the *carbonaria* allele is dominant and the *typica* allele recessive. From what we have learnt about the molecular basis of dominance and recessiveness, this result will not be surprising. We would anticipate that an allele specifying synthesis of a dark pigment would dominate over one specifying either synthesis of a light-colored pigment or synthesis of no pigment at all.

If we denote the pigment synthesis gene as M, for melanin, and its alleles *M* and *m*, then *carbonaria* moths would have the *MM* and *Mm* genotypes, and *typica* moths would be *mm*. Selection for the *carbonaria* form in industrial areas therefore increases the frequencies of the *MM* and *Mm* genotypes in the population and decreases the frequency of *mm*. This, in turn, means that the *M* allele becomes more frequent, and the *m* allele less so (Figure 16.16).

Continuing selection will eventually drive one allele to fixation and the other to extinction. Exactly how long this takes depends on the relative fitness of the different phenotypes specified by the alleles (Figure 16.17). Fitness is very difficult to measure with any degree of accuracy, especially in natural populations, as it requires a knowledge of the average number of offspring produced by individuals of each genotype. Most studies of the

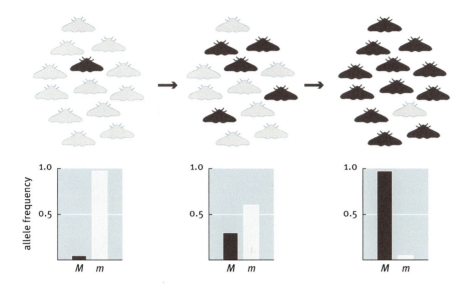

FIGURE 16.16 **The impact of natural selection on allele frequencies.** The melanin biosynthesis gene has two alleles *M* and *m*. Selection for the *carbonaria* phenotype of the peppered moth results in an increase in the frequency of the dominant allele *M*.

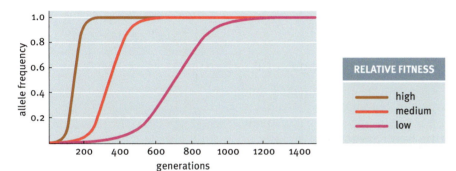

FIGURE 16.17 **The effect of relative fitness on fixation rate.** The rate at which the most advantageous of a pair of alleles becomes fixed in a population depends on the degree of relative fitness of the phenotype that it specifies compared with the phenotype specified by the less advantageous allele. The greater the degree of relative fitness, the more rapid the fixation of the advantageous allele.

influence of fitness on allele frequencies have therefore made use of computer simulations.

In these simulations, relative fitness is expressed as the **selection coefficient**, the most successful genotype being assigned a selection coefficient of 0, and the least successful given a value of 1. Individuals with the latter genotype would never produce progeny, perhaps because the genotype is lethal. Figures in between represent intermediate degrees of fitness. A selection coefficient of 0.05, for example, would be given to a genotype that confers a fitness level that is 95% that of the most successful genotype.

There are different types of natural selection, with different outcomes

The prevalence of *carbonaria* moths in an industrial setting that confers an advantage on insects with dark coloration is one example of natural selection. The outcome is a population shift in the light-to-dark direction. This is called **directional selection** and occurs when a single extreme phenotype is favored over all others (Figure 16.18a).

Directional selection is a common response to an environmental change but it is not the only way in which natural selection can act on a population. The phenotypes displayed by natural populations are often intermediate between the possible extremes (Figure 16.18b). One example is the gray coat color displayed by many rodents and other small ground-dwelling animals. In this environment, the gray color provides a degree of protection against predators as it allows the animal to blend in with the vegetation and soil. In contrast, animals with very light or very dark coats would stand out more and be at a greater risk of predation. This situation, where both extreme phenotypes are selected against, is called **stabilizing selection**.

The opposite of stabilizing selection is **disruptive selection**. Now the intermediate phenotype is absent and the natural population is divided into two groups, each displaying one of the two extreme phenotypes (Figure 16.18c). Disruptive selection was once thought to be relatively rare but we now look on it as important in many wild populations, especially ones inhabiting environments that display what are referred to as ecological discontinuities. An example is provided by stick insects that live on plants that have light green leaves and darkly colored stems. The stick insects themselves are either light green or brown, occupying the leaves and stems, respectively (Figure 16.19). Insects of intermediate color, which have no place to hide, have a lower fitness compared with green or brown insects and so are at a selective disadvantage. The population as a whole therefore occupies the two extreme phenotypes. It has been argued that the division of a population into two groups by disruptive selection can be a prelude to speciation, an issue we will explore later in this chapter.

Heterozygotes sometimes have a selective advantage

The link between selection and genotype can be quite complex, and there are examples where the most successful genotype is not one of the homozygous forms, but the heterozygote. This is variously called **overdominance**, **heterozygote superiority**, **heterozygote advantage** or **heterosis**.

Human sickle-cell anemia provides an excellent illustration of overdominance. This disease is associated with allele *S* of the β-globin gene, which differs from the normal allele *A* by a single point mutation. This mutation converts the 6th codon in the gene from one that specifies glutamic acid in the normal allele to one that codes for valine in the sickle-cell version (Figure 16.20). The resulting change in the electrical properties of the hemoglobin molecule causes the protein to form fibers when the oxygen tension is low. In the homozygous form *SS*, the red blood cells take on a

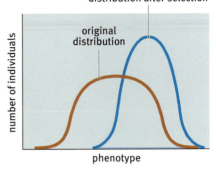

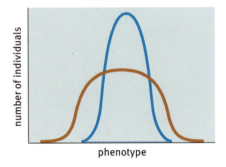

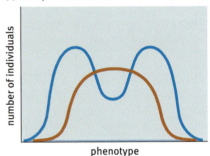

FIGURE 16.18 Three types of natural selection. (a) Directional selection, which favors a single extreme phenotype. (b) Stabilizing selection, where both extreme phenotypes are selected against. (c) Disruptive selection, where either of two extreme phenotypes are favored.

FIGURE 16.19 An example of disruptive selection. This stick insect population comprises light green insects that live on the leaves of the host plant, and brown ones that live on the darker stems.

sickle shape and tend to become blocked in capillaries, where they break down. The resulting sickle-cell anemia is fatal without treatment.

Heterozygotes, with the AS genotype, suffer less extensive breakdown of red blood cells and a milder form of anemia. We might therefore anticipate that the normal homozygous genotype AA would be the most successful and would have a selection coefficient of zero, with the SS genotype displaying a coefficient close to one, and AS at some intermediate value. This is indeed the situation in some parts of the world, but not in those areas where there is a high incidence of malaria. In those regions, the most successful genotype is AS. In malarial parts of Africa, for example, the selection coefficients are 0 for AS, 0.1 for AA and 0.8 for SS. This means that the fitness of AA (a homozygote with normal hemoglobin) is nine-tenths that of AS (a heterozygote with mild anemia).

The explanation of this unexpected result is that the malaria parasite *Plasmodium falciparum* finds it less easy to multiply within the bloodstream of an AS person, possibly because the low red blood cell count simply reduces the capacity of the parasite to reproduce. Individuals with the AS genotype are therefore more resistant to malaria than AA homozygotes. The increase in fitness conferred by malarial resistance is greater than the decrease in fitness due to the mild anemia associated with the AS genotype. Heterozygotes therefore have a selective advantage compared with either of the homozygous genotypes.

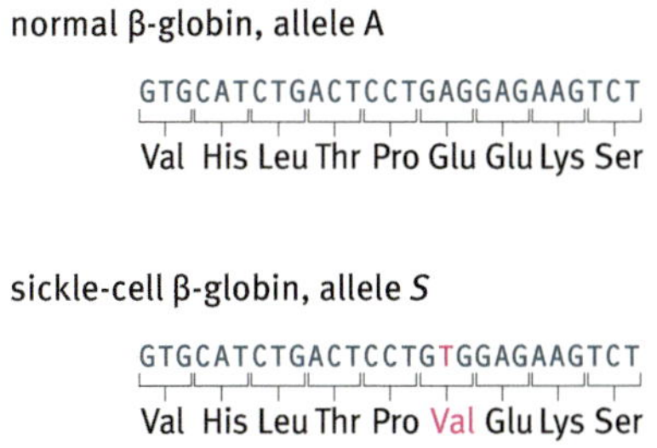

Alleles not subject to selection can hitchhike with ones that are

We must not forget that genes are contained within DNA molecules. This means that selection results not simply in a change in the frequency of an allele, but a change in the frequency of the segment of DNA containing that allele. This is an important point and we will spend a few minutes examining its implications.

Imagine that a mutation occurs that creates a new allele that has a selective advantage. Selection will result in the DNA molecule containing the advantageous mutation increasing in frequency in the population (Figure 16.21a). This increase in frequency will be accompanied by crossing-over events that occur during the production of gametes in individuals who are heterozygous for the allele that is under selection. Some of these crossovers will replace parts of the segment containing the allele with the equivalent stretches of DNA from chromosomes that lack the mutation (Figure 16.21b). In effect, the DNA segment that is increasing in frequency in the population will decrease in length.

Now let us imagine that the advantageous mutation becomes fixed in the population. The important point is that it is not just the mutated nucleotide that becomes fixed, but the DNA to either side. The length of this DNA segment will depend on how many crossovers have taken place since the mutation occurred (Figure 16.21c).

All of this would merely be of academic interest were it not for the possibility that other genes are located within the segment of DNA that becomes fixed. Imagine three adjacent genes, A, B and C. Genes A and C have no impact on fitness and their allele frequencies are changing solely as a result of genetic drift. In one particular DNA molecule, they are present as alleles *a* and *c* which, by chance, are relatively uncommon in the population as a whole. Now imagine that, within this DNA molecule, gene B undergoes mutation, creating a new allele, which we call *b**, which

normal β-globin, allele A

GTGCATCTGACTCCTGAGGAGAAGTCT
Val His Leu Thr Pro Glu Glu Lys Ser

sickle-cell β-globin, allele S

GTGCATCTGACTCCTGTGGAGAAGTCT
Val His Leu Thr Pro Val Glu Lys Ser

FIGURE 16.20 The mutation responsible for sickle-cell anemia. The first nine amino acids of the A and S variants of human β-globin are shown, along with the corresponding nucleotide sequences. Note that the β-globin gene begins with an ATG initiation codon, but the methionine is removed from the protein by post-translational processing, and so is not shown here.

(a) the impact of selection

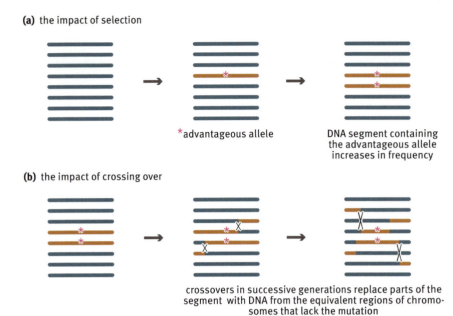

*advantageous allele   DNA segment containing the advantageous allele increases in frequency

(b) the impact of crossing over

crossovers in successive generations replace parts of the segment with DNA from the equivalent regions of chromosomes that lack the mutation

(c) the impact of selection and crossing over

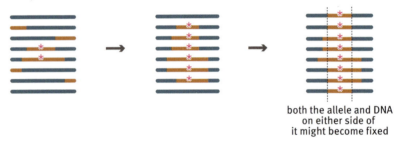

both the allele and DNA on either side of it might become fixed

FIGURE 16.21 **Selection acts on alleles, but those alleles are contained in DNA molecules.** (a) An advantageous mutation appears in a population. Selection acts not just on the new allele that is formed, but on the segment of DNA containing that allele. (b) Crossing over events that occur in successive generations will gradually reduce the length of the DNA segment that is increasing in frequency. (c) Selection may eventually result in the fixation of the advantageous allele. The length of the segment of DNA that is fixed along with the allele depends on how many crossovers have taken place since the mutation occurred.

confers an increase in fitness and so becomes subject to positive selection. The resulting increase in the frequency of allele $b*$ in the population will be accompanied by equivalent increases in the frequencies of a and c, until crossing over occurs in the regions separating gene A from B and/or gene B from C. If allele $b*$ becomes fixed before such crossings over occur, then alleles a and c will also become fixed, even though they themselves confer no increase in fitness (Figure 16.22). This process is called **hitchhiking** – alleles a and c are looked on as having 'hitchhiked' with allele $b*$ to fixation.

Hitchhiking therefore means that positive selection can cause a disproportionate reduction in the genetic diversity of a population. Not only do we lose variability at the gene under selection, but also at adjacent genes. We refer to this as a **selective sweep**.

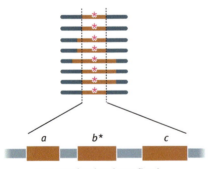

segment that has been fixed contains three genes

FIGURE 16.22 **An example of hitchhiking.** Alleles a and c have become fixed, even though they themselves confer no increase in fitness.

What are the relative impacts of drift and selection on allele frequencies?

Geneticists have long debated the impact of natural selection on populations, in particular, the extent to which changes in allele frequencies are influenced by the directional effects of selection as opposed to the random effects of genetic drift. The answer to this question clearly depends on

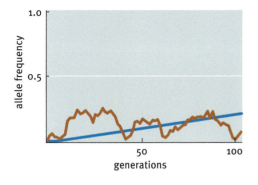
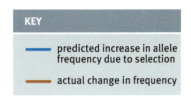

FIGURE 16.23 **The effect of weak selection can be masked by genetic drift.** In this graph, the gradual increase in allele frequency that is predicted to occur over 100 generations is not seen, because this increase is lesser in extent than the fluctuations in allele frequency occurring as a result of drift.

the selection coefficients of the different genotypes specifying an advantageous or disadvantageous trait. If the selection coefficients of the most and least successful genotypes are not hugely different, then the selection pressure will be small and the allele frequencies will change only slowly. If the frequency of a pair of alleles is changing only slowly as a result of selection, then those changes might be masked by the effects of drift (Figure 16.23).

For a new allele created by mutation, the period immediately after it appears in the population is critical. Initially, the allele will have a very low frequency, but if it contributes to a genotype that greatly increases the fitness of the individuals who possess it then we would expect its frequency to increase rapidly as a result of positive selection. But during the period when its frequency is low, a relatively small change resulting from drift could result in the elimination of the allele from the population (Figure 16.24). The accidental deaths of just a few individuals before reaching reproductive age could have such an effect. It is believed that most new advantageous alleles that occur in humans are lost in this way.

Although this area of research is still controversial, many population geneticists favor the **neutral theory**, which postulates that genetic drift is overwhelmingly the most important factor in determining the allele frequencies within a population. Selection has a much more minor impact, with negative selection eliminating the small number of disadvantageous alleles that arise by mutation, but positive selection only rarely resulting in the fixation of an advantageous allele.

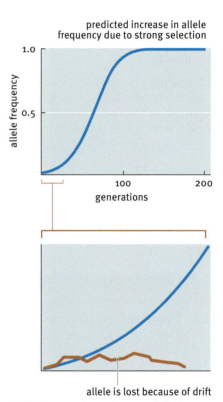

FIGURE 16.24 **Loss of a new allele due to drift.** Even if a new allele is subject to strong selection, it might be lost after a few generations because of the effects of drift.

16.2 THE EFFECTS OF POPULATION BIOLOGY ON ALLELE FREQUENCIES

Our understanding of 'population' has gradually developed as we have progressed through this chapter. We began with the infinitely large population on which the Hardy–Weinberg equilibrium is based, but quickly realized that all real populations are of finite size and this affects the way in which alleles are inherited within populations. When we considered selection for different variants of the peppered moth we became aware that populations do not occupy static ecological niches and that the environmental factors they are exposed to can change over time. From our study of sickle-cell anemia, we learnt that the selective pressures encountered by a population might not be the same throughout the entire geographical range occupied by that population and that different parts of the population might therefore display different allele frequencies. We must now look more closely at some of the complexities that the biological features of populations introduce into population genetics.

Environmental variations can result in gradations in allele frequencies

Many natural populations are large and occupy sizeable geographical regions. The environmental conditions throughout the entire range of a large population are unlikely to be uniform, which means that different parts of the population will be subject to different selective pressures. Those genotypes that provide the greatest fitness in one part of the population might therefore not be the genotypes that provide the greatest fitness in other parts (Figure 16.25). How do these factors affect allele frequencies?

An example is provided by cultivated barley, which is grown in many parts of the world and hence is exposed to many different environmental regimes. In Europe, for example, barley is grown throughout the continent, from the hot, dry areas on the Mediterranean coast all the way to the northern regions where the summers are much cooler and wetter.

In southern Europe, it is advantageous for barley to flower early in the season, from April to May, as this means that seeds are produced before the vegetative parts of the plant begin to wither and die when the temperature rises and the soil becomes dry. In northern Europe, on the other hand, it is better for the plants to continue growing during the cool, wet summer months, as this means that more nutrients can be taken up and a greater number of seeds produced when the plants eventually flower in August and September (Figure 16.26).

The genetic control over flowering time in plants is quite complicated, but plant biologists have discovered that one of the central genes involved in this phenotype is photoperiod-1. This gene enables the plant to respond to daylength, daylength being the trigger that causes the plant to flower. There are two alleles of photoperiod-1, called P and p. The first of these specifies the daylength response that results in early flowering in April and May. Allele p has an altered daylength response, such that flowering is delayed until later in the year.

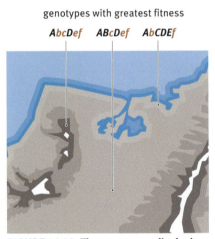

FIGURE 16.25 **The genotypes displaying the greatest fitness might vary across the geographical range of a species.** Here, we consider six genes, A–F, in a population with a varied geographical range. In the mountainous area, the greatest fitness is conferred by genotype *AbcDef*, on the inland plain by *ABcDef*, and in the coastal region by *AbCDEf*.

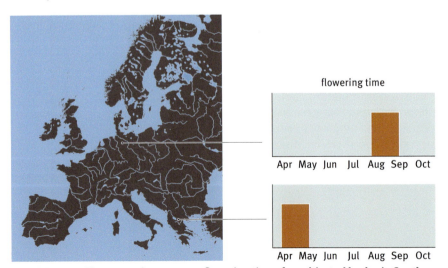

FIGURE 16.26 **The most advantageous flowering times for cultivated barley in Southern and Northern Europe.**

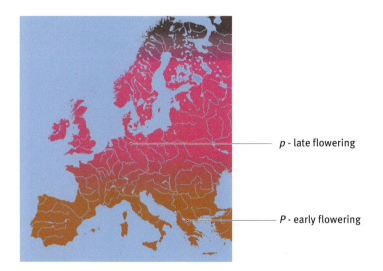

p - late flowering

P - early flowering

FIGURE 16.27 A cline of allele frequencies. The frequencies of the two flowering time alleles in barley cultivars growing in different parts of Europe make up a north-south cline. Allele *p*, associated with late flowering, is more frequent in Northern Europe, and allele *P*, for early flowering, is more frequent in Southern Europe.

(a) small animals initially inhabit a broad plain

When we examine the frequencies of these two alleles in barley cultivars grown in different parts of Europe, we see a gradation from north to south, the frequency of *p* being greatest in the north and that of *P* being greatest in the south (Figure 16.27). This is called a **cline** and is a feature of many alleles subject to environmental selection pressure in populations that have a large geographical range.

There may be internal barriers to the movement of alleles within a population

Real populations often contain subdivisions that are brought about and maintained by geographical or other barriers that reduce the amount of interbreeding that can take place between the subpopulations. To take a simple example, imagine a population of small animals inhabiting a broad plain (Figure 16.28a). Initially, there are no barriers to interbreeding, so any female in the population could mate with any male and vice versa. As time passes, a river begins to run across the plain, dividing the population into two groups (Figure 16.28b). These small animals cannot swim and the population therefore becomes divided into two components that do not interbreed. A barrier to interbreeding has sprung up between them.

(b) a river that cannot be crossed begins to run through the plain

subpopulation 1

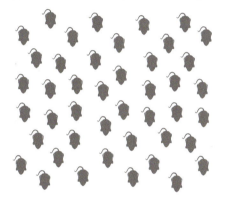

river that cannot be crossed

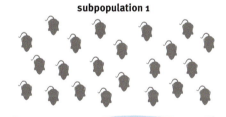

What will be the effects of this population subdivision on the genetic features of the two subpopulations? The process of genetic drift will still operate within the subpopulations, but its effect on allele frequencies will not be the same in both. By chance, some alleles that become more frequent in one subpopulation will become less so in the other (Figure 16.29a). There may even be cases where one allele of a pair becomes fixed in one subpopulation, and the other allele becomes fixed in the second subpopulation (Figure 16.29b).

If the effective sizes of each subpopulation are similar, then this differential allele fixation would result in the unusual situation where the two alleles have approximately equal frequencies in the population as a whole, but there are no heterozygotes present (Figure 16.30). This pattern would never arise in a single interbreeding population. If there is no barrier to interbreeding within a population then, for any pair of alleles of approximately equal frequency, there will always be many heterozygotes.

The example we have used is an extreme one where the barrier to interbreeding is clearly visible (it is a large river) and it is absolute (the animals cannot swim). In nature, subpopulations might be much less easy to discern, because the barriers between them are less obvious geographical

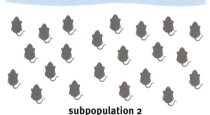

subpopulation 2

FIGURE 16.28 A population subdivision. (a) A population of small animals inhabits a broad plain. Initially, all individuals within the population have access to one another for breeding purposes. (b) If a river begins to flow through the plain, and if the animals are unable to swim, two subpopulations will soon develop on either side of the river that are unable to interbreed with each other.

(a) allele frequencies change independently in the two subpopulations

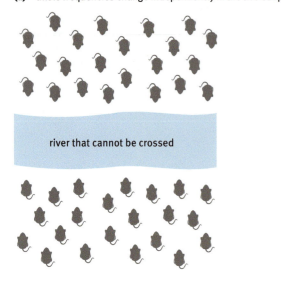
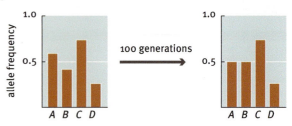
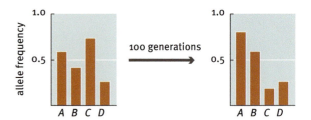

(b) an allele might be fixed in one subpopulation but lost in the other

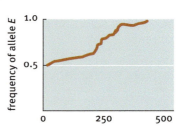
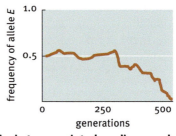

FIGURE 16.29 Drift will have different effects on allele frequencies in two non-interbreeding populations. (a) The changes in frequency for four alleles, *A–D*, are shown. After 100 generations, the allele frequencies are different in the two subpopulations. (b) The frequency of allele *E* is followed in the two subpopulations. In one subpopulation, the allele becomes fixed, but in the other it is lost.

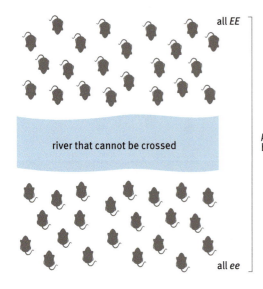

all *EE*

$pE = 0.5$, $pe = 0.5$ but no heterozygotes

all *ee*

FIGURE 16.30 Two alleles have equal frequencies but there are no heterozygotes. In this divided population, the two alleles *E* and *e* have the same frequency when the population as a whole is considered, but there are no heterozygotes.

features or because the barriers have a more subtle ecological context. In addition, most natural barriers are not absolute, and there is usually at least a limited amount of interbreeding between members of different subpopulations. How then can we detect the presence of subpopulations?

Even if there is some interbreeding between subpopulations, allele frequencies are still expected to change in different ways because of the random nature of genetic drift. The differences between the predicted and observed numbers of heterozygotes can therefore be used to determine if population subdivisions exist. This is the basis of the various **fixation indices**, which compare the heterozygosity – the proportion of heterozygotes – within different components of a population. The most commonly used fixation index is the F_{ST} **value**, calculated as

$$F_{ST} = \frac{H_T - H_S}{H_T}$$

In this equation, H_T is the heterozygosity that is predicted from the allele frequency in the population as a whole, and H_S is the average heterozygosity of the suspected subpopulations. If the F_{ST} value is less than 0.05, then the population as a whole is looked on as having little substructure, but if the value is over 0.25 then there is high genetic differentiation and relatively well-defined subpopulations are present.

Transient reductions in effective population size can result in sharp changes in allele frequencies

Another common feature of real populations is that their sizes might change over time. In particular, a change in the environment or a reduction in the available food supply can result in a rapid and severe drop in the number of individuals in a population. If the cause of the population crash is removed – the environmental conditions return to a more friendly state or the food supply increases – the population can grow again, possibly back to its original size. The period between the decline and recovery is called a **bottleneck** (Figure 16.31).

A number of species are thought to have gone through bottlenecks during the last few thousand years, including the elephant seal and African cheetah (Figure 16.32). The human population is thought to have declined to just 1,280 breeding individuals some 850,000 years ago, during a period of global cooling. Additionally, we are aware of many species whose population numbers are currently declining and for which the best future scenario is a severe bottleneck, the alternative being extinction.

A bottleneck has an important influence on the genetic features of a population because it can result in a change in allele frequencies. We are familiar with the principle from our daily experiences. Imagine there are

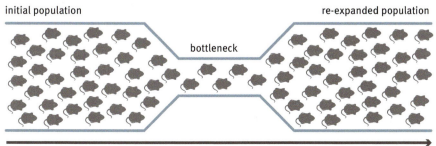

FIGURE 16.31 A population bottleneck.

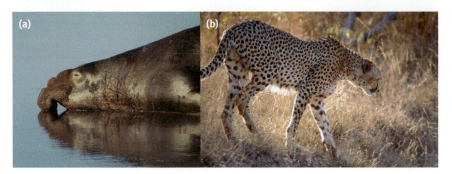

FIGURE 16.32 Two species that have gone through a population bottleneck during the last few thousand years. (a) The northern elephant seal (*Mirounga angustirostris*). (b) The African cheetah (*Acinonyx jubatus*). The elephant seal population is known to have declined to about 30 during the nineteenth century due to hunting, and although the population has now recovered it displays low genetic variability. The African cheetah population also has low genetic variability, but the bottleneck that caused this occurred in the more distant past. (a, courtesy of Robert Schwemmer, National Oceanic and Atmospheric Administration; b, courtesy of Thomas A. Hermann)

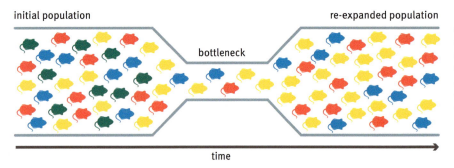

FIGURE 16.33 A bottleneck can cause a change in allele frequencies. Four alleles, each represented here by an animal of a different color, have approximately equal frequencies in the initial population. After the bottleneck, the frequencies are different. Note that the green allele is completely eliminated in the bottleneck.

120 balls in a bag, 30 blue ones, 30 red, 30 green and 30 yellow. Twelve balls are sampled at random from the bag. We do not expect those twelve always to comprise three of each color, instead we expect some variation and quite possibly the sample will not include any balls of one particular color. Exactly, the same principles apply to the sampling that occurs when a population goes through a bottleneck (Figure 16.33). The allele frequencies within the bottleneck might be quite different from those displayed by the population before its numbers began to decline, and some alleles, especially rarer ones, might be completely lost. The new expanded population that develops after the bottleneck therefore has different genetic features compared with the population that existed before the bottleneck. The allele frequencies will have shifted and there will probably have been an overall decline in variability due to allele loss. A bottleneck therefore causes a rapid change in the genetic features of a population.

Founder effects are similar to bottlenecks but take place in different contexts. A founder effect occurs when a few members of a population split away and initiate a new population (Figure 16.34). Typically, this happens when part of the population moves to new territory, previously unoccupied by that species, an event we call **colonization**. Again, there is a sampling effect as the small number of individuals who act as founders might not possess all the alleles present in the parent population and might display different allele frequencies for those that they do possess. Founder events have been important in human prehistory, as the gradual movement of our species from our original homeland in Africa into Asia, Europe and the New World is thought to have involved a series of

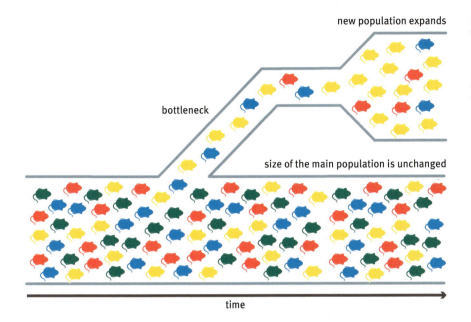

FIGURE 16.34 A founder effect. The founder effect shown here involves a tight bottleneck, as occurs when only a few individuals split from the main population. As in Figure 16.33, the four alleles have approximately equal frequencies in the main population.

colonizations, each involving a relatively small number of people. The change in allele frequency that occurred during each of these founder events is discernible, to some extent, in modern human populations. This enables the patterns of these past colonizations, and the approximate time when each one took place, to be deduced.

A population split can be the precursor to speciation

How do the creation of new alleles, and changes in the frequencies of alleles, relate to the evolution of new species? Our modern concept of species is based on reproductive incompatibility. Two members of the same species, of opposite sexes, are able to reproduce and give rise to viable offspring, whereas members of different species cannot interbreed, or if they do the progeny are sterile. To take a simple example, male and female horses produce viable baby horses even if the male and female parents are taken from different parts of the world. This is because all horses are members of the single species *Equus caballus* (Figure 16.35). In contrast, donkeys form a different species, *Equus asinus*. Donkeys and horses can produce offspring – a mule if the horse is the female parent, or hinny if the horse is the father – but these offspring are sterile. This is an example of **postzygotic isolation**, where the barrier to reproduction occurs after a productive mating. This contrasts with **prezygotic isolation**, where there is no mating or mating produces no offspring, not even sterile ones. To understand the genetic basis of **speciation,** we must therefore make a link between the microevolutionary events that we have studied in this chapter, involving the factors that influence allele frequencies within populations, and the processes that result in reproductive incompatibility.

A population split is often looked on as the starting point for speciation. Evolutionary biologists recognize two ways in which a split can occur, giving rise to two different modes of evolution. The first involves a population being split into two subpopulations by a geographical barrier, such as the river we envisaged in Figure 16.28. The split itself would not constitute speciation, because at this stage members of the two populations, if brought together, could still mate productively. However, the split does mean that there is no interbreeding between the two populations, which are now free to pursue their own microevolutionary pathways. These

FIGURE 16.35 **Horses and donkeys are different species.** Three members of the species *Equus caballus*: (a) American quarterhorse, (b) Shetland pony, (c) Mongolian horse; and (d) one member of *Equus asinus*, the donkey. (d, courtesy of BLM/Arizona.)

might eventually involve differential genetic changes that confer post- or prezygotic isolation. This is called **allopatric speciation**.

In contrast, **sympatric speciation** does not involve a geographical split. Instead, non-interbreeding subpopulations arise within a larger population that occupies a single geographical region. This can occur if disruptive selection has split the population into two groups with different genotypes (see Figure 16.18c). Again, the initial split does not constitute speciation as at this stage the members of the two subpopulations will still be able to interbreed. However, the nature of the phenotype might mean that interbreeding is rare. An example would be when a plant has split into two populations with different flowering times (Figure 16.36). The resulting absence of interbreeding means that, as with a geographical split, the two subpopulations follow different microevolutionary pathways.

The processes by which a population can split into two reproductively isolated groups are therefore well understood. A greater challenge is presented by the nature of the genetic changes that eventually lead to post- or pre-zygotic isolation. Several studies have looked at the possible role that chromosome pairing has on hybrid sterility. Sterility of mules and hinnies is thought to be due to horses and donkeys having different numbers of chromosomes – 64 for horses and 62 for donkeys. This difference means that although horse and donkey gametes can fuse to form a hybrid embryo that develops into a mule or hinny, the hybrid cannot

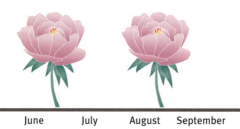

FIGURE 16.36 **Sympatric speciation.** A plant has split into two populations that do not interbreed because they flower at different times of the year.

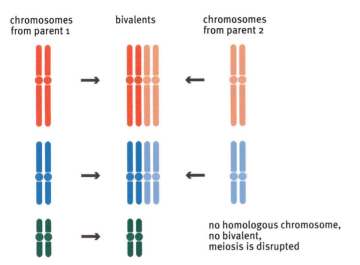

FIGURE 16.37 Hybrid sterility can arise if meiosis is disrupted. In this example, meiosis is disrupted because one parent has an additional chromosome not possessed by the second parent.

itself produce gametes because the chromosomes cannot pair effectively during meiosis (Figure 16.37). Difficulties in pairing can also arise if the two sets of chromosomes display rearrangements such as inversions and translocations. This type of genetic change might therefore be one of the factors that drives reproductive isolation and resulting speciation.

Other researchers are exploring the possibility that prezygotic isolation could be caused by a more simple interaction between pairs of genes. The hypothesis is that genes A and B have alleles, *a* and *b*, that cannot exist in the same cell because the proteins coded by these alleles interact in some deleterious way. In other words, the genotypes *AaBb*, *aabb*, *Aabb* and *aaBb* are lethal and all individuals in a population are *AABB*, *AaBB*, *aaBB*, *AABb* or *AAbb* (Figure 16.38). However, after a population split, allele *a* becomes fixed in one subpopulation and allele *b* in the other subpopulation. This would mean that members of the two populations could no longer interbreed because the offspring would inherit *a* and *b* alleles and hence be unable to survive. This is called **hybrid incompatibility**.

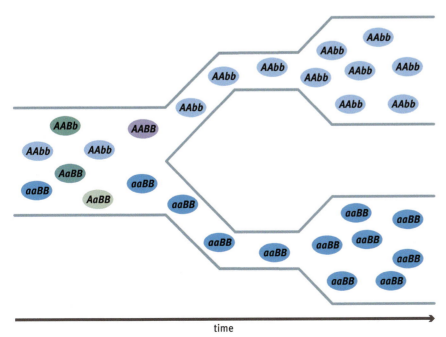

FIGURE 16.38 Hybrid incompatibility. Alleles *a* and *b* are incompatible, so individuals who possess both cannot survive. The starting population comprises individuals with genotypes *AABB*, *AaBB*, *aaBB*, *AABb* and *AAbb*. During a population split, genotypes *AABB*, *AaBB* and *AABb* die out, and *AAbb* becomes fixed in one subpopulation and *aaBB* in the other subpopulation. The two subpopulations cannot now interbreed because the F₁ would all have the incompatible genotype *AaBb*.

KEY CONCEPTS

- A population is a group of organisms whose members are likely to breed with one another.
- The Hardy–Weinberg equilibrium shows that mating, by itself, has no effect on the allele frequencies present in a population.
- Within a population, the number of individuals who possess a particular allele can change over time. These changes in allele frequencies are referred to as microevolution. They occur as a result of random processes called genetic drift and more directed processes called selection.
- New alleles are continually created by mutation.
- Natural selection favors individuals with high fitness for their environment. Natural selection does not act directly on genes, but the changes that it brings about within a population have important effects on allele frequencies.
- The neutral theory postulates that genetic drift is overwhelmingly the most important factor in determining the allele frequencies within a population. Selection has a much more minor impact, with negative selection eliminating the small number of disadvantageous alleles that arise by mutation, but positive selection only rarely resulting in fixation of an advantageous allele.
- Environmental variations can result in gradations in allele frequencies within a population.
- A transient reduction in effective population size can result in sharp changes in allele frequencies. This might occur if the population goes through a bottleneck or a few members leave the population and colonize a new geographical region.
- A population split is often looked on as the starting point for speciation. Different microevolutionary pathways within the two subpopulations might eventually result in differential genetic changes that confer post- or prezygotic isolation.

QUESTIONS AND PROBLEMS

Key terms

Write short definitions of the following terms:

allopatric speciation
bottleneck
cline
colonization
directional selection
disruptive selection
effective population size
F_{ST} value
fitness
fixation
fixation index

founder effect
genetic drift
Hardy–Weinberg equilibrium
heterosis
heterozygote advantage
heterozygote superiority
hitchhiking
hybrid incompatibility
microevolution
natural selection
neutral theory

overdominance
postzygotic isolation
prezygotic isolation
selection
selection coefficient
selective sweep
speciation
stabilizing selection
stochastic effect
sympatric speciation

Self-study questions

16.1 Describe the basis of the misconception that recessive alleles should gradually be lost from a population.

16.2 Explain how the gradual loss of recessive alleles from a population was shown to be a fallacy.

16.3 Why does microevolution occur despite the predictions of the Hardy–Weinberg equilibrium?

16.4 Explain how stochastic effects can lead to genetic drift.

16.5 What is the impact of effective population size on the rate at which alleles become fixed in a population?

16.6 Why is it important to understand the effective population size of an endangered species?

16.7 Explain why most new alleles are rapidly eliminated from a population. Under what circumstances is a new allele likely to increase in frequency?

16.8 Describe how the peppered moth illustrates the principles of natural selection.

16.9 Outline the key features of directional, stabilizing and disruptive selection.

16.10 Giving an example, explain why a heterozygote can be the most successful of a set of genotypes.

16.11 Distinguish between the terms 'hitchhiking' and 'selective sweep'.

16.12 According to the neutral theory, what are the relative impacts of genetic drift and natural selection on microevolution?

16.13 Describe how environmental factors can result in a gradation in allele frequencies.

16.14 How might an internal barrier to interbreeding arise in a population? How might such a barrier result in a situation where a pair of alleles have similar frequencies in the population as a whole, but there are no heterozygotes?

16.15 How are F_{ST} values used to study the internal structure of a population?

16.16 Distinguish between a bottleneck and a founder effect.

16.17 Give examples of species thought to have been through a recent population bottleneck.

16.18 What effect does a bottleneck have on the frequency of alleles in a population?

16.19 Describe the differences between prezygotic and postzygotic isolation.

16.20 Outline two ways in which genetic changes could result in prezygotic and/or postzygotic isolation.

Discussion topics

16.21 Is it possible for a real population ever to display the Hardy–Weinberg equilibrium?

16.22 To what extent do studies of microevolution provide an argument against those individuals who contend that evolution is a fallacy?

16.23 Discuss how a knowledge of the extent of genetic diversity displayed by an endangered species can be used to aid the planning of a breeding program aimed at avoiding that species becoming extinct.

16.24 Can you devise an analysis that would enable the length of time that has elapsed since a bottleneck occurred to be deduced from examination of the allele frequencies in a population?

16.25 Evaluate the hypothesis that prezygotic isolation can be achieved by hybrid incompatibility.

The answers to, or hints on how to answer, the self-study questions and discussion topic questions can be downloaded from the book's product page at this link: www.routledge.com/9781032743530

FURTHER READING

Charlesworth B (2009) Effective population size and patterns of molecular evolution and variation. *Nature Reviews Genetics* 10, 195–205. *A good description of the importance of effective population size in evolutionary studies.*

Cook L & Saccheri I (2013) The peppered moth and industrial melanism: evolution of a natural selection case study. *Heredity* 110, 207–212.

Hardy GH (1908) Mendelian proportions in a mixed population. *Science* 28, 49–50. *First description of the Hardy-Weinberg equilibrium.*

Hartl DL & Clark AC (1997) *Principles of Population Genetics*, 4th ed. Oxford: Oxford University Press.

Holsinger KE & Weir BS (2009) Genetics in geographically structured populations: defining, estimating and interpreting *F*st. *Nature Reviews Genetics* 10, 639–650.

Hu W, Hao Z, Du P et al. (2023) Genomic inference of a severe human bottleneck during the Early to Middle Pleistocene transition. *Science* 381, 979–984.

Jones H, Leigh FJ, Bower MA et al. (2008) Population-based resequencing reveals that the flowering time adaptation of cultivated barley originated east of the Fertile Crescent. *Molecular Biology and Evolution* 25, 2211–2219. *The barley photoperiod-1 allele gradation across Europe.*

Kimura M (1983) *The Neutral Theory of Molecular Evolution*. Cambridge: Cambridge University Press.

Nielsen R (2005) Molecular signatures of natural selection. *Annual Review of Genetics* 39, 197–218. *Explains how examination of DNA sequences and genotypes can reveal genes and larger regions of a genome that are under selection.*

Strasser BJ (1999) 'Sickle cell anemia: a molecular disease'. *Science* 286, 1488–1490. *An account of the discovery of the changes in hemoglobin structure that underlie sickle-cell anemia.*

Templeton RA (1980) The theory of speciation via the founder principle. *Genetics* 94, 1011–1038. *Founder effects.*

HOW GENES ARE STUDIED

17 Mapping the Positions of Genes in Chromosomes

18 Sequencing Genes and Genomes

CHAPTER 17
Mapping the Positions of Genes in Chromosomes

The previous 16 chapters have described our current understanding of the gene as a unit of biological information and as a unit of inheritance. Our focus has been on what we know about genes, and we have paid less attention to the way in which that knowledge has been acquired. In the next two chapters, we will begin to redress the balance by examining some of the most important research methods that are used to study DNA, genes and genomes. In this chapter, we will look at the methods used to map the positions of genes in chromosomes, and in Chapter 18, we will explore how genes and entire genomes are sequenced.

Gene mapping techniques were devised during the early days of genetics and for many years were the only way of studying the distribution and organization of genes in chromosomes. Even after DNA sequencing was invented, gene mapping remained important. As we will see in Chapter 18, it is relatively easy to find the genes in a DNA sequence, if we search the sequence by eye or with the help of a computer. It is much more difficult to assign a function to a gene if all we know about it is its DNA sequence. Gene mapping, on the other hand, locates the positions in chromosomes of genes responsible for known phenotypes, such as flower color or seed shape if plants are being studied. The information from gene mapping therefore enables functions to be assigned to genes that have been identified in a DNA sequence. This is particularly important in human genetics, where the gene being studied is often one responsible for an inherited disease (see Chapter 21).

We begin this chapter by studying the basic methodology for gene mapping in eukaryotes, including the techniques used to map genes in the human genome. We will then examine the rather different methodology used with bacteria. Finally, we will look at a set of techniques that use quite different approaches to map the positions of various sequence features, including but not only genes, in DNA molecules.

17.1 PARTIAL LINKAGE PROVIDES THE BASIS TO GENE MAPPING WITH EUKARYOTES

In Chapter 15, when we studied Mendel's work, we acknowledged that one of his conclusions – that alleles of different genes segregate independently of one another – is only partially correct. This became clear once it was established that genes reside on chromosomes. It was immediately obvious that the number of chromosomes possessed by an organism is considerably fewer than the total number of inheritable traits displayed by that organism. This being so then a single chromosome must carry a large number of genes. Different genes on the same chromosome would not be inherited independently, but instead would be transmitted together. The technical term for this is linkage.

Having established that genes on the same chromosome should display linkage, the early geneticists then complicated the issue by demonstrating that different genes on the same chromosome *can* segregate independently. It was this discovery that led to the methods used in gene mapping.

DOI: 10.1201/9781003473862-20

Linkage is not absolute, it is only partial

One of the first experimental demonstrations of linkage was provided in 1905 by William Bateson, Edith Saunders and Reginald Punnett. They carried out a series of dihybrid crosses along the same lines as Mendel, although with a different plant, the sweet pea. In their experiments, plants with purple flowers and long pollen grains were crossed with plants with red flowers and round pollen grains (Figure 17.1). In the F_1 generation, the plants, as expected, were all of the same phenotype, with purple flowers and long pollen grains. These two traits are therefore dominant. The F_1 plants were then allowed to self-fertilize and 6,950 F_2 plants were obtained. Using our understanding of dominance and recessiveness, we can predict the ratios of phenotypes that these 6,950 plants should display. If the genes for flower color and pollen grain shape lie on different chromosomes, then we would expect the 9:3:3:1 ratio of a typical dihybrid cross. This would give us:

- 3,909 purple, long
- 1,303 purple, round
- 1,303 red, long
- 435 red, round

But what if the two genes are on the same chromosome? They would segregate together, and in effect, we would have carried out a monohybrid cross between the alleles 'purple-flowers-and-long-pollen-grains' and 'red-flowers-and-round-pollen-grains'. We would therefore expect a 3:1 ratio:

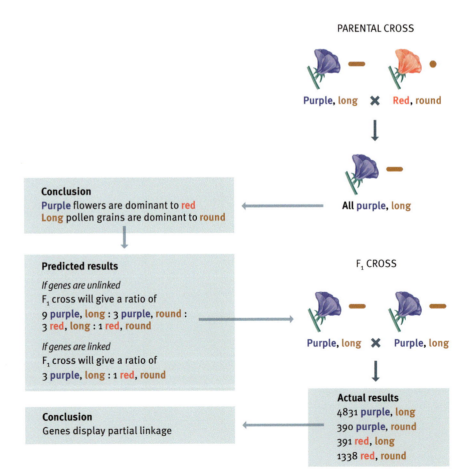

FIGURE 17.1 The discovery of partial linkage. Partial linkage was discovered in the early 20th century. The cross shown here was carried out by Bateson, Saunders and Punnett in 1905 with sweet peas. The parental cross gives the expected result with all the F_1 plants displaying the same phenotype, indicating that the dominant alleles are purple flowers and long pollen grains. The F_1 cross gives unexpected results, as the progeny show neither a 9:3:3:1 ratio (expected for genes on different chromosomes) nor a 3:1 ratio (expected if the genes are completely linked). An unusual ratio is typical of partial linkage.

5,212 purple, long

1,738 red, round

The important thing is that we would not expect to see any nonparental combinations (purple, round and red, long) in the F₂ generation, as these would require independent segregation of two genes that are linked together on the same chromosome.

Which of our predictions is correct? Surprisingly, it is neither. The results obtained by Bateson, Saunders and Punnett were:

4,831 purple, long

3,90 purple, round

391 red, long

1,338 red, round

This is neither a 9:3:3:1 nor a 3:1 ratio. Although the majority of the F₂ plants have the parental combination of characteristics, there are nonetheless a few with the nonparental features, although not as many as we would expect if the genes were unlinked. It is a half-way house: the genes display only **partial linkage**.

Partial linkage is explained by crossing over

The critical breakthrough in understanding partial linkage was achieved by Thomas Hunt Morgan, who made the conceptual leap between partial linkage and the crossing over that can occur between pairs of homologous chromosomes during meiosis.

To understand this, let us consider two genes, each of which has two alleles. We will call the first gene A and its alleles *A* and *a*, and the second gene B with alleles *B* and *b*. Imagine that the two genes are located on chromosome number 1 of the fruit fly *Drosophila melanogaster*. We will use fruit flies in this example because the techniques used to map genes were first applied to them, by Morgan and his students back in the 1910s.

We are going to follow the meiosis of a diploid nucleus in which one copy of chromosome 1 has alleles *A* and *B*, and the second has *a* and *b*. This situation is illustrated in Figure 17.2. There are two different scenarios. In the first, a crossover does not occur between genes A and B. If this is what happens, then two of the resulting gametes will contain copies of chromosome 1 with alleles *A* and *B*, and the other two will contain *a* and *b*. In other words, two of the gametes have the genotype *AB* and two have the genotype *ab*.

The second scenario is that a crossover does occur between genes A and B. This leads to segments of DNA containing gene A being exchanged between homologous chromosomes. The eventual result is that each gamete has a different genotype: 1 *AB*, 1 *aB*, 1 *Ab* and 1 *ab*.

Now think about what would happen if we looked at the results of meiosis in a hundred identical cells (Figure 17.3a). If crossovers never occur, then 200 of the resulting gametes will have the genotypes *AB*, and the other 200 will be *ab*. This is what we expect if the two genes are linked and therefore behave as a single unit during meiosis.

Now consider what will happen if crossovers occur between A and B in some of the meioses. Now the allele pairs will not be inherited as single units. Let us say that crossovers occur during 40 of the 100 meioses (Figure 17.3b). This would give rise to the following gametes: 160 *AB*, 160 *ab*, 40 *Ab* and 40 *aB*. The linkage is not complete, it is only partial. In

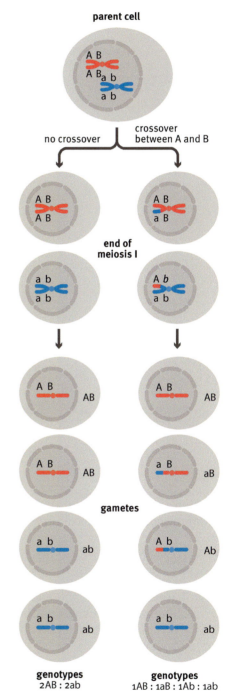

FIGURE 17.2 The effect of crossing over on linked genes. The drawing shows a pair of homologous chromosomes, one red and the other blue. A and B are linked genes with alleles *A*, *a*, *B* and *b*. On the left is meiosis with no crossover between A and B: two of the resulting gametes have the genotype *AB* and the other two are *ab*. On the right, a crossover occurs between A and B: the four gametes display all of the possible genotypes – *AB*, *aB*, *Ab* and *ab*.

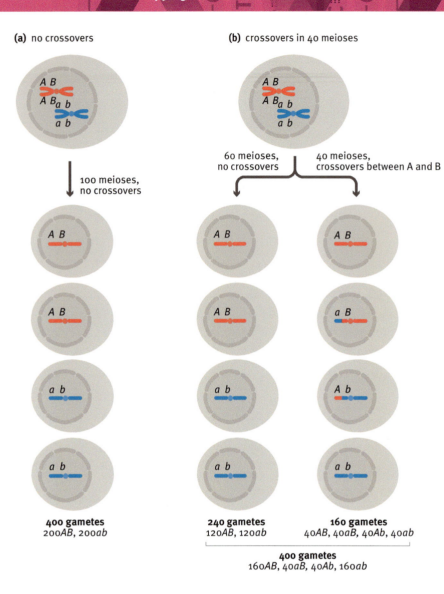

FIGURE 17.3 **The effect of partial linkage on genotype frequencies.** The gametes resulting from 100 meioses are examined. (a) Genes A and B display complete linkage, as no crossovers have occurred between them. (b) The genes display partial linkage, as crossovers have occurred between them during some of the meioses.

addition to the two **parental** genotypes (*AB*, *ab*), we see gametes with **recombinant** genotypes (*Ab*, *aB*).

The frequency of recombinants enables the map positions of genes to be worked out

Now that we appreciate the effect that crossing over has on the genotypes of the gametes resulting from meiosis, we only have to take a small step in order to understand how the positions of genes are mapped on a chromosome.

Let us assume that crossing over is a random event, there being an equal chance of it occurring at any position along a pair of lined-up chromosomes. If this assumption is correct, then two genes that are close together will be separated by crossovers less frequently than two genes that are more distant from one another. Furthermore, the frequency with which the genes are unlinked by crossovers will be directly proportional to how far apart they are on their chromosome. The **recombination frequency** is therefore a measure of the distance between two genes. If you work out the recombination frequencies for different pairs of genes, you can construct a map of their relative positions on the chromosome (Figure 17.4).

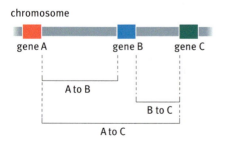

FIGURE 17.4 **The principle behind the use of recombination frequencies to map the positions of genes on a chromosome.**

The very first genetic map was constructed for chromosome 1 of the fruit fly by Arthur Sturtevant, an undergraduate who was working in Morgan's laboratory in 1913. Two of the genes that he studied specified eye color, gene V for vermilion eyes and gene W for ones that were white. A third, gene M, gave rise to miniature wings, and the fourth, Y, gave a yellow body. Sturtevant worked out the recombination frequencies between each pair of genes, obtaining values between 1.3% and 33.7%. From the data, he was able to draw the map shown in **Figure 17.5**.

Morgan's group then set about mapping as many fruit fly genes as possible and by 1915 had assigned locations for 85 of them. These genes fall into four **linkage groups**, corresponding to the four pairs of chromosomes seen in the fruit fly nucleus. The distances between genes are expressed in **map units** or **centiMorgans** (**cM**), with one map unit or cM being the distance between two genes that recombine with a frequency of 1%. According to this notation, the distance between the genes for white eyes and yellow body, which recombine with a frequency of 1.3%, is 1.3 map units.

When we compare Sturtevant and Morgan's map, which was put together a century ago, with the actual locations of these genes, taken from the *D. melanogaster* genome sequence, we see that the positions are more or less correct, but not precisely so (**Figure 17.6**). This is because the assumption about the randomness of crossovers is not entirely justified. Crossovers are less frequent near to the centromere of a chromosome, and some regions, called **recombination hotspots**, are more likely to be involved in crossovers than others. This means that a genetic map distance does not precisely indicate the actual distance between two genes. Also, we now realize that a single chromosome can participate in more than one crossover at the same time, but that there are limitations on how close together these crossovers can be, leading to more inaccuracies in the mapping procedure. Despite these qualifications, **linkage analysis** usually makes correct deductions about gene order, and distance estimates are sufficiently accurate for a genetic map to be valuable when searching for genes in a genome sequence.

The test cross is central to gene mapping when breeding experiments are possible

Most gene mapping projects make use of the procedures first devised by Morgan. These involve analysis of the progeny of experimental crosses set up between parents of known genotypes. This approach is applicable, at least in theory, to all eukaryotes, though practical considerations sometimes make it difficult if not impossible to carry out. The organism must have a relatively short life cycle in order for progeny to be obtained in a reasonable amount of time. It is therefore appropriate for fruit flies, mice and the like, but less so for elephants.

The key to gene mapping is being able to determine the genotypes of the gametes resulting from meiosis (see Figure 17.3). Then it is possible to deduce the crossover events that have occurred, and from these to calculate the recombination frequency for a pair of genes. The complication with a genetic cross is that the resulting diploid progeny are the product not of one meiosis but of two – one in each parent. In most organisms, crossover events are equally likely to occur during the production of both

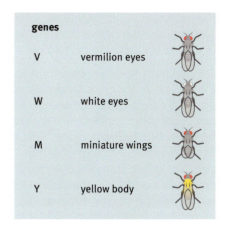

genes		
V	vermilion eyes	
W	white eyes	
M	miniature wings	
Y	yellow body	

recombination frequencies		
between M and V	=	3.0%
between M and Y	=	33.7%
between V and W	=	29.4%
between W and Y	=	1.3%

deduced map positions

Y W   V   M
0 1.3   30.7 33.7

FIGURE 17.5 Sturtevant's map showing the positions of four genes on chromosome 1 of the fruit fly. Recombination frequencies between the genes are shown, along with their deduced map positions.

Y W   V   M   Sturtevant's map

Y   W   V   M   genome sequence

FIGURE 17.6 Genetic maps have inaccuracies. Comparison between Sturtevant's map and the positions of the four genes in the fruit fly genome sequence.

the male and female gametes. Somehow we have to be able to disentangle from the genotypes of the diploid progeny the crossover events that occurred in each of these two meioses.

The solution to this conundrum is to use a **test cross**, as illustrated in Figure 17.7. Here we have set up a test cross to map the positions of genes A (alleles *A* and *a*) and B (alleles *B* and *b*). The critical feature of a test cross is the genotypes of the two parents. One of these parents has to be a **double heterozygote**. This means that all four alleles are present in this parent. Its genotype is therefore *AB/ab*. This notation indicates that one pair of the homologous chromosomes has alleles *A* and *B*, and the other has *a* and *b*. Double heterozygotes can be obtained by crossing two pure-breeding strains, for example, *AB/AB* × *ab/ab*.

The second parent has to be a **double homozygote**. In this parent, both homologous copies of the chromosome are the same. The alleles must be the recessive ones, which means that in the example shown in Figure 17.7, both homologous chromosomes have alleles *a* and *b*, and the genotype of the parent is *ab/ab*.

The double heterozygote has the same genotype as the cell whose meiosis we followed in Figure 17.2. Our objective is therefore to infer the genotypes of the gametes produced by this parent and to calculate the fraction that are recombinants. Note that all the gametes from the second parent (the double homozygote) will have the genotype *ab* regardless of whether crossovers occur during their production. Alleles *a* and *b* are both recessive, so meiosis in this parent is, in effect, invisible when the phenotypes of the progeny are examined. This means that, as shown in Figure 17.7, the phenotypes of the diploid progeny can be unambiguously converted into the genotypes of the gametes from the double heterozygous parent. The test cross therefore enables us to make a direct examination of a single meiosis and hence to calculate a recombination frequency and map distance for the two genes being studied.

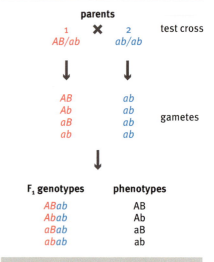

FIGURE 17.7 A test cross. Parent 1 can give rise to gametes with either of four possible genotypes, depending on whether or not crossing over takes place between genes A and B. Parent 2 can only give gametes with the double recessive genotype, regardless of whether there is crossing over. Parent 2 therefore makes no contribution to the phenotypes of the progeny. The phenotype of each individual in the F_1 generation is the same as the genotype of the gamete from Parent 1 that gave rise to that individual.

Multipoint crosses – mapping the positions of more than two genes at once

The power of gene mapping is enhanced if more than two genes are followed in a single cross. Not only does this generate recombination frequencies more quickly, it also enables the relative order of genes on a chromosome to be determined by simple inspection of the data. This is because two crossover events are required to unlink the central gene from the two outer genes in a series of three, whereas either of the two outer genes can be unlinked by just a single crossover (Figure 17.8). Two crossover events are less likely than a single one, so unlinking of the central gene will occur relatively infrequently.

If more than two genes are mapped at once then the breeding experiment is called a **multipoint cross**. A set of typical data from a three-point cross is shown in Table 17.1. A test cross has been set up between a triple heterozygote (*ABC/abc*) and a triple homozygote (*abc/abc*). The most frequent progeny are those with one of the two parental genotypes, resulting from an absence of recombination events in the region containing the genes A, B and C. Two other classes of progeny are relatively frequent, comprising 51 and 63 progeny, respectively, in the example shown. Both of these are presumed to arise from a single recombination. Inspection of their genotypes shows that in the first of these two classes, gene A has become unlinked from B and C, and in the second class, gene B has become unlinked from A and C. The implication is that A and B are the outer genes. This is confirmed by the number of progeny in which marker

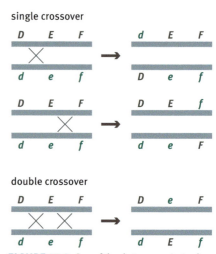

FIGURE 17.8 A multipoint cross. A single crossover is enough to unlink one of the outer genes, but two crossovers are needed to unlink the central gene.

PARTIAL LINKAGE PROVIDES THE BASIS TO GENE MAPPING WITH EUKARYOTES

TABLE 17.1 SET OF TYPICAL DATA FOR A MULTIPOINT CROSS BETWEEN A TRIPLE HETEROZYGOTE (*ABC/abc*) AND A TRIPLE HOMOZYGOTE (*abc/abc*)

Genotypes of Progeny	Number of Progeny	Inferred Recombination Events
ABC/abc, abc/abc	987	None (parental genotypes)
aBC/abc, Abc/abc	51	One, between A and B/C
AbC/abc, aBc/abc	63	One, between B and A/C
ABc/abc, abC/abc	2	Two, one between C and A and one between C and B

C has become unlinked from A and B. There are only two of these, showing that a double recombination is needed to produce this genotype. Gene C is therefore between A and B.

Having established the gene order we can use the frequencies of the genotypes to infer the distances between genes A and C, and C and B. Table 17.1 shows that there are 51 recombinants generated by a single recombination between genes A and C. We must also include the progeny generated by double recombination events because with each of these one of the events was located between genes A and C. So the recombination frequency for genes A and C is calculated as:

$$RF = \frac{sr(A \leftrightarrow C) + dr}{t} \times 100$$

In this equation, RF is the recombination frequency, $sr(A \leftrightarrow C)$ is the number of progeny resulting from single recombination events between genes A and C, dr is the number of progeny resulting from double recombination events and t is the total number of progeny. Putting in the appropriate numbers, we discover that the recombination frequency for genes A and C is

$$RF = \frac{51 + 2}{1,103} \times 100 = 4.8\%$$

and for genes C and B the recombination frequency is

$$RF = \frac{63 + 2}{1,103} \times 100 = 5.9\%$$

The deduced map is therefore as shown in **Figure 17.9**.

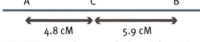

FIGURE 17.9 The map deduced from the multipoint cross.

Gene mapping in humans is carried out by pedigree analysis

With humans, it is, of course, impossible to preselect the genotypes of parents and set up crosses designed specifically for mapping purposes. But we can still map genes in humans. Indeed, of all the genomes in existence, our own is the one for which we would like to have the most detailed gene maps, as this is one of the ways of identifying genes that are involved in inherited disorders such as breast cancer and muscular dystrophy. Identification of the causative gene is often the first step in designing a cure or treatment for a disorder.

With humans, rather than carrying out planned experiments, data for the calculation of recombination frequencies have to be obtained by examining the genotypes of the members of successive generations of existing families. This means that only limited data are available, and their interpretation is often difficult. The limitation arises because genotypes of one or more family members are often unobtainable because those individuals are dead or unwilling to cooperate.

We will follow through a typical human **pedigree analysis** in order to explore what this method can achieve and to become familiar with some of the problems that can arise along the way. Imagine that we are studying a genetic disorder present in a family of two parents and six children (Figure 17.10a). Genetic disorders are pathological conditions such as breast cancer or cystic fibrosis (Section 21.1), which are frequently used as gene markers in humans. The affected state is one allele of the gene for the disorder and the unaffected state is the second allele. In our pedigree, the mother is affected by the disorder, as are four of her children. We know from family accounts that the maternal grandmother also suffered from this disorder, but both she and her husband – the maternal grandfather – are now dead. We can include them in the pedigree, with slashes indicating that they are dead, but we cannot obtain any further

(a) the pedigree

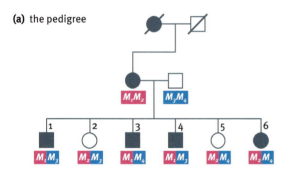
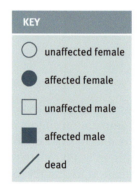

(b) possible interpretations of the pedigree

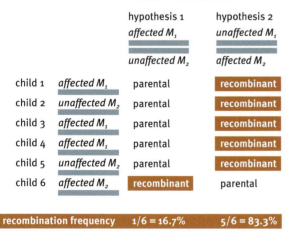

(c) resurrection of the maternal grandmother

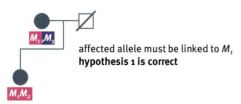

affected allele must be linked to M_1
hypothesis 1 is correct

FIGURE 17.10 An example of human pedigree analysis. (a) The pedigree shows inheritance of the gene for a genetic disorder in a family of two living parents and six children, with information about the maternal grandparents available from family records. The affected allele (closed symbols) is dominant over the unaffected allele (open symbols). (b) The pedigree can be interpreted in two different ways. Hypothesis 1 gives a low recombination frequency and indicates that the gene for the disorder is tightly linked to gene M. Hypothesis 2 suggests that the gene for the disorder and gene M are much less closely linked. In (c), the issue is resolved by the reappearance of the maternal grandmother, whose genotype is consistent only with hypothesis 1.

information on their genotypes. We know that the gene for the genetic disorder is present on the same chromosome as a multiallelic gene M, four alleles of which – M_1, M_2, M_3 and M_4 – are present in the living family members. Our aim is to map the position of the gene for the disorder relative to gene M.

To establish a recombination frequency between the gene for the disorder and gene M, we must determine how many of the children are recombinants. If we look at the genotypes of the six children, we see that numbers 1, 3 and 4 have the affected allele for the genetic disorder and allele M_1. Numbers 2 and 5 have the unaffected allele and M_2, and number 6 has the affected allele and M_2. We can therefore construct two alternative hypotheses. The first is that the two copies of the relevant homologous chromosomes in the mother have the genotypes *Affected-M_1* and *Unaffected-M_2*. This would mean that children 1, 2, 3, 4 and 5 have parental genotypes and child 6 is the one and only recombinant (Figure 17.10b). This would suggest that the gene for the disorder and gene M are relatively closely linked and crossovers between them occur infrequently.

The alternative hypothesis is that the mother's chromosomes have the genotypes *Unaffected-M_1* and *Affected-M_2*. According to this hypothesis, children 1 through 5 are recombinants, and child 6 has the parental genotype. This would mean that the two genes are relatively far apart on the chromosome. We cannot determine which of these hypotheses is correct. The data are frustratingly ambiguous.

The most satisfying solution to the problem posed by the pedigree in Figure 17.10 would be to know the genotype of the grandmother. Let us pretend that this is a soap opera family and it turns out that the grandmother is not really dead. She is written back into the story to boost the ratings and save the show from cancellation. Her genotype for gene M is found to be M_1M_5 (Figure 17.10c). This tells us that the chromosome inherited from her by the mother must have had the genotype *Affected-M_1*. It could not have been *Affected-M_2* because the grandmother does not possess the M_2 allele. We can now distinguish between our two hypotheses and conclude with certainty that only child 6 is a recombinant.

Practicalities of pedigree analysis

Data from human pedigrees are analyzed statistically, using a measure called the **lod score**. This stands for logarithm of the odds that the genes are linked and is used primarily to determine whether the two genes being studied lie on the same chromosome. A lod score of 3 or more corresponds to odds of 1,000:1 and is usually taken as the minimum for confidently concluding that this is the case. If it can be established that the two genes are on the same chromosome, then lod scores can be calculated for each of a range of recombination frequencies, in order to identify the frequency most likely to have given rise to the data obtained by pedigree analysis.

Lod scores are substantially more reliable if data are available for more than one pedigree, and the analysis is less ambiguous for families with larger numbers of children, and, as we saw above, it is important that the members of at least three generations can be genotyped. For this reason, family collections have been established, such as the one maintained by the Centre d'Etudes du Polymorphisme Humaine (CEPH) in Paris. The CEPH collection contains cultured cell lines from families in which all four grandparents as well as at least eight grandchildren could be sampled. This collection is available for gene mapping by any researcher who agrees to submit the resulting data to the central CEPH database.

17.2 GENE MAPPING WITH BACTERIA

The main difficulty that geneticists faced when trying to develop genetic mapping techniques for bacteria is that these organisms are normally haploid, and so do not undergo meiosis. Some other way therefore had to be devised to induce crossovers between homologous segments of bacterial DNA. The answer was to make use of the three natural methods that exist for transferring pieces of DNA from one bacterium to another – conjugation, transduction and transformation – each of which we met in Chapter 13. We will look at how these processes are used in gene mapping in a moment. First, we must consider the basic principles of this type of work when applied to bacteria.

Basic features of gene mapping in bacteria

The first issue that we must consider is how to distinguish between the alternative alleles of a gene whose position we wish to map in a bacterial chromosome. Bacteria have very few visual characteristics, so for the vast majority of genes we cannot distinguish alleles simply by observing the cells. Instead, alleles are distinguished by biochemical analysis. This may sound complicated, but the alternative alleles of many genes can be scored simply by determining whether the bacteria can grow on a **selective medium**. *Escherichia coli*, for example, has several genes that code for enzymes involved in the synthesis of the amino acid called threonine. Bacteria with functional **wild-type** versions of these gene can divide and form colonies on a selective medium that lacks threonine, whereas a bacterium that contains a threonine-synthesis gene that has been inactivated by mutation, and hence makes no threonine, will not be able to survive (Figure 17.11a). This particular type of selective medium, which lacks nutrients that wild-type bacteria can synthesize for themselves, is called a **minimal medium**. Other types of selective medium can be used to distinguish bacteria with alleles that confer sensitivity or resistance to toxic agents (Figure 17.11b).

The basic procedure for gene mapping in bacteria is shown in Figure 17.12. Two strains of the bacterium are required, one to act as a donor and one as a recipient. The donor must possess the wild-type versions of the genes under study, and the recipient bacterium must possess the non-wild-type versions.

A cross is then set up so that a part of the DNA molecule of the donor bacterium is transferred into the recipient bacterium. Exactly how this is achieved is the key distinction between the conjugation, transduction and transformation mapping techniques.

After DNA transfer, recombination can take place between the DNA from the donor bacterium and the chromosome of the recipient cell. Usually, to be of value for gene mapping, two crossover events must occur in such a way as to insert a piece of the donor cell DNA, containing one or more of the genes under study, into the recipient cell's chromosome (see Figure 13.24a). This insertion results in the copies of the relevant genes present in the recipient cell being replaced by the wild-type versions obtained from the donor cell. The resulting change in genotype of the recipient bacterium is then assessed by spreading samples onto the appropriate selective media.

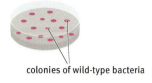

(a) minimal medium

colonies of wild-type bacteria

(b) medium containing a toxic agent

colonies of resistant bacteria

FIGURE 17.11 Selective media. (a) A minimal medium is used to distinguish between wild-type bacteria and ones that require a nutrient such as threonine. A minimal medium lacks threonine, so only wild-type bacteria can produce colonies. (b) A medium containing a toxic agent is used to select for resistant bacteria. Wild-type bacteria, which are sensitive to toxic agents, cannot grow on this medium.

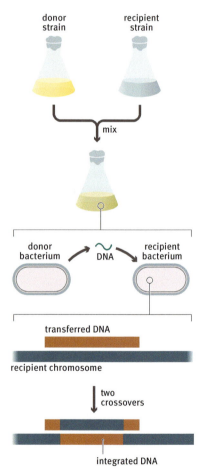

FIGURE 17.12 The basic features of a gene mapping experiment with a bacterium. The donor and recipient strains are mixed or some other manipulation carried out in order to achieve DNA transfer. After transfer, a pair of crossovers insert the DNA from the donor cell into the recipient cell's chromosome.

Mapping by conjugation – the interrupted mating technique

The discovery of bacterial conjugation (Section 13.3) was the key breakthrough that enabled the first gene mapping experiments to be carried out with *E. coli*. During conjugation between an Hfr and an F⁻ strain, DNA is transferred from the donor Hfr strain to recipient F⁻ strain in the same way that a string is pulled through a tube (see Figure 13.20). The relative positions of genes on the DNA molecule can be mapped by determining the times at which each gene appears in the recipient cell. To use conjugation as a gene mapping technique, we therefore need an Hfr strain that possesses the wild-type copies of two or more genes, and an F⁻ strain that carries the non-wild-type alleles of these genes. Transfer of each wild-type allele from the Hfr cell to the F⁻ cell will be signaled by acquisition by the recipient cells of the characteristic specified by that allele (Figure 17.13).

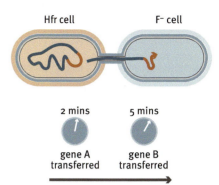

FIGURE 17.13 The basis to conjugation mapping. The time at which each wild-type allele is acquired by the F⁻ cell provides the map data.

One of the first experiments of this type was carried out by Elie Wollman and François Jacob at the Pasteur Institute in Paris in 1955. They studied the six genes listed in **Table 17.2.** The experiment was carried out by mixing the Hfr and F⁻ cells together and then removing samples from the culture at various time intervals (Figure 17.14). Each sample was immediately agitated to disrupt the pili linking Hfr and F⁻ cells and so interrupt the mating. The cells were then spread onto an agar medium containing streptomycin. Only the recipient cells would be able to grow on this medium because the original F⁻ cells were streptomycin-resistant, whereas the Hfr bacteria were sensitive to this antibiotic. The recipient cells therefore gave rise to colonies. The biochemical characteristics of the bacteria in each colony were then tested by transferring them to a series of selective media.

After about 8 minutes of mating, recipient cells that had acquired the ability to synthesize threonine and leucine began to appear (Figure 17.15). Next, after about 10 minutes, azide sensitivity began to appear in the recipient cells. Eventually, each of the wild-type characteristics had been transferred to the recipient cells.

It takes about 100 minutes for the entire *E. coli* chromosome to be transferred from the donor to the recipient cell during conjugation. The *E. coli* chromosome can therefore be divided into 100 map units or 'minutes', and the genes positioned on the chromosome according to the time at which they appear in the recipient cell (Figure 17.16). In fact, the transfer of the entire chromosome is not very likely during a single conjugation. Many bacteria naturally abandon mating well before 100 minutes have passed. To map the entire *E. coli* DNA, it is necessary to carry out a series of experiments, with a variety of Hfr strains, each with the F plasmid inserted at a different position. The parts of the *E. coli* chromosome that can be studied with each of these strains overlap, enabling the complete map to be built

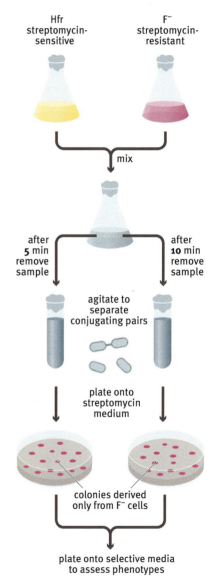

FIGURE 17.14 The interrupted mating experiment.

TABLE 17.2 THE GENES STUDIED BY WOLLMAN AND JACOB	
Gene	Characteristic
thr	Ability to synthesize threonine
leu	Ability to synthesize leucine
azi	Sensitivity to azide
ton	Resistance to colicin and T1 bacteriophage
lac	Ability to use lactose as an energy source
gal	Ability to use galactose as an energy source

up. These experiments were the first to show that the *E. coli* chromosome is a circular molecule.

Transduction and transformation can provide more detailed bacterial gene maps

During a conjugation mapping experiment, transfer of DNA from donor to recipient cells is not absolutely synchronous. This means that a gene will be transferred slightly earlier in some matings compared with others. For this reason, conjugation mapping is not completely accurate and is unable to distinguish the relative positions of two genes less than about 2 minutes apart, equivalent to about 90 kb of the chromosome.

Transduction and transformation mapping enable more detailed maps to be constructed. Both methods involve transfer into the recipient bacterium of a small segment of DNA, up to about 50 kb in length. The difference is that with transduction the transfer is mediated by a bacteriophage, whereas transformation involves the recipient cell taking up DNA from its environment.

The best studied transduction system is the P22 bacteriophage of *Salmonella typhimurium*. This bacteriophage follows a lysogenic infection cycle, which means that its genome can be inserted into the host DNA to form a prophage (Section 12.1). Every now and then, a P22 prophage is induced and enters the lytic stage of its infection cycle. During the lead up to lysis, the bacterial DNA molecule is broken into fragments, some of which might be about the same length as the P22 DNA molecule. If this is the case, then a segment of bacterial DNA could be packaged into a P22 particle instead of the bacteriophage DNA (Figure 17.17). The resulting P22 particle is still infective, as infection is a function purely of the protein components of the bacteriophage coat. The P22 particle will therefore transfer the bacterial DNA fragment into a new cell where, as a result of recombination, it can replace the equivalent stretch of DNA in the host chromosome. This means that gene transfer can occur. The donor cell is the one that was lysed by the bacteriophage, and the recipient cell is the one that is infected by the transducing bacteriophage particle.

A P22 bacteriophage can accommodate about 40 kb of DNA. Cotransduction, the transfer of two or more genes from donor to recipient, will occur if the genes are relatively close together on the *S. typhimurium* DNA molecule and hence can be contained in a single 40-kb fragment (Figure 17.18). A transduction mapping experiment therefore uses a donor bacterium that has the wild-type alleles of the genes that are being mapped, and a recipient that has the alternative versions of the genes. Following transduction, the recipient bacteria are tested for the acquisition of one or both of the wild-type alleles.

time after mixing (min)	genes transferred
8	thr, leu
10	thr, leu, azi
15	thr, leu, azi, ton
20	thr, leu, azi, ton, lac
25	thr, leu, azi, ton, lac, gal

FIGURE 17.15 The results of the interrupted mating experiment.

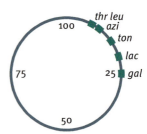

FIGURE 17.16 Map of the genes studied by Wollman and Jacob on the *E. coli* chromosome. The *E. coli* chromosome is divided into 100 map units called 'minutes'.

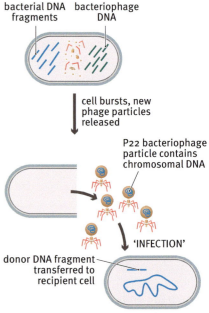

FIGURE 17.17 Transduction of DNA by a P22 bacteriophage. Replication of P22 bacteriophage DNA is accompanied by breakdown of the host bacterial DNA. Some bacterial DNA fragments might be about the same length as the P22 DNA molecule and could be packaged into a P22 particle instead of the bacteriophage DNA. The bacterial DNA fragment will then be transferred to a recipient cell by the transducing bacteriophage particle.

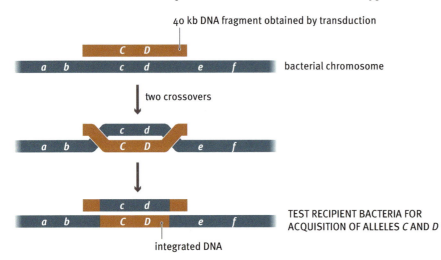

FIGURE 17.18 Cotransduction. The fragment of DNA transferred by transduction contains wild-type alleles for genes C and D. The recipient bacterium has alleles c and d. Cotransduction is detected by testing the recipient bacteria for acquisition of the wild-type alleles. If two alleles are cotransduced, then we can conclude that their genes are less than 40 kb apart on the bacterial chromosome.

The same principle applies to transformation mapping. The difference is that with transformation, DNA molecules are taken up by the recipient cell without the aid of a bacteriophage. A sample of DNA from the donor strain, carrying wild-type versions of the genes under study, is simply mixed with recipient cells in a high-salt solution, which promotes the binding of the DNA to the outer surface of the cells. A mild heat treatment is then applied to induce movement of the DNA into the bacterium.

17.3 PHYSICAL MAPPING OF DNA MOLECULES

In recent years, the gene mapping methods that were developed during the first half of the 20th century have been supplemented with what we call **physical mapping** techniques. These techniques are designed to identify the positions of sequence features by direct examination of chromosomes and purified DNA molecules. The sequence features can be genes, but are often much shorter sequences, perhaps just a few nucleotides in length, that recur at various positions in a DNA molecule.

What is the purpose of this type of mapping? We stated earlier that gene mapping enables the positions of genes responsible for known phenotypes to be identified in a chromosome. Physical mapping is less useful in this regard, but is very important as an aid to DNA sequencing. In the next chapter, we will learn that a limitation of the most commonly used sequencing method is that it is not possible to obtain more than about 500 bp of sequence in a single experiment. Overlaps between these short **sequence reads** must be identified in order to assemble a longer sequence (Figure 17.19). This is a massive task. The shortest human chromosome is 46,710,000 bp, so over 90,000 overlapping reads would be needed to build up its entire sequence. Errors can occur in identifying overlaps, leading to segments being omitted from the sequence or placed in the wrong position. A physical map of the chromosome is therefore valuable because it helps to ensure that the sequence has been assembled correctly (Figure 17.20).

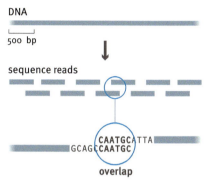

FIGURE 17.19 Assembling sequence reads into a full-length sequence. DNA sequencing results in short reads of about 500 bp. To assemble the sequence of a longer DNA molecule, a series of overlapping reads must be identified.

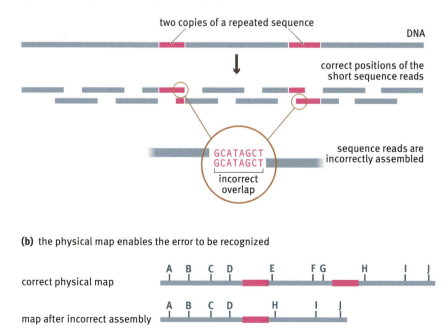

FIGURE 17.20 A physical map is an aid for correct sequence assembly. (a) In this example, the DNA molecule contains two copies of a repeat sequence. When the sequence reads are examined, two appear to overlap, but in fact one read contains the left-hand part of one repeat and the other read has the right-hand part of the second repeat. Failure to recognize this assembly error would lead to the segment of DNA between the two repeats being left out of the assembled sequence. If the two repeats were on different chromosomes, then the sequences of these chromosomes would mistakenly be linked together. (b) The error in sequence assembly is recognized because the relative positions of mapped features (A, B, C etc.) in the assembled sequence do not correspond with the correct positions of these features in the physical map.

Optical mapping can locate short sequence motifs in longer DNA molecules

The most useful physical mapping technique is **optical mapping**. This was first used to map the positions of **restriction sites** in DNA molecules. Restriction sites are short sequences, most between 4 bp and 8 bp in length, that are the recognition sequences for a **restriction endonuclease**. This is an enzyme that cuts DNA at a specific sequence. An example is *Bam*HI, which cuts only at the sequence 5′–GGATCC–3′.

The first step in optical mapping is to purify a set of DNA molecules and then extend them so they take up a linear configuration. This enables the locations of the cuts made by a restriction enzyme to be visualized by observation of the molecule. Linearization can be achieved by **molecular combing**. A silicone-coated cover slip is dipped into a solution of DNA and then slowly removed (Figure 17.21). The force required to pull the DNA molecules through the meniscus causes them to line up. Once in the air, the surface of the cover slip dries, retaining the DNA molecules as an array of parallel fibers. The restriction endonuclease is then added, along with a fluorescent dye which stains the DNA so that the fibers can be seen when the slide is examined with a high-power fluorescence microscope. The restriction sites in the extended molecules gradually become gaps as the degree of fiber extension is reduced by the natural springiness of the DNA, enabling the relative positions of the cuts to be recorded. Restriction sites less than 800 bp apart can be visualized (Figure 17.22).

Recording the results of optical mapping by observing the cut molecules under the microscope is labor-intensive, and accurate measurement of the distances between cuts can be difficult. Recent work has therefore focused on automating the procedure. One approach uses a microfluidic device that contains a nanochannel that is only just wide enough for a linear DNA molecule to squeeze through (Figure 17.23). This avoids the need for the combing step, as the DNA molecule becomes extended as it passes through the device. The microfluidic architecture is designed in such a way that a magnesium ion gradient is established within the nanochannel. This means that the restriction endonuclease, which is active only in the presence of magnesium, is activated when it enters the channel along with the DNA fragment. The resulting gaps in the DNA molecule are recorded as the molecule flows past a detection system. Using these automated approaches, optical maps have been obtained for a range of plant and animal genomes.

Optical mapping with fluorescent probes

The realization during the 2000s that optical mapping is a feasible method for detecting restriction sites in an extended DNA molecule led to the development of innovative versions of the technique that allow other sequence features to be mapped. One of these new approaches makes use of **hybridization probing**. The basis of this technique is the ability of any two complementary polynucleotides to form base pairs with

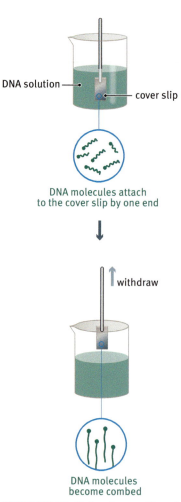

FIGURE 17.21 Molecular combing. A cover slip is dipped into a solution of DNA. The DNA molecules attach to the cover slip by their ends, and the slip is withdrawn from the solution at a rate of 0.3 mm s^{-1}, which produces a 'comb' of parallel molecules.

FIGURE 17.22 Observing the cut sites in combed DNA molecules. The natural springiness of a DNA molecule means that once a cut has been made, a gap appears in the DNA as the two cut ends are drawn apart.

FIGURE 17.23 A microfluidic device for optical mapping of restriction sites. The DNA molecule becomes partially extended by passage through the grid of electrodes and is then fully extended when it enters the nanochannel, which is only slightly wider than the double helix. The DNA is cut within the nanochannel, where there is a magnesium ion gradient, and the positions of the cuts recorded as the molecule passes the detector.

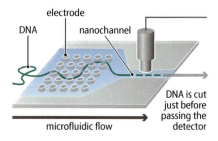

one another. The formation of base pairs between two polynucleotides is called **nucleic acid hybridization**.

Hybridization probing is combined with optical mapping in the following way. A short single-stranded DNA molecule is synthesized whose sequence is complementary to one or more segments of the DNA molecule that is being mapped. This probe will hybridize to its complementary target sites and thereby indicate their positions in this molecule (Figure 17.24). In order that these positions can be visualized, the probe DNA is modified prior to use by attachment of fluorescent chemical groups. The positions at which hybridization occurs can then be detected using a microfluidic device similar to the one described above, but with a fluorescent detector. This method is also called **fiber-FISH** (**fiber-fluorescent *in situ* hybridization**).

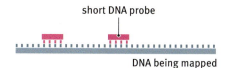

FIGURE 17.24 Hybridization of a probe DNA to complementary sequences in a longer DNA molecule.

The main challenge presented by this type of optical mapping is ensuring that the probe remains attached to the DNA molecule that is being mapped. The latter is, of course, double-stranded. A heat treatment is used to break the base pairs in this molecule, opening up single-stranded regions to which the probe can attach when the temperature is lowered. The difficulty is that the second DNA strand will compete with the probe and possibly displace it, reforming the double-stranded molecule. One solution to this problem is to use a **peptide nucleic acid** (**PNA**) as the probe. This is a polynucleotide analog in which the sugar–phosphate backbone is replaced by amide bonds (Figure 17.25). The hybrid between a PNA probe and its target on a DNA molecule is more stable than the normal DNA–DNA interaction, so attachment of the probe is favored over reattachment of the second DNA strand.

The probe used in this type of optical mapping could be a short sequence that is repeated multiple times in the DNA molecule that is being mapped. The information obtained would be similar to that obtained when a restriction endonuclease is used, but sequences other than those recognized by these enzymes could also be mapped. Another possibility is that probe is designed so that it is specific for the sequence of a particular gene, in which case the position of that gene can be identified.

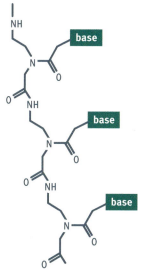

FIGURE 17.25 A short stretch of peptide nucleic acid. A peptide nucleic acid has an amide backbone instead of the sugar-phosphate structure found in a standard nucleic acid.

KEY CONCEPTS

- Gene mapping locates the positions in chromosomes of genes responsible for known phenotypes.
- Genes on the same chromosome can segregate independently if a crossover occurs between them during meiosis. This is called partial linkage.
- The relative positions of genes on chromosomes are determined by linkage analysis, which enables the recombination frequency between a pair of genes to be calculated.
- For many organisms, linkage analysis is carried out by following the inheritance of genes in planned breeding experiments. The necessary data are obtained from the phenotypes resulting from test crosses between double heterozygote and double homozygote parents.
- Multipoint crosses enable three or more genes to be mapped in one experiment.

- Planned breeding experiments are not possible with humans. Instead, gene mapping of humans depends on examination of inheritance in large families, a procedure called pedigree analysis.
- Genes specifying biochemical characteristics can be mapped in bacteria following their transfer into an F⁻ cell during conjugation with an Hfr donor bacterium.
- Bacterial genes can also be mapped by transduction and transformation, which result in short pieces of DNA being transferred from donor to recipient cells.
- Physical maps of short sequence motifs are an important aid for assembling a long DNA sequence from a set of sequence reads.
- Optical mapping enables the positions of restriction sites and other sequence features to be directly visualized in a long DNA molecule.

QUESTIONS AND PROBLEMS

Key terms

Write short definitions of the following terms:

centiMorgan	minimal medium	recombinant
double heterozygote	molecular combing	recombination frequency
double homozygote	multipoint cross	recombination hotspot
fiber-FISH	nucleic acid hybridization	restriction endonuclease
gene mapping	optical mapping	restriction site
hybridization probing	parental	selective medium
linkage analysis	partial linkage	sequence read
linkage group	pedigree analysis	test cross
lod score	peptide nucleic acid	wild-type
map unit	physical mapping	

Self-study questions

17.1 Clearly explain the differences between a gene map and a physical map of a DNA molecule.

17.2 Describe how studies of inheritance in sweet peas led to discovery of partial linkage.

17.3 Distinguish between a parental and a recombinant genotype. How do gametes with recombinant genotypes arise during meiosis?

17.4 Why does the frequency of recombinant gametes resulting from meiosis enable the map positions of genes to be worked out?

17.5 How were genes first mapped on chromosome 1 of the fruit fly?

17.6 Explain why the positions of genes deduced by linkage analysis are slightly different from their actual locations on a chromosome.

17.7 Explain why a double homozygote is used for a test cross in linkage analysis experiments. Why is it preferable that the homozygote alleles be recessive for the traits being tested?

17.8 Outline the limitations of human pedigree analysis and describe how the impact of these limitations is minimized in actual pedigree studies.

17.9 What is a lod score, and what information does it provide about the relative positions of two genes in the human genome?

17.10 Outline the basic features of an experiment to map genes in a bacterium.

17.11 Distinguish between the ways by which gene transfer is effected in conjugation mapping, transduction mapping and transformation mapping.

17.12 Describe how Wollman and Jacob carried out the first interrupted mating experiment.

17.13   A series of interrupted mating experiments with different Hfr strains produces the following linear gene maps:

   A J F C

   D H E B

   I B E H

   J F C I

   A G D H

   Draw a circular map of the bacterial DNA molecule showing the positions of these genes.

17.14   Explain how transduction mapping can be used in gene mapping.

17.15   In a series of transduction experiments, involving genes *A*, *B*, *C* and *D*, the following combinations were found to cotransduce relatively frequently:

   A and *B*

   C and *D*

   A and *D*

   Deduce the gene order.

17.16   Describe the basis to optical mapping of restriction sites.

17.17   How are microfluidic devices used in optical mapping?

17.18   Describe how a hybridization probe is used in an optical mapping experiment.

Discussion topics

17.19   Vermilion eyes and rudimentary wings are specified by two genes on chromosome 1 of the fruit fly. A cross between *VVRR* (red eyes and normal wings) and *vvrr* (vermilion eyes and rudimentary wings) produced the following F_2 generation:

   359 flies with red eyes and normal wings

   381 flies with vermilion eyes and rudimentary wings

   131 flies with red eyes and rudimentary wings

   139 flies with vermilion eyes and normal wings

   What is the map distance between the genes for vermilion eyes and rudimentary wings?

17.20   Genes A, B, C and D lie on the same fruit fly chromosome. In a series of crosses, the following recombination frequencies were observed:

Genes	Recombination Frequency (%)
A, C	40
A, D	25
B, D	5
B, C	10

   Draw a map of the chromosome showing the positions of these genes.

17.21   Deduce the methods used by Jacob and Wollman to test for transfer of each of the markers they studied in the first interrupted mating experiment. (The markers are listed in Table 17.2)

17.22   Discuss the reasons why gene mapping is still important today even though it is now relatively easy to obtain a complete DNA sequence for an organism's genome.

The answers to, or hints on how to answer, the self-study questions and discussion topic questions can be downloaded from the book's product page at this link: www.routledge.com/9781032743530

FURTHER READING

Cummings M (2015) *Human Heredity: Principles and Issues*, 11th ed. Boston: Cengage Learning. *Chapter 4 covers pedigree analysis.*

Michalet X, Ekong R, Fougerousse F et al. (1997) Dynamic molecular combing: stretching the whole human genome for high-resolution studies. *Science* 277, 1518–1523.

Morton NE (1955) Sequential tests for the detection of linkage. *American Journal of Human Genetics* 7, 277–318. *The use of lod scores in human pedigree analysis.*

Ott J, Wang J & Leal SM (2015) Genetic linkage analysis in the age of whole-genome sequencing. *Nature Reviews Genetics* 16, 275–284.

Sturtevant AH (1913) The linear arrangement of six sex-linked factors in *Drosophila*, as shown by their mode of association. *Journal of Experimental Zoology* 14, 43–59. *Construction of the first linkage map for the fruit fly.*

Yuan Y, Chung CYL & Chan TF (2020) Advances in optical mapping for genomic research. *Computational and Structural Biotechnology Journal* 18, 2051–2062.

CHAPTER 18

Sequencing Genes and Genomes

The most important technique available to the modern geneticist is **DNA sequencing**. By working out the sequence of nucleotides, we can gain access to the biological information contained in a DNA molecule. From the sequence, it might be possible to identify the genes present in the DNA molecule and possibly to deduce the functions of those genes.

We will begin this chapter by studying the methodology for DNA sequencing. We will discover that the most commonly used method has a limitation in that it provides individual sequence reads no greater than 500 bp in length. This means that a longer sequence has to be built up by searching for overlaps between individual sequence reads. We addressed this issue in Section 17.3 when we asked why physical maps of DNA molecules are important. We realized that we would need more than 90,000 overlapping reads to build up the entire sequence of the shortest human chromosome. In the second part of this chapter, we will examine the challenges that have to be met in order to assemble short sequence reads into the correct sequence of a longer DNA molecule such as chromosome or entire genome.

Finally, we will address the fact that a DNA sequence is simply a string of As, Cs, Gs and Ts and additional, possibly extensive work, is needed to annotate a DNA sequence. This is the term we use to describe the analyses, many carried out with a computer, that are used to locate the genes in a DNA sequence and to assign functions to those genes.

18.1 METHODS FOR DNA SEQUENCING

Before studying the methodology for DNA sequencing, we will make a brief overview of how sequencing technology has developed over the last 50 years. Over this time period, improvements in our ability to sequence DNA molecules have driven advances in our knowledge of the way in which biological information is stored in genes and genomes. Step-changes in DNA sequencing technology have resulted in step-changes in the applications of genetics in areas such as medicine, industry and agriculture. DNA sequencing is therefore largely responsible for the scope and reach of modern genetics, and future improvements in the technology will increase that scope and reach even further.

DNA sequencing methods have gradually become more powerful

The first efficient DNA sequencing technique to be developed was the chain termination method, which was invented by Frederick Sanger and colleagues at the University of Cambridge, United Kingdom, in 1977 (Figure 18.1). All of the important DNA sequencing projects up to 2005 used this method (see Chapter 1). Unfortunately, it had two important limitations. First, each fragment of DNA had to be purified before it could be sequenced. This meant that a certain amount of preparatory work had to be carried out, using either DNA cloning (Chapter 23) or the polymerase chain reaction (Chapter 21), before any sequence information could be obtained. The second limitation was that only about 800 bp of sequence could be read in a single experiment. Larger molecules had to be broken into fragments and each one sequenced individually to generate the master sequence. Automated systems were developed to maximize

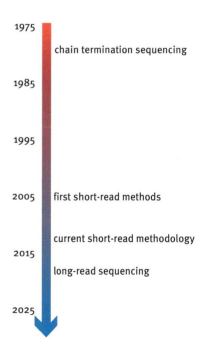

FIGURE 18.1 The timeline for DNA sequencing methodology. The diagram shows the approximate dates at which the various sequencing methodologies first became routinely available to genetics researchers.

DOI: 10.1201/9781003473862-21

throughput, but even in the most well-organized labs, it was rarely possible to obtain more than about 7 Mb of DNA sequence per day. Despite these drawbacks, the chain termination method was used to sequence many bacterial genomes and even some large eukaryotic ones. The latter included the human genome, which was completed in 2000, after 20 years of work and at an estimated cost of $100,000,000.

In order to make real progress with sequencing larger DNA molecules such as eukaryotic chromosomes and entire genomes, it was necessary to develop a less laborious and cheaper type of sequencing. This was achieved in the mid-2000s when **short-read sequencing** was introduced. Short-read sequencing enables more than 10^9 sequence reads to be obtained per experiment, with much less preparatory work prior to the actual sequencing stage. The individual reads are 500 bp or less in length, but with so many generated, it is possible to obtain up to 10^{11} bp (100 **gigabase pairs** or **Gb**) of sequence data in a single run of the sequencing machine. This is 250 times the length of the human genome. Short-read sequencing was used in almost all of the major sequencing projects carried out between 2005 and 2020, resulting in complete genome sequences for more than 30,000 prokaryotic species as well as for many eukaryotes.

The next logical improvement was to increase the length of the sequence reads. The longer the individual sequencing reads, the fewer that will be needed to cover an entire genome. **Long-read sequencing**, with read lengths of tens or even thousands of kb, first became widely available to genetics researchers in the early 2020s. Adoption of these methods decreases the amount of time and work needed to complete a genome sequencing project. Using a combination of short- and long-read sequencing, it is now possible to sequence a human genome for less than $1,000. This makes **personal medicine** a possibility – the use of an individual's genome sequence to diagnose their risk of developing a disease and to plan therapies and treatment regimes that are compatible with their personal genetic characteristics. It also makes possible ambitious ventures such as the **Earth BioGenome Project**, whose aim is to sequence the genome of every eukaryotic species. There is estimated to be 1.35 million of these, but with continued improvements in DNA sequencing methodology, it is hoped that the project can be completed by 2028.

Reversible terminator sequencing is the most popular short-read method

The first step in short-read sequencing is the preparation of a **sequencing library**. This is a collection of DNA fragments that have been immobilized on a solid support in such a way that multiple sequencing reactions can be carried out side by side (Figure 18.2). Library preparation begins with the purification of the DNA to be sequenced. The DNA is then broken into fragments between 200 bp and 500 bp in length, usually by **sonication**, a technique that uses high-frequency sound waves to make random cuts in DNA molecules.

After breakage, the DNA fragments are immobilized within a **flow cell** (Figure 18.3), a reaction chamber into which the reagents used in the sequencing experiment can be added and washed away one after the other. Immobilization occurs because the internal surface of the flow cell is coated with multiple copies of an **oligonucleotide**. These are single-stranded DNA molecules that are synthesized in the test tube, and all of which have the same sequence. **Adaptors**, which are short pieces of double-stranded DNA whose sequence matches that of the oligonucleotides, are attached to the ends of the DNA fragments. The DNA fragments therefore base pair to the oligonucleotides and become immobilized in the flow cell.

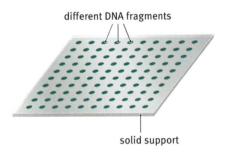

FIGURE 18.2 A sequencing library immobilized on a solid support. In reality, a single array will have millions of immobilized DNA fragments.

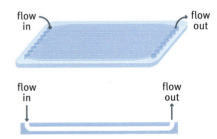

FIGURE 18.3 A typical flow cell used in DNA sequencing. The sequencing library is immobilized within the channels of the flow cell. To carry out the sequencing reactions, the necessary reagents flow through the cell, one after the other.

The most popular short-read method is based on **reversible terminator sequencing**. In this method, each DNA fragment acts as the template for the synthesis of a new polynucleotide by a DNA polymerase. The polynucleotide is, of course, complementary to the DNA fragment, so identification of each nucleotide that is added to its growing 3'-end enables the sequence of the fragment to be worked out (Figure 18.4).

How do we identify the nucleotides that are added? These are labeled with a fluorescent chemical group, a different one for each of the four nucleotides. The label is attached to the 3'-carbon of the nucleotide (Figure 18.5a). It therefore acts as a 'blocking group', preventing the next nucleotide from being added. Strand synthesis therefore pauses, giving time for an optical detector to identify the label and hence distinguish if an A, C, G or T is added. A reagent is then passed through the flow cell to remove the label, creating a 3'-OH group that can participate in the next nucleotide addition (Figure 18.5b). The cycle repeats, so one by one the nucleotides added to the growing end of the polynucleotide are identified and the sequence of the DNA fragment is revealed.

The same reactions occur in parallel with each of the fragments that are immobilized in the flow cell, monitored by a series of nanoscale optical detectors. Every one of the millions of fragments in the library is therefore sequenced at the same time. The maximum read lengths are about 250 bp. However, a single fragment can be sequenced from both ends. This gives two **paired-end reads** which, if they overlap, enable the entire sequence of a 500 bp fragment to be obtained.

There are different approaches to long-read sequencing

The main factor limiting read length when the reversible terminator method is used is the delay caused by the need to remove the blocking group after each nucleotide addition. If this delay can be avoided, then longer sequences are possible. This is what happens during **single-molecule real-time (SMRT) sequencing**, in which a sophisticated optical system called a **zero-mode waveguide** is used to observe the copying of each DNA molecule. The nucleotides that are used are still labeled with fluorescent markers, but the optical system is so precise that there is no

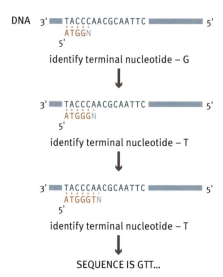

FIGURE 18.4 The basis to short-read sequencing. The DNA fragment being sequenced acts as the template for synthesis of a new polynucleotide by a DNA polymerase. Each nucleotide that is added to the growing 3' end of the polynucleotide is identified. Note that the sequence that is read is the *complement* of the DNA fragment that is being sequenced.

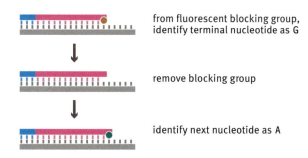

FIGURE 18.5 Reversible terminator sequencing. (a) The structure of a reversible terminator nucleotide with a removable fluorescent blocking group attached to the 3'-carbon. (b) After each nucleotide addition, there is a pause while the blocking group is removed. During this pause, the fluorescent label is detected and the terminal nucleotide identified.

need to use a blocking group to delay the polymerization process. Instead, the label is removed immediately after nucleotide incorporation, so DNA synthesis progresses without interruption. Read lengths of up to 200 kb are possible, with up to 160 Gb of sequence data obtained per flow cell. A problem with the early versions of SMRT sequencing was that on average one in ten nucleotides were read incorrectly – an error rate of 10%. However, with the last version of the technology, the error rate is <1%, which is comparable to short-read sequencing.

The next logical step is to dispense with the strand synthesis step and read the sequence of a DNA molecule directly without copying the molecule in any way. This has been made possible by **nanopore sequencing**. The principle of the method is that the flow of ions through a small pore in a membrane will be impeded when a single-stranded DNA molecule threads its way through the pore (Figure 18.6). Each of the four nucleotides has a different shape and so occludes the nanopore in a different way, resulting in a slightly different perturbation of the ion flow. These perturbations result in changes in the electrical current across the membrane, which can be measured in order to deduce the sequence of the DNA molecule as it threads its way through the pore.

Reading the sequence is not straightforward. A nanopore is so large that four or five nucleotides are present in the pore at the same time. The perturbations in ion flow do not therefore correspond to individual nucleotides, but to short series of nucleotides. Sophisticated computation systems are therefore needed to convert the plot of the electrical changes across the membrane into a nucleotide sequence. Even with the current technology, error rates of 5%–10% are expected. The great advantage of the method is that no DNA synthesis is involved so the length of the sequence is not limited by the capability of the DNA polymerase. If a very gentle extraction method is used, so the DNA molecules undergo the minimum amount of fragmentation, then reads of up to 4 Mb can be obtained.

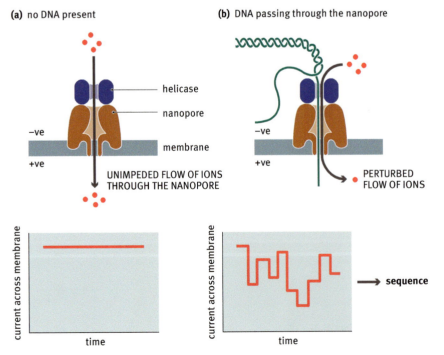

FIGURE 18.6 Nanopore sequencing. (a) In the absence of DNA, the flow of ions through a nanopore is unimpeded and the electrical current across the membrane is constant. (b) Passage of a polynucleotide through the nanopore perturbs the ion flow. The DNA is double-stranded, but the presence of a helicase enzyme in the vicinity of the nanopore breaks the base pairs, so the DNA unwinds and just one strand passes through the pore. Each nucleotide, or combination of adjacent nucleotides, perturbs the ion flow in a different way, resulting in fluctuations in the current from which the DNA sequence can be deduced.

18.2 ASSEMBLING A GENOME SEQUENCE

Most DNA sequencing projects involve more than just the generation of sequence reads. If a short-read method has been used, then the reads must be assembled into a contiguous sequence covering the entire length of the DNA molecule that is being studied. Even long reads will usually require some assembly to produce the master sequence. This is particularly important if an entire genome is being sequenced. Genome sequencing is a central component of modern genetics and will be the focus of the remainder of this chapter.

The simplest approach to **sequence assembly** is to build up the master sequence by identifying overlaps between the sequence reads (Figure 18.7). This is called the **shotgun method**.

Small bacterial genomes can be sequenced by the shotgun method

The first genomes were sequenced in the 1990s, when the chain termination method was the only technology that was available. As individual chain termination sequences are a maximum of 800 bp in length, a considerable amount of work is needed to identify overlaps among the thousands of sequence reads that would be needed to build up an entire genome. Initially, there was some doubt about whether this would be possible with the capabilities of the computer systems that were used at that time. These doubts were laid to rest in 1995 when the sequence of the 1,830 kb genome of the bacterium *Haemophilus influenzae* was assembled from 24,304 reads. It soon became accepted that the shotgun method was capable of assembling the genome sequences of most if not all prokaryotic species.

The 24,304 sequence reads used to assemble the *H. influenza* genome made up, in total, 11.6 Mb of sequence. This is 6.5× the length of the *H. influenzae* genome. Why was so much data generated? The answer is that no sequencing method is entirely accurate, and individual reads might contain one or more errors. It is therefore necessary to sequence each region of a genome multiple times in order to ensure that every nucleotide is correct (Figure 18.8). With the chain termination method, to ensure that all errors are identified, at least 5× **sequence depth** or **coverage** is needed, meaning that every nucleotide is present in five different reads. A similar degree of coverage is also required if a modern short-read method is used.

Assembly of genome simply by looking for overlaps between sequence reads is called ***de novo* sequencing**. The genomes of more than 80,000 bacterial species have been sequenced in this way. There are, however, almost 400,000 bacterial genome sequences in the databases. This is because some species have been sequenced multiple times in order to understand the amount of variation that exists in their genomes (Section 13.1). These resequencing projects make use of a second, less computer-intensive approach to assembly, in which the existing genome sequence is used as a **reference sequence**. Sequence reads

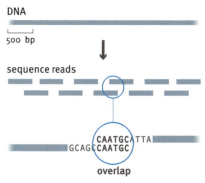

FIGURE 18.7 The shotgun method for sequence assembly. The master sequence is assembled by searching for overlaps between the sequence reads for individual fragments.

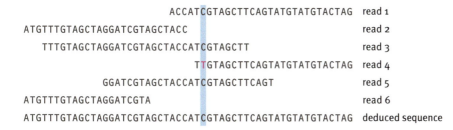

FIGURE 18.8 Read depth is needed in order to identify sequencing errors. Each region of a genome must be sequenced multiple times, in order to identify errors present in individual sequence reads. In this example, the discrepancy in Read 4 in the highlighted column can be ascribed to a sequencing error, the correct nucleotide at this position being C.

FIGURE 18.9 **Using a reference sequence during a resequencing project.** The reference sequence is used to place the reads from the second genome at their correct relative positions.

for the genome that is being assembled are compared with the reference sequence. Some will have an exact match with the reference sequence and can be placed at their correct position in the new genome sequence (Figure 18.9). Only in those regions where the new genome sequence differs significantly from the reference, will it be necessary to assemble reads solely by searching for overlaps.

The reference sequence approach can also be used to assemble the genome of a species that has not been studied before, so long as that species is related to one already in the genome database. The rationale is that the two sequences will be sufficiently similar for the known genome to direct the assembly of the sequence for the new species.

Repetitive DNA complicates shotgun assembly of eukaryotic genomes

The shotgun approach is also being applied to eukaryotic genomes, but with these organisms sequence assembly is more complicated. This is because eukaryotic genomes contain **repetitive DNA** sequences. These are sequences, up to several kilobases in length, that are repeated at two or more places in a genome. Almost half of the human genome is made up of repetitive DNA (Section 20.2) and for some species, especially plants, the amount is even higher. Most bacterial genomes, in contrast, contain little repetitive DNA.

When a genome containing repetitive DNA is broken into fragments in order to make a sequencing library, some of the resulting pieces will contain the same sequence motifs (Figure 18.10). It would be easy to reassemble the sequence reads from these fragments in such a way that a portion of a repetitive region is left out. It would even be possible to connect together two quite separate pieces of the same or different chromosomes.

Repetitive DNA also causes difficulties when a eukaryotic genome is used as a reference sequence for the assembly of additional versions of that genome. This is because the genomes of many eukaryotes, including humans, have **structural variants**. These are natural variations in the numbers and locations of repeat units, the presence or absence of certain sequences of 50–2,000 bp in length and the positioning of some sequences (Figure 18.11). Because of these variations, the genomes of

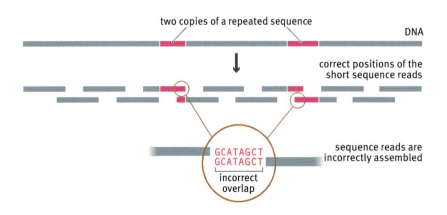

FIGURE 18.10 **A possible error in sequence assembly caused by repetitive DNA.** The DNA molecule contains two copies of a repeat sequence. When the sequence reads are examined, two fragments appear to overlap, but one fragment contains the left-hand part of one repeat and the other fragment has the right-hand part of the second repeat. Failure to recognize this assembly error would lead to the segment of DNA between the two repeats being left out of the master sequence. If the two repeats were on different chromosomes, then the sequences of these chromosomes would mistakenly be linked together.

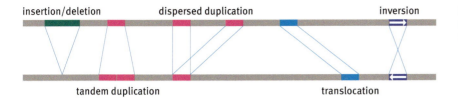

FIGURE 18.11 Structural variants that can complicate assemble of sequence reads on to a reference genome.

two individuals of the same species are different from one another. Some of these variations might be missed if a reference genome is followed too closely when assembling the reads from the second example of the genome. If that is the case, then the assembly that is obtained will be biased toward the structure of the reference genome.

For several years, the difficulties raised by the repetitive DNA content of eukaryotic genomes held back the assembly of complete and accurate eukaryotic genome sequences. Since 2020 the problems have largely been solved by the introduction of optical mapping and long-read sequencing. It is now relatively easy to use optical mapping to obtain a physical map of the restriction sites in a chromosome. If the assembled sequence and the map do not match, then a mistake has been made and the assembly must be revised (see Figure 17.20).

Similarly, the accuracy of a sequence assembled from short reads can be checked by comparing it with long sequence reads from the same part of the genome. Until recently, because of the high error rate, long-read sequencing on its own has not routinely been used to generate data for genome assembly. This is still a problem with nanopore sequencing and was also issue with the earlier versions of the SMRT method. However, even if a long sequence read contains a few errors, it should still be possible to use it to identify any regions where short sequence reads have been assembled incorrectly (**Figure 18.12**).

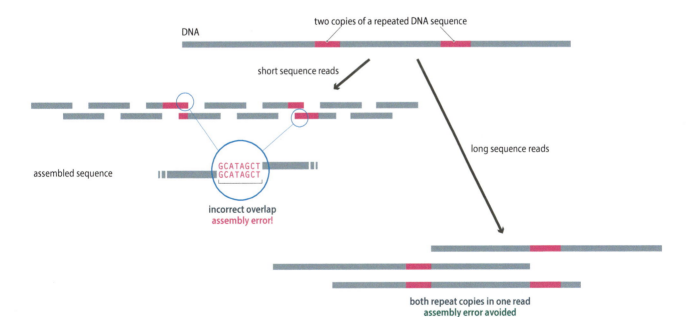

FIGURE 18.12 Long reads reduce assembly errors due to repetitive DNA. The DNA molecule contains two copies of a repeat sequence. On the left, two of the short reads appear to overlap, but one read contains the left-hand part of one repeat and the other read has the right-hand part of the second repeat. Failure to recognize this problem results in an assembly error, with the segment of DNA between the two repeats left out of the genome sequence. On the right, long reads have been obtained. Both copies of the repeat sequence are present in one of the reads, ensuring that the assembly error is avoided.

What is a 'genome sequence'

A 'complete' sequence – one in which every nucleotide is known, with no errors, and every segment placed at its correct position – has only recently become feasible for any genome. The less-complete sequence referred to as the **finished genome** will typically still have some unsequenced gaps and up to one error per 1,000 nucleotides. In contrast, a **draft genome sequence** has a greater error rate, more gaps and possibly some uncertainty about the order and/or orientation of some parts of the sequence. Moving down the scale, we have partially assembled genomes which might be 'work in progress' and in extreme cases unassembled collections of sequence reads.

The human genome illustrates these points (Figure 18.13). The outcome of the Human Genome Project, which was completed in 2000, was a draft sequence that covered just 90% of the genome, the missing 310 Mb being made up mainly of constitutive heterochromatin (Section 14.1). Within the 90% of the genome that was covered, each part had been sequenced at least four times, providing an 'acceptable' level of accuracy, but only 25% had been sequenced the 8–10 times that is necessary before the work is considered to be 'finished'. Furthermore, this draft sequence had approximately 150,000 gaps, and it was recognized that some segments had probably not been ordered correctly. The International Human Genome Sequencing Consortium, which managed the next phase of the project, set as its goal a 'finished' sequence of at least 95% of the euchromatin, with an error rate of less than one in 1,000 nucleotides, and all except the most difficult gaps filled. The chromosome sequences from this phase of the project began to appear in 2004, with the entire genome sequence being considered complete a year later.

The human genome sequence then stagnated for some time until long-read sequencing became available. These new methods made it possible to obtain a genuinely complete human genome sequence. When this degree of accuracy is sought, we have to take account of the fact that most human cells are diploid, and so contain two copies of each chromosome, one inherited from the mother and one from the father. The maternal and paternal chromosomes will not be identical. Some genes will be heterozygous, in which case the two chromosomal copies will carry different alleles with different sequences. If diploid cells are used as the source of DNA for a sequencing project, then the resulting sequence will contain a mixture of the male and female features (Figure 18.14): it will not be a genuine sequence of a single genome.

To avoid this problem, this new phase of genome sequencing made use of DNA from individual chromosomes and from **hydatidiform moles**. The latter derive from a defective egg cell that lacks a nucleus. After fertilization with a sperm, the cell implants in the uterus wall and forms a small growth that contains only the paternal chromosomes. Sequencing by the SMRT and nanopore methods was used to generate long reads with a coverage of >30×. The accuracy of the assembly, especially in difficult parts of the genome, was checked by reference to optical maps and by aligning short reads onto the long-read assemblies to ensure that all sequence errors have been accounted for (Figure 18.15). The first 'end-to-end' sequence, for the human X chromosome, was published in 2020, and the complete genome sequence in 2022.

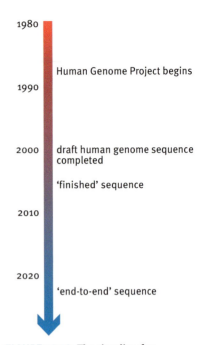

FIGURE 18.13 The timeline for sequencing the human genome.

...ATGAGCATCGATGCA$\overset{C}{T}$CAGCAGATTGAGCTAC...

FIGURE 18.14 The effect of heterozygosity on a sequence assembly. In this region of the assembly, some of the reads covering a particular position identify the nucleotide as C, whereas others identify the nucleotide as T. This is because of heterozygosity: one member of the pair of homologous chromosomes has a C at this position, and the other has a T. The nucleotide that is placed in the genome sequence will depend on which of the alternative reads is more frequent in the dataset.

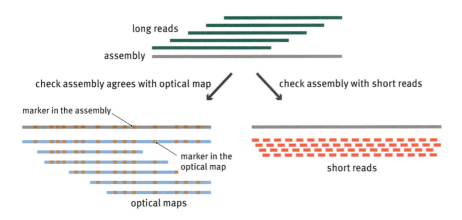

FIGURE 18.15 **Checking a genome assembly constructed with long reads.** The accuracy of the assembly is checked against the optical map (shown here as a series of overlapping maps that together cover the assembled sequence) and by ensuring that short reads can be aligned onto the assembly without ambiguities.

18.3 ANNOTATING A DNA SEQUENCE

Completion of a fully assembled genome sequence with sufficient sequence depth to be highly accurate is a time for celebration. This is not, however, the end of a genome sequencing project. A great deal of work is still needed to locate the genes and other interesting features within the sequence and to assign functions to those genes whose roles are unknown. In fact, this phase of the project may never be completed. Even today geneticists are still finding new features in the human genome and the functions of approximately one-fifth of the protein-coding genes are unknown or uncertain.

The location of genes in a genome sequence and identification of the functions of those genes is called **genome annotation**. There are two approaches to genome annotation. The first approach involves examination of the sequence with a computer, which enables sequences that might be genes to be located and can give some insights into the possible functions of at least some of those genes. The second approach makes use of experimental methods to confirm and extend the information obtained by computer analysis. The computer techniques form part of the methodology called **bioinformatics**, and it is with these that we will begin.

Genes can be located by searching for open reading frames

The DNA sequence of a protein-coding gene is an **open reading frame (ORF)**, consisting of a series of nucleotide triplets beginning with an initiation codon, which is usually but not always ATG, and ending with one of the termination codons, TAA, TAG or TGA (Figure 18.16). **ORF scanning**, which involves searching a DNA sequence for ORFs that begin with an ATG and end with a termination triplet, is therefore one way of looking for genes.

When carrying out an ORF scan, we must remember that a DNA molecule comprises two polynucleotides which run in opposite directions. Each molecule therefore has six **reading frames**, three of these on one polynucleotide and three in the opposite direction on the complementary strand (Figure 18.17). A gene could be present on any one of these six reading frames.

The key to ORF scanning is the frequency with which possible initiation and termination codons occur in the DNA sequence. If a DNA molecule contains equal amounts of each nucleotide (we refer to this as a **GC content** of 50%) and has a random sequence, then in each reading frame every one of the 64 possible triplets will appear, on average, once every 192 nucleotides. This means that there will be, on average, one ATG triplet every 192 nucleotides, and one TAA, TAG or TGA triplet every 64

FIGURE 18.16 **A protein-coding gene is an open reading frame of triplet codons.** The first four and last two codons of the gene are shown. The initiation and termination codons are highlighted in red.

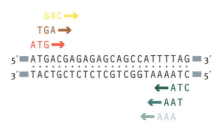

FIGURE 18.17 **A double-stranded DNA molecule has six reading frames.** Both strands are read in the 5'→3' direction. Each strand has three reading frames, depending on which nucleotide is chosen as the starting position.

GC<u>TAG</u>AGCAGGGCTCT<u>TAA</u>AATTCGGAGTCGTTG<u>ATG</u>CTCAATACTCCAATCGGTTTTTCGTGCACCACCGCGAGTGGC<u>TGA</u>CAAGGGTT<u>TGA</u>CATT<u>GA</u>

FIGURE 18.18 Random DNA contains short open reading frames. This 100-bp sequence has a GC content of 50% and was created using the University of California, Riverside, Random DNA Generator (https://faculty.ucr.edu/~mmaduro/random.htm). The sequence contains one ATG triplet (highlighted in green) and five potential termination triplets (red). The open reading frame that begins with the ATG triplet contains 19 'codons' before terminating at the second of the three TGA triplets.

nucleotides (Figure 18.18). If the DNA sequence has a GC content that is greater than 50%, then these four triplets, being AT-rich, will occur less frequently, but we would still expect a potential termination codon to be present in every 100–200 bp of sequence.

What this means is that in random DNA there will be many short ORFs that begin with an ATG and end in TAA, TAG or TGA. Most real genes are much longer than this. The average bacterial gene is 300–350 codons in length and so will be seen in the DNA sequence as an ORF of 900–1,050 nucleotides. Eukaryotic genes are even longer with an average size of 450 codons in humans. ORF scanning, in its simplest form, therefore takes a figure of, say, 250 nucleotides as the shortest length of a possible gene and records positive hits for all ORFs longer than this.

How well does this strategy work in practice? With bacterial genomes, simple ORF scanning is an effective way of locating most of the genes in a DNA sequence. This is illustrated by Figure 18.19, which shows a segment of the *E. coli* genome with all ORFs longer than 150 nucleotides highlighted. The two real genes in the sequence cannot be mistaken because they are much longer than this. With bacteria the analysis is further simplified by the fact that with many species there is relatively little intergenic DNA. If we assume that the real genes do not overlap, which is true for most bacterial genes, then a short, spurious ORF would have to be located in an intergenic region in order for it to be mistaken for a real gene. This means that if the intergenic component of a genome is small, then there is a reduced chance of making mistakes in interpreting the results of a simple ORF scan.

ORF scans are less effective with eukaryotic genomes

Although ORF scans work well with bacterial DNA, they are less effective for locating genes in eukaryotic genomes. There are two reasons for this. The first is that there is a lot more space between the real genes in a eukaryotic genome, increasing the chances of finding spurious ORFs in the intergenic regions.

The second, and more difficult, problem is the presence of introns in many eukaryotic genes (see Figure 3.2). If a gene contains one or more introns, then it will not appear as a continuous ORF in a DNA sequence. Starting from the initiation codon, the codon sequence for the first exon will be read correctly, but continuing this reading frame into an intron usually leads to a termination sequence that appears to close the ORF (Figure 18.20).

FIGURE 18.19 An ORF scan of part of the *Escherichia coli* DNA sequence. A 4,500-bp segment of the *E. coli* genome has been scanned for ORFs longer than 150 nucleotides. The sequence contains two genes, shown in blue. These real genes cannot be mistaken because they are much longer than the spurious ORFs, shown in red. The genes are lacZ and lacY, the first two genes of the lactose operon (Section 9.2). Arrows indicate the direction in which each ORF runs in the DNA sequence.

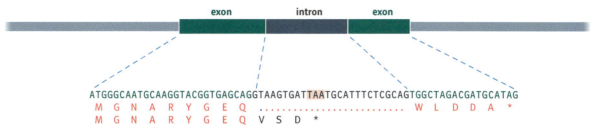

FIGURE 18.20 ORF scans are complicated by introns. The nucleotide sequence of a short gene containing a single intron is shown. The correct amino acid sequence of the protein translated from the gene is given immediately below the nucleotide sequence. In this sequence, the intron has been left out because it is removed from the transcript before the mRNA is translated into protein. In the lower line, the sequence has been translated without realizing that an intron is present. As a result of this error, the amino acid sequence appears to terminate within the intron. The amino acid sequences have been written using the one-letter abbreviations (see Table 7.1). Asterisks indicate the positions of termination codons.

To use ORF scanning with eukaryotic DNA, it is necessary to find a way of identifying the introns in the sequence. The simplest approach is to search for the distinctive sequence features of the 5′ and 3′ splice sites. In Section 5.2 we learnt that in vertebrate genes the two splice sites have the following consensus sequences:

5′ splice site   5′–AG↓GUAAGU–3′

3′ splice site   5′–PyPyPyPyPyPyNCAG↓–3′

where 'Py' is one of the two pyrimidine nucleotides (U or C), 'N' is any nucleotide, and the arrow indicates the exon–intron boundary. Other eukaryotes have similar consensus sequences at their splice sites. The difficulty is that the sequences are not precise and quite a lot of variation can occur with different introns. Some genuine splice sites may be missed when a sequence is scanned, and other sequences that are not associated with introns might be mistaken for splice sites.

A second possibility is to take account of **codon bias**. Not all codons are used equally frequently in the genes of a particular organism. For example, there are six codons for leucine (TTA, TTG, CTT, CTC, CTA and CTG; see Figure 7.20). In human genes, leucine is most frequently coded by CTG, and hardly ever by TTA or CTA. All organisms have a bias, which is different in different species. Real exons will display the codon bias whereas random series of triplets do not. The codon bias of the organism being studied can therefore be written into the ORF scanning software.

Despite these innovations, ORF scanning is an inefficient way of locating eukaryotic genes. For most eukaryotic genomes, it is possible to identify the starts and ends of genes with almost 100% accuracy, but the accuracy in identifying exon–intron boundaries is much lower, usually only 60%–70%. These figures assume that codon bias is taken into account, which in turns requires that a substantial number of genes in the genome have already been annotated correctly so that the bias can be quantified. If the genome is completely unstudied then the accuracy of gene prediction will be lower.

Locating genes by RNA sequencing

A rigorous way of distinguishing a genuine gene from a spurious ORF would be to check if the sequence is transcribed into an mRNA. This used to be a laborious procedure that required purifying the gene by DNA cloning and then searching RNA preparations from different tissues to see if any contained a transcript that hybridized to the gene. Today, this approach to genome annotation is much more routine, thanks to the development of short-and long-read sequencing techniques.

All DNA sequencing methods, by definition, are designed to work with DNA and not directly with RNA molecules. However, a double-stranded DNA copy of an mRNA can easily be made by **complementary DNA (cDNA)** synthesis. This is achieved by first copying the RNA strand into DNA, with the RNA-dependent DNA polymerase called **reverse transcriptase**. The new DNA strand is then copied with a DNA-dependent DNA polymerase to give double-stranded DNA (Figure 18.21). The resulting DNA can now be sequenced using a short-read method. This is called **RNA-seq**.

The sequence reads obtained by RNA-seq will correspond to fragments of the transcripts in the original RNA sample. Those reads can be mapped directly onto a genome sequence, in a manner identical to the way in which DNA sequence reads are mapped onto a reference sequence during a genome resequencing project (Figure 18.22a). An alternative strategy that achieves the same result is to apply a *de novo* assembly method to the collection of RNA-seq reads, and then map the assembled sequences, each of which will represent a different exon, onto the reference genome (Figure 18.22b). The advantage with the latter approach is that it gives greater accuracy with the members of a multigene family. Because these genes display sequence similarity, some short RNA-seq reads might be identical to segments of two or more members of the family. This is less likely to be the case if the reads are assembled into longer sequences before being mapped onto the reference genome.

Deducing the function of genes identified in a DNA sequence

Now we move on to the second question that is asked during the annotation phase of a genome sequencing project. What are the functions of the genes that have been located in the DNA sequence?

With some species, information regarding the approximate location of particular genes might be available from mapping experiments similar to the ones we studied in Chapter 17. However, when the region of DNA sequence thought to contain a mapped gene is examined, it is quite possible that more than one gene will be present. We therefore need methods for identifying which of these candidate genes is the one responsible for the phenotype that has been mapped. For many species, there are no mapping data, and gene identification has to proceed without any hints as to where the genes for particular phenotypes might be located.

The first step in functional annotation of a genome sequence is usually to perform a **homology search**. In this analysis, the sequence of

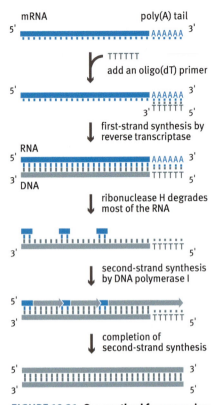

FIGURE 18.21 One method for preparing a double-stranded cDNA copy of an mRNA. The poly(A) tail at the 3′ end of the mRNA is used as the priming site for the first stage of cDNA synthesis, performed by reverse transcriptase. The oligo(dT) primer is a short single-stranded DNA molecule made up entirely of T nucleotides. When first-strand synthesis has been completed, the preparation is treated with ribonuclease H, which degrades part of the RNA molecule. The short RNA segments that remain prime the synthesis of the second DNA strand, catalyzed by DNA polymerase I.

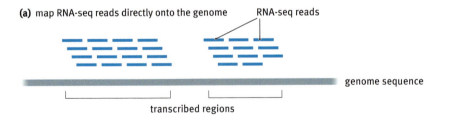

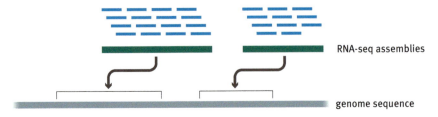

FIGURE 18.22 Two approaches to mapping RNA-seq reads onto a reference genome. (a) Direct mapping of reads onto the genome sequence. (b) Initial assembly of RNA-seq contigs, followed by mapping of the assembled sequences onto the genome.

a gene whose function is unknown is compared with the sequences of all the genes, from all species, that have already been sequenced. If the test sequence turns out to be similar to any known gene then there is a chance that the two are **homologous**, meaning that they are evolutionarily related. If this is the case, then they are likely to have similar biological roles, and the function of the known gene can be used to assign a function to the gene that has just been sequenced.

Usually, the DNA sequence of the unknown gene is converted into an amino acid sequence before the search is carried out. This is because there are 20 different amino acids in proteins but only four nucleotides in DNA, so genes that are unrelated usually appear to be more different from one another when their amino acid sequences are compared (Figure 18.23). A homology search is therefore less likely to give spurious results if the amino acid sequence is used.

The practicalities of homology searching are not at all daunting. Several software programs exist for this type of analysis, the most popular being **BLAST** (**Basic Local Alignment Search Tool**). The analysis can be carried out simply by logging on to the website for one of the DNA databases and entering the sequence of the unknown gene into the online search tool. The search program begins by making alignments between the query sequence and sequences from the databases. For each alignment, a score is calculated that indicates the likelihood that the query and test sequences are homologs. The simplest programs count the number of positions at which the same amino acid is present in both sequences. This number, when converted into a percentage, gives the degree of *identity* between two sequences.

More sophisticated programs use the chemical relatedness between nonidentical amino acids to assign a score to each position in the alignment. A higher score is given for identical or closely related amino acids (e.g. leucine and isoleucine, or aspartic acid and asparagine) and a lower score for less-related amino acids (e.g. cysteine and tyrosine, or phenylalanine and serine). This analysis determines the degree of *similarity* between a pair of sequences.

Gene editing can be used to assign functions to genes

Homology searching is becoming increasingly sophisticated, but the method has limitations and can rarely identify the function of every new gene in a DNA sequence. Experimental methods are therefore needed to complement and extend the results of computer analysis.

The function of an unknown gene might be revealed if it were possible to inactivate the gene and then assess the change in phenotype that results. Several methods for achieving this objective have been devised over the years. The most efficient of these is **gene editing**.

Gene editing makes use of a **programmable nuclease**. This is a nuclease that can be directed to a specific site in a genome where it makes a

```
            G  A  P  G  M  W  L  R  L  A  A  G  S  F  E  H  A  G
sequence 1 GGTGCACCCGGTATGTGACTGCGATTAGCAGCGGGATCATTTCAGCATGCAGGG
           *  * ***** **** **** ** *** **** ***** *** ** **** ** *
sequence 2 GATACACCCGTATTTGACAGCAATTTGCAGGGGATGATTGCACCATGGAGCG
            D  T  P  R  I  W  E  E  F  A  G  G  W  L  H  H  G  A
```

FIGURE 18.23 Lack of homology between two sequences is often more apparent when comparisons are made at the amino acid level. Two nucleotide sequences are shown, with nucleotides that are identical in the two sequences given in green and nonidentities given in red. The two nucleotide sequences are 76% identical, as indicated by the asterisks. This might be taken as evidence that the sequences are homologous. However, when the sequences are translated into amino acids, the identity decreases to 28%. Identical amino acids are shown in gold, and nonidentities in red. The comparison between the amino acid sequences suggests that the genes are not homologous, and that the similarity at the nucleotide level was fortuitous. The amino acid sequences have been written using the one-letter abbreviations (see Table 7.1).

double-stranded cut (Figure 18.24). The cut stimulates the natural repair process for double-strand breaks, called nonhomologous end-joining (NHEJ) in eukaryotes (see Figure 11.21). NHEJ is error-prone and usually results in a short insertion or deletion occurring at the repair site. If the repair is within a gene, then the change in nucleotide sequence will inactivate the gene.

The most popular gene editing system makes use of the **Cas9 endonuclease**. This enzyme makes a cut at a target site to which it is directed by a 20-nucleotide **guide RNA** (Figure 18.25). Performing the gene edit is straightforward. It simply requires that the gene for the Cas9 endonuclease, along with a DNA sequence specifying the guide RNA, is introduced into the cells whose genomes are being edited. Within the cell, the Cas9 gene is expressed so that the endonuclease is synthesized, and the guide RNA is transcribed from the DNA sequence.

The guide RNA sequence matches part of the sequence of the gene that we wish to inactivate. A critical issue is therefore the uniqueness of this sequence in the genome as a whole. The expected frequency of a 20-nucleotide motif in a DNA sequence is once every $4^{20} = 1.1 \times 10^{12}$ bp, which is 350 times the length of the human genome. This means that it is unlikely that there will be a second exact version of the target sequence. However, the Cas9 system does not require complete base-pairing between the guide RNA and genomic DNA. In early versions of the methodology, introduced in 2012, **off-target editing** was possible at sites where the DNA–RNA attachment had as many as five mismatched positions (Figure 18.26). Mismatching greatly increases the number of target sequences that can form an attachment with the guide RNA. As a result, the guide RNA might not be specific for the single site that we wish to edit.

Various ways of improving specificity have been explored, including the use of modified endonucleases that bind less strongly to the DNA. This might seem a counterintuitive way to increase specificity, but the rationale is that the protein-DNA interaction interferes with base pairing between the two strands of the double helix at the target site. Weaker attachment of the endonuclease results in a more stable double helix, preventing the guide RNA from hybridizing unless it too can make a perfectly base-paired attachment.

Genome browsers

The results of genome annotation must be displayed in some way. Usually, this is done in the form of a graph, with the DNA sequence forming the *x*-axis and the positions of genes and other interesting features marked at their appropriate positions.

A **genome browser** is a software package that enables genome annotation data to be displayed in this way. The display might simply indicate the positions of ORFs whose status as genes has been confirmed by RNA-seq and whose functions might or might not have been deduced. Usually, however, the display is more complex, also including short ORFs of questionable authenticity, noncoding RNA genes, the starts and ends of transcripts and the locations of repetitive DNA sequences. The software enables the annotation to be displayed at different degrees of resolution, so the entire length of a chromosome can be viewed, or the operator can zoom in to a level at which individual nucleotides are distinguished. Most browsers also incorporate a search facility so that particular genes can be quickly located.

Many genome browsers are used online, so that the annotations of newly sequenced genomes are accessible to researchers in other labs. Online

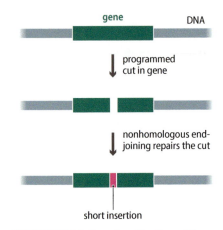

FIGURE 18.24 Gene inactivation with a programmable nuclease. The cut made by the nuclease is repaired by nonhomologous end-joining, which is error-prone and likely to insert or delete a few base pairs of DNA at the repair site, disrupting the target gene.

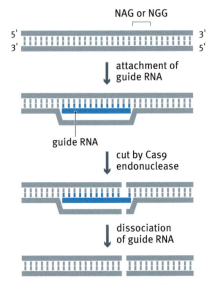

FIGURE 18.25 Cleavage of DNA by the Cas9 endonuclease. The cut position is specified by the 20-nucleotide guide RNA, which must be designed to base pair to a target site immediately upstream of a 5′–NGG–3′ or 5′–NAG–3′ sequence (where 'N' is any nucleotide).

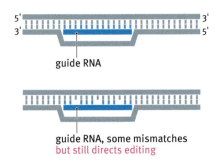

FIGURE 18.26 Off-target editing. The guide RNA can still initiate editing even if it is not completely base-paired to the target DNA.

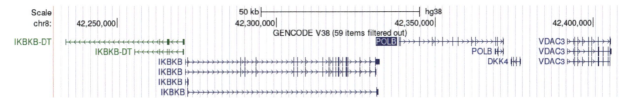

FIGURE 18.27 An example of the information provided by an online genome browser. This example shows the annotation for the 180 kb region of human chromosome 8 between nucleotide positions 42,250,000 and 42,410,000, as displayed by the UCSC Genome Browser. The user can configure the browser to show many different features of the genome annotation. This configuration shows the locations of genes. Protein-coding genes are in blue and noncoding RNA genes in green, with exons shown as boxes or vertical lines, and arrows indicating the DNA strand on which the gene is located. The different entries for an individual gene indicate variable start or end points and/or different splicing patterns. The genes are: IRBKB-DT, long noncoding RNA gene active mainly in testes; IRBKB, inhibitor of NF-κB kinase subunit β; POLB, DNA polymerase β; DKK4, WNT signalling pathway inhibitor and VDAC3, mitochondrial porin.

browsers include the UCSC Genome Browser developed by the University of California Santa Cruz, and Ensembl, which is run jointly by the European Bioinformatics Institute and the UK Sanger Centre. A typical example of the information provided by a genome browser is shown in Figure 18.27.

KEY CONCEPTS

- DNA sequencing is the most important technique available to the geneticist.
- Short-read sequencing enables more than 10^9 sequence reads to be obtained per experiment. The individual reads are 500 bp or less in length, but it is possible to obtain up to 100 Gb of sequence data in a single run.
- The first step in short-read sequencing is the preparation of a sequencing library which is immobilized in a flow cell.
- The most popular short-read method is reversible terminator sequencing.
- Long-read sequencing by the SMRT method provides read lengths of 200 kb. With nanopore sequencing read lengths up to 4 Mb can be obtained. Until recently, both methods were less accurate than short-read sequencing.
- The shotgun method of sequence assembly involves building up the master sequence by identifying overlaps between the sequence reads.
- The shotgun method is applicable to bacterial genomes, but is complicated by the presence of repetitive DNA when a eukaryotic genome is assembled.
- Only recently has it been possible to obtain end-to-end sequences of human chromosomes, by a combination of long- and short-read sequencing with assembly checked by reference to optical maps.
- Genes can be located in a genome sequence by searching for ORFs. With eukaryotic genomes, ORF searching is complicated by the presence of introns in most genes.
- Genes can also be located by mapping RNA-seq reads onto a genome sequence.
- The functions of some genes can be deduced by a homology search, where the sequence of the gene is compared to all other genes in the databases.

- Gene functions can also be assigned by inactivating a gene and assessing the resulting change in phenotype. This can be achieved by gene editing.

QUESTIONS AND PROBLEMS

Key terms

Write short definitions of the following terms:

adaptor
BLAST
Cas9 endonuclease
codon bias
complementary DNA synthesis
de novo sequencing
DNA sequencing
draft genome sequence
Earth BioGenome Project
finished genome
flow cell
GC content
gene editing
genome annotation
genome browser

gigabase pair
guide RNA
homologous
homology search
hydatidiform mole
long-read sequencing
nanopore sequencing
off-target editing
oligonucleotide
open reading frame
ORF scanning
paired-end reads
personal medicine
programmable nuclease
reading frame

reference sequence
repetitive DNA
reverse transcriptase
reversible terminator sequencing
RNA-seq
sequence assembly
sequence depth or coverage
sequencing library
short-read sequencing
shotgun method
single-molecule real-time sequencing
sonication
structural variant
zero-mode waveguide

Self-study questions

18.1 Summarize the strengths and weaknesses of chain termination, short-read and long-read sequencing methods.

18.2 Explain how a sequencing library for short-read sequencing is prepared and immobilized.

18.3 Describe how the DNA sequence is read by reversible terminator sequencing.

18.4 Outline the methodology used in SMRT sequencing.

18.5 Describe how a nanopore is used in long-read sequencing.

18.6 How is a DNA sequence assembled by the shotgun method?

18.7 Why is 5× sequence depth needed when a DNA sequence is assembled from short reads?

18.8 Distinguish between *de novo* sequencing and the use of a reference sequence.

18.9 What factors complicate the application of shotgun sequencing to a eukaryotic genome?

18.10 What are structural variants and why are these sometimes missed when reference genome is used to assemble a DNA sequence?

18.11 Distinguish between the terms 'draft' and 'finished' when referring to a genome sequence.

18.12 What methods have been used to obtain complete sequences of human chromosomes?

18.13 What is a hydatidiform mole and why are these useful in genome sequencing?

18.14 Why is it relatively easy to identify ORFs in prokaryotic genomes by computer analysis?

18.15 Describe how ORF scans are used to search for genes in eukaryotic genome sequences.

18.16 What is meant by the term 'codon bias'?

18.17 Describe how RNA-seq is carried out and why this method is useful in genome annotation.

18.18 Define the term 'homologous', as used when comparing gene sequences.

18.19 Describe how a BLAST search is carried out.

18.20 Explain how a gene can be inactivated by gene editing.

18.21 What is a genome browser?

Discussion topics

18.22 You have isolated a new species of bacterium whose genome is a single DNA molecule of approximately 2.6 Mb. Write a detailed project plan to show how you would obtain the genome sequence for this bacterium.

18.23 A 122-bp DNA molecule is randomly broken into overlapping fragments and those fragments are sequenced. The following sequence reads were obtained:

CGTAGCTAGCTAGCGATT

GATTTAGTTCGCCCATTCG

GCTGTAGCATGTTTTCGC

TTCGCTCAGCATCGGATTT

AGCTAGCTAGCGATTTAGT

TAGCATGTTTTCGCTCAGC

TTTCGCTCAGCATCGGATT

ATTTAGTTTAGCTGTAGCA

CATTCGCGATGCTATCTCT

GTTGACGCATACGGCGGG

TCGTAGCTAGCTAGCGAT

ATGCTATCTCATCTGATTT

ATTTAGTTCGCCCATTCGC

ATTTAGTTGACGCATACGG

ATGCATCGTAGCTAGCTAG

CTCAGCATCGGATTTAGTT

CGATGCTATCTCATCTGAT

CGCATACGGCGGGGGGAT

Is it possible to reconstruct the sequence of the original molecule by searching for overlaps between pairs of reads? If not, then what problem has arisen and how might this problem be solved?

18.24 To what extent do you believe it will be possible in future years to use computer analysis to obtain a complete description of the locations and functions of the protein-coding genes in a eukaryotic genome sequence?

18.25 Devise a hypothesis to explain the codon biases that occur in the genomes of various organisms. Can your hypothesis be tested?

18.26 Use the Ensembl Bacteria genome browser (http://bacteria.ensembl.org/index.html) to locate the position of the β-galactosidase gene in the *Escherichia coli* genome. Draw (or export) a map of the genes in the 30-kb region centered on the β-galactosidase gene.

18.27 Perform a BLAST search (https://blast.ncbi.nlm.nih.gov/Blast.cgi) with the following amino acid sequence:

GLSDGEWQLVLNVWGKVEADLAG
HGQEVLIRLFKGHPETLEKFDKFKH
LKSEKGSEDLKKHGNTVETALEGIL
KKKALELFKNDIAAKTKELGFLG

What protein has this amino acid sequence?

The answers to, or hints on how to answer, the self-study questions and discussion topic questions can be downloaded from the book's product page at this link: www.routledge.com/9781032743530

FURTHER READING

Adli M (2018) The CRISPR tool kit for genome editing and beyond. *Nature Communications* 9, 1911. *Gene editing.*

Giani AM, Gallo GR, Gianfranceshi L et al. (2020) Long walk to genomics: history and current approaches to genome sequencing and assembly. *Computational and Structural Biotechnology Journal* 18, 9–19.

Heather JM & Chain B (2016) The sequence of sequencers: the history of sequencing DNA. *Genomics* 107, 1–8.

Loman N & Pallen M (2015) Twenty years of bacterial genome sequencing. *Nature Reviews Microbiology* 13, 787–794.

Nurk S, Koren S, Rhie A et al. (2022) The complete sequence of the human genome. *Science* 376, 44–53. *End-to-end sequences of the 22 human autosomes and the X chromosome.*

Shendure J, Balasubramanian S, Church GM et al. (2017) DNA sequencing at 40: past, present and future. *Nature* 550, 345–353.

Stark R, Grzelak M & Hadfield J (2019) RNA sequencing: the teenage years. *Nature Reviews Genetics* 20, 631–656.

Wang J, Kong L, Gao G et al. (2012) A brief introduction to web-based genome browsers. *Briefings in Bioinformatics* 14, 131–143.

Yandell M & Ence D (2012) A beginner's guide to eukaryotic genome annotation. *Nature Reviews Genetics* 13, 329–342.

GENETICS IN OUR MODERN WORLD

19 Genes in Differentiation and Development
20 The Human Genome
21 Genes and Medicine
22 DNA in Forensic Genetics and Archaeology
23 Genes in Industry and Agriculture
24 The Ethical Issues Raised by Modern Genetics

CHAPTER 19

Genes in Differentiation and Development

We have completed our study of the two complementary roles of genes as units of biological information and as units of inheritance, and we have examined the methods used to map and sequence chromosomes and genomes. Now we will look more broadly at the role of genetics in the 21st century.

The topics that we will study are not an exhaustive list of all the important areas of modern genetics. It would require several books of this size to cover everything. The choice is based on three factors. The first is the importance of understanding how genetics works hand in hand with other areas of biology in research programs aimed at advancing our knowledge of cells and organisms. To this end, in this chapter, we will study differentiation and development, an area of research that has made great strides forward in recent years thanks to the combined endeavors of geneticists, cell biologists, physiologists and biochemists. We will examine how the genetic changes that occur when a cell takes on a specialized role are retained through many rounds of mitosis, and how changes in gene activity are regulated and coordinated in space and time during development.

The second factor influencing our choice of topics is our perception that, of all the species on the planet, *Homo sapiens* is the most important. Humans are probably in a minority of one in holding this view, but nevertheless we will devote three chapters to the genetics of humans. We will study the human genome, the way in which genetics is being used to understand human disease, and the use of genetics in forensic science and studies of human history.

The third factor is the important role that genetics now plays in industry and agriculture. We will investigate the use of genetic engineering as a means of producing drugs and vaccines, and we will also look at the applications of genetic manipulation in agriculture. In exploring these issues, we will encounter areas of modern genetics that are high in the public perception, not just because of the excitement of the new discoveries, but because of concerns that genetic technology might be used unwisely. As students of genetics, we cannot ignore these issues. The final chapter of this book will therefore be devoted to the ethical issues raised by modern genetics.

19.1 CHANGES IN GENE EXPRESSION DURING CELLULAR DIFFERENTIATION

Differentiation and development are inextricably linked but we must not confuse one with the other. Differentiation is the process by which an individual cell acquires a specialized function. Differentiation therefore requires a change in the pattern of gene expression in a cell, the resulting specialized expression pattern usually remaining in place for the remainder of the cell's lifetime, and often being inherited by the daughter cells, and the granddaughters and so on for many cell divisions. Development, on the other hand, is the pathway that begins with a fertilized egg cell and ends with an adult. A developmental pathway comprises a complex series of genetic, cellular and physiological events that must occur in the correct order, in the correct cells and at the appropriate times if the pathway is to reach a successful culmination. In humans, the pathway involves 10^{17} mitotic cell divisions and results in an adult containing 3×10^{13} cells differentiated into over 400 specialized types, the activity of each individual

DOI: 10.1201/9781003473862-23

cell coordinated with that of every other cell. Understanding how genes specify and regulate the differentiation and development pathways of multicellular eukaryotes such as humans is one of the biggest challenges in all of biology.

Differentiation requires permanent and heritable changes in gene activity

Fewer than 3,000 of the 20,000 protein-coding genes in the human genome are active in all cells. The other genes are differentially expressed, with up to 15,000 active in any particular cell type at any one time (Figure 19.1). These differential patterns of gene expression underlie the specialized biochemistries and physiologies of the 400 cell types that make up the human body plan.

In Chapter 9, we encountered processes by which the expression of individual genes or groups of genes can be modulated or even switched off in response to external and internal stimuli. Most, possibly all, of the processes that we studied result in transient changes in gene activity, changes that occur within the lifetime of a cell and might easily be reversed later in that cell's lifetime (Figure 19.2a). Cellular differentiation requires more permanent changes in gene expression patterns, changes that will last the lifetime of the cell and be maintained even when the stimulus that originally induced them has disappeared. Importantly, the gene expression patterns must be inherited by the descendants of that cell after mitosis

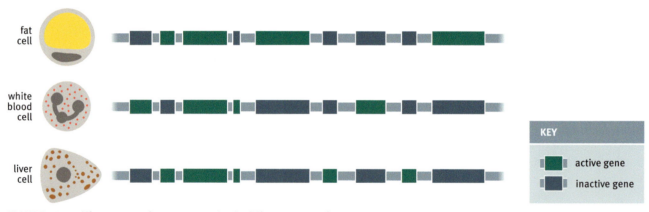

FIGURE 19.1 Different sets of genes are active in different types of specialized cell.

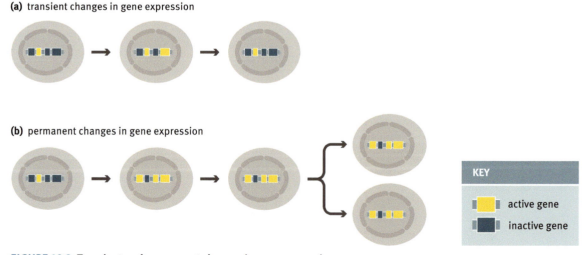

FIGURE 19.2 Transient and permanent changes in gene expression.

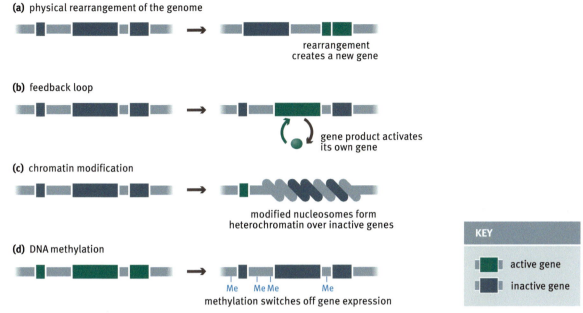

FIGURE 19.3 Four ways in which permanent changes in gene expression can be brought about. (a) Physical rearrangement of the genome. In this example, a new gene is formed, which will be inherited by the descendants of the cell in which the rearrangement occurred. (b) A genetic feedback loop can maintain a change in gene expression. (c) Chromatin modification involves the chemical alteration of nucleosomes so heterochromatin is formed over inactive genes. (d) DNA methylation switches off expression of adjacent genes.

(Figure 19.2b). We must therefore broaden our view beyond the types of gene regulation that we have studied so far.

Geneticists are aware of four different ways in which permanent changes in gene activity can be brought about (Figure 19.3). The first involves physical rearrangement of the genome, to bring together segments of DNA that previously were separated from one another. The new structure is then inherited when the cell divides. Second, it is possible for a genetic feedback loop to be established which will maintain a change in gene expression in a cell and its descendants. Third and fourth, there is **chromatin modification**, which involves chemical alteration of nucleosomes, and **DNA methylation**. Both of these processes are thought to have a general role in maintaining gene expression patterns in differentiated cells.

Increasingly, the term **epigenesis** is being used to describe these permanent changes in gene expression patterns. Epigenesis is defined as a heritable change to the genome that does not involve mutation.

Immunoglobulin and T-cell diversity in vertebrates results from genome rearrangement

First, we will examine an example of the way in which cellular differentiation can be brought about by physical rearrangement of the genome. This example concerns the differentiation of lymphocytes into cells that are specialized for the production of specific components of the mammalian immune system.

Immunoglobulins and T-cell receptors are related proteins that are synthesized by B and T lymphocytes, respectively. Both types of protein become attached to the outer surfaces of their cells, and immunoglobulins are also released into the bloodstream. The proteins help to protect the body against invasion by bacteria, viruses and other unwanted substances by binding to these **antigens**.

During its lifetime, an organism might be exposed to any number of different antigens, which means that the immune system must be able to synthesize an equally vast range of immunoglobulin and T-cell receptor proteins. Humans are able to make approximately 10^8 different immunoglobulin and T-cell receptor proteins, but there are only 20,000 protein-coding genes in the human genome, so where do all these proteins come from?

To understand the answer, we will look at the structure of a typical immunoglobulin protein. Each immunoglobulin is a tetramer of four polypeptides – two long 'heavy' chains and two short 'light' chains – linked by disulfide bonds (Figure 19.4). The heavy chains of immunoglobulins specific for different antigens differ mainly in their N-terminal regions, the C-terminal parts being similar, or 'constant', in all heavy chains. The same is true for the light chains, except that two families, κ and λ, can be distinguished.

In vertebrate genomes, there are no complete genes for the immunoglobulin heavy and light chain polypeptides. Instead, these proteins are specified by gene segments. The heavy chain segments are located on chromosome 14 in humans. There are 11 C_H gene segments, coding for different versions of the constant region, each with a slightly different amino acid sequence. The C_H segments are preceded by 123–129 V_H gene segments, 27 D_H gene segments and 9 J_H gene segments, coding for different versions of the V (variable), D (diverse) and J (joining) components of the variable part of the heavy chain (Figure 19.5). The entire heavy-chain region stretches over several Mb. A similar arrangement is seen with the light-chain loci on chromosomes 2 (κ locus) and 22 (λ locus), the only difference being that the light chains do not have D segments.

During the early stage of development of a B lymphocyte, the immunoglobulin region of the genome rearranges. Within the heavy-chain region, these rearrangements link one of the D_H gene segments with one of J_H gene segments and then link this D–J combination with a V_H gene segment (Figure 19.6). The end result is an exon that contains the complete open reading frame specifying the V, D and J segments of the immunoglobulin protein. This exon becomes linked to a second exon, containing the C segment sequence, by splicing during the transcription process. This creates a complete heavy-chain mRNA that can be translated into an immunoglobulin protein that is specific for just that one lymphocyte.

A similar series of DNA rearrangements results in the lymphocyte's light-chain V–J exon being constructed at either the κ or λ locus, splicing once again attaching an exon containing the C segment when the mRNA is synthesized. Each lymphocyte therefore makes its own specific immunoglobulin and passes the information for synthesis of that immunoglobulin, in the form of its rearranged genome, to all of its progeny.

Diversity of T-cell receptors is based on similar rearrangements that link V, D, J and C gene segments in different combinations to produce cell-specific genes. Each receptor comprises a pair of β molecules, which are similar to

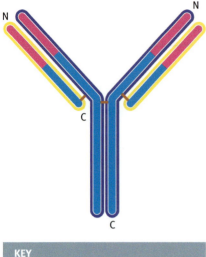

FIGURE 19.4 The structure of a typical immunoglobulin protein. Each heavy chain is 446 amino acids in length and consists of a variable region spanning amino acids 1–108 followed by a constant region. Each light chain is 214 amino acids, again with an N-terminal variable region of 108 amino acids. Additional disulfide bridges form between different parts of individual chains. These and other interactions fold the protein into a more complex three-dimensional structure.

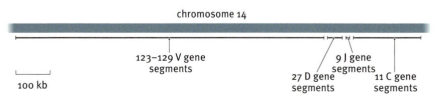

FIGURE 19.5 Organization of the human *IGH* (heavy-chain) region on chromosome 14.

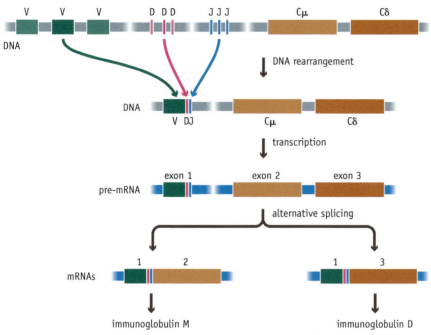

FIGURE 19.6 Synthesis of a specific immunoglobulin protein. DNA rearrangement links V, D and J segments, which are then linked to a C segment by splicing of the mRNA. In immature B cells, the V–D–J exon always becomes linked to the Cμ exon (exon 2) to produce an mRNA specifying a class M immunoglobulin. Later in the development of the B cell, some immunoglobulin D proteins are also produced by alternative splicing that links the V–D–J exon to the Cδ exon. Both types of immunoglobulin become bound to the cell membrane.

the immunoglobulin heavy chain, and two α molecules, which resemble the immunoglobulin κ light chains. As with immunoglobulins, the T-cell receptors become embedded in the cell membrane and enable each lymphocyte to recognize and respond to its own specific extracellular antigen.

A genetic feedback loop can ensure permanent differentiation of a cell lineage

The second mechanism for bringing about a permanent change in the pattern of gene expression involves the use of a feedback loop. In this system, a regulatory protein activates its own transcription so that once its gene has been switched on, it is expressed continuously (**Figure 19.7**).

One of the best-known examples of feedback regulation concerns the MyoD protein of vertebrates, which is involved in muscle development. A cell becomes committed to becoming a muscle cell when it begins to express the *myoD* gene. This gene codes for a transcription activator that targets a number of other genes involved in the differentiation of muscle cells. The MyoD protein also binds upstream of *myoD*, ensuring that its own gene is continuously expressed. The result of this positive feedback loop is that the cell continues to synthesize the MyoD protein and remains a muscle cell. The differentiated state is heritable because cell division is accompanied by transmission of MyoD to the daughter cells, ensuring that these are also muscle cells.

A second well-studied example of a genetic feedback loop involves the fruit fly protein called Deformed (Dfd). This protein plays an important role during *Drosophila* development as without it the fly's head fails to develop correctly. To perform its function, Dfd must be continuously expressed in the head progenitor cells. This is achieved by a feedback system, Dfd binding to an enhancer located upstream of the *Dfd* gene. Feedback

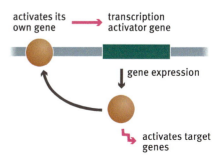

FIGURE 19.7 Feedback regulation of gene expression.

autoregulation also controls the expression of at least some of the important developmental genes of vertebrates.

Chemical modification of nucleosomes leads to heritable changes in chromatin structure

Permanent changes in gene expression can also result from the process that geneticists call chromatin modification. In Section 14.1, we learnt that the DNA in a eukaryotic chromosome is associated with nucleosomes, one nucleosome for every 200 bp or so of DNA. Interactions between nucleosomes are responsible for condensing the DNA into heterochromatin, which contains genes that are inactive in a particular cell. The packaging of regions of the genome into heterochromatin therefore influences which genes are expressed and is thought to be one of the primary determinants of cellular differentiation (Figure 19.8).

Chemical modification of the histone proteins present in the core octamer of the nucleosome appears to be the main factor that influences the formation of heterochromatin. The best studied of these modifications is acetylation. This is the attachment of an acetyl group to one or more of the lysine amino acids in the N-terminal regions of the histone proteins (Figure 19.9). These N termini form tails that protrude from the nucleosome (Figure 19.10). Their acetylation reduces the affinity of the histones for DNA and possibly also reduces the interaction between individual nucleosomes. The histones in heterochromatin are generally unacetylated, whereas those in areas containing active genes are acetylated, a clear indication that this type of modification is linked to DNA packaging. Whether or not a histone is acetylated depends on the balance between the activities of two types of enzyme, the **histone acetyltransferases** (**HATs**), which add acetyl groups to histones, and the **histone deacetylases** (**HDACs**), which remove these groups.

Acetylation is not the only type of histone modification that we know about. Methylation of lysine and arginine residues and phosphorylation of serines in the N-terminal regions of histones are also possible, as is the

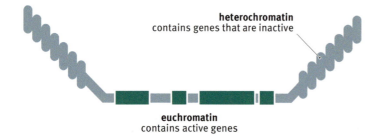

FIGURE 19.8 Euchromatin and heterochromatin. Genes present in heterochromatin cannot be expressed because RNA polymerase and other transcription proteins are unable to gain access to them.

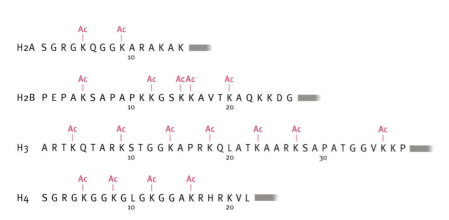

FIGURE 19.9 The positions at which acetyl groups can be attached within the N-terminal regions of the four core histones. The sequences shown are those of the human histones. Each sequence begins with the N-terminal amino acid.

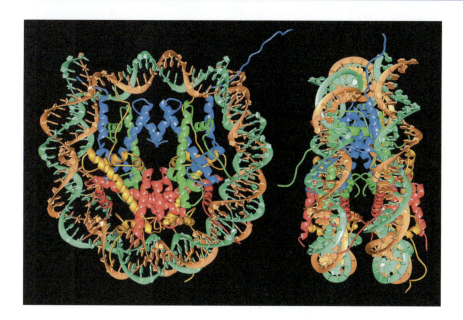

FIGURE 19.10 **Two views of the nucleosome core octamer.** The view on the left is downward from the top of the barrel-shaped octamer, and the view on the right is from the side. The two strands of the DNA double helix wrapped around the octamer are shown in brown and green. The octamer comprises a central tetramer of two histone H3 (blue) and two histone H4 (bright green) subunits plus a pair of H2A (yellow) – H2B (red) dimers, one above and one below the central tetramer. The N-terminal tails of the histone proteins can be seen protruding from the core octamer. (From K. Luger et al., *Nature* 389: 251–260, 1997. With permission from Springer Nature.)

addition of the small, common ('ubiquitous') protein called **ubiquitin** to lysines in the C-terminal regions. Altogether, at least 80 sites in the four core histones are known to be subject to covalent modification of one type or another (see Figure 8.19).

The modifications at these sites interact with one another to determine the degree of chromatin packaging taken up by a particular stretch of DNA. For example, methylation of lysine-9 (the lysine nine amino acids from the N terminus) of histone H3 forms a binding site for the HP1 protein which induces the formation of heterochromatin and silences gene expression, but this event is blocked by the presence of two or three methyl groups attached to lysine-4 (**Figure 19.11**). Methylation of lysine-4 therefore promotes an open chromatin structure and is associated with active genes.

Our growing awareness of the variety of possible histone modifications, and of the way in which different modifications work together, has led to the suggestion that there is a **histone code**, by which the pattern of chemical modifications specifies which regions of the genome are expressed at a particular time. We know of several examples where histone modification is directly linked to cellular differentiation (one such example is described below), but what is still difficult to understand is how these patterns of histone modification are inherited during cell division.

The fruit fly proteins called Polycomb and trithorax influence chromatin packaging

The example of differentiation via chromatin modification that we will look at involves the Polycomb and trithorax proteins of *Drosophila*. The *Polycomb* multigene family codes for proteins that bind to DNA sequences, called Polycomb response elements, and induce the formation of heterochromatin. Polycomb does this by methylating lysines 9 and 27 of histone H3, thereby inducing chromatin packaging.

FIGURE 19.11 **The differential effects of methylation of lysines 4 and 9 of histone H3.**

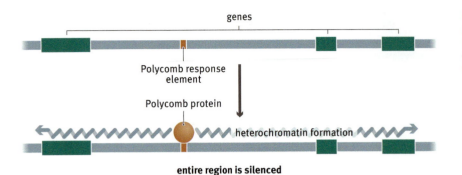

FIGURE 19.12 **The role of Polycomb proteins.** Polycomb proteins maintain silencing in regions of the *Drosophila* genome by initiating heterochromatin formation.

Each response element is approximately 10 kb in length. Exactly how it is recognized by Polycomb proteins is unknown, but the outcome is the nucleation of heterochromatin around the Polycomb proteins, the heterochromatin then propagating along the DNA for tens of kb in either direction (Figure 19.12). The regions that become silenced contain genes that control the development of the individual body parts of the fly – the head, legs and suchlike. Those genes that are not needed in any particular body part are permanently switched off by the action of the Polycomb proteins.

Polycomb proteins do not actually determine which genes will be silenced. Expression of these genes is already repressed before the Polycomb proteins bind to their response elements. The role of Polycomb is therefore to *maintain* rather than *initiate* gene silencing. The important point is that the heterochromatin induced by Polycomb is heritable. After division, the two new cells retain the heterochromatin established in the parent cell. This type of regulation of genome activity is therefore permanent not only in a single cell but also in the cell lineage.

The trithorax proteins have the opposite effect to Polycomb. They maintain an open chromatin state in the regions of active genes, the targets including the same genes as those that are silenced in different body parts by Polycomb proteins. The human equivalent of trithorax achieves this by trimethylating lysine-4 of histone 3, promoting an open chromatin structure. In *Drosophila*, trithorax proteins appear to have a different effect, called **nucleosome remodeling**, which results in the repositioning of nucleosomes within the target region of the genome (Figure 19.13). This does not involve chemical alterations to histone molecules, but instead is induced by an energy-dependent process that weakens the contact between the nucleosome and the DNA with which it is associated.

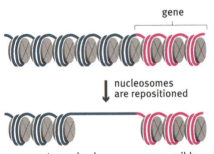

FIGURE 19.13 **Nucleosome remodeling.** Repositioning of nucleosomes in the region upstream of a gene enables the RNA polymerase and other transcription proteins to gain access to the promoter. This is thought to help maintain a gene in an active state.

DNA methylation can silence regions of a genome

Important alterations in gene activity can also be achieved by making chemical changes to the DNA itself. These alterations involve DNA methylation and are associated with the permanent silencing of regions of the genome, possibly entire chromosomes.

In eukaryotes, some cytosine bases in chromosomal DNA molecules are changed to 5-methylcytosine by the addition of methyl groups by enzymes called **DNA methyltransferases** (Figure 19.14). Cytosine methylation is relatively rare in lower eukaryotes but in vertebrates up to 10% of the total

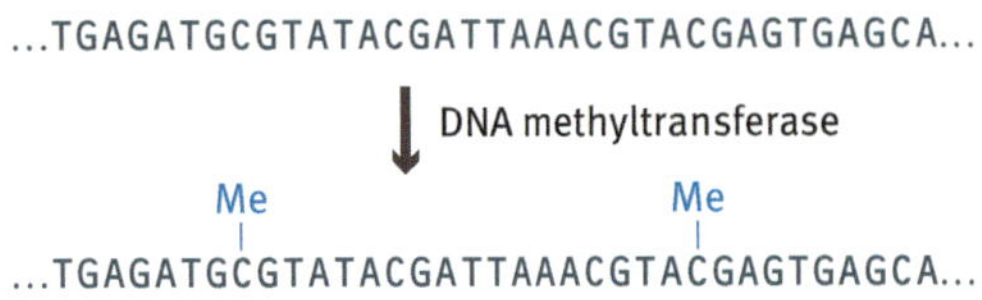

```
...TGAGATGCGTATACGATTAAACGTACGAGTGAGCA...
                    │ DNA methyltransferase
                    ▼
         Me                    Me
         │                     │
...TGAGATGCGTATACGATTAAACGTACGAGTGAGCA...
```

FIGURE 19.14 **Methylation of cytosines in the sequence 5′–CG–3′ in eukaryotic DNA.**

number of cytosines in a genome are methylated, and in plants the figure can be as high as 30%. The methylation pattern is not random, instead being limited to the cytosine in some copies of the sequences 5'–CG–3' and, in plants, 5'–CNG–3', where 'N' is any nucleotide.

A link between DNA methylation and gene expression becomes apparent when the methylation patterns in chromosomal DNAs are examined. These patterns show that active genes are located in unmethylated regions. For example, in humans, 40%–50% of all genes are located close to sequences of approximately 1 kb in which the GC content is greater than the average for the genome as a whole. These are called **CpG islands**. The methylation status of the CpG island reflects the expression pattern of the adjacent gene (Figure 19.15). Housekeeping genes – those that are expressed in all tissues – have unmethylated CpG islands, whereas the CpG islands associated with tissue-specific genes are unmethylated only in those tissues in which the gene is expressed.

With DNA methylation, unlike histone modification, we understand how the methylation pattern is inherited when cells divide. Following DNA replication, each daughter double helix comprises one parent strand and one newly synthesized strand. The former is methylated but the latter is not. This situation does not last long because a methyltransferase enzyme called Dnmt1 scans along the new strand, adding methyl groups to cytosines in the CpG sequences that are methylated in the parent strand (Figure 19.16). This **maintenance methylation** ensures that the two daughter double helices retain the methylation pattern of the parent molecule, which means that the pattern is inherited after cell division.

With methylation, the puzzle has been how the modification results in the repression of gene activity. A possible clue has been provided by the discovery of proteins called **methyl CpG-binding proteins** (**MeCPs**), which attach to methylated CpG islands. These work in conjunction with HDACs, which remove acetyl groups from histones, thereby inducing heterochromatin formation over the adjacent genes (Figure 19.17).

Genomic imprinting and X inactivation result from DNA methylation

Interesting examples of the role of DNA methylation in gene silencing are provided by **genomic imprinting** and **X inactivation**.

FIGURE 19.15 A CpG island.

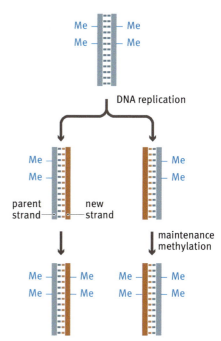

FIGURE 19.16 Maintenance methylation. Maintenance methylation ensures that, after DNA replication, the two daughter helices acquire the methylation pattern of the parent molecule.

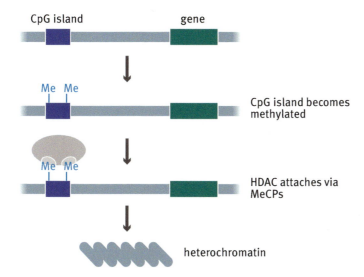

FIGURE 19.17 A model for the link between DNA methylation and genome expression. Methylation of the CpG island upstream of a gene provides recognition signals for the methyl-CpG-binding protein (MeCP) components of a histone deacetylase complex (HDAC). The HDAC modifies the chromatin in the region of the CpG island and hence inactivates the gene.

Genomic imprinting is a relatively uncommon but important feature of mammalian genomes in which one of a pair of genes present on homologous chromosomes is silenced by methylation (Figure 19.18). It also occurs in some insects and some plants. It is always the same member of a pair of genes that is imprinted and hence inactive. For some genes, this is the copy inherited from the mother, and for other genes, it is the paternal copy. Almost 200 genes in humans and mice have been shown to display imprinting, including both protein-coding and noncoding RNA genes. Imprinted genes are distributed around the genome but tend to occur in clusters. For example, in humans there is a 2.2-Mb segment of chromosome 15 within which there are at least ten imprinted genes, and a smaller, 1-Mb region of chromosome 11 which contains eight imprinted genes.

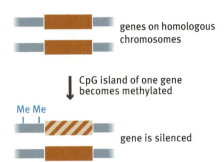

FIGURE 19.18 Silencing of one of a pair of genes by methylation of its CpG island.

An example of an imprinted gene is *Igf2*, which codes for a growth factor, a protein involved in signaling between cells. In mice, only the paternal gene is active (Figure 19.19). On the chromosome inherited from the mother, various segments of DNA in the region of *Igf2* are methylated, preventing the expression of this copy of the gene. A second imprinted gene, *H19*, is located some 90 kb away from *Igf2*, but the imprinting is the other way round – the maternal version of *H19* is active and the paternal version is silent.

Imprinting is controlled by **imprint control elements**. These are DNA sequences that are found within a few kb of clusters of imprinted genes. These elements mediate the methylation of the imprinted regions, but how they do this is unknown. The function of imprinting is also a mystery, but it must be important in development because genetically engineered mice that have two copies of the maternal genome fail to develop properly.

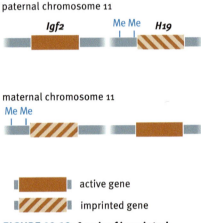

FIGURE 19.19 A pair of imprinted genes on human chromosome 11. *H19* is imprinted on the chromosome inherited from the father, and *Igf2* is imprinted on the maternal chromosome.

X inactivation is a special form of imprinting that leads to the silencing of 80% of the genes on one of the X chromosomes in a female mammalian cell. It occurs because females have two X chromosomes, whereas males have only one. If both of the female X chromosomes were active then proteins coded by genes on the X chromosome might be synthesized at twice the rate in females compared with males. To avoid this undesirable state of affairs, one of the female X chromosomes is silenced and is seen in the nucleus as a condensed structure called the **Barr body** (Figure 19.20), which is made up entirely of heterochromatin.

Silencing occurs early in embryo development and is controlled by the X inactivation center (*Xic*), a region present on each X chromosome (Figure 19.21). In a cell undergoing X inactivation, the inactivation center

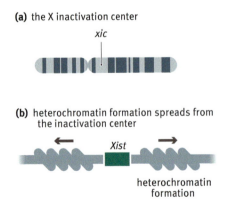

FIGURE 19.21 X inactivation. (a) The location of the inactivation center *xic* on the X chromosome. (b) During X inactivation, heterochromatin spreads from the inactivation center. The gene called *Xist*, located in the inactivation center, is transcribed into RNA which coats the chromosome.

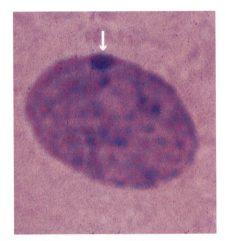

FIGURE 19.20 The Barr body. In this micrograph, the Barr body is seen as a dense object adjacent to the nuclear membrane. (Courtesy of Malcolm Ferguson-Smith, University of Cambridge.)

on one of the X chromosomes initiates the formation of heterochromatin, which spreads out from the nucleation point until the entire chromosome is affected, with the exception of a few short segments. The process takes several days to complete. The process involves a gene called *Xist*, located in the inactivation center, which is transcribed into a 17-kb noncoding RNA, copies of which coat the chromosome as the heterochromatin is formed. At the same time, various histone modifications occur and certain DNA sequences become methylated, although methylation appears to occur after the inactive state has been set up. X inactivation is heritable and is displayed by all cells descended from the initial one within which the inactivation took place.

19.2 COORDINATION OF GENE ACTIVITY DURING DEVELOPMENT

Genome rearrangements, feedback loops, chromatin modifications and DNA methylation, as described above, are the genetic processes thought to be responsible for cellular differentiation. The end products of differentiation – the specialized cell types present in a multicellular organism – must be formed at the correct times during a developmental pathway, and at the appropriate places within the developing organism. To understand the genetic basis for development, we must therefore examine how the gene expression patterns within individual cells can be controlled in time and space.

Research into developmental genetics is underpinned by the use of model organisms

Developmental processes, especially in higher eukaryotes such as humans, are extremely complex, but remarkably good progress toward understanding them has been made in recent years. One of the guiding principles of this research has been the assumption that there should be similarities and parallels between developmental processes in different organisms, reflecting their common evolutionary origins. This means that information relevant to human development can be obtained from studies of **model organisms** chosen for the relative simplicity of their developmental pathways.

The use of model organisms is not new. It has underpinned biological research for decades and explains why laboratory animals such as mice and rats have been used so extensively. For many biological questions, such as the causes of cancer, progress can only be made through studying a model organism that is closely related to humans. Developmental genetics is no different in this regard, but a great deal of information directly relevant to humans and other vertebrates has also been obtained from studies of species much lower down the evolutionary scale. Two species in particular are important as models in developmental genetics, the microscopic nematode worm *Caenorhabditis elegans* and the fruit fly *Drosophila melanogaster*.

Research with *C. elegans* (**Figure 19.22**) was initiated by Sydney Brenner in the 1960s, who chose the organism specifically to act as a model for development. Like all good model organisms, *C. elegans* is easy to grow in the laboratory and has a short generation time, measured in days but still convenient for genetic analysis. The worm is transparent at all stages of its life cycle, so internal examination is possible without killing the animal. This is an important point because it has enabled researchers to follow

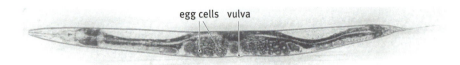

FIGURE 19.22 **The nematode worm *Caenorhabditis elegans*.** The micrograph shows an adult hermaphrodite worm, about 1 mm in length. Egg cells can be seen inside the worm's body in the region on each side of the vulva. The latter is the small projection located on the underside of the animal, about halfway along. (From J. Kendrew (ed.) *Encyclopedia of Molecular Biology*. Oxford: Blackwell, 1994. With permission from Wiley-Blackwell.)

the entire developmental process of the worm at the cellular level. Every cell division in the pathway from fertilized egg to adult worm has been charted, and every point at which a cell adopts a specialized role has been identified (Figure 19.23).

Unlike *C. elegans* the fruit fly was not chosen specifically as a model for developmental studies. It was first used by Thomas Hunt Morgan in 1910 as a convenient organism for studying gene inheritance. For Morgan, the advantages were its small size, enabling large numbers to be studied in a single experiment, its minimal nutritional requirements (the flies like bananas) and the presence in natural populations of occasional variants with easily recognized genetic characteristics such as unusual eye colors. Morgan was not initially aware of the potential of *Drosophila* in developmental research, but in 1915 one of his students, Calvin Bridges, isolated a bizarre mutant that he called *Antennapedia*, which has legs where its antennae ought to be (Figure 19.24). This was the first indication that mutation can affect not just phenotypes such as eye color and wing shape but also the underlying body plan of an organism. We will return to *Antennapedia* later as its study since 1915 has provided some

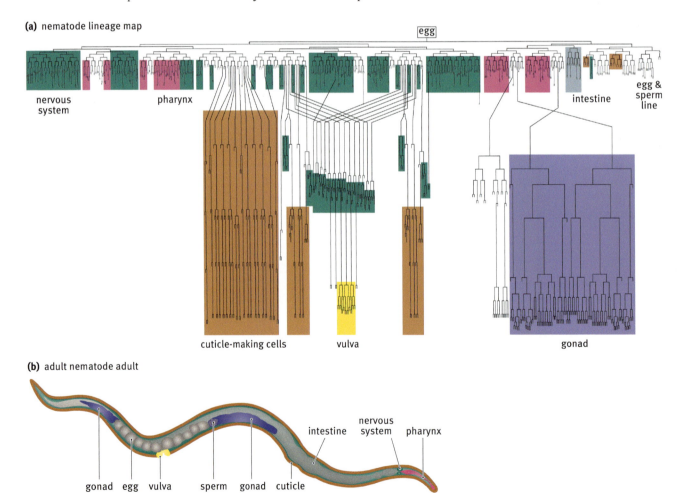

FIGURE 19.23 **Cell divisions on the pathway from fertilized egg to adult *C. elegans*.** The chart begins with the fertilized egg and shows how a series of cell divisions leads to construction of the different body parts of the worm.

of the foundations for our understanding of the spatial control over gene expression.

Our ability to use less complex organisms such as worms and insects as models for development in vertebrates indicates that the fundamental processes that underlie development evolved at a very early period, possibly soon after the first multicellular animals appeared. This means that although vertebrates have much more complex body plans than insects, the genetic processes responsible for the specification of those body plans are similar in both types of organism. We will therefore begin by examining the insights into developmental genetics provided by model organisms and then explore how this information has enabled us to understand the equivalent processes in vertebrates.

C. elegans reveals how cell-to-cell signaling can confer positional information

For development to proceed correctly, cells present at particular positions in an embryo must follow their individual differentiation pathways, so that the correct tissue or structure appears at the appropriate place in the adult organism (Figure 19.25). In order for this to happen, the cells in the embryo must acquire **positional information**. In simple terms, each cell must know where it is located in order to know which genes to express. Research with *C. elegans* has illustrated that cell-to-cell signaling is critical for the establishment of positional information within a developing embryo.

Most *C. elegans* worms are hermaphrodites, meaning that they have both male and female sex organs. The vulva is part of the female sex apparatus, being the tube through which sperm enter and fertilized eggs are laid. The adult vulva is derived from three vulva progenitor cells, called P5.p, P6.p and P7.p, which to begin with are located in a row on the undersurface of the developing worm (Figure 19.26). Each of these progenitor cells becomes committed to the differentiation pathway that leads to the production of vulva cells. The central cell, called P6.p, adopts the 'primary vulva cell fate' and divides to produce eight new cells. The other two cells – P5.p and P7.p – take on the 'secondary vulva cell fate' and divide into seven cells each. These 22 cells then reorganize their positions to construct the vulva.

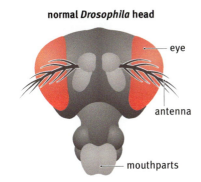
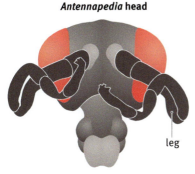

FIGURE 19.24 The *Antennapedia* mutation.

FIGURE 19.25 A 6-day-old human embryo implanting into the wall of its mother's uterus. At this stage of development, the embryo has already acquired the positional information ensuring that the correct structures will appear at the correct positions in the adult. (Courtesy of Yorgos Nikas/Wellcome Images.)

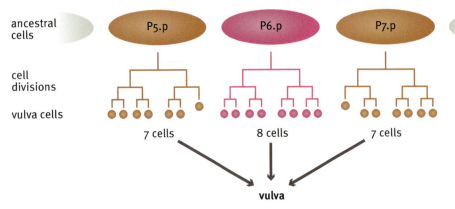

FIGURE 19.26 Cell divisions resulting in production of the vulva cells of *C. elegans*. Three ancestral cells divide in a programmed manner to produce 22 progeny cells, which reorganize their positions relative to one another to construct the vulva.

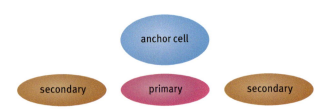

FIGURE 19.27 The relative positions of the anchor cell and the primary and secondary vulva cells.

Vulva development must occur in the correct position relative to the gonad, the structure containing the egg cells. If the vulva develops in the wrong place, then the gonad will not receive sperm and the egg cells will never be fertilized. The positional information needed by the progenitor cells is provided by a cell within the gonad called the anchor cell (**Figure 19.27**). The importance of the anchor cell has been demonstrated by experiments in which it is artificially destroyed in the embryonic worm. In the absence of the anchor cell, a vulva does not develop. This is because the anchor cell secretes an extracellular signaling compound that induces P5.p, P6.p and P7.p to differentiate. This signaling compound is a protein called LIN-3.

Why does P6.p adopt the primary cell fate, whereas P5.p and P7.p take on secondary cell fates? The LIN-3 protein forms a concentration gradient which enables it to have different effects on P6.p, the cell which is closest to the anchor cell, and P5.p and P7.p, which are more distant. The cell surface receptor for LIN-3 is a protein called LET-23 which, when activated by binding LIN-3, initiates a series of intracellular reactions that switches on a variety of genes. The identity of the genes that are switched on depends on the number of LET-23 receptors that are activated, which in turn depends on the extracellular LIN-3 concentration. This means that one set of genes is activated in P6.p, which is exposed to the highest concentration of LIN-3, and a different set is activated in P5.p and P7.p, which are exposed to lower amounts of the signaling protein (**Figure 19.28**). The LIN-3 concentration gradient therefore provides P5.p, P6.p and P7.p with the positional information that these cells need in order to follow their specific differentiation pathways.

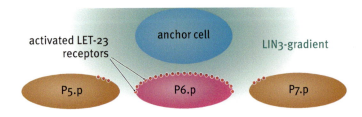

FIGURE 19.28 A LIN-3 concentration gradient establishes the primary and secondary cell fates. The anchor cell secretes LIN-3 protein which forms a concentration gradient. Because of this gradient, a greater number of LET-23 receptors are activated on the P6.p cell than on P5.p or P7.p. This results in P6.p becoming the primary vulva cell, and P5.p and P7.p becoming the secondary cells.

Positional information in the fruit fly embryo is established by a cascade of gene activity

The *C. elegans* vulva provides us with an understanding of the way in which cell-to-cell signaling underlies the establishment of positional information during development. Studies of *Drosophila* have taken this understanding much further.

The body plan of the adult fly is built up as a series of segments, each with a different structural role. This is clearest in the thorax, which has three segments, each carrying one pair of legs, and the abdomen, which is made up of eight segments (Figure 19.29). It is also true for the head, even though in the head the segmented structure is less visible. The early embryo, on the other hand, is a single **syncytium** comprising a mass of cytoplasm and multiple nuclei (Figure 19.30). The developmental process must therefore convert the undifferentiated embryo into a young larva with the correct segmentation pattern.

Initially, the positional information that the embryo needs is the ability to distinguish front (anterior) from back (posterior), as well as similar information relating to up (dorsal) and down (ventral). This information is provided by concentration gradients of proteins that become established in the syncytium. Most of these proteins are not synthesized from genes in the embryo, but are translated from mRNAs injected into the embryo by the mother. One of these **maternal effect genes** is called *bicoid*. The *bicoid* gene is transcribed in the maternal nurse cells, which are in contact with the egg cells, and its mRNA is injected into the anterior end of the unfertilized egg (Figure 19.31). This position is defined by the orientation of the egg cell in the egg chamber. The *bicoid* mRNA remains in the anterior region of the egg cell, attached to the cell's cytoskeleton. Bicoid proteins, translated from the mRNA, diffuse through the syncytium, setting up a concentration gradient, highest at the anterior end and lowest at the posterior end. Additional maternal effect proteins, such as Hunchback, Nanos, Caudal and Torso, contribute in a similar way to the anterior–posterior axis, while Dorsal and others set the dorsal–ventral axis. As a result, each point in the syncytium acquires its own unique chemical signature defined by the relative amounts of the various maternal effect proteins.

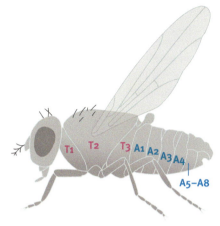

FIGURE 19.29 The segmentation pattern of the adult *Drosophila melanogaster*. The head is also segmented, but the pattern is not easily discernible from the morphology of the adult fly. T1–T3 are segments of thorax and A1–A8 make up the abdomen.

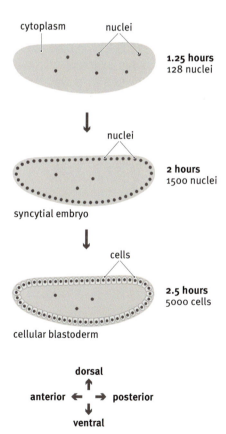

FIGURE 19.30 Early development of the *Drosophila* embryo. To begin with, the embryo is a single syncytium containing a gradually increasing number of nuclei. These nuclei migrate to the periphery of the embryo after about 2 hours, and within another 30 minutes cells begin to be constructed. The embryo is approximately 500 μm in length and 170 μm in diameter.

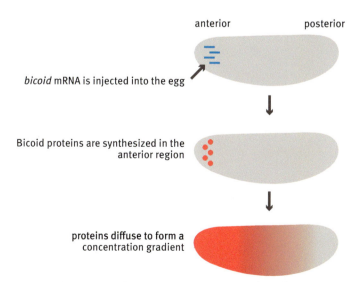

FIGURE 19.31 Establishment of the Bicoid gradient in a *Drosophila* egg cell. The *bicoid* mRNA is injected into the anterior end of the egg. Bicoid proteins, translated from the mRNA, diffuse through the syncytium setting up a concentration gradient.

This positional information is made more precise by the expression of the **gap genes**. Three of the anterior–posterior gradient proteins – Bicoid, Hunchback and Caudal – are transcription activators that target the gap genes in the nuclei that now line the inside of the embryo (Figure 19.32). The identities of the gap genes expressed in a particular nucleus depend on the relative concentrations of the maternal effect proteins and hence on the position of the nucleus along the anterior–posterior axis. The gap proteins therefore form their own, more detailed, gradients within the embryo (Figure 19.33).

The next set of genes to be activated, the **pair rule genes**, establishes the basic segmentation pattern. Transcription of these genes responds to the relative concentrations of the gap proteins. The embryo can now be looked upon as comprising a series of stripes, each stripe consisting of a set of cells expressing a particular combination of pair rule genes (Figure 19.34a). In a further round of genes activation, the **segment polarity genes** become switched on, providing greater definition to the stripes by setting the sizes and precise locations of what will eventually be the segments of the larval fly (Figure 19.34b). Gradually, the positional information of the maternal effect gradients has been converted into a sharply defined segmentation pattern.

Segment identity is determined by the homeotic selector genes

The pair rule and segment polarity genes establish the segmentation pattern of the fruit fly embryo but do not themselves determine the identities of the individual segments. This is the role of the **homeotic selector genes**. These were first discovered because of the extravagant effects that mutations in these genes have on the appearance of the adult fly. An example is the *Antennapedia* mutation, which we met earlier, which transforms the head segment that usually produces an antenna into one that makes a leg, so the mutant fly has a pair of legs where its antennae should be (see Figure 19.24). The early geneticists were fascinated by

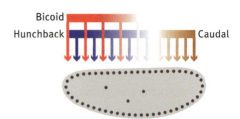

FIGURE 19.32 Activation of the gap genes in the *Drosophila* embryo. Bicoid, Hunchback and Caudal are transcription activators that target gap genes in the nuclei lining the inside of the *Drosophila* embryo. Because each of these three proteins has formed a concentration gradient, a differential expression pattern for the gap genes is established along the anterior–posterior axis.

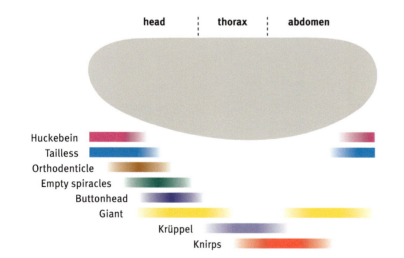

FIGURE 19.33 The role of the gap proteins in conferring positional information in the *Drosophila* embryo. The concentration gradient for each of the gap proteins is denoted by the colored bars. These gradients are established by differential activation of the gap genes by Bicoid, Hunchback and Caudal. The parts of the embryo that give rise to the head, thorax and abdomen of the adult fly are indicated.

FIGURE 19.34 Later stages in development of the Drosophila embryo. The positions at which particular genes are expressed have been revealed by staining the embryo with a labeled antibody that binds specifically to (a) the pair rule protein Even-skipped and (b) the segment polarity protein Engrailed. (a, Courtesy of Michael Levine, University of California, Berkeley, and Stephen Small, New York University. b, From C. Hama, Z. Ali & T.B. Kornberg, Genes Dev. 4: 1079-1093, 1990. With permission from Cold Spring Harbor Laboratory Press.)

these monstrous **homeotic mutants** and many were collected during the first few decades of the 20th century.

Genetic mapping of homeotic mutations has revealed that the selector genes are clustered in two groups on chromosome 3. These clusters are called the Antennapedia complex (ANT-C), which contains genes involved in determination of the head and thorax segments, and the Bithorax complex (BX-C), which contains genes for the abdomen segments (Figure 19.35). The order of genes corresponds to the order of the segments in the fly, the first gene in ANT-C being *labial palps*, which controls the most anterior segment of the fly, and the last gene in BX-C being *Abdominal B*, which specifies the most posterior segment.

The correct selector gene is expressed in each segment because the activation of each one is responsive to the positional information provided by the distributions of the gap and pair rule proteins. The selector gene products are themselves transcription activators that switch on the sets of genes needed to initiate differentiation of the specified segment. Maintenance of the differentiated state is ensured partly by the repressive effect that each homeotic selector gene product has on the expression of the other selector genes, and partly by the work of Polycomb which, as we saw above, constructs inactive heterochromatin over the selector genes that are not expressed in a particular cell.

FIGURE 19.35 The Antennapedia and Bithorax gene complexes of *D. melanogaster*. The full gene names are as follows: *lab*, labial palps; *pb*, proboscipedia; *Dfd*, Deformed; *Scr*, Sex combs reduced; *Antp*, Antennapedia; *Ubx*, Ultrabithorax; *abdA*, abdominal A and *AbdB*, Abdominal B.

Homeotic selector genes are also involved in vertebrate development

Up until the mid-1980s, it was not realized that studies of development in the fruit fly were directly relevant to development in vertebrates. This remarkable discovery was made after the *Drosophila* homeotic selector genes were sequenced and it was found that each one contains a segment, 180 bp in length, that is very similar in the different genes (Figure 19.36). We now know that this sequence, called the **homeobox**, codes for a DNA-binding structure, the **homeodomain**, that enables these proteins to attach to DNA and hence perform their roles as transcription activators.

The similarities between the homeoboxes of the various *Drosophila* selector genes led researchers in the 1980s to search for other genes containing this sequence. First, the *Drosophila* genome was examined, resulting in the isolation of several previously unknown homeobox-containing genes. These have turned out not to be selector genes but other types of gene coding for transcription activators involved in development. Examples include the pair rule genes *even-skipped* and *fushi tarazu*, and the segment polarity gene *engrailed*.

The presence of additional homeobox genes in *Drosophila* was not particularly surprising. The real excitement came when the genomes of other

FIGURE 19.36 The homeobox is a nucleotide sequence that is similar in different homeotic selector genes.

organisms were searched and it was realized that homeoboxes are present in genes in a wide variety of animals, including humans. Some of the homeobox genes in these other organisms are homeotic selector genes organized into clusters similar to ANT-C and BX-C, these genes having equivalent functions to the *Drosophila* versions, specifying the construction of the body plan. This is illustrated by the HoxC8 gene of mouse, mutations in this gene resulting in an animal that has an extra pair of ribs, due to the conversion of a lumbar vertebra (normally in the lower back) into a thoracic vertebra (from which the ribs emerge). Other Hox mutations in animals lead to limb deformations, such as the absence of the lower arm, or extra digits on the hands or feet.

We now look on the ANT-C and BX-C clusters of selector genes in *Drosophila* as two parts of a single complex, the homeotic gene complex or HOM-C. In vertebrates, there are four homeotic gene clusters, called HoxA to HoxD. As in *Drosophila*, the order of genes in the vertebrate clusters reflects the order of the structures specified by the genes in the adult body. This is clearly seen with the mouse HoxB cluster, which controls development of the nervous system (Figure 19.37).

When the four vertebrate clusters are aligned with one another and with the *Drosophila* HOM-C cluster, similarities are seen between the genes at equivalent positions (Figure 19.38). This suggests that they have evolutionary relationships. The implication is that in the vertebrate lineage, there were two duplications of the original Hox cluster. The first duplication converted the single cluster seen today in fruit flies into a pair of Hox clusters, and the second duplication resulted in the four clusters seen in vertebrates (Figure 19.39).

Not all of the vertebrate Hox genes have been ascribed functions, but we believe that the additional versions possessed by vertebrates relate to the added complexity of the vertebrate body plan. From our anthropocentric viewpoint, we might therefore assume that four is the maximum number of Hox clusters possessed by any organism on the planet. We would be sadly disillusioned. Several types of fish have more. Teleosts, a group of

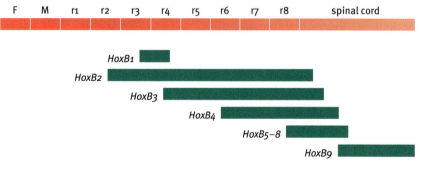

FIGURE 19.37 Specification of the mouse nervous system by selector genes of the HoxB cluster. The nervous system is shown schematically and the positions specified by the individual HoxB genes (*HoxB1* to *HoxB9*) indicated by the green bars. The components of the nervous system are as follows: F, forebrain; M, midbrain; r1–8, rhombomeres 1–8; followed by the spinal cord. Rhombomeres are segments of the hindbrain seen during development.

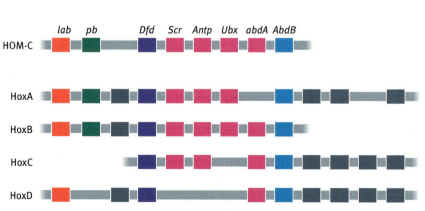

FIGURE 19.38 Comparison between the *Drosophila* HOM-C gene complex and the four Hox clusters of vertebrates. Genes that code for proteins with related structures and functions are indicated by the same colors.

ray-finned fish, have seven or eight Hox clusters, presumed to have arisen by duplication of the vertebrate set, followed by the loss of one cluster in the ancestor of those fish that have just seven clusters. Their extra copies of the Hox genes are thought to underlie the vast range of different variations of the basic body plan displayed by these fish (Figure 19.40). A further duplication and loss of cluster in the salmonid lineage has resulted in 13 Hox clusters in the Atlantic salmon *Salmo salar* and the rainbow trout *Oncorhynchus mykiss*.

Homeotic genes also underlie plant development

The power of *Drosophila* as a model system for development extends even beyond vertebrates. Developmental processes in plants are, in many respects, very different from those of fruit flies and other animals, but at the genetic level, there are similarities. These are sufficient for knowledge gained from *Drosophila* to be of value in interpreting developmental processes in plants. In particular, the recognition that a limited number of homeotic selector genes control the *Drosophila* body plan has led to a model for plant development which postulates that the structure of the flower is also determined by a small number of homeotic genes.

All flowers are constructed along similar lines, made up of four concentric whorls, each comprising a different floral organ (Figure 19.41). The outer whorl, number 1, contains sepals, which are modified leaves that envelop the bud during its early development. The next whorl, number 2, contains

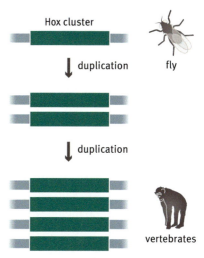

FIGURE 19.39 Duplications in the Hox gene cluster during the evolutionary lineage leading to vertebrates.

FIGURE 19.40 Teleost fish, probably the most diverse group of organisms to have evolved. These examples were painted by the French naturalist, François-Louis Laporte, comte de Castelnau, from specimens observed in the Amazon and La Plata river systems during 1842–1847. The species are (left to right, top to bottom): bluespotted cornetfish (*Fistularia tabacaria*), silver mylossoma (*Mylossoma duriventre*), *Mesonauta acora* (a type of chiclid), planet catfish (*Corydoras splendens*), armored catfish (*Pseudacanthicus spinosus*), blue tang surgeonfish (*Acanthurus coeruleus*) and yellowtip damselfish (*Stegastes pictus*).

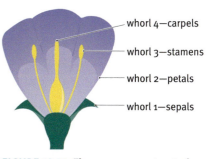

FIGURE 19.41 Flowers are constructed from four concentric whorls.

the distinctive petals, and within these are whorls 3 (stamens, the male reproductive organs) and 4 (carpels, the female reproductive organs).

Most of the research on plant development has been carried out with *Antirrhinum* (the snapdragon) and *Arabidopsis thaliana*, a type of cress that has been adopted as a model species, partly because it has a genome of only 125 Mb, one of the smallest known among flowering plants (Figure 19.42). Although these plants do not appear to contain homeodomain proteins, they do have genes which, when mutated, lead to homeotic changes in the floral architecture, such as the replacement of sepals by carpels. Analysis of these mutants has led to the **ABC model**, which states that there are three types of homeotic genes – A, B and C – which control flower development. According to the ABC model, whorl 1 is specified by A-type genes (which include *apetala1* and *apetala2* in *Arabidopsis*), whorl 2 by A genes acting with B genes (such as *apetala3*), whorl 3 by the B genes plus the single C gene (*agamous*) and whorl 4 by the C gene acting on its own (Figure 19.43).

As anticipated from the work with *Drosophila*, the A, B and C homeotic gene products are transcription activators. All except the APETALA2 protein contain the same DNA-binding domain, the **MADS box**, which is also found in other proteins involved in plant development. These include SEPALLATA1, 2 and 3, which work with the A, B and C proteins in defining the detailed structure of the flower. Other similarities with *Drosophila* include the presence of Polycomb group proteins, such as CURLY LEAF, which methylate histones in order to silence parts of the genome containing those homeotic genes that are inactive in a particular whorl.

FIGURE 19.42 *Arabidopsis thaliana,* **a model organism for plant genetics.** (Courtesy of Dr Jeremy Burgess/Science Photo Library.)

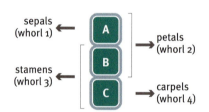

FIGURE 19.43 The ABC model for flower development. According to this model, whorl 1 is specified by A-type genes, whorl 2 by A genes acting with B genes, whorl 3 by the B and C genes and whorl 4 by the C gene on its own.

KEY CONCEPTS

- Differentiation is the process by which an individual cell acquires a specialized function. Differentiation requires a change in the pattern of gene expression in a cell, the resulting specialized expression pattern usually remaining in place for the remainder of the cell's lifetime, and often being inherited by the daughter cells, and the granddaughters and so on for many cell divisions.

- Development is the process that begins with a fertilized egg cell and ends with an adult. A developmental pathway comprises a complex series of genetic, cellular and physiological events that must occur in the correct order, in the correct cells, and at the appropriate times if the pathway is to reach a successful culmination. These events include the differentiation of many cells into many different specialized types.

- There are four ways in which permanent changes in gene activity can be brought about. Physical rearrangement of the genome can result in a change in the structure of the genome inherited by a cell lineage. A genetic feedback loop can be established in order to maintain a change in gene expression in a cell and its descendants. Chromatin modification of nucleosomes and DNA methylation result in gene silencing.

- Epigenesis is the term used to describe any heritable change to the genome that does not involve mutation.

- One of the guiding principles of developmental genetics is the assumption that there should be similarities and parallels between developmental processes in different organisms, reflecting their common evolutionary origins. This means that information relevant to human development can be obtained from studies of model organisms chosen for the relative simplicity of their developmental pathways.

- Studies of *C. elegans* have shown how cell-to-cell signaling can confer positional information during embryo development.
- The establishment of the segmentation pattern of a fruit fly embryo involves the sequential expression of different groups of genes. Positional information initially contained in the gradient concentrations of the maternal effect proteins is converted into a sharply defined segmentation pattern.
- The activities of homeotic selector genes define the identities of the fruit fly segments.
- Homeotic selector genes are present in all multicellular animals including humans.
- Homeotic genes also control developmental processes in plants such as the construction of the flower.

QUESTIONS AND PROBLEMS

Key terms

Write short definitions of the following terms:

ABC model
antigen
Barr body
chromatin modification
CpG island
DNA methylation
DNA methyltransferase
epigenesis
gap gene
genomic imprinting

histone acetyltransferase
histone code
histone deacetylase
homeobox
homeodomain
homeotic mutant
homeotic selector gene
imprint control element
MADS box
maintenance methylation

maternal effect gene
methyl-CpG-binding protein
model organism
nucleosome remodeling
pair rule gene
positional information
segment polarity gene
syncytium
ubiquitin
X inactivation

Self-study questions

19.1 Distinguish between the terms 'differentiation' and 'development'.

19.2 Describe how immunoglobulin diversity is brought about by physical rearrangement of the human genome.

19.3 Give two examples of genetic feedback loops.

19.4 Distinguish between the activities of histone acetyltransferases and histone deacetylases.

19.5 What types of chemical modification are known to be made to histone proteins? What effects do these modifications have on expression of the genome?

19.6 Explain what is meant by the term 'histone code'.

19.7 Describe the role of the fruit fly *Polycomb* gene family.

19.8 Using an example, explain how nucleosome remodeling can influence expression of a gene.

19.9 Explain the link between DNA methylation and silencing of the genome.

19.10 Give two examples that illustrate the principles of genomic imprinting.

19.11 What is the purpose of X inactivation, and how is X inactivation brought about?

19.12 Explain why model organisms are important in studies of development.

19.13 Outline the features of *Caenorhabditis elegans* and *Drosophila melanogaster* that make these species good model organisms for the study of development.

19.14 Describe how studies of vulva development in *C. elegans* illustrate how positional information is generated and used during a developmental pathway.

19.15 Explain how positional information in the fruit fly embryo is initially established through the action of the maternal effect genes.

19.16 Distinguish between the roles of the gap genes, pair rule genes and segment polarity genes during the development of the fruit fly embryo.

19.17 What is the genetic basis of the *Antennapedia* mutation of the fruit fly?

19.18 Describe the role of homeotic selector genes during development in the fruit fly.

19.19 Explain why the discovery of homeotic selector genes in the fruit fly is relevant to development in vertebrates.

19.20 What is the ABC model, and how does it explain the development of a flower?

Discussion topics

19.21 Explore and assess the histone code hypothesis.

19.22 In many areas of biology, it is difficult to distinguish between cause and effect. Evaluate this issue with regard to nucleosome remodeling and genome expression – does nucleosome remodeling cause changes in genome expression, or is it the effect of these expression changes?

19.23 Maintenance methylation ensures that the pattern of DNA methylation on two daughter DNA molecules is the same as the pattern on the parent molecule. In other words, the methylation pattern, and the information on gene expression that it conveys, is inherited. Other aspects of chromatin structure might also be inherited in a similar way. How do these phenomena affect the principle that inheritance is specified by genes?

19.24 In a normal diploid female, one X chromosome is inactivated and the other remains active. Remarkably, in diploid females with unusual sex chromosome constitutions, the process still results in just a single X chromosome remaining active. For example, in those rare individuals that possess just a single X chromosome no inactivation occurs, and in those individuals with an XXX genotype two of the three X chromosomes are inactivated. What might be the means by which the numbers of X chromosomes in a nucleus are counted so that the appropriate number of X chromosomes can be inactivated?

19.25 What would be the key features of an ideal model organism for the study of development in vertebrates?

19.26 Are *C. elegans* and *D. melanogaster* good model organisms for development in vertebrates?

The answers to, or hints on how to answer, the self-study questions and discussion topic questions can be downloaded from the book's product page at this link: www.routledge.com/9781032743530

FURTHER READING

Alt FW, Blackwell TK & Yancopoulos GD (1987) Development of the primary antibody repertoire. *Science* 238, 1079–1087. *Generation of immunoglobulin diversity.*

Amores A, Force A, Yan Y-L et al. (1998) Zebrafish *hox* clusters and vertebrate genome evolution. *Science* 282, 1711–1714. *Hox genes in fish.*

Barlow DP & Bartolomei MS (2016) Genomic imprinting in mammals. *Cold Spring Harbor Perspectives in Biology* 6, a018382.

Blackledge NP & Klose RJ (2021) The molecular principles of gene regulation by Polycomb repressive complexes. *Nature Reviews Molecular Cell Biology* 22, 815–833.

Du Q, Luu P-L, Stirzaker C et al. (2015) Methyl-CpG-binding domain proteins: readers of the epigenome. *Epigenomics* 7, 1051–1073.

Gebelein B & Ma J (2016) Regulation in the early *Drosophila* embryo. *Reviews in Cell Biology and Molecular Medicine* 2, 140–167.

Heard E, Clerc P & Avner P (1997) X-chromosome inactivation in mammals. *Annual Review of Genetics* 31, 571–610.

Irish V (2017) The ABC model of plant development. *Current Biology* 27, R887–R890.

Kouzarides T (2007) Chromatin modifications and their function. *Cell* 126, 693–705. *Histone modifications.*

Lawrence M, Daujat S & Schneider R (2016) Lateral thinking: how histone modifications regulate gene expression. *Trends in Genetics* 32, 42–56.

Margueron R, Trojer P & Reinberg D (2005) The key to development: interpreting the histone code? *Current Opinion in Genetics and Development* 15, 163–176.

Schindler AJ & Sherwood DR (2013) Morphogenesis of the *Caenorhabditis elegans* vulva. *Wiley Interdisciplinary Reviews: Developmental Biology* 2, 75–95.

Shin H & Reiner DJ (2018) The signaling network controlling *C. elegans* vulval cell fate patterning. *Journal of Developmental Biology* 6, 30.

Suzuki MM & Bird A (2008) DNA methylation landscapes: provocative insights from epigenomics. *Nature Reviews Genetics* 9, 465–476. *The role of methylation in controlling gene activity.*

Zakany J & Duboule D (2007) The role of *Hox* genes during vertebrate limb development. *Current Opinion in Genetics and Development* 17, 359–366.

CHAPTER 20

The Human Genome

The human genome is a typical example of a mammalian genome, but it also illustrates many of the general features of the genomes of all animals and plants. By devoting a chapter to the human genome, we can therefore learn more about the genetics of our own species while, at the same time, gaining an understanding of how genomes are organized in many other species.

In this chapter, we will study three important topics concerning the human genome. The first is what the genome contains. We will look at the genes that are present and how they are arranged. In doing so, we will learn that the genes take up relatively little space, and a greater fraction of the genome is made up of repetitive DNA sequences that have no obvious function.

The second topic we will examine is the relationship between the human genome and those of other primates. We will explore the extent to which the information contained in these genomes helps us to understand what it is that makes us human.

The final topic in this chapter is the human mitochondrial genome. In addition to the DNA in the nucleus, humans and almost all other eukaryotes also have smaller DNA molecules in their mitochondria, the energy-generating organelles present in the cytoplasm (Figure 20.1). These mitochondrial molecules contain genes, and we must investigate the functions of these genes and ask why they are located in the mitochondria.

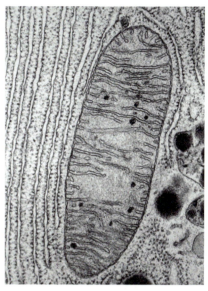

FIGURE 20.1 An electron micrograph of a cross section of a mitochondrion. (Courtesy of Don Fawcett/Science Photo Library.)

20.1 THE GENETIC ORGANIZATION OF THE HUMAN NUCLEAR GENOME

The human nuclear genome is made up of approximately 3,100 Mb of DNA, split into 24 chromosomes. Each chromosome contains a single DNA molecule, the shortest 47 Mb and the longest 249 Mb. About 39% of the genome has, or at one time had, a function that we understand. This fraction includes all the genes, pseudogenes and other types of gene relic, as well as the promoter regions upstream of genes and the introns contained within discontinuous genes (Figure 20.2). The remainder, a little more than 60% of the genome, is made up of intergenic DNA.

The human genome has fewer protein-coding genes than we expected

The most recent estimates for the number of protein-coding genes in a range of eukaryotic genomes are given in **Table 20.1**. These figures indicate that the simplest eukaryotes, typified by the single-celled yeasts, have approximately 5,000–7,000 protein-coding genes. As we move up the scale, we see that it takes almost 14,000 genes to make a fruit fly and just under 20,000 to make a *Caenorhabditis elegans* worm, a surprising result as, intuitively, we might expect a microscopic worm to be less complex than an insect.

We also see that, based on these comparisons, humans are only slightly more complex than *C. elegans*, as our genome has only 20,449 protein-coding genes. This number is much lower than originally expected, as a 'best guess' of 80,000–100,000 was still in vogue up to a few months before the human genome sequence was completed in 2000. These early estimates were high because they were based on the supposition that,

DOI: 10.1201/9781003473862-24

FIGURE 20.2 The composition of the human genome.

TABLE 20.1 GENE NUMBERS FOR VARIOUS EUKARYOTES	
Species	Number of Protein-Coding Genes
Schizosaccharomyces pombe (fission yeast)	5,145
Saccharomyces cerevisiae (budding yeast)	6,600
Drosophila melanogaster (fruit fly)	13,968
Caenorhabditis elegans (nematode worm)	19,985
Homo sapiens (human)	20,449
Arabidopsis thaliana (thale cress)	27,655
Zea mays (maize)	39,756

in most cases, a single gene specifies a single protein. According to this model, the number of protein-coding genes in the human genome should be similar to the number of different types of proteins in human cells, leading to the estimates of 80,000–100,000. The discovery that the number of genes is much lower than this indicates that alternative splicing, the process by which exons are assembled in different combinations so that more than one protein can be produced from a single gene, is more prevalent than was originally appreciated.

The comparisons between species should not therefore be on the basis of the number of genes, but on the number of different proteins that can be produced from those genes. Now humans do exhibit a more satisfying degree of additional complexity compared to microscopic worms, because 75% of all human protein-coding genes, representing 95% of those with two or more introns, undergo alternative splicing, with an average of four different spliced mRNAs per gene. This means that the 20,449 human genes specify a total of 78,120 proteins. Alternative splicing also occurs in lower eukaryotes, but it is less prevalent. In *C. elegans*, only about 25% of the protein-coding genes have alternative splicing pathways, with an average of 2.2 variants per gene.

The functions of about 70% of the 20,449 human protein-coding genes are known or can be inferred with a reasonable degree of certainty. Almost a quarter of these genes are involved in expression, replication and maintenance of the genome. Enzymes responsible for the general biochemical functions of the cell account for another 20% of the known genes, and the remainder are involved in activities such as the transport of compounds into and out of cells, the folding of proteins into their correct three-dimensional structures, the immune response and synthesis of structural proteins such as those found in the cytoskeleton and in muscles.

The arrangement of genes in the human genome is largely random

Human genes cover a spectrum of sizes from less than 100 bp to more than 2,000 kb. The smallest genes specify noncoding RNAs such as tRNAs, snRNAs and snoRNAs. Some of these are as short as 65–75 bp. The smallest protein-coding genes are slightly longer, for example, 189 bp for the *KRTAP6-2* gene, which specifies a protein found in hair. At the other end of the size spectrum are genes whose bulk is made up not of coding sequence but of introns. The longest human gene is the one coding for Rbfox1, a protein involved in the regulation of alternative splicing, which contains 15 introns in its 2,474 kb length. The coding sequence of the *RBFOX1* gene (i.e. all of the exons added together) is only 1,191 bp, less than 0.05% of the length of the gene. A point worth noting is that this single gene is over one-half the length of the entire *E. coli* chromosome.

The genes are not spread evenly throughout the human genome. Some chromosomes are relatively gene-rich, such as chromosome 13 which has 41.2 genes per Mb of sequence. Others are gene-poor, including the Y chromosome which has only 3.0 genes per Mb. Within a single chromosome, the gene-rich areas are interspersed with relatively empty stretches of DNA, the latter including **gene deserts** in which the density is very low over regions as long as several Mb (Figure 20.3). These less dense regions are often adjacent to, or spanning, the centromere, which is made up mainly of repetitive DNA (Section 14.2).

The randomness of the processes that shaped chromosome architecture is also reflected in the distribution of the members of multigene families (Figure 20.4). With some families, the individual genes are clustered together at a single position in the genome. Examples are the growth hormone gene family, whose five members are clustered on chromosome 17, and the 5S rRNA gene family, comprising 2,000 genes in a tandem array on the long arm of chromosome 1. With other families, the genes are interspersed around the genome. For example, the three members of the aldolase gene family are located on chromosomes 9, 16 and 17. Some large gene families are both clustered and interspersed at the same time. There are about 280 copies of the rRNA transcription unit, grouped into five clusters of 50–70 units each, the clusters located on the short arms of chromosomes 13, 14, 15, 21 and 22. The implication is that, in most cases, chance events during the past evolution of the genome have been responsible for placing the genes in the positions in which we find them today.

The genes make up only a small part of the human genome

The variations in gene density that occur along the length of a human chromosome mean that it is difficult to identify regions in which the organization of the genes can be looked on as 'typical' of the genome as a whole. Bearing this in mind, we will take a closer look at a 200-kb segment of chromosome 1 in order to understand the detailed organization of the genome (Figure 20.5).

This segment contains all or part of three genes. There is no obvious relationship between them, emphasizing the apparently random nature of the distribution of genes in the human genome. First, we have the end

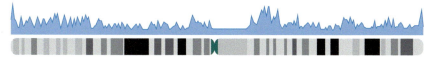

FIGURE 20.3 Gene density along human chromosome 1.

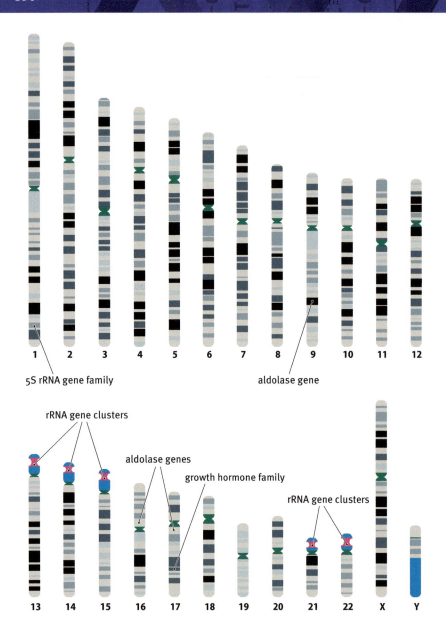

FIGURE 20.4 Locations of some genes and gene families in the human genome.

of the *BSND* gene, which codes for a chloride channel protein. This is a membrane-bound protein that forms a pore through which various ions, including chloride, can enter and leave the cell. The second gene is *PCSK9*, specifying an enzyme called proprotein convertase subtilisin/kexin type 9, which is involved in the breakdown of low-density lipoproteins, thereby playing an important role in the metabolism of cholesterol. Finally, there is the start of *USP24*, which codes for ubiquitin-specific peptidase 24, a protease that removes ubiquitin side chains from proteins that have been modified by ubiquitination.

We can see from Figure 20.5 how little of the human genome actually contains biological information. Each of the three genes is discontinuous, the number of introns ranging from three for *BSND* to 67 for *USP64*. When added together, the total length of the exons is just 10,664 bp, equivalent to 5.33% of the 200-kb segment. Throughout the genome as a whole, the amount of coding DNA is rather less than this. All the exons in the human genome make up only 48 Mb, just 1.5% of the total (see Figure 20.2).

In addition to the genes, there are multiple repeat sequences dispersed at various positions in this segment of chromosome 1. These are examples

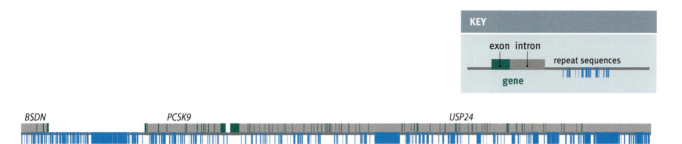

FIGURE 20.5 A 200-kb segment of the human genome. The map shows the locations of genes and repeat sequences in a 200 kb segment of human chromosome 1.

of sequences that recur at many places in the genome. Most of them are located in the intergenic regions of our segment, but several lie within introns. The distribution of these repeats is typical of the genome as a whole, as we will see later in this chapter when we examine the repetitive DNA in more detail.

Genes are more densely packed in the genomes of lower eukaryotes

How extensive are the differences in gene organization among eukaryotes? To explore this question, we will compare our 200-kb segment of the human genome with 'typical' segments of the same length, from the genomes of the yeast *Saccharomyces cerevisiae*, the fruit fly *Drosophila melanogaster* and maize (Figure 20.6).

The yeast genome segment comes from chromosome IV. The first striking feature is that it contains more genes than the human segment, 106 genes thought to code for proteins, four that specify transfer RNAs and one small nucleolar RNA gene. Relatively few of these genes are discontinuous, reflecting the fact that in the entire yeast genome, there are only 287 discontinuous genes and the vast majority of these genes have just one intron each. There are also fewer repeat sequences. This part of chromosome IV contains just two repeats, both truncated long terminal repeat (LTR) retrotransposons (Section 12.3). When all 16 yeast chromosomes are considered, the total amount of sequence taken up by repetitive DNA is only 3.3% of the total.

The picture that emerges is that the genetic organization of the yeast genome is much more economical than that of the human version. The genes themselves are more compact, having fewer introns, and the spaces between the genes are relatively short, with much less space taken up by repetitive DNA and other noncoding sequences.

The hypothesis that less complex eukaryotes have more compact genomes holds true when other species are examined. Figure 20.6 also shows a 200-kb segment of the fruit fly genome. If we agree that a fruit fly is more complex than a yeast but less complex than a human, then we would expect the organization of the fruit fly genome to be intermediate between that of yeast and humans. This is what we see, this 200-kb segment of the fruit fly genome having nine genes, more than in the human segment but fewer than in the yeast sequence. Eight of these genes are discontinuous, with some of the introns similar in length to those in human genes. Again there are just two repeat sequences. The picture is similar when the entire genome sequences of the three organisms are compared (Table 20.2). The gene density in the fruit fly genome is intermediate between that of yeast and humans, and the average fruit fly gene has many more introns than the average yeast gene, but only one-third the number in the average human gene.

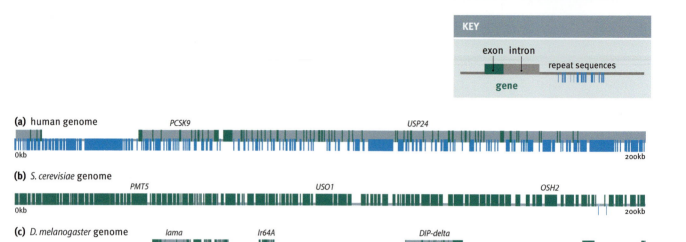

FIGURE 20.6 Genome organization in different eukaryotes. Comparison between 200kb segments of the (a) human, (b) yeast, (c) fruit fly and (d) maize genomes. In (b)–(d), names are given for only a few of the genes.

TABLE 20.2 COMPACTNESS OF THE YEAST, FRUIT FLY AND HUMAN GENOMES			
Feature	Yeast	Fruit Fly	Human
Gene density (average number per Mb)	543	97	7
Introns per gene (average)	0.05	3.5	10
Amount of the genome that is taken up by repetitive DNA	3.3%	20.5%	48.5%

The comparison between the yeast, fruit fly and human genomes also reveals the differing repetitive DNA contents of these genomes. Repeat sequences make up 3.3% of the yeast genome, about 20% of the fruit fly genome and 49% of the human genome. It is beginning to become clear that repeat sequences play an intriguing role in dictating the compactness or otherwise of a genome. This is strikingly illustrated by the maize genome, which at 2,500 Mb is relatively small for a flowering plant. The maize genome is dominated by repetitive elements. In the segment shown in Figure 20.6, there are nine genes, eight of which contain one or more short introns. Instead of genes, the dominant feature of this genome segment is the repeat sequences. The majority of these are LTR retrotransposons, which comprise virtually all of the noncoding part of the segment, and on their own are estimated to make up approximately 50% of the maize genome.

20.2 THE REPETITIVE DNA CONTENT OF THE HUMAN NUCLEAR GENOME

The genome segments shown in Figure 20.6 show us that repetitive DNA is an important component of most eukaryotic genomes. We must therefore look more closely at these repeat sequences.

Repetitive DNA can be divided into two categories (Figure 20.7). The first is genome-wide or **interspersed repeats**, whose individual repeat units are distributed around the genome in an apparently random fashion.

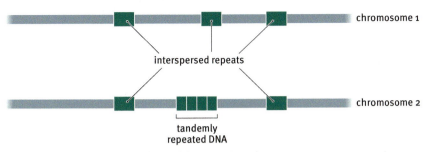

FIGURE 20.7 The two types of repetitive DNA. Two chromosome segments are shown. Interspersed repeats are distributed around the genome in an apparently random fashion. Tandemly repeated units are located next to each other in an array.

Then there is **tandemly repeated DNA**, whose repeat units are placed next to each other in an array. Repetitive DNA, and the other parts of the genome of no known function, used to be called **junk DNA**. This term is falling out of favor, partly because the number of surprises resulting from genome research over the last few years has meant that geneticists have become less confident in asserting that any part of the genome is unimportant simply because currently we do not know what it might do.

Interspersed repeats include RNA transposons of various types

We are already familiar with many of the most important types of interspersed repeat in the human and other eukaryotic genomes, as these are the cellular RNA transposons, related to viral retroelements, that we studied in Section 12.3. LTR retrotransposons of the *Ty1/copia* and *Ty3/gypsy* families are present in many eukaryotic genomes, but humans and other mammals are slightly unusual in this regard. The bulk of their LTR transposons are decayed endogenous retroviruses rather than the types of LTR retrotransposon that predominate in other eukaryotes. They are all members of the same family of elements so this is only a small difference.

Not all types of retrotransposon have LTR elements. In mammals, the most important types of non-LTR retroelements, or **retroposons**, are the **LINEs (long interspersed nuclear elements)** and **SINEs (short interspersed nuclear elements)**. SINEs have the highest copy number for any type of interspersed repetitive DNA in the human genome, with more than 1.7 million copies comprising 14% of the genome as a whole. LINEs are less frequent, with about 850,000 copies, but as they are longer they make up a larger fraction of the genome, more than 20%. Most of the interspersed repeats in the segment of the human genome shown in Figure 20.5 are LINEs and SINEs.

There are three families of LINEs in the human genome, of which one group, LINE-1, is both the most frequent and the only type that is still able to transpose. The LINE-2 and LINE-3 families are made up of inactive relics. A full-length LINE-1 element is 6.1 kb and has two genes. One of these genes codes for a polyprotein similar to the product of the *pol* gene of viral retroelements, this polyprotein including a reverse transcriptase enzyme (Figure 20.8a). There are no LTRs, but the 3' end of the LINE is marked by a series of A–T base pairs, giving what is usually referred to as a poly(A) sequence, though of course it is a poly(T) sequence on the other strand of the DNA. Only 80–100 of the 500,000 LINE-1 elements in the human genome are full-length versions, with the average size of all the copies being just 900 bp. Most are therefore inactive, and LINE-1 transposition is a rare event, although it has been observed in cultured cells. It also appears to be responsible for hemophilia in some patients, whose factor VIII gene contains a LINE-1 sequence that prevents synthesis of the important factor VIII blood clotting protein.

SINEs are much shorter than LINEs, being just 100–400 bp and not containing any genes (Figure 20.8b). Instead SINEs 'borrow' enzymes that have been synthesized by LINEs in order to transpose. The commonest SINE in primate genomes is **Alu**, which has a copy number of approximately 1.2 million in humans. Some Alu elements are actively copied into RNA, providing the opportunity for the proliferation of the element.

Alu is derived from the gene for the 7SL RNA, a noncoding RNA involved in the movement of proteins around the cell. The first Alu element may have arisen by the accidental reverse transcription of a 7SL RNA molecule and integration of the DNA copy into the genome. Other SINEs are derived from tRNA genes which, like the gene for the 7SL RNA, are transcribed by RNA polymerase III in eukaryotic cells, suggesting that some feature of the transcripts synthesized by this polymerase make these molecules prone to occasional conversion into retroposons.

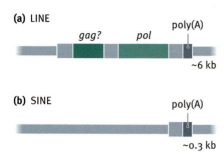

FIGURE 20.8 Non-LTR retroelements in the human genome.

Tandemly repeated DNA forms satellite bands in density gradients

Tandemly repeated DNA is also called **satellite DNA** because DNA fragments containing tandemly repeated sequences form 'satellite' bands when genomic DNA is centrifuged in a density gradient. This is the type of centrifugation that was used in the Meselson–Stahl experiment, which proved that DNA replication is semiconservative (Section 10.1).

The buoyant density of a DNA molecule, which determines the position it takes up in a density gradient, depends on its GC content. Human DNA has GC content of 40.3% and a buoyant density of 1.701 g cm^{-3}. Human DNA therefore forms a band at the 1.701 g cm^{-3} position in a density gradient. This is not, however, the only band that appears. With human DNA, there are also three 'satellite' bands, at 1.687, 1.693 and 1.697 g cm^{-3} (Figure 20.9). These additional bands contain repetitive DNA. They form because the long chromosomal molecules become cleaved into fragments, 50 kb or so in length, when the cell is broken open. Fragments containing predominantly single-copy DNA have GC contents close to the 'standard' human value of 40.3%, and so move to the 1.701 g cm^{-3} position in the density gradient. However, fragments containing large amounts of repetitive DNA behave differently. For instance, a fragment made up entirely of ATTAC repeats has a GC content of just 20%, lower than the standard value, and so has a buoyant density somewhat less than 1.701 g cm^{-3}. These repetitive DNA fragments therefore migrate to a satellite position, above the main band in a density gradient. A single genome can contain several different types of satellite DNA, each with a different repeat unit, these units being anything from <5 to >200 bp. The three satellite bands in human DNA include at least four different repeat types.

We have already encountered one type of human satellite DNA, the alphoid DNA repeats found in the centromere regions of chromosomes (Section 14.2). Although some satellite DNA is scattered around the genome, most is located in the centromeres, where it may play a structural role, possibly as binding sites for one or more of the special centromere proteins.

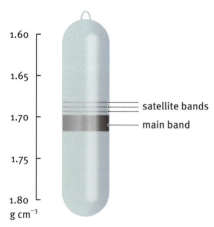

FIGURE 20.9 Satellite DNA from the human genome. Human DNA has an average GC content of 40.3% and average buoyant density of 1.701 g cm^{-3}. Fragments made up mainly of single-copy DNA have a GC content close to this average and are contained in the main band in the density gradient. The satellite bands at 1.687, 1.693 and 1.697 g cm^{-3} consist of fragments containing repetitive DNA. The GC contents of these fragments depend on their repeat motif sequences and are different from the genome average, meaning that these fragments have different buoyant densities to single-copy DNA and migrate to different positions in the density gradient.

tRNAs, translate the mRNAs into protein. There are only 22 tRNAs, fewer than the minimum usually needed to read the genetic code. This means that the standard codon–anticodon pairing rules have to be supplemented by 'superwobble', which allows some tRNAs to recognize all four codons of a single family (Figure 20.18). The genetic code that is used is slightly different to the one that operates in the rest of the cell (see Table 7.2).

The various genes account for 15,368 bp of the mitochondrial DNA, with most of the remainder taken up by the replication origin. Very little of the genome, only 87 bp or so, is genetically unimportant. In fact, the human mitochondrial genome seems to have gone to great lengths to become as compact as possible. The rRNA genes are among the smallest known, coding for a large subunit rRNA with a sedimentation coefficient of only 16S and a small subunit RNA of 12S. Compare these figures with the typical values for eukaryotic rRNAs given in Figure 6.4. There are no intergenic regions, so the mRNAs just comprise the coding sequence of the gene. Several mRNAs are so truncated that the termination codons are incomplete. Five mRNAs end with just a U or UA, with the rest of the termination codon provided by polyadenylation after transcription (Figure 20.19). The small number of tRNAs also appears to result from a need to keep the genome as small as possible. Why the genome has to be so small is not understood.

Mitochondrial genomes as a whole display a great deal of variability

Almost all eukaryotes have mitochondrial genomes. Initially, it was thought that virtually all of these were circular DNA molecules, like the human mitochondrial genome. We still believe this to be the case for most organisms, but we now recognize that there is a great deal of variability. In many eukaryotes, the circular molecules coexist with linear versions, and in some microbial eukaryotes (e.g. *Paramecium*, *Chlamydomonas* and several yeasts), the mitochondrial genome is always linear.

Mitochondrial genomes vary in size and gene content (Table 20.3), with neither feature related to the complexity of the organism. Most multicellular animals have small mitochondrial genomes with a compact genetic organization, the genes being close together with little space between them. The human mitochondrial genome is typical of this type. In contrast, most lower eukaryotes such as *S. cerevisiae*, as well as flowering

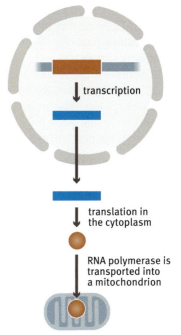

FIGURE 20.17 **The mitochondrial RNA polymerase is coded by a nuclear gene.** The mRNA is translated in the cytoplasm and the protein transported into the mitochondrion.

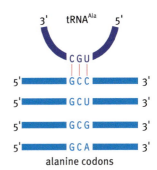

FIGURE 20.18 **An example of superwobble occurring during translation of human mitochondrial mRNAs.** Superwobble enables a single mitochondrial tRNA to recognize all four codons in a four-codon family.

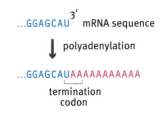

FIGURE 20.19 **Some human mitochondrial mRNAs lack a complete termination codon.** The codon sequence is completed by polyadenylation.

TABLE 20.3	FEATURES OF MITOCHONDRIAL GENOMES			
Species	Genome Size	Number of Genes		
		Protein-coding	Noncoding	Total
Plasmodium falciparum (protozoan)	6 kb	3	0	3
Chlamydomonas leiostraca (green alga)	14 kb	7	12	19
Homo sapiens (human)	17 kb	13	24	37
Saccharomyces cerevisiae (yeast)	86 kb	19	27	46
Reclinomonas americana (protozoa)	69 kb	67	30	97
Arabidopsis thaliana (plant)	367 kb	33	25	58

plants, have larger and less compact mitochondrial genomes, with several of the genes containing introns. The number of genes varies from three for the malaria parasite *Plasmodium falciparum* to 97 for the protozoan *Reclinomonas americana*. All but the smallest mitochondrial genomes contain genes for the rRNAs and at least some of the protein components of the respiratory chain, the latter being the main biochemical feature of the mitochondrion. The more gene-rich genomes also code for tRNAs, ribosomal proteins, RNA polymerase and proteins involved in transport of other proteins into the mitochondrion from the surrounding cytoplasm.

A general feature of mitochondrial genomes emerges from Table 20.3. These genomes specify some of the proteins found in mitochondria, but not all of them. The other proteins are coded by nuclear genes, synthesized in the cytoplasm and transported into the mitochondria. If the cell has mechanisms for transporting proteins into mitochondria, then why not have all the mitochondrial proteins specified by the nuclear genome? We do not yet have a convincing answer to this question, although it has been suggested that at least some of the proteins coded by the mitochondrial genome are extremely hydrophobic and cannot be transported through the membranes that surround the mitochondrion, and so simply cannot be moved into the organelle from the cytoplasm. The only way in which the cell can get them into the mitochondrion is to make them there in the first place.

Mitochondria and chloroplasts were once free-living prokaryotes

The discovery of genomes in mitochondria and chloroplasts led to many speculations about their origins. Today, most biologists accept that the **endosymbiont theory** is correct, at least in outline, even though it was considered quite unorthodox when first proposed in the 1960s.

The endosymbiont theory is based on the observation that the gene expression processes occurring in mitochondria and chloroplasts are similar in many respects to the equivalent processes in bacteria. In addition, when nucleotide sequences are compared, the genes in these organelles are found to be more similar to equivalent genes from bacteria than they are to eukaryotic nuclear genes. The endosymbiont theory therefore holds that mitochondria and chloroplasts are the relics of free-living bacteria that formed a symbiotic association with the precursor of the eukaryotic cell, way back at the very earliest stages of evolution (Figure 20.20).

Support for the endosymbiont theory has come from the discovery of organisms which appear to exhibit stages of endosymbiosis that are less advanced than those seen with mitochondria and chloroplasts. For example, an early stage in endosymbiosis is displayed by the group of algae called glaucophytes, which possess photosynthetic **cyanelles** that are different from chloroplasts and instead resemble ingested cyanobacteria (Figure 20.21).

If mitochondria and chloroplasts were once free-living bacteria, then since the endosymbiosis was set up, there must have been a transfer of genes from the organelle into the nucleus. We do not understand how this occurred, or indeed whether there was a mass transfer of many genes at once, or a gradual trickle from one site to the other. But we do know that DNA transfer from organelle to nucleus, and indeed between organelles, still occurs. This was discovered in the early 1980s, when the first partial sequences of chloroplast genomes were obtained. It was found that in some plants the chloroplast genome contains segments of DNA, often including entire genes, that are copies of parts of the mitochondrial

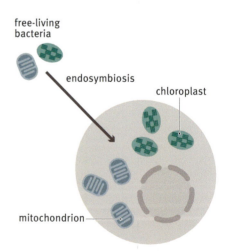

FIGURE 20.20 The endosymbiont theory. According to this theory, mitochondria and chloroplasts are the relics of free-living bacteria that formed a symbiotic association with the precursor of the eukaryotic cell.

genome. The implication is that this so-called **promiscuous DNA** has been transferred from one organelle to the other.

We now know that the transfer of DNA between genomes has occurred in the evolutionary histories of many species. This point is illustrated by the plant *Arabidopsis thaliana*. Its mitochondrial genome contains various segments of nuclear DNA as well as 16 fragments of the chloroplast genome, including six tRNA genes that have retained their activity after transfer to the mitochondrion. The nuclear genome of *Arabidopsis* includes several short segments of the chloroplast and mitochondrial genomes as well as a 270-kb piece of mitochondrial DNA located within the centromere region of chromosome 2. The transfer of mitochondrial DNA to vertebrate nuclear genomes has also been documented.

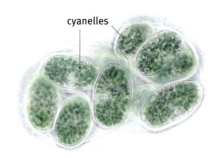

FIGURE 20.21 A possible living clue to the origin of chloroplasts. Cyanelles in cells of the glaucophyte alga *Cyanophora*. (Courtesy of Michael Abbey/Science Photo Library.)

KEY CONCEPTS

- The human genome has approximately 20,500 protein-coding genes but specifies 80,000–100,000 proteins. Because of alternative splicing, many genes code for more than one protein.

- The genes are not spread evenly throughout the human genome and there is little obvious order or logic to the locations of individual genes.

- As a general rule, the more complex types of eukaryotes have less compact genomes, with a greater number of introns, longer intergenic regions and a greater proportion of repetitive DNA.

- Repetitive DNA can be divided into two categories. The first is genome-wide or interspersed repeats, whose individual repeat units are distributed around the genome in an apparently random fashion. The second category is tandemly repeated DNA, whose repeat units are placed next to each other in an array.

- The human and chimpanzee genomes display 0.9% nucleotide dissimilarity with more than 30% of the genes coding for proteins that are identical in the two species.

- Structural variations between the human and chimpanzee genomes affect some genes involved in brain development, including the *SRGAP2* gene, which is a single copy in the chimpanzee genome but multicopy in human DNA. The human *FOXP2* protein has two amino acid differences compared to the chimpanzee version which might be important in specifying the degree of synaptic plasticity in the human nervous system.

- Almost all eukaryotes have mitochondrial genomes and photosynthetic species also have genomes in their chloroplasts.

- The endosymbiont theory for the origins of organelle genomes holds that mitochondria and chloroplasts are the relics of free-living bacteria that formed a symbiotic association with the precursor of the eukaryotic cell.

QUESTIONS AND PROBLEMS

Key terms

Write short definitions of the following terms:

Alu
cyanelle
DNA profile
endosymbiont theory
extrachromosomal gene
gene desert
hominid
interspersed repeat

junk DNA
LINE
microsatellite
minisatellite
molecular phylogenetics
promiscuous DNA
retroposon
satellite DNA

short tandem repeat
simple sequence length
  polymorphism
simple sequence repeat
SINE
tandemly repeated DNA
variable number tandem repeats

Self-study questions

20.1 Why is the human genome able to specify many more types of protein than are present in *Caenorhabditis elegans*, even though the two species have similar gene numbers?

20.2 Give examples that illustrate the different lengths of the genes present in the human genome.

20.3 Describe the differences between the organizations of the human growth hormone and aldolase gene families.

20.4 How are genes for the rRNA molecules arranged in the human genome?

20.5 What is a 'gene desert'?

20.6 Describe the key differences in gene organization in humans, yeast and the fruit fly.

20.7 What is unusual about the repetitive DNA content of the maize genome?

20.8 Distinguish between interspersed and tandemly repeated DNA.

20.9 Describe the difference between a LINE and SINE.

20.10 What are the key features of the LINE-1 family of sequences in the human genome?

20.11 Explain how Alu sequences are thought to have arisen.

20.12 Describe how the buoyant density of a DNA molecule is measured.

20.13 Give examples of satellite and minisatellite DNA sequences in the human genome.

20.14 What are the special features of microsatellite DNA?

20.15 Outline the evolutionary relationships between humans and chimpanzees.

20.16 Describe the similarities and differences between the human and chimpanzee genomes.

20.17 Describe the important features of the human mitochondrial genome.

20.18 Outline how the human mitochondrial genome is similar to or different from the genomes of other organisms.

20.19 According to the endosymbiont theory, how did organelle genomes originate?

20.20 Explain how the features of glaucophyte algae lend support to the endosymbiont theory.

Discussion topics

20.21 Devise a research project whose objective is to identify all parts of the human genome that are transcribed into RNA.

20.22 Discuss the theory that repetitive DNA is simply 'junk'.

20.23 Explain how the study of microsatellites might be used to establish whether a male and female adult and three children are all members of a single family.

20.24 How might the comparisons between the human and chimpanzee genomes be supplemented with other types of study in order to understand what makes us human?

20.25 How do we know that human chromosome 2 arose by fusion of two chromosomes that are separate in chimpanzees, rather than the two chimpanzee chromosomes arising from a single human chromosome that split into two?

20.26 What reasons might there be for having some proteins coded by genes in the mitochondrion and others coded in the nucleus?

 The answers to, or hints on how to answer, the self-study questions and discussion topic questions can be downloaded from the book's product page at this link: www.routledge.com/9781032743530

FURTHER READING

Anderson S, Bankier AT, Barrell BG et al. (1981) Sequence and organization of the human mitochondrial genome. *Nature* 290, 457–465.

Csink AK & Henikoff S (1998) Something from nothing: the evolution and utility of satellite repeats. *Trends in Genetics* 14, 200–204.

Enard W, Gehre S, Hammerschmidt K et al. (2009) A humanized version of Foxp2 affects cortico-basal ganglia circuits in mice. *Cell* 137, 961–971.

Hatje K, Mühlahausen S, Simm D et al. (2019) The protein-coding human genome: annotating high-hanging fruits. *BioEssays* 41, 1900066. *Describes changes in the estimated numbers of human genes and the difficulties in identifying the exact number.*

Ostertag EM & Kazazian HH (2005) LINEs in mind. *Nature* 435, 890–891. *Brief review of recent research into LINEs.*

Payseur BA, Jing P & Haasl RJ (2011) A genomic portrait of human microsatellite variation. *Molecular Biology and Evolution* 28, 303–312.

Roger AJ, Muñoz-Gómez SA & Kamikawa R (2017) The origin and diversification of mitochondria. *Current Biology* 27, R1177–R1192.

Sánchez-Baracaldo P, Raven JA, Pisani D et al. (2017) Early photosynthetic eukaryotes inhabited low-salinity habitats. *Proceedings of the National Academy of Sciences USA* 114, E7727–E7745. *Identifying the bacterial ancestor of chloroplasts.*

Suntsova MV & Buzdin AA (2020) Differences between human and chimpanzee genomes and their implications in gene expression, protein functions and biochemical properties of the two species. *BMC Genomics* 21, 535.

CHAPTER 21

Genes and Medicine

A great deal of biological research is directly or indirectly aimed at medical questions. This is equally as true with genetics as it is in any other area of biology. Throughout this book, we have touched on the importance of genetics in the understanding human health and disease. In this chapter, we will pull several of these threads together by considering some of the links between genetics and medicine. Our focus will be on the genetic basis to **inherited disorders** and to cancer, and the genetic approaches, such as **gene therapy**, which are being explored as possible ways of curing or managing these illnesses.

21.1 THE GENETIC BASIS TO INHERITED DISORDERS

An inherited disorder is one that is caused by the sequence of the genome and which, like other genetic features, can be passed from parents to offspring. Sometimes these are called **genetic disorders** but the term must, however, be used with caution because there are some genetic disorders that are not inherited, but instead arise due to mutations in somatic cells that occur during an individual's lifetime.

There are more than 6,000 inherited **monogenic** disorders, ones that result from a mutation in a single gene. Approximately one in every 200 births is of a child with one or other of these disorders (**Table 21.1**). The commonest include the lung disease cystic fibrosis, Huntington's disease, which is a neurological disorder characterized by uncoordinated body movements, and Marfan syndrome, which affects connective tissue and leads to various problems including heart disease. Other disorders are much rarer, with just a few cases reported worldwide, possibly affecting just a few families. Inherited diseases also affect other animals, in particular those subject to inbreeding such as some pedigree dogs.

TABLE 21.1 SOME OF THE COMMONEST MONOGENIC DISORDERS

Disorder	Symptoms	Frequency (UK Births per Year)
Inherited breast cancer	Cancer	1 in 300 females
Cystic fibrosis	Lung disease	1 in 2,000
Huntington's disease	Neurodegeneration	1 in 2,000
Duchenne muscular dystrophy	Progressive muscle weakness	1 in 3,000 males
Hemophilia A	Blood disorder	1 in 4,000 males
Sickle-cell anemia	Blood disorder	1 in 10,000
Phenylketonuria	Mental retardation	1 in 12,000
β-Thalassemia	Blood disorder	1 in 20,000
Retinoblastoma	Cancer of the eye	1 in 20,000
Hemophilia B	Blood disorder	1 in 25,000 males
Tay-Sachs disease	Blindness, loss of motor control	1 in 200,000

Many inherited disorders are caused by loss-of-function mutations

In Chapter 11, we learnt that mutations in a multicellular organism can result in either a loss of function or a gain of function, with the former being much more common. This is as true with inherited disorders as with any other phenotype, and the vast majority of human monogenic illnesses are due to a loss-of-function mutation that inactivates a particular gene. The protein coded by that gene is absent, or modified in some way that prevents it from functioning correctly, leading to the inherited disorder (Figure 21.1).

A loss of function leading to an inherited disorder can arise in a number of different ways (Figure 21.2). The loss of function could be caused by a point mutation, such as a nonsense mutation that creates in an internal termination codon within a gene. It could equally well be caused by a small deletion or insertion within the gene, or by a large deletion that removes the entire gene. The mutation might also be in a regulatory sequence upstream of the gene so that no transcription initiation occurs, or within one of the internal sequences that control splicing.

A single disorder is often caused by more than one mutation, with different individuals carrying different mutations but still suffering from the same disease. We addressed this issue in Section 3.2 when we examined *CFTR*, the causative gene of cystic fibrosis, as an example of a gene with many different variants (see Figure 3.17). Geneticists place these 2,500 variants in five groups depending on the effect that the mutation has on the CFTR protein (Table 21.2). The commonest variant is called ΔF508. The underlying mutation is the deletion of nucleotides 1,514–1,516 from the *CFTR* gene. The notation ΔF508 indicates that this mutation results in the deletion (Δ) of a phenylalanine (F) from position 508 of the CFTR protein (Figure 21.3a). ΔF508 is class II mutation: the protein is synthesized by ribosomes in the cytoplasm, but the absence of the phenylalanine affects the ability of the protein to be transported to the cell membrane. The resulting loss of function leads to inflammation and mucus accumulation in the lungs, the primary symptoms of cystic fibrosis.

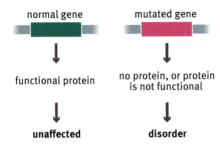

FIGURE 21.1 The genetic basis of an inherited disorder.

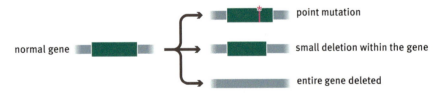

FIGURE 21.2 Three ways in which a loss of function can give rise to an inherited disorder.

TABLE 21.2 CLASSES OF CYSTIC FIBROSIS MUTATION		
Class	Effect on CFTR Protein	Prevalence*
I	No protein synthesis	10%
II	Protein is synthesized but fails to attach to the cell membrane	88%
III	Regulation of protein activity is defective	4%
IV	Reduced rate of chloride transport	<2%
V	Reduced amount of protein	rare
*Some individuals have more than one mutation in the CFTR gene, which is why the percentages for prevalence add up to more than 100		

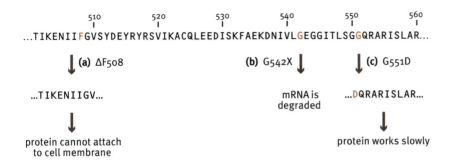

FIGURE 21.3 **The three commonest cystic fibrosis mutations.** The amino acid sequence of the CFTR protein from positions 501 to 560 is shown. (a) The ΔF508 mutation results in deletion of the phenylalanine at position 508, and gives rise to a protein that cannot attach to the cell membrane. (b) The G542X mutation replaces the codon for G452 with a termination codon. The mRNA transcribed from this mutant gene is degraded before any protein is made. (c) The G551D mutation replaces the glycine at position 551 with an aspartic acid, giving rise to a protein with only 4% the activity of the unmutated version.

The next most common cystic fibrosis mutation, G542X, is a class I variant. It is a mutation at position 1,624 in the gene that converts the codon for a glycine (G), normally present at position 542 in the protein, into a termination codon (X) (**Figure 21.3b**). The presence of this internal termination codon in the mRNA transcribed from the mutant *CFTR* gene switches on the surveillance system that degrades mutated mRNAs. This means that the CFTR protein is not synthesized, again resulting in the loss of function.

The third most common mutation, G551D, has a class IV effect. Replacing glycine-551 with aspartic acid does not prevent the protein from being synthesized, but changes its kinetic properties so it now transports chloride ions at only 4% of the rate displayed by the unmutated protein (**Figure 21.3c**). Again, the mutation results in a loss of function. Although the protein is still able to transport chloride ions, it does so at such a slow rate that the physiological function of the protein is lost.

Inherited disorders resulting from a gain of function are much less common

Gain of function is much less common among inherited disorders, simply because there are relatively few types of underlying mutation that can cause a gain of function. One possibility is overexpression of a gene, so that its protein product accumulates in the cell in greater than normal quantities. Overexpression could be due to a mutation in one of the regulatory sequences that control transcription initiation, but more frequently arises from gene duplication. This is the cause of Charcot-Marie-Tooth disease, one of the commonest inherited neurological disorders, which affects the peripheral nerves and muscle strength so that actions such as walking become difficult. The increase from two to three copies of the myelin protein gene *PMP22* is sufficient, in some unknown way, to cause the symptoms of this disease. Duplication of *PMP22* arises by recombination between repeat sequences either side of the 1.5-Mb region of chromosome 17 that contains this gene (**Figure 21.4**). The recombination can also lead to the deletion of *PMP22* on one of the chromosome copies, reducing the gene number from two to one. This leads to a *loss* of function, manifested as a completely different disease called tomaculous

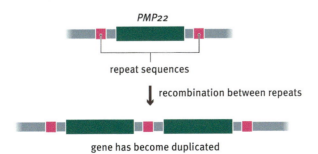

FIGURE 21.4 **Duplication of the *PMP22* gene.**

neuropathy, again affecting the peripheral nerves, but in this case causing numbness and unusual sensitivity to pressure.

A second way of acquiring a gain of function is by mutation of a protein that acts as a cell surface receptor and which relays a message into the cell when it binds an external signaling compound, such as a hormone. The signal passed on by the surface receptor might have various effects inside the cell, including changes in gene expression patterns. A normal cell will respond to the external stimulus in an appropriate way, but some mutations can affect a cell surface receptor so that it remains active even when its external signaling compound is absent (see Figure 11.29). The cell therefore acts as if it is continuously being affected by the signaling compound, resulting in the gain in function. Early onset of male puberty can be caused in this way by mutation of the luteinizing hormone receptor, and the very rare Jansen's disease, with only 20 known cases in the entire world, is due to this kind of mutation in the parathyroid hormone receptor, leading to growth issues.

Although rare among inherited disorders, gain-of-function mutations are relatively common causes of cancer. We will therefore return to them later in the chapter.

Trinucleotide repeat expansions can cause inherited disorders

One particular type of DNA mutation that often leads to an inherited disorder deserves a special mention. This is trinucleotide repeat expansion, in which a relatively short series of trinucleotide repeats become elongated to two or more times its normal length. These expansions are associated with a number of neurodegenerative diseases, including Huntington's disease (HTT), one of the commonest inherited disorders (see Table 21.1). The normal *HTT* gene contains the sequence 5'–CAG–3' repeated between 6 and 29 times in tandem, coding for a series of glutamines in the protein product. In Huntington's disease, this repeat expands to a copy number of 38–180, increasing the length of the polyglutamine tract and resulting in loss of function of the HTT protein (see Figure 11.8). The exact biochemical role of this protein has not yet been discovered but it is thought act within the nerve cells of the brain.

Several other inherited disorders are also caused by expansions of polyglutamine codons (**Table 21.3**). There are also examples of diseases resulting from expansion of different trinucleotides that lie outside of the coding region of a gene. In Friedreich's ataxia, a 5'–GAA–3' expansion within the first intron of the frataxin gene decreases the synthesis of the mature mRNA (**Figure 21.5**). This results in a loss of function that leads to a decrease in the number of nerve cells in the spinal cord. Similarly, a 5'–CTG–3' expansion in the trailer region of the myotonic dystrophy protein kinase gene leads to premature degradation of the mRNA, resulting in muscle wasting.

Some trinucleotide expansions in the 5' untranslated region of a gene give rise to a **fragile site**, a position where the chromosome is likely to break. Examples of these trinucleotide expansions are an expansion of CGG triplets upstream of the *FMR1* gene, and a GCC expansion upstream of *AFF2*. Both of these genes are on the X chromosome and the syndromes are called fragile X and fragile XE, respectively. Fragile site syndromes are associated with intellectual disability, but chromosome breakage is not the underlying cause of the disorder. Instead, the expansion interferes with methylation of a CpG island upstream of the affected gene, which is silenced leading to loss of the protein product.

TABLE 21.3 EXAMPLES OF TRINUCLEOTIDE REPEAT EXPANSION DISORDERS

Gene	Repeat Sequence		Associated Disease
	Normal	Mutated	
Polyglutamine expansions (within the coding region of the gene)			
HTT	$(CAG)_{6-29}$	$(CAG)_{38-180}$	Huntington's disease
AR	$(CAG)_{13-31}$	$(CAG)_{40}$	Spinomuscular bulbar atrophy
ATN1	$(CAG)_{6-35}$	$(CAG)_{49-88}$	Dentatorubral-pallidoluysian atrophy
ATXN1	$(CAG)_{6-39}$	$(CAG)_{41-83}$	Spinocerebellar ataxia type 1
ATXN3	$(CAG)_{12-40}$	$(CAG)_{52-86}$	Machado–Joseph disease
Other expansions (outside the coding region of the gene)			
FXN	$(GAA)_{5-30}$	$(GAA)_{70-1,000}$	Friedreich's ataxia
DMPK	$(CTG)_{5-37}$	$(CTG)_{40-50}$	Myotonic dystrophy
FMR1	$(CGG)_{6-50}$	$(CGG)_{200-4,000}$	Fragile X syndrome
AFF2	$(GCC)_{4-39}$	$(GCC)_{>200}$	Fragile XE syndrome
EPM1	$(CCCCGCCCCGCG)_{2-3}$	$(CCCCGCCCCGCG)_{>30}$	Progressive myoclonus epilepsy

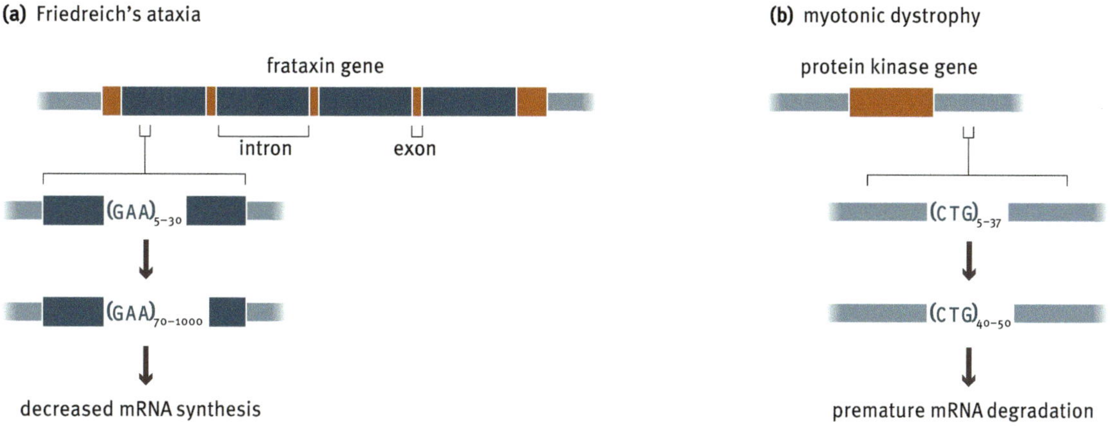

FIGURE 21.5 The genetic basis of two trinucleotide repeat expansion disorders. (a) Friedreich's ataxia and (b) myotonic dystrophy.

A few disease-causing mutations involve expansions of longer sequences. An example is progressive myoclonus epilepsy. This is caused by the copy number of the sequence 5′–CCCCGCCCCGCG–3′ being increased from 2–3 to more than 30. This sequence lies in the promoter region of the cystatin B gene, the product of which is involved in regulating the activities of proteases within human cells.

Exactly how trinucleotide expansions are generated is not understood. At first, it was assumed that they result from errors during DNA replication, but expansions can also occur in nondividing cells that are not replicating their genomes. Current thinking is that an expansion can be generated during nucleotide excision repair of damaged segments of DNA. This process involves the removal of part of a polynucleotide containing a mutation, followed by resynthesis of the DNA strand to fill in the single-stranded gap that is formed (Section 11.2). If the segment that is cut out includes part or all of a trinucleotide region, then slippage during the DNA synthesis step of the repair process could result in the triplet expansion.

There are dominant and recessive inherited disorders

The mutated version of a gene that leads to an inherited disorder is an allele, and like all other alleles, it will have a dominant–recessive relationship with the other alleles of the gene. Those inherited disorders that are dominant are displayed by individuals who are either homozygous for the mutated allele or heterozygotes with one mutated and one normal allele (Figure 21.6a). At least one of the parents of an affected child must also have the disorder, which in turn tells us that the disorder, however unfortunate its symptoms, is not so severe as to prevent at least some afflicted individuals reaching reproductive age and having children. It is therefore not surprising that many dominant inherited disorders have incomplete penetrance so not all individuals with the affected genotype actually develop the symptoms. Examples are Huntington's disease and Marfan syndrome. Note, however, that a mutation resulting in a dominant disorder can also arise spontaneously at an early stage in embryo development. In this case, neither parent will display the disorder and it is quite possible that the symptoms will be so severe as to prevent the affected individual from reaching reproductive age.

If the mutated allele is recessive, as is the case for cystic fibrosis, then only homozygotes will suffer from the disorder (Figure 21.6b). With recessive disorders, it is therefore quite possible for an affected child to have two unaffected parents. Both parents would be heterozygotes, **carriers** who have one normal and one mutated allele. Recessive disorders often have more severe symptoms than dominant ones and it is quite possible that most affected people die before being able to have their own children. The disorder persists in the population because its allele survives in the pool of unaffected carriers.

In the examples shown in Figure 21.6, the relevant gene lies on one of the autosomes. The inheritance pattern will be different if the gene is on one of the sex chromosomes. If the gene is on the X chromosome, then it will always be expressed by males who possess the mutated allele. This is because males have just one copy of the X chromosome, making it immaterial if the mutated allele is dominant or recessive (Figure 21.7). If the disorder is dominant then all females with a mutated allele will also display the disorder, but if it is recessive then the disorder will be expressed only by homozygous females. X-linked dominant disorders are in fact quite rare. Most X-linked disorders are recessive, the best known being hemophilia A. This disease gained notoriety in the late 19th and early 20th centuries as it affected some of the descendants of Queen Victoria, leading to various succession crises among the ruling families of Europe as male heirs suffering from the disease died young (Figure 21.8). Y-linked disorders are uncommon because the Y chromosome has very

(a) dominant inherited disorder

DD	Dd	dd
affected		unaffected

(b) recessive inherited disorder

DD	Dd	dd
unaffected		affected

FIGURE 21.6 Genotypes and phenotypes for autosomal inherited disorders. (a) A dominant autosomal disorder and (b) a recessive autosomal disorder. In both cases, the disorder gene D has a dominant allele *D* and a recessive allele *d*.

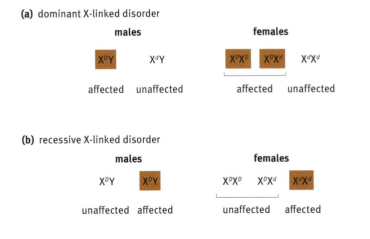

FIGURE 21.7 Genotypes and phenotypes for X-linked inherited disorders. (a) A dominant X-linked disorder and (b) a recessive X-linked disorder. In both cases, the disorder gene D has a dominant allele *D* and a recessive allele *d*.

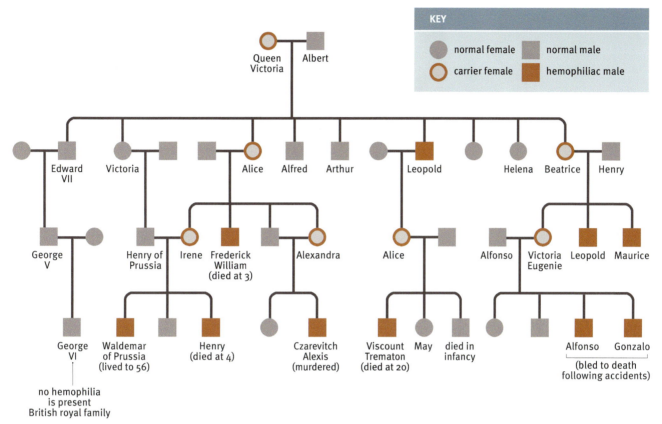

FIGURE 21.8 Partial pedigree of the descendants of Queen Victoria. The symbols indicate those females who were carriers for hemophilia, and those males who had the disease. For clarity, not all descendants are shown, and siblings are not always positioned according to their order of birth.

few genes – fewer than 100 – so the scope for mutation is limited. Most Y-linked disorders affect fertility and all, of course, are restricted to males.

The dominant–recessive relationship can tell us something about the basis of a disorder

Understanding the basis of an inherited disorder is clearly important, so how can dominance and recessiveness help us in this regard? A simple principle is that a disorder that results from a gain-of-function mutation is likely to be dominant and one due to a loss-of-function mutation is likely to be recessive. The rationale behind this statement is that in a heterozygote a new activity resulting from a gain-of-function mutation is likely to mask the effect of the normal allele, and so the gain of function is dominant. A loss of function, on the other hand, can often be compensated for by the presence of the single unmutated allele, so a loss of function is recessive.

Exceptions to this principle are likely to be interesting for one reason or another. For example, many of the disorders that result from trinucleotide expansions within the coding region of a gene give rise to a loss of function, but the expanded alleles are dominant and the normal alleles recessive. The reason for this is not known, but it might relate to the tendency of the abnormal protein products coded by the expanded genes to form insoluble aggregates within nerve cells (Figure 21.9). Dominance would be explained if these aggregates have any role in the manifestation of the disorder, as the presence of the normal allele would not prevent or mask the pathological effect of the aggregate.

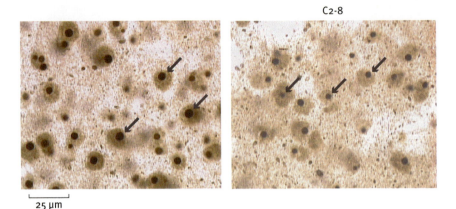

FIGURE 21.9 Insoluble aggregates of the Huntingtin protein in neurons from a mouse brain. The aggregates are visible as brown nuclear inclusions in neurons from the brain of a mouse with the murine equivalent of Huntington's disease. (Courtesy of Dr Vanita Chopra, Massachusetts General Hospital.)

A few other loss-of-function mutations are neither dominant nor recessive, but instead display haploinsufficiency. This is the situation where the phenotype displayed by the heterozygote is affected by the approximately 50% reduction in protein activity caused by the presence of the mutated allele (Figure 21.10a). This is not a common situation with inherited disorders because the single normal allele possessed by a heterozygote is usually able to direct synthesis of enough protein to maintain the unaffected state. A gene could display haploinsufficiency simply because large amounts of its protein are needed in the cell, and a single allele cannot satisfy this need, but disorders of this type are uncommon. Supravalvular aortic stenosis, an inherited narrowing of the aorta brought about by mutation of the gene for the connective tissue protein elastin, is thought to be an example.

Most other examples of haploinsufficiency are more subtle and involve a gene whose protein product must be present in the cell in a carefully controlled amount, possibly because it is a signaling protein whose oversynthesis would disrupt its signaling pathway (Figure 21.10b). As the protein is normally synthesized at just the correct level, a 50% reduction would cause a significant loss of activity. This is probably the explanation of the haploinsufficiency displayed by Alagille syndrome, sufferers of which have heart and liver problems. The syndrome results from mutation of the gene for the CD339 protein, leading to disruption of a cell-to-cell signaling system that plays an important role in embryo development.

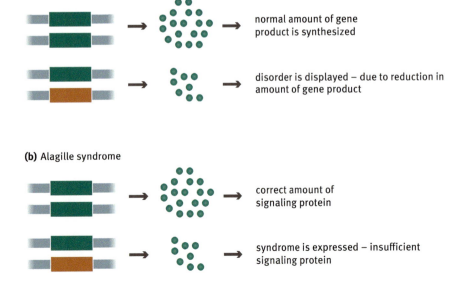

FIGURE 21.10 Haploinsufficiency and its role in Alagille syndrome.

Inherited disorders can also be caused by large deletions and chromosome abnormalities

Not all inherited disorders result from the mutation of a single gene. Many have more complex origins that involve, for example, the deletion of regions of the genome containing several genes or duplication of a chromosome.

On average, there are seven human genes per megabase of the genome, so deletions of more than 1 Mb are likely to result in the loss of at least a few genes. Often this will be so disastrous that the fertilized egg will not develop and the deletion will never appear in the population. On occasions, however, deletions of a few Mb can be tolerated, albeit associated with some kind of genetic disorder. Many of these involve the deletion of a single important gene, possibly along with a few adjacent genes whose loss can be tolerated. Often, such a deletion will give rise to symptoms indistinguishable from those resulting from point mutation within the gene responsible for the disorder (Figure 21.11). Several of the disorders mentioned so far in this chapter are known to arise both by simple mutation and by gene deletion. This is the case with Alagille syndrome, one of the examples of haploinsufficiency, which results from a deletion of the CD339 gene in 93% of cases and a point mutation in the same gene in the other 7%.

Occasionally, a deletion removes a segment of the genome containing two or more important genes that are individually associated with inherited disorders. The unfortunate person will then display a range of symptoms characteristic of different syndromes. There has been one recorded case where deletion of a large segment of the X chromosome led to a combination of retinitis pigmentosa, chronic granulomatous disease and Duchenne muscular dystrophy, the genes for these disorders occupying adjacent positions on the chromosome (Figure 21.12). Such multiple conditions are exceedingly rare.

More common are genetic disorders arising from duplication of a chromosome, resulting in a cell that contains three copies of one chromosome and two copies of all the others. This condition is called **trisomy** and is probably lethal for most chromosomes because for several it is never seen. Trisomy of chromosomes 21 or 18 are, however, relatively common, resulting in, respectively, Down syndrome (approximately one in 800 births) and Edwards syndrome (one in 6,000 births). Probably, the resulting increase in copy number for the genes on the duplicated chromosome leads to an imbalance of their gene products and disruption of the cellular biochemistry, giving rise to the developmental abnormalities that characterize trisomy disorders.

Trisomies are genetic but not inherited disorders. They result from aberrations during the meiosis that produces the affected reproductive cell and occur spontaneously in the population as a whole (Figure 21.13). The same is true of the syndromes resulting from nonstandard numbers of the sex chromosomes. These include Klinefelter syndrome, displayed by individuals with an extra X chromosome (XXY), and the self-explanatory XYY syndrome.

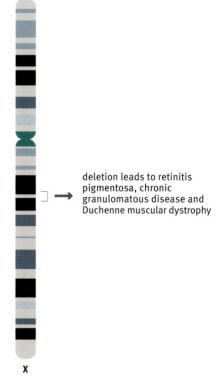

FIGURE 21.12 **A single deletion can give rise to a multifactorial syndrome.** Location of the single deletion that gives rise to a combination of retinitis pigmentosa, chronic granulomatous disease and Duchenne muscular dystrophy.

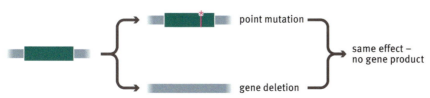

FIGURE 21.11 **A point mutation and a gene deletion can give rise to the same inherited disorder.**

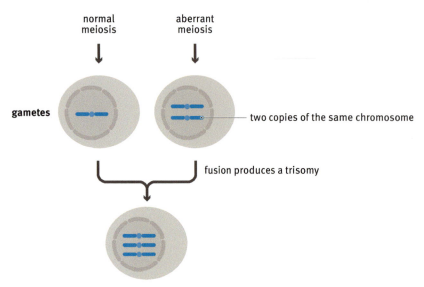

FIGURE 21.13 Production of trisomy by the fusion of one normal and one abnormal gamete.

21.2 IDENTIFYING A GENE FOR AN INHERITED DISORDER

For any inherited disorder, identification of the causative gene is a very important objective. Once the gene has been identified, the biochemical basis of the disease can be studied, possibly leading to ways of treating the disease by conventional means or by gene therapy. A description of the mutations that result in the disease is equally important because it enables a screening program to be devised. The mutated gene can then be detected in individuals who have not yet developed the disease or who are carriers. So how do we go about identifying the gene for an inherited disease?

We answered this question, at least in outline, in Section 17.1, when we studied how pedigree analysis is used to map human genes. The example we used to illustrate the technique involved a gene for a genetic disorder, which we mapped relative to a second gene, whose position in the genome was already known (see Figure 17.10). Once the approximate map position of a gene is known, the DNA sequence can be searched to see what genes are actually present in that region.

Although we have already studied the basic principles of pedigree analysis, there are some important details that we must examine in order to understand how the procedure is used in the real world. Once we have understood those details, we will follow through the research project that identified the gene for inherited breast cancer.

DNA markers are often used in pedigree analysis

In our previous example of pedigree analysis, we made use of a gene, which we called M, whose position in the genome was known. In reality, genes are rarely used as the fixed markers when a human pedigree analysis is carried out. This is because there are only a few human genes that exist in allelic forms that can be distinguished conveniently, and these are widely spaced out in the genome. This means that mapping data based entirely on genes is not very accurate.

Pedigree analysis therefore makes use of **DNA markers**. A DNA marker is a sequence that does not form part of a gene but is variable and so exists

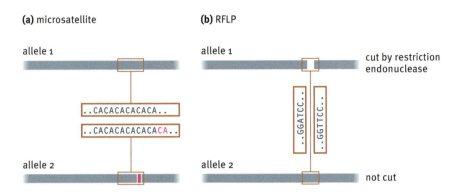

FIGURE 21.14 Two alleles of (a) a microsatellite, and (b) a RFLP.

as two or more alleles. Microsatellites are an example. These are short sequences made up of tandem repeats of a motif such as the dinucleotide 5′–CA–3′ (Section 20.2). Many microsatellites are multiallelic because of variations in the number of repeat units in the array (Figure 21.14a). There are more than four million microsatellites with repeat units of 2–6 bp in the human genome, on average one every 775 bp, so there is high chance that one or more will be located close to the gene for an inherited disorder.

A second type of DNA marker is a **restriction fragment length polymorphism** (**RFLP**). An RFLP results from a mutation that changes the sequence of a recognition site for a restriction endonuclease. This is an enzyme that cuts DNA at a specific sequence (Section 17.3). An example is *Bam*HI, which cuts only at the sequence 5′–GGATCC–3′. If a mutation alters this sequence to, say, 5′–GGTTCC–3′, then it will not be recognized by *Bam*HI and the site will not be cut if the DNA is treated with this enzyme. The two versions of the sequence – the one that is cut by the enzyme and the one that is not – are the two alleles of the RFLP (Figure 21.14b).

Microsatellite and RFLP alleles can be typed by the **polymerase chain reaction** (**PCR**). This important technique enables any short segment of a DNA molecule to be copied repeatedly so that large quantities are obtained (Figure 21.15). The amplified segment can then be examined to identify which allele of the DNA marker is present. For a microsatellite, the alleles are distinguished because each one gives rise to a PCR product of a slightly different length, the length indicating the number of repeat motifs that are contained in the microsatellite. The length of each product is therefore measured by **capillary gel electrophoresis** (Figure 21.16). A capillary gel is a long, thin tube of polyacrylamide contained in a plastic sleeve. The PCR product is loaded at one end of the capillary and an electric current is applied. The migration rate of each DNA molecule through the capillary gel depends on its electric charge. Each phosphodiester bond in a DNA molecule carries a single negative electric charge (see Figure 2.12), so the overall negative charge of a polynucleotide is directly proportional to its length. By the time the DNA has reached the bottom of the capillary, molecules differing in length by just one nucleotide become separated. The PCR products resulting from microsatellites of different lengths can therefore be distinguished.

Gel electrophoresis is also used to type RFLP alleles, but in a slightly different way. The two alleles are distinguished by treating the PCR product with the appropriate restriction endonuclease (Figure 21.17). If the non-mutated allele is present then the PCR product will be cut into two pieces, but if the RFLP is mutated then the PCR product will remain intact. These intact and cut pieces are likely to differ in length by tens of nucleotides, so the resolving power of a capillary gel is not needed. Instead, electrophoresis is carried out in a slab of agarose gel.

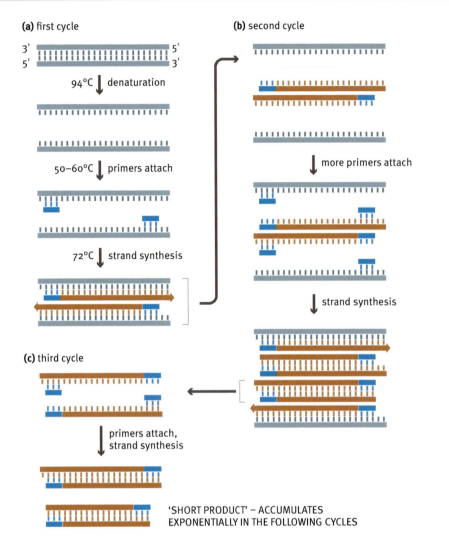

FIGURE 21.15 The polymerase chain reaction. The procedure requires two short oligonucleotides whose sequences match the DNA sequence either side of the segment that we wish to amplify. The oligonucleotides act as primers for repeated cycles of DNA synthesis carried out by a DNA polymerase. (a) To begin the procedure, the DNA is heated to 94°C so the strands separate – this is called 'denaturation'. The DNA is then cooled to 50°C–60°C which enables the primers to attach, and heated again to 72°C, which is the optimum temperature for strand synthesis by the type of DNA polymerase that is used in the reaction. (b) The reaction cycle is repeated: the new DNA molecules made in the first cycle are denatured, more primers attach and another round of DNA synthesis takes place. Products of varying lengths are synthesized. (c) In the third cycle, another variety of products is formed. In the diagram, we focus on just one of these, the 'short product', which appears for the first time. This is a product whose 5' and 3' ends are set by the positions at which the primers attach to the DNA. In subsequent cycles, the short products accumulate in an exponential fashion. This means that after 30 cycles, there will be more than 130 million short products derived from each starting molecule.

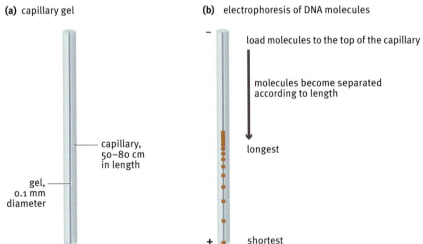

FIGURE 21.16 Separating DNA molecules of different lengths by capillary gel electrophoresis. (a) The capillary gel is typically 50–80 cm in length with a diameter of 0.1 mm, contained in a plastic sleeve. (b) When an electric current is applied, DNA molecules migrate toward the positive electrode at rates depending on their lengths. Here we see a series of molecules of different lengths becoming separated. The PCR product from a microsatellite would give one band if both alleles had the same number of repeats, or two bands if the alleles are different. The size of a PCR product can be identified by comparing its migration rate with the migration rates of DNA molecules of known length, and the number of repeats in the microsatellite can then be calculated from the length of the PCR product.

LANES

1. DNA size markers
2. Unmutated RFLP – cut by the restriction endonuclease
3. Mutated RFLP – not cut

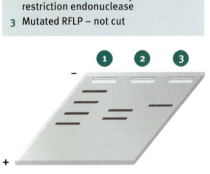

FIGURE 21.17 Identifying RFLP alleles by agarose gel electrophoresis. The DNA samples are loaded into the wells at the top of the slab of gel. The gel is then submerged in buffer and the electric charge applied. As in capillary electrophoresis, the DNA molecules migrate toward the positive electrode, the shorter molecules more quickly than the longer ones. In lane 2, there are two bands because the RFLP was cut by the restriction endonuclease. In lane 3 there is just one band, showing that this sample contains the mutated version of the RFLP, which is not cut by the endonuclease.

How the human breast cancer susceptibility gene *BRCA1* was identified

We have seen how DNA markers are typed and we also know, from Section 17.1, how to perform a pedigree analysis. Now we will bring these strands of knowledge together and examine how pedigree analysis and DNA markers have been used in the identification of the gene for an inherited disorder. The example we will study concerns breast cancer, one of the most prevalent genetic disorders (see Table 21.1). The gene we will identify if *BRCA1*, which is one of two genes closely associated with the inheritance of the disease.

Pedigree studies carried out in the late 1980s showed that in families with a high incidence of breast cancer, a significant number of the women who suffered from the disease possessed the same allele of the RFLP called *D17S74*. This RFLP had previously been mapped to the long arm of chromosome 17. It could therefore be concluded that *BRCA1* must also be located on the long arm of chromosome 17, probably within the chromosomal region designated q21 (**Figure 21.18**).

More than 1000 genes are located in the 20 Mb of DNA contained within this particular stretch of chromosome 17. The next objective was therefore to carry out pedigree studies with a greater number of families to try to pinpoint *BRCA1* more accurately. These studies made use of microsatellites located in the q21 region, and placed *BRCA1* between two markers called *D17S1321* and *D17S1325*, which are approximately 600 kb apart.

There are 60 genes located between the *D17S1321* and *D17S1325* markers. Any one of these **candidate genes** could be *BRCA1*. Further pedigree analysis, with increasing numbers of families, could conceivably have narrowed down the search even further, but at this stage of a gene identification project, more direct strategies are usually employed. One of these is to examine the expression patterns of the candidate genes, to identify ones that are active in tissues that express symptoms associated with the disorder. Inherited breast cancer is frequently associated with ovarian cancer, so in this case attention was focused on those candidate genes that were expressed in these two tissues.

A second approach is based on the assumption that any gene that is important enough to cause a serious disorder when mutated is likely to be present not just in humans but also in related species. Today we would use a homology search of the DNA databases (Section 18.3) to identify which of the candidate genes have homologs in other species. However, *BRCA1* was identified in the 1990s, before the databases were large enough for efficient homology searching. Instead, the presence of each candidate gene in other mammals had to be tested by purifying DNA from those species and using the hybridization technique called **zoo blotting** (**Figure 21.19**).

When these analyses were complete, the most likely candidate for *BRCA1* was an approximately 100-kb gene, comprising 22 exons and coding for a 1,863-amino acid protein. Transcripts of the gene were detectable in breast and ovary tissues, and homologs were present in mice, rats, rabbits, sheep and pigs. Most importantly, the genes from five susceptible families contained mutations likely to lead to a nonfunctioning protein. Subsequent studies have shown that that *BRCA1* is involved in transcription regulation and DNA repair, also acts as a tumor suppressor gene which, as we will see in the next section, means that it can prevent abnormal cell division.

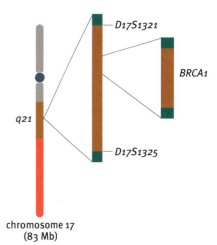

FIGURE 21.18 Mapping the breast cancer susceptibility gene *BRCA1*. Initially, the gene was mapped to a segment of the long arm of chromosome 17, within the region designated q21 (highlighted in the left-hand drawing). Additional mapping experiments narrowed this down to a 600-kb region flanked by two microsatellite markers, *D17S1321* and *D17S1325* (middle drawing). From the 60 genes in this region, a strong candidate for *BRCA1* was eventually identified (right-hand drawing).

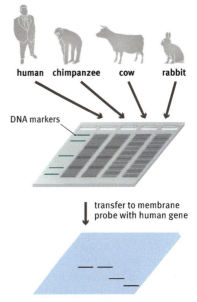

FIGURE 21.19 Zoo blotting. In this method, samples of DNA are prepared from different species and cut with a restriction endonuclease. Each sample gives thousands or millions of fragments that form a smear when the DNA is electrophoresed in an agarose gel. The DNA smears are transferred to a nylon membrane and tested by hybridization probing (Section 17.3) with the human gene as a probe. In the example shown, a positive hybridization signal is obtained with each of the animal DNAs, showing that each of these species possesses a homolog of the human gene.

Genome-wide association studies can also identify genes for genetic disorders

Although pedigree analysis can be used to identify the gene responsible for a monogenic disorder, it is less successful with those syndromes that have more complex genetic backgrounds. Several types of cancer, as well as disorders such as coronary heart disease and osteoporosis, are polygenic, meaning that they are controlled not just by one gene but by many genes working together. One difficulty here is that two members of an affected family can have different combinations of alleles for the causative genes. Under these circumstances, the inheritance pattern of the disorder is likely to be complicated, and it is impossible to use pedigree data to link the disorder to DNA markers with any degree of certainty.

A **genome-wide association study** (**GWAS**) is an alternative approach to gene identification that can work with polygenic traits. Rather than linking DNA markers with individual genes, a GWAS attempts to identify all of the markers, from all over the genome, that are associated with the disorder. The positions of these markers will reveal the positions of candidate genes that might form part of the polygenic trait.

Microsatellites and RFLPs can be used as markers, but more often these are **single-nucleotide polymorphisms** (**SNPs**). An SNP is a position in the genome where some individuals have one nucleotide (e.g. G) and others have a different nucleotide (e.g. C) (Figure 21.20). Some SNPs also give rise to RFLPs, but many do not because the sequence in which they lie is not recognized by any restriction endonuclease.

SNPs are popular in GWAS studies because many can be typed at once by **oligonucleotide hybridization analysis**. An oligonucleotide will only hybridize with a DNA molecule if its sequence is an exact complement of the target site (Figure 21.21). If there is a single mismatch – as there will be if the target contains the alternative allele of an SNP – then there will be no hybridization. To use this method to type multiple SNPs, many different oligonucleotides are attached to thin wafer of silicon called a **DNA chip**. The DNA to be tested is labeled with a fluorescent marker and applied to the surface of the chip. Hybridization is then detected by examining the chip with a fluorescence microscope. This reveals the positions where the test DNA has hybridized to an oligonucleotide (Figure 21.22). Those oligonucleotides contain the version of the SNP present in the test DNA.

One of the first GWAS projects studied age-related macular degeneration, which causes vision problems in elderly people. More than 225,000 SNPs were typed in 96 people who were affected by the disorder and 50 others who were unaffected. Two SNPs were identified whose alleles were significantly associated with the disorder. By this, we mean that most of the affected individuals had one allele of the SNP, and most of the unaffected ones had the second allele. For all of the other SNPs, the allele distributions did not correlate with the presence or absence of the disorder. Both of the associated SNPs were located in an intron within the gene for complement factor B, a protein involved in control of the inflammatory response. This is one of five genes, on three different chromosomes, now thought to be involved in age-related macular degeneration.

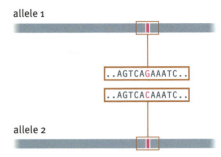

FIGURE 21.20 A single-nucleotide polymorphism.

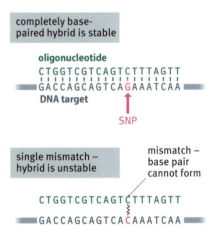

FIGURE 21.21 The basis to oligonucleotide hybridization analysis. A stable hybrid can form only if the oligonucleotide is exactly complementary with the target DNA sequence. If there is a single mismatch, then hybridization does not occur.

21.3 THE GENETIC BASIS TO CANCER

Cancer is a group of disorders characterized by uncontrolled division of a somatic cell. Cancer is a *genetic* disorder in the sense that the inheritance of certain mutations gives a predisposition to cancer. But cancer itself is not an *inherited* disorder as the cancerous state is not directly passed on from parent to offspring.

A full appreciation of cancer requires an understanding of not only its genetic basis but also aspects of cell biology, biochemistry, physiology and developmental biology, as well as a consideration of the myriad different types of cancer that are known to occur. Of necessity, therefore, our discussion of cancer will be limited in extent. We will consider just two central questions, concerning the nature of the genetic changes that can give rise to cancer and the multistep model for the development of a cancer.

Cancers begin with the activation of proto-oncogenes

One of the most important breakthroughs in understanding cancer came in the 1960s when the transforming viruses were discovered. As we learnt in Chapter 12, these are retroviruses that carry a copy of a human gene and which, when they infect a cell, can transform that cell into a cancerous state. It was realized that this type of transformation is brought about by the uncontrolled expression of the gene carried by the retrovirus, which overrides the strictly controlled expression pattern of the cellular version of the gene (Figure 21.23). The transformation process is called oncogenesis, and the gene carried by the virus is called an **oncogene**.

A second breakthrough was made in the early 1980s when the normal cellular function of a viral oncogene was identified for the first time. It was discovered that the cellular version of the v-*sis* oncogene is the gene for the platelet-derived growth factor B protein, which plays a central role in the control of cell growth and division. Other viral oncogenes were subsequently found to be versions of cellular genes involved in activities such as intracellular signaling, regulation of transcription and control of the cell cycle.

The terminology we now use is to refer to the normal cellular versions of these genes as **proto-oncogenes**. In its normal state, a proto-oncogene is not harmful to the cell – quite the reverse – but activation of a proto-oncogene converts it into the oncogene capable of initiating cancer. One way of activating a proto-oncogene is, as we have seen, to place it in a retrovirus genome where its normal expression pattern is lost and the gene becomes overexpressed.

Activation can, however, occur without the intervention of the retrovirus. Mutations are common activating factors for proto-oncogenes such as the *RAS* genes, which code for cell surface receptors that transmit external signals into the cell. Typically, the mutation results in the receptor sending signals into the cell when it should not, one possible consequence being

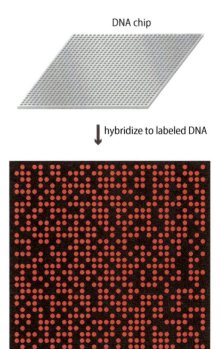

FIGURE 21.22 SNP typing with a DNA chip. Oligonucleotides are immobilized in an array on the surface of the chip. Labeled DNA is applied and the positions at which hybridization occurs are determined by fluorescence microscopy. More than 500,000 oligonucleotides can be immobilized on a chip with an area of just 2 cm².

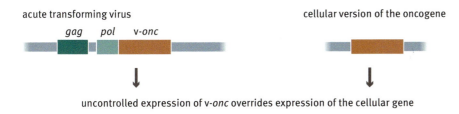

FIGURE 21.23 Uncontrolled expression of a viral oncogene (v-*onc*) can override the controlled expression of the equivalent cellular gene.

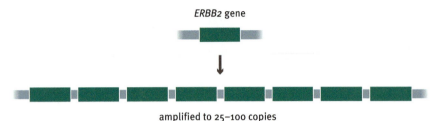

FIGURE 21.24 Amplification of the *ERBB2* gene, one of the causes of breast cancer.

that the cell divides even though there has been no external stimulus for it to do so.

Activation can also occur by duplication and further increases in the copy number of the proto-oncogene, possibly resulting from a DNA replication error, so that greater than normal amounts of the protein product are synthesized. Many breast cancers derive from the amplification of the *ERBB2* gene, which codes for a cell surface receptor. In the activated state, multiple copies of *ERBB2* are found side by side on chromosome 17 (Figure 21.24).

Chromosome translocation, which results in one segment of DNA being exchanged between chromosomes, can also activate a proto-oncogene. For reasons that are unknown, some translocations are relatively common, including one between human chromosomes 9 and 22, resulting in the abnormal product called the **Philadelphia chromosome** (Figure 21.25). The breakpoint in chromosome 9 lies within the *ABL* gene, which codes for a tyrosine kinase enzyme that participates in intracellular signaling. In the Philadelphia chromosome, the *ABL* segment is fused with part of a second gene normally resident on chromosome 22. This gene is called the 'breakpoint cluster region' as its precise function is unknown, though it appears also to be involved in protein phosphorylation. The product of the hybrid gene created by the translocation resembles the *ABL* tyrosine kinase, but has altered properties that result in cell transformation.

Note that the above activation processes all lead to a gain of function. They are therefore dominant and need to affect only one of a pair of alleles for cell transformation to occur.

Tumor suppressor genes inhibit cell transformation

The **tumor suppressors** are a second type of gene which, when mutated, can contribute to the development of a cancer. This class of genes was first discovered as a result of studies of retinoblastoma, an unpleasant cancer of the eye that affects children and can lead to blindness. Sporadic retinoblastoma occurs occasionally but rarely throughout the population, in contrast to familial retinoblastoma, which is much more prevalent but is only found in certain families. The members of these families clearly have an inherited predisposition to this type of cancer.

The 'two-hit' model was devised to explain the difference between sporadic and familial retinoblastoma. According to this model, the retinoblast cells in the eye, being rapidly dividing, have a predisposition to becoming cancerous. The cancer is prevented, however, by the retinoblastoma gene *Rb* (Figure 21.26). Because it can prevent the cancer from arising, *Rb* is called a tumor suppressor gene. Most people have two functioning copies of *Rb*. It is very unlikely that inactivating mutations will occur in both genes of a single cell, so these homozygotes virtually never develop the disease. The sporadic version of retinoblastoma is therefore rare. Heterozygotes, on the other hand, are more vulnerable to retinoblastoma. They have just

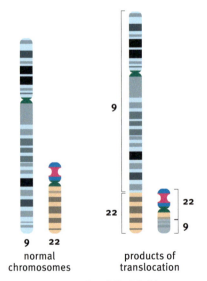

FIGURE 21.25 The Philadelphia chromosome. Details of the reciprocal translocation that results in the production of the Philadelphia chromosome. The breakpoints lie within the *ABL* gene on chromosome 9 and the breakpoint cluster region (*BCR*) of chromosome 22. The Philadelphia chromosome is the smaller of the two products.

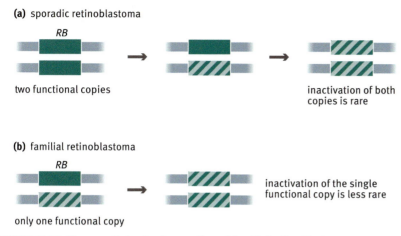

FIGURE 21.26 The genetic basis of sporadic and familial retinoblastoma.

one functional copy of *Rb* per cell, and so a single inactivating mutation in any retinal cell could lead to the disease. Familial retinoblastoma therefore occurs in families whose members include heterozygotes who have inherited one nonfunctional allele from a parent and may pass that allele on to their offspring.

More than 1,000 tumor suppressor genes are known. Most of these are involved in regulation of the cell cycle, especially the detection of and prevention of aberrant cell division. If a problem of some kind is detected then division is halted, and if the problem cannot be corrected, then the cell might be induced to undergo apoptosis (Figure 21.27). One role of tumor suppressors is therefore to counter the transforming effects of an oncogene by causing the death of a cell in which an oncogene has been activated. Mutation of a tumor suppressor gene leads to at least partial loss of this surveillance process, enabling a rogue cell to escape and initiate tumor formation.

Inactivation of tumor suppressor genes often occurs by point mutation or deletion of some or part of the coding sequence. Additionally, tumor suppressor genes can become inactive not as a result of a standard mutation

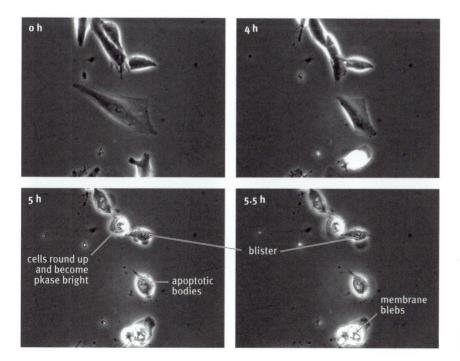

FIGURE 21.27 **Human trophoblast cells undergoing apoptosis induced by treatment with spermine.** After 5 hours, the cells have become rounded and appear bright in these phase contrast micrographs. The membranes start to blister but remain intact. Cell fragments called apoptotic bodies may bud off and are seen as tiny dark spots around some dying cells. Apoptosis is looked on as a 'clean' form of cell death, compared with necrosis, which gives rise to broken cells that often cannot be disposed of by macrophages. (Courtesy of Guy Whitley, St George's University of London.)

but because of DNA methylation. In some transformed cells, the CpG island upstream of a tumor suppressor gene is heavily methylated, and occasionally this is the only reason why the gene is nonfunctional – the coding region remains intact and lacks point mutations. We do not yet understand if this methylation occurs by chance or if these particular CpG islands have an abnormal predisposition toward methylation.

The development of cancer is a multistep process

If every mutation in a proto-oncogene or pair of mutations in a tumor suppressor gene gave rise to a mature cancer, then the frequency of deaths in the population would be much higher than it is. Most cancers develop via a multistep process that involves a number of mutations and other genome modifications that occur over a lengthy period of time. Often, detection of the developing cancer at an early stage in the process enables more effective treatment, and better survival rates, than is possible if detection is delayed until the cancer is mature.

The multistep development of a tumor has been most clearly described for the colon cancer called familial adenomatous polyposis (**Figure 21.28**). The morphological changes occurring during this process involve the initial formation of an adenoma, a small bulge in the epithelial tissue in the wall of the colon. This is followed by the growth of the adenoma into a polyp a centimeter or so in diameter, giving rise finally to a tumor called a carcinoma which contains metastatic cells, ones that can spread to other places in the body and initiate new tumors. The genetic events accompanying progression through these stages are not the same in all cases, but usually they begin with inactivation of the tumor suppressor gene called *APC*. As described above for retinoblastoma, there are sporadic and familial versions of colon cancer, the latter more frequent because it occurs in a person who has inherited an inactive *APC* allele from one parent and so needs just one additional mutation in order to completely inactivate the *APC* gene.

Subsequent events are variable but often there is a decrease in the overall degree of DNA methylation within the adenoma, followed by activation of the *KRAS* proto-oncogene, which codes for an intracellular signaling protein. Progression to this stage gives rise to a structure intermediate between the adenoma and polyp. Inactivation of the tumor suppressor gene *SMAD4*, coding for another signaling protein, gives rise the polyp, and the carcinoma is produced if the *p53* tumor suppressor gene is inactivated. Metastasis might occur now or following other genetic events. The multistep process reflects the number of controls that exist within the cell to prevent aberrant division, controls that must be broken down one by one in order for a metastatic tumor to develop.

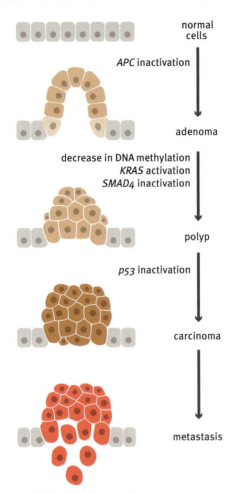

FIGURE 21.28 The multistep pathway leading to tumor formation in colon cancer. The pathway begins with the inactivation of the tumor suppressor gene *APC* and the formation of an adenoma. The adenoma develops into a polyp and then a carcinoma. These changes are associated with genetic events such as a decrease in DNA methylation, activation of the *KRAS* proto-oncogene and inactivation of the tumor suppressor genes *SMAD4* and *p53*. After the carcinoma has formed, metastasis might occur.

21.4 GENE THERAPY

A knowledge of the genetic basis of an inherited disorder is important for two reasons. First, it provides the information needed to devise a screening program so that the mutated gene can be identified in individuals who are carriers or who have not yet developed symptoms of the disorder. Carriers can receive counseling regarding the chances of their children inheriting the disorder. Early identification in individuals who have not yet developed symptoms allows appropriate precautions to be taken to reduce the risk of the disorder becoming expressed.

Genetics also can lead to strategies for treating an inherited disorder. Once the gene responsible for the disorder has been identified, it might be possible to work out its normal function in unaffected individuals, thereby indicating the biochemical basis of the disorder. This might enable drugs to be designed that will reverse or alleviate the disorder.

Treatment is, however, different to cure. The only way of curing an inherited disorder is to replace the mutated gene with a functional version. This is the objective of gene therapy.

There are two approaches to gene therapy

There are two basic approaches to gene therapy. These are **germline therapy** and **somatic cell therapy**. In germline therapy, a fertilized egg is provided with a copy of the correct version of the mutated gene and reimplanted into the mother. The new gene will be present in all cells of the individual that develops from that egg. Germline therapy is usually carried out by **microinjection** of the gene into a somatic cell followed by transfer of the nucleus of that cell into an oocyte whose own nucleus has been removed (Figure 21.29). The nuclear transfer is needed because successful microinjection directly into a fertilized egg is very difficult to achieve. Theoretically, germline therapy could be used to treat any

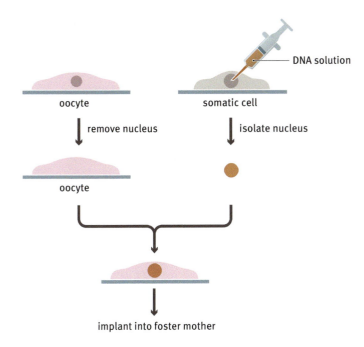

FIGURE 21.29 **The nuclear transfer process for the introduction of new DNA into an oocyte.**

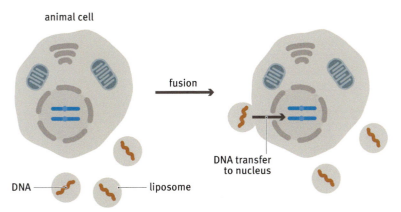

FIGURE 21.30 **The use of liposomes to introduce new DNA into an animal cell.**

inherited disease caused by a point mutation, but it is considered ethically unacceptable to manipulate the human germline in this way, and at present, this technology is used only to produce genetically engineered animals.

The transfer of new genes into somatic cells is less controversial because it only results in changes to the treated individual: the new genes are not passed on to offspring. Two ways of carrying out this type of transfer have been explored. The first is to enclose DNA containing the new gene in membrane-bound vesicles called **liposomes**. These can then be fused with the membrane of the target cell (Figure 21.30). Usually, DNA that is transported into a cell in this way is not stable, though the genes it contains might be expressed for a few days or weeks. The method has been tried, with limited success, as a way of treating lung diseases such as cystic fibrosis, with the DNA-containing liposomes introduced into the respiratory tract via an inhaler.

The second transfer strategy makes us of a virus which, as part of its natural infection strategy, transports its genome into the nucleus of the host cell. **Adenoviruses** (Figure 21.31) have been used for this purpose, because they can take up semi-permanent residence within the nucleus of an infected cell and are passed to daughter cells when the infected cell divides. However, many people are naturally exposed to adenoviruses during their lifetime and hence build up an immunity to these viruses. This means that an adenovirus used in a gene therapy procedure might be destroyed by the patient's immune system before it is able to deliver its DNA to the host cells. Attention is now focused on **adeno-associated viruses** (**AAVs**), which despite their name are unrelated to adenoviruses. AAVs are much less likely to trigger the patient's immune response.

With both liposome and AAV transfer, the new DNA that is introduced will eventually be degraded. The therapeutic effect is therefore only temporary. This problem could be solved by gene editing (Section 18.3), which enables a permanent change to be made to a lineage of somatic cells.

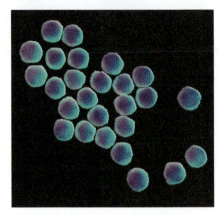

FIGURE 21.31 Adenoviruses, sometimes used as a vector for gene therapy. (Courtesy of Dr Hans Gelderblom, Visuals Unlimited/Science Photo Library.)

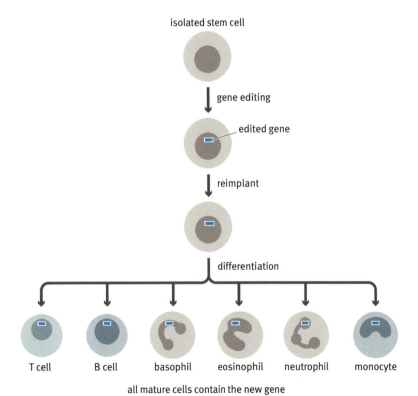

FIGURE 21.32 Gene therapy directed at a blood disorder. The disorder is treated by editing stem cells in the bone marrow. These cells give rise to mature blood cells, all of which will carry the edit.

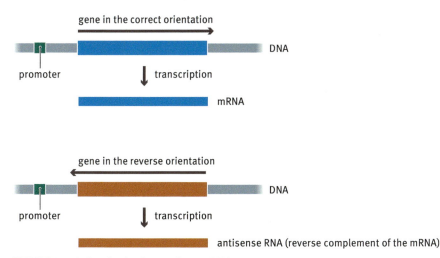

FIGURE 21.33 Synthesis of an antisense RNA.

Trials are being carried out in which liposomes or AAVs are used to transfer a programmable nuclease plus a guide RNA into host cells. Once in the nucleus the guide RNA directs an edit that corrects the mutation carried by the target gene. This change is permanent and will be inherited by all the cells that descend from the one in which the edit was performed. In 2023, the United Kingdom became the first country in the world to approve the use of gene editing in this way, as a treatment for sickle-cell anemia and β-thalassemia. To treat these blood disorders, stem cells from the bone marrow are edited, because these cells give rise to all the specialized cell types in the blood (Figure 21.32). The replication of the treated stem cells therefore results in the presence of the edited gene throughout the blood system.

Gene therapy can also be used to treat cancer

The clinical uses of gene therapy are not limited to the treatment of inherited disorders. There have also been attempts to use this technique as a treatment for cancer.

Inactivation of a tumor suppressor gene could be reversed by the introduction of the active version of the gene by liposome or AAV transfer, or by gene editing. Inactivation of an oncogene, on the other hand, cannot be achieved by the introduction of the nonmutated version. This is because the oncogenic copy is the dominant allele and as this allele would still be present the pathological condition would not be affected.

Gene editing could be used to inactivate those oncogenes that differ from the normal gene by point mutation. Editing the mutation would convert the oncogene back to the harmless version. As editing works directly on the oncogene, the pathological condition will be reversed. However, gene editing would be unable to correct viral oncogenes or cellular ones that are activated by duplication or chromosome translocation.

Because of these problems, gene therapy has been directed not at the DNA copies of oncogenes, but at their mRNA transcripts. This approach makes use of an **antisense RNA** version of the mRNA transcribed from the oncogene (Figure 21.33). An antisense RNA is the reverse complement of a normal RNA and can prevent synthesis of the protein coded by the gene it is directed against. It probably does this by hybridizing to the mRNA producing a double-stranded RNA molecule that is rapidly degraded by cellular ribonucleases (Figure 21.34). Introduction of the antisense RNA into a tumor can therefore inactivate a targeted oncogene.

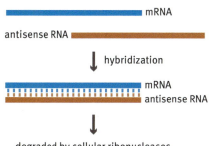

FIGURE 21.34 The probable mechanism by which an antisense RNA prevents the expression of its target gene.

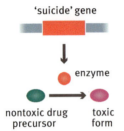
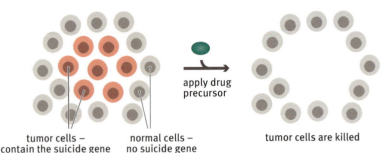

FIGURE 21.35 Suicide gene therapy. (a) The mode of action of the suicide gene. (b) When the drug precursor is applied, the tumor cells are killed.

An alternative is to introduce a gene that selectively kills cancer cells or promotes their destruction by drugs administered in a conventional fashion. This is called **suicide gene therapy** and is looked on as an effective general approach to cancer treatment, because it does not require a detailed understanding of the genetic basis of the particular disease being treated. Many genes that code for toxic proteins are known, and there are also examples of enzymes that convert nontoxic precursors of drugs into the toxic form (Figure 21.35). Introduction of the gene for one of these toxic proteins or enzymes into a tumor should result in the death of the cancer cells, either immediately or after drug administration. It is obviously important that the introduced gene is targeted accurately at the cancer cells, so that healthy cells are not killed. This requires a very precise delivery system, such as direct inoculation into the tumor, or some other means of ensuring that the gene is expressed only in the cancer cells.

A third approach involves engineering the patient's lymphocytes so that they are better able to recognize and kill tumor cells. The lymphocytes are engineered, either by introducing new DNA or by gene editing, so that they synthesize T-cell receptors (Section 19.1) that target proteins that are present on the surfaces of cancer cells but absent from the cells of normal tissues. The engineered lymphocytes are called **chimeric antigen receptor T (CAR-T) cells**.

KEY CONCEPTS

- An inherited disorder is one that is caused by a mutation and which, like other genetic features, can be passed from parents to offspring.

- Most human inherited disorders are caused by a loss-of-function mutation that inactivates a particular gene. The protein coded by that gene is absent or modified in some way that prevents it from functioning correctly, giving rise to the symptoms of the inherited disorder.

- There are dominant and recessive types of inherited disorder. Understanding whether a disorder is dominant or recessive can help in understanding the basis of the disorder.

- Identification of the causative gene for a genetic disorder can lead to ways of treating the disease and enables screening programs to be designed to identify individuals who have not yet developed the disease or who are carriers.

- Pedigree analysis with DNA markers such as microsatellites or RFLPs is used to locate the chromosomal position of a gene for an inherited disorder.
- Candidate genes for a disorder are identified by examining expression patterns, the presence of gene homologs in related species and the presence in affected families of mutations that might result in a non-functioning gene product.
- Cancer is a group of disorders characterized by uncontrolled division of a somatic cell. Cancer is looked on as a genetic disorder because the inheritance of certain mutations gives a predisposition to cancer.
- Cancer begins by activation of a proto-oncogene, a normal gene that can give rise to uncontrolled cell division when mutated.
- Tumor suppressor genes are involved in the regulation of the cell cycle, especially in the detection and prevention of aberrant cell division. Mutation of a tumor suppressor gene leads to at least partial loss of this surveillance process, increasing the likelihood of a cell becoming cancerous.
- In germline gene therapy, a fertilized egg is provided with a copy of the correct version of a mutated gene and reimplanted into the mother. The new gene will be present in all cells of the individual that develops from that egg. Theoretically, germline therapy could be used to treat any inherited disease caused by a point mutation, but it is considered ethically unacceptable to manipulate the human germline in this way, and at present, this technology is used only to produce genetically engineered animals.
- Somatic gene therapy is less controversial because it only results in changes to the treated individual, the new genes not being passed on to offspring.

QUESTIONS AND PROBLEMS

Key terms

Write short definitions of the following terms:

adeno-associated virus
adenovirus
antisense RNA
candidate gene
capillary gel electrophoresis
carrier
chimeric antigen receptor T cells
DNA chip
DNA marker
fragile site
gene therapy
genetic disorder
genome-wide association study
germline therapy
inherited disorder
liposome
microinjection
monogenic
oncogene
oligonucleotide hybridization
 analysis
Philadelphia chromosome
polymerase chain reaction
proto-oncogene
restriction fragment length
 polymorphism
single-nucleotide polymorphism
somatic cell therapy
suicide gene therapy
trisomy
tumor suppressor
zoo blotting

Self-study questions

21.1 Describe the ways in which a loss-of-function mutation could lead to an inherited disorder.

21.2 Using cystic fibrosis as an example, explain how a single genetic disorder can be caused by more than one mutation.

21.3 Why are inherited disorders that result from a gain-of-function mutation relatively uncommon?

21.4 Describe the special features of trinucleotide repeat expansion disorders.

21.5 Give examples of (a) a dominant inherited disorder and (b) a recessive inherited disorder.

21.6 How will a disorder whose causative gene is on the X chromosome be inherited? What will be the inheritance pattern if the gene is on the Y chromosome?

21.7 Describe how a knowledge of whether a disorder is dominant or recessive can be used to make inferences about the molecular basis of the disorder. Why are exceptions to a simple dominant–recessive relationship particularly informative?

21.8 Give examples of inherited disorders caused by large deletions.

21.9 Why is trisomy not a type of inherited disorder?

21.10 Why are DNA markers used when a gene for an inherited disorder is being mapped?

21.11 Distinguish between the different alleles of (a) a microsatellite, (b) an RFLP and (c) an SNP.

21.12 Describe how microsatellites, RFLPs and SNPs are typed.

21.13 Outline the research project that led to the identification of *BRCA1*.

21.14 Describe how a genome-wide association study is carried out.

21.15 Explain how activation of a proto-oncogene can lead to cancer.

21.16 What is the Philadelphia chromosome and why is its possession linked with cancer?

21.17 Describe the role of tumor suppressor genes in preventing cancer.

21.18 Outline the 'two-hit' model for the onset of retinoblastoma.

21.19 Describe the multistep process that leads to colon cancer.

21.20 Distinguish between germline therapy and somatic cell therapy.

21.21 Give examples of the way in which new DNA is introduced into an organism to perform gene therapy.

21.22 What are the advantages of gene editing when used as a form of gene therapy?

21.23 Describe how gene therapy might be used to treat cancer.

21.24 Explain what is meant by the term 'suicide gene therapy'.

Discussion topics

21.25 Why are some inherited disorders more common than others?

21.26 Explore the current knowledge concerning trinucleotide repeat expansion disorders, including hypotheses that attempt to explain why triplet expansion in these genes leads to the disorder.

21.27 If two parents are both carriers of an inherited disorder, what are the chances that their next child will suffer from the disorder? What are the chances that the next child but one will have the disorder?

21.28 A pharmaceutical company has invested a great deal of time and money to isolate the gene for a genetic disorder. The company is studying the gene and its protein product and is working to develop drugs to treat the disorder. Does the company have the right (in your opinion) to patent the gene? Justify your answer.

21.29 Is gene therapy a realistic approach to the treatment of inherited disorders?

21.30 Discuss the ethical issues that are raised by the development of gene therapy techniques. In your discussion, make sure that you distinguish between somatic cell therapy and germline therapy.

The answers to, or hints on how to answer, the self-study questions and discussion topic questions can be downloaded from the book's product page at this link: www.routledge.com/9781032743530

FURTHER READING

Anguela XM & High KA (2019) Entering the modern era of gene therapy. *Annual Review of Medicine* 70, 273–288.

Bobadilla JL, Macek M, Fine JP & Farrell PM (2002) Cystic fibrosis: a worldwide analysis of *CFTR* mutations – correlation with incidence data and application to screening. *Human Mutation* 19, 575–606.

Caplan A (2019) Getting serious about the challenges of regulating germline gene therapy. *PLoS Biology* 17, e3000223.

Hassold TJ & Jacobs PA (1984) Trisomy in man. *Annual Review of Genetics* 18, 69–97.

Karjoo Z, Chen X & Hatefi A. (2016) Progress and problems with the use of suicide genes for targeted cancer therapy. *Advanced Drug Delivery Reviews* 99, 113–128.

Kontomanolis EN, Koutras A, Syllaios A et al. (2020) Role of oncogenes and tumor-suppressor genes in carcinogenesis: a review. *Anticancer Research* 40, 6009–6015.

Koretzky GA (2007) The legacy of the Philadelphia chromosome. *Journal of Clinical Investigation* 117, 2030–2032. *Describes the discovery of the Philadelphia chromosome and what its study has told us about cancer.*

Lupski JR, de Oca-Luna RM, Slaugenhaupt S et al. (1991) A duplication associated with Charcot-Marie-Tooth disease type 1A. *Cell* 66, 219–232.

Miki Y, Swensen J, Shattuck-Eidens D et al. (1994) A strong candidate for the breast and ovarian cancer susceptibility gene *BRCA1*. *Science* 266, 66–71. *The identification of the gene for inherited breast cancer.*

Mirkin AM (2007) Expandable DNA repeats and human disease. *Nature* 447, 932–940. *Trinucleotide repeat expansion diseases.*

Papanikolaou E & Bosio A (2021) The promise and the hope of gene therapy. *Frontiers in Genome Editing* 3, 618346.

Smith KR (2003) Gene therapy: theoretical and bioethical concepts. *Archives of Medical Research* 34, 247–268.

Tam V, Patel N, Turcotte M et al. (2019) Benefits and limitations of genome-wide association studies. *Nature Reviews Genetics* 20, 467–484.

Vogelstein B & Kinzler KW (1993) The multistep nature of cancer. *Trends in Genetics* 9, 138–141.

Zhu H, Zhang H, Xu Y et al. (2020) PCR past, present and future. *Biotechniques* 69, 317–325. *A review of all aspects of this important technique.*

CHAPTER 22

DNA in Forensic Genetics and Archaeology

In the public perception, one of the most important applications of genetics is its use in forensic science. Hardly a week goes by without a report in the media of a murder or rape that has been solved thanks to DNA. The increasing sophistication of the technology means that not only are recent crimes being solved, but also many 'cold cases' dating back tens of years. In addition to the reports of real events on the front pages of newspapers, the TV guides are also full of dramas in which glamorous geneticists use DNA to fight the forces of evil.

We will begin this chapter by exploring the science behind the real-life crime stories and the fictional TV dramas. In doing so, we will discover that there is an unbroken progression from the use of DNA in criminal cases to its application in archaeology. This is because the methods used to identify individuals from DNA samples taken from crime scenes can also be used with **ancient DNA** – DNA extracted from fossil or archaeological remains.

22.1 DNA AND FORENSIC GENETICS

Although every copy of the human genome has the same genes in the same order, and the same stretches of intergenic DNA between those genes, the genomes possessed by different people are by no means identical. This because the human genome, as well as those of other organisms, contains many **polymorphisms**, positions where the nucleotide sequence is not the same in every member of the population (Figure 22.1). Because of these polymorphisms, there is a high statistical probability that no two people alive today have exactly the same genome sequences. The only possible exception would be monozygotic siblings – identical twins, triplets etc. Even these individuals will not have absolutely identical genomes because different mutations will have occurred in their genomes since their birth. The variability of the human genome is the key to the use of DNA in forensic science.

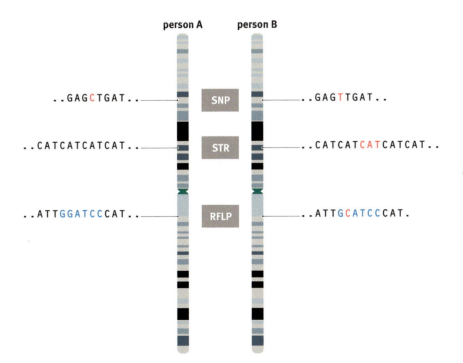

FIGURE 22.1 Polymorphisms in the human genome. Two copies of chromosome 1 are shown, from two different people. The DNA sequences differ because of variations in a single-nucleotide polymorphism (SNP), short tandem repeat (STR) and restriction fragment length polymorphism (RFLP). The variable nucleotides are shown in red, and in the RFLP the recognition sequence for the restriction endonuclease *Bam*HI is shown in green.

The first genetic fingerprints were based on minisatellite variability

The first DNA method used in forensic science was called **genetic fingerprinting**. It was invented by Alec Jeffreys of the University of Leicester, United Kingdom, and made use of minisatellite variability as the means of distinguishing between individuals. Each minisatellite is made up of several hundred repeats of a motif up to 60 bp in length, the repeats arranged one after the other in an array (Figure 22.2). The number of repeats in an array is variable, contributing in part to the overall polymorphism of the genome.

The sequence of one type of minisatellite contains the core motif 5′–GGGCAGGANG–3′, where 'N' is any nucleotide. Minisatellites of this type are located at various places in the genome and are sometimes called **hypervariable dispersed repetitive sequences**. To construct a genetic fingerprint, DNA is prepared and cut into fragments with a restriction endonuclease, and the fragments are separated into a smear of DNA by agarose gel electrophoresis (Figure 22.3). The DNA is then transferred to a nylon membrane and a labeled hybridization probe applied. The probe is designed so that it will base pair to the sequence 5′–GGGCAGGANG–3′, thereby revealing those restriction fragments that contain a copy of the minisatellite. The length of each fragment will depend on the number of repeat units in the minisatellite that it contains. The fragment pattern – the genetic fingerprint – will therefore be slightly different in different people.

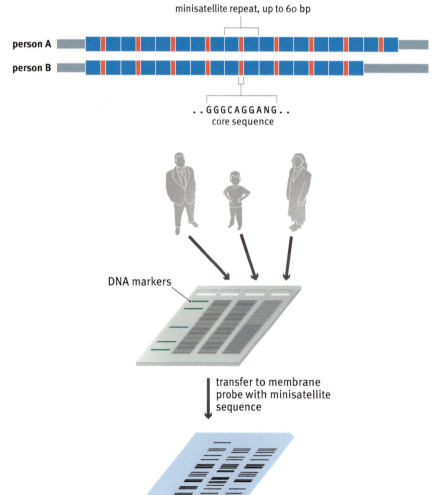

FIGURE 22.2 Minisatellites. A minisatellite is made up of multiple repeats of a motif up to 60 bp in length. The repeat motifs have small sequence variations, but the minisatellites used in genetic fingerprinting have a core sequence 5′–GGGCAGGANG–3′ (where 'N' is any nucleotide) that is present in every copy of the repeat. In this diagram, the same minisatellite is compared in two people. In person A the minisatellite has nine repeat units, but in person B there are only eight repeats. Note that real minisatellites are much longer than this, with several hundred copies of the repeat unit.

FIGURE 22.3 Genetic fingerprinting. The DNA from each person is cut with a restriction endonuclease and the fragments are separated into a smear by agarose gel electrophoresis. The DNA is then transferred to a membrane and a hybridization probe that detects the minisatellite core sequence is applied. The probe therefore reveals those restriction fragments that contain a minisatellite, with the length of the fragment indicating how many repeat units are present.

Genetic fingerprinting was first used in 1986 to identify the person responsible for the rape and murder of two teenage girls in Leicestershire, United Kingdom. Traces of semen were used as the source of the DNA that was fingerprinted. This fingerprint did not match that of the person initially suspected of the crime, who was therefore shown to be innocent. More than 4,000 men living in the region volunteered blood samples that were tested but still no match was identified. However, one of these volunteers was overheard in his local pub saying that when he gave his blood sample he used the name of work colleague, who had asked him to take his place as he did not want to be involved with the police. The work colleague was arrested and found to have a genetic fingerprint identical to that from the semen sample. He admitted his crimes and after trial was sentenced to life imprisonment.

Short tandem repeats are used to obtain DNA profiles

Genetic fingerprinting was a major new development in forensic science, but the method has two important weaknesses. The first is that a relatively large amount of DNA is needed to obtain a genetic fingerprint, which means that the method is not applicable at many crime scenes. Semen or blood samples often yield enough DNA for a fingerprint, but dried bloodstains usually do not. The second problem is that minisatellites have limited variability. This means that there is a relatively high possibility that two individuals could have the same, or very similar, fingerprints. A genetic fingerprint could therefore be discredited as evidence on the basis that although it matches that of the suspect, the criminal could in fact be somebody else whose DNA has not been tested.

DNA profiling is a more powerful technique that circumvents these problems. In a DNA profile, the polymorphic sequences called microsatellites or short tandem repeats (STRs) are typed. These are similar to minisatellites, but an STR is rarely longer than 150 bp and the repeat motifs are less than 15 bp in length. In many countries, DNA profiles are obtained by typing the alleles present at just 20 STRs (Figure 22.4). These are called the **CODIS (Combined DNA Index System) set**. Other systems, for example, the European Standard Set, use a similar set of STRs. Nineteen of the STRs in the CODIS set have tetranucleotide repeats (e.g. 5′–AATG–3′ for the STR called TPOX, which is located on chromosome 2, in the tenth intron of the gene for thyroid peroxidase). The exception is D22S1045, which has a trinucleotide repeat. Each of the STRs has at least ten alleles, which means that together they have sufficient variability that there is only a one in 10^{18} chance that two individuals, other than monozygotic siblings, have the same profile. As the world population is around 7.7×10^9, the statistical likelihood of two individuals on the planet, other than monozygotic siblings, sharing the same profile is so low as to be considered implausible when DNA evidence is assessed.

A match between a profile obtained from a crime scene and that of a suspect is therefore looked on as highly suggestive. In most countries, a match is not considered enough to secure a conviction on its own, but if the prosecution can provide another form of evidence linking the suspect with the crime, then often the DNA evidence is taken as conclusive. Equally importantly, the absence of a match can exclude suspects from an enquiry, and in particular has resulted in the release from prison of people who were wrongly convicted at a trial held before the days of DNA testing.

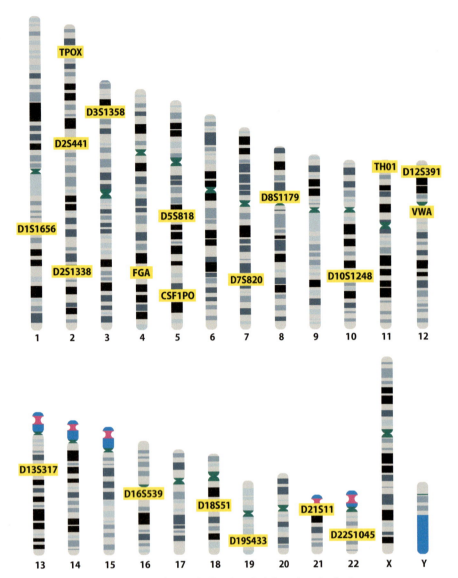

FIGURE 22.4 The CODIS set of STRs, indicating their locations in the human genome.

A DNA profile is obtained by multiplex PCR

How do we identify the alleles present at an STR sequence? The answer is to use the polymerase chain reaction (PCR – see Figure 21.15). The primers for the PCR are designed so that they anneal either side of the STR. This means that the length of the molecule that is produced during the PCR will indicate how many repeats are present and therefore identify the allele. One of the primers used in the PCR is labeled with a fluorescent marker so that the length of the PCR product can be measured by capillary gel electrophoresis (Figure 22.5). One band, indicating a single allele, will be detected if the person whose DNA is being tested is homozygous for the STR. Two bands will be seen if the individual is a heterozygote.

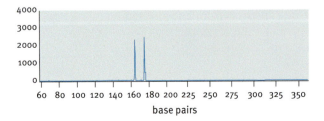

FIGURE 22.5 Identifying the alleles present at an STR by capillary gel electrophoresis. This STR is heterozygous with different repeat numbers on the two copies of its chromosome. PCR therefore gives a mix of two products of different lengths. These lengths are measured by capillary gel electrophoresis and displayed as a pair of peaks on the readout.

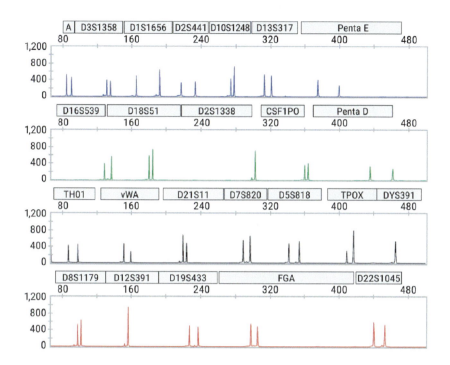

FIGURE 22.6 **A DNA profile.** This profile has been performed as a multiplex PCR with four different fluorescent labels. Each row shows the sizes of the products obtained with a different label. The profile includes the twenty CODIS STRs, a Y chromosome STR (DYS391) and an amelogenin PCR (A) for sex identification, as well as two additional STRs, Penta D and Penta E, which are highly variable though not included in the official CODIS set. (Adapted from figure courtesy of Promega Corporation.)

Carrying out 20 PCRs in parallel is time-consuming, an important consideration in a forensic laboratory that might have to process hundreds of samples a week. To save time, the profile is usually constructed by performing all the PCRs together in a single reaction. This is called a **multiplex PCR**. The products of multiple PCRs can easily be separated in the capillary gel, but a means is needed for identifying which molecules come from which STR. One possibility would be to use a different fluorescent label for each individual PCR in the multiplex, but this makes the detection process overly complicated. Instead, just four different labels can be used, the results of PCRs with the same label being distinguished by ensuring that the product sizes do not overlap (Figure 22.6).

The multiplex PCR is often designed in such a way that DNA profiling is accompanied by a DNA-based sex identification. Two approaches to sex identification are used. The first is simply to include a PCR that targets an STR located on the Y chromosome. If a PCR product is detected, then it can be concluded that Y chromosomal DNA is present in the DNA sample. If there is no product then the Y chromosome is absent.

The second approach to sex identification does not involve an STR. Although the X and Y chromosomes are largely different from one another, there are two short segments, called the **pseudoautosomal regions**, where the sequences of the two chromosomes are very similar (Figure 22.7). Several genes are located in these regions, including one for the tooth enamel protein called **amelogenin**. At one position in this gene, the version on the Y chromosome has a 6-bp insertion that is not present in the gene on the X chromosome (Figure 22.8). A PCR that is designed to span this insertion therefore gives two products of different size if both X and Y chromosomes are present, and just a single product, the smaller of the two, if the Y chromosome is absent. The results of this PCR can be seen in Figure 22.6, in this case revealing that the sample was female.

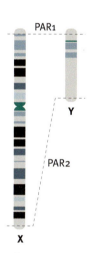

FIGURE 22.7 **The two pseudoautosomal regions, PAR1 and PAR2, of the human sex chromosomes.**

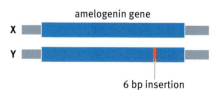

FIGURE 22.8 **The amelogenin gene.** The Y version of the amelogenin gene has a 6 bp insertion that is not present in the X version.

DNA profiling can solve cold cases

PCR has great sensitivity and can yield amplification products from the DNA content of just a single cell. DNA profiling is therefore a much more powerful technique than genetic fingerprinting, which required a sample of semen or blood as the starting material. Gradually, as methods have improved, DNA profiling has been applied to smaller samples, including bloodstains, the roots of hairs and even from skin cells shed into clothing or caught under the fingernails of assault victims.

The expansion of the range of material that can be used to produce a result has meant that DNA profiling is increasingly being applied to 'cold cases', ones from years or decades in the past that have remained unsolved. Evidence collected at the time and stored since then may still contain traces of DNA that can be used to construct a profile. This work is challenging, because DNA is an unstable molecule that will gradually break down after the death of a cell. In particular, even if evidence such as clothing is kept dry, there will be sufficient water present to induce breakage of the polynucleotides (see Figure 11.14). Over time the DNA will degrade into fragments, which eventually will become too short to act a targets for PCR (Figure 22.9). Special PCR systems that position the primers in such a way as to reduce the lengths of the PCR products may help if the DNA is only slightly fragmented, but it is still possible that **allelic dropout** will occur, where one or more STR alleles that are present fail to amplify (Figure 22.10a). A second problem with degraded DNA is **stutter**, where the authentic PCR product from an STR is accompanied by products differing in length by one or more repeat units (Figure 22.10b). Allelic dropout and stutter both affect the accuracy of the DNA profile, which can be a critical factor when a profile is used during a trial.

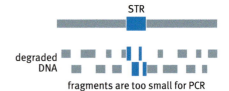

FIGURE 22.9 Degradation of DNA over time will prevent STR typing. Over time the DNA will degrade into fragments. A PCR directed at the STR will not work with the fragmented DNA, because the entire STR is not present in any of the fragments that remain.

Among the cold cases that have been solved in this way is the murder of a woman on Wimbledon Common, London, in July 1992. At the time, investigators could not obtain sufficient DNA to produce a profile, but when the case was reopened 10 years later, the forensic geneticists were able to identify a male sample from the clothing of the murder victim. This sample had been left on the clothing in the form of skin cells that had rubbed off of the hands of the murderer. The profile that was obtained matched a profile held by the UK National DNA Database. This database holds DNA profiles taken from individuals who have been charged with a criminal

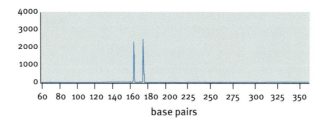

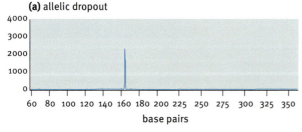
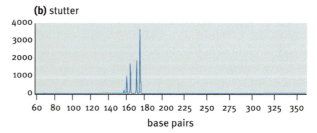

FIGURE 22.10 Two artifacts that occur when STRs are amplified from degraded DNA. The correct STR allele profile is shown at the top. (a) Allelic dropout, when one of the STR alleles fails to amplify. (b) Stutter, where additional peaks are seen.

offence. In this case, the match was with a man who had been convicted in 1995 of the murder of another woman and her daughter. This person subsequently confessed to the Wimbledon Common murder.

Familial DNA testing extends the range of DNA profiling

A DNA profile is simply a description of the alleles present at a set of STRs in a human genome. Like all alleles, the profile has been inherited partly from the mother and partly from the father (Figure 22.11). Parents and their offspring, brothers and sisters and close relatives, will all have profiles that reflect these inherited similarities.

These relationships raise the possibility that if even if a sample taken from a crime scene cannot be linked directly with the perpetrator of the crime, it might still be possible to identify the criminal by recognizing a similarity between the DNA profile and that of a relative. This is called **familial DNA testing**. It is currently a controversial area of forensic genetics and we will return to it in that context in Chapter 24. Here we will look at the ways in which familial DNA testing is carried out.

Familial DNA testing involves using a computer to compare a test DNA profile with all the profiles in a database. The computer assigns a degree of similarity to each database profile depending on the number of alleles shared with the test profile and the frequency of those alleles in the population as a whole. A profile assigned a high degree of similarity might be from a relative of the person who left the sample at the crime scene, but this cannot be assumed. Further DNA testing is therefore needed to confirm, or discount, the match. For this testing to be possible there must be DNA available from the individual whose profile was identified in the database. This is often a terminal problem as in many countries it is illegal for the authorities to store the actual DNA sample after the profile has been obtained and entered in the database.

If further comparisons are possible, then often these focus on the Y chromosome, presuming the database hit is a male sample. This approach reflects the fact that the majority of perpetrators of violent crime are male. STRs, and also single-nucleotide polymorphisms (SNPs), on the Y chromosome will be typed in the crime scene and database samples. If there is sufficient DNA, then the analysis could be extended to include STRs and SNPs from the autosomes. The resulting information might increase the likelihood that the similarity between the profiles is due to a family relationship, providing investigators with leads regarding possible suspects.

The accuracy of familial DNA testing is increased greatly if actual family members of the suspect can be persuaded to give DNA samples. This has been a feasible approach with those cold cases where the suspect is deceased. If living family members of the suspect are willing to donate DNA samples, then it might be possible for that suspect to be linked to DNA taken from the crime scene. Some of the oldest cold cases have been solved in this way, including the murder of a teenage boy and girl in Montana in 1956. The main suspect had died in 2007, but his children volunteered to give DNA samples in 2021. These were similar enough to a sample of semen stored from the crime scene for the suspect to be associated with the crime.

DNA testing is also used to identify human remains

In addition to the forensic applications, DNA profiling and other methods for DNA testing are also used to identify human remains, in those situations where more conventional means of identification are not possible. These

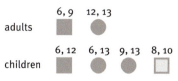

FIGURE 22.11 DNA profiles are inherited. Squares represent males and circles represent females. A single STR has been typed in each individual, and the alleles are indicated by the numbers. Three of the children could be siblings and the two adults could be their parents. The fourth child, shown as the open square, cannot be a member of this family.

include the victims of disasters such as the World Trade Center attacks and airplane crashes. This work pushes DNA analysis to its very limits as events of this nature are often associated with incineration, so DNA yields are low and whatever DNA can be recovered is highly degraded. If profiles or other DNA data can be obtained, then these can be compared with the victim's own DNA, taken from sources such as a toothbrush, or a familial analysis can be performed with DNA donated by relatives.

Similar challenges occur in the identification of fallen service personnel, especially from past conflicts where in the heat of battle victims were buried without identification or were listed as missing in combat. The US Armed Forces DNA Identification Laboratory has a commitment to identify and repatriate the remains of missing service personnel from conflicts back to the mid-20th century. Many of these bodies suffered trauma at the time of death and since then have been exposed to harsh environmental conditions that have resulted in extensive DNA decay. There are stored blood samples for all US military personnel who enlisted since 1992, aiding their identification as DNA data from remains can be compared with the blood samples. For remains predating 1992, familial DNA analysis is the only possible approach.

A famous example of DNA identification of a lost serviceman concerns James 'Earthquake McGoon' McGovern, an American pilot who crashed close to Dien Bien Phu, Laos, attempting to fly ammunition to French troops in 1954, during the First Indochina War. The garrison surrendered the next day and the remains thought to be those of McGovern and his crewmen were not recovered until 2002. DNA samples from bone material yielded a partial CODIS STR profile and data was also obtained for additional Y-STRs. These were compared with the profiles of various of McGovern's living relatives. A critical result was the identity between the Y-STR alleles obtained from the bone sample and those of McGovern's paternal nephew – the son of his brother. This, and other correlations between the test sample and those of the relatives, was sufficient to confirm the remains as belonging to McGovern. Similar work enabled his crew members to be distinguished. Following his identification, McGovern was buried at Arlington National Cemetery in 2007.

DNA testing can be used in kinship analysis

Our exploration of familial DNA testing as an aid in identification of forensic DNA samples and those obtained from human remains leads us directly to the final application of DNA testing that we will study. This is the use of DNA profiling and other genetic tests in kinship analysis.

The objective of kinship analysis is to identify individuals who are related to one another and to establish the exact nature of those relationships. DNA profiling can contribute to kinship analysis because, as we acknowledged above, the STR alleles that make up a profile are inherited in the same way as other genetic alleles. Full siblings – children with the same parents – will therefore have allele combinations that are similar to one other and to those of their parents (see Figure 22.11). The most common application of this type of genetic profiling is in paternity testing, when there is uncertainty regarding the identity of the father of a child.

The 20 STRs of the CODIS set are all located on autosomes (see Figure 22.4). Sometimes it is an advantage to follow just the male or female line in a family. The male line can be followed by typing one or more STRs located on the Y chromosome, as the Y chromosome is passed from a father to his sons. This means that male siblings possess the same alleles for a set of Y-STRs, these being the alleles that their father has (Figure 22.12a).

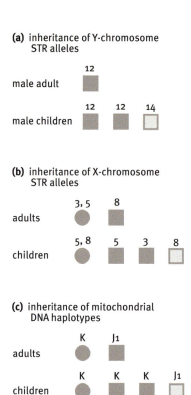

FIGURE 22.12 Following the male or female lineages within a family. Squares represent males and circles represent females. (a) The male line can be followed by typing STRs on the Y chromosome. In this example, a single Y chromosome STR has been typed. Two of the children have the same allele as the male adult and could be his sons. The third child, shown as an open circle, cannot be his son. (b) The maternal line cannot be precisely followed by typing STRs on the X chromosomes. The combinations shown here are compatible with the hypothesis that the female is the parent of three of the children, but the inheritance pattern is not strictly maternal because the female child also inherits an X chromosome from her father. The fourth child cannot be the son of either parent. Although he has the same X chromosome allele as the male adult, he cannot have inherited it from that person. (c) The female line can be followed by typing the mitochondrial genome. The first three children have the same mitochondrial DNA haplotype as the female adult and so could be her daughter and sons. The fourth child cannot be a member of this family group. Although his mitochondrial haplotype is the same as that of the male adult, he cannot have inherited it from that person.

How can we follow the female line in a family? We could type STRs located on the X chromosome but this would provide only part of the answer. Male children always inherit a X chromosome from their mother, but one of the X chromosomes inherited by a female child comes from the father (Figure 22.12b).

The female line is more precisely traced by studying mitochondrial DNA. Unlike the nuclear DNA, the mitochondrial genome is inherited in a **uniparental** manner, with only the mother contributing her genome to the offspring. The paternal mitochondrial genome is present in the male sperm but does not penetrate the egg cell during fertilization, or if it does it is degraded soon afterward. Each of a set of full siblings therefore has the same version of the mitochondrial genome, this being the version possessed by their mother (Figure 22.12c). There are no STRs in the human mitochondrial DNA, but there are many SNPs, especially in the region either side of the replication origin, the only part of the genome that lacks genes (Figure 22.13). The sequence of this **hypervariable region** defines the mitochondrial DNA haplotype, and it is this haplotype that is shared by a set of siblings and their mother.

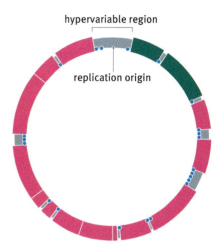

FIGURE 22.13 The human mitochondrial genome, showing the position of the hypervariable region. The hypervariable region is 747 bp in length and contains the replication origin for the mitochondrial DNA. See Figure 20.16 for a description of the genes present in the mitochondrial genome.

22.2 ANCIENT DNA EXTENDS GENETICS INTO ARCHAEOLOGY

We have learnt that DNA testing is being applied to the remains of fallen service personnel from the conflicts of the mid-20th century. If DNA is still present 70 years after the death of an individual, then might it be possible to obtain DNA from older material, including much older human remains from archaeological periods?

The answer is yes. Ancient DNA is now known to be present in many archaeological remains and even in some paleontological samples dating as far back as 1.2 million years ago. The discovery that ancient DNA is preserved in this type of material has led to the new discipline called **archaeogenetics**.

Ancient DNA is degraded and prone to contamination with modern DNA

The first indication that DNA might be preserved in very old material came in 1984 when DNA sequences were obtained from a 140-year-old museum skin of a quagga, an extinct type of zebra that had once been common in parts of southern Africa (Figure 22.14). Five years later, in 1989, the first successful of extraction of ancient DNA from archaeological bone samples was reported.

There is never very much DNA in an archaeological specimen, possibly only a few picograms in a gram of bone, but this is not a problem because PCR can amplify these tiny amounts into larger quantities that can be typed by DNA sequencing and STR analysis. It was soon realized that the decay processes that cause polynucleotides to fragment over time mean that ancient DNA molecules are rarely longer than 100 bp, and some nucleotide positions might be unreadable due to reaction with chemicals in the burial environment. Nevertheless, in the early 1990s sequences from samples going back millions of years were obtained.

Unfortunately, the claims for multi-million-year-old ancient DNA proved to be misguided. It was realized that extreme care must be taken to avoid contaminating archaeological material with modern DNA. Contamination

FIGURE 22.14 A quagga. This mare was photographed at London Zoo in 1870. Quaggas were hunted to extinction in the wild by the late 1870s and the last captive quagga died in 1883. (Courtesy of Frederick York and the Biodiversity Heritage Library.)

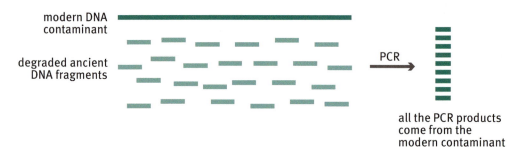

FIGURE 22.15 The problem that occurs when an ancient DNA sample is contaminated with modern DNA. The single modern DNA molecule acts as a better template for PCR than any of the degraded ancient DNA fragments. As a result, all of the PCR products are copies of the modern DNA contaminant and none come from the ancient DNA.

could be human DNA from the hands of archaeologists and geneticists who handle the samples, or DNA from PCR products generated by other experiments carried out in the same laboratory, and which had escaped from their test tubes in the form of small aerosol droplets. Because modern DNA is intact and lacks chemical damage, it acts as a better template for PCR than a fragmented, partially degraded sample of ancient DNA. The presence of a single modern DNA molecule in an ancient DNA preparation could, therefore, result in sequences derived entirely from that contaminant (Figure 22.15). This, sadly, was the explanation of the highly publicized claims in the early 1990s of sequences from dinosaur bones (these were contaminating human DNA sequences from parts of the genome that at that time were uncharacterized) and from insects preserved in amber. It did, however, make a good movie.

Ancient DNA researchers now adopt stringent precautions to prevent contamination of samples, including wearing full-body protective clothing while preparing extracts and setting up PCRs. The study of ancient DNA has also been greatly boosted by the development of short-read DNA sequencing methods (Section 18.1). Originally, a different PCR had to be set up for each small segment of ancient DNA that was being studied. This meant that DNA data from an archaeological sample could be acquired only very slowly. Short-read sequencing provides data from all the ancient DNA in a sample in a single experiment, making it much easier to obtain information on sequence polymorphisms that might be used to address archaeological questions. We will now move on and examine exactly what kinds of archaeological questions can be asked.

Ancient DNA typing provides a means of studying kinship in past societies

It is very difficult to assess genetic relatedness simply by examining the physical structure of a pair of skeletons. Osteologists recognize more than 400 features of the human skeleton that vary among different people, and like all biological characteristics those variations are inherited. They include the positions of holes in the base of the cranium where nerves enter and leave the head, variations in the structure of the teeth, and differences in the patterns formed by the fusion of the bones that form the skull. Providing the skull is intact and well-preserved – which is not always the case with archaeological skeletons – careful measurement of these various structures can give an indication of kinship between a pair of skeletons if these come from members of a single family. Whatever conclusions are drawn have to be treated with care because the traits can be affected not just by genetic factors but also diet and environmental conditions.

Ancient DNA provides a possible solution to these problems because genetic relationships inferred from DNA data are much more accurate than those obtained from osteological measurements. The key issue is whether it is possible to obtain enough data from an ancient DNA sample to type a sufficient number of STRs or other polymorphisms. Until recently, most of this work focused entirely on mitochondrial DNA. There are about 8000 identical copies of the mitochondrial genome per cell, compared to just two copies of the nuclear genome. It should therefore be easier to type polymorphisms in the mitochondrial DNA simply because there should be more of these molecules in an ancient DNA sample.

There is a limit to the information that can be obtained from mitochondrial DNA typing. Even if the entire hypervariable region is sequenced, the variations that exist are only enough to distinguish 275 different haplotypes. This means that many people who are unrelated have the same haplotype. Also, mitochondrial DNA can only reveal maternal relationships (see Figure 22.12c). It can identify a mother and her children but cannot provide information on the male parent. Note, however, that mitochondrial DNA can provide definite information on nonrelatedness: if two ancient DNA samples yield different mitochondrial DNA haplotypes, then we can be certain that the individuals they come from were not full siblings. This aspect of mitochondrial DNA typing was important in DNA analysis of a female skeleton discovered at a site in South Uist, one of the Outer Hebridean islands off the northwest coast of Scotland. This skeleton was buried underneath a dwelling about 1100 BC, during the Late Bronze Age. Mitochondrial DNA typing led to the remarkable discovery that the skeleton was made up of the bones of different people (Figure 22.16). A tooth from the lower jaw belonged to haplogroup U5, the right femur was typed as T1a and the right humerus as haplogroup H2a2a1. Archaeologists are still speculating about the origin and meaning of this composite skeleton.

Ancient DNA has identified family groups among archaeological skeletons

An example of how ancient DNA typing can provide kinship information in an archaeological project was provided by research carried out on skeletons from a 4,600-year-old cemetery at Eulau, in central Germany. Four graves were excavated, these containing a total of 13 skeletons, a mixture of adult males and females as well as children. The arrangement of the skeletons in the graves suggested that each was a family group comprising one or both parents and their children. For two of the graves, this hypothesis was supported by ancient DNA results. In grave 99, for example, the female adult had mitochondrial haplotype K, and the male adult possessed the combination of Y-STR alleles that are designated haplotype R1a. The two children, both males, also possessed haplotypes K and R1a (Figure 22.17). The children therefore had the combination of alleles that would be expected if the two adults were their parents.

The skeletons at Eulau date from the period when people in Northern Europe were abandoning their traditional lifestyle of hunting wild animals and gathering wild plants, and becoming farmers. Hunting and gathering requires a great deal of mobility as the animals must be followed as they migrate from place to place. Farming, on the other hand, enables a more settled lifestyle, with a community staying in one place for years on end. The DNA results show that in these early farming communities, people were organizing themselves into family groups similar to ones that we are familiar with today.

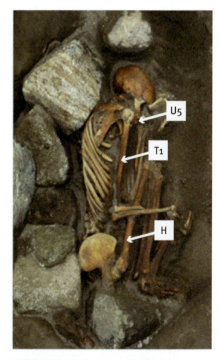

FIGURE 22.16 A composite Bronze Age skeleton from Cladh Hallan. This female skeleton is made up of bones from at least three people. The mitochondrial DNA haplogroups for the different skeletal parts are indicated. (From J. Hanna et al., *J. Archaeol. Sci.* 39: 2774–2779, 2012. With permission from Elsevier.)

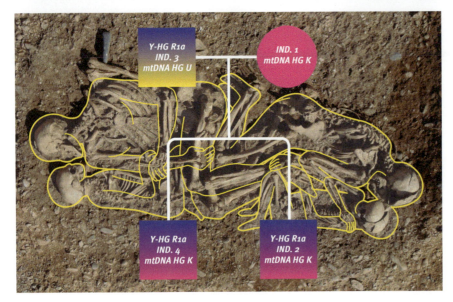

FIGURE 22.17 **The genetic relationships between the four skeletons in grave 99 at Eulau.** The outline drawing shows the two adult skeletons at the top, with the female on the right and the male on the left. The two children, both boys, are below. In the family tree, squares represent males and the circle represents the female adult. The mitochondrial and Y chromosome haplotypes that were identified for each individual are shown. Abbreviations: HG, haplogroup; IND., individual. (Courtesy of Andrea Hörentrup, Landesamt für Denkmalpflege und Archäologie Sachsen-Anhalt, State Office for Heritage Management and Archaeology Saxony-Anhalt.)

An interestingly sideline was revealed by analyzing the strontium isotope composition of the teeth of some of the skeletons. The strontium in teeth reflects the geological isotope composition of the area where a person is living when their teeth form. The results with the Eulau skeletons suggested that although the adult males had been born and raised in the area surrounding Eulau, the females – their wives in our modern parlance – came from the Harz mountains some 40 km away. Anthropologists refer to this as an exogamous and patrilocal marriage system. Exogamy is where marriage is between members of two different social groups, and is often practised as a way of forging alliances and reducing conflict. Patrilocal marriage is where the female partner moves and the new couple live close to the man's family. In modern societies, patrilocal marriage typically occurs when there is patrilineal inheritance, a married son staying close to his parents so he can take over the homestead when his father dies.

Ancient DNA can reveal more distant relationships between groups of prehistoric people

Genetic relationships are present not just within family groups. The alleles possessed by an individual are inherited from a vast web of grandparents, great-grandparents and earlier ancestors, back into the distant past. Many people who have no knowledge of their distant relationship share alleles from a common ancestor who may have lived centuries ago (Figure 22.18).

To identify these more distant relationships, thousands of SNPs and other polymorphisms must be typed in multiple individuals. Until recently, this was beyond the capability of ancient DNA research, but advances since 2010 have made this type of analysis possible. One of these advances is the discovery that ancient DNA samples prepared from certain parts of the skeleton are less degraded than samples from other parts. In particular, the temporal bones, two of which are located at the base of each skull, include a segment called the petrous, which has a very high bone density. This provides a protective environment that retards the decay processes that affect DNA in other bones. The high yields of ancient DNA that can be obtained from the petrous, combined with modern short-read sequencing methods, mean that it is now possible to obtain a complete genome sequence for a person who lived thousands of years ago.

FIGURE 22.18 Some of your relatives. The tree shows relationships among individuals who are all descended from your great-great-great-grandparents. Each category in the tree can, of course, be occupied by more than one person.

How are genome sequences analyzed to reveal distant genetic relationships? The method is called **identity-by-descent** (**IBD**). It involves searching for segments of DNA that contain groups of SNPs that display identical alleles in different individuals. The allele identity indicates that recombination has not occurred within such a segment during the period since the common ancestor of those two people (Figure 22.19).

The lengths of the DNA segments shared by IBD indicate how far back in time we have to go to reach the common ancestor. This is because recombination events that occur every generation break up the shared

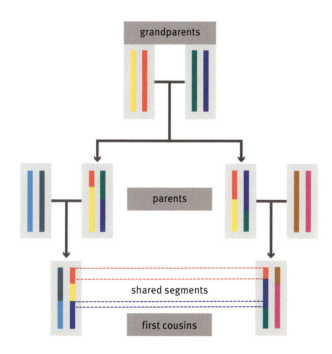

FIGURE 22.19 The basis to identity-by-descent. The diagram shows the inheritance of a pair of homologous chromosomes. Because of recombination, the chromosomes inherited by offspring are mosaics of the parental versions. The diagram shows that after two generations quite large shared segments can be identified when the chromosomes of first cousins are compared. With additional generations and additional recombination, the lengths of the shared segments will decrease, but are still identifiable in 10th cousins.

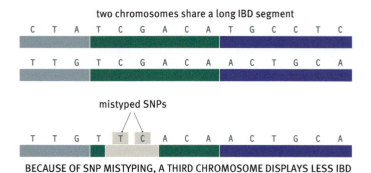

FIGURE 22.20 SNP mistyping reduces the amount of IBD that is detected. The SNPs that are typed in a chromosome from three individuals are shown. The results show that the top two chromosomes share a lengthy segment, shown in green. In the lower chromosome, two SNPs have been mistyped. The shared segment is split into three shorter pieces, reducing the amount of IBD that is detected.

segments, so these gradually become shorter and shorter. The genomes of closely related individuals such as siblings have long segments of shared DNA, but first cousins, who share a grandparent, display a lesser degree of IBD. When modern DNA databases are analyzed, IBD can be used to identify relatives as distant as 10th cousins – ones who share a great-great-great-great-great-great-great-great-grandparent. If we take 25 years as the average for a human generation, that ancestor lived 250 years ago.

To use IBD to identify relationships as distant as 10th cousins, it is necessary for the DNA data to be absolutely accurate. If an SNP is mistyped then the shared segment in which it is located will appear to be broken into two segments (Figure 22.20) reducing the amount of IBD that is detected. This is problem with ancient DNA because the data are often low quality. Recall that it is necessary to sequence each part of the genome multiple times to ensure that errors made by the DNA sequencer can be identified (see Figure 18.8). Often the amount of ancient DNA that can be obtained from a skeleton is too little to enable this checking process to be performed, and sometimes the sequence will have gaps where SNPs cannot be typed. As a result, it has been estimated that relationships only up to 6th cousins can be detected when IBD is applied to typical ancient DNA datasets. Despite this limitation, use of the method has led to interesting discoveries about prehistoric populations. In one study, it was shown that two groups of people from approximately 3000 BC, referred to as the Yamnaya and Afanasievo cultures, have close genetic links, even though the Yamnaya lived on the border of modern Asia and Europe and the Afanasievo in south Siberia (Figure 22.21). The extent to which people moved around during that period is illustrated by a pair of 5th cousins belonging to the Afanasievo cultures, who were buried 1,400 km apart, in Mongolia and southern Russia, respectively.

The Neanderthal genome has been sequenced

Providing the environmental conditions are not too harsh, ancient DNA can be preserved long enough to be used to obtain sequences not just from *Homo sapiens*, but also from our extinct ancestors. Neanderthals (Figure 22.22) are an extinct type of human who lived in Europe and parts of Asia between 200,000 and 30,000 years ago. Many paleontologists classify them as a distinct species, *Homo neanderthalensis*, while others argue that their skeletons are sufficiently similar to our own for Neanderthals to be classified as a subspecies of *H. sapiens*. In either case, they are our closest extinct relatives.

The first Neanderthal genome sequence was obtained from a 50,000-year-old toe bone preserved in Denisova cave in the Altai Mountains of Siberia. Enough sequences were generated to achieve 52× coverage of the genome, sufficient to ensure a high degree of accuracy. This amount of coverage is necessary because of the possibility that individual fragments

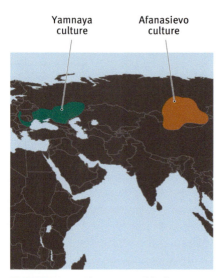

FIGURE 22.21 The geographical ranges of the Yamnaya and Afanasievo cultures, c. 3000 BC.

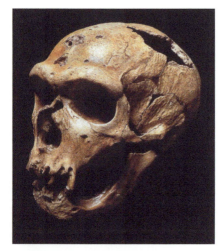

FIGURE 22.22 Skull of a male Neanderthal. This individual was 40–50 years old when he died about 50,000 years ago in what is now La Chapelle-aux-Saints, France. (Courtesy of John Reader/Science Photo Library.)

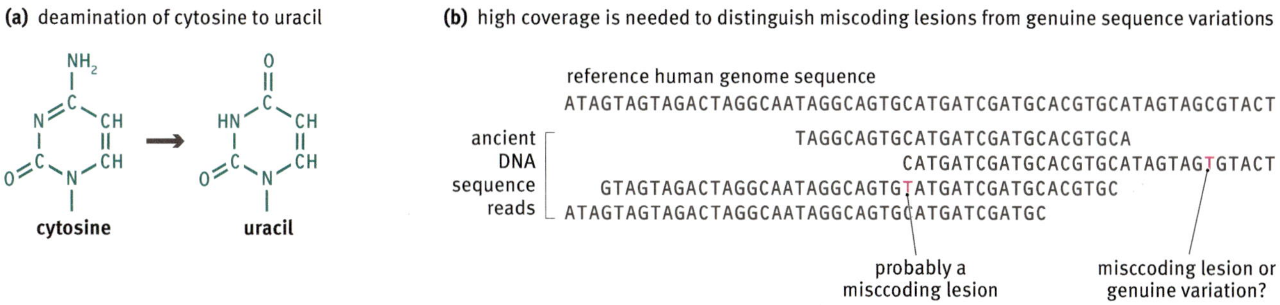

(a) deamination of cytosine to uracil

(b) high coverage is needed to distinguish miscoding lesions from genuine sequence variations

```
                          reference human genome sequence
                          ATAGTAGTAGACTAGGCAATAGGCAGTGCATGATCGATGCACGTGCATAGTAGCGTACT
              ancient  ┌                      TAGGCAGTGCATGATCGATGCACGTGCA
              DNA      │                                 CATGATCGATGCACGTGCATAGTAGTGTACT
              sequence │         GTAGTAGACTAGGCAATAGGCAGTGTATGATCGATGCACGTGC
              reads    └ ATAGTAGTAGACTAGGCAATAGGCAGTGCATGATCGATGC
```

probably a miscoding lesion

miscoding lesion or genuine variation?

FIGURE 22.23 Miscoding lesions cause problems when sequencing ancient DNA. (a) Deamination of cytosine gives uracil. When a DNA sequence is read, this miscoding lesion results in the C being read as a T. (b) High coverage is needed to distinguish miscoding lesions from genuine sequence variations. In this example, there are two possible miscoding lesions. One can be identified with confidence as it is present in just one of four sequences. The second possible miscoding lesion is present in a part of the genome that is covered by just a single ancient DNA sequence read. It is impossible to tell if this C→T anomaly is a miscoding lesion or a genuine sequence variation.

of Neanderthal DNA contain **miscoding lesions.** These are positions where the degradation process results in the sequence being misread. An example is the conversion of a cytosine to uracil (Figure 22.23a), which occurs by spontaneous deamination in both cellular DNA molecules and in ancient DNA. In cellular DNA the damage is corrected by the base excision repair process, but after death the damage accumulates. The uracils that result will be read as Ts when the ancient fragment is sequenced. A high coverage ensures that these errors are not carried over into the completed genome sequence (Figure 22.23b).

A second bone sample from the Altai cave proved to be even more exciting. This came from the most incomplete of skeletons: all that was found was a single bone from a little finger. The mitochondrial DNA sequence from this bone was quite unlike any of the mitochondrial haplotypes known in *H. sapiens*, and also different from the Neanderthal sequence. At first, it was thought that the finger bone might come from an entirely unknown species of extinct human, which were dubbed the Denisovans. However, the nuclear genome sequence showed a greater degree of similarity with Neanderthals, and it is now uncertain if Denisovans were a genuinely different kind of human or a subgroup of Neanderthals.

Our ancestors interbred with both Neanderthals and Denisovans

We share about 99.7% sequence identity with Neanderthals and Denisovans. As with the comparison between the human and chimpanzee genomes (Section 20.3), examination of the 0.3% of difference does not provide any obvious clues regarding the genetic basis to the special biological characteristics of *H. sapiens*.

Instead, the most striking insight into recent human evolution provided by the genome sequences is a clear indication of interbreeding, or **admixture**, between *H. sapiens* and both Neanderthals and Denisovans. This was first established from studies of the Neanderthal genome. If there has been no admixture, then we would expect the degree of difference between the Neanderthal genome and that of modern Europeans to be about the same as the difference between Neanderthals and people from other parts of the world. This is not the case: there is less difference between Neanderthals and Europeans than there is between Neanderthals and other modern populations, suggesting that some Neanderthal DNA is present in the genomes of modern Europeans (Figure 22.24). Our species originated in

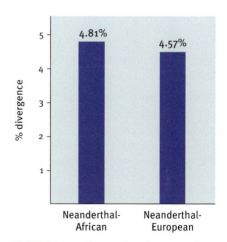

FIGURE 22.24 Comparison between the Neanderthal genome and the genomes of modern-day Africans and Europeans. There is less divergence between Neanderthal and European genomes, suggesting that the ancestors of modern Europeans interbred with Neanderthals.

Africa but small groups left that continent around 70,000 years ago, starting a migration that eventually reached those parts of Europe and Asia populated by Neanderthals some 45,000 years ago. As Neanderthals did not become extinct until 30,000 years ago, there was a window of opportunity of 15,000 years when interbreeding must have occurred.

Admixture can also be detected between Denisovans and *H. sapiens*. The latest estimates are that approximately 2.0% of the DNA of present-day humans outside of Africa derives from admixture with Neanderthals, and 1.0% of the genomes of inhabitants of Oceania comes from prehistoric interbreeding with Denisovans. And it does not stop there. There was also admixture between Neanderthals and Denisovans, and between Denisovans and an extinct, unidentified type of human (Figure 22.25).

What impact did interbreeding have on our genome? Researchers are still exploring this question but one interesting conclusion is that admixture with Denisovans has been responsible for the ability of people from Tibet to live at high altitudes, where the oxygen content of the atmosphere is particularly low. One of the genes that allows the human body to tolerate low oxygen conditions is *EPAS1*. This gene codes for a protein that regulates the expression of other genes that together provide some of the necessary biochemical and physiological adaptations. Tibetans possess a unique allele of *EPAS1* that is not seen in any other group of people today. Remarkably the same allele is present in the Denisovan genome, leading to the conclusion that this allele, and the adaptation to low oxygen that it provides, was transferred into the genome of the distant ancestor of Tibetans by admixture with Denisovans.

Ancient DNA can be obtained from many species other than humans

So far we have focused entirely on ancient DNA from human skeletons. Ancient DNA is also present in the remains of many other species. The study of nonhuman ancient DNA adds another dimension to archaeogenetics and is also an important tool in paleontology and environmental biology.

The oldest skeleton so far found to contain ancient DNA comes from a steppe mammoth (Figure 22.26a), a predecessor of the wooly mammoth, which lived about million years ago. The mammoth became buried in permafrost which, being a low and consistent temperature, reduced the rate of DNA degradation. Ancient DNA has also been isolated from the remains of more recent mammoths, enabling the genetic relationships between different types of mammoth to studied.

From the more recent past, ancient DNA from the skeletons of domesticated animals such as cattle, dogs and cats has allowed the relationships between the domestic and wild versions of these species to be studied. In particular, DNA sequences have been obtained from bones of the aurochs, the extinct progenitor of domesticated cattle, as well as from early versions of cattle from archaeological sites in Europe and Asia. We know that farming began in southwest Asia around 12,000 years ago, and that farmers, along with their animals and crops, migrated into Europe and south Asia during the following millennia. Ancient DNA analysis has shown that in Europe, domesticated cattle interbred with the local aurochs populations, similar to the admixing that occurred many thousands of years earlier between *H. sapiens* and Neanderthals. Additionally, during a period 4,200 years ago, when the climate became warmer and dryer, domesticated cattle from India, called the zebu, which were already adapted to dry conditions, were crossbred with the less well adapted cattle from the Near

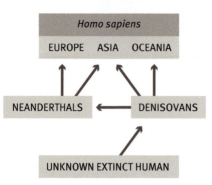

FIGURE 22.25 Interbreeding between *H. sapiens* and extinct types of human. The arrows indicate the direction in which DNA has been transferred.

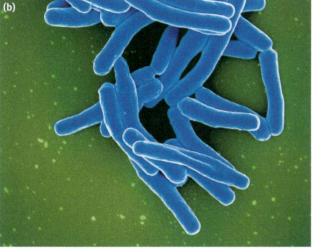

FIGURE 22.26 Two species that have been studied by ancient DNA sequencing. (a) The steppe mammoth, *Mammuthus trogontherii*, (b) *Mycobacterium tuberculosis*, the causative agent of tuberculosis. (a, From Roman Uchytel/Science Photo Library, with permission; b, From NIAID, published under CC BY 2.0 license.)

East. Prehistoric farmers therefore had to respond to challenges caused by climate change similar to those we face today.

Ancient DNA from nonhuman species can also be obtained from human skeletons. Several types of pathogenic bacteria leave their DNA traces in the skeleton, either because they enter bone tissue as part of their infective process or because they are present in the blood vessels that permeate hard tissue such as bone. *Mycobacterium tuberculosis* (Figure 22.26b), the causative agent of tuberculosis, is an example. Although looked on mainly as a lung disease, in about 4% of patients the tuberculosis bacteria move through blood and lymphatic systems and invade bones in various parts of the body. If this happens, then it might be possible to detect ancient DNA from *M. tuberculosis* in the skeleton thousands of years after death. In addition to indicating the presence of the disease in the individual concerned, the bacterial sequences reveal the differences between the ancient and modern versions of the pathogen. This work has indicated that previous ideas about the origins of tuberculosis are incorrect. It had been thought that humans originally contracted tuberculosis from domesticated cattle, but now it appears that it is a much older disease that has afflicted the human population since at least 70,000 years ago.

The million-year-old steppe mammoth is the oldest skeleton from which ancient DNA sequences have been obtained. But this is not the oldest DNA of all. The most ancient of DNA samples have been obtained from permafost sediments, dated to 2 million years ago, from North Greenland. Sedimentary DNA is believed to derive from the cells of animals and plants which are shed into the soil, where the DNA becomes bound to mineral particles. Immobilization in this way may help to slow down the breakdown of the DNA molecules. Ancient DNA from the Greenland sediments gave sequences from large mammals including caribou and mastodon, the latter an extinct type of elephant. Plant DNA included sequences from conifers such as poplar, as well as smaller plants that would have been browsed by herbivores. Two million years ago the climate was warmer than at present, and the sedimentary DNA analysis revealed that Northern Greenland, rather than being the icy wasteland that it is today, was a lush forest with trees, vegetation and many animals.

KEY CONCEPTS

- The presence of polymorphisms in the human genome means that no two people, except monozygotic siblings, have identical genome sequences.
- Genetic fingerprinting made use of minisatellite variability to distinguish between individuals.
- DNA profiling involves typing the 20 STR markers that make up the CODIS set. These have sufficient variability for there to be only a one in 10^{18} chance that two individuals, other than monozygotic siblings, have the same profile.
- DNA profiles are constructed by multiplex PCR with groups of STRs labeled with different fluorescent markers.
- Typing a Y-STR or the amelogenin gene is used for sex identification.
- DNA profiling of samples stored from past crime scenes has been used to solve cold cases.
- Familial DNA testing enables a DNA profile from a crime sample to be linked with individuals who may be related to the suspect.
- DNA testing is used to identify disaster victims.
- Kinship can be established by DNA testing. The Y chromosome is used to follow the male lineage, and mitochondrial DNA for the maternal lineage.
- Ancient DNA samples can be obtained from some archaeological samples. Ancient DNA is fragmentary and has suffered chemical damage, and is prone to contamination with modern DNA.
- Ancient DNA has been used to establish kinship between archaeological skeletons, and to identify broader relationships between groups of prehistoric people.
- The Neanderthal and Denisovan genomes have been sequenced. Comparison with the *H. sapiens* sequence has revealed traces of past interbreeding.
- Ancient DNA can also be obtained from the bones of extinct species, domesticated animals, pathogenic bacteria present in human bones, and from sedimentary samples taken from permafrost.

QUESTIONS AND PROBLEMS

Key terms

Write short definitions of the following terms:

admixture
allelic dropout
amelogenin gene
ancient DNA
archaeogenetics
CODIS set
DNA profiling
familial DNA testing
genetic fingerprinting
hypervariable dispersed repetitive sequence
hypervariable region
identity-by-descent
miscoding lesion
multiplex PCR
polymorphism
pseudoautosomal region
stutter
uniparental

Self-study questions

22.1 Describe the types of polymorphism present in the human genome.

22.2 Explain how a genetic fingerprint is obtained.

22.3 What are the key differences between genetic fingerprinting and DNA profiling? Why does DNA profiling result in a lower probability that two unrelated people have the same profile?

22.4 What are the features of the CODIS set of STRs?

22.5 Describe how multiplex PCR is used to obtain a DNA profile.

22.6 How is DNA testing used to identify sex?

22.7 What additional technical challenges occur when DNA is used to investigate cold cases?

22.8 Describe the basis to familial DNA testing.

22.9 Giving an example, explain how DNA testing is used to identify disaster victims.

22.10 Distinguish between the inheritance patterns of the Y chromosome and mitochondrial DNA. Why is X chromosome typing not a reliable way of following a maternal lineage?

22.11 Describe the problems that have been caused by contamination of ancient samples with modern DNA. How can these problems be avoided?

22.12 Why has mitochondrial DNA been a popular target for PCRs of ancient DNA?

22.13 Give two examples of archaeological projects that involved mitochondrial DNA typing.

22.14 What is meant by the term 'identity-by-descent'. When this method is used, how are close and more distant relatives distinguished?

22.15 What is a miscoding lesion? How were miscoding lesions identified when the Neanderthal genome was sequenced?

22.16 What patterns of interbreeding have been identified by comparisons of the Neanderthal, Denisovan and *H. sapiens* genomes?

22.17 Give one example of an allele thought to have been acquired by a modern human population through past interbreeding.

22.18 What information can be obtained by studies of ancient DNA from nonhuman vertebrates?

22.19 How has ancient DNA typing been used in studies of (A) ancient disease and (B) past ecosystems.

Discussion topics

22.20 The forensic use of DNA requires that the DNA profile of the criminal be obtained so that it can be matched with DNA from the crime scene. This process would be easier if the profiles of all members of the public were held in a national database. Discuss the ethical implications raised by the concept of a national DNA database.

22.21 Should a match between the DNA at a crime scene and that of a suspect ever be accepted as the sole type of evidence on which to base a conviction?

22.22 Are Neanderthal and *H. sapiens* the same or different species?

22.23 What is the potential of ancient DNA in studies of human evolution?

 The answers to, or hints on how to answer, the self-study questions and discussion topic questions can be downloaded from the book's product page at this link: www.routledge.com/9781032743530

FURTHER READING

Haak W, Brandt G, de Jong HN et al. (2008) Ancient DNA, strontium isotopes, and osteological analyses shed light on social and kinship organization of the Later Stone Age. *Proceedings of the National Academy of Sciences USA* 105, 18226–18231. *The family groups at Eulau.*

Hanna J, Bouwman AS, Brown KA et al. (2012) Ancient DNA typing shows that a Bronze Age mummy is a composite of different skeletons. *Journal of Archaeological Science* 39, 2774–2779.

Huerta-Sánchez E, Jin X, Asan BZ et al. (2014) Altitude adaptation in Tibetans caused by introgression of Denisovan-like DNA. *Nature* 512, 194–197.

Irwin JA, Edson SM, Loreille O et al. (2007) DNA identification of 'Earthquake McGoon' 50 years postmortem. *Journal of Forensic Science* 52, 1115–1118.

Jeffreys AJ, Wilson V & Thein LS (1985) Individual-specific fingerprints of human DNA. *Nature* 316, 76–79. *Genetic fingerprinting.*

Jobling MA & Gill P (2004) Encoded evidence: DNA in forensic analysis. *Nature Reviews Genetics* 5, 739–751.

Katsanis SH & Wagner JK (2013) Characterization of the standard and recommended CODIS markers. *Journal of Forensic Science* 58, S169–172.

Kjær KH, Winther Pedersen M, De Sanctis B et al. (2022) A 2-million-year-old ecosystem in Greenland uncovered by environmental DNA. *Nature* 612, 283–291.

Llamas B, Willerslev E & Orlando L (2017) Human evolution: a tale from ancient genomes. *Philosophical Transactions of the Royal Society of London, series B* 372, 20150484.

Nakahori Y, Hamano K, Iwaya M & Nakagome Y (1991) Sex identification by polymerase chain reaction using X–Y homologous primer. *American Journal of Medical Genetics* 39, 472–473. *The amelogenin method.*

Prüfer K, Racimo F, Patterson N et al. (2014) The complete genome sequence of a Neanderthal from the Altai Mountains. *Nature* 505, 43–49.

Ringbauer H, Huang Y, Akbari A et al. (2024) Accurate detection of identity-by-descent segments in human ancient DNA. *Nature Genetics* 56, 143–151.

Wolf AB & Akey JM (2018) Outstanding questions in the study of archaic hominin admixture. *PLoS Genetics* 14, e1007349.

Genes in Industry and Agriculture

CHAPTER 23

Some of the most important and controversial applications of genetics in our modern world lie in the areas of industry and agriculture. The industrial use of biology is called **biotechnology**. The common perception is that biotechnology is a new subject, but the industry has much more ancient roots. Fermentation processes that make use of living yeast cells to produce ale and mead have been exploited by humans for at least 4,000 years. Since the beginning of the 20th century, microorganisms have been used extensively for the industrial production of important organic compounds and enzymes such as glycerol and invertase. The discovery of penicillin by Alexander Fleming in 1929 led to a further expansion of the biotechnology industry driven by the large-scale production of antibiotics by fungi and bacteria.

The biotechnology industry was transformed in the 1970s by the first use of **DNA** or **gene cloning** to expand the repertoire of products that can be obtained from microorganisms. Cloning enables a gene for an important animal protein to be taken from its normal host and introduced into a bacterium. The transfer can be designed so that the animal gene remains active in the bacterium, which now synthesizes the protein that the gene specifies (Figure 23.1). In this way, large quantities of proteins with important pharmaceutical applications can be obtained relatively cheaply. We will examine this technology during the first part of this chapter.

More recently, gene cloning has been used with animals or plants as the host organisms. In some cases, the objective is once again the production of an important protein, and in others, it is to change the biological characteristics of the organism in some way that is beneficial for its use by humans. Examples include the development of **genetically modified (GM) plants** that are resistant to herbicides. We will study these aspects of biotechnology in the second part of this chapter.

23.1 PRODUCTION OF RECOMBINANT PROTEIN BY MICROORGANISMS

The term **recombinant protein** is used to describe a protein that is produced from a cloned gene. Microorganisms are looked on as ideal hosts for recombinant protein production because they can be grown at high density, enabling large amounts of the protein to be obtained. One possibility is simply to grow the engineered microbe in a large culture vessel from which the product is purified after the cells have been removed (Figure 23.2a). This **batch culture** method gives good yields but those yields are limited by the size of the culture vessel. Alternatively, the

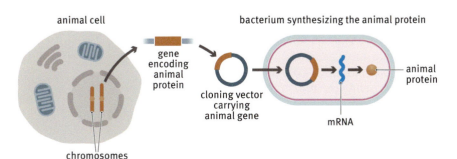

FIGURE 23.1 An outline of the way in which DNA cloning is used to synthesize an animal protein in a bacterium.

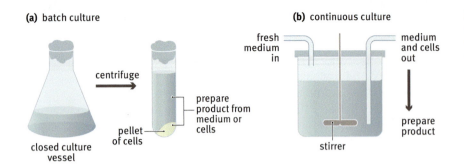

FIGURE 23.2 Two different methods for the large-scale culture of a microorganism. (a) In batch culture, the microorganism is grown in a large culture vessel from which the product is purified. (b) In continuous culture, a constant supply of product is obtained.

microbes can be grown in **continuous culture**, with a **fermenter**, from which samples are continuously drawn off (Figure 23.2b). This provides a nonstop supply of the product.

High yields of the product are therefore possible, but how is a microorganism engineered so that it synthesizes a recombinant protein? We will look first at the procedures that underlie this type of genetic engineering, and then at how these procedures are used to make recombinant proteins such as insulin and factor VIII, as well as vaccines that give protection against viruses.

DNA cloning is used to transfer foreign genes into a bacterium

DNA cloning enables any piece of DNA up to about 300 kb in length, from any species, to be transferred into the bacterium *Escherichia coli*. After transfer, the cloned DNA replicates and is passed on to daughter cells when the bacterium divides.

To illustrate how DNA cloning is carried out, we will follow through the procedure used to clone a human gene in *E. coli*. The first step is to prepare a fragment of human DNA containing the gene we wish to clone. There are various ways of doing this. We will use the simplest approach, where the gene has been amplified by PCR and is available as a short DNA molecule.

The next step is to insert the animal gene into a **cloning vector**. This is a DNA molecule that is able to replicate inside an *E. coli* cell. Often the vector is a modified version of a naturally occurring plasmid. In its replicative state, the plasmid is a circular molecule, but at the start of the cloning experiment, it is needed in a linear form. It is therefore cut with a restriction endonuclease, taking care to choose an enzyme that has just a single recognition sequence in the vector. The linear vector and PCR product are then mixed and a DNA ligase enzyme is added, which joins DNA molecules together end to end. Various combinations are produced, including some where a vector molecule has become linked to one of the DNA fragments and the two ends then joined together to produce a circular **recombinant DNA molecule** (Figure 23.3a).

In the next stage of the cloning procedure, the molecules resulting from ligation are mixed with *E. coli* cells, which have been chemically treated so that they are **competent**, a term used to describe cells that are able to take up DNA molecules from their environment. Linear molecules or circular molecules that lack the vector sequences, cannot be propagated inside a bacterium. Circular vectors, on the other hand, including the recombinant ones, are able to replicate, so that multiple copies are produced, exactly how many depending on the identity of the cloning vector, but usually several hundred per cell (Figure 23.3b). If a few cells containing recombinant vectors are used to start a batch or continuous culture, then each of the billions of bacteria present in the culture, when it reaches maximum density, will contain a copy of the cloned animal gene (Figure 23.3c).

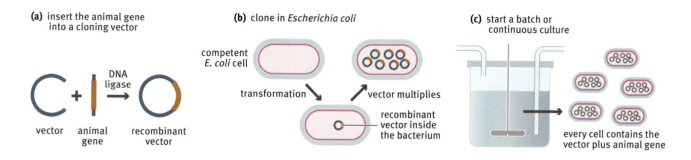

FIGURE 23.3 Cloning an animal gene in *E. coli*. (a) The animal gene is inserted into a cloning vector. (b) The recombinant vector is taken up by *E. coli* cells and replicates inside the bacteria. (c) A few transformed cells are used to start a batch or continuous culture. All of the bacteria in the culture will contain a copy of the animal gene.

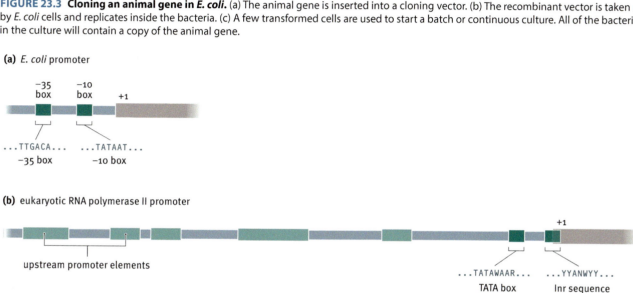

FIGURE 23.4 Differences between bacterial and eukaryotic promoters. Consensus sequences are shown for the promoters of (a) *E. coli* and (b) a eukaryotic protein-coding gene transcribed by RNA polymerase II. 'R' is a purine, 'W' is A or T, 'Y, is a pyrimidine and 'N' is any nucleotide.

Special cloning vectors are needed for the synthesis of recombinant protein

In order to produce recombinant protein, it is necessary that the cloned animal gene is expressed within *E. coli*. This requires a special type of cloning vector because an animal gene will not itself possess the correct signals for expression in a bacterium. To be transcribed and translated, the gene must be preceded by an *E. coli* promoter and a ribosome binding sequence, and followed by a sequence able to act as a transcription termination site. Eukaryotic genes possess all of these signals but the sequences are different from the *E. coli* versions. This is illustrated by comparing the consensus sequences of the promoters from protein-coding genes in *E. coli* and eukaryotes. There are similarities, but it is unlikely that an *E. coli* RNA polymerase would recognize and attach to a eukaryotic promoter (Figure 23.4). Most animal genes are therefore inactive in *E. coli*.

To solve this problem, a cloning vector that will be used for recombinant protein production must contain a set of *E. coli* expression signals. These are positioned in such a way that when the animal gene is inserted into the vector it is placed downstream of a promoter and ribosome binding site and upstream of a termination sequence (Figure 23.5). Plasmids that provide these signals are called **expression vectors**.

Of these three signals, the promoter is the one that is most important. This is for two reasons. First, the promoter determines how rapidly a gene is expressed. The strongest *E. coli* promoters direct 1,000 times as many productive initiations of transcription as the weakest ones. For recombinant

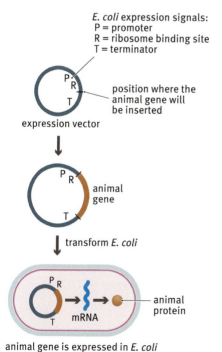

FIGURE 23.5 An expression vector and the way in which it is used.

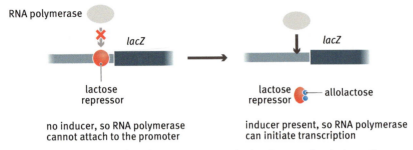

FIGURE 23.6 Induction of the lactose promoter. In the absence of an inducer, the repressor prevents the RNA polymerase from accessing the promoter.

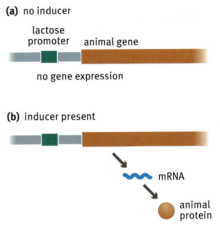

FIGURE 23.7 The use of the lactose promoter in the synthesis of animal protein in *E. coli*.

protein synthesis, where high yields are important, the promoter must be as strong as possible.

The second reason why the promoter is important is because it provides a means of controlling gene expression. This can be an advantage in the production of recombinant protein, especially if the protein is harmful to the bacteria, in which case its synthesis must be carefully regulated to prevent the protein accumulating to lethal levels. Even if the recombinant protein has no harmful effects, it is still useful to be able to regulate the expression of the cloned gene. This is because a continuously high level of transcription might interfere with plasmid replication and the partitioning of plasmids during the division of the host bacterium. The recombinant plasmid might therefore be lost gradually from the culture.

The ideal promoter is therefore one that directs a high rate of transcription initiation and can be controlled by the addition of a regulatory chemical to the culture medium. There are a number of possible choices, but the most popular is the lactose promoter. This is the promoter that, in normal *E. coli* cells, controls the transcription of the lactose operon. It is a strong promoter and so directs a large amount of protein synthesis, and it is regulated by the lactose repressor. In the absence of an inducer, the repressor prevents the RNA polymerase from accessing the promoter (Figure 23.6). In the natural system, the inducer is allolactose, which binds to the repressor, changing its conformation so it detaches from the DNA. The polymerase is now able to gain access to the promoter and gene expression is switched on.

These regulatory events still occur even after the segment of DNA containing the promoter has been removed from the lactose operon and placed in a cloning vector. An animal gene that is inserted just downstream of the lactose promoter will be expressed at a high level, with expression switched on by adding an inducer to the culture medium (Figure 23.7). We could use allolactose as the inducer, but this compound is not particularly stable. It would be necessary to continually add fresh allolactose in order to prevent the animal gene from being switched off. Instead, an artificial inducer, such isopropylthiogalactoside (IPTG), is used. IPTG is an analog of allolactose (Figure 23.8), and so can bind to the repressor. It is much more stable than allolactose so does not need continual replenishment in order to maintain expression of the animal gene.

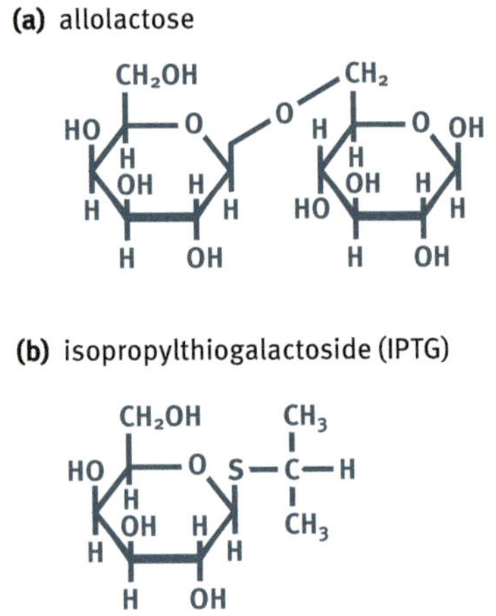

Insulin was one of the first human proteins to be synthesized in *E. coli*

The initial goal of 'recombinant' biotechnology was to use *E. coli* to produce large quantities of important therapeutic proteins. Insulin, needed for the treatment of diabetes mellitus, provides a good example. Insulin is synthesized by the β-cells of the islets of Langerhans in the pancreas,

FIGURE 23.8 Inducers of the lactose promoter. Structures of (a) allolactose, the natural inducer of the lactose promoter, and (b) isopropylthiogalactoside (IPTG), which is often used to induce the promoter when recombinant protein is being synthesized. Allolactose is less stable than IPTG because *E. coli* can break it down into its constituent monosaccharides.

and controls the level of glucose in the blood. An insulin deficiency leads to diabetes, which can be fatal if untreated. Fortunately, many forms of diabetes can be controlled by insulin injections, supplementing the limited amount of hormone synthesized by the patient's pancreas.

The insulin used in this treatment was originally obtained from the pancreas of pigs and cows slaughtered for meat production. Pig and cow insulins have very similar amino acid sequences to the human version (one and three amino acid differences, respectively) and both are able to control blood sugar levels after injection into humans. Although this is an effective way of treating diabetes, purification of insulin from pancreas tissue is not straightforward, and carryover of contaminants occasionally leads to allergic reactions. Recombinant insulin would avoid this problem because purification from a bacterial culture is a much simpler process and less prone to contamination.

Insulin displays two features that facilitate its production by recombinant methods. The first is that the human protein does not undergo extensive post-translational modification. Bacteria lack many of the enzymes needed to carry out these modifications and, in particular, are unable to attach the sugar side chains that are possessed by those eukaryotic proteins that are modified by glycosylation. Because insulin lacks these modifications, the recombinant version synthesized by *E. coli* should therefore be identical to the version made by the human pancreas.

The second important feature is the size of the molecule. Insulin is a relatively small protein, made up of two polypeptides, the A chain which is 21 amino acids in length and the B chain of 30 amino acids (**Figure 23.9**). In the pancreas, these chains are synthesized as a precursor called preproinsulin, which contains the A and B segments linked by a third chain (the C chain) and preceded by a signal peptide. The signal peptide is removed after synthesis and the C chain is excised, leaving the A and B polypeptides linked to each other by two disulfide bonds.

Several strategies have been used to obtain recombinant insulin. In one of the first projects, carried out in the 1970s, two artificial genes were synthesized, one specifying the A chain and the other the B chain. Two recombinant plasmids were constructed, one carrying the artificial

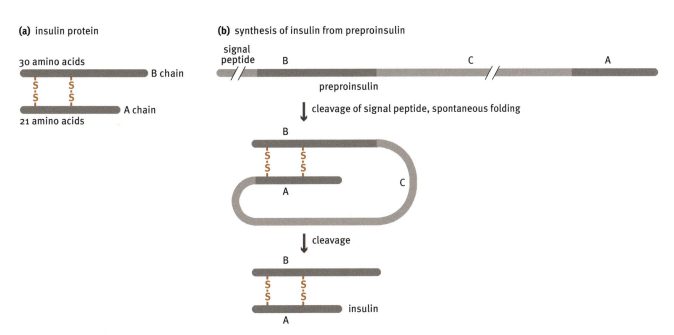

FIGURE 23.9 The structure of insulin and its processing pathway, beginning with preproinsulin.

gene for the A chain and one the gene for the B chain. In each case, the artificial gene was inserted into an expression vector so that its reading frame became contiguous with the first few codons of the gene for the *E. coli* β-galactosidase (Figure 23.10). This is the first gene in the lactose operon and is immediately downstream of the lactose promoter. The insulin genes were therefore placed under the control of the lactose promoter, and the A and B chains were synthesized as **fusion proteins**, each consisting of the first few amino acids of β-galactosidase followed by the A or B polypeptides. Each fusion gene was designed so that the β-galactosidase and insulin segments of the protein were separated by a methionine amino acid. Methionine is attacked by cyanogen bromide, so after purification from the culture each fusion protein was treated with this chemical, to cut apart the insulin polypeptide and the β-galactosidase fragment. The β-galactosidase fragment was discarded, and the purified A and B chains were attached to each other by treating with an oxidizing agent, resulting in the formation of the interchain disulfide bonds.

E. coli is not an ideal host for recombinant protein production

Despite the development of sophisticated expression vectors, recombinant protein synthesis in *E. coli* is a difficult and not always successful procedure. There are two types of problem, the first relating to the differences

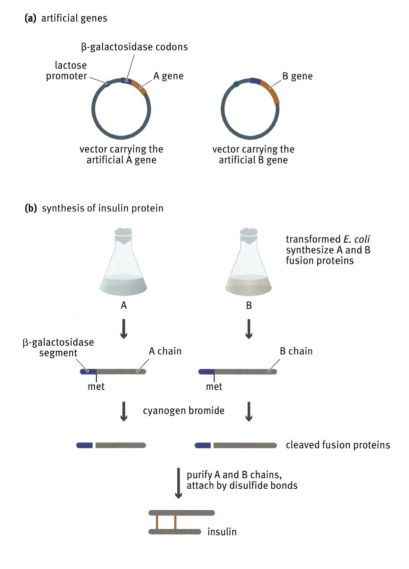

FIGURE 23.10 Synthesis of insulin as a recombinant protein in *E. coli*. (a) Two recombinant vectors were constructed, one carrying an artificial gene for the A chain and the other carrying a gene for the B chain. Each of these artificial genes was fused with the first few codons of the *E. coli* β-galactosidase gene. (b) *E. coli* was transformed with the recombinant vectors and the fusion proteins were purified. The proteins were cleaved with cyanogen bromide, the β-galactosidase fragments discarded, and the A and B chains attached to each other by the formation of the disulfide bonds.

between animal and *E. coli* genes, and the second concerning the biochemical and physiological limitations of bacteria.

We have already seen how it is necessary to place the animal gene under the control of an *E. coli* promoter, ribosome binding site and transcription termination sequence. This is not a major problem and is solved by the design of the expression vector. But there are three other ways in which the sequence of an animal gene might hinder its expression in *E. coli*. First, the animal gene might contain introns. This is a problem because *E. coli* genes do not contain introns and therefore the bacterium does not possess the necessary machinery for removing introns from mRNA molecules. The second problem is that the animal gene might contain sequences that act as transcription termination signals in *E. coli*. These sequences would have no function when the gene was inside its animal cell, but after transfer to the bacterium might result in premature termination of transcription and a loss of gene expression (Figure 23.11).

A final possibility is that the codon bias of the gene may not be ideal for translation in *E. coli*. Virtually, all organisms use the same genetic code, but each species has a bias toward preferred codons. For example, leucine is specified by six codons (TTA, TTG, CTT, CTC, CTA and CTG), but in human genes leucine is most frequently coded by CTG and is only rarely specified by TTA or CTA. The biological reason for codon bias is not understood, but all organisms have a bias, which is different in different species. If an animal gene contains a high proportion of codons that are relatively rare in *E. coli*, then recombinant protein synthesis might only occur at a low rate. The reason why codon bias should affect the rate of protein synthesis is not understood, but it probably relates to the efficiency with which *E. coli* tRNAs are able to recognize the codons in the animal gene.

None of these problems are insoluble. If the animal gene contains introns, then an intron-free version can be obtained by converting its mRNA into complementary DNA (cDNA) (see Figure 18.21). As the mRNA version of a gene lacks introns, the double-stranded cDNA will also be intron-free. If the gene still contains termination sequences, or if codon bias is a problem, then **site-directed mutagenesis** techniques can be used to alter the nucleotide sequence (Figure 23.12). Alternatively, if the gene is less than

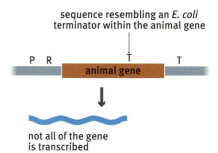

FIGURE 23.11 The problem caused by a termination signal within a cloned animal gene. Abbreviations: P, promoter; R, ribosome binding site; T, termination signal.

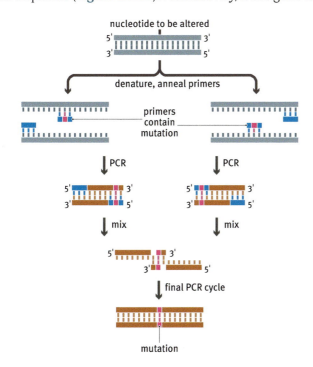

FIGURE 23.12 One method for site-directed mutagenesis. This method can be used to alter one or more nucleotides in a DNA molecule such as an animal gene. Two PCRs are performed. Each reaction uses one normal primer, which forms a fully base-paired hybrid with the template DNA, and a 'mutagenic primer', which contains a single base-pair mismatch corresponding to the alteration that we wish to introduce into the animal gene. There are two mutagenic primers, which base pair to identical positions in the DNA, but on opposite strands. After PCR, the two products are mixed together and a final PCR cycle is performed to construct the full-length, mutated DNA molecule.

1 kb in length then it might be possible to synthesize an entirely artificial version in the test tube, designed to ensure that introns and termination sequences are absent and that only the preferred *E. coli* codons are used.

The limitations that are placed on recombinant protein synthesis by the biochemical and physiological properties of bacteria are more difficult to solve. The protein folding processes occurring in bacterial cells are less sophisticated than those in eukaryotes. An animal protein synthesized in *E. coli* might therefore not be folded correctly, in which case it usually becomes insoluble and forms an **inclusion body** within the bacterium (**Figure 23.13**). It is not difficult to recover the protein from an inclusion body, but subsequently converting it into its correctly folded form in the test tube might be impossible. An incorrectly folded protein is, of course, inactive.

The inability of *E. coli* to carry out some of the complex chemical modifications that occur with animal proteins is an equally severe problem. In particular, glycosylation is extremely uncommon in bacteria, and recombinant proteins synthesized in *E. coli* are not glycosylated correctly. Glycosylation is not always essential for the function of the animal protein, but incorrect or deficient glycosylation might significantly reduce the stability of the protein when injected into a patient, and might also induce an allergic reaction.

Recombinant protein can also be synthesized in eukaryotic cells

DNA can be cloned in many different organisms, not just *E. coli*. Cloning vectors and other methods of propagating foreign DNA in host cells have been developed for other bacteria as well as for a variety of eukaryotic species. The problems associated with obtaining high yields of active recombinant proteins from animal genes cloned in *E. coli* can therefore be solved by using expression systems designed for eukaryotic cells. Microbial eukaryotes, such as yeast and filamentous fungi, can be grown almost as easily as bacteria in continuous culture, and being more closely related to animals might be able to deal with recombinant protein synthesis more efficiently than *E. coli*.

Expression vectors are still required because the promoters and other expression signals for animal genes do not in general work efficiently in these lower eukaryotes. *Saccharomyces cerevisiae* is currently the most popular microbial eukaryote for recombinant protein synthesis, partly because it is accepted as a safe organism for the production of proteins for use in medicines or in foods, and partly because of the wealth of knowledge built up over the years regarding its biochemistry and genetics. Many yeast expression vectors carry the *GAL* promoter, from the gene coding for galactose epimerase, an enzyme involved in the metabolism of galactose. The *GAL* promoter is induced by galactose, providing a convenient system for regulating the expression of a cloned animal gene (**Figure 23.14**).

Although yields of recombinant protein are relatively high, *S. cerevisiae* does not glycosylate animal proteins correctly, often hyperglycosylating – adding too many sugar units (**Figure 23.15**). This problem can be countered, to some extent, by using mutant strains of *S. cerevisiae* with altered glycosylation activity, or by using a different species of yeast, such as *Pichia pastoris* (now called *Komagataella phaffii*), as the host. Even then it might not be possible to synthesize an animal protein efficiently. For proteins with complex and essential glycosylation structures, an animal cell might be the only type of host within which the active protein can be synthesized.

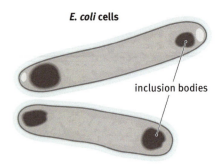

FIGURE 23.13 Inclusion bodies of recombinant protein in *E. coli*.

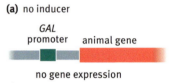

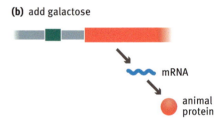

FIGURE 23.14 Induction of the *S. cerevisiae GAL* promoter. (a) When no inducer is present there is no gene expression. (b) Expression occurs after addition of galactose.

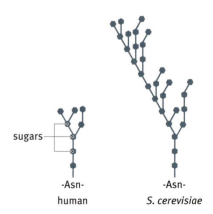

FIGURE 23.15 Hyperglycosylation. Comparison of a typical N-linked glycosylation structure synthesized by human cells and the hyperglycosylated structure made by *S. cerevisiae*.

Culture methods for animal cells have gradually been developed over the last 50 years, to the stage where today it is possible to grow at least some types of cell in continuous systems at relatively high cell densities. Recombinant protein yields are much lower than is possible with microorganisms, but this can be tolerated if it is the only way of obtaining the active protein. Mammalian cell lines derived from humans or hamsters are often used, which means that there are few problems with gene expression and protein synthesis, and the proteins that are made are indistinguishable from the nonrecombinant versions. The expression vector is needed only to maximize yields and enable protein synthesis to be regulated. The promoter is often from a virus such as simian virus 40 (SV40), cytomegalovirus or Rous sarcoma virus. Although this is the most reliable approach to the synthesis of active proteins, it is also the most expensive, especially as the possible presence of viruses in the cell lines means that rigorous quality control procedures must be employed to ensure that the purified protein is safe.

Factor VIII is an example of a protein that is difficult to produce by recombinant means

Synthesis of factor VIII as a recombinant protein has been one of the major goals of modern biotechnology. Factor VIII is the protein that is nonfunctional in the commonest form of hemophilia, leading to a breakdown in the blood clotting pathway and the well-known symptoms associated with the disorder. Hemophilia can be treated by regular injections of purified factor VIII, but until recently the only source of the protein was human blood provided by donors. Purification of factor VIII directly from the blood is a complex procedure that is beset with difficulties, in particular the need to remove viruses that might be present in the donated blood. Sadly, both hepatitis and immunodeficiency viruses have been passed on to hemophiliacs via factor VIII injections. Recombinant factor VIII, free from contamination problems, would be a significant achievement for biotechnology.

Factor VIII has proved difficult to obtain by recombinant methods. This is mainly because of the complexity of the pathway that converts the initial translation product of the factor VIII gene into the active protein. The translation product is a polypeptide of 2,351 amino acids, which is cut into three segments (Figure 23.16). The central segment is discarded and the two external ones joined by 17 disulfide bonds. While these events are taking place, the protein becomes extensively glycosylated. As might be anticipated for a protein with such a complex processing pathway, it has proved impossible to synthesize an active version of recombinant factor VIII in E. coli or yeast. Attention has therefore focused on mammalian cells.

Initially, a cDNA version of the full-length factor VIII gene was cloned in hamster cells. Yields were disappointingly low, probably because the post-translational modifications, although carried out correctly in hamster cells, did not convert all of the initial translation product into an active form. To try to improve yields, two fragments from the cDNA were cloned, the aim being to synthesize the A and C subunits of the protein separately. Each cDNA fragment was inserted into an expression vector, downstream of a promoter and upstream of a polyadenylation signal (Figure 23.17). The two recombinant plasmids were then introduced into a hamster cell line. Within the hamster cells, the A and C protein segments were synthesized separately and then joined together by the normal processing pathway. The yields were over ten times greater than those from cells containing the complete cDNA, and the resulting factor VIII protein was indistinguishable in terms of function from the nonrecombinant version.

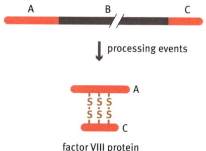

FIGURE 23.16 Processing of the human factor VIII protein.

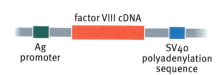

FIGURE 23.17 The construct used to obtain the expression of a factor VIII cDNA in hamster cells. The Ag promoter is a hybrid of the promoter sequences of the chicken β-actin and rabbit β-globin genes. The polyadenylation sequence is from SV40 virus.

To produce sufficient amounts of recombinant factor VIII for commercial needs, the recombinant hamster cells are grown in continuous culture. The cells secrete the recombinant factor VIII into the culture medium from which it is purified. The cell debris is removed, and various chromatography and filtration procedures carried out to remove impurities, and to stabilize the protein so it can be stored prior to use by patients.

Vaccines can also be produced by recombinant technology

In addition to proteins used to treat diseases, recombinant techniques have also been used to synthesize vaccines. After inoculation into the bloodstream, a vaccine stimulates the immune system to synthesize antibodies that protect the body against subsequent infection with the virus, bacterium or other pathogenic organism against which the vaccine is targeted. Most vaccines contain an inactivated form of the infectious organism, such as heat-killed virus particles. The large amounts of virus that are needed to prepare these vaccines are usually obtained from infected tissue cultures. Some viruses, such as the one that causes hepatitis B, cannot be grown in tissue culture, so obtaining enough material to make a vaccine is difficult. A second problem with these conventional vaccines is the possibility that a few viruses survive the heat treatment. If a live virus is present in a vaccine, then there is the risk that inoculation will cause the disease, rather than providing protection. This has happened with vaccines for the cattle disease foot-and-mouth.

Vaccines can be effective even if they do not contain intact virus particles. Isolated components of a virus, in particular purified preparations of the capsid proteins, will also stimulate the immune system and give protection against a later infection (Figure 23.18). This provides the basis to recombinant vaccine production. The gene coding for a virus capsid protein is transferred to a microbial or animal cell and the recombinant protein then used as the vaccine. A vaccine prepared in this way is entirely free from intact virus particles and can be obtained in large quantities.

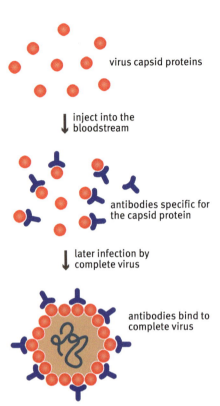

FIGURE 23.18 The rationale underlying the use of virus capsid proteins as a vaccine.

This approach has been most successful with hepatitis B virus. Hepatitis B is endemic in many tropical parts of the world and causes liver disease. If the infection persists then the patient might develop cancer of the liver, which has a very low survival rate – over half a million people die every year from primary liver cancer and 80% of these contracted cancer as a result of hepatitis B infection.

Fortunately, most people recover from hepatitis B infection with no long-term effects. These individuals are immune to future infection because their blood contains antibodies to the capsid protein called the hepatitis B surface antigen (HBsAg) (Figure 23.19). Recombinant HBsAg has been synthesized in *S. cerevisiae* and hamster cells, and in both cases was obtained in reasonably high quantities. When injected into test animals the recombinant proteins provided protection against hepatitis B. Both the yeast and hamster cell vaccines have been approved for use in humans, and the yeast version is being produced commercially by two companies.

A second strategy for recombinant vaccine production makes use of engineered vaccinia viruses. Vaccinia is the cowpox virus, which is harmless to humans but which stimulates immunity to smallpox when inoculated as live virus particles. Recombinant vaccinia viruses have been constructed by inserting new genes, such as the gene for HBsAg, into the vaccinia genome (Figure 23.20). The idea is that after injection into the bloodstream, replication of the recombinant virus will result not only in new vaccinia particles but also in significant quantities of HBsAg. Although this approach has not yet been approved for use with humans, it has been

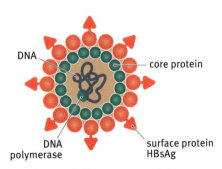

FIGURE 23.19 The structure of the hepatitis B virus. The HBsAg proteins are located on the surface of the virus.

shown to work with animals. A vaccinia virus expressing a coat protein from the rabies virus is being used as a vaccine to control rabies in Europe and North America. It is also possible to place a number of foreign genes in a single vaccinia genome, to produce a recombinant vaccine that is active against a variety of diseases. The potential of this approach has been demonstrated with a single vaccine that confers immunity to influenza, hepatitis B and herpes simplex in monkeys.

23.2 GENETIC ENGINEERING OF PLANTS AND ANIMALS

So far in this chapter, we have only considered examples of DNA cloning where a new gene is introduced into a microorganism or into eukaryotic cells that are being grown in culture. It is also possible to introduce new genes into animals and plants.

The objective of this type of genetic engineering is to change the phenotype of an animal or plant in a way that is considered beneficial for the applied use of the organism. Genetic engineering is being extensively used to breed improved varieties of domesticated plants and farm animals as one possible means of helping to address the food shortages facing the world in decades to come. If the engineering involves the introduction of a new gene then the product is referred to as a **transgenic** plant or animal. In the case of plants, the popular term **genetically modified** (**GM**) is widely used.

A number of projects are being carried out around the world, many by biotechnology companies, aimed at exploiting the potential of genetic engineering in crop and animal improvement. In this section, we will investigate a representative selection of these projects.

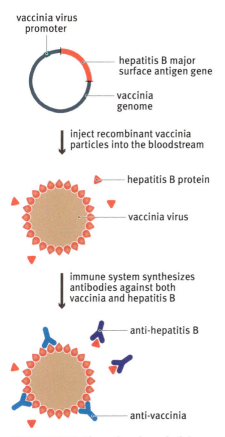

FIGURE 23.20 The rationale underlying the use of a recombinant vaccinia virus that synthesizes HBsAg proteins.

Crops can be engineered to become resistant to herbicides

In commercial terms, the most important genetically engineered organisms are crop plants that have been modified so that they are resistant to the herbicide called glyphosate. This herbicide is widely used because it is nontoxic to animals and insects and has a short residence time in soil, breaking down after a few days into harmless products. It is therefore less damaging to the environment than some other types of herbicide that have been used in the past. The problem for the farmer is that glyphosate is a broad-spectrum herbicide. It kills all plants, not just weeds, and so has to be applied to fields very carefully to avoid harming the crop plants. If crop plants could be engineered so that they are resistant to the effects of glyphosate, then a field could be treated with herbicide simply by spraying the entire area, not worrying about avoiding contact with the crop. This less rigorous application process would save time and money.

Glyphosate kills plants by inhibiting the enzyme called enolpyruvylshikimate-3-phosphate synthase (EPSP synthase). Enolpyruvylshikimate-3-phosphate is a substrate for the synthesis of the aromatic amino acids tryptophan, tyrosine and phenylalanine (Figure 23.21). Inhibition of EPSP synthase therefore leads to a deficiency in these amino acids and the subsequent death of the plant. The first crops to be engineered for glyphosate resistance were not, strictly speaking, transgenic as they did not contain any genes from other species. The aim of the genetic modification was simply to increase the amount of EPSP synthase that was made by the crop plant, in the expectation that plants overproducing the enzyme would be able to withstand higher levels of glyphosate than the nonengineered weeds. As it turned out, this approach was unsuccessful.

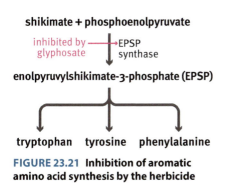

FIGURE 23.21 Inhibition of aromatic amino acid synthesis by the herbicide glyphosate.

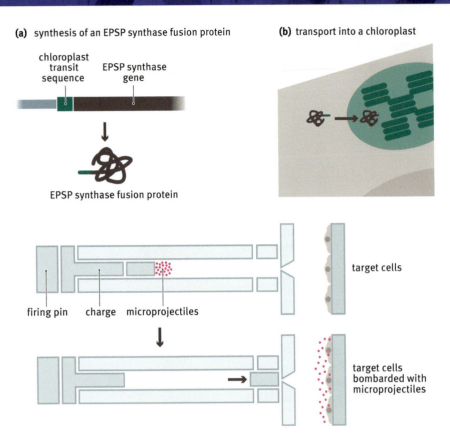

FIGURE 23.22 The transgenic strategy used to create glyphosate-resistant soybean plants. (a) The EPSP synthase gene was fused with a leader segment coding for a chloroplast transit sequence. (b) After synthesis, the transit sequence directs the fusion protein into the chloroplast. Note that the transit sequence is detached during the transport process.

FIGURE 23.23 Biolistics. DNA is fired into plant cells on gold microprojectiles.

Engineered plants that made up to 80 times the normal amount of EPSP synthase were obtained, but the resulting increase in glyphosate tolerance was not sufficient to protect these plants from herbicide application in the field.

A transgenic strategy was therefore adopted, making us of organisms whose EPSP synthase enzymes are naturally resistant to glyphosate. Transfer of the EPSP synthase gene from such an organism into a crop plant should confer resistance to the herbicide. The gene that was chosen was not from a plant but from a soil bacterium, called *Agrobacterium* strain CP4. EPSP synthase is located in the plant chloroplasts, so the *Agrobacterium* gene was inserted into a cloning vector in such a way that it became fused with a set of codons specifying a chloroplast transit sequence. This is an amino acid sequence that, after translation, directs a protein across the chloroplast membrane and into the organelle (Figure 23.22). The recombinant plasmid was introduced into soybean cells by **biolistics**. In this method, small gold microprojectiles are coated with the plasmid and literally fired into the cells with a gun (Figure 23.23). If done carefully, this does not cause terminal damage to the cells. Plant cells are totipotent, so a single transformed cell can give rise to a mature plant, all of whose cells will contain the new gene (Figure 23.24). These plants were three times more resistant to glyphosate than nonengineered soybean. Similar experiments have more recently been carried out with maize.

Various GM crops have been produced by gene addition

Herbicide resistance is just one of the ways in which the addition of new genes into plant genomes has been used to breed crops with advantageous properties. One of the first examples of this **gene addition** approach to genetic engineering involved the introduction into plants of genes for insect resistance. The conventional way of protecting crops

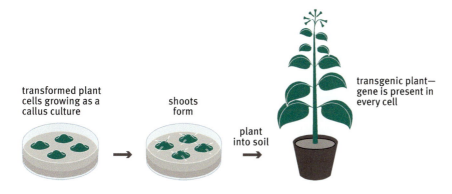

FIGURE 23.24 **Regeneration of a transgenic plant from transformed plant cells.** The transformed cells are initially plated onto solid agar and grown into small callus cultures. The addition of plant growth compounds, such as 6-benzylaminopurine and naphthalene acetic acid, to the agar stimulates the calluses to produce shoots that can be propagated to give the mature plant.

from the potentially devastating effects of insect attack is to spray regularly with insecticides. Most conventional insecticides, such as pyrethroids and organophosphates, are nonspecific and kill many different types of insects, not just the ones eating the crop. Several of these insecticides also have potentially harmful side effects for other species, including humans. To make matters worse, the treatment is sometimes ineffective because insects that live within the plant, or on the undersurfaces of leaves, can sometimes avoid the insecticide altogether.

The search for more environment-friendly insecticides has focused largely on the δ-endotoxins of the soil bacterium *Bacillus thuringiensis*. The δ-endotoxins are proteins that the bacteria make to protect themselves from being eaten by insects. The proteins are some 80,000 times more poisonous than chemical insecticides and are synthesized initially as a harmless precursor that is activated into the toxic form only after ingestion by the insect. They are selective, different strains of *B. thuringiensis* synthesizing proteins effective against the larvae of different groups of insects. All of these properties make the δ-endotoxins very attractive as environment-friendly insecticides. It is no surprise that the first patent for their use was taken out as far back in 1904.

In 1904, nobody had any idea that it might be possible to use DNA cloning to engineer plants to make their own δ-endotoxins. Instead, up to the 1980s, they were used as conventional insecticides and were not particularly popular because their instability meant that they had to be sprayed repeatedly onto crops. Transfer of δ-endotoxin genes into plants was first achieved in 1993, resulting in GM maize plants with significant levels of resistance to the European corn borer, the larva of the moth *Ostrinia nubilalis* (Figure 23.25). This caterpillar tunnels into the plant from eggs laid on the undersurfaces of the leaves, thereby evading the effects of insecticides applied by spraying. The average length of the tunnels made by larvae in the GM plants was just 6.3 cm, compared with 40.7 cm for the unmodified controls. In addition to maize, 'Bt' versions of potato and cotton have also been produced, displaying resistance to a range of insects and hence improving yields of these important crops.

Gene addition has also been used to improve the nutritional content of crop plants. Rice, for example, has been engineered so that the grains produce β-carotene, which humans use as a precursor for vitamin A synthesis. The objective is to produce a GM variety of rice for those parts of the world where there is a high incidence of vitamin A deficiency. Almost half a million children every year become blind because they do not have enough vitamin A in their diet. Engineering rice grains to produce β-carotene is a more complicated project than the ones we have studied so far, because two genes must be transferred, one from maize and one from the bacterium *Erwinia uredovora* (now called *Pantoea ananas*). The enzymes specified by these two genes work together to convert geranylgeranyl

FIGURE 23.25 **The European corn borer.** The European corn borer is the larva of the moth *Ostrinia nubilalis*. (Courtesy of Keith Weller/USDA.)

diphosphate into lycopene. This compound can then be converted into β-carotene by an enzyme that is already present in rice (Figure 23.26). The higher β-carotene content gives the grains a yellow color, and the GM variety is called 'golden rice' (Figure 23.27).

Other examples of GM plants produced by gene addition are listed in **Table 23.1**. These projects include a second approach to insect resistance, using genes coding for proteinase inhibitors. These are small proteins that interfere with enzymes in the insect gut, preventing the insect from feeding on the GM plants. Proteinase inhibitors are produced naturally by legumes such as cowpeas and common beans, and genes from these species have successfully been transferred to other crops which do not normally make the proteins in significant amounts. Proteinase inhibitors are particularly effective against beetle larvae that feed on stored seeds. In a different sphere of commercial activity, ornamental plants with novel flower colors are being produced by transferring genes involved in pigment synthesis from one species to another.

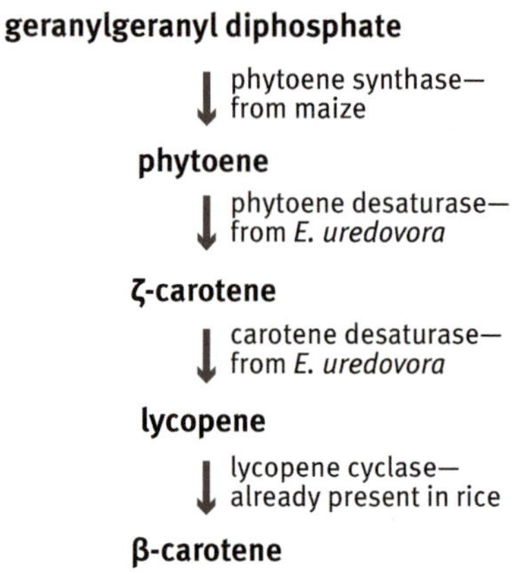

Plant genes can also be silenced by genetic modification

Plants can also be genetically engineered so that targeted genes are partially or completely suppressed. This is sometimes called **gene subtraction**, but the term is inaccurate as the gene is inactivated not removed. There are several possible ways of inactivating a single, chosen gene in a living plant. One method is to use antisense RNA technology, which we met in Section 21.4 as a potential means of silencing oncogenes by gene therapy.

Antisense RNA technology has been used to inactivate ethylene synthesis in plants that are likely to spoil due to overripening during storage. Ethylene is a regulatory compound that switches on the genes involved in the later stages of fruit ripening in plants such as tomato. Fruit ripening

geranylgeranyl diphosphate
↓ phytoene synthase— from maize
phytoene
↓ phytoene desaturase— from *E. uredovora*
ζ-carotene
↓ carotene desaturase— from *E. uredovora*
lycopene
↓ lycopene cyclase— already present in rice
β-carotene

FIGURE 23.26 The pathway leading from geranylgeranyl diphosphate to β-carotene. The sources of the enzymes used to synthesize β-carotene in transgenic rice are indicated. Note that the gene transferred from *E. uredovora* has a dual function: the enzyme it specifies acts as both a phytoene desaturase and a carotene desaturase.

TABLE 23.1 EXAMPLES OF GENE ADDITION PROJECTS WITH PLANTS

Added Gene	Source Organism	Characteristic Conferred on the Engineered Plants
Ornithine carbamyltransferase	*Pseudomonas syringae*	Bacterial resistance
Phosphatidylinositol specific phospholipase C	Maize	Drought tolerance
Chitinase	Rice	Fungal resistance
Enolpyruvylshikimate-3-phosphate synthase	*Agrobacterium* spp.	Herbicide resistance
Glyphosate N-acetyltransferase	*Bacillus licheniformis*	Herbicide resistance
Phytoene synthase + phytoene/carotene desaturase	Various plants and bacteria	Improved nutritional content
Methionine-rich protein	Brazil nuts	Improved sulfur content
δ-Endotoxin	*Bacillus thuringiensis*	Insect resistance
Proteinase inhibitors	Various legumes	Insect resistance
Acyl carrier protein thioesterase	California bay laurel tree	Modified fat and oil content
Δ12-Desaturase	Soybean	Modified fat and oil content
Dihydroflavanol reductase	Various flowering plants	Modified flower color
Monellin, thaumatin	Sweet prayers plant	Sweetness
Virus coat proteins	Various viruses	Virus resistance

FIGURE 23.27 **Comparison of golden rice (right) with white rice (left).** (Courtesy of Golden Rice Humanitarian Board [www.goldenrice.org].)

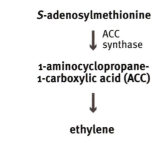

FIGURE 23.28 **The ethylene biosynthesis pathway.**

can therefore be delayed by inactivating one or more of the genes involved in ethylene synthesis. Fruits on these plants initially develop as normal, but are unable to complete the ripening process, and can be transported to the marketplace with less danger of the crop spoiling. Before selling to the consumer, artificial ripening is induced by spraying the fruits with ethylene.

Ethylene is synthesized from a precursor called 1-aminocyclopropane-1-carboxylic acid (ACC), which in turn is made from S-adenosylmethionine (SAM) (**Figure 23.28**). In tomato, the strategy for delaying fruit ripening involved the inactivation of the gene for ACC synthase, the enzyme that converts SAM into ACC. This gene was inserted into a cloning vector in the reverse orientation, so that the cloned sequence directed synthesis of an antisense version of the ACC synthase mRNA (**Figure 23.29**). The modified plants made only 2% of the amount of ethylene produced by non-engineered ones, and their fruits were unable to complete the ripening process until treated with ethylene.

The applications of gene subtraction in plant genetic engineering are less broad than those of gene addition. This is simply because there are relatively few genes whose inactivation would be advantageous to the farmer or consumer. Most gene subtraction projects have been aimed at reducing crop spoilage, but a few ways of using this approach to improve nutritional value have also been devised (**Table 23.2**).

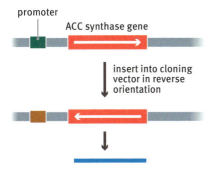

FIGURE 23.29 **Synthesis of an antisense version of the ACC synthase mRNA.**

TABLE 23.2	EXAMPLES OF GENE SUBTRACTION PROJECTS WITH PLANTS
Inactivated Gene	**Modified Characteristic**
1-Aminocyclopropane-1-carboxylic acid synthase	Modified fruit ripening in tomato
1-D-myo-Inositol 3-phosphate synthase	Reduction of indigestible phosphorus content of rice grains
Chalcone synthase	Modification of flower color in various decorative plants
Δ12-Desaturase	High oleic acid content in soybean
Polygalacturonase	Delay of fruit spoilage in tomato
Polyphenol oxidase	Prevention of discoloration in fruits and vegetables
Starch synthase	Reduction of starch content in vegetables

Gene editing has many applications in plant genetic engineering

The methods described above for the production of GM crops have been supplemented during the last few years by the introduction of gene editing. In Section 18.3, we studied how this method can be used as an aid to assigning functions to unidentified genes. This type of gene editing, called **error-prone editing**, introduces a deletion into the target gene (see Figure 18.25), which is therefore silenced. Error-prone editing is therefore another way of achieving gene subtraction. The method is quick and easy, which means that it is possible to make multiple edits in a single plant. This approach was used to increase the grain yield of rice, by inactivating a set of three genes, affecting the size, weight and number of grains per plant.

Other versions of gene editing take plant genetic engineering into new realms by making it possible to change the sequence of a target gene. The objective is not to inactivate the gene, but to change the activity of the protein product in some defined way. In one method, called **homology-directed repair**, the sequence alteration is directed by a short template DNA molecule that is introduced into the cell along with the DNA encoding the Cas9 endonuclease gene and the guide RNA sequence. The template DNA contains the sequence alteration that we wish to introduce into the target gene. The endonuclease and guide RNA direct a recombination event that replaces the equivalent segment of the gene with the template DNA (Figure 23.30). The sequence alterations are therefore introduced into the gene.

Several plant genetic engineering projects have made use of homology-directed repair, including one in which a nitrate transporter gene was edited in order to improve the grain yield of rice. Unfortunately, homology-directed repair is an inefficient method. Editing often fails to occur even though the endonuclease, guide RNA and template DNA have been successfully introduced into the plant cell. Modifications of the process are therefore being examined to try to improve the success rate.

One of these new approaches dispenses with the template DNA and instead uses a more direct approach to make the sequence alterations. The alterations are made by a **base editor**, which is an enzyme whose natural function is to convert one nucleotide into another. An example is adenine deaminase, which converts adenine nucleotides into inosines (Figure 23.31). An artificial gene that codes for a fusion between the Cas9 endonuclease and a base editor is introduced into the cell along with the guide RNA. After editing, the inosine base pairs with cytosine when the gene is transcribed or replicated, so in effect an A to G edit has been made. Base editors that convert C to T have also been designed.

Although the base editor only makes edits in the region specified by the guide RNA, it might make more than one edit if multiple copies of the target nucleotide are present (Figure 23.32). In some projects, this is not a problem, but often the objective is to make just one nucleotide alteration. If that is the case then these **bystander mutations** must be

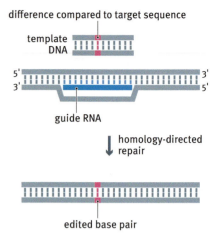

FIGURE 23.30 Homology-directed repair. When a short template DNA is present, the Cas9 endonuclease directs a recombination event that inserts the template into the target DNA. If the sequence of the template is different to that of the target, then the target DNA becomes edited.

FIGURE 23.31 Adenine deaminase converts adenine into inosine. Inosine base pairs with C rather than T.

FIGURE 23.32 Base editing. Fusion between the Cas9 endonuclease and adenine deaminase creates the base editor, which carries out the adenine to inosine conversion in the strand displaced by the guide RNA. In this example, two edits have been made.

avoided. Current research is therefore aimed at identifying ways of altering the structure of the base editor so it edits just a single nucleotide in the region covered by the guide RNA. Despite the problems with bystander mutations, A to G and C to T base editing has been used with some success with crops including wheat, rice and potato.

Engineered plants can be used for the synthesis of recombinant proteins, including vaccines

Now we return to the subject with which we started this chapter, the use of cloning to transfer genes into new hosts in order to obtain recombinant protein. We did not discuss the possibility that plants might be used for this purpose. Plants would be a good choice because plants and animals have similar protein processing activities, so most animal proteins produced in plants would be expected to undergo the correct post-translational modifications and to be fully functional.

This approach to recombinant protein production has been used with a variety of crops, such as maize, tobacco, rice and sugarcane. By inserting the new gene downstream of a promoter that is active only in developing seeds, large amounts of recombinant protein can be obtained in a form that is easily harvested and processed. Recombinant proteins have also been synthesized in the leaves of tobacco and alfalfa and the tubers of potatoes.

Whichever production system is used, plants offer a low-technology means of mass-producing recombinant proteins. One promising possibility is that plants could be used to synthesize vaccines, providing the basis for a cheap and efficient vaccination program. If the recombinant vaccine is effective after oral administration, then immunity could be acquired simply by eating part or all of the transgenic plant. The feasibility of this approach has been demonstrated by trials with vaccines such as HbsAg and the capsid proteins of measles virus and respiratory syncytial virus. With each of these, immunity has been conferred by feeding the transgenic plant to test animals. The main difficulty is ensuring that the amount of recombinant protein synthesized by the plant is sufficient to stimulate complete immunity against the target disease. This requires vaccine yields of 8–10% of the soluble protein content of the part of the plant which is eaten, but in practice yields achieved so far are much less than this, usually not more than 0.5%. One possibility is to design the cloning system so the protein is synthesized in the chloroplasts, as this generally results in a higher yield. In experimental trials, vaccines active against anthrax, plague, smallpox and the malaria parasite *Plasmodium falciparum* have all been produced in this way.

Recombinant proteins can be produced in animals by pharming

It is also possible to obtain transgenic animals by genetic engineering, using the technology that we encountered when we considered the means by which germline gene therapy might be carried out (Section 21.4). With mice, it is possible to produce a transgenic animal by direct microinjection of the gene to be cloned into a fertilized egg cell, but this is not possible with many other mammals. With these, the more sophisticated nuclear transfer procedure is required, in which the gene is first injected into the nucleus of a somatic cell (see Figure 21.29).

Recombinant proteins have been produced in the blood of transgenic animals, and in the eggs of transgenic chickens, but the most successful approach has been to obtain the protein from the milk of farm animals such as sheep or pigs. This is possible if the gene is attached to a promoter that is active in mammary tissue (Figure 23.33). Sheep and pigs are, of course, mammals and most human proteins that have been produced in this way undergo the correct post-translational modifications and are fully active. The process is called **pharming**. The first recombinant protein obtained by pharming and approved for medical use was antithrombin III, from goat's milk. This is an anticoagulant sometimes used to prevent blood clotting during heart surgery. Other blood proteins, including factor VIII, have also been synthesized in the milk of transgenic animals.

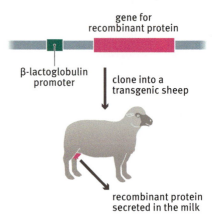

FIGURE 23.33 Production of recombinant protein in the milk of a sheep. In this example, the gene for the recombinant protein is attached to the β-lactoglobulin promoter, which is active only in mammary tissue.

KEY CONCEPTS

- DNA cloning enables a gene for an important animal protein to be taken from its normal host and introduced into a bacterium. The transfer can be designed so that the animal gene remains active. The term 'recombinant protein' is used to describe a protein that is produced in this way from a cloned gene.

- Microorganisms are looked on as ideal hosts for recombinant protein production because they can be grown at high density, enabling large amounts of the protein to be obtained.

- An expression vector contains a bacterial promoter, ribosome binding site and termination sequence, and is used to ensure that a cloned animal gene is active after transfer to a bacterium.

- Simple proteins such as insulin can be synthesized in *E. coli* but more complex ones might not be processed correctly.

- Yeasts such as *S. cerevisiae* and *P. pastoris* are used as hosts for recombinant protein production because they are able to carry out post-translational modifications such as glycosylation, but some complex proteins, such as human factor VIII, can only be produced in animal cells.

- Virus coat proteins synthesized as recombinant proteins have been used as vaccines.

- The objective of genetic engineering is to change the phenotype of an animal or plant in a way that is considered beneficial for the applied use of the organism.

- In commercial terms, the most important genetically engineered organisms are crop plants that have been modified so that they are resistant to the herbicide called glyphosate.

- Plant genes can be inactivated by genetic engineering techniques. This approach has been used to delay fruit ripening by inactivating genes involved in ethylene synthesis.

- Gene editing provides another dimension to plant genetic engineering by making it possible to change the sequence of a target gene.

- Recombinant proteins, including vaccines, have been produced in transgenic plants, and others have been obtained from the milk of transgenic animals.

QUESTIONS AND PROBLEMS

Key terms

Write short definitions of the following terms:

- base editor
- batch culture
- biolistics
- biotechnology
- bystander mutation
- cloning vector
- competent
- continuous culture
- DNA cloning
- error-prone editing
- expression vector
- fermenter
- fusion protein
- gene addition
- gene subtraction
- genetically modified plant
- homology-directed repair
- inclusion body
- pharming
- recombinant DNA molecule
- recombinant protein
- site-directed mutagenesis
- transgenic

Self-study questions

23.1 Distinguish between the batch and continuous methods for culture of a microorganism.

23.2 Outline the steps in a DNA cloning experiment designed to clone an animal gene in *E. coli*.

23.3 Describe the key features of an expression vector and explain why this type of vector is needed when an animal gene is cloned in *E. coli*.

23.4 What are the desirable features of the promoter carried by an expression vector? What specific features of the lactose promoter make it a good choice for a vector used for expressing animal genes in *E. coli*?

23.5 Outline the procedure that was used to obtain recombinant insulin from *E. coli*.

23.6 Why is codon bias sometimes a problem when an animal gene is being expressed in *E. coli*?

23.7 Describe the possible solutions to the problems caused by introns, internal termination sequences and codon bias during expression of an animal gene in *E. coli*.

23.8 What is an inclusion body and why is the formation of inclusion bodies sometimes a problem when *E. coli* is used as a host for recombinant protein synthesis?

23.9 Explain why glycosylation can be a problem when an animal protein is synthesized in *E. coli*. To what extent is this problem solved by using *S. cerevisiae* or *P. pastoris* as the host?

23.10 Outline the procedure that was used to obtain human factor VIII from hamster cells. Why was the production of factor VIII a challenge for recombinant technology?

23.11 Describe how recombinant protein synthesis is used to prepare a vaccine.

23.12 Describe how a recombinant vaccinia virus is constructed and give an example of the successful use of this technology.

23.13 Outline the advantages of a glyphosate-resistant crop compared with a nonresistant crop, and describe how glyphosate-resistant soybean was produced by genetic engineering.

23.14 Describe how plants that synthesize δ-endotoxins have been obtained by genetic engineering.

23.15 List the various characteristics that have been conferred on genetically modified plants by gene addition.

23.16 Giving examples, describe how antisense RNA technology has been used to inactivate plant genes.

23.17 Describe how gene editing is being used in plant genetic engineering.

23.18 Give examples of the production of recombinant protein in plants.

23.19 Outline how a transgenic animal is obtained.

23.20 Explain what is meant by the term 'pharming', and give examples of proteins that have been obtained by pharming.

Discussion topics

23.21 Most cloning vectors carry one or more genes for antibiotic resistance. Suggest a way in which the presence of such genes could be used to distinguish bacteria that have taken up a vector molecule from those that have not been transformed.

23.22 With most cloning vectors, insertion of a new piece of DNA disrupts a gene carried by the vector. Suggest a way in which this insertional inactivation could be used to distinguish bacteria that have taken up a recombinant vector molecule from those that have taken up a vector that contains no inserted DNA.

23.23 Speculate on the advantages of obtaining a recombinant protein as a fusion with an endogenous bacterial protein such as β-galactosidase.

23.24 What strategies might be used to prevent or reduce the development among insect populations of resistance to the δ-endotoxin produced by a genetically modified crop plant?

23.25 Discuss the concerns that have been raised about the possibility that GM crops might harm the environment.

23.26 Discuss the ethical issues raised by the development of pharming.

 The answers to, or hints on how to answer, the self-study questions and discussion topic questions can be downloaded from the book's product page at this link: www.routledge.com/9781032743530

FURTHER READING

Brocher B, Kieny MP, Costy F et al. (1991) Large-scale eradication of rabies using recombinant vaccinia-rabies vaccine. *Nature* 354, 520–523.

Feitelson JS, Payne J & Kim L (1992) *Bacillus thuringiensis*: insects and beyond. *Biotechnology* 10, 271–275. *Details of endotoxins and their potential as conventional insecticides and in genetic engineering.*

Goeddel DV, Kleid DG, Bolivar F et al. (1979) Expression in Escherichia coli of chemically synthesized genes for human insulin. *Proceedings of the National Academy of Sciences USA* 76, 106–110.

Gomes AMV, Carmo TS, Carvalho LS et al. (2018) Comparison of yeasts as hosts for recombinant protein production. *Microorganisms* 6, 38.

Houdebine L-M (2009) Production of pharmaceutical proteins by transgenic animals. *Comparative Immunology, Microbiology and Infectious Diseases* 32, 107–121.

Kind A & Schnieke A (2008) Animal pharming, two decades on. *Transgenic Research* 17, 1025–1033. *Reviews the progress and controversies with animal pharming.*

Liu MA (1998) Vaccine developments. *Nature Medicine* 4, 515–519. *Describes the development of recombinant vaccines.*

Matas AJ, Gapper NE, Chung M-Y et al. (2009) Biology and genetic engineering of fruit maturation for enhanced quality and shelf-life. *Current Opinion in Biotechnology* 20, 197–203.

Padgette SR, Kolacz KH, Delannay X et al. (1995) Development, identification, and characterization of a glyphosate-tolerant soybean line. *Crop Science* 35, 1451–1461. *Transfer of the Agrobacterium EPSP synthase gene into soybean.*

Robinson M, Lilley R, Little S et al. (1984) Codon usage can affect efficiency of translation of genes in *Escherichia coli*. *Nucleic Acids Research* 12, 6663–6671.

Rosano GL & Ceccarelli EA (2014) Recombinant protein expression in *Escherichia coli*: advances and challenges. *Frontiers in Microbiology* 5, 172.

Schillberg S, Raven N, Spiegel H et al. (2019) Critical analysis of the commercial potential of plants for the production of recombinant proteins. *Frontiers in Plant Science* 10, 720.

Tiwari S, Verma PC, Sing PK & Tuli R (2009) Plants as bioreactors for the production of vaccine antigens. *Biotechnology Advances* 27, 449–467.

Yonemura H, Sugawara K, Nakashima K et al. (1993) Efficient production of recombinant human factor VIII by co-expression of the heavy and light chains. *Protein Engineering* 6, 669–674.

Zhu H, Li C & Gao C (2020) Applications of CRISPR–Cas in agriculture and plant biotechnology. *Nature Reviews Molecular Cell Biology* 21, 661–677. *Applications of gene editing.*

Zhu J (2012) Mammalian cell protein expression for biopharmaceutical production. *Biotechnology Advances* 30, 1158–1170.

CHAPTER 24
The Ethical Issues Raised by Modern Genetics

It is no longer possible for a student of genetics to consider only the academic aspects of the subject. The ethical issues raised by the applications of genetics in our modern world must also be examined. To many people, some of the recent advances in genetics are alarming, with the potential to do great harm if the discoveries are not used wisely. For some, any form of science is a threat and opposition to genetics is merely part of a broader philosophy that science is bad. But for others, concerns about genetics are based on well-reasoned arguments. These views are important: they must be, as they are held by many geneticists.

We must therefore examine the ethical issues raised by modern genetics. We have already begun to do this in the discussion topics for the last three chapters, where you were asked to think about the broader implications of some of the medical, forensic and industrial applications of genetics. Discovering these issues for yourself is important, because every person, whether or not they are studying genetics, must make up their own mind about the possible dangers. Simply following the lead of others is not an option, especially for a student of genetics who is aware of the underlying science and so is able to form an opinion based on an expert knowledge of the subject. Bear in mind that you have a much better understanding of genetics than that possessed by most politicians and many leaders of public thought.

In this chapter, we will look at four areas of modern genetics that are of particular concern to the general public (**Table 24.1**). These are the development of GM crops, the use of personal DNA data, the possible misuse of gene therapy, and pharming. The last two topics are related in that the public concerns center on the underlying technology, which is the same for both germline gene therapy and the production of pharmed animals. As far as possible, we will attempt to discuss these issues this in a nonjudgmental manner. In the second part of the chapter, we will explore the more general framework within which the ethical aspects of genetics should be debated.

TABLE 24.1 SOME OF THE AREAS OF CONCERN IN MODERN GENETICS	
Topic	**Specific Issues**
GM crops	Harmful effects on human health Harmful effects on the environment Use of genetic and legal methods to prevent crops being grown for a second season from seeds collected from the previous harvest
Personal DNA data	Retention of genetic profiles of individuals who have not committed any crime Use of genetic data by insurance companies, and the more general discrimination against individuals with 'bad' alleles of particular genes Personal freedom to know, or not know, one's genetic susceptibility to delayed-onset diseases
Gene therapy	Use of germline gene therapy to create 'designer babies'
Pharming	Creation of animals specifically for use as producers of recombinant protein Suffering associated with the manipulations used to produce transgenic animals

24.1 AREAS OF CONCERN RELATING TO GM CROPS

Plant genetic engineering is by no means the only aspect of modern genetics that raises ethical issues, but it is the topic that is most prominent in the public perception. This is partly because the introduction of GM crops in the 1990s was, for many people, their first encounter with the applications of genetic technology. GM crops therefore became the arena within which many of the concerns about genetic engineering have been played out. The key questions that have been debated surround the possible hazards that GM crops might have on human health and on the environment, and the possibility that GM techniques could be used to exploit local farmers for the benefit of the companies that market the crops. We will look at each of these issues in turn.

There have been concerns that GM crops might be harmful to human health

The first issue that we will consider is the perceived hazard to human health that might result from the consumption of GM crops. The first GM crop to be approved for sale to the public was a type of tomato that had been engineered so that its ripening was delayed, enabling the fruits to be transported from grower to consumer with less spoilage. These 'FlavrSavr' tomatoes appeared in supermarkets in 1994 (**Figure 24.1**). They raised public awareness about genetics because, up to that point, it was not widely appreciated that plants and animals can be modified by genetic engineering. Not surprising, when faced with GM tomatoes, many people asked the question 'are these safe to eat?'.

FIGURE 24.1 Three overripe tomatoes (left) compared with three FlavrSavr tomatoes (right). The FlavrSavr tomato was the first type of GM crop to be approved for sale to the public. (Courtesy of Martyn F. Chillmaid/Science Photo Library.)

FlavrSavr tomatoes had been engineered by the introduction of an antisense version of the gene for the polygalacturonase enzyme. This enzyme slowly breaks down the pectin molecules present in the cell walls in the fruit pericarp (**Figure 24.2**). This results in the gradual softening that makes the fruit palatable but which, if allowed to proceed too far, results in spoilage. By inhibiting the synthesis of polygalacturonase, fruit ripening is delayed. The rationale is exactly the same as that behind the use of antisense technology to inhibit ethylene production in developing fruits, the only difference being that another part of the fruit ripening pathway is targeted.

It seems inconceivable that the introduction of the polygalacturonase antisense RNA into these tomatoes would cause a health hazard. But this was not the point. Of much greater concern was the fact that these tomatoes

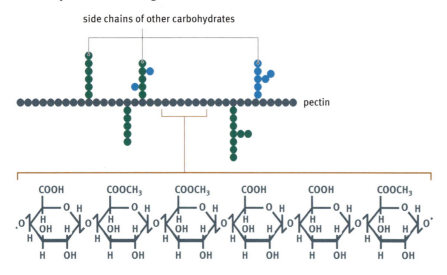

FIGURE 24.2 The structure of pectin. This component of the plant cell wall is a carbohydrate polymer made up of galacturonic acid units, with side chains of other carbohydrates. Some of the galacturonic acid units are also modified by methylation. Polygalacturonase breaks the pectin chains in the cell wall of the tomato pericarp, leading to softening of the fruit.

also contained a gene for the enzyme called aminoglycoside 3-phosphotranferase or APH(3′), which confers resistance to aminoglycoside antibiotics such as kanamycin. The gene acts as a **selectable marker** for plant cells that had taken a copy of the recombinant plasmid carrying the antisense polygalacturonase gene. Methods used to transport cloning vectors into plant cells, such as biolistics (see Figure 23.23), are inefficient and not all cells take up a copy of the recombinant plasmid. To ensure that only an recombinant cells give rise to calluses, the cells are initially plated onto an agar medium that contains kanamycin (Figure 24.3). The nonrecombinant cells, lacking the plasmid and so having no APH(3′) gene, cannot withstand the antibiotic and die.

After the selection process, the APH(3′) gene remains in the tomato cells, and will be present in every cell of the engineered plant, along with the gene for the polygalacturonase antisense RNA. In addition to kanamycin, the APH(3′) gene confers resistance to aminoglycoside antibiotics that are used to treat bacterial infections. The concern, therefore, is that the APH(3′) gene might possibly be transferred, after ingestion, from the tomato fruit to the bacteria present in the human gut. Further movement of the gene between different bacteria might lead to antibiotic resistance spreading to the types of harmful bacteria that these antibiotics are currently used against.

Geneticists have therefore devised ways of deleting the APH(3′) gene after the recombinant cells have been selected. One of these strategies makes use of an enzyme called Cre, which comes from bacteriophage P1. The Cre enzyme catalyzes a type of recombination event that removes a segment of DNA that is flanked by 34-bp recognition sequences (Figure 24.4). Two recombinant plasmids are introduced into a single plant cell. One carries the gene being added to the plant along with its APH(3′) selectable marker, the latter being flanked by the Cre target sequences. The second plasmid carries the Cre gene. Once inside the plant cell, the activity of the Cre enzyme results in the excision of the APH(3′) gene from the plant DNA (Figure 24.5). There is still time, however, to test the cells for kanamycin resistance so that recombinant ones can be selected.

What if the Cre gene is itself hazardous in some way? This is not an issue as the two recombinant plasmids would probably become inserted into different chromosomes. Random segregation during sexual reproduction will therefore result in a new generation of plants that contain one inserted plasmid but not the other. It is, therefore, possible to obtain a plant that does not contain either the Cre or APH(3′) gene, but does contain the important gene that we wish to add to the plant's genome.

There are also concerns that GM crops might be harmful to the environment

The second area of concern regarding GM plants is that they might be harmful to the environment in some way. The development of maize, soybean and oilseed rape engineered for increased herbicide resistance,

(a) a typical cloning vector for introducing a new gene into a plant

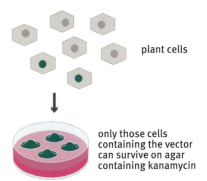

(b) the APH(3′) gene enables plant cells containing the vector to be selected

FIGURE 24.3 The role of the APH(3′) gene as a selectable marker during a DNA cloning experiment with plant cells. (a) The structure of a typical plant cloning vector. (b) The APH(3′) gene enables plant cells containing the vector to be selected.

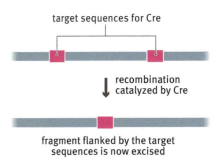

FIGURE 24.4 The mode of action of the Cre enzyme.

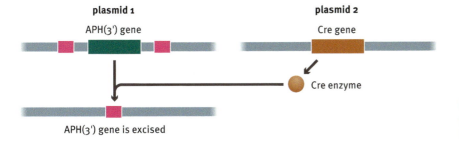

FIGURE 24.5 Excision of the APH(3′) gene by the Cre enzyme in a recombinant plant cell.

and cotton and potato varieties that make δ-endotoxin insecticides, has resulted in large increases in the amounts of land given over to growth of GM plants. What impact might these plants have on the ecology of the environment in which they are growing?

Two aspects of the environmental impact of GM crops must be considered. The first is whether the introduced gene can move from the crop into the natural vegetation, possibly leading to insect and herbicide-resistant versions of wild plants. For this to happen, pollen from the GM crop must be able to fertilize the wild species. This type of cross-hybridization is rare, but can occur between a crop and closely related wild plants. Examples are oil seed rape (*Brassica napus*) and the weed species called charlock (*Sinapis arvensis*) (**Figure 24.6**). Transfer of herbicide resistance from rape to charlock has been observed in experimental cultivations.

Movement of a gene from a GM crop to a wild plant must be distinguished from a second, and much commoner way, in which so-called superweeds can arise. This is through the natural process of evolution. Experience has shown that, when cultivating a herbicide-resistant GM crop, some farmers are tempted to use larger amounts of herbicide than in the past. This ensures that as many weeds as possible are killed, without any risk of harming the crop. The problem is that increased application of herbicide places a selective pressure on the environment that encourages propagation of those rare, naturally resistant weed varieties that already exist, or which evolve through the generation of new alleles by mutation events. Once these weeds become common in a field, the advantage of growing the engineered crop is lost. This is an issue specific to herbicide resistance and, although it affects farming practices, it is not obvious that these weeds will have any longer term impact on the natural environment. Outside of an area in which the herbicide is being used, resistance confers no selective advantage, and will probably be lost from the plant population.

The evolution of superweeds in response to the overuse of herbicides leads us to the second potential environmental impact of GM crops. It is possible that the cultivation of these plants might be accompanied by changes in farming practices that will have indirect impacts on the local environment and affect the abundance and diversity of farmland wildlife. This is, of course, an issue for any change in the way in which land is managed, but because of the controversies much more work has been carried out on the effects of growing GM crops than on other farming practices. One such study was commissioned by the UK Government in 1999 and completed in 2003. An independent investigation was carried out into how herbicide-resistant crops, whose growth in the United Kingdom was not at that time permitted, might affect the environment. The study involved 273 field trials throughout England, Wales and Scotland, and

FIGURE 24.6 (a) Oilseed rape and (b) its wild relative, charlock.
(a, courtesy of James King-Holmes/Science Photo Library; b, courtesy of Bjorn Rorslett/Science Photo Library.)

included glyphosate-resistant sugar beet as well as maize and spring rape engineered for resistance to a second herbicide, glufosinate-ammonium. The summary of the results, as described in the official report, illustrates the difficulties in understanding the impact that GM crops have on the environment:

'The team found that there were differences in the abundance of wildlife between GM crop fields and conventional crop fields. Growing conventional beet and spring rape was better for many groups of wildlife than growing GM beet and spring rape. There were more insects, such as butterflies and bees, in and around the conventional crops because there were more weeds to provide food and cover. There were also more weed seeds in conventional beet and spring rape crops than in their GM counterparts. Such seeds are important in the diets of some animals, particularly some birds. In contrast, growing GM maize was better for many groups of wildlife than conventional maize. There were more weeds in and around the GM crops, more butterflies and bees around at certain times of the year, and more weed seeds. The researchers stress that the differences they found do not arise just because the crops have been genetically modified. They arise because these GM crops give farmers new options for weed control. That is, they use different herbicides and apply them differently. The results of this study suggest that growing such GM crops could have implications for wider farmland biodiversity. However, other issues will affect the medium- and long-term impacts, such as the areas and distribution of land involved, how the land is cultivated and how crop rotations are managed. These make it hard for researchers to predict the medium- and large-scale effects of GM cropping with any certainty. In addition, other management decisions taken by farmers growing conventional crops will continue to impact on wildlife'.

The safety or otherwise of GM crops remains a contentious subject

There is still no consensus regarding the safety or otherwise of GM crops, with divergent opinions held not just by individuals but by government agencies. In 2019, GM crops accounted for 177 million acres of cultivation in the United States, representing approximately 40% of the total area of the United States devoted to crops. There was also 130 million acres of GM cultivation in Brazil and 59 million acres in Argentina, as well as significant acreages in other parts of North and South America and in parts of Asia (Figure 24.7). To give context, the 177 million acres of GM cultivation in the United States are about the same as the acreage devoted to organic farming worldwide, as calculated in 2018 by the Research Institute for Organic Farming.

Not surprisingly, the government agencies for those countries where GM crops are cultivated contend that these crops are hazard free and even beneficial for the environment. This view is supported by the majority of studies reported in the scientific literature. For example, a global analysis published in 2020 estimated that the cultivation of GM crops has reduced pesticide use by 8.3% and decreased the environmental impact of herbicide and pesticide application by 18.5%. The latter figure was estimated using the Environmental Impact Quotient, which is an established risk indicator used by weed scientists. The 2020 study also concluded that GM crops had an indirect benefit because of changes in the way farmland is prepared prior to and during cultivation, which resulted in a reduction in greenhouse gas emissions equivalent to 15.27 million automobiles.

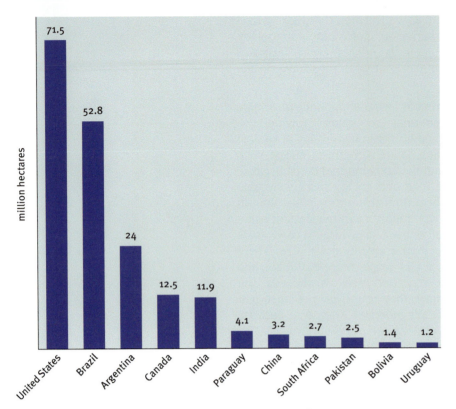

FIGURE 24.7 **Cultivation of GM crops in 2019.** Data are taken from www.statista.com.

Elsewhere there is greater caution regarding GM crops. Virtually, no GM crops are cultivated in countries of the European Union, either because cultivation is banned in those countries or because cultivators have not attempted to obtain the necessary approvals which, although issued by the member state, follows a procedure laid down by the European Commission. This caution reflects more negative opinions of GM safety expressed by some campaign groups. For example, a 2023 factsheet issued by the Canadian Biotechnology Action Network states that herbicide sales increased by 234% since the first cultivation of GM crops in Canada in 1995, and that 57 weed species have become resistant to glyphosate, including 17 present in the United States and eight in Canada. Other pressure groups object both to the cultivation of GM crops and their use in animal or human feedstuffs. The Greens/European Free Alliance group of the European Parliament contend, in a 2019 report, that 'health risks from GM crops imported into the EU for food and feed cannot be ruled out for humans and farm animals' and, referring to plants that synthesize δ-endotoxins from *Bacillus thuringiensis*, 'studies show that Bt toxins may reinforce the allergenic properties of other foodstuffs'.

The development of gene editing (see Section 23.2) has changed the landscape for ethical debates concerning GM crops. Gene editing alters the phenotype of a plant not by introducing new genes but by making directed alterations in genes that are natural components of the crop genome. Most of the current gene editing methods do require that new DNA be introduced into the plant, coding for the Cas9 endonuclease and guide RNA that perform the edit, but even this addition can be avoided with more recent methods that involve the introduction of just a ribonucleoprotein complex comprising the Cas9 protein and the guide RNA. With this approach, the only change made to the plant is an alteration in one or a few nucleotides. This is equivalent to the outcome of mutation events that occur in the natural environment, the only difference being that with gene editing the resulting alterations are not random and their positions in the genome are known. Current debate centers on whether gene edited crops should be designated as GM and hence subject to the regulations of other GM crops, or as non-GM and hence permitted for cultivation after the much less rigorous procedures used to assess the safety of crops obtained from conventional breeding programs.

GM technology could be used to ensure that farmers have to buy new seed every year

Now we turn our attention to a different type of ethical issue that is raised by our ability to introduce new genes into crop plants. In addition to the development of plants with properties such as insect or herbicide resistance, genetic engineering has also been used to introduce genes that are of benefit not to the farmer or the consumer, but to the company that produces the seed.

An example of this type of engineering is the **genetic use restriction technology** (**GURT**). This is a process that enables a company that develops and markets GM crops to protect its financial investment by ensuring that farmers buy new seeds every year. In some parts of the world, in particular poorer regions, it is quite common for a farmer to save some seed from the crop and to sow this second-generation seed the following year. This is, of course, cheaper than buying new seed from the plant breeding company every year.

The intellectual property rights of plant breeding companies have been debated for many years. Some argue that a new variety of crop should be looked on as an 'invention' and so covered by patent laws that would give the company control over the use of its invention. Currently, plant breeder's rights are slightly different, and in many countries 'saved seed' is specifically excluded from those rights with certain conditions, such as the payment of a royalty if more than a specified amount of seed is produced. This applies to any type of new crop variety, including ones produced by conventional breeding. The controversy arises because exclusion of saved seed from plant breeders' rights does not necessarily mean that a company is obliged to market crops that produce viable seeds. If the seeds are nonviable then the farmer cannot grow them the following year and so must keep on buying new stocks from the breeding company. GURT is a suite of methods that can be used to engineer crops so that their seeds are nonviable.

There are different versions of GURT. One of these is called the **terminator technology** and makes use of the gene for the **ribosome inactivating protein** (**RIP**). This protein blocks the translation stage of gene expression by cutting one of the ribosomal RNA molecules into two pieces (Figure 24.8a). Any cell in which the RIP is active will die. Genetic engineering has been used to place the RIP gene under control of a promoter that is active only during embryo development. These GM plants grow normally and will produce seeds, but the seeds will not be able to germinate because they do not contain embryos – the part that gives rise to the new plant.

Why are the first-generation seeds, the ones sold to farmers, not sterile? The RIP gene is nonfunctional in the plants that produce these seeds because it is disrupted by a segment of DNA (Figure 24.8b). This DNA is flanked by the 34-bp recognition sequences for the Cre enzyme. The plants are also engineered to contain the gene for the Cre enzyme, attached to a promoter that is switched on by tetracycline. The plants grow normally and produce seeds with viable embryos. Once these seeds have been

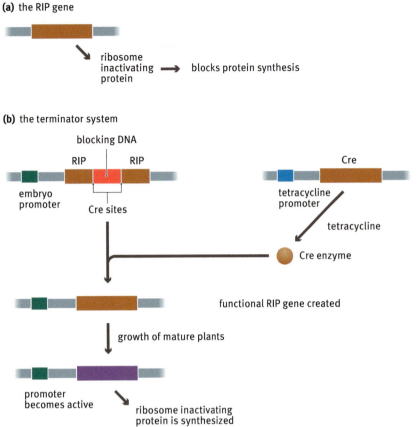

FIGURE 24.8 The terminator technology. (a) The RIP gene codes for the ribosome inactivating protein, which prevents protein synthesis by cutting one of the ribosomal RNA molecules into two pieces. (b) The RIP gene is modified so it contains a piece of blocking DNA flanked by Cre recognition sites, and is preceded by an embryo-specific promoter. The Cre gene is under control of a promoter that is switched on by tetracycline. When tetracycline is added, the Cre enzyme is synthesized and the blocking DNA is removed from the RIP gene. Later, when the plants begin to produce seed, the embryo promoter is switched on and the RIP protein is made.

obtained, the supplier activates the Cre enzyme by placing the seeds in a tetracycline solution. The blocking DNA is deleted from the RIP gene. The seeds are still able to germinate, because the RIP gene is not expressed until its promoter becomes active during the next round of embryogenesis, when the next generation of seeds are being produced by the adult plants. These are seeds that the farmer would normally save.

The terminator technology was initially developed as a means of preventing GM crops from cross-hybridizing with wild plants. The RIP protein will kill the embryos in seeds resulting from cross-pollination between a GM plant and a wild relative, so the new genes carried by the GM plant will never be transferred to the natural environment. But today it is looked on more as a means of preventing farmers from saving seeds, and although there are moves to introduce the technology, it is currently banned in most countries.

24.2 ETHICAL ISSUES RELATING TO THE USE OF PERSONAL DNA DATA

The second area of modern genetics whose controversial aspects we will examine is the use of **personal DNA data**. By 'personal DNA data', we mean any information about an individual's genome, including DNA profiles, complete genome sequences and catalogs of single-nucleotide polymorphisms present in genes that might be associated with health or disease susceptibility.

DNA profiling alerted the public to the concept of personal DNA data

The first use of genetic fingerprinting in criminal investigation in 1986, followed by DNA profiling a few years later (Section 22.1), raised public awareness regarding the individuality of human genome sequences. The understanding that some features of the genome are unique to a single individual led to the concept of personal DNA data. A logical progression from this concept is that a genome sequence is a personal possession. As such there should be restrictions on how it can be used by other people and by companies and government agencies.

This debate centered initially on the collection and retention of DNA profiles. In most countries, the authorities are allowed to retain DNA profiles that are obtained as part of a criminal investigation. National legislations differ on issues such as the circumstances under which a DNA sample must be given, the length of time that a profile can be kept on record and whether profiles of individuals who are eventually cleared of having committed any offence can be kept.

The argument for having as large a database as possible is that it increases the chances of a new crime being solved quickly. If a DNA profile from the scene of a crime matches a profile in the database, then that person can be interviewed without delay. If there is no match then the investigation will clearly take longer while suspects are tested, and quite possibly a conviction will never be made. There have been several reports of criminals only being apprehended after a period of years, often because they have been forced to give a DNA sample after committing a relatively minor offence such as a traffic violation. Their sample when tested matches the profile from an unsolved crime. In some cases, it has emerged that the individual concerned committed other crimes in the interim, some of a serious nature. Prevention of a rapist from reoffending on multiple occasions is a powerful argument in favor of large databases of DNA profiles.

These considerations have led to suggestions that there should be universal DNA databases, with every person in a country or region required to provide a profile. Public opinion regarding this idea is variable but as many as 60% of participants in some surveys have supported a universal database for their country of residence. However, attempts by governments to set up such schemes have generally been unsuccessful. For example, legislation introduced in Kuwait in 2015, in response to a terrorist attack, that DNA profiles would be obtained from all citizens, residents and visitors was repealed 2 years later as contrary to that country's constitution.

A different concern relating to universal DNA databases is whether the information they contain could be used to discriminate against an individual whose genome contains polymorphisms associated with genetic disorders, especially ones with late onset which might be expressed later in a person's lifetime. Should this information be available to insurance companies who wish to judge the risk that they are taking by offering a life insurance policy? Life insurance can be refused if a regular health check discovers ailments that reduce life expectancy, such as heart disease, and smokers may be asked to pay higher premiums in the expectation that they will die from lung cancer at a relatively young age. Is there a difference between this type of health information and the information on health is obtained from a genome sequence?

A DNA profile, being simply a list of short tandem repeat alleles, contains no information on an individual's biological characteristics and therefore cannot be used to deduce an individual's health status or susceptibility to genetic disorders. The question is still relevant, however, because a DNA sample is required in order to construct the profile, and if retained by the authorities that sample could be used for additional testing that did reveal an individual's health status. To provide reassurance that this scenario could not happen, in most countries DNA samples collected as part of a criminal investigation must be destroyed once the DNA profile has been obtained, whether or not the person from which the sample was taken is subsequently convicted of a crime.

Nonforensic DNA databases complicate issues regarding personal DNA data

Any debate about the ethical issues relating to the collection of DNA profiles by crime prevention agencies must take account of the willingness of many people to donate DNA samples to nonforensic genetic databases, many of which contain information on personal polymorphisms associated with disease susceptibility. These databases include ones set up for research purposes and others that are described as **direct-to-consumer** (**DTC**) DNA testing.

The primary aim of the research databases is to establish large enough collections of DNA polymorphisms to conduct genome-wide association studies (GWAS) aimed at identifying genes involved in complex genetic disorders (Section 21.2). The accuracy and sensitivity with which such genes can be identified are proportional to the number of DNA samples that are tested. One project in the United Kingdom has collected whole genome sequences for 500,000 individuals who had donated biological samples, along with their personal health details, to the **UK biobank**. Being a national research project, the biobank is subject to stringent regulations to protect the anonymity of participants, so that it is impossible for users of the data to discover the identity of any person providing a sample. Samples are donated after informed consent which includes face-to-face meetings with healthcare professionals. Similar initiatives are underway in other countries. One issue that has been raised is the possible bias among volunteers toward or against particular social groups. If a DNA database used in a GWAS project lacks representation from a particular group, then polymorphisms specific to genetic disorders in that group will be missed, and future healthcare provision for that group will be negatively impacted.

DTC DNA testing works on a rather different basis. This type of testing is operated by companies that charge a fee to provide information based on an individual's DNA data. The first of these companies dates back to the 2000s and used mitochondrial DNA sequencing to construct narratives regarding a person's ancestry. Since then, ancestry typing has become more sophisticated with companies typing genome-wide polymorphisms in order to determine genetic affiliations and also offering identity-by-descent (IBD) analysis to identify possible relatives (see Figure 22.19). For other companies the main focus is on polymorphisms associated with health and genetic susceptibility. Some of the resulting databases are huge: one company had 14 million customers by 2023.

Participation in DTC testing is a free choice that has been made by millions of individuals who are willing to pay a fee for the privilege, and as such it could be argued that ethical considerations must be minor or irrelevant. Nevertheless, three issues have been raised. The first is the impact that the discovery of genetic susceptibility to a disorder such as Alzheimer's or Parkinson's disease, which at present are not preventable, has on the individual concerned. In the past, when such tests could only be carried out by trained professionals, the results would be presented to the patient with a careful explanation of their implications. Many people, for example, are unsure what a phrase such as '70% lifetime risk' actually means. A personal test carried out voluntarily but without access to professional counseling might be unnecessarily deleterious to a person's quality of life.

A second issue to consider is the impact that the results of DNA testing might have on relatives of the individual who has been tested. Not all people wish to know their susceptibility to disorders that they might or might not experience in their later lives. The discovery that a family member has susceptibilities of different degrees to various disorders, in the absence of counseling to inform that person how likely it is that close relatives also possess those susceptibilities, could have a negative impact on the other members of the family.

The third issue is the use to which DNA data held by DTC companies can be used. Most companies have restrictions on sharing data with other bodies, but this does not always exclude partnerships with pharmaceutical companies that might mine the data for their own ends. There are also questions about what would happen to the database if a company becomes insolvent. However, the most controversial of these issues is the possible use of a DTC database for familial DNA testing in a crime investigation (Section 22.1). If a DNA sample from a crime scene is submitted to an ancestry site that uses IBD to search for potential relatives, then any matches that are identified could be relatives of the perpetrator of the crime. This approach was taken in the search for the 'Golden State Killer', who raped 51 people and murdered at least 13 others in California during 1974–1986. The cold case was reopened in 2016 when archived DNA samples were entered into a DTC database, revealing several possible third and fourth cousins, which enabled the authorities to construct a family tree from which the criminal was eventually identified.

24.3 ETHICAL ISSUES RAISED BY GENE THERAPY AND PHARMING

Gene therapy and pharming are two additional areas of modern genetics that raise ethical concerns. The underlying issues are very similar and we can therefore deal with them together.

Germline gene therapy has the potential to make permanent changes to a human lineage

When considering gene therapy, it is important to distinguish between the somatic and germline versions. Somatic cell therapy is less controversial because in this technique the genetic manipulation is not inherited. It is difficult to sustain an argument against the application via a respiratory inhaler of correct versions of the cystic fibrosis gene as a means of managing this disorder. The new gene will be active only in the lungs of the patient, and for only a short period of time before another administration is needed. Other than the fact that the active ingredient is DNA, this kind of treatment appears no different from the use of any other type of drug.

Germline therapy is more controversial because with this procedure the change that is made to the genome is heritable and so will be passed to all the descendants of the person who is treated. The concerns do not center so much on the use of germline therapy to eradicate terrible genetic disorders from the human population. The problem is that the techniques used for germline correction of inherited disorders are exactly the same techniques that could be used for germline manipulation of other inherited characteristics. This leads to fears that methods developed for germline gene therapy could be used to create 'designer babies' whose features match the preferences of the parents, or which are produced for less benign purposes.

The critical step in germline gene therapy or any other method that aims to manipulate the germline is achieving successful division and subsequent development of the cell whose DNA is altered. Various approaches have been attempted but by far the major focus has been on **somatic cell nuclear transfer** (**SCNT**). This is the method that involves injection of DNA into the nucleus of a somatic cell followed by transfer of the nucleus to an oocyte whose own nucleus has been removed (see Figure 21.29). Although SCNT can be carried out with reasonable efficiency with most animals, it has proved difficult to induce the engineered oocyte to divide to produce a **blastocyst**. A blastocyst is the structure, comprising 100 or so cells, that attaches to the wall of the uterus during a normal pregnancy (Figure 24.9). The difficulties in blastocyst formation have been greatest with primates, and for many years this problem provided an effective block to the possibility of human germline therapy. The problem has still not been entirely solved, but low efficiency implantation of human blastocysts has now been achieved. That success, combined with the ease with which human DNA can be altered by gene editing, makes the ethical issues associated with this work much more pertinent than they were just 10 years ago.

FIGURE 24.9 A 5-day-old human blastocyst. The technical difficulty in getting a human oocyte created by somatic nuclear transfer to develop into a blastocyst was for some time a block to the development of human germline therapy.

Most countries have legislation that bans germline gene manipulation for reproductive purposes, although research studies that do not involve actual reproduction are often permitted subject to licensing. Despite these controls at least one case of reproductive germline editing involving humans has been carried out. In 2018, a Chinese researcher announced the birth of twin girls whose genomes had been engineered to provide resistance to HIV infection. The work attracted widespread condemnation both from within China and abroad, and the researcher concerned was given a three-year jail sentence.

The use of SCNT in pharming raises additional issues

What distinction should be made between humans and other animals? The techniques that could be used to produce designer babies are already being applied to farmed animals and fish to make 'improvements' such as greater meat yield (Figure 24.10). They also underlie the creation of transgenic animals for pharming. Here, concerns have been raised that the procedures that are used might cause suffering to the animals. Animals produced by SCNT often have defects that result in them being stillborn, or dying soon after birth. Some that are healthy do not display the desired phenotype and are destroyed. This type of genetics is therefore accompanied by a high 'wastage'.

FIGURE 24.10 A genetically modified salmon. The larger of these two salmon has been engineered to produce a greater body mass than an unmodified salmon. (Courtesy of AquaBounty Technologies.)

Those animals that survive to adulthood sometimes appear to undergo premature aging. This is thought to have been the case with 'Dolly the sheep' who, although not transgenic, was the first animal to be produced by nuclear transfer (Figure 24.11). Dolly was a Finn Dorset, a breed that usually has a life expectancy of up to 12 years. Dolly had to be put down when she was six because she was found to be suffering from terminal lung disease, which is normally found only in much older sheep. The previous year, when she was just five, she had developed arthritis. It has been suggested that her premature aging might have been related to the age of the somatic cell that was used as the source of the nucleus that gave rise to Dolly. This cell came from a 6-year-old sheep, so it could be argued that Dolly was already 6 years old when she was born, although the scientists who created Dolly contend that this interpretation is incorrect.

The technology for the creation of engineered animals by SCNT has moved on dramatically since 1997, when Dolly was born, but the welfare issues regarding transgenic animals have not been resolved. This application of modern genetics remains at the forefront of public awareness.

FIGURE 24.11 Dolly the sheep. Dolly was the first animal to be produced by somatic nuclear transfer. (Courtesy of Ph. Plailly/ Eurelios/Science Photo Library.)

24.4 REACHING A CONSENSUS ON ETHICAL ISSUES RELATING TO MODERN GENETICS

How should the ethical issues that are raised by the applications of modern genetics be resolved? There is no easy answer to this question. Indeed, the problem is so complex that **bioethics** has emerged as a distinct sub-discipline of biology. Entire textbooks and university departments are now devoted to academic discussion of ethical issues and how they should be dealt with. This raises the danger that 'procedures' might be devised for resolving ethical concerns. Policy makers tend to like 'procedures' because they allow a decision to be justified not on the basis of whether it is correct or not, but because it was reached by the agreed procedure.

We will not attempt here to devise a procedure for dealing with ethical issues. We will limit ourselves to a consideration of the fundamental requirements that must be met by any debate into the ethical problems raised by the applications of modern genetics.

Ethical debates must be based on science fact not science fiction

The principal requirement when the ethical implications of genetics are debated is that the discussion is based in the real world and does not stray into the realms of science fiction. It is, of course, important to be forward looking and to recognize the ways in which genetics is likely to develop in the immediate future, and to identify any dangers that those new developments will present. The relevant ethical issues should be debated now so that when the subject has developed, there is already an understanding of the rules and regulations that ought to be enforced to prevent the dangers from becoming reality.

On the other hand, there is no point in considering the ethical implications of areas of science that are not yet plausible and which have little prospect of becoming so in the near future, if at all. It might be amusing to debate, for example, whether there should be any exceptions to the *Star Trek* prime directive that humans in interstellar spaceships must not interfere with alien cultures who have not yet developed spaceflight. Websites are devoted to the subject. But this is not a debate on scientific ethics, it is a debate on science fiction.

The *Star Trek* analogy might seem obvious and trivial but a search of the internet reveals many other web pages devoted to debates on scientific issues that an expert will immediately recognize as being fictional, but which the debaters passionately believe to be realistic. A good way to find these sites is to google emotive terms such as 'Frankenstein food'. It is easy to be critical of the lack of scientific understanding of people who believe that any kind of GM crop has the potential to cause brain damage. One might consider such views to be extremist, but it is important to be aware that the other end of the spectrum is the equally extreme view that a particular application of genetics, such as GM crops, is harmless and benign simply because that is the way we would like things to be. A focus on science fact is not an invitation to minimize or ignore potential issues.

The line of reasoning that we have followed leads us to a key conclusion. Debates on the ethical issues surrounding any area of science must be driven by facts. Scientists therefore have a critical role to play in these debates. The issues surrounding GM crops, personal DNA data, germline gene therapy and pharming must be resolved by the community as a whole, but geneticists must take the lead in these discussions. It is ethically unsound for a geneticist to live their life in a research lab and not engage in the wider debates that their subject generates.

Few geneticists, probably none at all, are mad or evil

If geneticists must lead the debate on the ethical issues raised by their research, then the public must trust geneticists. The public perception of scientists is an interesting subject in its own right. For much of the last 20 years, geneticists have benefitted from the assumption, driven by popular TV dramas, that although many of us are geeks most of us are also rather glamorous once we take off our glasses. The truth is that geneticists are just ordinary people and are as concerned about the ethical outcomes of their work as are any other members of the public.

Geneticists, in fact, have a good record in self-regulation and in open discussion of the possible dangers of their subject. This has not always been the case, but it is true for the period beginning with the discovery of the double helix structure of DNA. We have already considered the potential problems that might arise if a gene for antibiotic resistance was transferred from a GM foodstuff into bacteria living in the human gut. This is not just an issue for GM crops, because antibiotic resistance genes are used as selectable markers in virtually all cloning experiments, including those used to introduce genes into bacteria such as *Escherichia coli*. As *E. coli* is a component of the gut microflora, the transfer of an antibiotic resistance gene, or any other type of gene, from an engineered *E. coli* to bacteria in the gut is a much greater possibility than transfer from a foodstuff. Steps must therefore be taken to prevent it from happening. This problem was appreciated by the geneticists who designed the first cloning vectors and who developed the techniques for their use. So concerned were these geneticists that they voluntarily halted their work until a scientific conference could be held to discuss the hazards of gene cloning and how these ought to be avoided. That conference, held in Asilomar, California, in February 1975, established a set of guidelines for ensuring that cloned genes could not be transferred from *E. coli* to other bacteria. These guidelines enabled the new technology to continue to develop in a safe way while the hazards were explored more comprehensively. In a similar vein, following the announcement in 2018 that germline gene manipulation had been used in the conception of twin girls engineered to display resistance to HIV, the scientific community led the efforts to achieve a global consensus on the limitations that should be placed on this type of work.

The potential problems resulting from modern genetics were also a theme of the Human Genome Project. When this project was first conceived, in the 1980s, it was agreed that funds totaling 3–5% of the total annual budget would be set aside every year for studying the **ethical, legal and social issues** (**ELSI**) that might arise as a result of knowing the complete sequence of the human genome (**Table 24.2**). This funding continues today, under the auspices of the National Human Genome Research Institute, part of the US National Institutes of Health. The principles established by the Human Genome Project now underlie the ethical frameworks to most large genetics research projects.

The implication is that geneticists can be trusted to lead discussions on the ethical issues raised by their work. But there must be a caution. The Asilomar and ELSI initiatives are positives in the recent history of genetics, but there have also been negatives. Genetics has become big business, not just because of the commercial applications, but also because of the prestige that academic researchers acquire from large research grants and publications in high-impact journals. The recent history of science is littered with cases where the temptations of prestige and fame have become too great for a researcher, who has made up results, or suppressed others, in order to achieve the objective that they are searching for. And there are many cases where a scientist has become so passionately convinced that their own favorite hypothesis is correct that they fail to recognize when other research proves it to be wrong. Geneticists should therefore lead the debate on the ethical implications of genetics, but they should not be the sole arbiters of that debate. In the end, these issues must be resolved by the public at large, with all members of the community contributing, and all views given a balanced hearing based on their merits.

TABLE 24.2 THE ETHICAL, LEGAL AND SOCIAL GOALS OF THE HUMAN GENOME PROJECT AS SET OUT IN 1988	
1.	Examine the issues surrounding the completion of the human DNA sequence and the study of human genetic variation
2.	Examine issues raised by the integration of genetic technologies and information into health care and public health activities
3.	Examine issues raised by the integration of knowledge about genomics and gene–environment interactions into nonclinical settings
4.	Explore ways in which new genetic knowledge may interact with a variety of philosophical, theological and ethical perspectives
5.	Explore how socioeconomic factors and concepts of race and ethnicity influence the use, understanding and interpretation of genetic information, the utilization of genetic services and the development of policy

CHAPTER 24: The Ethical Issues Raised by Modern Genetics

KEY CONCEPTS

- Students of genetics must be aware of, and form opinions on, the ethical issues arising from the applications of genetics in the modern world.
- Geneticists have developed methods for ensuring that the antibiotic resistance genes carried by the cloning vectors used in the production of GM foodstuffs cannot be transferred to bacteria in the human gut.
- The environmental impacts of GM crops have been extensively studied, but are complex.
- The terminator technology has been proposed as a method for enabling a plant breeding company to protect the financial investment it makes in the development of a new variety of crop produced by GM or conventional methods.
- The ease with which personal DNA data can be acquired raises issues concerning privacy, use of data by insurance companies and the right to know, or not know, one's susceptibility to genetic disorders.
- Germline gene therapy and pharming raise issues concerning the manipulation of human characteristics for cosmetic purposes, and the welfare of transgenic animals.
- Any debate on the ethical implications of genetics must be based on facts and not fiction.
- Geneticists should lead debates on the ethical implications of genetics, but should not be sole arbiters of those debates.

QUESTIONS AND PROBLEMS

Key terms

Write short definitions of the following terms:

bioethics
blastocyst
direct-to-consumer
ethical, legal and social issues
genetic use restriction technology
personal DNA data
ribosome inactivating protein
selectable marker
somatic cell nuclear transfer
terminator technology
UK biobank

Self-study questions

24.1 Describe the genetic features of FlavrSavr tomatoes.

24.2 How is the Cre recombinase used to delete a selectable marker gene from a plant genome?

24.3 Outline two ways in which superweeds could arise following cultivation of a herbicide-resistant crop.

24.4 Describe how the introduction of gene editing has altered the basis to ethical discussions regarding genetic engineering of crop plants.

24.5 What is 'genetic use restriction technology'?

24.6 Describe how the terminator technology could be used to prevent farmers using saved seed to grow the following year's crop.

24.7 Define the term 'personal DNA data'.

24.8 What would be the advantages of a universal database of DNA profiles? Give an example where attempts to establish such a database were unsuccessful.

24.9 What is 'direct-to-consumer' DNA testing?

24.10 What is the main purpose of large scale DNA testing projects such as the one conducted by the UK biobank?

24.11 Explain how a DTC database could be used for familial DNA testing as part of a criminal investigation.

24.12 Why has somatic cell nuclear transfer restricted the use of germline manipulation with human oocytes?

24.13 What are the ethical issues relating to the use of somatic cell nuclear transfer in pharming?

24.14 Describe how geneticists dealt with the potential hazards of gene cloning when this technology was first invented in the 1970s.

24.15 Outline the ethical, legal and social issues that were studied as part of the Human Genome Project.

Discussion topics

24.16 Is it possible to argue that GM crops pose a danger to human health?

24.17 Should there be greater restrictions on the cultivation of GM crops bearing in mind possible harmful effects on the environment?

24.18 Is there a distinction between gene edited and GM crops?

24.19 What are the arguments for and against genetic use restriction technology?

24.20 Would a universal database of DNA profiles represent a violation of personal DNA data?

24.21 If you were the chief executive of a direct-to-consumer DNA testing company, what restrictions would you place on the use of the data your company collects?

24.22 Is it possible to place an international ban on technology such as human germline manipulation? Would such a ban be desirable?

24.23 Do the ethical issues relating to pharming outweigh the benefits of the technology?

24.24 Is the public perception of genetics and geneticists overly influenced by fictional TV dramas? Is this good or bad, and if the latter then what should real geneticists do to correct the view of genetics held by the public?

24.25 Can geneticists be trusted to self-regulate the use of new technology to ensure this is used solely for purposes that are in the public interest?

 The answers to, or hints on how to answer, the self-study questions and discussion topic questions can be downloaded from the book's product page at this link: www.routledge.com/9781032743530

FURTHER READING

Berg P (2008) Meetings that changed the world: Asilomar 1975: DNA modification secured. *Nature* 455, 290–291.

Brookes G & Barfoot P (2020) Environmental impacts of genetically modified (GM) crop use 1996–2018: impacts on pesticide use and carbon emissions. *GM Crops and Food* 11, 215–241.

Buchholzer M & Frommer WB (2033) An increasing number of countries regulate genome editing in crops. *New Phytologist* 237, 12–15.

Dedrickson K (2018) Universal DNA databases: a way to *improve* privacy? *Journal of Law and the Biosciences* 22, 637–647.

Joseph AM, Karas M, Ramadan Y et al. (2022) Ethical perspectives of therapeutic human genome editing from multiple and diverse viewpoints: a scoping review. *Cureus* 14, e31927.

Loi P, Czernik M, Zacchini F et al. (2013) Sheep: the first large animal model in nuclear transfer research. *Cellular Reprogramming* 15, 367–373.

Lombardo L (2014) Genetic use restriction technologies: a review. *Plant Biotechnology Journal* 12, 995–1005.

Martins MF, Murry LT, Telford L et al. (2022) Direct-to-consumer genetic testing: an updated systematic review of healthcare professionals' knowledge and views, and ethical and legal concerns. *European Journal of Human Genetics* 30, 1331–1343.

McEwen JE, Boyer JT, Sun KY et al. (2014) The ethical, legal, and social implications program of the National Human Genome Research Institute: reflections on an ongoing experiment. *Annual Review of Genomics and Human Genetics* 15, 481–505.

Miki B & McHugh S (2003) Selectable marker genes in transgenic plants: applications, alternatives and biosafety. *Journal of Biotechnology* 107, 193–232.

Shelton AM, Zhao JZ & Roush RT (2002) Economic, ecological, food safety, and social consequences of the deployment of BT transgenic plants. *Annual Review of Entomology* 47, 845–81. *Various issues relating to endotoxin plants.*

Various authors (2003) Theme issue: the Farm Scale Evaluations of spring-sown genetically modified crops. *Philosophical Transactions of the Royal Society of London series B* 358, 1775–1889. *A series of papers on the environmental impact of GM crops.*

GLOSSARY

α helix One of the commonest secondary structural conformations taken up by segments of polypeptides.

β-N-glycosidic bond The linkage between the base and sugar of a nucleotide.

β sheet One of the commonest secondary structural conformations taken up by segments of polypeptides.

β turn A sequence of four amino acids, the second usually glycine, which causes a polypeptide to change direction.

γ complex A component of DNA polymerase III comprising subunit γ in association with δ, δ′, χ and ψ.

(6–4) lesion A dimer between two adjacent pyrimidine bases in a polynucleotide, formed by ultraviolet irradiation.

(6–4) photoproduct photolyase An enzyme involved in photoreactivation repair.

10 nm fiber An unpacked form of chromatin consisting of nucleosome beads on a string of DNA.

2–aminopurine A base analog that can cause mutations by replacing adenine in a DNA molecule.

3″–OH terminus The end of a polynucleotide that terminates with a hydroxyl group attached to the 3′-carbon of the sugar.

3′→5′ exonuclease An enzyme that sequentially removes nucleotides in the 3′→5′ direction from the end of a nucleic acid molecule.

30 nm fiber A relatively unpacked form of chromatin consisting of a possibly helical array of nucleosomes in a fiber approximately 30 nm in diameter.

5-bromouracil (5-bU) A base analog that can cause mutations by replacing thymine in a DNA molecule.

5′-P terminus The end of a polynucleotide that terminates with a mono-, di- or triphosphate attached to the 5′-carbon of the sugar.

5′→3′ exonuclease An enzyme that sequentially removes nucleotides in the 5′→3′ direction from the end of a nucleic acid molecule.

ABC model A model for flower development based on three groups of homeotic genes that work together to specify the structures of the flower.

acceptor arm Part of the structure of a tRNA molecule.

acceptor site The splice site at the 3′ end of an intron.

accessory genome The component of a prokaryotic genome comprising all those genes not present in the core genome.

acridine dye A chemical compound that causes a frameshift mutation by intercalating between adjacent base pairs of the double helix.

acrocentric A chromosome whose centromere is positioned toward one end.

acylation The attachment of a lipid side chain to a polypeptide.

Ada enzyme An *Escherichia coli* enzyme that is involved in the direct repair of alkylation mutations.

adaptor A synthetic, double-stranded oligonucleotide used to attach sticky ends to a blunt-ended molecule.

adenine A purine base found in DNA and RNA.

adeno-associated virus (AAV) A virus that is unrelated to adenovirus but which is often found in the same infected tissues, because AAV makes use of some of the proteins synthesized by adenovirus to complete its replication cycle.

adenosine deaminase acting on RNA (ADAR) An enzyme that edits various eukaryotic mRNAs by deaminating adenosine to inosine.

adenovirus An animal virus, derivatives of which have been used to clone genes in mammalian cells.

adenylate cyclase The enzyme that converts ATP to cyclic AMP.

admixture Interbreeding between two individuals from different genetic populations.

A-form of DNA A structural configuration of the double helix, present but not common in cellular DNA.

alkylating agent A mutagen that acts by adding alkyl groups to nucleotide bases.

allele One of two or more alternative forms of a gene.

allelic dropout When only one of a pair of alleles is detected by PCR directed at, for example, a short tandem repeat.

allopatric speciation Speciation that occurs due to two populations becoming separated by a geographical barrier.

alphoid DNA The tandemly repeated nucleotide sequences located in the centromeric regions of human chromosomes.

alternative exons An alternative splicing scenario where the mRNA contains either of a pair of exons, but not both at the same time.

alternative polyadenylation The use of two or more different sites for polyadenylation of an mRNA.

alternative promoter One of two or more different promoters acting on the same gene.

alternative splice site selection An alternative splicing scenario where the usual donor or acceptor site is ignored and a second site used in its place.

alternative splicing The production of two or more mRNAs from a single pre-mRNA by joining together different combinations of exons.

Alu A type of SINE found in the genomes of humans and related mammals.

amelogenin gene A gene present in the pseudoautosomal region of the X and Y chromosomes, used in DNA-based sex identification.

amino acid One of the monomeric units of a protein molecule.

amino terminus The end of a polypeptide that has a free amino group.

aminoacyl or A site The site in the ribosome occupied by the aminoacyl-tRNA during translation.

aminoacylation Attachment of an amino acid to the acceptor arm of a tRNA.

aminoacyl-tRNA synthetase An enzyme that catalyzes the aminoacylation of one or more tRNAs.

anaphase The period during the division of a cell when the pairs of replicated chromosomes separate.

ancient DNA DNA preserved in ancient biological material.

anticodon The triplet of nucleotides, at positions 34–36 in a tRNA molecule, that base pairs with a codon in an mRNA molecule.

anticodon arm Part of the structure of a tRNA molecule.

antigen A substance that elicits an immune response.

antiparallel Refers to the arrangement of polynucleotides in the double helix running in opposite directions.

antisense RNA An RNA molecule that is the reverse complement of a naturally occurring mRNA, which can be used to prevent translation of that mRNA in a transformed cell.

antitermination A bacterial mechanism for regulating the termination of transcription.

antiterminator protein A protein that attaches to bacterial DNA and mediates antitermination.

AP endonuclease An enzyme involved in base excision repair.

apoptosis Programmed cell death.

applied science Any area of science that addresses practical issues.

AP (apurinic/apyrimidinic) site A position in a DNA molecule where the base component of the nucleotide is missing.

archaea One of the two main groups of prokaryotes, mostly found in extreme environments.

archaeogenetics The use of DNA analysis to study the human past.

attenuation A process used by some bacteria to regulate the expression of an amino acid biosynthetic operon in accordance with the levels of the amino acid in the cell.

autoradiography The detection of radioactively labeled molecules by exposure of an X-ray-sensitive photographic film.

autosome A chromosome that is not a sex chromosome.

back mutation A mutation that reverses the effect of a previous mutation by restoring the original nucleotide sequence.

backtracking The reversal of an RNA polymerase a short distance along its DNA template strand in order to correct an error in transcription.

bacteria One of the two main groups of prokaryotes.

bacterial chromosome One of the DNA molecules (possibly the only one) in a bacterial cell that carries essential genes and is located in the nucleoid.

bacteriophage A virus that infects a bacterium.

Barr body The highly condensed chromatin structure taken on by an inactivated X chromosome.

basal rate of transcript initiation The number of productive initiations of transcription occurring per unit time at a particular promoter.

base analog A compound whose structural similarity to one of the bases in DNA enables it to act as a mutagen.

base editor An enzyme capable of changing one nucleotide to another within a DNA molecule.

base excision A DNA repair process that involves the excision and replacement of an abnormal base.

baseless site A position in a DNA molecule where the base component of the nucleotide is missing.

base pair The hydrogen-bonded structure formed by two complementary nucleotides. The shortest unit of length for a double-stranded DNA molecule.

base pairing The attachment of one polynucleotide to another, or one part of a polynucleotide to another part of the same polynucleotide, by base pairs.

base ratio The ratio of A to T, or G to C, in a double-stranded DNA molecule. Chargaff showed that the base ratios are always close to 1.0.

batch culture Growth of bacteria in a fixed volume of liquid medium in a closed vessel, with no additions or removals made during the period of incubation.

B-form of DNA The commonest structural conformation of the DNA double helix in living cells.

bioethics Study of the ethical issues raised by the applications of biology.

biolistics A means of introducing DNA into cells that involves bombardment with high-velocity microprojectiles coated with DNA.

biological information The information that is contained in the genome of an organism and which directs the development and maintenance of that organism.

biotechnology The use of living organisms, often (but not always) microbes, in industrial processes.

bivalent The structure formed when a pair of homologous chromosomes lines up during meiosis.

BLAST (Basic Local Alignment Search Tool) An algorithm frequently used in homology searching.

blastocyst The embryonic structure that attaches to the wall of the uterus during a normal pregnancy in a mammal.

blastoderm The structure that forms when individual cells start to appear around the outside of the syncytium of a developing fruit fly embryo.

bottleneck The period between the decline and recovery in the size of a population.

branch migration A step during homologous recombination, involving exchange of polynucleotides between a pair of recombining double-stranded DNA molecules.

break repair A process for the repair of single- or double-strand breaks in a DNA molecule.

buoyant density The density possessed by a molecule or particle when suspended in an aqueous salt or sugar solution.

bystander mutation Unwanted mutations occurring adjacent to the target nucleotide when base editing is carried out.

CAAT box A regulatory sequence that controls the basal rate of transcription of a gene.

candidate gene A gene, identified by any of various means, that might be responsible for a genetic disorder.

cap binding complex The complex that makes the initial attachment to the cap structure at the beginning of the scanning phase of eukaryotic translation.

capillary gel electrophoresis Polyacrylamide gel electrophoresis carried out in a thin capillary tube, providing high resolution.

capsid The protein coat that surrounds the DNA or RNA genome of a virus.

cap structure The chemical modification at the 5' end of most eukaryotic mRNA molecules.

capsule A coating of polysaccharide that surrounds some bacterial cells.

carboxyl terminus The end of a polypeptide that has a free carboxyl group.

carcinogen An environmental agent that causes cancer.

carrier An individual who has one normal and one mutated allele for a gene responsible for an inherited disorder, but does not have the disorder because the mutated allele is recessive.

Cas9 endonuclease A programmable nuclease that is directed to its target site by a 20-nucleotide guide RNA.

cascade system A system in which the completion of one event triggers the initiation of a second event. In genetics, usually a system for controlling the order in which genes are expressed.

catabolite activator protein A regulatory protein that binds to various sites in a bacterial genome and activates transcription initiation at downstream promoters.

cell cycle The series of events occurring in a cell between one division and the next.

cell cycle checkpoint A period before entry into S or M phase of the cell cycle, a key point at which regulation is exerted.

cell-free protein synthesizing system A cell extract containing all the components needed for protein synthesis and able to translate added mRNA molecules.

cell surface receptor A protein located in the cell membrane that responds to an external signal by causing a biochemical change within the cell.

cell transformation The alteration in morphological and biochemical properties that occurs when an animal cell is infected by an oncogenic virus.

centimorgan (cM) The unit used to describe the distance between two genes on a chromosome. 1 cM is the distance that corresponds to a 1% probability of recombination in a single meiosis.

centromere The constricted region of a chromosome that is the position at which the pair of chromatids are held together.

centrosome The structures on opposite sides of the mitotic spindle from which microtubules radiate during cell division.

chain termination sequencing A DNA sequencing method that involves enzymatic synthesis of polynucleotide chains that terminate at specific nucleotide positions.

chaperonin A multisubunit protein that forms a structure which aids the folding of other proteins.

charging Aminoacylation of a tRNA.

chiasmata The crossover points visible when a pair of recombining chromosomes are observed by microscopy.

chi form An intermediate structure seen during recombination between DNA molecules.

chimeric antigen receptor T cells Lymphocytes that are engineered to recognize and kill tumor cells.

chi (crossover hotspot initiator) site A repeated nucleotide sequence in the *E. coli* genome that is involved in the initiation of homologous recombination.

chloroplast genome The multicopy DNA molecule located in the chloroplast of photosynthetic organisms.

chromatid The arm of a chromosome.

chromatin The complex of DNA and histone proteins found in chromosomes.

chromatin modification Regulation of genome expression by chemical alteration of nucleosomes.

chromatosome A subcomponent of chromatin made up of a nucleosome core octamer with associated DNA and a linker histone.

chromid A bacterial DNA molecule that has the characteristic features of a plasmid but which carries essential genes.

chromosomal interaction domain A loop of DNA, attached to a protein core, which forms a structural component of a prokaryotic nucleoid.

chromosome One of the DNA-protein structures that contains part of the nuclear genome of a eukaryote. Less accurately, the DNA molecule(s) containing a bacterial genome.

clastogen An environmental agent that causes breaks in DNA molecules.

cleavage and polyadenylation specificity factor (CPSF) A protein that has an ancillary role during polyadenylation of eukaryotic mRNAs.

cleavage stimulation factor (CstF) A protein that plays an ancillary role during polyadenylation of eukaryotic mRNAs.

cline A geographical gradation in the allele frequencies within a population.

cloning vector A DNA molecule that is able to replicate inside a host cell and therefore can be used to clone other fragments of DNA.

closed promoter complex The structure formed during the initial step in assembly of the transcription initiation complex. The closed promoter complex consists of the RNA polymerase and/or accessory proteins attached to the promoter, before the DNA has been opened up by breakage of base pairs.

cloverleaf A two-dimensional representation of the structure of a tRNA molecule.

CODIS set A standard set of short tandem repeats used for the construction of DNA profiles.

codominance The relationship between a pair of alleles that both contribute to the phenotype of a heterozygote.

codon A triplet of nucleotides coding for a single amino acid.

codon–anticodon recognition The interaction between a codon on an mRNA molecule and the corresponding anticodon on a tRNA.

codon bias Refers to the fact that not all codons are used equally frequently in the genes of a particular organism.

Col plasmid A plasmid that carries one or more genes coding for colicins.

colicin A toxic bacterial protein often coded for by a gene carried by a plasmid.

colonization Movement of a population to new territory previously unoccupied by that species.

compatible Refers to two plasmids that can coexist in the same cell.

competent Refers to a culture of bacteria that have been treated, for example, by soaking in calcium chloride, so that their ability to take up DNA molecules is enhanced.

complementary Refers to two nucleotides or nucleotide sequences that are able to base pair with one another.

complementary DNA (cDNA) synthesis Synthesis of a double-stranded DNA copy of an mRNA molecule.

complementary gene action The situation in which a particular combination of alleles at two separate genes is needed in order to produce a phenotype.

complex multigene family A type of multigene family in which the genes have similar nucleotide sequences but are sufficiently different to code for proteins with distinctive properties.

concatemer A DNA molecule made up of linear genomes or other DNA units linked head to tail.

conjugation Transfer of DNA between two bacteria that come into physical contact with one another.

consensus sequence A nucleotide sequence that represents an 'average' of a number of related but nonidentical sequences.

conservative replication A hypothetical mode of DNA replication in which one daughter double helix is made up of the two parental polynucleotides and the other is made up of two newly synthesized polynucleotides.

constitutive heterochromatin Chromatin that is permanently in a compact organization.

context-dependent codon reassignment Refers to the situation whereby the DNA sequence surrounding a codon changes the meaning of that codon.

continuous culture The culture of microorganisms in liquid medium under controlled conditions, with additions to and removals from the medium over a lengthy period.

COOH- or C terminus The end of a polypeptide that has a free carboxyl group.

copy number The number of molecules of a plasmid contained in a single cell.

core enzyme The version of *E. coli* RNA polymerase, subunit composition $\alpha_2\beta\beta'\omega$, that performs RNA synthesis but is unable to locate promoters efficiently.

core genome The component of a bacterial pan-genome that contains the set of genes possessed by all members of the species.

core octamer The central component of a nucleosome, made up of two subunits each of histones H2A, H2B, H3 and H4, around which DNA is wound.

co-repressor A molecule that must bind to a bacterial repressor before the repressor can attach to its operator site.

core promoter The position within a eukaryotic promoter where the initiation complex is assembled.

***cos* site** One of the cohesive, single-stranded extensions present at the ends of the DNA molecules of certain strains of λ phage.

CpG island A GC-rich DNA region located upstream of approximately 56% of the genes in the human genome.

CREB transcription factor A protein that controls the expression of genes involved in cell growth and division.

cross-fertilization Fertilization of a female gamete with a male gamete derived from a different individual.

crossing over The exchange of DNA between chromosomes during meiosis.

cryptic splice site A site whose sequence resembles an authentic splice site and which might be selected instead of the authentic site during aberrant splicing.

C-terminal domain (CTD) A component of the largest subunit of RNA polymerase II, important in activation of the polymerase.

C- or COOH terminus The end of a polypeptide that has a free carboxyl group.

cyanelle Photosynthetic structures, resembling ingested cyanobacteria, inside the cells of *Cyanophora* protozoa.

cyclic AMP (cAMP) A modified version of AMP in which an intramolecular phosphodiester bond links the 5′ and 3′ carbons.

cyclin A regulatory protein whose abundance varies during the cell cycle and which regulates biochemical events in a cell-cycle-specific manner.

cyclin-dependent kinase A protein that activates enzymes and other proteins that have specific functions during the cell cycle.

cyclobutyl dimer A dimer between two adjacent pyrimidine bases in a polynucleotide, formed by ultraviolet irradiation.

cytokinesis The period during the division of a cell when the two daughter cells form around the divided nuclei.

cytosine One of the pyrimidine bases found in DNA and RNA.

dark repair A type of nucleotide excision repair process that corrects cyclobutyl dimers.

D arm Part of the structure of a tRNA molecule.

deadenylation-dependent decapping A process for degradation of eukaryotic mRNAs that is initiated by removal of the poly(A) tail.

deaminating agent A mutagen that acts by removing amino groups from nucleotide bases.

defective retrovirus A viral genome that has lost essential nucleotide sequences and so cannot replicate on its own.

degenerate Refers to the fact that the genetic code has more than one codon for most amino acids.

degradative plasmid A plasmid whose genes allow the host bacterium to metabolize unusual molecules such as toluene and salicylic acid.

degradosome A multienzyme complex responsible for degradation of bacterial mRNAs.

delayed onset Refers to a mutation whose effect is not apparent until a relatively late stage in the life of the mutant organism.

deletion A mutation resulting from deletion of one or more nucleotides from a DNA sequence.

denaturation Breakdown by chemical or physical means of the noncovalent interactions, such as hydrogen bonding, that maintain the secondary and higher levels of structure of proteins and nucleic acids.

***de novo* sequencing** A strategy in which a genome sequence is assembled solely by finding overlaps between individual sequence reads.

density gradient centrifugation A technique in which a cell fraction is centrifuged through a dense solution, in the form of a gradient, so that individual components are separated.

deoxyribonuclease An enzyme that cleaves phosphodiester bonds in a DNA molecule.

diauxie The phenomenon whereby a bacterium, when provided with a mixture of sugars, uses up one sugar before beginning to metabolize the second sugar.

Dicer The ribonuclease that has a central role in RNA interference.

dihybrid cross A sexual cross in which the inheritance of two pairs of alleles is followed.

diploid A nucleus that has two copies of each chromosome.

directional selection A population shift that occurs when a single extreme phenotype is favored over all others.

direct repair A DNA repair system that acts directly on a damaged nucleotide.

direct-to-consumer A commercial DNA testing service.

discontinuous gene A gene that is split into exons and introns.

dispersive replication A hypothetical mode of DNA replication in which both polynucleotides of each daughter double helix are made up partly of parental DNA and partly of newly synthesized DNA.

displacement replication A mode of replication that involves continuous copying of one strand of the helix, the second strand being displaced and subsequently copied after the synthesis of the first daughter strand has been completed.

disruptive selection Selection that results in loss of an intermediate phenotype.

disulfide bridge A covalent bond linking cysteine amino acids on different polypeptides or at different positions on the same polypeptide.

D loop The structure formed when a DNA double helix is invaded by a single-stranded DNA or RNA molecule, which forms a region of base pairing with one of the polynucleotides of the helix.

DNA One of the two forms of nucleic acid in living cells; the genetic material for all cellular life forms and many viruses.

DNA adenine methylase (Dam) An enzyme involved in methylation of *E. coli* DNA.

DNA chip A high-density array of DNA molecules used for parallel hybridization analyses.

DNA cloning Insertion of a fragment of DNA into a cloning vector and subsequent propagation of the recombinant DNA molecule in a host organism.

DNA cytosine methylase (Dcm) An enzyme involved in methylation of *E. coli* DNA.

DNA-dependent DNA polymerase An enzyme that makes a DNA copy of a DNA template.

DNA-dependent RNA polymerase An enzyme that makes an RNA copy of a DNA template.

DNA glycosylase An enzyme that cleaves the β-*N*-glycosidic bond between a base and the sugar component of a nucleotide as part of the base excision and mismatch repair processes.

DNA gyrase A type II topoisomerase of *E. coli*.

DNA ligase An enzyme that synthesizes phosphodiester bonds as part of DNA replication, repair and recombination processes.

DNA marker A DNA sequence that exists as two or more readily distinguished versions and which can therefore be used to mark a map position on a genetic, physical or integrated genome map.

DNA methylation Refers to the chemical modification of DNA by the attachment of methyl groups.

DNA methyltransferase An enzyme that attaches methyl groups to a DNA molecule.

DNA photolyase A bacterial enzyme involved in photoreactivation repair.

DNA polymerase An enzyme that synthesizes DNA.

DNA polymerase I The bacterial enzyme that completes the synthesis of Okazaki fragments during DNA replication.

DNA polymerase II A bacterial DNA polymerase involved in DNA repair.

DNA polymerase III The main DNA replicating enzyme of bacteria.

DNA polymerase α The enzyme that primes DNA replication in eukaryotes.

DNA polymerase γ The enzyme responsible for replication of the mitochondrial genome.

DNA polymerase δ The enzyme responsible for replication of the lagging DNA strand in eukaryotes.

DNA polymerase ε The enzyme responsible for replication of the leading DNA strand in eukaryotes.

DNA profile The alleles present at a defined set of short tandem repeat loci.

DNA profiling The use of multiplex PCR to construct a DNA profile.

DNA repair The biochemical processes that correct mutations arising from replication errors and the effects of mutagenic agents.

DNA sequence The order of nucleotides in a DNA molecule.

DNA sequencing The technique for determining the order of nucleotides in a DNA molecule.

DNA topoisomerase An enzyme that introduces or removes turns from the double helix by the breakage and reunion of one or both polynucleotides.

DNA unwinding element (DUE) The AT-rich component of a bacterial origin of replication; the position at which helix melting occurs.

dominant Refers to the allele that is expressed in a heterozygote.

donor site The splice site at the 5' end of an intron.

double helix The base-paired double-stranded structure that is the natural form of DNA in the cell.

double heterozygote A nucleus that is heterozygous for two genes.

double homozygote A nucleus that is homozygous for two genes.

draft genome sequence An incomplete chromosome or genome sequence, typically containing some errors, gaps and ambiguity about the order and/or orientation of some parts of the sequence.

early genes Bacteriophage genes that are expressed during the early stages of infection of a bacterial cell; the products of early genes are usually involved in replication of the phage genome.

Earth BioGenome Project A project that aims to sequence the genome of every eukaryotic species.

effective population size The size of that part of the population that participates in reproduction.

elongation factor A protein that has an ancillary role in the elongation step of transcription or translation.

endogenous retrovirus (ERV) An active or inactive retroviral genome integrated into a host chromosome.

endonuclease An enzyme that breaks phosphodiester bonds within a nucleic acid molecule.

endosymbiont theory A theory that states that the mitochondria and chloroplasts of eukaryotic cells are derived from symbiotic prokaryotes.

enhancer A regulatory sequence that increases the rate of transcription of a gene or genes located some distance away in either direction.

enzyme A protein that catalyzes one or more biochemical reactions.

epigenesis A heritable change to the genome that does not involve mutation.

episome A plasmid capable of integration into the host cell's chromosome.

epistasis The situation in which the alleles present at one gene mask or cancel those at a second gene.

error-prone editing A version of genome editing that stimulates nonhomologous end joining and results in a short insertion or deletion at the target site.

ethical, legal and social issues (ELSI) An ethical framework for genetic research, first developed as part of the Human Genome Project.

ethidium bromide A type of intercalating agent that causes mutations by inserting between adjacent base pairs in a double-stranded DNA molecule.

ethylmethane sulfonate (EMS) A mutagen that acts by adding alkyl groups to nucleotide bases.

euchromatin Regions of a eukaryotic chromosome that are relatively uncondensed, thought to contain active genes.

eukaryote An organism whose cells contain membrane-bound nuclei.

excision repair A DNA repair process that corrects various types of DNA damage by excising and resynthesizing a region of polynucleotide.

exit or E site A position within a bacterial ribosome to which a tRNA moves immediately after deacylation.

exon A coding region within a discontinuous gene.

exon skipping Aberrant splicing, or an alternative splicing event, in which one or more or exons are omitted from the spliced RNA.

exonuclease An enzyme that sequentially removes nucleotides from the ends of a nucleic acid molecule.

exosome A multiprotein complex involved in degradation of mRNA in eukaryotes.

expression vector A special type of cloning vector designed for the synthesis of recombinant protein.

extra, optional or variable loop A structural component of tRNAs.

extrachromosomal gene A gene in a mitochondrial or chloroplast genome.

F' A bacterium that carries a modified F plasmid, one carrying a small piece of DNA derived from the host bacterial DNA molecule.

F+ A bacterium that carries an F plasmid.

F− A bacterium that does not carry an F plasmid.

facultative heterochromatin Chromatin that has a compact organization in some but not all cells, thought to contain genes that are inactive in some cells or at some periods of the cell cycle.

familial DNA testing Comparison of DNA profiles to identify possible relatives of a person whose profile has been recovered from a crime scene.

fermenter A vessel used for the large-scale culture of microorganisms.

fertility or F plasmid A fertility plasmid that directs conjugal transfer of DNA between bacteria.

fiber-FISH (fiber-fluorescent *in situ* hybridization) A hybridization method used to map the positions of DNA sequence features within a long DNA molecule.

filamentous bacteriophage A bacteriophage or virus capsid in which the protomers are arranged in a helix, producing a rod-shaped structure.

finished genome A genome sequence that is almost complete, but typically still has some unsequenced gaps between contigs and an average of up to one error per 10^4 nucleotides.

fitness The ability of an organism or allele to survive and reproduce.

fixation Refers to the situation that occurs when a single allele reaches a frequency of 100% in a population.

fixation index A statistic indicating the degree of difference between the allele frequencies present in two populations.

flap endonuclease (FEN1) An enzyme involved in replication of the lagging strand in eukaryotes.

flow cell A reaction chamber ion which DNA sequencing, or some other biochemical reaction, is carried out.

foldback RNA The precursor RNA molecules that are cleaved to produce microRNAs.

folding pathway The series of events, involving partly folded intermediates, that results in an unfolded protein attaining its correct three-dimensional structure.

founder effect The situation that occurs when a few members of a population split away and initiate a new population.

F plasmid A fertility plasmid that directs conjugal transfer of DNA between bacteria.

fragile site A position in a chromosome that is prone to breakage, possibly because it contains an expanded trinucleotide repeat sequence.

frameshift mutation A mutation resulting from the insertion or deletion of a group of nucleotides that is not a multiple of three and which therefore changes the frame in which translation occurs.

F_{ST} **value** A statistic indicating the degree of difference between the allele frequencies present in two populations.

functional RNA RNA that has a functional role in the cell, i.e. RNAs other than mRNA.

fusion protein A protein that consists of a fusion of two polypeptides, or parts of polypeptides, normally coded for by separate genes.

G1 phase The first gap period of the cell cycle.

G1-S checkpoint A cell cycle checkpoint that a cell must pass before it is able to replicate its DNA.

G2 phase The second gap period of the cell cycle.

G2-M checkpoint A cell cycle checkpoint that can only be passed when a cell is ready to enter mitosis.

gain of function A mutation that results in an organism acquiring a new function.

gamete A reproductive cell, usually haploid, that fuses with a second gamete to produce a new cell during sexual reproduction.

gap gene Developmental gene that has a role in establishing positional information within the fruit fly embryo.

Gaussian distribution The distribution pattern commonly called a bell-shaped curve.

GC box A regulatory sequence that controls the basal rate of transcription of a gene.

GC content The percentage of nucleotides in a genome that are G or C.

gene A DNA segment containing biological information and hence coding for an RNA and/or polypeptide molecule.

gene addition A genetic engineering strategy that involves the introduction of a new gene or group of genes into an organism.

gene amplification The production of multiple copies of a DNA segment so as to increase the rate of expression of a gene carried by the segment.

gene desert A region of genome in which here are few if any genes.

gene editing A method that enables directed changes to be made in a target gene.

gene expression The series of events by which the biological information carried by a gene is released and made available to the cell.

gene mapping Any of various methods used to identify the positions of genes in DNA molecules.

gene subtraction A genetic engineering strategy that involves the inactivation of one or more of an organism's genes.

gene therapy A clinical procedure in which a gene or other DNA sequence is used to treat a disorder.

genetically modified (GM) plant A plant whose phenotype has been altered by genetic engineering.

genetic analysis Methods, based on examination of the progeny resulting from breeding experiments, that enable interactions between the alleles of the same and different genes to be studied.

genetic code The rules that determine which triplet of nucleotides codes for which amino acid during protein synthesis.

genetic disorder A disorder caused by the sequence of the genome.

genetic drift The changes in allele frequencies that occur in a population over time due to random processes.

genetic engineering The use of experimental techniques to produce DNA molecules containing new genes or new combinations of genes.

genetic fingerprinting A method by which a genetic profile is constructed from minisatellite variability.

genetic material The chemical material of which genes are made, now known to be DNA in most organisms, and RNA in a few viruses.

genetics The branch of biology devoted to the study of genes.

genetic use restriction technology (GURT) A process that enables a company that develops and markets GM crops to protect its financial investment by ensuring that farmers must buy new seed every year.

genome The entire genetic complement of a living organism.

genome annotation The process by which the genes and other interesting features are identified in a genome sequence.

genome browser A website that allows the user to examine a genome for the locations of features such as genes, transcripts and repeated sequences.

genome-wide association study (GWAS) A method that attempts to identify all of the markers, from all over the genome, that are associated with a disease or other phenotype.

genomic imprinting Inactivation by methylation of a gene on one of a pair of homologous chromosomes.

genotype A description of the genetic composition of an organism.

germline therapy A type of gene therapy in which a fertilized egg is provided with a copy of the correct version of the mutated gene and reimplanted into the mother.

gigabase pair (Gb) 1,000,000 kb; 1,000,000,000 bp.

glycosylation The attachment of sugar units to a polypeptide.

guanine One of the purine nucleotides found in DNA and RNA.

guanine methyltransferase The enzyme that attaches a methyl group to the 5′ end of a eukaryotic mRNA during the capping reaction.

guanylyl transferase The enzyme that attaches a GTP to the 5′ end of a eukaryotic mRNA at the start of the capping reaction.

guide RNA An RNA that attaches to a DNA or RNA target site as acts as guide for the enzyme involved in an editing process.

half-life A measure of the rate of degradation of a molecule or the rate of decay of an atom.

haploid A nucleus that has a single copy of each chromosome.

haploinsufficiency The situation in which the inactivation of a gene on one of a pair of homologous chromosomes results in a change in the phenotype of the mutant organism.

haplotype A sequence variant of a gene, used as an alternative to 'allele' when there are several variants that confer the same phenotype.

Hardy-Weinberg equilibrium The equilibrium between allele frequencies that occurs in a population of infinite size for genes that are not subject to selection.

head-and-tail A bacteriophage capsid made up of an icosahedral head, containing the nucleic acid, and a filamentous tail that facilitates entry of the nucleic acid into the host cell.

heat shock module A regulatory sequence upstream of genes involved in the protection of a cell from heat damage.

helicase An enzyme that breaks base pairs in a double-stranded DNA molecule.

helix-turn-helix A common structural motif for attachment of a protein to a DNA molecule.

heterochromatin Chromatin that is relatively condensed and is thought to contain DNA that is not being transcribed.

heteroduplex An intermediate in recombination in which two double-stranded DNA molecules are linked together by a pair of Holliday structures.

heterogeneous nuclear ribonucleoprotein (hnRNP) One of a broad group of RNA-protein complexes, which play several roles in the nucleus, most of which involve binding to RNAs.

heterogenous nuclear RNA (hnRNA) The nuclear RNA fraction that comprises unprocessed transcripts synthesized by RNA polymerase II.

heterosis The situation in which the heterozygote is the most successful genotype in a population.

heterozygote A diploid cell or organism that contains two different alleles for a particular gene.

heterozygote advantage or superiority The situation in which the heterozygote is the most successful genotype in a population.

Hfr A bacterium whose DNA molecule contains an integrated copy of the F plasmid.

histone One of the basic proteins found in nucleosomes.

histone acetyltransferase (HAT) An enzyme that attaches acetyl groups to core histones.

histone code The hypothesis that the pattern of chemical modification on histone proteins influences various cellular activities.

histone deacetylase (HDAC) An enzyme that removes acetyl groups from core histones.

hitchhiking The process by which alleles that are not subject to selection become fixed or lost because they are adjacent to an allele that is under selection.

Holliday structure An intermediate structure formed during recombination between two DNA molecules.

holoenzyme The version of the *E. coli* RNA polymerase, subunit composition $\alpha_2\beta\beta'\sigma\omega$, that is able to recognize promoter sequences.

homeobox A conserved sequence element, coding for a DNA-binding motif called the homeodomain, found in several genes believed to be involved in the development of eukaryotic organisms.

homeodomain A DNA-binding motif found in many proteins involved in the developmental regulation of gene expression.

homeotic mutant An organism that has a mutation that results in the transformation of one body part into another.

homeotic selector gene A gene that establishes the identity of a body part such as a segment of the fruit fly embryo.

hominid A member of the taxonomic family called the Hominidae, comprising humans, chimpanzees, gorillas and orangutans.

homologous Two genes or other sequences that share a common evolutionary origin.

homologous chromosomes Two or more identical chromosomes present in a single nucleus.

homologous recombination Recombination between two homologous double-stranded DNA molecules, i.e. ones which share extensive nucleotide sequence similarity.

homology-directed repair The version of gene editing that enables individual nucleotides in a target gene to be changed.

homology search A technique in which genes with sequences similar to that of an unknown gene are sought, the objective being to gain an insight into the function of the unknown gene.

homozygote A diploid cell or organism that contains two identical alleles for a particular gene.

horizontal gene transfer Transfer of a gene from one species to another.

hormone response element A nucleotide sequence upstream of a gene that mediates the regulatory effect of a steroid hormone.

housekeeping gene A gene that is continually expressed in all or at least most cells of a multicellular organism.

Hsp70 chaperone A family of proteins that bind to hydrophobic regions in other proteins so as to aid their folding.

HU family A group of proteins found in bacterial nucleoids.

hybrid incompatibility A process that would enable prezygotic isolation to be achieved by the interaction between pairs of genes.

hybridization probing A technique that uses a labeled nucleic acid molecule as a probe to identify complementary or homologous molecules to which it base pairs.

hydatidiform mole A small growth that results when an egg cell that lacks a nucleus is fertilized by a sperm and implants in the uterus wall. The cells in a hydatidiform mole contain only the paternal chromosomes.

hydrogen bond A weak electrostatic attraction between an electronegative atom such as oxygen or nitrogen and a hydrogen atom attached to a second electronegative atom.

hyperediting The extensive conversion of adenosines to inosines in an RNA.

hypervariable dispersed repetitive sequence A type of minisatellite used in genetic fingerprinting.

hypervariable region One of the two very variable regions of the noncoding part of the mitochondrial genome.

icosahedral A bacteriophage or virus capsid in which the protomers are arranged into a three-dimensional geometric structure that surrounds the nucleic acid.

identity-by-descent A method that identifies possible relatives by searching for blocks of shared DNA in their genome sequences.

immunoelectron microscopy An electron microscopy technique that uses antibody labeling to identify the positions of specific proteins on the surface of a structure such as a ribosome.

imprint control element A DNA sequence, found within a few kilobases of clusters of imprinted genes, which mediates the methylation of the imprinted regions.

inclusion body A crystalline or paracrystalline deposit within a cell, often containing substantial quantities of insoluble protein.

incompatibility group Comprises several different types of plasmid, often related to each other, that are unable to coexist in the same cell.

incomplete dominance Refers to a pair of alleles, neither of which displays dominance, the phenotype of a heterozygote being intermediate between the phenotypes of the two homozygotes.

incomplete penetrance A mutation that provides a predisposition to a phenotype, which appears only in response to a later trigger.

inducer A molecule that induces the expression of a gene or operon by binding to a repressor protein and preventing the repressor from attaching to the operator.

inducible operon An operon that is switched on by an inducer molecule that prevents the repressor from attaching to its DNA-binding site.

induction The excision of the integrated form of λ and accompanying switch to the lytic mode of infection, in response to a chemical or other stimulus.

inherited disorder A disorder caused by the sequence of the genome.

initiation codon The codon, usually but not exclusively 5′-AUG-3′, found at the start of the coding region of a gene.

initiation complex The complex of proteins that initiates transcription. Also the complex that initiates translation.

initiation factor A protein that has an ancillary role during initiation of translation.

initiation region A region of eukaryotic chromosomal DNA within which replication initiates at positions that are not clearly defined.

initiator (Inr) sequence A component of the RNA polymerase II core promoter.

insertion A mutation that arises by insertion of one or more nucleotides into a DNA sequence.

integron A set of genes and other DNA sequences that enable plasmids to capture genes from bacteriophages and other plasmids.

intercalating agent A compound that can enter the space between adjacent base pairs of a double-stranded DNA molecule, often causing mutations.

intergenic DNA The regions of a genome that do not contain genes.

internal ribosome entry site (IRES) A nucleotide sequence that enables the ribosome to assemble at an internal position in some eukaryotic mRNAs.

interspersed repeat A sequence that recurs at many dispersed positions within a genome.

intrinsic terminator A position in bacterial DNA where termination of transcription occurs without the involvement of Rho.

intron A noncoding region within a discontinuous gene.

inversion A mutation that involves the excision of a portion of a DNA molecule followed by its reinsertion at the same position but in the reverse orientation.

inverted repeat Two identical nucleotide sequences repeated in opposite orientations in a DNA molecule.

iron-response element A regulatory sequence upstream of a gene involved in iron uptake or storage.

isoaccepting tRNAs Two or more tRNAs that are aminoacylated with the same amino acid.

isoform One of set of mRNAs that result from alternative splicing of the same transcript.

junk DNA One interpretation of the intergenic DNA content of a genome.

karyogram The entire chromosome complement of a cell, with each chromosome described in terms of its appearance at metaphase.

kilobase pair (kb) 1,000 base pairs.

kinase An enzyme that adds phosphate groups to other proteins, often as a means of activating those proteins.

kinetochore The part of the centromere to which spindle microtubules attach.

Kozak consensus The nucleotide sequence surrounding the initiation codon of a eukaryotic mRNA.

Ku70-Ku80 heterodimer A key component of the protein complex responsible for nonhomologous end-joining.

lagging strand The strand of the double helix that is copied in a discontinuous fashion during DNA replication.

late gene A bacteriophage gene that is expressed during the later stages of the infection cycle. Late genes usually code for proteins needed for the synthesis of new phage particles.

latent period The period between injection of a phage genome into a bacterial cell and the time at which cell lysis occurs.

leading strand The strand of the double helix that is copied in a continuous fashion during DNA replication.

lethal allele An allele that contains a lethal mutation and so causes the death of an organism at a very early stage in its development.

lethal mutation A mutation that results in the death of the cell or organism.

LINE (long interspersed nuclear element) A type of interspersed repeat, often with transposable activity.

linkage The physical association between two genes that are on the same chromosome.

linkage analysis The procedure used to assign map positions to genes by genetic crosses.

linkage group A group of genes that display linkage. With eukaryotes, a single linkage group usually corresponds to a single chromosome.

linker DNA The DNA that links nucleosomes.

linker histone A histone, such as H1, that is located outside the nucleosome core octamer.

liposome A lipid vesicle sometimes used to introduce DNA into an animal or plant cell.

lod score A statistical measure of linkage as revealed by pedigree analysis.

long noncoding RNA (lncRNA) Noncoding RNAs longer than 200 nucleotides in length.

long patch repair A nucleotide excision repair process of *E. coli* that results in the excision and resynthesis of up to 2 kb of DNA.

long-read sequencing A sequencing method that generates reads tens or thousands of kb in length.

long terminal repeat (LTR) Repeat sequences present at the ends of certain types of transposable element.

loss of function A mutation that reduces or abolishes a protein's activity.

lysogenic infection cycle The type of bacteriophage infection that involves integration of the phage genome into the host DNA molecule.

lytic infection cycle The type of bacteriophage infection that involves lysis of the host cell immediately after the initial infection, with no integration of the phage DNA molecule into the host genome.

m^{6}A writer An enzyme that methylates adenosines in an mRNA.

MADS box A DNA-binding domain found in several transcription factors involved in plant development.

maintenance methylation Addition of methyl groups to positions on newly synthesized DNA strands that correspond to the positions of methylation on the parent strand.

major groove The larger of the two grooves that spiral around the surface of the B-form of DNA.

MAP kinase system A signal transduction pathway.

map unit A unit used to describe the distance between two genes on a chromosome, now superseded by centimorgan.

marker A nucleotide that emits a radioactive, fluorescent or other detectable signal, used to label DNA or RNA molecules.

maternal effect gene A fruit fly gene that is expressed in the parent and whose mRNA is subsequently injected into the egg, after which it influences development of the embryo.

mediator A protein complex that forms a contact between various activators and the C-terminal domain of the largest subunit of RNA polymerase II.

megabase pair (Mb) 1,000 kb; 1,000,000 bp.

meiosis The series of events, involving two nuclear divisions, by which diploid nuclei are converted to haploid gametes.

messenger RNA (mRNA) The transcript of a protein-coding gene.

metacentric A chromosome whose centromere is positioned in the middle.

metaphase The period during the division of a cell when the chromosomes line up in the middle of the nuclear region.

metaphase chromosome A chromosome at the metaphase stage of cell division, when the chromatin takes on its most condensed structure and features such as the banding pattern can be visualized.

metastasis The process that results in cells from one cancer spreading to other places in the body and initiating new tumors.

methyl-CpG-binding protein (MeCP) A protein that binds to methylated CpG islands and may influence the acetylation of nearby histones.

MGMT (O^6-methylguanine-DNA methyltransferase) An enzyme involved in the direct repair of alkylation mutations.

microevolution The changes in allele frequencies that occur in a population over time.

microinjection A method of introducing new DNA into a cell by injecting it directly into the nucleus.

microRNA (miRNA) A class of short RNAs involved in regulation of gene expression in eukaryotes, which act by a pathway similar to RNA interference.

microsatellite A type of repeat sequence comprising tandem copies of, usually, di-, tri- or tetranucleotide repeat units. Also called a simple tandem repeat (STR).

minimal medium A medium that provides only the minimum number of different nutrients needed for the growth of a particular bacterium.

minisatellite A type of repetitive DNA comprising tandem copies of repeats that are a few tens of nucleotides in length.

minor groove The smaller of the two grooves that spiral around the surface of the B-form of DNA.

miscoding lesion A sequence error caused by chemical alteration of a nucleotide, resulting from the partial degradation of an ancient DNA molecule.

mismatch A position in a double-stranded DNA molecule where base pairing does not occur because the nucleotides are not complementary; in particular, a non-base-paired position resulting from an error in replication.

mismatch repair A DNA repair process that corrects mismatched nucleotide pairs by replacing the incorrect nucleotide in the daughter polynucleotide.

mitochondrial genome The genome present in the mitochondria of a eukaryotic cell.

mitosis The series of events that results in nuclear division.

mitotic spindle The microtubular arrangement that, during cell division, occupies the region of the cell previously taken up by the nucleus.

model organism An organism that is relatively easy to study and hence can be used to obtain information that is relevant to the biology of a second organism that is more difficult to study.

Modern Synthesis A framework that links together Mendelian genetics, natural selection and population variation into a single consistent explanation of evolutionary genetics.

molecular approach The approach to the study of genetics that takes as its starting point the function of DNA rather than the inheritance of genes.

molecular chaperone A protein that helps other proteins to fold.

molecular clock A device based on the inferred mutation rate that enables times to be assigned to gene duplication events and to the branch points in a phylogenetic tree.

molecular combing A technique for preparing restricted DNA molecules for optical mapping.

molecular phylogenetics A set of techniques that enable the evolutionary relationships between DNA sequences to be inferred by making comparisons between those sequences.

monogenic An inherited disorder that is caused by a mutation in a single gene.

monohybrid cross A sexual cross in which the inheritance of one pair of alleles is followed.

motor protein A protein involved in cell motility.

M phase The stage of the cell cycle when mitosis or meiosis occurs.

mRNA surveillance A RNA degradation process in eukaryotes.

multigene family A group of genes, clustered or dispersed, with related nucleotide sequences.

multiplex PCR PCR with two or more primer pairs in a single reaction.

multipoint cross A genetic cross in which the inheritance of three or more markers is followed.

mutagen A chemical or physical agent that can cause a mutation in a DNA molecule.

mutasome A protein complex that is constructed during the SOS response of *E. coli*.

mutation An alteration in the nucleotide sequence of a DNA molecule.

N-linked glycosylation The attachment of sugar units to an asparagine residue in a polypeptide.

nanopore sequencing A long-read DNA sequencing method.

natural selection The preservation of favorable alleles and the rejection of injurious ones.

neutral theory The theory that genetic drift is the most important factor in determining the allele frequencies within a population.

N- or NH_2 terminus The end of a polypeptide that has a free amino group.

noncoding RNA An RNA molecule that does not code for a protein.

nonhomologous end-joining (NHEJ) A process for repairing a double-strand break in a DNA molecule.

nonpolar A hydrophobic (water-hating) chemical group.

nonsense mutation An alteration in a nucleotide sequence that changes a triplet coding for an amino acid into a termination codon.

nonsynonymous mutation A mutation that converts a codon for one amino acid into a codon for a different amino acid.

normal distribution The distribution pattern commonly called a bell-shaped curve.

nuclear genome The DNA molecules present in the nucleus of a eukaryotic cell.

nuclease protection A technique that uses nuclease digestion to determine the positions of proteins on DNA or RNA molecules.

nucleic acid hybridization Formation of a double-stranded hybrid by base pairing between complementary polynucleotides.

nucleoid The DNA-containing region of a prokaryotic cell.

nucleoid-associated protein A protein component of a bacterial nucleoid.

nucleolus The region of the eukaryotic nucleus in which rRNA transcription occurs.

nucleoplasm A general name for the complex mixture of molecules that comprises the ground substance of the nucleus of a living cell.

nucleoside A purine or pyrimidine base attached to a five-carbon sugar.

nucleosome remodeling A change in the positioning of a nucleosome, associated with a change in access to the DNA to which the nucleosome is attached.

nucleosome The complex of histones and DNA that is the basic structural unit in chromatin.

nucleotide A purine or pyrimidine base attached to a five-carbon sugar, to which a mono-, di- or triphosphate is also attached. The monomeric unit of DNA and RNA.

nucleotide excision A repair process that corrects various types of DNA damage by excising and resynthesizing a region of a polynucleotide.

octamer sequence A regulatory sequence that controls the basal rate of transcription of a gene.

off-target editing Editing of sites other than the target site during a gene editing experiment.

Okazaki fragment One of the short segments of RNA-primed DNA synthesized during replication of the lagging strand of the double helix.

oligonucleotide A short synthetic single-stranded DNA molecule.

oligonucleotide hybridization analysis The use of an oligonucleotide as a hybridization probe.

O-linked glycosylation The attachment of sugar units to a serine or threonine in a polypeptide.

oncogen An environmental agent that causes tumor formation.

oncogene A gene that when carried by a retrovirus is able to cause cell transformation.

one-step growth curve A single infection cycle for a lytic bacteriophage.

open promoter complex A structure formed during assembly of the transcription initiation complex.

open reading frame (ORF) A series of codons starting with an initiation codon and ending with a termination codon. The part of a protein-coding gene that is translated into protein.

operator The nucleotide sequence to which a repressor protein binds to prevent the transcription of a gene or operon.

operon A set of adjacent genes in a bacterial genome, transcribed from a single promoter and subject to the same regulatory regime.

optical mapping A technique for the direct visual examination of restricted DNA molecules.

optional loop A structural component of tRNAs.

ORF scanning Examination of a DNA sequence for open reading frames in order to locate the genes.

origin licensing The construction of prereplication complexes on replication origins.

origin of replication A site on a DNA molecule where replication initiates.

origin recognition complex (ORC) A set of proteins that bind to the origin recognition sequence.

origin recognition sequence A component of a eukaryotic origin of replication.

overdominance The situation in which the heterozygote is the most successful genotype in a population.

overlapping genes Two genes whose coding regions overlap.

paired-end reads Mini-sequences from the two ends of a single DNA fragment.

pair rule gene Developmental genes that establish the basic segmentation pattern of the fruit fly embryo.

pan-genome concept The concept that views a bacterial genome as a combination of a core and accessory genome.

ParABS A bacterial partitioning system.

pararetrovirus A viral retroelement whose capsid contains the DNA version of the viral genome.

parental The genotype possessed by one or both of the parents in a genetic cross.

partial linkage The type of linkage usually displayed by a pair of genetic and/or physical markers on the same chromosome, the markers not always being inherited together because of the possibility of recombination between them.

partitioning The process by which copies of the bacterial chromosome are passed to daughter cells when the bacterium divides.

pedigree analysis The process used to map human genes by examining inheritance patterns in families.

pentose A sugar containing five carbon atoms.

peptide bond The chemical link between adjacent amino acids in a polypeptide.

peptide nucleic acid (PNA) A polynucleotide analog in which the sugar-phosphate backbone is replaced by amide bonds.

peptidyl or P site The site in the ribosome occupied by the tRNA attached to the growing polypeptide during translation.

peptidyl transferase The enzyme activity that synthesizes peptide bonds during translation.

pericentromeric The region of the chromosome either side of the centromere.

personal DNA data Information about an individual's genome, including DNA profiles, complete genome sequences and catalogs of single-nucleotide polymorphisms.

personal medicine The use of an individual's genome sequence to diagnose their risk of developing a disease and to plan therapies and treatment regimes that are compatible with their personal genetic characteristics.

phage An abbreviation of 'bacteriophage'.

pharming Genetic modification of a farm animal so that the animal synthesizes a recombinant pharmaceutical protein, often in its milk.

phenotype The observable characteristics displayed by a cell or organism.

Philadelphia chromosome An abnormal chromosome resulting from a translocation between human chromosomes 9 and 22, a common cause of chronic myeloid leukemia.

phosphodiesterase A type of enzyme that can break phosphodiester bonds.

phosphodiester bond The chemical link between adjacent nucleotides in a polynucleotide.

photo 51 An X-ray diffraction pattern obtained by Franklin and Gosling in 1952, which was important in elucidation of the double helix structure of DNA.

photoproduct A modified nucleotide resulting from the treatment of DNA with ultraviolet radiation.

photoreactivation A DNA repair process in which cyclobutyl dimers and (6–4) photoproducts are corrected by a light-activated enzyme.

physical mapping Location of a gene or other sequence feature by direct examination of a DNA molecule.

pilus A structure involved in bringing a pair of bacteria together during conjugation, and possibly the tube through which DNA is transferred.

plasmid A usually circular piece of DNA often found in bacteria and some other types of cell.

plectonemic Refers to a helix whose strands can only be separated by unwinding.

pleiotropy The situation that arises when a mutation in a single gene affects more than one trait.

point mutation A mutation that results from a single nucleotide change in a DNA molecule.

polar A hydrophilic (water-loving) chemical group.

polyadenylation site The position within a eukaryotic mRNA to which the poly(A) tail is attached.

poly(A) polymerase The enzyme that attaches a poly(A) tail to the 3' end of a eukaryotic mRNA.

poly(A) tail A series of A nucleotides attached to the 3' end of a eukaryotic mRNA.

polymer A compound made up of a long chain of identical or similar units.

polymerase chain reaction (PCR) A technique that results in exponential amplification of a selected region of a DNA molecule.

polymorphism Refers to a gene or other variable sequence that is represented by several different alleles or haplotypes in the population as a whole.

polynucleotide A single-stranded DNA or RNA molecule.

polynucleotide phosphorylase An enzyme that degrades RNA in the cell, but which, in the test tube, will catalyze the synthesis of RNA.

polypeptide A polymer of amino acids.

polyprotein A translation product consisting of a series of linked proteins that are processed by proteolytic cleavage to release the mature proteins.

polypyrimidine tract A pyrimidine-rich region near the 3' end of a GU-AG intron.

polysome An mRNA molecule that is being translated by more than one ribosome at the same time.

positional information Information that enables a cell to follow the differentiation pathway that is appropriate for its particular location in a developing organism.

post-translational chemical modification Chemical modification of a protein that occurs after that protein has been synthesized by translation of an mRNA.

postzygotic isolation The situation in which members of two species can interbreed but the offspring are sterile.

pre-initiation complex The structure comprising the small subunit of the ribosome, the initiator tRNA plus ancillary factors that forms the initial association with the mRNA during protein synthesis in eukaryotes. Also the structure that forms at the core promoter of a gene transcribed by RNA polymerase II.

prepriming complex A protein complex, initially comprising six copies of DnaB and six copies of DnaC, which initiates the construction of a replication fork on a DNA molecule.

pre-replication complex (pre-RC) A protein complex that is constructed at a eukaryotic origin of replication and enables the initiation of replication to occur.

pre-rRNA The primary transcript of a gene or group of genes specifying rRNA molecules.

prezygotic isolation The situation in which interbreeding between members of two species produces no offspring.

Pribnow box A component of the bacterial promoter.

primary structure The sequence of amino acids in a polypeptide.

primary transcript The initial transcript of a gene prior to splicing or other modification events.

primase The RNA polymerase enzyme that synthesizes RNA primers during bacterial DNA replication.

primosome A protein complex involved in DNA replication.

prion An unusual infectious agent that consists purely of protein.

processing Events that change the chemical or physical structure of an RNA or protein molecule.

proflavin A chemical mutagen, one of the acridine dyes, that is frequently used to induce mutations for experimental purposes.

programmable nuclease A nuclease that can be directed to a specific site in a genome.

prokaryote An organism whose cells lack a distinct nucleus.

proliferating-cell nuclear antigen (PCNA) An accessory protein involved in DNA replication in eukaryotes.

prometaphase The period during the division of a cell when the chromosomes begin to attach to microtubules.

promiscuous DNA DNA that has been transferred from one organelle genome to another.

promoter The nucleotide sequence, upstream of a gene, to which RNA polymerase binds to initiate transcription.

proofreading The 3'→5' exonuclease activity possessed by some DNA polymerases, enabling the enzyme to replace a misincorporated nucleotide.

prophage The integrated form of the genome of a lysogenic bacteriophage.

prophase The period during the division of a cell when the chromosomes condense.

protease An enzyme that degrades protein.

proteasome A multisubunit protein structure that is involved in the degradation of other proteins.

protein The polymeric compound made of amino acid monomers.

proteome The collection of proteins in a living cell.

protomer One of the individual polypeptide subunits that combine to make the protein coat of a virus.

proto-oncogene The normal cellular version of an oncogene.

pseudoautosomal region A segment of the X and Y chromosomes where the sequences of the two chromosomes are very similar.

pseudogene An inactivated and hence nonfunctional copy of a gene.

punctuation codon A codon that specifies either the start or the end of a gene.

Punnett square A tabular analysis for predicting the genotypes of the progeny resulting from a genetic cross.

pure-breeding Refers to a population of homozygous plants that, when self-fertilized, give rise to progeny whose phenotypes are identical to those of the parents.

purine One of the two types of nitrogenous base found in nucleotides.

pyrimidine One of the two types of nitrogenous base found in nucleotides.

quantitative trait A phenotype, such as height, that has a continuous distribution pattern within a population, which is typically determined by the combined effects of several genes.

quaternary structure The structure resulting from the association of two or more polypeptides.

radiolabeling Attachment of a radioactive marker to a DNA or RNA molecule.

reading frame A series of triplet codons in a DNA sequence.

readthrough mutation A mutation that changes a termination codon into a codon specifying an amino acid and hence results in readthrough of the termination codon.

RecBCD enzyme An enzyme complex involved in homologous recombination in *E. coli*.

recessive The allele that is not expressed in a heterozygote.

reciprocal strand exchange The exchange of DNA between two double-stranded molecules, occurring as a result of recombination, such that the end of one molecule is exchanged for the end of the other molecule.

recognition helix An α-helix in a DNA-binding protein, one that is responsible for recognition of the target nucleotide sequence.

recombinant A progeny member that possesses neither of the combinations of alleles displayed by the parents.

recombinant DNA molecule A DNA molecule created in the test tube by ligating pieces of DNA that are not normally joined together.

recombinant protein A protein synthesized in a recombinant cell as the result of the expression of a cloned gene.

recombination The outcome of crossing over between pairs of homologous chromosomes, and the physical process involving breakage and reunion of DNA molecules that underlies crossing over.

recombination frequency The proportion of recombinant progeny arising from a genetic cross.

recombination hotspot A region of a chromosome where crossovers occur at a higher frequency than the average for the chromosome as a whole.

reference sequence An existing genome sequence that is used to aid assembly of the reads obtained by sequencing of a related genome.

regulatory gene A gene that codes for a protein, such as a repressor, involved in regulation of the expression of other genes.

regulatory protein A protein that controls one or more cellular activities.

relaxed Refers to a plasmid whose replication is not linked to replication of the host genome, and which therefore can exist as multiple copies within the cell.

release factor A protein that has an ancillary role during the termination of translation.

renaturation The return of a denatured molecule to its natural state.

repetitive DNA A DNA sequence that is repeated two or more times in a DNA molecule or genome.

replication fork The region of a double-stranded DNA molecule that is being opened up to enable DNA replication to occur.

replication fork barrier A sequence in a eukaryotic genome that impedes the progress of a replication form and might be involved in termination of replication at a few sites.

replication protein A (RPA) The main single-strand binding protein involved in the replication of eukaryotic DNA.

replication slippage An error in replication that leads to an increase or decrease in the number of repeat units in a tandem repeat such as a microsatellite.

replisome A complex of proteins involved in DNA replication.

repressible operon An operon that is switched off by the repressor working in conjunction with a co-repressor molecule.

repressor A regulatory protein involved in the control of expression of a gene or operon in a bacterium.

resistance or R plasmid A plasmid that carries genes conferring resistance to an antibiotic or other toxic agent.

resolution Separation of a pair of recombining double-stranded DNA molecules.

restriction endonuclease An enzyme that cuts DNA molecules at a limited number of specific nucleotide sequences.

restriction fragment length polymorphism (RFLP) A restriction fragment whose length is variable because of the presence of a polymorphic restriction site.

restriction site The recognition sequence for a restriction endonuclease.

retroposon A retroelement that does not have LTRs.

retrotransposition Transposition via an RNA intermediate.

retrovirus A viral retroelement whose capsid contains the RNA version of the genome.

reverse transcriptase A polymerase that synthesizes DNA on an RNA template.

reversible terminator sequencing A DNA sequencing method in which the sequence is read by detection of the fluorescent label attached to each nucleotide that is added to a growing polynucleotide.

R group The variable group in the structure of an amino acid.

Rho-dependent terminator A position in bacterial DNA where termination of transcription occurs with the involvement of the protein called Rho.

ribonuclease An enzyme that degrades RNA.

ribosomal protein One of the protein components of a ribosome.

ribosomal RNA (rRNA) The RNA molecules that are components of ribosomes.

ribosome One of the protein-RNA assemblies on which translation occurs.

ribosome binding site The nucleotide sequence that acts as the attachment site for the small subunit of the ribosome during the initiation of translation in bacteria.

ribosome inactivating protein (RIP) A protein that blocks the translation stage of gene expression by cutting one of the ribosomal RNA molecules into two pieces.

ribosome recycling factor (RRF) A protein responsible for disassembly of the ribosome at the end of protein synthesis in bacteria.

ribozyme An RNA molecule that has catalytic activity.

RNA Ribonucleic acid, one of the two forms of nucleic acid in living cells; the genetic material for some viruses.

RNA-dependent DNA polymerase An enzyme that makes a DNA copy of an RNA template; a reverse transcriptase.

RNA-dependent RNA polymerase An enzyme that makes an RNA copy of an RNA template.

RNA editing A process by which nucleotides not encoded by a gene are introduced at specific positions in an RNA molecule after transcription.

RNA-induced silencing complex (RISC) A complex of proteins that cleaves and hence silences an mRNA as part of the RNA interference pathway.

RNA interference (RNAi) An RNA degradation process in eukaryotes.

RNA polymerase An enzyme that synthesizes RNA on a DNA or RNA template.

RNA polymerase I The eukaryotic RNA polymerase that transcribes ribosomal RNA genes.

RNA polymerase II The eukaryotic RNA polymerase that transcribes protein-coding and snRNA genes.

RNA polymerase III The eukaryotic RNA polymerase that transcribes tRNA and other short genes.

RNA-seq Short- or long-read sequencing of RNA.

RNA silencing An RNA degradation process in eukaryotes.

RNA splicing The process by which introns are removed from RNA molecules.

RNA transposon A transposable element that transposes via an RNA intermediate.

rolling-circle replication A replication process that involves continual synthesis of a polynucleotide that is 'rolled off' a circular template molecule.

satellite DNA Repetitive DNA that forms a satellite band in a density gradient.

satellite RNA An infective RNA molecule some 320–400 nucleotides in length that does not encode its own capsid proteins, instead moving from cell to cell within the capsid of a helper virus.

scanning A system used during the initiation of eukaryotic translation, in which the pre-initiation complex attaches to the 5′-terminal cap structure of the mRNA and then scans along the molecule until it reaches an initiation codon.

secondary structure The conformations, such as α helix and β sheet, taken on by a polypeptide.

second-site reversion A second mutation that reverses the effect of a previous mutation in the same gene but without restoring the original nucleotide sequence.

sedimentation coefficient The value used to express the velocity at which a molecule or structure sediments when centrifuged in a dense solution.

segmented genome A virus genome that is split into two or more DNA or RNA molecules.

segment polarity gene A developmental gene that provides greater definition to the segmentation pattern of the fruit fly embryo established by the action of the pair rule genes.

segregation The separation of homologous chromosomes, or members of allele pairs, into different gametes during meiosis.

selectable marker A gene carried by a vector and conferring a recognizable characteristic on a cell containing the vector or a recombinant DNA molecule derived from the vector.

selection A process that brings about the preservation of favorable alleles and the rejection of injurious ones.

selection coefficient A measure of the relative fitness of a genotype.

selective medium A medium that permits growth only of cells with certain genetic features.

selective sweep A disproportionate decrease in the genetic diversity of a population caused by the loss of variability of genes adjacent to one that is under selection.

self-fertilization Fertilization of a female gamete with a male gamete derived from the same individual.

semiconservative replication The mode of DNA replication in which each daughter double helix is made up of one polynucleotide from the parent and one newly synthesized polynucleotide.

sequence assembly Assembly of the many short reads obtained by a sequencing experiment into a contiguous DNA sequence.

sequence depth or coverage The average number of reads that cover each nucleotide position in a DNA sequence assembly.

sequence read A single sequence from the output of a DNA sequencing experiment.

sequencing library A set of DNA fragments that have been immobilized on a solid support in such a way that multiple sequencing reactions can be carried out side by side in an massively parallel array format.

serotype A bacterial strain distinguished because of its immunological properties.

sex cell A reproductive cell; a cell that divides by meiosis.

shelterin A structure, comprising telomere binding proteins, that protects the telomeres from degradation by nuclease enzymes and mediates the enzymatic activity that maintains the length of each telomere during DNA replication.

Shine–Dalgarno sequence The ribosome binding site upstream of an *E. coli* gene.

short interfering RNA (siRNA) An intermediate in the RNA interference pathway.

short noncoding RNA (sncRNA) Noncoding RNAs less than 200 nucleotides in length.

short patch repair A nucleotide excision repair process of *E. coli* that results in the excision and resynthesis of about 12 nucleotides of DNA.

short-read sequencing A sequencing method that generates millions of sequence reads <600 bp in length.

short tandem repeat (STR) A type of simple sequence length polymorphism comprising tandem copies of, usually, di-, tri- or tetranucleotide repeat units. Also called a microsatellite.

shotgun method A genome sequencing strategy in which the molecules to be sequenced are randomly broken into fragments which are then sequenced individually.

signal peptide A short sequence at the N terminus of some proteins that directs the protein across a membrane.

signal transduction Control of cellular activity, including gene expression, via a cell surface receptor that responds to an external signal.

silent mutation A change in a DNA sequence that has no effect on the expression of any gene or gene product.

simple multigene family A multigene family in which all of the genes are the same.

simple sequence length polymorphism (SSLP) Another name for a short tandem repeat.

simple sequence repeat (SSR) Another name for a short tandem repeat.

SINE (short interspersed nuclear element) A type of interspersed repeat, typified by the Alu sequences found in the human genome.

single-molecule real-time (SMRT) sequencing A DNA sequencing method which uses an advanced optical system to observe the addition of individual nucleotides to a growing polynucleotide.

single-nucleotide polymorphism (SNP) A point mutation that is carried by some individuals of a population.

single-strand binding protein (SSB) One of the proteins that attach to single-stranded DNA in the region of the replication fork, preventing base pairs from forming between the two parent strands before they have been copied.

site-directed mutagenesis Techniques used to produce a specified mutation at a predetermined position in a DNA molecule.

small interfering RNA (siRNA) A type of short eukaryotic RNA molecule involved in the control of gene expression.

small nuclear ribonucleoprotein (snRNP) A structure involved in RNA splicing and other RNA processing events, comprising one or two snRNA molecules complexed with proteins.

small nuclear RNA (snRNA) A type of short eukaryotic RNA molecule involved in RNA splicing and other RNA processing events.

small nucleolar RNA (snoRNA) A type of short eukaryotic RNA molecule involved in the chemical modification of rRNA.

somatic cell A nonreproductive cell; a cell that divides by mitosis.

somatic cell nuclear transfer (SCNT) A method that involves injection of DNA into the nucleus of a somatic cell followed by transfer of the nucleus to an oocyte whose own nucleus has been removed.

somatic cell therapy A type of gene therapy in which the correct version of a gene is introduced into a somatic cell.

sonication A procedure that uses ultrasound to cause random breaks in DNA molecules.

SOS response A series of biochemical changes that occur in *E. coli* in response to damage to the genome and other stimuli.

speciation The process that gives rise to new species.

S phase The stage of the cell cycle at which DNA synthesis occurs.

spliceosome The protein-RNA complex involved in RNA splicing.

splicing The removal of introns from the primary transcript of a discontinuous gene.

splicing code A hypothetical code that would explain the impact on a splicing pathway of the various interactions that can occur between enhancers, silencers and their binding proteins.

splicing enhancer A nucleotide sequence that has a positive regulatory role during splicing of GU–AG introns.

splicing silencer A nucleotide sequence that has a negative regulatory role during splicing of GU–AG introns.

SR protein A protein that has a role in splice-site selection during the RNA splicing.

stabilizing selection Selection against the two extremes of a phenotype.

STAT (signal transducer and activator of transcription) A type of protein that responds to the binding of an extracellular signaling compound to a cell surface receptor by activating a transcription factor.

steroid hormone A type of extracellular signaling compound.

steroid receptor A protein that binds a steroid hormone after the latter has entered the cell and then moves to the nucleus where it activates the expression of target genes.

stochastic effect A deviation from statistical perfection that occurs with real events in the real world.

stringent Refers to a plasmid with a low copy number of perhaps just one or two per cell.

strong promoter A promoter that directs a relatively large number of productive initiations per unit time.

structural gene A gene that codes for an RNA molecule or protein other than a regulatory protein.

structural protein A protein that has a structural role, for example, by giving rigidity to the framework of an organism.

structural variant Variations that result in the genomes of two individuals of the same species differing in the numbers and locations of repeat units, the presence or absence of certain sequences of 50–2,000 bp in length and the positioning of some sequences.

stutter A PCR artifact where the authentic product from amplification of an STR is accompanied by products differing in length by one or more repeat units.

submetacentric A chromosome whose centromere is positioned a little off-center.

substitution A mutation that escapes the repair processes and results in a permanent change in a DNA sequence.

suicide gene therapy A type of gene therapy aimed at cancerous cells, involving the introduction of a gene that selectively kills the cell or promotes its destruction by drugs administered in a conventional fashion.

supercoiling A conformational state in which a double helix is overwound or underwound so that superhelical coiling occurs.

suppression Refers to a mutation in one gene that reverses the effect of a mutation in a second gene.

sympatric speciation Speciation that occurs when non-interbreeding populations arise within a single geographical region.

syncytium A cell-like structure comprising a mass of cytoplasm and many nuclei.

synonymous mutation A mutation that changes a codon into a second codon that specifies the same amino acid.

TΨC arm Part of the structure of a tRNA molecule.

TAF- and initiator-dependent cofactor (TIC) A type of protein involved in the initiation of transcription by RNA polymerase II.

tandem array A set of identical or very similar genes that are arranged one after the other in a group.

tandemly repeated DNA Direct repeats that are adjacent to each other.

TATA-binding protein (TBP) A component of the transcription factor TFIID that makes contact with the TATA box of the RNA polymerase II promoter.

TATA box A component of the RNA polymerase II core promoter.

tautomers Structural isomers that are in dynamic equilibrium.

TBP-associated factor (TAF) One of several components of the transcription factor TFIID, with ancillary roles in the recognition of the TATA box.

telocentric A chromosome whose centromere is positioned very close to one end.

telomerase The enzyme that maintains the ends of eukaryotic chromosomes by synthesizing telomeric repeat sequences.

telomere The end of a eukaryotic chromosome.

telophase The period during the division of a cell when the new nuclear membranes are formed.

temperate bacteriophage A bacteriophage that is able to follow a lysogenic mode of infection.

template-dependent DNA synthesis Synthesis of a DNA molecule on a DNA or RNA template.

teratogen An environmental agent that causes developmental abnormalities.

termination codon One of the three codons that mark the position where translation of an mRNA should stop.

terminator sequence One of several sequences on a bacterial genome involved in the termination of genome replication.

terminator technology A recombinant DNA process that results in the synthesis of the ribosome inactivating protein in plant embryos, used to prevent GM crops from producing seeds.

terminator utilization substance (Tus) The protein that binds to a bacterial terminator sequence and mediates the termination of genome replication.

tertiary structure The structure resulting from folding the secondary structural units of a polypeptide.

test cross A genetic cross between a double heterozygote and a double homozygote that carries recessive alleles of the two genes being studied.

thymine One of the pyrimidine bases found in DNA.

transcription The synthesis of an RNA copy of a gene.

transcription bubble The non-base-paired region of the double helix, maintained by RNA polymerase, within which transcription occurs.

transcription factor IID (TFIID) The protein complex, including the TATA-binding protein, that recognizes the core promoter of a gene transcribed by RNA polymerase II.

transcription initiation complex The complex of proteins, including the enzyme that performs transcription, that is assembled on the DNA adjacent to a eukaryotic gene that is going to be expressed.

transcriptome The entire RNA content of a cell.

transduction Transfer of bacterial genes from one cell to another by packaging in a phage particle.

transfer RNA (tRNA) A small RNA molecule that acts as an adapter during translation and is responsible for decoding the genetic code.

transformation The acquisition by a cell of new genes by the uptake of naked DNA.

transforming principle The compound, now known to be DNA, responsible for the transformation of an avirulent *Streptococcus pneumoniae* bacterium into a virulent form.

transgenic Refers to an organism that carries a cloned gene.

transition A point mutation that replaces a purine with another purine, or a pyrimidine with another pyrimidine.

translation The synthesis of a polypeptide whose amino acid sequence is determined by the nucleotide sequence of an mRNA in accordance with the rules of the genetic code.

translocation The movement of a ribosome along an mRNA molecule during translation.

transversion A point mutation that involves the replacement of a purine by a pyrimidine, or vice versa.

trinucleotide repeat expansion disorder A disorder that results from the expansion of an array of trinucleotide repeats in or near a gene.

triplet binding assay A technique for determining the coding specificity of a triplet of nucleotides.

triplex helix A DNA structure comprising three polynucleotides.

trisomy The presence of three copies of a homologous chromosome in a nucleus that is otherwise diploid.

tRNA nucleotidyltransferase The enzyme responsible for the post-transcriptional attachment of the triplet 5'-CCA-3' to the 3' end of a tRNA molecule.

***trp* RNA-binding attenuation protein (TRAP)** A protein involved in the attenuation regulation of some operons in bacteria such as *Bacillus subtilis*.

tumor suppressor A gene coding for a tumor suppressor protein, the inactivation of which can lead to tumor formation.

turnover Complete or partial degradation of a set of RNA or protein molecules.

type 0 cap The basic cap structure, consisting of 7-methylguanosine attached to the 5' end of an mRNA.

type 1 cap A cap structure comprising the basic 5'-terminal cap plus an additional methylation of the ribose of the second nucleotide.

type 2 cap A cap structure comprising the basic 5'-terminal cap plus methylation of the riboses of the second and third nucleotides.

type I topoisomerase A topoisomerase that makes a single-strand break in a double-stranded DNA molecule.

type II topoisomerase A topoisomerase that makes a double-strand break in a double-stranded DNA molecule.

ubiquitin A 76-amino-acid protein that, when attached to a second protein, acts as a tag directing that protein for degradation.

UK biobank A collection of biological material, including genome sequences, donated by volunteers,

uniparental The inheritance pattern displayed by mitochondrial DNA, in which only the mother contributes her genome to the offspring.

unit factor Mendel's term for a gene.

unit of biological information The concept of the gene as containing information for a biological process.

unit of inheritance The concept of the gene as a controlling factor in inheritance.

upstream Toward the 5' end of a polynucleotide.

upstream control element A component of an RNA polymerase I promoter.

UvrABC endonuclease A multienzyme complex involved in the shortpatch repair process of *E. coli*.

variable loop Part of the structure of a tRNA molecule.

variable number tandem repeats Another name for minisatellites and microsatellites.

viral retroelement A virus whose genome replication process involves reverse transcription.

viroid An infectious RNA molecule 240–375 nucleotides in length that contains no genes and never becomes encapsidated, spreading from cell to cell as naked RNA.

virulence plasmid A plasmid whose genes confer pathogenicity on the host bacterium.

virulent bacteriophage A bacteriophage that follows the lytic mode of infection.

virus An infective particle, composed of protein and nucleic acid, that must parasitize a host cell to replicate.

virusoid An infectious RNA molecule some 320–400 nucleotides in length that does not encode its own capsid proteins, instead moving from cell to cell within the capsid of a helper virus.

wild-type A gene, cell or organism that displays the typical phenotype and/or genotype for the species and is therefore adopted as a standard.

wobble hypothesis The process by which a single tRNA can decode more than one codon.

X inactivation Inactivation by methylation of most of the genes on one copy of the X chromosome in a female nucleus.

X-ray diffraction The diffraction of X-rays through a crystal.

Z-DNA A conformation of DNA in which the two polynucleotides are wound into a left-handed helix.

zero-mode waveguide A nanostructure that enables individual molecules to be observed.

zoo blotting A technique that uses hybridization probing to determine if a gene present in one species is also present in the genomes of other species.

INDEX

Note: **Bold** page numbers refer to tables; *italics* page numbers refer to figures.

α-globin 47
α helix 113, 118–119
α-LPH *143*
α-MSH *143*
α subunits 57
β-carotene 463–464, *464*
β-ENDO *143*
β-galactosidase 154, 456
β-galactoside transacetylase 154
β-globin 47
β-globin gene 310
β-globin proteins 11–12, *12*
β-MSH *143*
β-*N*-glycosidic bond 24, 29, *29*, 201, *202*
β sheet 113, 114, *114*
β-thalassemia 425
β turn 115
β-type globins 47
γ complex 185
δ-endotoxins 463
Δ12-desaturase **464**
σ subunit 61–62, 163–164
−10 box 58, 62
−35 box 58, 62
1-aminocyclopropane-1-carboxylic acid (ACC) 465
1-D-*myo*-inositol 3-phosphate synthase **465**
2-aminopurine 200
2′-deoxyribose 25
2′-deoxyadenosine 5′-triphosphate 25
2′deoxycytidine 5′-triphosphate 25
2′-deoxyguanosine 5′-triphosphate 25
2′-deoxythymidine 5′-triphosphate 25
2′-*O*-methylation 103, *103*
3′→5′ exonuclease 180
3′ splice sites 83
3″–OH terminus 26
3′-terminus 26
5′-P terminus 26
5′-terminus 26
5′→3′ exonuclease 180
5′ splice site 83
(6–4) lesion 201, 204
(6–4) photoproduct photolyase 204
7SL RNA 394
10 nm fiber 258, *258*, 260, *260*
    globular chromatin structures 259, *260*
30 nm fiber 260, *260*

ABC model 384, *384*
Abdominal B gene 381
ABL gene 420
ABO blood groups 291, *291*
acceptor
    arm 97
    site 83
accessory genome 242
ACC synthase mRNA 465, *465*
Acinonyx jubatus 318
acridine dye 117
acrocentric 261
ACTH *143*
acute transforming virus 419
acylation 141
acyl carrier protein thioesterase **464**
Ada enzyme 204
adaptor 346
adenine 24
adenine deaminase 466, *466*
adeno-associated virus (AAV) 424
adenoma 422, *422*
adenosine deaminase acting on RNA (ADAR) 88
adenovirus **228**, 424
adenylate cyclase 158
admixture 445
adrenocorticotropic hormone *143*
A-form of DNA 27, 29
agamous gene 384
Agrobacterium strain CP4 462
Agrobacterium tumefaciens 238
AIDS 219, 229
Alagille syndrome 412, *412*
alanine **112**, *113*
alcohol oxidase promoter 458
aldolase genes 389
alkylating agents 201
alleles
    bell-shaped curve 293, *293*, 296, *296*
    carotenoid pigment synthesis 295, *296*
    codominance 290–291, *291*
    complementary gene action 294–295, *295*
    creation by mutation 301, 306
    definition 43–44
    epistasis 294, *294*
    Hemerocallis genes 292, 292–293
    incomplete dominance 288, 288–289

    interacting synthesis genes 292–293, **293**
    lethal 289–290, *290*
    mesocarp 295, *295*
    multiple 44–46
    in populations 302–321
    recessive genes 286–288, *287*
    round pea phenotype 287, *287*
    white-flowered sweet peas 294, *294*
allelic dropout 436
allolactose 156, 454
allopatric speciation 320
alphoid DNA 261, 394
alternative exons 86, *86*
alternative polyadenylation 70
alternative promoters 60–61, *61*
alternative splice site selection 86, *86*
alternative splicing 75, 388
alternative σ subunits 164
Alu 394
amelogenin gene 435, *435*
amino acid 20, 111
amino acid breakdown 397
amino acid sequences 357, *357*
aminoacyl 134
aminoacylation 122–125, *123*, *124*
aminoacyl site 134
aminoacyl-tRNA synthetase 122–123
amino-adenine *197*
amino terminus 112, 113
amphibians 46
Anabaena 53
anaphase 266
ancient DNA 13
ANT-C 381
Antennapedia complex 381, *381*
Antennapedia gene *381*
Antennapedia mutation *377*, 380
anti-adenosine 29
antibodies 39
anticodon 124
anticodon arm 97
antigens 367–368
antiparallel 28
Antirrhinum 384
antisense RNA 43, 425, *425*, 464–465
antitermination 164–166, *165–166*
antithrombin III 468
APC gene 422

AP endonuclease 204
apetala genes 384
apocytochrome b gene *398*
apoptosis 420, 421, *421*
applied science 1
AP (apurinic/apyrimidinic) site 203
Arabidopsis thaliana 384, **388, 399**
arabinose operon 158
archaea 54–55, 250
archaeogenetics 13
Ardipithecus ramidis 397
AR gene **409**
arginine **112,** *113*
A site 134
asparagine **112,** *113*
aspartic acid **112,** *113*
ataxia telangiectasia 209
ATPase subunit genes *398*
attenuation 166–168
ATX gene 209
Australopithecus africanus 397
autosome 256

Bacillus licheniformis **464**
Bacillus species 248
Bacillus subtilis 164, 168, *250*
Bacillus thuringiensis 463
back mutation 212
backtracking 66
bacteria 5, 54, 336, *336; see also individual species*
bacterial chromosome 237, 238
bacterial ribosomes 94, *95, 96, 97*
bacterial transformation 21, *21*
bacteriophage 5
    φ **221**
    φX174 220
    features 222
    genes *22,* 22–23
    genomes 222–223
    lytic infection cycle 222–224
    M13 220, 226–227
    MS2 220
    PM2 220
    SPO1 **221**
    structures *220,* 220–222
    T2 220
    T4 117, 220–223
    T6 220
    T7 **221**
    transduction 249
    unusual life cycles 226–227
bacteriophage λ
    antitermination 165
    Cro protein 118–119
    features **221**
    lysogenic infection cycle 224–227
barley 314

Barr body 374, 375
basal promoter element 160
basal rate of transcript initiation 59
basal rate of transcription 58
base analog 198, 200
base editing 466, *466*
base editor 466
base excision 204–205
base excision repair 203, 204
baseless site 203
base pairing 28, 29
base ratio 27
Basic Local Alignment Search Tool (BLAST) 357
batch culture 451
BCR-ABL 420
bell-shaped curve 293, *293,* 296, *296*
B-form of DNA 27, 29
Bicoid protein 379, *379*
bioethics 11
biolistics 462
biological information 1, 6–8, 37–46
biotechnology 10, 451
biotinylation **141**
Biston betularia 307
Bithorax complex 381, *381*
bivalent 268–271, *270–272*
blastocyst 482
blood clotting 39
Bloom syndrome 209
bone 13, 39, 77
bone marrow 77, *424,* 425
Borrelia burgdorferi 238
bottleneck 317–319, *317–319*
bovine spongiform encephalopathy (BSE) 232
branch migration 272, *272*
Brassica oleracea 474
brazil nuts **464**
BRCA1 gene 417, *417*
breakpoint cluster region 420
break repair 203
breast cancer 78, 209, **405,** 420, *420*
Brenner, Sydney 376
5-bromouracil (5-bU) 200, *200*
Bronze Age skeleton 440–441, *441*
buoyant density 175, 394
BX-C 381
bystander mutation 466

CAAT box 160
Caenorhabditis elegans 12, *12*
    development 375, *376*
    model organism 376–377
    number of genes 387
California bay laurel tree **464**
California State Route 128 13, *13*

cAMP 158
cancer 77–78, 229–230, 419–422, 425–426
Candida rugosa **121**
candidate gene 417
Canis lupus 54
CAP 157
cap binding complex 131–133
capillary gel electrophoresis 415
capping 66
capsid 220
cap structure 66, 80, 131–133
capsule 20
carbonaria peppered moth 307
carboxylase enzymes **141**
carboxyl terminus 113
carcinogen 199
carcinoma 422, *422*
carotene desaturase *464*
carotenoid pigment synthesis 295, *296*
carpels 384
carriers 410
cartilage 39
cascade system 223
Cas9 endonuclease 358, *358*
catabolite activator protein (CAP) 157
Caudal protein 379–380, *380*
cDNA 457
CD4 protein 229
CD339 protein 412
cell cycle 190
    DNA replication 190, *190, 191*
cell cycle checkpoint 190
cell division 3
cell-free protein synthesizing system 118, *118*
cell specialization 152
cell structure 49
cell surface receptor 161–163, 212, 408, 419–420
cell transformation 229–230, 419–422
centimorgan (cM) 331
centromere 261
centrosome 265
CFTR gene 44–46, 406–407
chain termination sequencing 11
chalcone synthase **465**
chaperonins 138–140
Chargaff, Erwin 27
charging 122
cheetah 317
chemical modification of proteins 370–371
chiasmata 270

chi form 272, *273*
chimeric antigen receptor T cells 426
chimpanzee 396–398
chi (crossover hotspot initiator) site 274
chitinase **464**
Chlamydomonas 399
chloride ions 44
chloroplast 462
chloroplast genome 255, 398, 400–401
cholera 54
chromatid 261
chromatin 256
chromatin modification 367, 370–372
chromatosome 259
chromid 241
chromosomal interaction domain 244
chromosomes 1–3, *3, 19,* 19–20
    abnormalities 413–414, *413–414*
    human 3, 387
    partitioning systems *244,* 244–245
chronic granulomatous disease 413
cilia 39
cI repressor 225–226
citrus exocortis viroid 232
CJD 232
clamp loader 185
clastogen 199
cleavage and polyadenylation specificity factor (CPSF) 70
cleavage stimulation factor (CstF) 71
cline 315
CLIP *143*
cloning vector 452–454
closed promoter complex 61
cloverleaf 97, *98*
CMT disease 407
Cockayne syndrome 209
codominance 290–291, *291*
codons 117, **121**
    bias 355, 457
codon-anticodon recognition 123–125
cold cases
    DNA 436–437
colicin 240
colinearity 117
collagen 39, *39,* **141**
colon cancer 422
colonization 318

Colorado potato beetle *54*
Col plasmid 240
Combined DNA Index System (CODIS) set 433
common bean 464
compatible 240
competent 452
complementary 29
complementary base pairing 31–32
complementary DNA (cDNA) synthesis 356
complementary gene action 294–295, *295*
complex multigene family 46–47
concatemers 224
conjugation 240, 246–248
conjugation mapping 337, *337*
consensus sequences 58
conservative DNA replication 173
conservative replication 173
constitutive heterochromatin 259
context-dependent codon reassignment 120–122
continuous culture 452
COOH-terminus 113
copy number 240
core enzyme 57
core genome 242
core octamer 259
co-repressor 159
core promoter 59
corticotropin-like intermediate lobe protein *143*
cos site 224
cotransduction 338, *338*
cotton 463
cowpeas 464
CpG islands 373–374, 422
CREB transcription factor 77
Cre protein *473,* 473–475
Creutzfeld-Jakob disease (CJD) 232
Crick, Francis 27–28
cro protein 225–226
cross-fertilization 283
crossing over 270
C-terminal domain (CTD) 62–63
C terminus 113
curly leaf gene 384
cyanelles 400
cyanogen bromide 456
Cyanophora paradoxa 401
cyclic AMP (cAMP) 158
cyclin 190
cyclin-dependent kinase (CDK) 190
cyclobutyl dimer 201, 204
cysteine **112,** *113*
cystic fibrosis 15, 44–46, 405–407, 424
cystic fibrosis gene 15, 44–46, 406–407
cytochrome c oxidase subunit genes *398*
cytoglobin 49
cytokine 162
cytokinesis 266, *267*
cytomegalovirus 459
cytosine 24
cytosine methylation 373

Dam 207
dark repair 205
D arm 97
Darwin's evolutionary theory 306
deadenylation-dependent decapping 80
Dead Sea 54
deaminating agents 201
defective retrovirus 230, *230*
Deformed gene *381*
Deformed protein 369–370
degeneracy 119
degenerate 119
degradative plasmid 240
degradosome 80
Deinococcus radiodurans 241
delayed onset 290, 298
delayed onset phenotypes 298, *298*
deletion 196
deletion mutation 196–198
denaturation 114, *138*
de novo sequencing 349
density gradient centrifugation 93–95, *94,* 175, *175,* 394
dentatorubral-pallidoluysian atrophy **409**
deoxyribonuclease 21
deoxyribonucleic acid (DNA) 5, 5–6, *6,* 19, 440, *440*
    amelogenin 435, *435*
    archaeological skeletons 441–442, *442*
    bacterial transformation 21, *21*
    bacteriophage genes *22,* 22–23
    biological role 31–32
    Bronze Age skeleton 440–441, *441*
    chromosomes *19,* 19–20
    cold cases 436–437
    familial DNA testing 437, *437*
    genetic fingerprinting *432,* 432–433
    glycosylase 204
    gyrase 178
    Hershey–Chase experiment 22, *23*
    IBD 444, *444*
    identity-by-descent (IBD) 443, *443*

deoxyribonucleic acid (DNA) (cont.)
   minisatellite 432, *432*
   miscoding lesions 445, *445*
   Neanderthal genome *444*, 444–446, *445*, *446*
   polymerase chain reaction (PCR) 434–435
   polymorphisms 431, *431*
   profile 1, 395, 435, *435*
   pseudoautosomal regions 435, *435*
   quagga 439, *439*
   radiolabeling 22, *22*
   short tandem repeats (STRs) 433, 434, *434*, 436, *436*
   Streptococcus pneumonia 20, *20*
   structure 23–30
   supercoiling 243, *243*, 244
   testing 437–439
   transforming principle 20–21, *21*
   tree shows relationships 442–443, *443*
deoxyribonucleotide *see* nucleotide
development
   Caenorhabditis elegans 375, *376*
   Drosophila melanogaster 375, 379–384, *379–380*, *382*
   model organisms 376–377
   plants 383–384
   role of gene regulation 152–154
   vertebrates 381–383
Dfd protein 369
diabetes 454
diauxie 157
Dicer 81–82
Dicer protein 81–82
differentiation 365–375
dihybrid crosses 284–285, *284*
dimethylnitrosamine 201
diploid 256
directional selection 309
direct repair systems 203, 204
direct-to-consumer (DTC) 480
discontinuous genes 38
dispersive DNA replication 173
dispersive replication 173
displacement replication 178
disruptive selection 310, *310*
disulfide bridge 114
DKPLA gene **409**
D loop 178, 272

DMPK gene **409**
DNA *see* deoxyribonucleic acid (DNA)
DNA adenine methylase (Dam) 207
DnaA protein 184
DNA-binding proteins 29, 118–119
DnaB protein 184–185
DNA break repair 203, 208
DNA chip 418, *419*
DNA cloning 451–454, *453*
DnaC protein 184
DNA cytosine methylase (Dcm) 207
DNA-dependent DNA polymerase 180
DNA-dependent RNA polymerase 56
DNA helicase II 206
dnaJ 7
dnaK 7
DNA ligase 186
DNA markers 414
   pedigree analysis 414–416, *415–416*
DNA methylation 208, 367, 373–375, 422
DNA methyltransferase 372, 373
DNA molecules
   physical mapping techniques 339, *339*
DNA photolyase 204
DNA polymerase(s) 8, 32, **39**, 180–184
   accuracy 197
   proofreading 197
DNA polymerase I **181**, 182, 186
DNA polymerase II 182
DNA polymerase III **181**, 182, 183, 185–186
DNA polymerase V 209
DNA polymerase α **181**, 182, 183, 188–189
DNA polymerase γ **181**, 182, 183
DNA polymerase δ **181**, 182, 188–189
DNA repair 196
   base excision repair 203–205, *205*
   defects leading to disease 209
   direct repair 203, 204
   DNA breaks 203, 208
   mismatch repair 203, 206–208
   nucleotide excision repair 203, 205–206
   types 203
DNA replication
   accuracy 196–197, 203
   bacteria 184–187

   cell cycle 190, *190*, 191
   DNA polymerases 180–184
   errors 197–198
   eukaryotes 187–191, *189*
   events at replication fork 185–186, 188–189
   flap endonuclease (FEN1) 188–189, *189*
   origins 187–188, *189*
   origins of replication 184–185, 187–188
   role of complementary base pairing 31–32
   semiconservative replication 173
   termination 186–187, 189
   topological problem 173–180
   variations on semiconservative replication 178–180
DNA sequence *12*, 31
DNA sequencing 11, 12, *13*, 345–346, *345–346*, 446–448, *447*
   gene editing 357–358
   genes function 356–357
   genome browser 358–359, *359*
   genome sequence 349–350, *349*, *350*, 352, *352*, 353
   open reading frame (ORF) 353–355, *353–355*
   RNA sequencing 355–356
   shotgun method 346–347, *347*, 349, 349–351
DNA structure
   double helix 28–30
   nucleotides 23
   polynucleotide 25–26
DNA topoisomerase 176–178
DNA unwinding element (DUE) 184
dogs 405
Dolly the sheep 483, *483*
dominance 410–412
dominant 283
donkey 319
donor site 83
Dorsal protein 379
double helix 23
   different forms 29–30
   structure 28–30
double heterozygote 332
double homozygote 332
double-stranded DNA 356, *356*
double-stranded RNA 100, *100*
Down syndrome 413
draft genome sequence 352
Drosophila melanogaster 3, 4

development 379–384, *379–380, 382*
    feedback loop 369
    gene number **388**
    gene organization 391
    LTR retrotransposons 231
    model organism 3, *4*, 375, *379*, 383
    Polycomb proteins 372
    trithorax proteins 373
Duchenne muscular dystrophy 60, **405,** 413
dynein 39
dystrophin gene 60–61, 68

early genes 223
early onset of male puberty 408
Earth BioGenome Project 346
E. coli chromosome 337, *338*
edge of life 231–233
Edwards syndrome 413
eEF-1 134
eEF-2 135
EF-2 135
EF-1A 134
EF-1B 134
effective population size 305
egg white 39
eIF-1 132
eIF-1A 132
eIF-2 132, 133
eIF-2B 132
eIF-3 132
eIF-5 133
eIF-6 133
eIF-4A 132–133
eIF-4B 133
eIF-4E 132
eIF-4F 133
eIF-4G 132
eIF-4H **132**
elastin protein 412
elephant seal 317
elongation factors 68, 134–135
EMS 201
ENDO *143*
endogenous retrovirus (ERV) 232
endonuclease 79
endorphin *143*
endosymbiont theory 400–401
engrailed gene 381
enhancer 85–87, 160
enol-guanine 197
enolpyruvl-shikimate-3-phosphate (EPSP) synthase 461–462
enol-thymine 197
env gene 229, 231
enzymes 38, 39, *40*
    introns 104–105

epigenesis 367
episome 239
epistasis 294, *294,* 295
EPM1 gene **409**
EPSP synthase 461–462
Equus asinus 319
Equus cabalus 319
ERBB2 gene 420, *420*
eRF-1 136
eRF-2 136
error-prone editing 466
ERV 232
Erwinia uredovora 463
Escherichia coli (*E. coli*)
    antitermination 164–166, 165–166
    attenuation 166–168
    bacteriophages 220–227
    chromosome 237
    conjugation 246–248
    discovery 237, 238
    DNA cloning 31–32
    DNA polymerases 182
    DNA replication 184–187
    error rate in DNA replication 203
    example of a prokaryote 54
    gene regulation 149–150, 154–159, 163–169
    genome 238
    growth in medium 237
    heat shock response 7, *7,* 163–164
    horizontal gene transfer 245–250
    insulin production 7
    lactose operon 154–159
    model organism 237
    nucleoid 243–244
    origin of replication 184–185
    promoter 58
    recombinant protein 452–458
    RNA polymerase 57, 66, 70, 163–164
    rRNA genes 46
    SOS response 209
    transcription 61, 63, *63,* 68
    translation 130–131, 134–136
    tryptophan operon 158–159, 166–168
E site 135
estrogens 161
ethical, legal and social issues (ELSI) 485
ethics 15
ethidium bromide 201
ethylene 464–465
ethylene biosynthesis pathway 465, *465*

ethylmethane sulfonate (EMS) 201
euchromatin 259, *259*
eukaryotes 53, *54*
    DNA replication 189, *189*
    gene mapping techniques *see* gene mapping techniques
eukaryotic pre-tRNAs 105–107, *106*
eukaryotic proteasome 144, *144*
Euplotes **121**
European corn borer 463
even-skipped gene 381
evolution 8–9, 303–313, 320
excision repair 203
exit site 135
exon 38
exon skipping 86
exonuclease 79, 180–182
exoskeleton 39
exosome 80
expression vectors 452–454
extrachromosomal genes 398
extra loop 98

F' 247
F^+ 246
F^- 246
factor VIII 459–460
facultative heterochromatin 259
familial adenomatous polyposis 422
familial DNA testing 437, *437*
fatty acids 39
F^- cell 246–247
F´ cell 246
F^+ cell 246–247
F_1 cross produces *283,* 283–284
feathers 39
feedback loop 367, 369–370
FEN1 protein 189
fermenter 452
ferritin 39, 169
fertility 239
fertility plasmid 246
fibrous proteins 39
filamentous bacteriophages 220
filamentous fungi 458
Fischer structure 24
fish 382–383, *383*
fitness 307
fixation 304–306
fixation indices 317
flagella 39
flap endonuclease (FEN1) 188–189, *189*
'FlavrSavr' tomatoes 472, *472*
flower development 383–384
flowering time 314, *314–315*

foldback RNA 81
folding pathway 138
foot-and-mouth 460
forensic science
    genetic profiling 15
formate dehydrogenase **121**
formyl-methanofuran dehydrogenase **121**
founder effect 318
FOXP2 protein 398
F plasmid 239, 246–248
fragile site 408
frameshift mutation 211
frataxin gene 408
Friedreich's ataxia 408, *409*
fruit fly *see Drosophila melanogaster*
fruit ripening 464–465
F_{ST} value 317
functional RNA 41
fushi tarazu gene 381
fusion protein 454–456

gag gene 229, 231
gain of function 211
galactose 154
galactose epimerase 458
galactose operon 158
galE gene **130**
GAL promoter 458
gamete 256
gap genes 380
Gaussian distribution 293
GC box 160
GC content 353
gene(s) 1
    discontinuous genes 38
    gene expression 6
    gene families 46–49
    human gene number 37
    instructions for making RNA and protein 38
    lengths 37
    segments of DNA molecules 37–38
    units of biological information 6–8
    units of inheritance 1, 15
    variations in information content 43–46
gene addition 462–464
gene amplification 46
gene desert 389
gene deserts 389, *389*
gene duplication 47–49
gene editing 11, 357–358
gene expression 6
    different types of organism 53–56
    DNA makes RNA makes protein 55
    importance 6–7, *7*
    outline 37
    role of protein synthesis 38–43
gene families 46–49
gene mapping 327
gene mapping techniques
    bacteria 336, *336*
    eukaryotes 327
    multipoint cross 332–333, *332*, **333**, *333*
    partial linkage 328–330, *328–330*
    pedigree analysis 333–335, *334*
    recombination frequency 330–331, *330–331*
    test cross 331–332, *332*
    interrupted mating technique 337–338, *337*, *338*
gene regulation
    alternative σ subunits 163–164, 224
    antitermination 164–166, *165–166*
    attenuation 166–168
    bacterial transcription initiation 154–159
    eukaryotic transcription initiation 159–163
    importance 7, 150–154
    initiation of translation 168–169
    lactose operon 154–159
    tryptophan operon 158–159, 166–168
genes function 356–357
gene subtraction 464–465, **465**
gene therapy 1, 405, 422–426
genetics 1
    chromosomes 1–3, *3*
    DNA *5*, 5–6, *6*
    Drosophila melanogaster 3, *4*
    engineering 9–11, *10*
    ethical issues 483–484
    evolution 8–9
    eye colors of fruit fly 3, *4*
    genomes 11–12
    Human Genome Project 485, *485*
    Mendel, Gregor 1–3
    organization of book 13–15, *14*
    origins 1–13
    unit of biological information 6–8
    variation 12–13
genetically modified (GM) crops 10, 451, 461–468
    Cre protein *473*, 473–475
    cultivation 475–477, *476*
    Dolly the sheep 483, *483*
    'FlavrSavr' tomatoes 472, *472*
    pectin 472, *472*
    SCNT 482–483
    selectable marker 473, *473*
    terminator technology 477–478, *478*
genetically modified (GM) plant 10
genetic analysis 280
genetic code 7, 118–119
    basic features 31, 38
    cell-free translation 118, *118*
    codon meanings 121
    colinearity 117
    degeneracy 119
    nonstandard codes 120–122
    punctuation codons 119–120
    random heteropolymers 119, *119*
    triplet nature 117
genetic disease *see* inherited disease
genetic disorder 405
genetic drift 302, 304
genetic engineering 1
    animals 467–468
    GM crops 10, 451, 461–468
genetic fingerprinting 13, *432*, 432–433
genetic material 9, 19
genetic profile 15, 395
genetic use restriction technology (GURT) 477
genome(s) 11–12, 237–242, 389–391, *390*, *391*
genome annotation 353
genome browser 358–359, *359*
genome rearrangement 367–369
genome sequence 349–350, *349*, *350*
genome-wide association study (GWAS) 418
genomic imprinting 373–375
genotype 283
geranylgeranyl diphosphate 463–464
germ-line gene therapy 423–425
germline therapy 423, 481–482, *482*
gfpE 7
gigabase pair (Gb) 346
globin genes 46–49, 151
globin mRNA 79
globular chromatin structures 10 nm fiber 259, *260*
glucocorticoid hormone 162
glucose 154, 156–158

glucose metabolism 39–40
glutamic acid **112**, *113*
glutamine **112**, *113*
glycine **112**, *113*
glycine reductase **121**
glycosylation 141, 455, 458
glyphosate 461–462
glyphosate *N*-acetyltransferase **464**
G2-M checkpoint 190
GM crops 451, 461–468
Golden rice *10*, 464
gorilla 396, *396*
G1 phase 190
G2 phase 190
gray wolf *54*
GroEL/GroES complex 139–140
growth factors 151
growth hormone 40
growth hormone genes 389
G1-S checkpoint 190
guanine 24
guanine methyltransferase 66
guanylyl transferase 66
G–U base pair 125, *125*
guide RNA 358

Haemophilus 248
hair 39
hairpin loop 68–70
half-life 78–79
hamster cells 459–460, *459*
haploid 256
haploinsufficiency 212, 412, 413
haplotype 45, 46
Hardy, Godfrey 303
Hardy-Weinberg equilibrium 303
HATs 370
Haworth structure 24
HBsAg 460
HDACs 370
HD protein *411*
head-and-tail 220
head-and-tail bacteriophages 220
heat as a mutagen 201, *202*
heat shock *7*, 163–164
heat shock genes *7*
heat shock module 160
helicase 62, 184
helix-turn-helix 115
helix-turn-helix motif 115
Hemerocallis genes *292*, 292–293
hemoglobin 39, *40*
hemophilia **405,** 410, *411,* 459
hepatitis B surface antigen (HBsAg) 460, 467
hepatitis B virus **228,** 460, *460*
herbicide resistance 461–462

hereditary nonpolyposis colorectal cancer 209
Hershey–Chase experiment 22, *23*
HERV-K group 232
heterochromatin 259, *259,* 370–372, 375
heteroduplex 272
heterogeneous nuclear 87
heterogeneous nuclear ribonucleoprotein (hnRNP) 87
heterogenous nuclear RNA (hnRNA) 82
heterosis 310
heterozygote 283
heterozygote advantage 310
heterozygote superiority 310
hexokinase **39**
hexose-1-phosphate uridyltransferase **130**
Hfr 247
Hfr cell 247
H19 gene 375
hinny 319, 320
histidine **112**, *113*
histone **141,** 257, 370–372
histone acetyltransferases (HATs) 370
histone code 371
histone deacetylases (HDACs) 370
histone H3 140, *370*, 371, *371*
histone H4 *370*, 371
histone H2A *370*
histone H2B *370*
hitchhiking 312, *312*
HIV-1 229
HIV-2 229
HNPCC 209
hnRNA 82
Holliday structure 272
holoenzyme 57
HOM-C 382
homeobox 381
homeodomain 381
homeotic mutant 381
homeotic selector genes 381–384
hominids 396
Homo erectus 397
Homo habilis 397
Homo heidelbergensis 397
homologous 267
homologous chromosomes 267, *268,* 268–269
 bivalent 270–271, *271, 272*
homologous recombination 271–273, *272*
homology-directed repair 466, *466*

homology search 356
Homo neanderthalensis 397
homozygote 283
horizontal gene transfer 237, *237,* 245–250
hormone 39, 151
hormone response element 162
horse 319, 320
housekeeping genes 158
HoxA cluster 382
HoxB cluster 382
HoxB genes 382
HoxC8 gene 382
HoxD cluster 382
HP1 protein 371
Hsp70 chaperones 139
HTT protein 408
HU family 244
human apolipoprotein 88, *88*
human chromosomes 3, 37
human embryo *377*
human evolution 1
human gene number 37
human genome
 basic features 387–388
 composition 387, 388
 gene density 389, 392
 gene functions 387–388
 genetic organization 387–392
 introns per gene 392
 mitochondrial genome 398–399
 multigene families 389
 number of active genes *367*
 number of genes 387
 relationship to other animals 396–398
 repetitive DNA 389, 391, 392–394
Human Genome Project 485, **485**
human globin genes 46–49
human immunodeficiency viruses 219, 229
human mitochondrial genome 439, *439*
Hunchback protein 379–380, *380*
huntingtin protein *412*
Huntington's disease 405, 408
hybrid incompatibility 321
hybridization probing 340–341, *341*
hybrid sterility 320, *321*
hydatidiform mole 352
hydrogenase **121**
hydrogen bonds 29, 113
hydrogen peroxide 169
hyperediting 89

hypervariable dispersed repetitive sequence 432
hypervariable region 439
hypoxanthine 201, *201*

IBD 444, *444*
icosahedral 220
icosahedral bacteriophages 220
identity-by-descent (IBD) 443, *443*
IF-1 131
IF-2 131
IF-3 131
Igf2 gene 375
IIA^{Glc} protein 158
imino-adenine 197
immunoelectron microscopy 96, *97*
immunoglobulin 39, *41*, 368–369
immunoglobulin diversity 367–369
imprint control elements 374, *375*
inclusion body 458
incompatibility group 240
incomplete dominance 288–289, *288*
incomplete penetrance 296–298, *297*, 410
inducer 156
inducible operons 158
induction 154, 224
industry 1
influenza virus **228**
inherited disease 405
    chromosome abnormalities 413–414, *413–414*
    commonest ones **405**
    dominant and recessive 410–412
    gain of function 407–408
    loss of function 406–407
    number and frequency 13, **405**
    trinucleotide repeat expansions 408–409
initiation codon 120, 130–133
initiation complex 131
initiation factors 131–133
initiation of translation 168–169
initiation region 188
initiator sequence 60
initiator (Inr) sequence 60
initiator tRNA 130–133
inosine 123–125
Inr sequence 62
insecticides 463
insertion 196
insertion mutation 196–198
insulin *10*, 39, *41*, 142–143, 454–456
insulin gene 160

integron 241
intercalating agent 201
intergenic DNA 37
internal ribosome entry site (IRES) 133
interrupted mating technique 337–338, *337*, *338*
interspersed repeats 392–394
intrinsic terminator 68, *69*
introns 38
    basic features 38
    discovery 82–83
    effect on colinearity 117
    enzymes 104–105
    eukaryotic pre-tRNAs 105–107, *106*
    in human genome 389
    in yeast and fruit fly genomes 391, *392*
inversion 196
inversion mutation 196
inverted repeat 68
iodothyronine deiodinase **121**
ionizing radiation 208
IPTG 454
IRES 133
iron 39, 169
iron-response element 169
Islets of Langerhans 142, 454
isoaccepting tRNAs 122
isoform 85
isoleucine **112**, *113*
isopropylthiogalactoside (IPTG) 454

Jacob, François 154
Jansen's disease 408
junk DNA 393

karyogram 261
Kenyanthropus platyops 397
keratin protein 39
keto-guanine *197*
keto-thymine 197
kilobase pair (kb) 37
kinase 161
kinetochores 263, 266, *267*
Klebsiella pneumoniae 164
Klinefelter's syndrome 413
Kozak consensus 132
KRAS proto-oncogene 422
Ku70-Ku80 heterodimer 208
Ku protein 208

labial palps gene 381
lacA gene 155
lacI gene 155–156
lactose 154–158
lactose operon **130**, 154–159, 454
lactose permease 154

lactose promoter **59**, 454, *454*
lactose repressor 156
lacY gene 155
lacZ gene 155
lagging strand 182–186, 188
late genes 223
latent period 222
leading strand 182–186, 188
Lederberg, Joshua 246
Leptinotarsa decemlineata 54
lethal alleles 289–290, *290*
lethal mutation 211
leucine **112**, *113*
LINE 393–394
linkage 285, *286*
linkage analysis 331
linkage group 331
linker DNA 259
linker histone 259
LIN-3 protein 378
liposome 424, *425*
lipotropin *143*
liver 39
lod score 335
long interspersed nuclear element (LINE) 393–394
long noncoding RNA (lncRNA) 42
long patch repair 206
long-read sequencing 346
long terminal repeats (LTRs) 231–232
loss of function 211
LPH *143*
LTR retrotransposon 231–232, 391, 393
LTRs 231–232
luminal A subtype breast cancer 78
luminal B subtype breast cancer 78
luteinizing hormone receptor 408
lycopene 464
lycopene cyclase *464*
lymphocytes 367–369
lysine **112**, *113*
lysogenic infection cycle 224–227, *224*
lytic infection cycle 222–225

Machado-Joseph disease **409**
MADS box 384
maintenance methylation 373, *374*
maize **388**, 391–392, 462, *463*
major groove 28, *28*
malaria 311, 400
male siblings possess 438–439, *438*
mammalian cell culture 459
Manchester 307

MAP kinase system 162
Marfan syndrome 405, 410
marker 22
maternal effect genes 379–380
m^6A writer 89
ME *143*
measles virus 467
mediator 161
megabase pair (Mb) 37
meiosis 265, 267, *268*, 269, *270, 271, 285*, 285–286
meiosis I 268, *268*
meiosis II 268, *269*
melanin 307
melanotropin *143*
melittin **141,** 142
Mendel, Gregor 1–3, *2, 9*, 12
Mendel's experiments 280–281
    characteristics 281–282, *281, 281*
    dihybrid crosses 284–285, *284*
    F1 cross produces 283–284, *283*
    linkage 285, *286*
    monohybrid crosses 282, *282, 282*, 283
    pea chromosomes 285, *286*
    Pisum sativum 280, *280*
    Punnett square 283, *284*
Meselson–Stahl experiment 175–176, *175, 176*
mesocarp 295, *295*
messenger RNA (mRNA) 38
    amount in cell 40
    chemical modification 87–89, *87*
    component 76–77, *76*
    composition 78–79, *79*
    function 38
    human apolipoprotein 88, *88*
    splicing 82–87
    synthesis 56–71
    translation 130–136
metacentric 261
metaphase 265–266
metaphase chromosomes 259–260, *260*
metastasis 78, *78*, 422
met-encephalin *143*
methanogens 54
methicillin-resistant *Staphylococcus aureus* (MRSA) 249
methionine **112,** *113*
methionine-rich protein **464**
6-methyladenosine 89, *89*
methyl-CpG-binding proteins (MeCPs) 373–374
5-methylcytosine 373

7-methylguanosine 66
methyl halides 201
MGMT (O^6-methylguanine-DNA methyltransferase) 204
microevolution 302–313
microinjection 423, 467
microRNA (miRNA) 42, 81
microsatellites 395, *395*
mineralocorticoid hormone 162
mine streams 54
minimal medium 336
minisatellites 395, 432, *432*
minor groove 28, *28*
Mirounga angustirostris 318
miscoding lesions 445, *445*
mismatch 82, 195
mismatch repair 203, 206–208
mitochondrial genome 255
    basic features 387
    endosymbiont theory 400–401
    human genome content 398–399
    nonstandard genetic code 120
    variability in different species 399–400
mitogen-activated protein 162
mitosis 265–266, *266, 269, 270*
mitotic spindle 266, *266*
model organisms 375–377
modern synthesis 9
molecular approach 19
molecular approach to genetics 19
molecular chaperones 138–140
molecular clock 48, 396
molecular combing 340, *340*
molecular phylogenetics 396
monellin **464**
Monod, Jacques 154, 156
monogenic 405
monogenic disorder 405
monohybrid crosses 282, **282,** *282*, 283
Morgan, Thomas Hunt 376
mother cell 164
motor proteins 39
M phase 190
mRNA surveillance 80
mRNA turnover 79–82
MRSA 249
mule 319, 320
multigene families 46–49
multipartite genomes 240–241
multiple promoters 61
multiplex PCR 435
multipoint cross 332–333, *332,* **333,** *333*
multistep cancer development 422, *422*

muscle 39
muscle cells 369
muscular dystrophy **405,** 413
mutagens 195, 196, 198–203
mutasome 209
mutation 47
    accumulation over time 47
    causes 196–203
    creation of new alleles 301, 306
    effect on genes 210–211
    effect on multicellular organisms 211–212
    importance 8
    mutagens 198–203
    repair 203–209
    replication errors 197–198
    reversal 213
Mut proteins 207
Mycobacterium tuberculosis 54, **238**
Mycoplasma genitalium **238**
myelin protein 407
MyoD protein 369
myoglobin 49
myosin 39
myotonic dystrophy 408

NADH dehydrogenase subunit genes *398*
nanopore sequencing 348, *348*
Nanos protein 379–380
natural selection 306–310, *310*
Neanderthal genome 444–446, *444–446*
Neanderthals 397
neuroglobin 49
neutral theory 313
N-formylation **141**
N-formylmethionine 123, 131
NH_2 terminus 113
nitrogen fixation 164
nitrous acid 201
N-linked glycosylation 141
N-myristoylation **141**
noncoding RNA 41–43
nonhomologous end-joining (NHEJ) 208
nonpolar 112
nonsense mutation 210
nonstandard genetic codes 120–122
nonsynonymous mutation 210
normal distribution 293
normal-like subtype breast cancer 78
N terminus 113
nuclear genome 255
nuclear transfer 423, *423*

nuclease protection technique 95, 96, *96*
nucleic acid hybridization 341
nucleoid 243–244, *243, 244*
nucleoid-associated protein 244
nucleolus 82
nucleoplasm 82
nucleoside 24
nucleosome(s) 370–372
nucleosome remodeling 372, 373
nucleotide(s) 5, 23
    bases 24
        components 24–25
        excision 205–206, *206*
        names 24
        polynucleotide 25–26
        structure 24–25
nucleotide excision repair 203, 205–206

octamer sequence 160
off-target editing 358
Okazaki fragments 182–186, 188
oligonucleotide 346
oligonucleotide hybridization analysis 418, *418*
O-linked glycosylation 141
O^6-methylguanine-DNAmethyltransferase 204
OmpC protein 222
oncogen 199
oncogene 230, 419–420
one-step growth curve 222
oocyte 423
open promoter complex 61
open reading frame (ORF) 353–355, *353–355*
operator 156
operon 154–159
optical mapping 340–341
optional loop 98
orangutan 396
ORC 187–188
ORF scanning 353
origin licensing 190
origin of replication 184
origin recognition complex (ORC) 187–188
origin recognition sequence 187
ornithine carbamyltransferase **464**
Ostrinia nubilalis 463
ovalbumin 39
ovarian cancer 209
overdominance 310
over-dominance 310
overlapping genes 220, *221*
OxyS 169

paired-end reads 347
pair-rule genes 380
pancreas 142, 454
pan-genome concept 242, *242*
ParABS system 244, *245*
Paramecium 399
pararetroviruses 229
parathyroid hormone receptor 408
parental genotypes 330
PARP1 proteins 208
partial linkage 328–330, *328–330*
partitioning systems 244–245, *244*
parvovirus **228**
P22 bacteriophage 338, *338*
PCNA 182
pea chromosomes 285, *286*
peas 43–44
pea shape 43–44
pectin 472, *472*
pedigree analysis 333–335, *334, 414–416*
pentose 24
peppered moth 307
peptide bond 112, *114*
peptide hormones 143
peptide nucleic acid (PNA) 341, *341*
peptidyl site 134–135
peptidyl transferase 134
pericentromeric 261
personal DNA data 478–481
    concept 479–480
    issues 480–481
personal medicine 346
petals 384
phage 22
pharming 468
phenotype 283
phenylalanine **112**, *113*
phenylalanine operon 159
phenylketonuria **405**
Philadelphia chromosome 420
phosphatidylinositol-specific phospholipase C **464**
phosphodiesterase 205
phosphodiester bond 25
photo 51 27
photoperiod-1 gene 314
photoproduct 201
photoreactivation 204
physical mapping techniques 339
    DNA molecules 339, *339*
    optical mapping 340–341
phytoene *464*
phytoene synthase *464*
Pichia pastoris 458
picornavirus 133
pilus 246

Pisum sativum 280, *280*
plasmid 239
Plasmodium falciparum 311, 400
plectonemic 174
pleiotropy 296–297, *297*
PMP22 gene 407
point mutation 196–197
polar 112
pol gene 229, 231
poliovirus 133, **228**
polyadenylation 70
polyadenylation site 70
Polycomb proteins 372, 381
Polycomb response elements 371
polygalacturonase **465**
polymer 5, 23
polymerase chain reaction (PCR) 13, 415, 416, 434–435
polymorphisms 431, *431*
polynucleotide 23, 25–26
polynucleotide phosphorylase 118
polyp 422
polypeptide 111, 113–114
polyphenol oxidase **465**
poly(A) polymerase 70
polyprotein 142, 229
polypyrimidine tract 83
polysome 135
poly(A) tail 70, 80, 132
population
    biology 313–321
    genetics 301–321
positional information 377–380
post-translational chemical modification 56
post-translational processing
    chemical modification 140–141
    folding 136–140
    outline 40–41
    proteolytic cleavage 142–143
postzygotic isolation 319
potato 467
P6.p cell 377–378
p53 protein 422
pre-initiation complex 131
prepriming complex 184
preproinsulin 142
pre-replication complex (pre-RC) 190
pre-rRNAs 99, *100,* 104–107
pre-tRNAs 101, *102,* 104–107
prezygotic isolation 319
Pribnow box 58
primary structure 113
primary structure of protein 113
primary transcript 75
primase 183

primers 183
primosome 185
prions 232–233
proboscipedia gene *381*
processing 56
proflavin 117
programmable nuclease 357, *358*
progressive myoclonus
epilepsy 409
proinsulin 142
prokaryotes 53–54, *54*
proliferating-cell nuclear antigen
(PCNA) 182
proline **112,** *113*
proline reductase **121**
promelittin 142
prometaphase 265
promiscuous DNA 401
promoter 58–61, 453–454
promoter clearance *61*
proofreading 180, 197
proopiomelanocortin 143, *143*
prophage 224, 249
prophase 265
protease 21, 137
proteasome 144, *144*
protective proteins 39
protein 19
  amino acid sequence
    114–116
  structure 111–116
  types 38–41
proteinase inhibitors 464
protein degradation 143–144
protein folding 136–140, 458
protein synthesis
  key to expression of biological
    information 38–41
  post-translational processing
    136–143
  role of ribosome 130–136
proteome 129
protomers 220
proto-oncogene 419–420
PrP^C protein 232–233
PrP^SC protein 232
pseudoautosomal regions
  435, *435*
pseudogene 49
Pseudomonas **221**
Pseudomonas aeruginosa 220
Pseudomonas syringae **464**
Psi+ prion 233
P site 134, 135
punctuation codons 120
Punnett square 283, *284*
pure-breeding 281
purine 24

pyrimidine 24
pyrophosphate 63
pyrrolysine 120–122, 140

quagga 439, *439*
quantitative trait 292
quaternary structure 114
quaternary structure of
  protein 114
Queen Victoria 410

rabies virus 461
radiolabeling 22
random heteropolymers 119, *119*
ray-finned fish 383
RB gene 421
reading frame 117
readthrough mutation 211
RecA protein 226
RecBCD enzyme 274
recessive 283
recessive genes 286–288, *287*
recessiveness 410–412
reciprocal strand exchange 272
Reclinomonas americana 400
recognition helix 115
recombinant 330
recombinant DNA molecule 452
recombinant insulin 454–456
recombinant protein 451
  Escherichia coli 452–458
  expression vectors 452–454
  mammalian cells 459
  microbial eukaryotes 458–459
  pharming 467–468
  plants 466–467
  vaccines 460–461
recombination 270
recombination frequency 330–331,
  330–331
recombination hotspot 331
red blood cells *40*
reference sequence 349
regulatory gene 155, 156
regulatory proteins 39–40
relaxed plasmids 240
release factor 136
renaturation 115, 138, *138*
reovirus **228**
repetitive DNA 350, 392–394
repetitive DNA sequences 350,
  350–351
replication fork 178, 185–186,
  188–189
replication fork barrier 189
replication protein A (RPA) 188
replication slippage 197, 395
replisome 185

repressible operons 158
repressor 156
resistance 239
resolution 272
respiratory syncytial virus 467
restriction endonuclease 340
restriction fragment length
  polymorphism
  (RFLP) 415
restriction sites 340, *340*
reticulocytes 79
retinitis pigmentosa 413
retinoblastoma **405,** 420
retinoic acid *162*
retroposons 393–394
retrotransposition 231
retrovirus 219, **228,** 228–230, 419
reverse transcriptase 229, 393
reversible terminator sequencing
  347, *347*
RF-1 136
RF-2 136
RF-3 136
R groups 111
rhinovirus 133
Rho 69
Rho-dependent terminator
  69, *70*
ribonuclease 21, 79, 138
ribonucleic acid (RNA) 21
  introns from pre-rRNAs and
    pre-tRNAs 104–107
  product of gene expression
    38, 40
  ribosomal 93–96
  rRNAs and tRNAs 99–104
  sequencing 355–356
  structure 38
  transfer 97–98
  types 41–43
ribonucleoproteins (hnRNPs) 87
ribonucleotide *38*
ribose 38
ribosomal protein 94
ribosomal protein genes 168
ribosomal protein L1 *168*
ribosomal protein L10 **130**
ribosomal protein L11 *168*
ribosomal RNA (rRNA) 93–96
  amount in cell 42
  chemical modification
    103–104
  density gradient centrifugation
    93–95, *94*
  fine structure 95–96, *95, 96*
  function 41–42
  genes 389, 399
  introns 104–105, *105*

ribosomal RNA (rRNA) (cont.)
  molecules 99–101, 99
  multigene family 46
  transcription units 101, 101
ribosome 93, 93
  number in cell 46
  role in attenuation 168
  role in protein synthesis 130–136
ribosome binding site 130
ribosome inactivating protein (RIP) 477
ribosome recycling factor (RRF) 136
ribozyme 104, 134
rice 463–464
RISC 81
RNA-dependent DNA polymerase 180
RNA-dependent RNA polymerase 56
RNA editing 87–89, **88**
RNA-induced silencing complex (RISC) 81
RNA interference (RNAi) 42, 81
RNA polymerase 56–58
RNA polymerase I 56, **58,** 60, 63
RNA polymerase II 57–58, 62–63
RNA polymerase III 57, 60, 63
RNA processing 56
RNA-seq 77, 356, *356*
RNA silencing 81
RNA splicing 42
  eukaryotic mRNAs 82–87
  importance 42
RNA synthesis 63, *63*
RNA transposons 231, 393–394
rolling-circle replication 178–180, *247*
round pea phenotype 287, *287*
round peas 43
Rous sarcoma virus 230, 459
RPA 188
R plasmid 239
rplJ gene **130**
rRNA genes 389, 399

Saccharomyces cerevisiae
  gene number **388**
  gene organization 391
  LTR retrotransposons 231
  mitochondrial genome 399–400
  prions 233
  recombinant protein 458, 460–461
S-adenosylmethionine (SAM) 465
SAM 465

sampling effect 318
satellite DNA 394
satellite RNA 232, 394
SBE1 gene 44
SCA genes **409**
scanning 131–132, *132,* 133
Schizosaccharomyces pombe **388**
SCNT 482–483
secondary structure 111
secondary structure of protein 113
second-site reversion 213
sedimentation coefficient 94
segmented genome 220
segment polarity genes 380, 381
segregation 283
selectable marker 473, *473*
selection 302
selection coefficient 309, 311
selective medium 336, *336*
selective sweep 312
selenocysteine 120–122, *122,* 140
self-fertilization 281
semiconservative DNA replication 173
semiconservative replication 173
SEPALLATA1 protein 384
sepals 384
sequence assembly 349
sequence coverage 349
sequence depth 349
sequence read 339
sequencing library 346
serine **112,** *113*
serotype 20
serum albumin 39
sex cell 256
Sex combs reduced gene *381*
sexual reproduction 15, 279–280, *279*
shelterin 263
Shine–Dalgarno sequence 130
short interfering RNA (siRNA) 81
short interspersed nuclear elements (SINEs) 393–394
short noncoding RNA (sncRNA) 42
short-patch repair 205–206
short-read sequencing 346
short tandem repeats (STRs) 395, 433, 434, *434,* 436, *436*
shotgun method 349–351, *349*
sickle-cell anemia 310–311, *311,* **405**
signal peptide 142, 455, *455*
signal transducer and activator of transcription (STAT) 162
signal transduction 162

silent mutation 210, *210*
simian virus 40 459
simple multigene family 46
simple mutation 196
SINEs 393–394
single-molecule real-time (SMRT) sequencing 347
single-nucleotide polymorphisms (SNPs) 45, 418, *418*
single-site mutation 196
single-strand binding proteins (SSBs) 185
siRNA 42, 81
site-directed mutagenesis 457
site selection 86
skin cancer 209
sliding clamp 185
SMAD4 gene 422
small interfering RNA (siRNA) 42
small nuclear ribonucleoprotein (snRNP) 84
small nucleolar RNAs (snoRNAs) 42, 103–104, *104*
smallpox 460
snapdragon 384
sodium bisulfite 201
somatic cell 256
somatic cell nuclear transfer (SCNT) 482
somatic cell therapy 423, 481
somatic nuclear transfer 467
somatostatin 40
somatotrophin 40
sonication 346
Sorangium cellulosum 238
SOS response 209, 226
soybean 462
speciation 319
speciation genes 321
S phase 190
spinal and bulbar muscular atrophy **409**
spinocerebellar ataxia 209, **409**
spliceosome 84
splicing 75
splicing code 86
splicing enhancer 86–87
splicing pathways 85–87, *86, 87*
splicing silencer 87
sporulation 164
SR protein 87
stabilizing selection 310
stamens 384
Staphylococcus aureus 249
starch branching enzyme type I 43
starch synthase **465**
starch synthesis 43–44
STAT 162

stem cells 425
steroid hormone 161, 162
steroid receptor 162
stochastic effects 303–304
stone tools 397
storage proteins 39
Streptococcus pneumonia 20, *20*
Streptomyces coelicolor 238
stress 133
stringent 240
strong promoter 59
Strophariaceae 54
structural gene 155
structural protein 39
structural variant 350
Sturtevant's map 331, *331*
stutter 436
submetacentric 261
substitution 196
subunit *224*
sucrose 43
sugar-utilizing genes 150
suicide gene therapy 425–426
supercoiling 243–244
superwobble 399
suppression 213
supravalvular aortic stenosis 412
SV40 virus 459
sweet prayers plant **464**
sympatric speciation 320
syn-adenosine *29*
syncytium 379
synonymous mutation 210

TAF- and initiator-dependent cofactor (TIC) 62
tandem array 46
tandemly repeated DNA 393, 394
TATA-binding protein 62–63
TATA box 60, 62
tautomers 197
Tay-Sachs disease **405**
TBP-associated factor (TAF) 62–63
T-cell diversity 367–369
telocentric 261
telomerase 265
telomere 263
telomeric DNA 395
telophase 266
temperate bacteriophage 222
template-dependent DNA synthesis 31–32
tendons 39
teratogen 199
termination codon 80, 120
terminator sequences 186–187
terminator technology 477

terminator utilization substance (Tus) 187
tertiary structure 113
tertiary structure of protein 113
test cross 331–332, *332*
Tetrahymena rRNA intron 104, *105*
TFIIB 62
TFIID 62
TFIIE 62
TFIIH 62
TFIIIB 63
thalassemia **405**
thaumatin **464**
Thermotoga maritima 250
thioredoxin reductase **121**
threonine **112,** *113*
thrombin 39
thymine 24
TICs 62
tobacco mosaic virus 114, **228**
tomato 465
transcription 38
    elongation 63–68
    enzymes 56–58
    importance 7, 38
    initiation 61–63
    recognition sequences 58–61
    termination 68–71
transcription bubble 64
transcription factor IID (TFIID) 62
transcription initiation complex 56
transcriptome 75–82
    cancer research 77–78
    mRNAs
    component 76, 76–77
    composition 78–79, *79*
transcriptome analysis, mRNA 76–82
transduction 248–249
    bacterial gene maps 338–339
transferrin 169
transfer RNAs (tRNAs) 97–98, *98*
    chemical modification 102–103, *103*
    cloverleaf structure 97, *98*
    function 42
    molecules 99, *99*
    role in protein synthesis 122–125, 130–136
    suppression 213
    three-dimensional structure 98, *98*
    transcription units 101
transformation 248–249
    bacterial gene maps 338–339
transforming principle 5
transgenic 461

transition 196
translation 7, 38
    elongation 134–136
    initiation in bacteria 130–131
    initiation in eukaryotes 131–133
    outline 38
    termination 136
translation pre-initiation complex 56
translocation 135
transport proteins 39
transversion 196
TRAP 168
tree shows relationships 442–443, *443*
trinucleotide repeat expansion disorder 198
trinucleotide repeat expansions 408–409, 411
triplet binding assay 119
triplex helix 274
trisomy 413
trithorax proteins 373
tRNA nucleotidyltransferase 101
trp RNA-binding attenuation protein (TRAP) 168
tryptophan **112,** *113*
tryptophan operon 158–159, 166–168
tryptophan promoter **59**
tryptophan synthetase 116
tuberculosis 54
tulip *54*
tumor suppressor genes 420–422
turnover 56
Tus proteins 186–187
two-hit model 420–421
TyA polyprotein 231
TyB polyprotein 231
Ty1/copia 231
Ty3/gypsy 231
type 0 cap 67, *67*
type 1 cap 67
type 2 cap 67
type II topoisomerase 177
type I topoisomerase 176
typica peppered moth 307, 308
tyrosine **112,** *113*
TΨC arm 98

ubiquitin 144, *144*, 371
UK biobank 480
Ultrabithorax gene *381*
ultraviolet radiation 201
UmuD′$_2$C complex 209
uniparental 439
unit factor 2

unit of biological information 1, 6–8
unit of inheritance 1
upstream 58
upstream control element 60
uracil 38
urea 138
Ure3 prion 233
UvrABC endonuclease 206

vaccine 460–461, 466–467
vaccinia virus **228,** 460–461
valine **112,** *113*
variable loop 98
variable number tandem repeats (VNTRs) 395
Vibrio cholerae 54, 241, *241*
viral retroelements 228–229
viroid 232
virulence plasmid 240
virulent bacteriophage 222

viruses
    bacteriophages 220–227
    genomes 227
    influenza virus **228**
    retroviruses 219, 228–230, **228,** 419
    structures 227–228
    viral retroelements 228–229
viruslike particles (VLPs) 231
virusoid 232
vitamin A deficiency 463
VLPs 231
v-*onc* 230
vulva development 377–378

Watson, James 27–28
weak promoter 59
Weinberg, Wilhelm 303
Werner syndrome 209
white-flowered sweet peas 294, *294*

wild-type 336
wobble hypothesis 123–125
wrinkled pea phenotypes 43, 287, *287*

xanthine 201
xeroderma pigmentosum 209
X inactivation 373–375, *376*
X inactivation center *(Xic)* 375
Xist gene 375
X-linked disease 410, *410*
X-ray crystallography 96
X-ray diffraction 5
XYY syndrome 413

Y-linked disease *410,* 410–411
Yule, Udny 302

Z-DNA 30
Zea mays **388**
zero-mode waveguide 347
zoo blotting 417, *417*

For Product Safety Concerns and Information, please contact our EU representative GPSR@taylorandfrancis.com Taylor & Francis Verlag GmbH, Kaufingerstraße 24, 80331 München, Germany

Printed by Integrated Books International, United States of America